Coral Reefs and Associated Marine Fauna around the Arabian Peninsula

Coral Reefs and Associated Marine Fauna around the Arabian Peninsula is a unique text that contains studies on a diverse range of topics related to the biology of the Red Sea and Arabian (Persian) Gulf region.

Containing invited and peer-reviewed chapters, this book is a compilation of the works of various experts in their respective fields. The authors delve into the marine fauna around the Arabian Peninsula, including marine reptiles and mammals, coral reefs, fish, invertebrates, algae and phytoplankton. They also explore the changes resulting from anthropogenic and climate effects.

This book will be a helpful resource for researchers in Biology and will also be a valuable reference for anyone interested in the biology of these two warm semi-isolated seas with their unique environments.

Coral Reefs and Associated Marine Fauna around the Arabian Peninsula

Edited by Najeeb M.A. Rasul and Ian C.F. Stewart

CRC Press
Taylor & Francis Group
Boca Raton London New York

CRC Press is an imprint of the
Taylor & Francis Group, an **informa** business
A BALKEMA BOOK

First published 2025
by CRC Press/Balkema
4 Park Square, Milton Park, Abingdon, Oxon, OX14 4RN

and by CRC Press/Balkema
2385 NW Executive Center Drive, Suite 320, Boca Raton FL 33431

CRC Press/Balkema is an imprint of the Taylor & Francis Group, an informa business

British Library Cataloguing-in-Publication Data
A catalogue record for this book is available from the British Library

Library of Congress Cataloging-in-Publication Data
Names: Rasul, Najeeb M. A., editor. | Stewart, Ian C. F. (Ian Charles Ferguson), editor.
Title: Coral reefs and associated marine fauna around the Arabian Peninsula / edited by Najeeb M.A. Rasul and Ian C.F. Stewart.
Description: Abingdon, Oxon ; Boca Raton, FL : CRC Press/Balkema, 2024. | Includes bibliographical references and index.
Identifiers: LCCN 2023058865 (print) | LCCN 2023058866 (ebook) | ISBN 9781032342900 (hardback) | ISBN 9781032342931 (paperback) | ISBN 9781003321392 (ebook)
Subjects: LCSH: Reef animals—Red Sea. | Reef animals—Persian Gulf. | Reef ecology—Red Sea. | Reef ecology—Persian Gulf. | Reef animals—Effect of human beings on—Red Sea. | Reef animals—Effect of human beings on—Persian Gulf. | Reef animals—Effect of global warming on—Red Sea. | Reef animals—Effect of global warming on—Persian Gulf.
Classification: LCC QL137 .C67 2024 (print) | LCC QL137 (ebook) | DDC 591.77/890953—dc23/eng/20240313
LC record available at https://lccn.loc.gov/2023058865
LC ebook record available at https://lccn.loc.gov/2023058866

ISBN: 9781032342900 (hbk)
ISBN: 9781032342931 (pbk)
ISBN: 9781003321392 (ebk)

DOI: 10.1201/9781003321392

Typeset in Times
by Apex CoVantage, LLC

Contents

Preface

The Red Sea is a young ocean basin in the process of formation that has a unique environment and biology. On the eastern side of the Arabian Peninsula, the Arabian (or Persian) Gulf is a region of extensive onshore and offshore development, largely associated with hydrocarbon production. Large areas of both the Red Sea and the Gulf are relatively shallow, and their semi-enclosed location with limited contact with the Indian Ocean within an arid region means that they also experience high evaporation rates that result in high temperatures and salinities. All of this has had a far-reaching impact on the marine life and ecology of the region, aspects of which are described in this volume.

A range of topics are examined in this volume, but it is not intended to be a comprehensive description of all aspects of the marine fauna. Nine chapters deal with corals and fish in the Red Sea and the Gulf, while a further ten chapters describe a range of species, from invertebrates such as oysters and sea slugs to crustaceans and plankton. Marine reptiles (turtles) and mammals (dugongs) are also covered in four chapters, including measures for their conservation. The chapters aim to present some of the current thinking and summaries of recent research in each field of study, including useful reference lists for further study.

The contributions of the many technical referees to improving the contents of the chapters are also greatly appreciated.

Editor biographies

Najeeb M.A. Rasul obtained a Ph.D. in Geological Oceanography from the University of Wales, Swansea, U.K. He was with the Faculty of Marine Science, King Abdulaziz University, Jeddah, before joining the Saudi Geological Survey, where he was a Technical Adviser until 2022, and Head of the Center for Marine Geology from 2003 to 2010. He has held key research positions at the Geological and Geophysical Research Systems in Canada, the Challenger Division for Seafloor Processes at the Southampton Oceanography Centre, currently NOC, Southampton, and Environment Canada. He has conducted research cruises with European Union research centres, including the Institute of Marine Sciences (ISMAR), Italy, to the Red Sea and the IOC-UNESCO (TTR programme) in the Atlantic Ocean and Mediterranean Sea. He was a EUROMARGINS and Saudi Scientific Coordinator. He has also been involved in research projects as chief scientist and principal investigator in the Western Basin of Lake Erie, the Indian and Atlantic Oceans, and the Arabian Sea, and currently the Red Sea. His research interests have concentrated on both shallow and deepwater sedimentary processes, depositional mechanisms and environmental aspects.

Ian C.F. Stewart received a Ph.D. in seismology from the University of Adelaide, South Australia. He is an independent geophysical consultant based in Adelaide, and has worked for numerous mining and oil companies in Australia, the Middle East, and elsewhere. From 1974 until 1982, he was with the Department of Earth Sciences at Memorial University of Newfoundland, Canada, where he taught geophysics. He then carried out research for Saudi Aramco in the Exploration Department in Dhahran, Saudi Arabia, from 1982 until 1996, and was in charge of the development and application of potential field methods throughout the Kingdom. Later he was involved in geophysical exploration in Australia. He was the Geophysical Technical Adviser to the Saudi Geological Survey (SGS), Jeddah, between 2003 and 2013, applying geophysical methods for mineral and groundwater exploration in the Kingdom, as well as carrying out earthquake studies. He has published numerous papers in addition to reports for the SGS, Saudi Aramco and other companies.

Contributors

Khadija Y.I. Abaker
Faculty of Marine Science and Fisheries
Red Sea University
Port Sudan, Red Sea, Sudan

Jeff B.R. Abrogueña
Environmental Protection and Control Department
Royal Commission for Jubail and Yanbu, Jazan City for Primary and Downstream Industries
Saudi Arabia

Greta S. Aeby
Department of Biological and Environmental Sciences
College of Arts and Sciences, Qatar University
Doha, Qatar

Sakinah Al-Haddad
Ecosystem-Based Management of Marine Resources Program, Environment and Life Sciences Research Center
Kuwait Institute for Scientific Research
Safat, Kuwait

Manal Al-Kandari
Ecosystem-Based Management of Marine Resources, Environment and Life Sciences Research Center
Kuwait Institute for Scientific Research
Salmiya, Kuwait

Abdulsalam Alkawri
Department of Marine Biology, Hodeidah University
Yemen

Reem AlMealla
Nuwat for Environmental Research and Education
Al Jasrah, Bahrain

Muaadh A. Alnuwairah
Applied Research Center for Environment and Marine Studies, Research and Innovation
King Fahd University of Petroleum & Minerals
Dhahran, Saudi Arabia

Hashim Al-Sayed
Department of Biology, College of Science
University of Bahrain
Kingdom of Bahrain

Abdulaziz Al-Suwailem
King Abdullah University of Science and Technology
Thuwal, Saudi Arabia

Zainab Al-Wazzan
Environment Public Authority
Kuwait

Faiza Al-Yamani
Ecosystem-Based Management of Marine Resources Program, Environment and Life Sciences Research Center
Kuwait Institute for Scientific Research
Salmiya, Kuwait

Ibrahim A. Alzoghiby
Applied Research Center for Environment and Marine Studies, Research and Innovation
King Fahd University of Petroleum and Minerals
Dhahran, Saudi Arabia

Mohamed Asharaf
Applied Research Center for Environment and Marine Studies, Research and Innovation
King Fahd University of Petroleum and Minerals
Dhahran, Saudi Arabia

Yasir Y. Asiri
Applied Research Center for Environment and Marine Studies, Research and Innovation
King Fahd University of Petroleum and Minerals
Dhahran, Saudi Arabia

Radhouane Ben-Hamadou
Department of Biological and Environmental Sciences, College of Arts and Sciences
Qatar University
Doha, Qatar

John A. Burt
Arabian Center for Climate and Environmental Sciences and Water Research Center
New York University at Abu Dhabi
Abu Dhabi, UAE

Susana Carvalho
King Abdullah University of Science and Technology
Red Sea Research Center
Thuwal, Saudi Arabia

Svenja Christiansen
Department of Biosciences
University of Oslo
Norway

Priya B. Das
Applied Research Center for Environment and Marine Studies, Research & Innovation
King Fahd University of Petroleum and Minerals
Dhahran, Saudi Arabia

Alessandro De Maddalena
Marine Sciences
University of Milano-Bicocca
Milan, Italy

Muhammad Y.A. Dosoky
Department of Marine Science, Faculty of Science
Port Said University
Port Said, Egypt

Sheikheldin M. Elamin
Faculty of Marine Science and Fisheries
Red Sea State University
Port Sudan, Red Sea, Sudan

Elfatih B.A. Eltaib
Faculty of Marine Science and Fisheries
Red Sea State University
Port Sudan, Red Sea, Sudan

Saja A. Fakhraldeen
Ecosystem-Based Management of Marine Resources Program, Environment and Life Sciences Research Center
Kuwait Institute for Scientific Research
Safat, Kuwait

Ace V.B. Flandez
Applied Research Center for Environment and Marine Studies, Research and Innovation
King Fahd University of Petroleum and Minerals
Dhahran, Saudi Arabia

Ahmed Ghallab
Egyptian Environmental Affairs Agency, Natural Conservation Sector
Red Sea Protectorates, Egypt

Jinoy V. Gopalan
Applied Research Center for Environment and Marine Studies, Research and Innovation
King Fahd University of Petroleum and Minerals
Dhahran, Saudi Arabia

Nazima Habibi
Biotechnology Program, Environment and Life Sciences Research Center
Kuwait Institute for Scientific Research
Safat, Kuwait

Mahmoud H. Hanafy
Department of Marine Science, Faculty of Science
Suez Canal University
Ismailia, Egypt

Layla J. Hazeem
Department of Biology, College of Science
University of Bahrain
Kingdom of Bahrain

Moritz Heinle
Applied Research Center for Environment and Marine Studies, Research & Innovation
King Fahd University of Petroleum and Minerals
Dhahran, Saudi Arabia

Leanne Hepburn
School of Life Sciences
University of Essex
UK

Javed A. Hussain
Applied Research Center for Environment and Marine Studies, Research & Innovation
King Fahd University of Petroleum and Minerals
Saudi Arabia

Hussein N.M. Hussein
National Institute of Oceanography and Fisheries
Red Sea Branch
Hurghada, Egypt

Amany Ismael
Oceanography Department, Faculty of Science
Alexandria University
Egypt

Thadickal V. Joydas
Applied Research Center for Environment and Marine Studies, Research & Innovation
King Fahd University of Petroleum and Minerals
Dhahran, Saudi Arabia

Stein Kaartvedt
Department of Biosciences
University of Oslo
Norway

Syed A. Khan
Applied Research Center for Environment and Marine Studies, Research & Innovation
King Fahd University of Petroleum and Minerals
Dhahran, Saudi Arabia

Rebecca Klaus
Equipe Cousteau
Paris, France

Diego F. Lozano-Cortés
Environmental Protection Department
Saudi Aramco
Dhahran, Saudi Arabia

Rakhesh Madhusoodhanan
Ecosystem-Based Management of Marine Resources, Environment and Life Sciences Research Center
Kuwait Institute for Scientific Research
Salmiya, Kuwait

Hashem A. M. Madkour
National Institute of Oceanography and Fisheries
Red Sea Branch
Hurghada, Egypt

Aldoushy Mahdy
Zoology Department, Faculty of Science
Al-Azhar University, Assiut Branch
Assiut, Egypt

Rommel H. Maneja
Applied Research Center for Environment and Marine Studies, Research and Innovation
King Fahd University of Petroleum and Minerals
Dhahran, Saudi Arabia

Karuppasamy P. Manikandan
Applied Research Center for Environment and Marine Studies, Research & Innovation
King Fahd University of Petroleum and Minerals
Dhahran, Saudi Arabia

Seerangan Manokaran
Applied Research Center for Environment and Marine Studies, Research & Innovation
King Fahd University of Petroleum and Minerals
Dhahran, Saudi Arabia

Gopikrishna Mantha
Ecosystem-Based Management of Marine Resources, Environment and Life Sciences Research Center
Kuwait Institute for Scientific Research
Salmiya, Kuwait

Jeffrey D. Miller
Biological Research and Education Consultants
Missoula
Montana, USA

Amin R. Mohamed
Biology Program
New York University
Abu Dhabi, Saadiyat Island, Abu Dhabi, United Arab Emirates

Ashraf S. Mohammad
National Institute of Oceanography and Fisheries
Red Sea Branch
Hurghada, Egypt

Even Moland
Institute of Marine Research
Nordnes
Bergen, Norway

Kirsty L. Nash
Centre for Marine Socioecology
University of Tasmania
Hobart, Australia

Portia J. Nillos-Kleiven
Institute of Marine Research
Nordnes, Bergen, Norway

Manickam Nithyanandan
Ecosystem-Based Management of Marine Resources, Environment and Life Sciences Research Center
Kuwait Institute for Scientific Research
Salmiya, Kuwait

Alaa Osman
Zoology Department, Faculty of Science
Al-Azhar University, Assiut Branch
Assiut, Egypt

Mauricio L. Padierna
Department of Biological and Environmental Sciences, College of Arts and Sciences
Qatar University
Doha, Qatar

Shamseer Pavithratha
Applied Research Center for Environment and Marine Studies, Research & Innovation
King Fahd University of Petroleum and Minerals
Dhahran, Saudi Arabia

John K. Pearman
Molecular and Algal Ecology Group
Cawthron Institute
Nelson, New Zealand

Igor Polikarpov
Ecosystem-Based Management of Marine Resources Program, Environment and Life Sciences Research Center
Kuwait Institute for Scientific Research
Salmiya, Kuwait

Ali Qasem
Environmental Protection Department
Saudi Aramco
Dhahran, Saudi Arabia

Mohammed A. Qurban
National Center for Wildlife
Riyadh, Saudi Arabia

Kolchar M. Rajaneesh
Applied Research Center for Environment and Marine Studies, Research & Innovation
King Fahd University of Petroleum and Minerals
Saudi Arabia

Madhusoodhanan Rakhesh
Ecosystem-Based Management of Marine Resources Programme, Environment, and Life Sciences Research Center
Kuwait Institute for Scientific Research
Salmiya, Kuwait

Pedro A.G. Range
Environmental Science Center
Qatar University
Doha, Qatar

Najeeb M.A. Rasul
Geological and Geophysical Research Systems
Mississauga, Ontario, Canada

Bernhard Riegl
Department of Marine and Environmental Sciences, Halmos College of Natural Sciences and Oceanography
Nova Southeastern University
Dania Beach, Florida, USA

Anders Røstad
King Abdullah University of Science and Technology
Saudi Arabia

Maria Saburova
Ecosystem-Based Management of Marine Resources Program, Environment and Life Sciences Research Center
Kuwait Institute for Scientific Research
Salmiya, Kuwait

Mahsa Alidoost Salimi
School of Geography, Earth and Atmospheric Sciences, Faculty of Science
The University of Melbourne
Victoria, Australia

Ahmed M. Shawky
Technical Office of the Minister of the Environment
30 Misr Helwan El-Zyrae Road
Maadi, Cairo, Egypt

Ian C.F. Stewart
Stewart Geophysical Consultants Pty. Ltd.
Adelaide, South Australia, Australia

Fathima Thuslim
Ecosystem-Based Management of Marine Resources, Environment and Life Sciences Research Center
Kuwait Institute for Scientific Research
Salmiya, Kuwait

Felipe De Oliveira Torquato
Department of Biological and Environmental Sciences, College of Arts and Sciences
Qatar University
Doha, Qatar

Peter J. Vine
Earth and Ocean Sciences, NUIG
University of Galway
Ireland

Nathalie Yonow
Department of Biosciences
Swansea University
Singleton Park, Swansea, Wales, United Kingdom

Mohamed Youssef
Department of Geology and Geophysics, College of Science
King Saud University
Riyadh, Saudi Arabia

Mohamed Yusuf
College of Science
University of Bahrain
Kingdom of Bahrain

Introduction to coral reefs and associated marine fauna around the Arabian Peninsula

1

Najeeb M.A. Rasul, Ian C.F. Stewart and Peter J. Vine

1.1 INTRODUCTION

The Red Sea and Arabian (or Persian) Gulf have unique environments and biology. Due to their semi-enclosed nature and locations within an arid region, their environments are affected by high evaporation rates that, together with limited contact with the Indian Ocean, result in high temperatures and salinities. In the Red Sea, lower sea levels in the past have also led to extensive evaporite deposition within its basin. All of this, together with anthropogenic effects due to rapid development in the area, has had a far-reaching impact on the marine and terrestrial life of the region and on its human inhabitants.

The Red Sea is an active spreading zone in the early stages of forming an ocean basin. It is about 2000 km long, with a surface area of 458,620 km^2 and a maximum width of 355 km. The narrow axial trough with depths greater than 1000 m underlies about 15% of the Red Sea, while depths of up to about 3000 m occur in the "Deeps"; the shelves are much shallower, with about 40% of the Red Sea area having depths of less than 100 m (Figure 1.1). Details of the Red Sea morphology and environment are given by Rasul and Stewart (2015, and the chapters and many references therein) and will not be repeated here.

On the eastern side of the Arabian Peninsula, the relatively shallow Arabian (or Persian) Gulf (PAG) covers about 251,00 km^2, is 989 km long, and is quite shallow, with a maximum depth of over 90 m, and an average depth of only 50 m. The Gulf lies over a stable Phanerozoic sedimentary platform, and the deepest sections lie towards the east of the Gulf due to subduction of the Arabian plate under the Eurasian Plate, forming the Zagros thrust and fold belt in Iran. The thick sedimentary column is host to extensive evaporite deposits and major hydrocarbon reservoirs, the exploitation of which has affected the biology and environment of the region. At the northwestern end, the Euphrates and the Tigris rivers converge in the delta of the Shatt al-Arab, which carries the waters of these rivers. The maximum width of the Gulf is 340 km whilst its narrowest point (56 km wide) is at the Strait of Hormuz, where the Gulf joins the Indian Ocean (Figure 1.2).

'Uniqueness' and 'biodiversity' are recurring words employed to describe the Red Sea and its neighbouring basin, the Arabian/Persian Gulf. They are justifiable adjectives from many different perspectives and hint at the fascinating features that attract the attention of scientists from a wide variety of disciplines. Competition, photosynthesis and symbiosis are key elements upon which healthy reefs develop, and the more we learn, the more complex and incredible the story becomes. It gives resonance to the familiar adage, 'variety is the spice of life', and these ocean basins represent two of the most biodiverse regions of the world, hosting a vast array of marine species.

The present volume follows on from the earlier volumes edited by Rasul and Stewart (2015, 2019) that dealt exclusively with the Red Sea, whereas here chapters are also included on topics related to the Persian-Arabian Gulf (PAG).

1.2 THIS VOLUME

Brief summaries of the various topics covered in this volume are presented in the following sections.

1.2.1 Coral biology

Corals illustrate the biodiversity, with over 400 recorded species in the Red Sea and more being discovered on a regular basis, especially within the reef framework, described in these pages as the coral reef 'cryptobiome' (Carvalho and Pearman). Such studies provide the foundations for integrated studies of coral reef communities, from both spatial and temporal perspectives, and this volume adds to our understanding of the complex interactions that make the PAG and Red Sea so rich in habitats and species. Several reef surveys are described as a basis for establishing baseline data that can be used to monitor

DOI: 10.1201/9781003321392-1

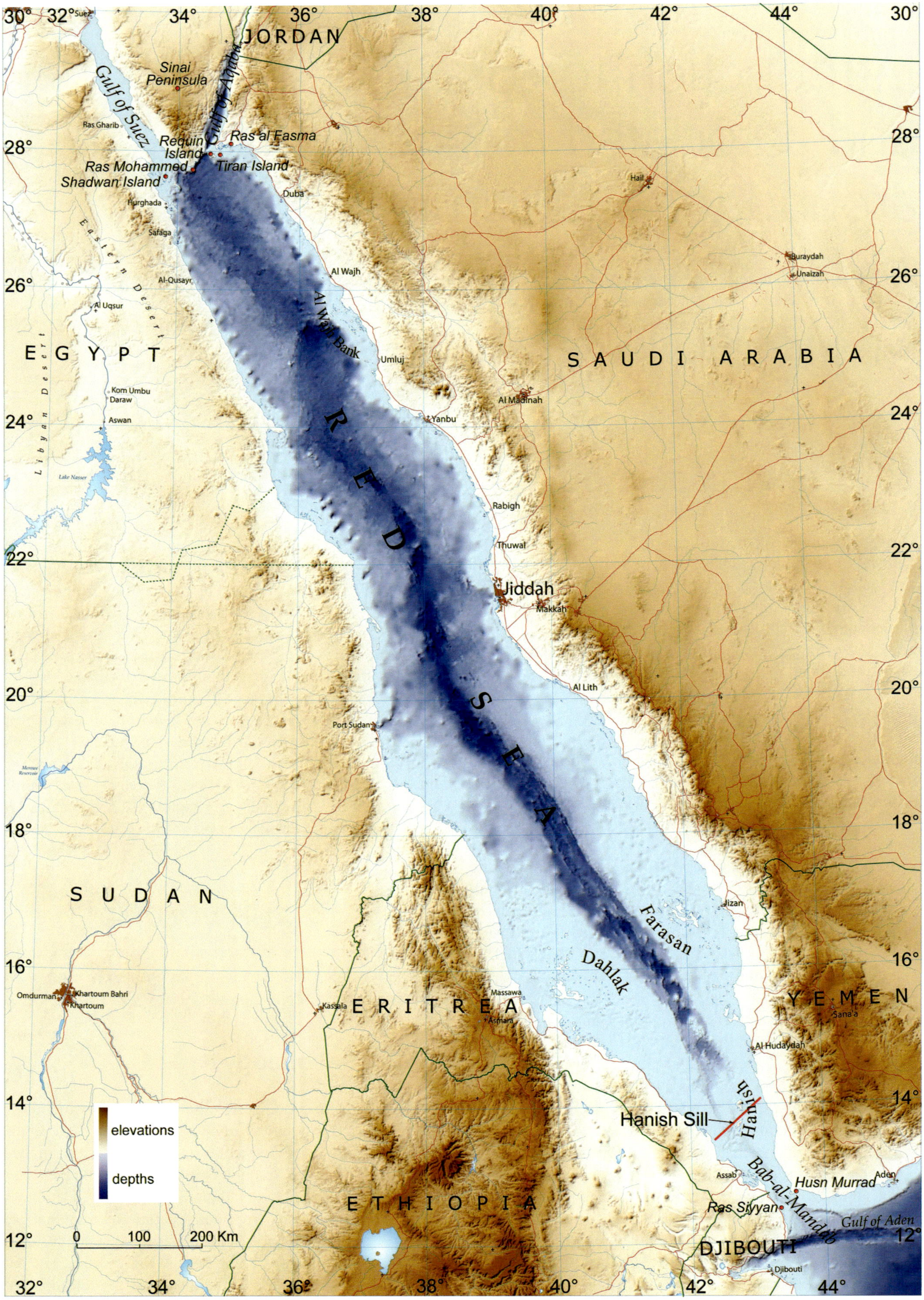

FIGURE 1.1 Map of the Red Sea area; darker colours indicate greater depths or higher elevations (after Rasul et al. 2015).

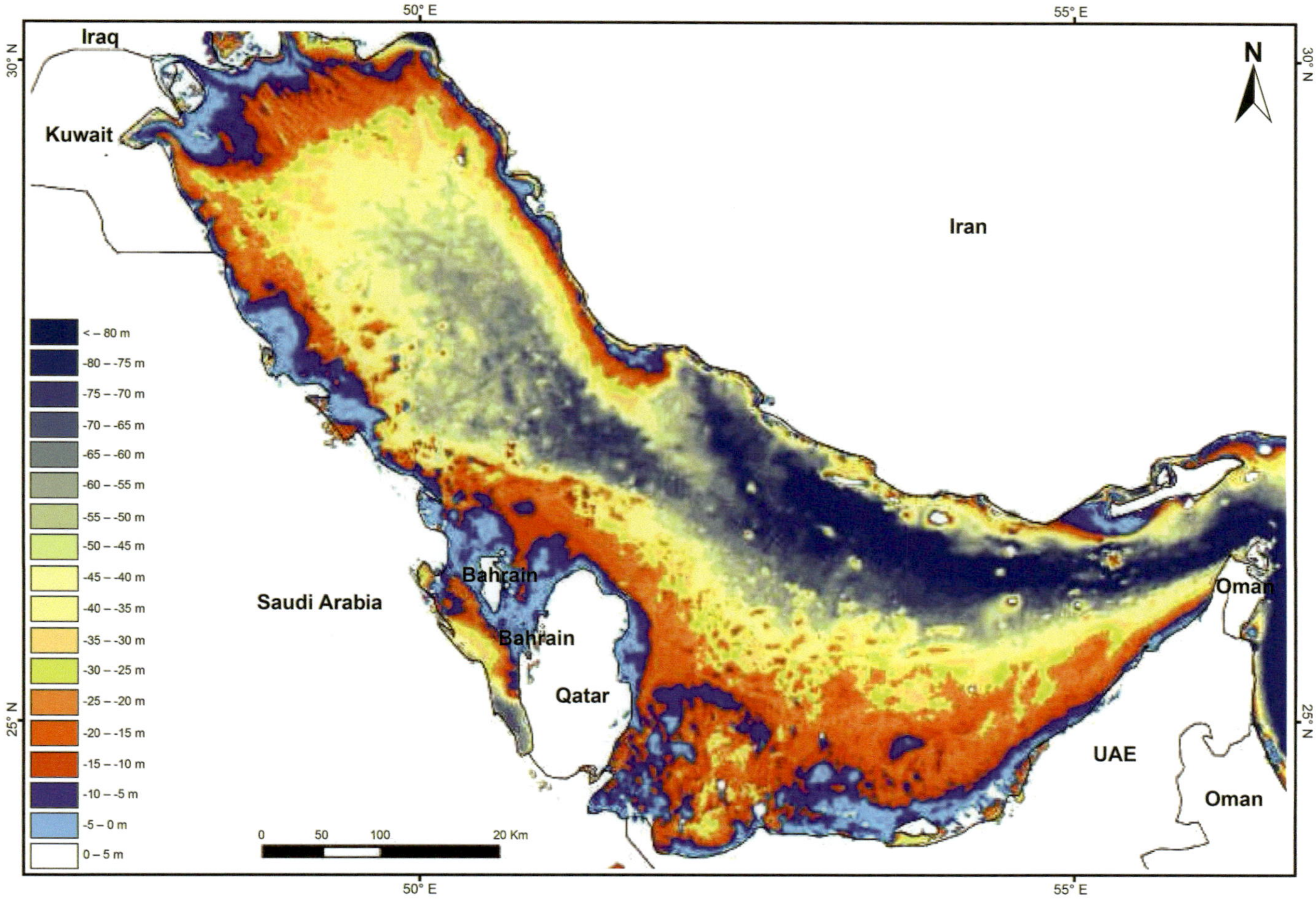

FIGURE 1.2 Arabian (Persian) Gulf (PAG) bathymetry map (after Hereher 2020). Depths are in metres.

the reefs in the future. Twenty-five reefs in the Sudanese Red Sea (Abaker et al.) with a mean hard coral cover of 35% lost up to 40% of live corals to bleaching in 2016 and 28% in 2017.

Our knowledge of the latter has been further developed by a survey of 19 Egyptian sites, demonstrating an increase in reef building substrata (corals, sponges, soft corals and algae) south of Hurghada (Ghallab et al.). Meanwhile, PAG reef surveys published in this volume are welcome additions to our understanding of the marine habitats of Kuwait and Bahrain – two areas that have been less well known than the Red Sea but are nevertheless of great importance, both commercially and biologically (AlMealla and Hepburn). Reefs have undergone severe degradation in the last four decades but signs of recovery documented on some PAG reefs may support a more positive outlook.

Several authors remind us of the impact of coral bleaching (Aeby et al.) and the unexpected resilience of some Red Sea and PAG corals in the face of extremes of temperature that would elsewhere trigger bleaching. Given the importance of corals to the wellbeing of marine resources throughout the world, there is great interest in how PAG conditions influence coral bleaching and disease. The impact of ocean warming and the level of thermal stress experienced by coral reefs along the Egyptian coast of the Red Sea are also investigated by Hanafy et al. Despite summer sea surface temperature anomalies occurring in the north, conditions may still be within the normal range of thermotolerance, whereas in the south they could be close to the bleaching threshold.

At the heart of these studies is a focus on the organisms that inhabit the cryptic spaces within the reef framework, known as the cryptobiome. Knowledge of these assemblages is in its infancy, as they have traditionally been overlooked in reef assessments. The coral holobiont encompasses the sum total of biological elements, such as microalgae and diseases, that form living corals. At a micro-level the territorial waters of Kuwait provide a unique niche for microbial communities given the environmental extremes to which they are exposed. Shotgun metagenomic sequencing was employed by Fakhraldeen and colleagues to identify key bacterial communities and their functions.

1.2.2 Fish

There are more than 1200 fish species in the Red Sea and more than 750 in the PAG. Of these, at least 165 are endemics (only found in the Red Sea) whilst a further 51 are restricted to the Red Sea and Gulf of Aden. This abundance of wildlife is repeated in most of the animal groups.

Fish larvae and their distribution in the western PAG are covered in terms of both spatial and temporal patterns, especially in relation to water column characteristics. A study undertaken in Saudi Arabian waters revealed 62 families, predominantly Gobiidae, Engraulidae and Clupeidae. Water temperature, salinity, dissolved oxygen and chlorophyl α were crucial background variables explaining the abundance and distribution patterns of fish larvae. Fisheries, including those for pearl oysters in Bahrain and other PAG waters, were a valuable cultural heritage. It is noted that Red Sea mesopelagic fishes spawn throughout the year, though apparently with minima in mid-summer. Typically occupying the twilight zone at around 200 m to 1000 m depth, the occupants of these waters make relatively brief visits to the surface at night, especially around dawn and dusk.

Shawky has continued his work of identifying important wildlife species with a new study on oceanic white-tipped sharks in the Egyptian Red Sea. *Carcharhinus longimanus* is one of the top predators and main ecotourism attractions for Red Sea divers. Individual sharks can be identified by markings on their first dorsal fin. Recording variations by underwater photography techniques is a valuable tool to determine data such as migratory patterns, and a photo-ID project at Elphinstone Reef in the Egyptian Red Sea has identified 57 individuals, of whom 50 were females. The research may help to avoid shark attacks on tourists and once again underlines the urgency of preserving the local population of this species, which is threatened by overfishing throughout its range.

1.2.3 Invertebrates

Yonow's contributions on sea slugs of the Red Sea and a co-authored review (led by Manickam Nithyanandan) on those from the PAG include the latest information on cryptic species, adding to records of both non-endemic and endemics while demonstrating how a more holistic overview of the reef system enables valuable data upon which to base faunal lists. In the case of heterobranchs of the PAG, the current study depends on published works, specimens examined and photographs from the region to delineate their taxonomic distinctness and zoogeographic affinities. One hundred and forty species of heterobranch molluscs were recorded from the PAG, of which 62 are new records.

Pearling in the PAG is also covered up to the 1930s, when Bahrain and the surrounding area relied heavily on the pearl trade, and Bahrain seas produced pearls of a higher quality than those found in other areas. The history and economy of the pearl oyster trade in the Kingdom of Bahrain are reviewed here.

1.2.4 Benthos

One hundred and thirty-five species from 28 diverse benthic crustacean taxa were found, comprising 60 species of amphipods, 37 species of cumaceans, 14 species of ostracods, and 24 others including mysid, nebaliacea, stomatopoda, euphausiacea, stenopodidea, penaeidea, caridea, scyllaridea, nephropsidea, thalassinidea, paguridea, hippidea, brachyura, tanaidacea, isopoda, hyperiidea, caprellidea, cladocera, copepoda, cirripedia, mystacocarida and pycnogonida. Amphipods were the dominant taxa, constituting 44% of the total density and 41% of the total crustacean species. There are around 200 documented crab species in the PAG, inhabiting a wide range of different coastal habitats. Al Wazzan summarises the available information about the biology and ecology of the brachyuran crabs in the PAG.

Meiofauna are an important metazoan assemblage of the marine realm, defined by their body size (those organisms which pass through 1000 μm and are retained in 42 μm mesh sieves). They are the most diversified and abundant microscopic faunal groups of marine biota, which accommodate around 24 of the 35 animal phyla present today (five of which are exclusively meiobenthic, i.e., Gastrotricha, Gnathostomulida, Kinorhyncha, Loricifera and Tardigrada).

1.2.5 Bacteria, algae and phytoplankton

A total of 490 phytoplankton species were identified, of which more than 65% were diatoms and nearly 30% dinoflagellates. There were peaks during early summer and troughs during late fall. The frequent occurrence of toxic blooms in both PAG and Red Sea waters can lead to mass mortalities that have serious ecological consequences. Microalgae and phytoplankton blooms may be both beneficial and damaging to the reefs, depending on their context. Ninety potentially harmful planktonic and benthic species were recorded from the entire Red Sea, causing severe fish mortalities and numerous ecological impacts in the region.

Mention is made of *Trichodesmium erythraeum*, which blooms in spring and early summer, causing the surface to turn red and, according to some observers, providing a justification for the 'Red' title of this basin. *Alexandrium monilatum* was also noted during summer in offshore waters. Temperature and nutrient concentrations significantly influenced the seasonal variation in the phytoplankton in the PAG and the Red Sea.

1.2.6 Marine turtles

Both the PAG and Red Sea have important breeding and nesting sites for marine turtles – primarily hawksbill and green turtles. On Zabargad Island in southern Egypt, a sandy beach 2.5 km long is used by green turtles as one of the most suitable and largest areas for turtle nesting in Egypt. Ecological studies of sea turtles add to our knowledge of these important species that come to lay their eggs on many other remote beaches along the coast and at low-lying islands.

1.2.7 Marine mammals

Despite their large size, and because of their shyness, nocturnal behaviour and scattered distribution, Red Sea dugongs remain something of a mystery. One way of gathering data on existing populations is to measure and monitor their feeding trails among the seagrass beds. Ahmed M. Shawky has made something of a speciality of this art and in so doing has revealed valuable new information on these charismatic creatures. First and foremost, feeding trails are relatively easy to identify among the seagrass beds upon which the dugongs feed. Dimensions of feeding trails, grazing rates, frequency of feeding trails and estimates of local populations were all possible using these methods – a valuable basis for conservation planning.

REFERENCES

Hereher, M.E. 2020. Assessment of climate change impacts on sea surface temperatures and sea level rise – The Arabian Gulf. *Climate* 8: 50. doi:10.3390/cli8040050.

Rasul, N.M.A., and I.C.F. Stewart (Eds.). 2015. *The Red Sea: The formation, morphology, oceanography and environment of a young ocean basin*, 633. Berlin Heidelberg: Springer Earth System Sciences. doi:10.1007/978-3-662-45201-1.

Rasul, N.M.A., and I.C.F. Stewart (Eds.). 2019. *Oceanographic and biological aspects of the Red Sea*, 550. Cham, Switzerland: Springer Nature. doi:10.1007/978-3-319-99417-8.

Rasul, N.M.A., Stewart, I.C.F., and Z.A. Nawab. 2015. Introduction to the Red Sea: Its origin, structure and environment. In: *The Red Sea: The formation, morphology, oceanography and environment of a young ocean basin*. Eds: N.M.A. Rasul, and I.C.F. Stewart, 1–28. Berlin Heidelberg: Springer Earth System Sciences. doi:10.1007/978-3-662-45201-1_1.

PART 1

Coral biology

The Red Sea coral reef cryptobiome

Spatial and temporal patterns

2

Susana Carvalho and John K. Pearman

2.1 INTRODUCTION

Coral reefs are essential sources of goods and services for human well-being (Morrison et al. 2019; Woodhead et al. 2019) but are also one of the biomes most affected by human pressures (Carpenter et al. 2008; Hoegh-Guldberg and Bruno 2010). Coral reefs worldwide are experiencing rapid degradation (Hughes et al. 2017b) due to a combination of direct and indirect effects resulting from local pressures (e.g., overfishing, eutrophication from agriculture and land development, tourism) (Hughes 1994; Jackson et al. 2001; Pandolfi et al. 2003), combined with the impacts from climate change (Hughes et al. 2003; Ateweberhan et al. 2013; Cheal et al. 2017; Magel et al. 2020).

Coral reefs are also well known as the most diverse marine ecosystems – the rainforests of the sea – yet their biodiversity is severely underestimated (Fisher et al. 2015). Traditionally, investigations of changes in reef biodiversity and biological composition have focused on large, conspicuous, and sometimes charismatic components of the fauna and flora, such as corals, sponges, macroalgae, and fish. Most of the reef biodiversity, however, inhabits small cryptic spaces provided by an intricate network of habitats and ecological niches within the three-dimensional framework of the coral reef. This structural complexity and the resulting array of niches that can vary considerably even within a reef outcrop undoubtedly contributes to the well-known high levels of biodiversity observed. However, this structural complexity also makes this biodiversity hard to sample in a non-destructive manner. As a result, it is estimated that approximately 74% of the species inhabiting coral reefs are yet to be discovered (Fisher et al. 2015).

Unfortunately, our understanding of the effects of multiple stressors on reef biodiversity is also very limited (Ellis et al. 2019), and many species may go extinct even before being discovered. Indeed, our knowledge of the most diverse biological component inhabiting coral reefs – the reef cryptobiome – is in its infancy compared to what we already know for corals and fish. The term cryptobiome was recently defined by Carvalho et al. (2019) to describe cryptic micro-, meio- and macro-organisms that reside within the three-dimensional structure of the reef, including those associated with coral colonies, sponges, or other foundation species. These small-size organisms are completely overlooked during traditional biodiversity reef surveys (e.g., photo transect and visual censuses); however, they most likely play critical roles in reef functioning (Brandl et al. 2019). A large proportion of cryptobenthic organisms are cavity-dwelling suspension feeders, including ascidians, sponges, and bryozoans that can play an important role in organic carbon (C) and nitrogen (N) (re)-cycling. Richter et al. (2001) showed that there was a high density of sponges in the cryptic spaces of coral reefs that were able to filter more than 60% of the phytoplankton passing through the crevices. Sponges are also important in the uptake of the dissolved organic matter (DOM) that has originated from coral reef primary producers, thereby preventing energy and nutrient loss to the open ocean (de Goeij et al. 2013; Rix et al. 2017). The uptake of DOM by sponges is the base of the sponge loop with subsequent release of particulate organic material which is available to detritivores, enabling the transfer of energy and nutrients to higher trophic levels within the coral reef (de Goeij et al. 2013; Rix et al. 2017). Cryptobenthic fish also contribute significantly to reef biodiversity (approximately 40% of reef fish biodiversity) and the trophodynamics of the reef ecosystem (Brandl et al. 2019).

2.2 AUTONOMOUS REEF MONITORING STRUCTURE – ARMS

Autonomous Reef Monitoring Structures (ARMS) emerged due to a need to discover and quantify the biodiversity of the organisms hidden in the reef matrix that due to their nature, behaviour, and size were not included in traditional benthic

DOI: 10.1201/9781003321392-3

surveys. ARMS resulted from a joint effort of researchers from different institutions such as the Smithsonian, National Oceanic and Atmospheric Administration (NOAA), and the Australian Institute of Marine Science during the Census of Marine Life initiative (www.oceanarms.org) that has now developed into a global program, with over 1500 ARMS having been deployed worldwide.

Each ARMS unit is a cube-like structure consisting of nine layers of grey Type I PVC plates (approximately 23 cm wide) with one inch (2.5 cm) of space between them, placed on a larger rectangular PVC base (Figure 2.1). Bars of PVC block the water flow between alternate PVC layers to provide different conditions for species preferring low water flow, whereas the open layers allow for water to freely flow, simulating the variability in natural conditions. The design of an ARMS intends to partially mimic the 3D structure of a reef (Zimmerman and Martin 2004; Leray and Knowlton 2015) while allowing for a non-destructive and quantitative assessment of both the sessile components growing on the plates as well as the mobile organisms that move into the spaces. Even though it was not initially incorporated into the concept of an ARMS unit, a research team at the King Abdullah University of Science and Technology has equipped one of the ARMS units deployed per reef with seawater temperature sensors, allowing for a continuous record of seawater temperature changes over the colonization period. This information may be very relevant, particularly in remote areas, especially considering that recent studies have identified a mismatch between *in situ* and satellite-derived seawater temperature products (DeCarlo et al. 2021; Steyaert et al. 2022).

The standardized nature of ARMS and the protocols for their assessment ensure that ARMS samples are comparable across multiple user groups (Ransome et al. 2017) and provide an adequate tool to make comparisons across a variety of factors irrespective of the benthic reef community nearby. This represents an advantage in comparison to methodologies previously used to investigate the cryptobiome such as coral deadheads (Plaisance et al. 2011). It should be noted that as ARMS are artificial structures, they may not entirely represent the natural communities within a coral reef, as observed previously for other artificial substrates in temperate reefs (Carvalho et al. 2013). Nevertheless, comparisons with coral deadheads showed the same average similarity as between deadhead communities (Plaisance et al. 2011). Further studies showed similar microphytobenthic communities on ARMS compared to natural substrates (Pennesi and Danovaro 2017). Thus, the standardized, non-destructive, and mobile nature of ARMS as a sampling tool offers a reproducible framework for long-term temporal monitoring of cryptobenthic reef assemblages, by retrieving and deploying new ARMS units at the same site on a reef over consistent colonization periods.

As part of the standardization of protocols (both regarding sampling and processing), the development team suggested that at least three replicates would be deployed per reef 2–5 m apart along the 10 m depth contour, and for periods ranging from 1 to 3 years of colonization, depending on the colonization rate of a certain area or specificities of a research project (Pearman et al. 2019). As the use of ARMS has increased, flexibility has been incorporated into the deployment period to account for increased differences in colonization rates, such as along a depth profile from shallow water to mesophotic reefs (Luiz A. Rocha, pers. comm.). Due to the time period of deployment, ARMS are of limited use for detecting short-term (e.g., seasonal) changes in the reef-associated communities. Although short-term submersions are possible, the associated costs (fabrication, processing) mean other alternatives such as settlement tiles might be preferable (Roth et al. 2020).

ARMS were designed to assess the eukaryotic metazoan diversity of the cryptobenthos, and an ARMS unit generates different samples, ranging from the sessile community that is scraped from all the PVC plates (except the base) to the large mobile organisms (>2 mm) (Figure 2.1). The mobile components of the cryptobenthic communities are fractionated based on their size using a sequence of three screens: 0.125 mm, 0.5 mm, and 2 mm. The ARMS can also be used for investigating compositional changes of other biological components such as microalgae, Fungi, and prokaryotes that are associated with biofilms in the reef matrix (Pearman et al. 2019). To our knowledge, only one study has investigated the reef-associated bacteria using ARMS (Pearman et al. 2019), while Pennesi and Danovaro (2017) have assessed whether microphytobenthos (and in particular diatoms) colonizing these artificial substrates could be used as a tool for detecting environmental changes. Both studies highlight the potential of ARMS as a non-destructive and standardized tool to provide more comprehensive information across a broader range of taxonomic groups that play vital roles in the functioning of reef communities (Roth et al. 2020, 2021).

The sessile component as well as the >2 mm fraction are ideally analysed by a combination of morphology and molecular methods, while the two smaller mobile fractions are analysed using high-throughput amplicon sequencing (i.e., metabarcoding). Morphology-based taxonomic assessments are time-consuming, present low biodiversity coverage, are highly dependent on the availability of expert taxonomists, are mainly focused on adults, often fail to distinguish closely related species, and are particularly challenging in regions where sound taxonomic studies are scarce. These constraints can hinder the advance of knowledge of biodiversity at multiple spatial and temporal scales. Indeed, the comparison between ARMS and traditional surveys showed an increased biodiversity depth provided by the inclusion of ARMS in reef biodiversity assessments (Pearman et al. 2016). ARMS provide an ideal platform, especially for the >2 mm organisms which can be picked individually, for the continued collaboration between taxonomists and molecular scientists to improve reference databases. Taxonomic classifications rely on comparisons to reference databases containing sequences from identified species. These databases (such as SILVA, BOLD and NCBI nr) are generally biased towards species of interest or more frequently investigated habitats (Carugati et al. 2015). Small cryptic species, from relatively understudied areas, such as the Red Sea, are often under-represented in these databases, making species-level classifications harder to achieve. Thus, collaboration between taxonomists and molecular scientists and efforts such as BIOSCAN are resulting in the enlargement of these

FIGURE 2.1 Schematic representation of the processing workflow of an ARMS unit. ARMS are positioned on the reef for 2–3 years before being retrieved. The cryptobiome is divided into a sessile fraction, which is attached to the ARMS plates and three different mobile size fractions. These components are then assessed using molecular techniques to assess the structure and composition of the cryptobiome. Photo by Morgan Bennett-Smith.

databases and thus improved taxonomical classifications (Hobern 2021).

With the constraints of morphological methods and improvements in molecular technologies, the use of environmental DNA has become increasingly prevalent over the last decade (Taberlet et al. 2018). High-throughput sequencing technologies result in large numbers of sequences and enable in-depth analysis of the species present in a habitat and are touted as being a cost-effective approach to assessing entire biological communities. These molecular approaches are especially beneficial for the smaller size fractions that are present within ARMS that would be difficult to morphologically identify and can also distinguish between juvenile organisms (e.g., fish larvae), which are often morphologically similar.

Molecular methods generally reveal a higher diversity than morphological methods (Pearman et al. 2016; Cahill et al. 2018) and highlight the potential of molecular approaches to reveal unrecognized diversity. It should, however, be noted that molecular approaches have the potential to overinflate diversity due to sequencing errors, the presence of pseudogenes, and the inherent variability within the species for the selected genes. Bioinformatic approaches have either clustered sequencing reads into operational taxonomic units (OTUs) based on a defined similarity threshold or more recently have inferred amplicon sequence variants (ASVs) via the correction of erroneous sequences (Callahan et al. 2016). Both of these approaches have consequences for identifying species diversity. The use of a defined similarity threshold is problematic as different taxa are known to have different thresholds for determining species, and this can change for different genes. Thus, using an arbitrary cutoff for clustering can either clump distinct species together and result in an underestimation of the diversity or split species apart causing an overestimation. On the other hand, ASVs can refine differences down to a single nucleotide difference and are thus likely to cause an overestimation of diversity as intra-species variation is determined as a unique ASV. Pseudogenes, non-functional duplications of functional genes, also have the potential to inflate diversity. For example, work on the snapping shrimp genus *Alpheus* showed several copies of the cytochrome oxidase I (COI) gene in genomic DNA, but only a single copy in the cDNA as non-functional copies are not transcribed (Williams and Knowlton 2001). These duplications thus have the potential to inflate the assessment of diversity, although methods which identify frameshifts, indels, and stop codons have been used to minimize the impacts of pseudogenes on diversity (Pearman et al. 2018). To identify a broad range of taxa morphologically, a variety of experts are required, which is time consuming and expensive. In molecular assessments, genes that are present in a wide variety of taxa can be selected (e.g., 18S rRNA and mitochondrial cytochrome oxidase I (COI)) and primers designed to amplify this broad set of taxa. However, it should be noted that primers are not universal and can fail to amplify specific taxa. To mitigate this issue, it has been suggested that multiple primer sets are used to ascertain the full diversity of taxa present in the community. It has been shown that the use of 18S rRNA and mt COI genes gave complementary information on the taxa present in the Red Sea cryptobiome (Pearman et al. 2018) as well as in Indonesia (Casey et al. 2021).

2.3 THE RED SEA CRYPTOBIOME

The Red Sea cryptobiome incorporates a plethora of small-sized organisms of varied Phyla (Carvalho et al. 2019). The first studies of cryptic organisms in the Red Sea focused on the prevalence of suspension feeders in coral reefs (Richter and Wunsch 1999; Wunsch et al. 2000; Richter et al. 2001). More recent research has focused on the use of artificial surfaces such as ARMS to investigate the cryptobiome community in the Red Sea due to their non-destructive nature and standard design. ARMS units were deployed across 22 reefs along the Saudi Arabian Red Sea between 2013 and 2017 (for periods of two years), with Carvalho et al. (2019) observing a phylogenetic-rich community with 20 Phyla, of which 14 were metazoans. To date, 30 metazoan Phyla have been reported as inhabiting coral reefs (Paulay 1997). Although the Red Sea coral reefs may not contain these metazoan Phyla, the disparity in our results may partially be related to the non-representation in the online DNA databases. Of the 16 Phyla that have been reported in coral reefs but were missing from the Saudi Arabian Red Sea ARMS, 11 do not have representation in the BOLD or NCBI databases of DNA. It is worth noting that 29% of the sequences could not be assigned to any eukaryotic phylum, representing 45% of the OTUs observed, highlighting the urgency of improving DNA databases by incorporating sequences of more species into the reference databases. This will require the collaboration of taxonomists (to properly identify and describe novel species) and molecular biologists. It is especially important for the cryptobiome of the Red Sea coral reefs, which has until recently received scant research attention compared with other regions (such as the Mediterranean), resulting in the fauna being under-represented in reference databases. This is further highlighted by the fact that classification improved seven-fold for organisms in ARMS in French Polynesia when a geographically local reference database was used (Ransome et al. 2017). Steps have been made to address this issue with the barcoding of the >2 mm fraction of the ARMS, with thousands of sequences recently being deposited in reference databases (Villalobos 2021).

Arthropoda and Annelida made up a great proportion of the cryptobiome, followed by Mollusca, which also contributed considerably to the diversity (Figure 2.2). Chordata were also prevalent in the ARMS communities. Although minor contributors when compared to the previous ones, Echinodermata, Porifera, Chlorophyta, Bryozoa, and Cnidaria were also consistently observed in the samples. Overall, Arthropoda were large contributors in both the sessile (26%) and the mobile (41%) assemblages. Porifera (34%) was clear in the sessile fraction, while annelids (24%) were major contributors to the mobile counterpart. In terms of OTUs, the study by Carvalho et al. (2019) reported a comparatively higher richness of the mobile component of the reef cryptobiome in comparison to the sessile fraction, with the former being twice as rich as the latter.

While the use of standardized protocols, such as ARMS, has improved the knowledge of the cryptobiome in the Red Sea, further research should be complemented with assessments characterizing the surrounding benthic habitats. Cryptobenthic species often develop symbiotic relationships

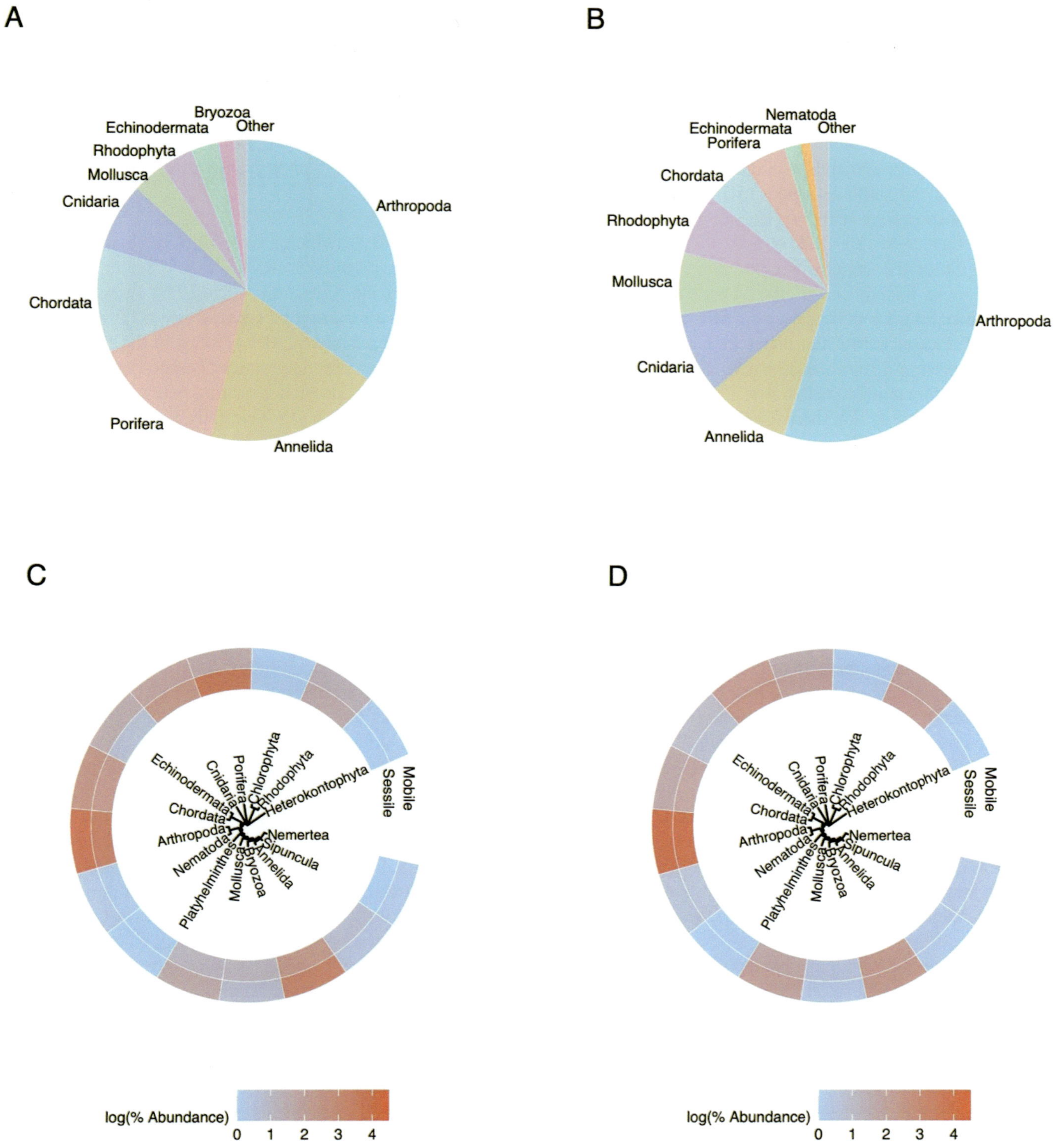

FIGURE 2.2 Overall composition and structure of the ARMS-related eukaryotic cryptobiome in the Red Sea. A) Relative abundance of eukaryotic phyla; B) Proportion of ASVs belonging to each phylum; C) Relative abundance of phyla observed in the mobile and sessile fractions of the ARMS; D) Proportion of ASVs belonging to each phylum in the different fractions.

with foundation species in coral reefs (e.g., corals, sponges, algae), but the effect of these associations on cryptobenthic colonization and succession is poorly understood (Roth et al. 2020). A pilot study is currently ongoing off Thuwal (central Red Sea) comparing the cryptobenthic communities associated with contrasting habitats (hard-coral dominated, algae-pavement dominated, rubble) at the pioneer (7 months) and mature (24 months) stages. This project will help shed light into the potential biodiversity and functional changes associated with changes in ecosystem state. The close associations between cryptobenthic organisms and the benthic habitat may magnify the shifts in community structure even if biodiversity increases (Timmers et al. 2021). Indeed, increases in biodiversity and density may not indicate better reef health, as previous studies have found that the number of species and abundance of cryptobenthic species tend to increase in coral rubble compared to live coral habitats (Enochs and Manzello 2012).

2.4 SMALL-SCALE PATTERNS OF DISTRIBUTION IN THE RED SEA

In the Gulf of Aqaba, distinct gradients were noted in the composition of the cryptobiome, with coralline algae being more abundant near the sunlit entrances to crevices, while sponges abounded in posterior sections (Kobluk 1988; Wunsch et al. 2000; Richter et al. 2001). David et al. (2019) used photo identification of the organisms colonizing the ARMS plates to show these differences in the abundance of taxonomic groups, including Annelida, Bryozoa, Mollusca, Porifera, and colonial Tunicata, across the different plates comprising an ARMS. There are differences in the light exposure, predation exposure, sedimentation, and water flow amongst the different plates, suggesting that there are distinct micro-niches that specific taxa occupy. This highlights the diversity of niches prevalent within coral reefs and that to get a more informed representation of the diversity of coral reefs, a wide diversity of these niches needs to be sampled.

On a within-reef scale in the Gulf of Aqaba, Al-Rshaidat et al. (2016) showed that replicates within a site were more similar in both the sessile and larger (0.5–2 mm) mobile size fraction of eukaryotes compared to those replicates of a site approximately 200 m away within the same reef. However, no fine-scale geographic clustering was observed for the >2 mm size fraction of the ARMS, indicating that there may be dispersal differences between the different size classes that comprise the cryptobiome; however, more work on fine-scale geographic patterns is required to be undertaken in the Red Sea to ascertain this with more certainty. On a larger spatial scale along the coastline of Saudi Arabia, Pearman et al. (2016) showed distinct differences in the cryptobiome community amongst three reefs spreading over approximately 40 km in the region of Jeddah in the central Red Sea. Although these two studies identified spatial differences in the cryptobiome across different sites, the drivers of these differences were not investigated. More work is required to identify the drivers of changes in community composition within and across reefs along small spatial scales of the coastline.

Cross-shelf habitats have been shown to have some of the sharpest gradients in marine ecosystems. For instance, nearshore areas are more influenced by terrestrial factors while outer regions are more exposed to oceanic currents and wave action. Ellis et al. (2017) also showed that nutrient availability varies across the shelf gradient in the southern Red Sea. These environmental differences can in turn affect the distribution of the organisms inhabiting the reefs across the shelf, with differences being shown across a variety of organisms globally, including macrobenthic organisms in the southern Red Sea (Ellis et al. 2017) and fish, scleractinians, and algal biofilms (Cleary et al. 2005, 2016). For the cryptobiome, there was no significant difference in the number of OTUs present on an ARMS in reefs that differed in shelf position in the central Red Sea (Pearman et al. 2018), although shared OTUs were higher between the near- and mid-shore reefs than those between the near- and offshore reefs. Indeed, the offshore sites showed a more unique community compared to those in the mid- and nearshore regions (Pearman et al. 2018), which could indicate that there are limitations in dispersal, such as the presence of boundary currents, that alongside selective pressures (e.g., nutrient availability) are preventing the homogenization of the cryptobiome across the shelf.

2.5 LARGE-SCALE PATTERNS OF DISTRIBUTION

Investigations of distribution patterns of the benthic cryptobiome across large spatial scales in the Red Sea are limited. On a basin-wide scale, Carvalho et al. (2019) showed that there was a small core eukaryotic community (0.1% mean relative abundance and present in 90% of sites) in the Red Sea cryptobiome, but this core community consisted of 12 phyla with Arthropoda, Annelida, and Chordata being especially prevalent. The majority of ASVs were only found in a single reef. A similar pattern was found for bacteria inhabiting ARMS units across the Red Sea, with less than 5% of ASVs being observed in all three regions (north, central, and south) of the Red Sea. Investigations of both bacteria (Pearman et al. 2019) and eukaryotes (Carvalho et al. 2019) showed that there were distinct regional differences in the composition and diversity of the cryptobiome across the Red Sea. The Red Sea has substantial environmental gradients along its length, with the southern region having higher sea surface temperatures and organic matter while the northern region is substantially cooler but with a higher salinity. Using ecological community modelling frameworks (Stegen et al. 2013, 2015), it was shown that differences in the environmental conditions experienced by the bacterial benthic cryptobiome community was the predominant driving force behind the differences in the composition in the community (Pearman et al. 2019). This result suggests a potential role of bacterial communities as indicators of environmental quality. Considering these communities have short generation times, they may be ideal candidates for monitoring programs aiming to inform decision making in a timely manner, i.e., before changes are no longer reversible.

Differences have been shown in the relative abundance of different bacterial families along the Red Sea, with Phyllobacteriaceae more prevalent in the southern Red Sea, being correlated with higher sea surface temperatures and organic matter, whilst the opposite pattern was seen for Flammeovirgaceae and environmental clades TK85 and Sva0996 (Pearman et al. 2019). Carvalho et al. (2019) identified bacterial indicator ASVs for the different regions of the Red Sea, with indicators of the northern region in general being negatively correlated with sea surface temperature while those of the southern region were positively correlated with sea surface temperature and organic matter. Further, organic matter and sea surface temperature were shown to be significant environmental factors affecting the alpha diversity of the eukaryotic component of the cryptobiome, with lower diversity observed in the southern region of the Red Sea, where temperatures and organic matter were higher. Currently, there is a global trend toward the use of molecular-based tools to assess changes in environmental quality. However, limited knowledge exists to allow the use of these tools alone. Our studies pave the way toward more specific research testing and validating future indicators of reef health. This information is critical in the Red Sea, where local communities are not only experiencing the effects of global warming but are also under increasing pressure due to intense coastal development.

The distinct nature of the Red Sea cryptobiome has been highlighted when contrasted with benthic communities found in several European seas using ARMS (David et al. 2019; Pearman et al. 2020) and artificial substrate units (ASUs; Cahill et al. 2018). Overall, our studies also highlight high levels of small- to large-scale variability, with communities mainly driven by turnover (species replacement), highlighting the need for conservation of multiple sites across the Red Sea basin.

2.6 TEMPORAL VARIABILITY IN BIODIVERSITY PATTERNS

Snapshot studies may provide relevant information regarding the composition and structure of marine communities. However, long-term studies are critical in providing knowledge regarding temporal dynamics in environmental (e.g., temperature, salinity, light availability) and ecological patterns (Ducklow et al. 2009; Hughes et al. 2017a). For example, time series provide fundamental data on natural ecological variability, which is essential to defining critical background levels above which significant and undesirable changes may be detected. Ultimately, these can be used as thresholds helping the early detection of habitat degradation and predicting future trajectories through modelling (Condon et al. 2013; Boero 2015). However, securing funding for sustaining time series is incredibly challenging (Hughes et al. 2017a), as most funding streams don't allow for long-term field data collection.

One of the few studies targeting inter-annual variability of the coral reef cryptobiome was conducted in the Red Sea for six years. In this study, Villalobos et al. (2022) investigated patterns of cryptobenthic coral reef communities in three time periods (2013–2015; 2015–2017; 2017–2019) across a shelf gradient initially described by Pearman et al. (2018). Over this period, reefs in the region experienced the global mass bleaching event in 2015/2016, which allowed the investigation of potential effects of bleaching in the cryptobenthic biodiversity patterns. The authors reported that cryptobenthic communities responded to bleaching events and that responses might be size-dependent. Alpha-diversity measures (i.e., the number of OTUs/ASVs) showed contrasting trends, increasing after the bleaching for larger organisms (>2 mm), while smaller organisms (<2 mm) showed a clear decrease in diversity two years after the bleaching, with numbers recovering to pre-bleaching levels in 2019. The cross-shelf gradient previously identified persisted through the six-year period, suggesting that the ecological signature was maintained despite the mass bleaching event in 2015 (Villalobos et al. 2022). The reefs surveyed were not as strongly affected by the 2015/2016 bleaching event as they were by previous bleaching events like the one in 2010 (DeCarlo et al. 2020), and the levels of bleaching registered did not result in a potential homogenization of the community composition that could attenuate the differences across the shelf gradient. Nevertheless, partitioning of beta-diversity (turnover and nestedness; Baselga 2010) showed higher turnover (i.e., changes in species identities) between pre-bleaching and post-bleaching samples when compared to samples collected post-bleaching, suggesting effects of bleaching in the cryptobiome. Specific differences in taxa were also noted, with an increase in richness of Annelida and the decrease in Chordata (mainly fish). Although annelids are known to prefer reef substrate not covered by living corals (Enochs and Manzello 2012), cryptobenthic fish, members of the Chordata, particularly those in close association with scleractinian corals, may experience clear declines in response to changes in their coral hosts (Froehlich et al. 2021). Bellwood et al. (2012) also reported that cryptobenthic fish assemblages may not recover to pre-bleaching levels, even when coral cover is restored, as shifts in coral communities may persist beyond its cover. In the case of more frequent and intense bleaching events, as predicted by Hughes et al. (2018), communities may not be able to recover fast enough to sustain their characteristics through time. Ultimately, this can result in changes in ecosystem functioning, yet our lack of knowledge on the ecological and functional roles of the cryptobiome is alarming, and we may be overlooking the potential impacts of environmental change resulting from changes in these reef communities at global scales.

2.7 MOVING FORWARD: ECOLOGICAL FUNCTIONS

There is a paucity of information regarding how changes in cryptobenthic biodiversity affect reef functioning. Therefore, the next step in understanding the coral reef cryptobiome is to investigate their ecological functions and get a better understanding of the biogeochemical processes to which the cryptobiome contribute. Methods have been developed to infer the metabolic functions of bacterial taxa from their 16S rRNA using software such as Paprica (Bowman and Ducklow 2015) or Picrust2 (Douglas et al. 2020). These approaches can give an indication of the functioning of the bacterial community without undertaking more expensive and computationally intensive metagenomic/metatranscriptomic analysis. Indeed, Pearman et al. (2019) showed that by using Paprica on 16S rRNA ARMS data, there were significant spatial differences in the functional potential of microbial communities colonizing ARMS units in the Red Sea. However, because this approach is only able to infer functions based on placement of 16S rRNA genes in phylogenetic trees of taxa that have their genomes sequenced, it has its limitations. To get a more complete assessment of how organisms are responding functionally, metagenomic or metatranscriptomic approaches would be required. Metagenomics is the sequencing of the DNA from an environmental sample and gives the functional potential of a community, while metatranscriptomics involves the RNA in a community and gives an indication of the functions that are being performed by the community at the time of sampling. These approaches would also be able to incorporate eukaryotes, which are currently not incorporated into the inference tools but make up a substantial proportion of the cryptobiome. Work in the western Indian Ocean (Wambua et al. 2021) has shown the use of metagenomics in assessing the effect of human perturbation on the functioning

FIGURE 2.3 ARMS incubation in the Red Sea (light incubation) following Roth et al. (2019). Photo by Joao Rosado.

of the coral reef microbial community and highlights the potential of the future use of this methodology in the Red Sea. This is especially important with the increased coastal development occurring along the Red Sea coastline and its potential to have a substantial impact on the coral reef cryptobiome, resulting in cascading changes up the trophic levels. Similar to their use for investigating changes in the taxonomic composition of cryptobiome, ARMS provide an ideal platform for the investigation of functional changes, providing a standardized substrate that will allow for the temporal investigation of functional changes in the community, and thus enable a better understanding of response to pressures such as climate change and nutrient inputs.

A complementary way of understanding the links between cryptobenthic biodiversity and function is using *in situ* incubation chambers (Roth et al. 2020). The exchange of energy and nutrients are integral components of reef functioning and are dependent upon *in situ* environmental conditions. *Ex situ* (e.g., laboratory experiments) studies do not incorporate the natural environmental complexity, leading to less accurate assessments of metabolic rates and biogeochemical fluxes (Roth et al. 2019). Recently developed incubation chambers allow the assessment of metabolic rates and biogeochemical fluxes integrating local environmental conditions (Roth et al. 2019), and in 2021 these chambers were used to incubate ARMS in the Red Sea coral reef habitats with different characteristics (e.g., hard-coral dominated, algae/pavement, rubble). Combined with photogrammetry tools, it is possible to estimate community budgets of photosynthesis and respiration of the cryptobenthic communities assessed with ARMS (Figure 2.3). In addition, when combined with specific measurements (e.g., total alkalinity, dissolved inorganic carbon), these incubations can also allow the estimation of calcification processes, which are vital to reef building. We hypothesized that varying biogeochemical characteristics – indicative of the functional role of these communities as C and N sources and/or sinks – occur among habitats, reflecting compositional changes in the pioneer communities, but results are still being analysed. Given the contribution of cryptobenthic communities to reef functioning, this kind of study will help to better understand the consequences of phase-shift scenarios to reef functioning.

2.8 CONCLUSION

With coral reefs in crisis, a more holistic approach to research is required to reverse the degradation that is occurring to reef systems. Traditionally, coral reef research in the Red Sea, and elsewhere, has predominantly focused on longer-lived and more conspicuous species. Changes in the patterns of these species often take time to manifest, meaning coral reefs may have reached a tipping point before problems are identified. By incorporating members of the cryptobiome – which are short lived and respond rapidly to environmental changes (Brandl et al. 2019) – into assessments, more informative conclusions can be ascertained in environmental impact assessment studies. Thus, the development and deployment of ARMS in the Red Sea has provided a step change in the understanding of the cryptobiome in this region and will provide a baseline status against which future health conditions can be assessed. This should be combined with the collaboration of molecular biologists and taxonomists to identify novel species and make sure their genetic information is present in reference databases. This is a critical step to improve molecular-based approaches to characterize biodiversity changes in the future, a method that is becoming increasingly used for environmental monitoring (Aylagas et al. 2020, 2021).

Even though there are increasing amounts of data (and knowledge) being collected about the cryptobiome in the Red Sea, there is still a lack of information on the species structure, composition, and the natural variability of the cryptobiota in the region. This limits our ability to fully disentangle the effects of local stressors or even bleaching events on the distribution patterns of this assemblage and prevents a better understanding of reef dynamics, which could help preserve coral reefs for future generations.

ACKNOWLEDGEMENTS

We would like to thank Eva Aylagas, Rodrigo Villalobos, Holger Anlauf, Darren Coker, Joao Cúrdia, Aislinn Dunne, Joao Rosado, Nauras Daraghmeh, Ute Langner, Gloria Gil Ramos, Micaela Justo, and Irene Salinas for their work deploying and processing the ARMS in the Red Sea. We are also indebted to many students in the Red Sea Research Center (RSRC) who contributed to the recovery and processing of ARMS over the last decade. We would also like to acknowledge the core laboratories at King Abdullah University of Science and Technology for the help they provided and Saudi Aramco for the funding to enable establishing and developing an ARMS research program in the Red Sea. The current version of this chapter benefited from the comments of three reviewers and the editor, to whom we are also grateful.

REFERENCES

Al-Rshaidat, M.M., Snider, A., Rosebraugh, S., et al. 2016. Deep COI sequencing of standardized benthic samples unveils overlooked diversity of Jordanian coral reefs in the northern Red Sea. *Genome* 59: 724–737.

Ateweberhan, M., Feary, D.A., Keshavmurthy, S., et al. 2013. Climate change impacts on coral reefs: Synergies with local effects, possibilities for acclimation, and management implications. *Mar Pollut Bull* 74: 526–539.

Aylagas, E., Atalah, J., Sánchez-Jerez, P., et al. 2021. A step towards the validation of bacteria biotic indices using DNA metabarcoding for benthic monitoring. *Mol Ecol Resour* 21: 1889–1903.

Aylagas, E., Borja, A., Pochon, X., et al. 2020. Translational molecular ecology in practice: Linking DNA-based methods to actionable marine environmental management. *Sci Total Environ* 744: 140780.

Baselga, A. 2010. Partitioning the turnover and nestedness components of beta diversity. *Glob Ecol Biogeogr* 19: 134–143.

Bellwood, D.R., Hoey, A.S., and T.P. Hughes. 2012. Human activity selectively impacts the ecosystem roles of parrotfishes on coral reefs. *Proc Royal Soc Biol Sci* 279: 1621–1629.

Boero, F. 2015. The future of the Mediterranean Sea ecosystem: Towards a different tomorrow. *Rend Fis Acc Lincei* 26: 3–12.

Bowman, J.S., and H.W. Ducklow. 2015. Microbial communities can be described by metabolic structure: A general framework and application to a seasonally variable, depth-stratified microbial community from the coastal West Antarctic Peninsula. *PLoS One* 10: e0135868.

Brandl, S.J., Tornabene, L., Goatley, C.H.R., et al. 2019. Demographic dynamics of the smallest marine vertebrates fuel coral reef ecosystem functioning. *Science* 364: 1189–1192.

Cahill, A.E., Pearman, J.K., Borja, A., et al. 2018. A comparative analysis of metabarcoding and morphology-based identification of benthic communities across different regional seas. *Ecol Evol* 8: 8908–8920.

Callahan, B.J., McMurdie, P.J., Rosen, M.J., et al. 2016. DADA2: High-resolution sample inference from Illumina amplicon data. *Nat Methods* 13: 581.

Carpenter, K.E., Abrar, M., Aeby, G., et al. 2008. One-third of reef-building corals face elevated extinction risk from climate change and local impacts. *Science* 321: 560–563.

Carugati, L., Corinaldesi, C., Dell'Anno, A., et al. 2015. Metagenetic tools for the census of marine meiofaunal biodiversity: An overview. *Mar Genomics* 24 Pt 1: 11–20.

Carvalho, S., Aylagas, E., Villalobos, R., et al. 2019. Beyond the visual: Using metabarcoding to characterize the hidden reef cryptobiome. *Proc Royal Soc Biol Sci* 286: 20182697.

Carvalho, S., Moura, A., Cúrdia, J., da Fonseca, L.C., and M.N. Santos. 2013. How complementary are epibenthic assemblages in artificial and nearby natural rocky reefs? *Mar Environ Res* 92: 170–177.

Casey, J.M., Ransome, E., Collins, A.G., et al. 2021. DNA metabarcoding marker choice skews perception of marine eukaryotic biodiversity. *Environ DNA* 3: 1229–1246.

Cheal, A.J., MacNeil, M.A., Emslie, M.J., et al. 2017. The threat to coral reefs from more intense cyclones under climate change. *Glob Change Biol* 23: 1511–1524.

Cleary, D.F.R., Becking, L.E., de Voogd, N.J., et al. 2005. Variation in the diversity and composition of benthic taxa as a function of distance offshore, depth and exposure in the Spermonde Archipelago, Indonesia. *Estuar Coast Shelf Sci* 65: 557–570.

Cleary, D.F.R., Polónia, A.R.M., Renema, W., et al. 2016. Variation in the composition of corals, fishes, sponges, echinoderms, ascidians, molluscs, foraminifera and macroalgae across a pronounced in-to-offshore environmental gradient in the Jakarta Bay – Thousand Islands coral reef complex. *Mar Pollut Bull* 110: 701–717.

Condon, R.H., Duarte, C.M., Pitt, K.A., et al. 2013. Recurrent jellyfish blooms are a consequence of global oscillations. *Proc Natl Acad Sci* 110: 1000–1005.

David, R., Uyarra, M.C., Carvalho, S., et al. 2019. Lessons from photo analyses of Autonomous Reef Monitoring Structures as tools to detect (bio-)geographical, spatial, and environmental effects. *Mar Pollut Bull* 141: 420–429.

DeCarlo, T.M., Carvalho, S., Gajdzik, L., et al. 2021. Patterns, drivers, and ecological implications of upwelling in coral reef habitats of the southern Red Sea. *J Geophys Res Oceans* 126: e2020JC016493.

DeCarlo, T.M., Gajdzik, L., Ellis, J., et al. 2020. Nutrient-supplying ocean currents modulate coral bleaching susceptibility. *Sci Adv* 6: eabc5493.

de Goeij, J.M., van Oevelen, D., Vermeij, M.J.A., et al. 2013. Surviving in a marine desert: The sponge loop retains resources within coral reefs. *Science* 342: 108–110.

Douglas, G.M., Maffei, V.J., Zaneveld, J.R., et al. 2020. PICRUSt2 for prediction of metagenome functions. *Nat Biotechnol* 38: 685–688.

Ducklow, H.W., Doney, S.C., and D.K. Steinberg. 2009. Contributions of long-term research and time-series observations to marine ecology and biogeochemistry. *Ann Rev Mar Sci* 1: 279–302.

Ellis, J., Anlauf, H., Kürten, S., et al. 2017. Cross shelf benthic biodiversity patterns in the southern Red Sea. *Sci Rep* 7: 437.

Ellis, J.I., Jamil, T., Anlauf, H., et al. 2019. Multiple stressor effects on coral reef ecosystems. *Glob Change Biol* 25: 4131–4146.

Enochs, I.C., and D.P. Manzello. 2012. Species richness of motile cryptofauna across a gradient of reef framework erosion. *Coral Reefs* 31: 653–661.

Fisher, R., O'Leary, R.A., Low-Choy, S., et al. 2015. Species richness on coral reefs and the pursuit of convergent global estimates. *Curr Biol* 25: 500–505.

Froehlich, C.Y.M., Klanten, O.S., Hing, M.L., et al. 2021. Uneven declines between corals and cryptobenthic fish symbionts from multiple disturbances. *Sci Rep* 11: 16420.

Hobern, D. 2021. BIOSCAN: DNA barcoding to accelerate taxonomy and biogeography for conservation and sustainability. *Genome* 64: 161–164.

Hoegh-Guldberg, O., and J.F. Bruno. 2010. The impact of climate change on the world's marine ecosystems. *Science* 328: 1523–1528.

Hughes, B.B., Beas-Luna, R., Barner, A.K., et al. 2017a. Long-term studies contribute disproportionately to ecology and policy. *BioSci* 67: 271–281.

Hughes, T.P. 1994. Catastrophes, phase shifts, and large-scale degradation of a Caribbean coral reef. *Science* 265: 1547–1551.

Hughes, T.P., Baird, A.H., Bellwood, D.R., et al. 2003. Climate change, human impacts, and the resilience of coral reefs. *Science* 301: 929–933.

Hughes, T.P., Barnes, M.L., Bellwood, D.R., et al. 2017b. Coral reefs in the anthropocene. *Nature* 546: 82.

Hughes, T.P., Kerry, J.T., Baird, A.H., et al. 2018. Global warming transforms coral reef assemblages. *Nature* 556: 492–496.

Jackson, J.B., Kirby, M.X., Berger, W.H., et al. 2001. Historical overfishing and the recent collapse of coastal ecosystems. *Science* 293: 629–637.

Kobluk, D.R. 1988. Cryptic faunas in reefs: Ecology and geologic importance. *PALAIOS* 3: 379–390.

Leray, M., and N. Knowlton. 2015. DNA barcoding and metabarcoding of standardized samples reveal patterns of marine benthic diversity. *Proc Natl Acad Sci* 112: 2076–2081.

Magel, J.M.T., Dimoff, S.A., and J.K. Baum. 2020. Direct and indirect effects of climate change-amplified pulse heat stress events on coral reef fish communities. *Ecol Appl* 30: e02124.

Morrison, T.H., Hughes, T.P., Adger, W.N., et al. 2019. Save reefs to rescue all ecosystems. *Nature* 573: 333–336.

Pandolfi, J.M., Bradbury, R.H., Sala, E., et al. 2003. Global trajectories of the long-term decline of coral reef ecosystems. *Science* 301: 955–958.

Paulay, G. 1997. Diversity and distribution of reef organisms. In: *Life and death of coral reefs*. Ed: C. Birkeland, 198–229. New York: Chapman & Hall.

Pearman, J.K., Anlauf, H., Irigoien, X., et al. 2016. Please mind the gap – Visual census and cryptic biodiversity assessment at central Red Sea coral reefs. *Mar Environ Res* 118: 20–30.

Pearman, J.K., Aylagas, E., Voolstra, C.R., et al. 2019. Disentangling the complex microbial community of coral reefs using standardized Autonomous Reef Monitoring Structures (ARMS). *Mol Ecol* 28: 3496–3507.

Pearman, J.K., Chust, G., Aylagas, E., et al. 2020. Pan-regional marine benthic cryptobiome biodiversity patterns revealed by metabarcoding Autonomous Reef Monitoring Structures. *Mol Ecol* 29: 4882–4897.

Pearman, J.K., Leray, M., Villalobos, R., et al. 2018. Cross-shelf investigation of coral reef cryptic benthic organisms reveals diversity patterns of the hidden majority. *Sci Rep* 8: 8090.

Pennesi, C., and R. Danovaro. 2017. Assessing marine environmental status through microphytobenthos assemblages colonizing the Autonomous Reef Monitoring Structures (ARMS) and their potential in coastal marine restoration. *Mar Pollut Bull* 125: 56–65.

Plaisance, L., Caley, M.J., Brainard, R.E., et al. 2011. The diversity of coral reefs: What are we missing? *PLoS One* 6: e25026.

Ransome, E., Geller, J.B., Timmers, M., et al. 2017. The importance of standardization for biodiversity comparisons: A case study using Autonomous Reef Monitoring Structures (ARMS) and metabarcoding to measure cryptic diversity on Mo'orea coral reefs, French Polynesia. *PLoS One* 12: e0175066.

Richter, C., and M. Wunsch. 1999. Cavity-dwelling suspension feeders in coral reefs – A new link in reef trophodynamics. *Mar Ecol Prog Ser* 1999: 105–116.

Richter, C., Wunsch, M., Rasheed, M., et al. 2001. Endoscopic exploration of Red Sea coral reefs reveals dense populations of cavity-dwelling sponges. *Nature* 413: 725–730.

Rix, L., de Goeij, J.M., van Oevelen, D., et al. 2017. Differential recycling of coral and algal dissolved organic matter via the sponge loop. *Funct Ecol* 31: 778–789.

Roth, F., Karcher, D.B., Rädecker, N., et al. 2020. High rates of carbon and dinitrogen fixation suggest a critical role of benthic pioneer communities in the energy and nutrient dynamics of coral reefs. *Funct Ecol* 34: 1991–2004.

Roth, F., Rädecker, N., Carvalho, S., et al. 2021. High summer temperatures amplify functional differences between coral-and algae-dominated reef communities. *Ecology* 102: e03226.

Roth, F., Wild, C., Carvalho, S., et al. 2019. An in situ approach for measuring biogeochemical fluxes in structurally complex benthic communities. *Methods Ecol Evol* 10: 712–725.

Stegen, J.C., Lin, X., Fredrickson, J.K., et al. 2013. Quantifying community assembly processes and identifying features that impose them. *ISME J* 7: 2069.

Stegen, J.C., Lin, X., Fredrickson, J.K., et al. 2015. Estimating and mapping ecological processes influencing microbial community assembly. *Front Microbiol* 6: 370.

Steyaert, M., Lindhart, M., Khrizman, A., et al. 2022. Remote reef cryptobenthic diversity: Integrating autonomous reef monitoring structures and in situ environmental parameters. *Front Mar Sci* 9: 932375. doi:10.3389/fmars.2022.932375.

Taberlet, P., Bonin, A., Zinger, L., et al. 2018. *Environmental DNA: For biodiversity research and monitoring*. Oxford, UK: Oxford University Press.

Timmers, M.A., Jury, C.P., Vicente, J., et al. 2021. Biodiversity of coral reef cryptobiota shuffles but does not decline under the combined stressors of ocean warming and acidification. *Proc Natl Acad Sci* 118: e2103275118.

Villalobos, R. 2021. *Spatial and temporal biodiversity patterns of coral reef cryptofauna on the Arabian Peninsula*. Ph.D. diss., King Abdullah University of Science and Technology, Saudi Arabia.

Villalobos, R., Aylagas, E., Pearman, J.K., et al. 2022. Inter-annual variability patterns of reef cryptobiota in the central Red Sea across a shelf gradient. *Sci Rep* 12: 16944.

Wambua, S., Gourlé, H., de Villiers, E.P., et al. 2021. Cross-sectional variations in structure and function of coral reef microbiome with local anthropogenic impacts on the Kenyan coast of the Indian Ocean. *Front Microbiol* 12: 673128.

Williams, S.T., and N. Knowlton. 2001. Mitochondrial pseudogenes are pervasive and often insidious in the snapping shrimp genus *Alpheus*. *Mol Biol Evol* 18: 1484–1493.

Woodhead, A.J., Hicks, C.C., Norström, A.V., et al. 2019. Coral reef ecosystem services in the anthropocene. *Funct Ecol* 33: 1023–1034.

Wunsch, M., Al-Moghrabi, S.M., and I. Kötter. 2000. *Communities of coral reef cavities in Jordan, Gulf of Aqaba (Red Sea)*. Proc 9th International Coral Reef Conf, Bali, Indonesia, 1, pp. 595–600.

Zimmerman, T.L., and J.W. Martin. 2004. Artificial Reef Matrix Structures (ARMS): An inexpensive and effective method for collecting coral reef-associated invertebrates. *Gulf Caribb Res* 16: 59–64.

Assessment of coral communities along Sudan's Red Sea coast

3

Khadija Y.I. Abaker, Sheikheldin M. Elamin, Kirsty L. Nash, Elfatih B.A. Eltaib, Portia J. Nillos-Kleiven, Rebecca Klaus and Even Moland

3.1 INTRODUCTION

Coral reefs cover approximately 284,300 km^2 of the globe (~ 0.02% of the ocean surface), adjoining the coastlines of 100 countries (Spalding et al. 2001). Coral reefs support substantial fisheries, both subsistence and commercial in nature, which in turn provide food, jobs, and livelihoods for an estimated 200 million people (Teh et al. 2013; Hoegh-Guldberg et al. 2019). The Red Sea, one of the Earth's youngest seas, is characterized by extensive coral reef ecosystems despite the warm, highly saline, and highly oligotrophic water that arises from high evaporation rates and little freshwater input from rivers (Rasul et al. 2015). The Red Sea is one of the most diverse tropical marine environments outside the coral triangle (Spalding et al. 2001), and it is a hotspot of coral reef endemism, with a pool of unique genetic lineages that are exported to other regions. A total of 346 species of scleractinian corals, 635 polychaetes, 211 echinoderms, 79 ascidians, 231 crustaceans, 137 molluscs, and 1071 shallow-water fishes have been recorded (DiBattista et al. 2016). Levels of endemism range from less than 10% for scleractinian corals to more than 12% for fish and 16% for ascidians (DiBattista et al. 2016).

Despite the high biodiversity and endemism, until recently comparatively little work was undertaken on Red Sea reefs. Berumen et al. (2013) carried out a quantitative assessment of research published from the Red Sea across a range of topics, including coral bleaching, coral reproductive biology, herbivory, and marine protected areas. They compared the resulting literature with the research done in other coral reef regions. They found that the coral reef–associated scientific literature from the Red Sea represented one-sixth and one-eighth of that undertaken on the Great Barrier Reef and the Caribbean, respectively.

This lack of focus on the Red Sea is changing, with increasing scientific attention in recent years, predominantly from active research institutions in Saudi Arabia, Israel, and Egypt. Global changes in coral reefs arising from accelerating anthropogenic global warming have made the heat-tolerant corals found in the central Red Sea the focus of high-impact studies (e.g., Savary et al. 2021). However, little of this recent work has targeted Sudanese waters. The Sudanese coast runs along >800 km of the western side of the Red Sea, extending from 18°N to 22°N. The coastline is bordered by fringing reefs 1–3 km wide, separated by deep channels from an offshore barrier complex 1–14 km wide, that has developed on offshore tectonically formed banks, rather than as the result of growth on ancient fringing reefs (Bemert and Ormond 1981). Three distinct depth zones are recognized: shallow reefs and reef-studded shelves less than 50 m in depth, deep shelves from 500 to 1000 m, and a central trench that reaches a maximum of 3000 m off the city of Port Sudan.

Historically, the long-term work conducted at Sanganeb Atoll is notable (Reinicke et al. 2003; Mergner and Schuhmacher 1981, 1985; Mergner et al. 1992). An early species inventory of Sanganeb Atoll and extensive descriptions of the reefs in the vicinity of Port Sudan including the Towartit reef complex were published in the 1980s (Schroeder and Nasr 1983; Head 1980; Vine and Vine 1980). These early studies reported greater hard coral cover at the offshore Towartit complex than on inshore reefs (Head 1980), and the offshore reefs of Suakin were reported as having high coral diversity with decreasing diversity closer towards the coast (Ormond 1976).

More recently, large-scale work exploring diversity and coral cover of the fringing reefs along a 300 km stretch of Sudan's northern coastline recorded a mean coral cover of over 40% between 2008 and 2011 (Ali 2011; Ali and Elhag 2016). The most comprehensive study prior to the 2010 and 2015–16 bleaching events reported hard coral cover of 11–65% along 200 km of the central Sudanese coast covering both fringing and barrier reef sites (Klaus et al. 2008).

DOI: 10.1201/9781003321392-4

In the last five years, international projects focused on capacity building and support for coral reef management have resulted in publications describing Sudanese reefs (Kessel et al. 2017; Olsen et al. 2021). However, this literature targets fish communities with no comprehensive description of the state of the reef-building coral community since significant bleaching events in 2010 and 2015–2016 (Furby et al. 2013; Monroe et al. 2018; Hughes et al. 2018). This chapter describes an extensive, coast-wide survey of the Red Sea State province of Sudan conducted in 2016 and 2017 as part of a UNIDO project funded by Norway. The study assesses the status of coral communities along the Sudanese Red Sea coast, describing patterns in coral cover and quantifying the presence and prevalence of coral bleaching to provide baseline information for future studies and monitoring.

3.2 MATERIALS AND METHODS

3.2.1 Study sites

The surveys reported here were conducted over two years during the periods 20 October to 3 December 2016, and 13 October to 25 November 2017 as part of the cruise plan of the UNIDO project mentioned previously. A total of 25 fringing and offshore coral reef sites were surveyed, spanning approximately 600 km of the coast running from the border with Egypt in the north to the border with Eritrea in the south (Figure 3.1, Table 3.1). Sampling was conducted at the same 25 sites in both years. These data are compared here.

The survey sites were chosen in relation to the frequency of use by the local small-scale fishers (Marine Fisheries Administration, pers. comm.), with the aim of sampling equally across regularly, occasionally, and less visited reefs. Some additional sites were chosen from the Equipe Cousteau studies (Klaus et al. 2008).

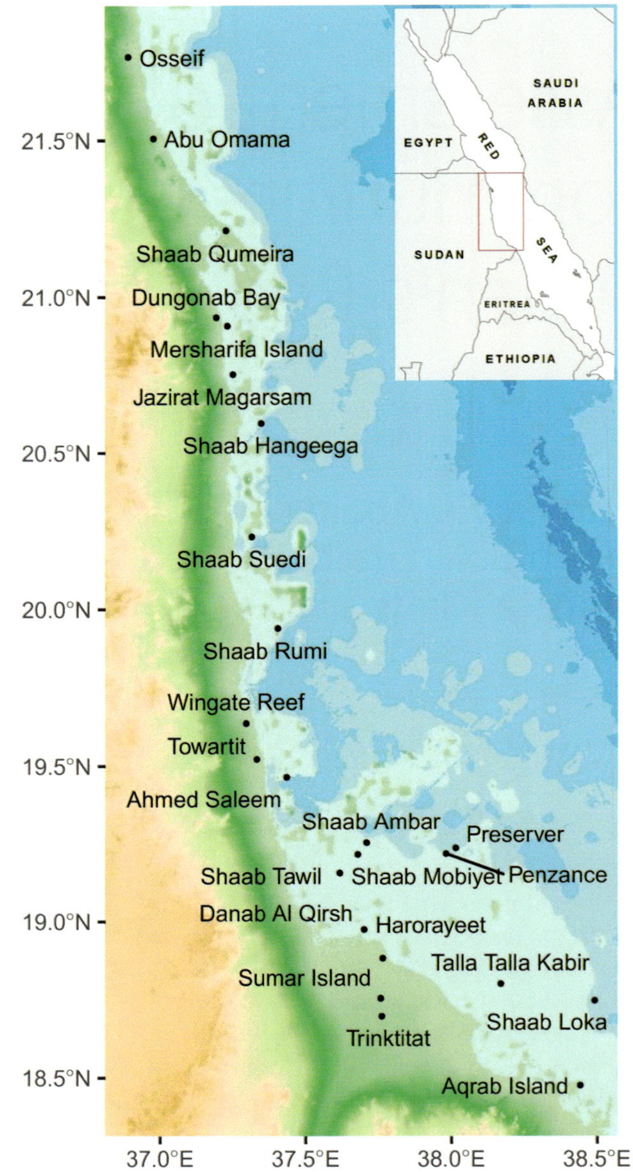

FIGURE 3.1 Location of the survey sites along the Sudanese Red Sea coast (Red Sea State) completed in 2016 and 2017. Map area corresponds to the red rectangle in the insert.

3.2.2 Survey

The surveys covered both ecological parameters (benthic composition and cover), which were recorded using Diver-Operated Video (DOV) consisting of two GoPro Hero4 cameras in a stereo configuration mounted in SeaGis® housings/DOV system. Physical parameters (temperature, salinity and dissolved oxygen), which were sampled with a SAIV® conductivity, temperature, and depth (CTD) sensor.

3.2.2.1 Benthic cover

Quantification of benthic cover and composition in this study used the recorded DOV transects (Figure 3.2A). Two 50-m transects were recorded at each site, with pairs of transects being located approximately 10 m apart. The transects were laid on the reef slope at depths ranging from 6–16 m (see Table 3.1). The videos were analysed as point intercept transects, with points separated by 0.5 m to give 100 points per 50-m video transect. To identify the specific frame associated with each of the 100 points, we assumed a constant swimming speed for the whole 50 m, and then divided the time length of the video by 100 (Figure 3.2B). In each of the sample video frames, the benthic cover located mid-frame and directly below the transect tape was recorded (Figure 3.2C). Benthic cover at each point was identified according to categories used in the standardized Reef Check protocol (Table 3.2; Hodgson et al. 2004). In addition, at each point (along point intercept transect) where a live coral was observed, the corals were identified to genus level according to Veron and Stafford-Smith (2000), and the degree of coral bleaching was recorded using a 4-point scale: no bleaching (0), partially bleached (1), white (2), and bleaching and partly dead (3). The sample frames were annotated in the SeaGIS video analysis software "EventMeasure," and the percentage cover of each benthic category was calculated and averaged (mean) across transects. These data were plotted for each site from the north (Osseif) to the south (Aqrab Island S).

TABLE 3.1 Global Positioning System (GPS) position (decimal degrees), depth, and percentage cover (± SE) of hard corals and algae at survey sites in 2016 and 2017, listed from north to south. Where no SE data are presented, only one transect was surveyed at a site.

SITE	LAT	LONG	DEPTH (M)	2016 HARD CORAL	2017 HARD CORAL	2016 ALGAE	2017 ALGAE
Osseif	21.76650	36.89112	10	16 ± 3.5	37.5 ± 2.5	0 ± 0	1.5 ± 0.4
Abu Omama	21.50572	36.97748	13	12.5 ± 1.8	9 ± 3.5	0 ± 0	0 ± 0
Shaab Qumeira	21.21212	37.22513	16	47 ± 0	45.5 ± 3.2	0 ± 0	0.5 ± 0.4
Dungonab Bay	20.93443	37.19152	7	38 ± 2.8	17.5 ± 0.4	0 ± 0	0 ± 0
Mersharifa Island	20.90748	37.22952	7	38.5 ± 9.5	20 ± 2.1	0 ± 0	0 ± 0
Jazirat Magarsam	20.90705	37.26998	6	13 ± 2.1	16 ± 2.8	3.5 ± 0.4	0 ± 0
Shaab Hangeega	20.75255	37.24847	15	61	25 ± 1.4	0	0 ± 0
Shaab Suedi	20.59522	37.34610	13	43	17.5 ± 5.3	1	0.5 ± 0.4
Shaab Rumi	20.23137	37.31307	15	41.5 ± 0.4	18	0 ± 0	0
Wingate Reef	19.63667	37.29335	10	60 ± 1.4	15.5 ± 2.5	0 ± 0	1.5 ± 1.1
Towartit	19.52183	37.32953	10	21 ± 1.4	12.5 ± 3.2	1.5 ± 0.4	0.5 ± 0.4
Ahmed Saleem	19.46453	37.43352	13	41	15.5 ± 4.6	0	0 ± 0
Shaab Ambar	19.25370	37.70698	13	42	18 ± 2.1	0	0.5 ± 0.4
Preserver	19.23637	38.01320	13	41	37.5 ± 0.4	2	0 ± 0
Penzance	19.21818	37.98008	13	47.5 ± 3.9	52.5 ± 2.5	2 ± 1.4	0 ± 0
Shaab Mobiyet	19.21588	37.67688	16	48	29.5 ± 3.2	5	0 ± 0
Shaab Tawil	19.15655	37.61511	14	63 ± 0.7	15 ± 3.5	1 ± 0.7	1 ± 0.7
Dhanab Al Qirsh	18.97450	37.69885	15	39 ± 1.4	28 ± 3.5	2.5 ± 1.1	0 ± 0
Harorayeet	18.88258	37.76238	7	30 ± 1.4	43 ± 9.9	3.5 ± 0.4	0.5 ± 0.4
Talla Talla Kebir	18.80091	38.16808	8	50 ± 2.1	34.5 ± 4.6	3.5 ± 1.1	0 ± 0
Sumar Island	18.75495	37.75550	7	67 ± 4.2	56.5 ± 6	3.5 ± 1.1	1.5 ± 0.4
Shaab Loka	18.74728	38.49030	14	50.5 ± 1.1	66 ± 8.5	0 ± 0	0 ± 0
Trinkitat	18.69680	37.75925	7	57.5 ± 0.4	56	3.5 ± 1.1	0
Aqrab Island N	18.47612	38.44140	12	32.5 ± 1.8	36.5 ± 6	0 ± 0	0 ± 0
Aqrab Island S	18.45463	38.44655	9	19.5 ± 3.2	22 ± 5.7	0 ± 0	0 ± 0

Between-year differences in live coral cover were assessed using a paired two-way t-test. Between-year differences in benthic cover were visualized using non-metric multidimensional scaling (nMDS; Clarke 1993) based on the Bray-Curtis measure of similarity using log transformed data in the "Vegan" package in R (Oksanen et al. 2022). Analysis of similarities (ANOSIM; Clarke 1993) was used to compare benthic composition between years. Between-year differences in prevalence of total bleaching were assessed using paired two-way t-tests. Between-year differences in bleaching categories were visualized using nMDS based on the Bray-Curtis measure of similarity using log transformed data. ANOSIM was used to compare bleaching categories among years. To understand potential unobserved bleaching between 2016 and 2017, the difference in prevalence of recently killed vs. live hard coral was assessed using a paired two-way t-test.

Scleractinian corals present beneath sampled points were identified to genus level, creating a genera list for the region.

3.2.2.2 CTD and oxygen measurements

Ocean temperature, salinity, and dissolved oxygen were measured in vertical profiles from the surface to near the bottom using a handheld CTD probe at the closest location to the survey sites that provided at least 40 m of water depth. Patterns in these parameters are important to understand in relation to the health of coral communities since high temperature, low salinity, and hypoxia have all been implicated in coral bleaching (Johnson et al. 2021; Gegner et al. 2017). The oceanographic data were then visualized using Ocean Data View to provide vertical profiles for each parameter across the 25 sites. The 2017 data are summarized and shown later.

3.3 RESULTS

3.3.1 Benthic cover

Thirty-three genera of scleractinian corals belonging to 12 families were identified (Appendix 3.1). In 2016, the highest value of coral cover was recorded in Sumar Island (67% ±4.2 SE) and the lowest at Abu Omama (12.5% ±1.8 SE) (Figure 3.3, Table 3.1). In 2017 the highest value of coral cover was recorded at Shaab Loka N (66.0% ±8.5 SE), and the lowest value was at Abu Omama (9% ±3.5 SE). Mean coral cover across all sites decreased from 2016 (41.8% ±15.5 SE) to 2017 (29.8% ±16 SE). The difference in coral cover among years was statistically significant ($t = 3.1008$, d.f. = 24, p-value = 0.005). In both years,

FIGURE 3.2 A) Image of Diver-Operated Video (DOV) recording the benthos at Shaab Hangeega; B) Screenshot of EventMeasure (left camera view), the SeaGIS video analysis software used to determine the percentage cover of each benthic category; C) Screen grab of *Porites* coral identified on Point Intercept Transect (PIT). Photo credit (A): Even Moland/IMR.

TABLE 3.2 Categories of substrate type (Hodgson et al. 2004) used to classify benthos in videos.

BENTHOS	*DESCRIPTION*
Hard coral (HC)	Scleractinia and fire coral (*Millepora*)
Soft coral (SC)	Including zoanthids
Recently killed coral (RKC)	Coral that died within the past year – skeletal structure of corallites still visible. The coral may be broken into pieces.
Rock (RC)	Any hard substrate and coral dead for more than one year – skeletal structure of corallites eroded.
Rubble (RB)	Rocks and coral pieces the size 0.5–15 cm in diameter
Sand (SD)	Rocks smaller than 0.5 cm in diameter that falls quickly to the bottom when dropped, i.e., this was a visual determination of sand.
Algae (ALG)	Various type algae, including both turf and macroalgae
Sponge (SP)	Sponges
Other (OT)	Sessile organisms including Bivalvia, *Strombus* etc.
Silt/Clay (SI)	Sediment that remains in suspension if disturbed, i.e., this was a visual determination of silt.

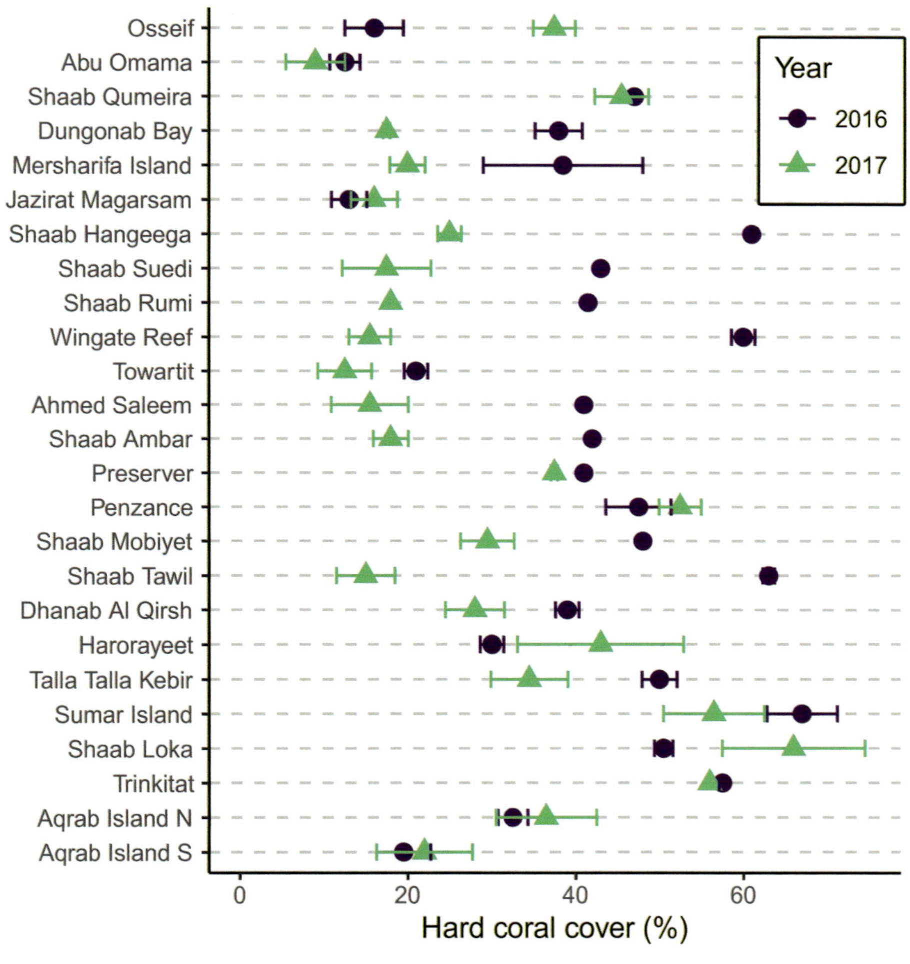

FIGURE 3.3 Live hard coral cover (± SE) across all sites (plotted north to south) for 2016 (purple shading) and 2017 (green shading).

algal cover was low (2016: 0–5% across sites; 2017: 0–1.5% across sites).

The nMDS plot of benthic cover across the two years differentiated sites primarily along axes of sand to rock (axis nMDS1) and soft coral to algae (axis nMDS2) (Figure 3.4). There was a significant difference in benthic communities between years, but the low R value suggests considerable variation in the data is unexplained by year (ANOSIM R= 0.17; $p < 0.001$).

Coral bleaching was observed on transects in both years, but most colonies were unbleached (Figure 3.5A). Forty percent (±0.0 SE) of colonies recorded on the transect were bleached in 2016. This percentage fell to 28% (±0.0 SE) in 2017. This difference was highly statistically significant ($t = 4.3647$, d.f. = 24, p-value <0.001). The nMDS (Figure 3.5B) of bleaching categories across the two years differentiated sites primarily along no bleaching (B_0) to partially bleached (B_1; axis nMDS1) and bleached and partly dead (B_3) to bleached (B_2; axis nMDS2) (Figure 3.5B). The low R value suggests considerable variation in these data is unexplained by year (ANOSIM R= 0.10; p =0.005); however, there was a significant difference in prevalence of bleaching

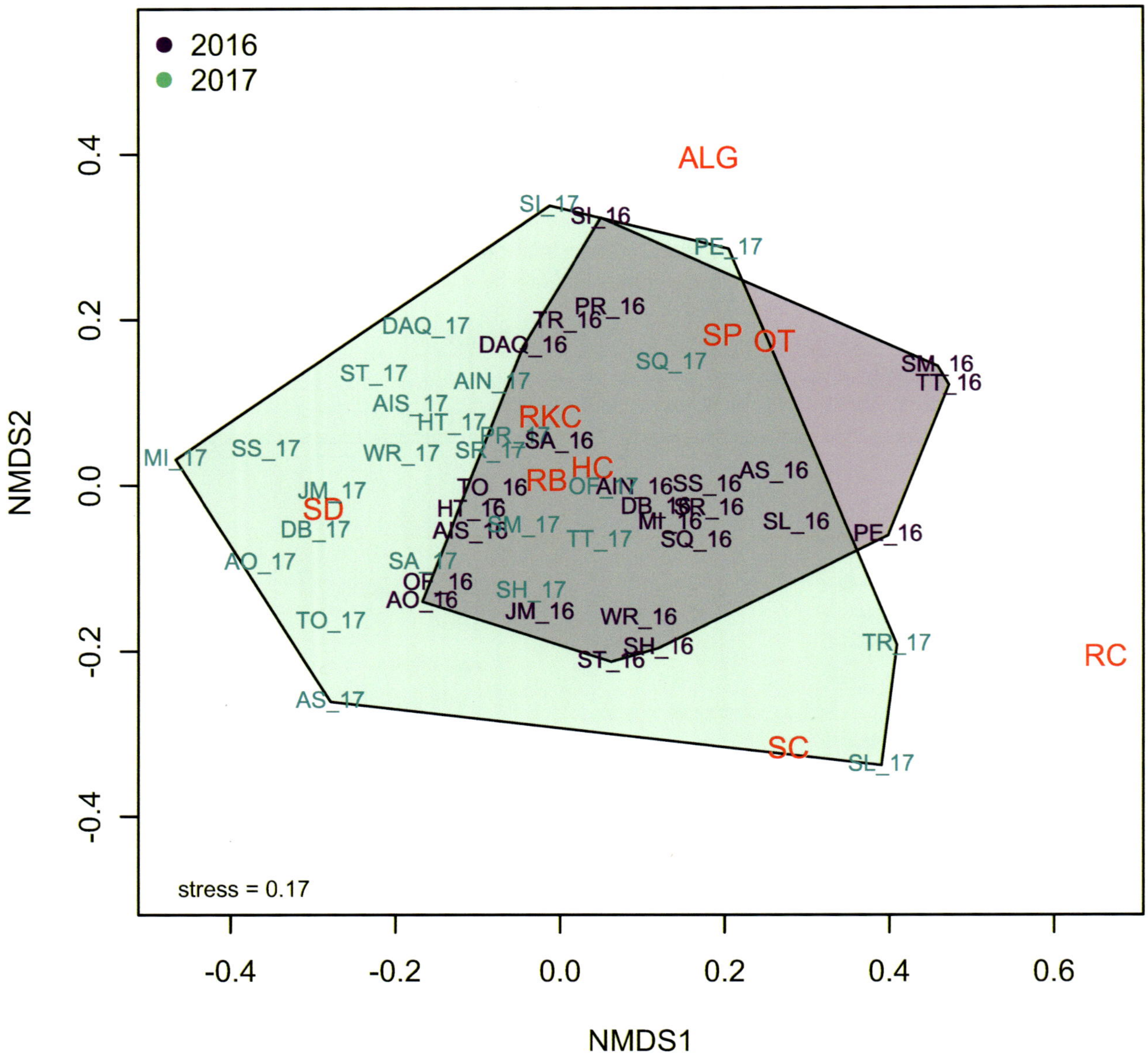

FIGURE 3.4 nMDS of benthic cover at all sites in 2016 (purple shading) and 2017 (green shading) based on benthic categories described in Table 3.1. Abu Omama (AO), Ahmed Saleem (AS), Aqrab Island N(AIN), Aqrab Island S (AIS), Dhanab Al Qirsh (DAQ), Dungonab Bay (DB), Harorayeet (HT), Jazirat Magarsam (JM), Mersharifa Island (MI), Osseif (OI), Penzance (PE), Preserver (PR), Shaab Ambar (SA), Shaab Hangeega (SH), Shaab Loka (SL), Shaab Mobiyet (SM), Shaab Qumeira (SQ), Shaab Rumi (SR), Shaab Suedi (SS), Shaab Tawil (ST), Sumar Island (SI), Talla Talla Kebir (TT), Towartit (TO), Trinkitat (TR), Wingate Reef (WR).

categories among years. The nMDS suggests that 2017 has greater proportions of unbleached (B_0) vs. bleached or partially bleached (B_1 to B_3) colonies as differentiated along nMDS axis 1. When comparing live vs. recently killed coral cover (Figure 3.6), mean live cover did not change significantly from 2016 (72.1% ±9.8 SE) to 2017 (64.6% ±1.6 SE) (t = 1.9488, d.f. = 24, p-value = 0.06).

3.3.2 CTD and oxygen measurements from 2017

Surface water temperature (top 20 m) increased from north to south (Table 3.3), ranging from 28.5°C at Osseif to 31.5°C at Sumar Island; however, south of Sumar Island, surface water temperatures fell to 26°C (Figure 3.7A). In contrast, surface salinity decreased from north to south (40 to 39 ppt; Figure 3.7B). Surface oxygen concentration showed no clear trend from north to south (4.5–5 mg/l across all sites; Figure 3.7C).

3.4 DISCUSSION

The earth's marine ecosystems are experiencing broad-scale and accelerating anthropogenic impacts (Halpern et al. 2017). The current and future health of coral reefs is of particular concern (Hughes et al. 2010). Coral reef ecosystems are extremely

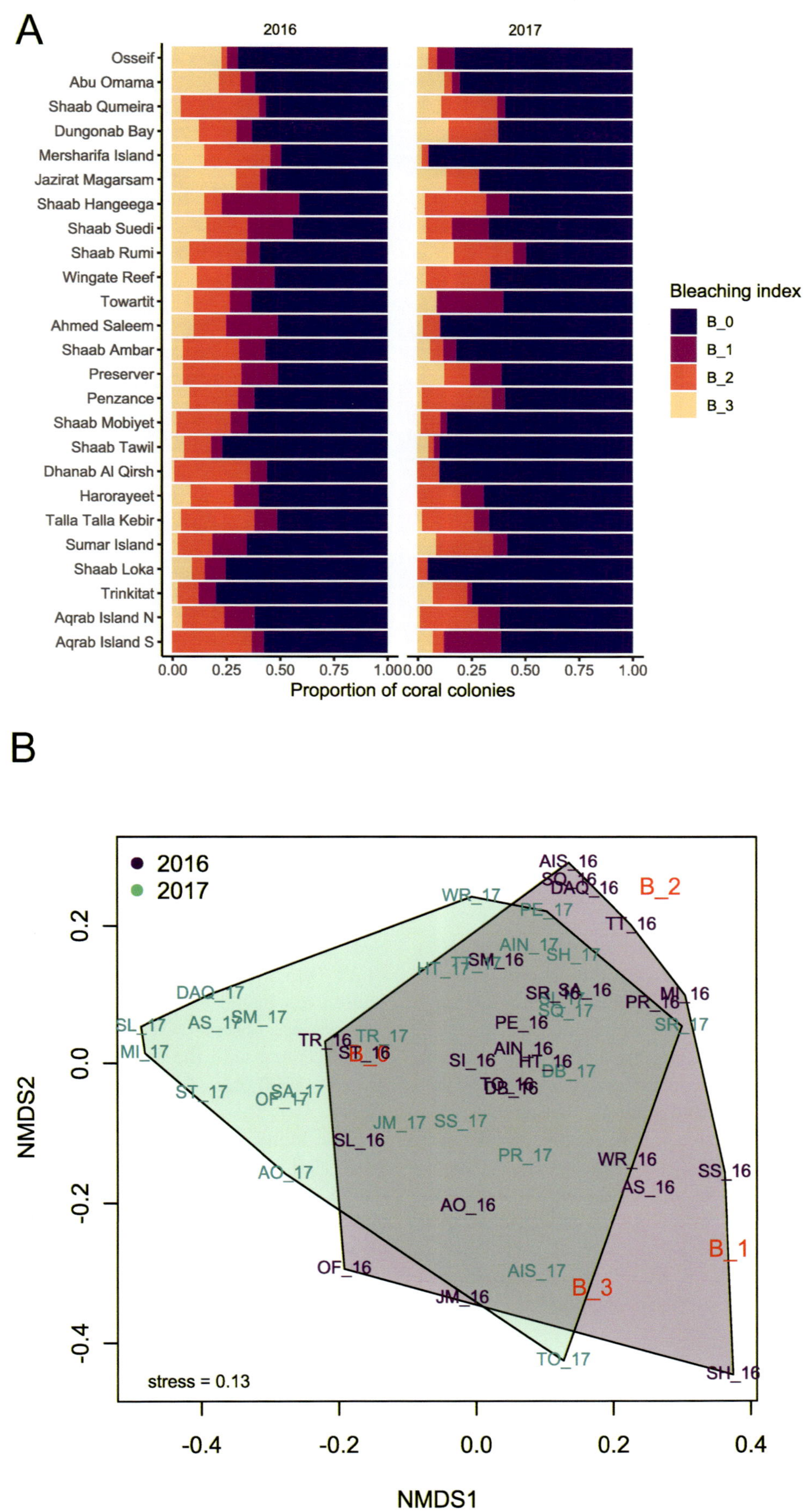

FIGURE 3.5 A) Coral bleaching across all sites (plotted north to south) for 2016 and 2017. No bleaching (0), partially bleached (1), bleached (2), bleached and partly dead (3). B) nMDS of bleaching at all sites in 2016 (purple shading) and 2017 (green shading) based on benthic categories described in Table 3.1. Abu Omama (AO), Ahmed Saleem (AS), Aqrab Island N(AIN), Aqrab Island S (AIS), Dhanab Al Qirsh (DAQ), Dungonab Bay (DB), Harorayeet (HT), Jazirat Magarsam (JM), Mersharifa Island (MI), Osseif (OI), Penzance (PE), Preserver (PR), Shaab Ambar (SA), Shaab Hangeega (SH), Shaab Loka (SL), Shaab Mobiyet (SM), Shaab Qumeira (SQ), Shaab Rumi (SR), Shaab Suedi (SS), Shaab Tawil (ST), Sumar Island (SI), Talla Talla Kebir (TT), Towartit (TO), Trinkitat (TR), Wingate Reef (WR).

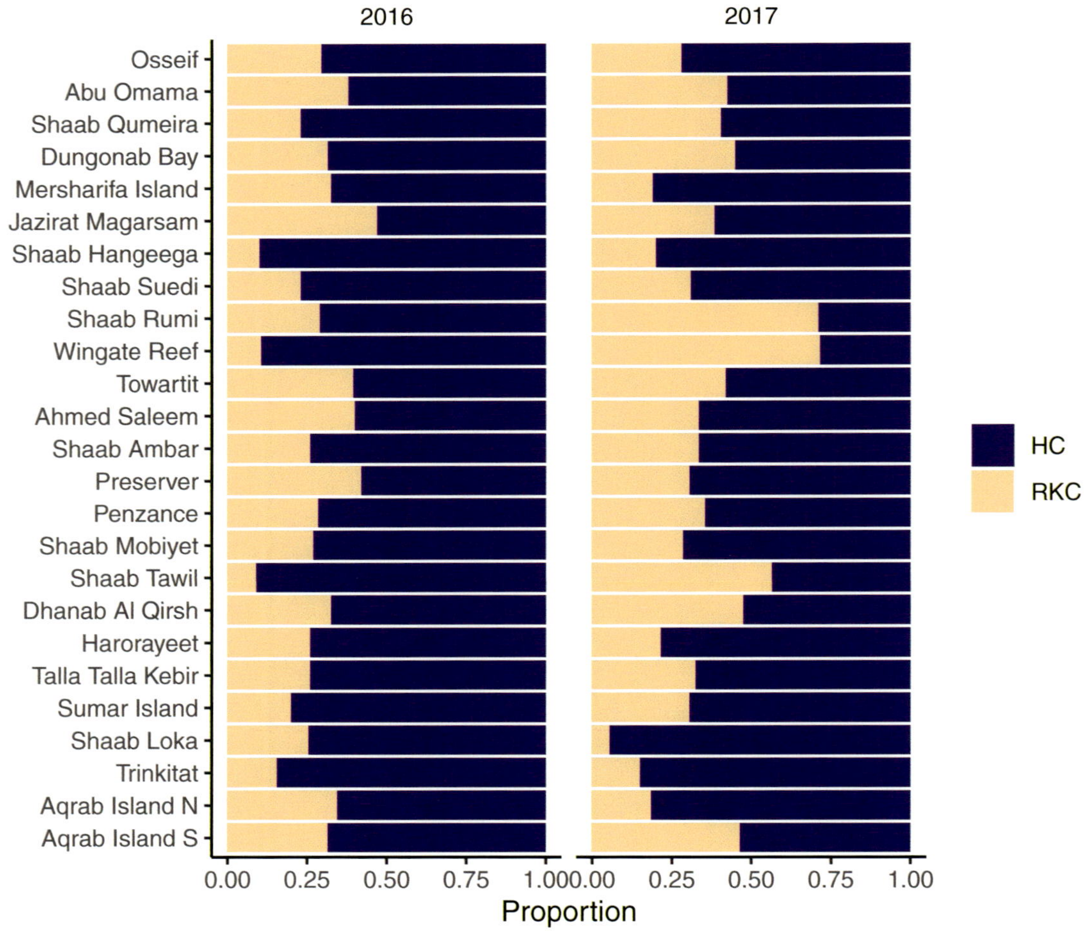

FIGURE 3.6 Recent coral mortality (plotted north to south) for 2016 and 2017. Live hard coral (HC; including live colonies that are bleached) and recently killed coral (RKC).

biodiverse, supporting 35% of marine life (Reaka-Kudla 1997). The ecosystem services provided by coral reefs are important economically, socially, and culturally for many nations (Hoegh et al. 2019). However, there is burgeoning evidence of a decline in the health of coral reefs, with an estimated average decrease from 36% to 19% live coral cover globally between 1998 and 2018 (Tebbett et al. 2023), and multiple widespread coral bleaching events that are increasing in frequency (Harrison et al. 2019; Hughes et al. 2017).

Considering these declines, it is critical to understand the current state of coral reefs, particularly in understudied locations such as the broader Red Sea in general, and Sudan in particular (Berumen et al. 2013; Klaus et al. 2008). Here we provide the results of recent surveys along 600 km of the Red Sea coast of Sudan. We recorded a mean coral cover in 2016 of 41.8% (±15.5 SE). These findings correspond with prior work undertaken in the early 2000s when a mean coral cover of 34–65% was observed (Hassan et al. 2002; Klaus et al. 2008; Ali 2011) and with the few published records of coral cover prior to 2000 (e.g., Reinicke et al. 2003; Braithwaite et al. 2000).

We found evidence of bleaching in both 2016 and 2017 (40% ±0.0 SE and 28% ±0.0 SE of hard coral colonies respectively), and ongoing bleaching was observed at both shallow and deeper sites during 2016 (E. Moland, pers. comm.). Mean bleaching prevalence was greater than that found in previous bleaching studies in Sudan, which observed that between 0% and 5% of coral colonies were bleached (PERSGA 2010; Ali

TABLE 3.3 Mean physical parameters for sea water (temperature, salinity, oxygen, density, and conductivity) for 2017, with latitude from south to north.

PARAMETERS	*UNIT*	*19° N (32 SITES)*	*20° N (6 SITES)*	*21° N (5 SITES)*
Temperature	(°C)	30	29.5	28.75
Salinity	(‰)	39.3	39.8	40.9
Oxygen	mg/l	3.8	4.1	4.2
Density	kg/m^3	25.27	25.68	26.60
Conductivity		63	64.1	64.9

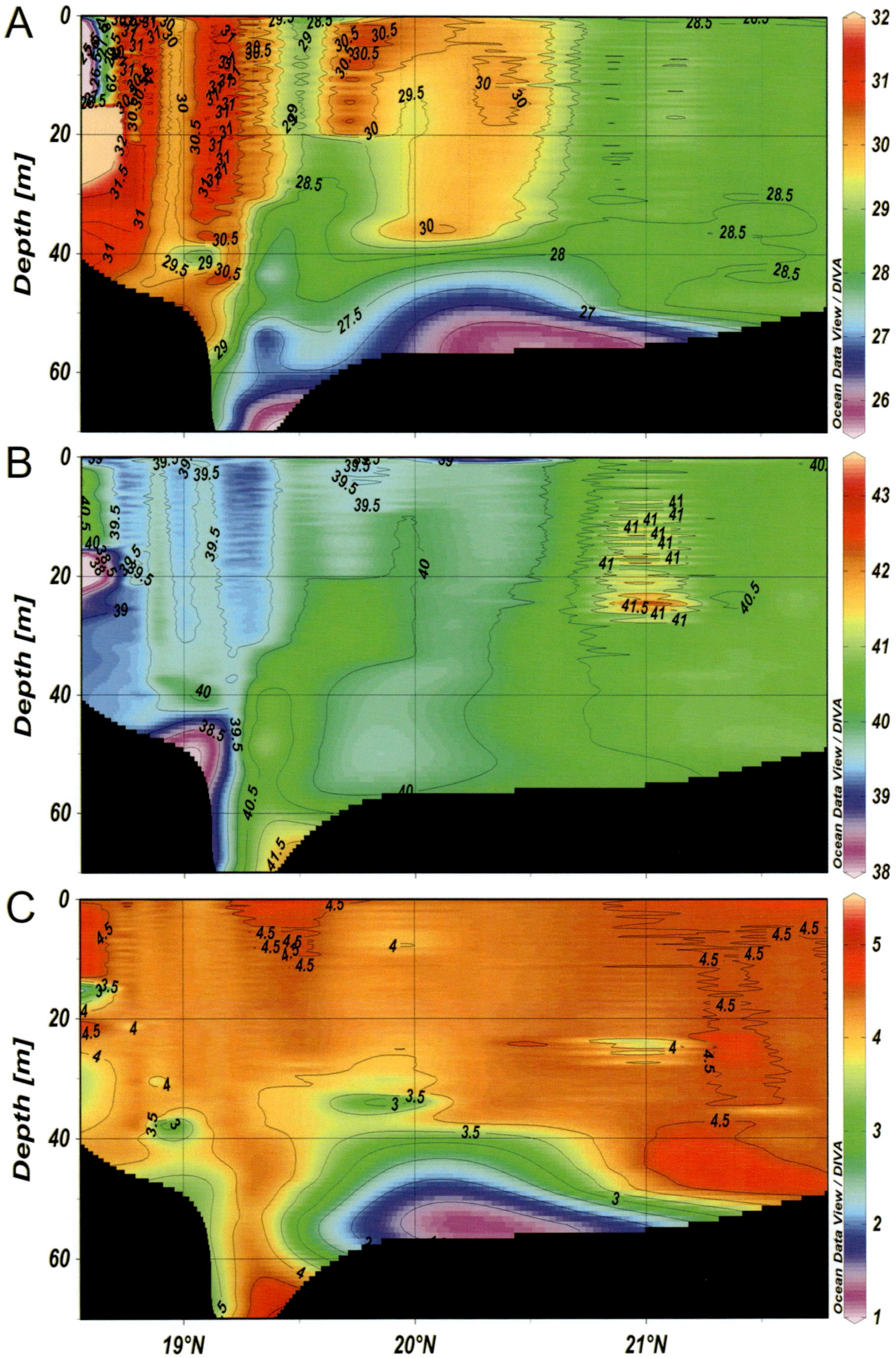

FIGURE 3.7 A) Water temperature (°C) of the upper 70 m of the water column measured in relation to location from the southernmost to the northernmost positions sampled in 2017. B) Salinity (‰) of the upper 70 m of the water column measured in relation to location from the southernmost to the northernmost positions sampled in 2017. C) Oxygen concentration (mg/l) in the upper 70 m of the water column measured in relation to location from the southernmost to the northernmost positions sampled in 2017.

2011). This greater prevalence of bleaching in 2016 and 2017 is not surprising considering the third global bleaching event that occurred throughout 2016 and 2017 (Eakin et al. 2019). Nonetheless, we found no latitudinal patterns in bleaching prevalence despite the latitudinal gradient in temperature. To explore the influence of temperature (or salinity and oxygen concentration) on coral cover and bleaching, there is now a need to move beyond snapshot data and bring together information on temperature trends and anomalies.

In 2017, we observed lower levels of coral cover (29.8% ±16 SE) than in 2016. The relatively high levels of bleaching observed in our surveys potentially indicate bleaching in 2016 that led to coral mortality and lower coral cover in 2017. However, there were considerable variations in the data, and we had low replication within sites. Furthermore, we saw no increase in recently killed coral from 2016 to 2017. As such, we suggest that there is currently insufficient evidence to assume widespread mortality of corals along the coast of Sudan between November 2016 and November 2017. This contrasts with the outcome of bleaching between April and September in the Arabian Gulf in 2017, which led to over 60% coral mortality (Burt et al. 2019). The potential resilience of Sudanese corals to bleaching-driven mortality is an exciting outcome of our study that needs further investigation but supports the burgeoning interest in research around the capacity for corals within the warm, saline Red Sea to adapt to a changing climate (Berumen et al. 2013; Gegner et al. 2017; Osman et al. 2018). Furthermore, relatively little research has focused on bleaching events in the Red Sea, compared to other coral reef regions (but see Osman et al. 2018). The combination of coral cover, bleaching, and oceanographic data along the whole Sudanese coast over two years thus provides an important addition to the literature.

Although bleaching may have contributed to a decline in coral cover between 2016 and 2017, low replication and the patchiness of reef benthic communities is likely driving sites exhibiting large increases in coral cover between 2016 and 2017. For example, we observed an increase in coral cover at Osseif reef from 16% (±3.5 SE) to 37.5% (±2.5 SE) between years. Expansion of the reef monitoring program to provide greater replication is an important first step for future work in the region.

Evaluation of Diver-Operated Video (DOV) methods have shown them to be robust and time-efficient for assessing benthic composition (Leujak and Ormond 2007; Lam et al. 2006). Here we used a stereo-video setup to quantify benthic communities. DOV methods allow for rapid completion of transects underwater, and comparison of percentage hard coral cover recorded in our surveys and prior work suggests this stereo-video system is useful for quantifying coral cover. However, the genera list (Appendix 3.1) generated through our study suggests the DOV method likely misses cryptic species or those with small corallites that are hard to distinguish on the videos. We only observed 33 genera compared to 42 (Ali and Elhag 2016) and 46 (Elmak 1988) recorded in prior work in Sudan, despite the larger-scale nature of our surveys. Genera that were not observed on our survey include *Leptastrea* and *Plesiastrea*, corals with small corallites and fine skeletal features. There is now a need to evaluate and test the stereo-video approach further to fully understand whether certain genera are poorly resolved on the videos. In addition, algal cover was low at all sites in both 2016 and 2017, even though our algal classification included both macroalgae and turf algae. We suggest that the video method may make it difficult to identify fine turfs, and thus the percentage algal cover (Table 3.2) is more likely representative of the macroalgae percentage cover rather than the total algal cover. Further work is needed to evaluate this proposition. The present work found that hard corals on reefs surveyed in Sudan's Red Sea State seem to be able to resist the effects of ocean warming, at least for the time being, and supports the notion of a 'Red Sea refuge' (Fine et al. 2013; Osman et al. 2018). However, with accelerating warming and other pending threats to Sudan's coral reefs – such as local pollution and unsustainable fishing practices – there is a continued need to survey and protect these unique coral reefs.

3.5 CONCLUSION

This study assessed the status of the coral reefs along the Sudanese Red Sea coast. Based on our findings of relatively high live coral cover, comparable to that of previous studies, it seems that the reef complexes along this part of the Red Sea are not yet seriously threatened by cumulative anthropogenic impacts. However, evidence of increased levels of coral bleaching suggests these reefs need careful and ongoing monitoring. Critically, this study constitutes an updated baseline for the included sites, and is an important source of information in an ongoing effort to develop a conservation and management plan for Sudan's unique coral reef ecosystems.

ACKNOWLEDGEMENTS

Coral reef surveys conducted in 2016 and 2017 were integral activities within the UNIDO administered project (SAP ID 130130) 'Building institutional capacities for the sustainable management of the marine fishery in the Red Sea State' (2015–2018), funded by the Norwegian Ministry of Foreign Affairs through its embassy in Khartoum. Khadija Y.I. Abaker's master's project at the Red Sea University was supported by the consecutive project phase (SAP ID 170230) 'Building institutional capacities for an ecosystem approach to management of the marine fisheries in the Red Sea State' (2018–2024) funded by the Norwegian Agency for Development Cooperation (NORAD). The content of this chapter was improved by insightful reviews provided by G. Banc-Prandi, A.R. Mohamed, R. Ormond, and P. Vine. We thank L. Segalini and the crew of the live-aboard M/Y Don Questo for their service during our expeditions.

REFERENCES

Ali, M.E. 2011. *Assessment of the health status of the Sudanese Red Sea fringing reefs*. Ph.D. Thesis, University of Khartoum, Sudan.

Ali, M.E., and E.A. Elhag. 2016. Coral diversity and similarity along Sudanese Red Sea fringing reef. *Int J Adv Res* 4: 720–727.

Bemert, G., and R. Ormond. 1981. *Red Sea coral reefs*. London, UK: Routledge.

Berumen, M.L., Hoey, A.S., Bass, W.H., Bouwmeester, J., Catania, D., Cochran, J.E.M., Khalil, M.T., Miyake, S., Mughal, M.R., Spaet, J.L.Y., and P. Saenz-Agudelo. 2013. The status of coral reef ecology research in the Red Sea. *Coral Reefs* 32, no. 3: 737–748. doi:10.1007/s00338-013-1055-8.

Braithwaite, C.J.R., Montaggioni, L.F., Camoin, G.F., Dalmasso, H., Dullo, W.C., and A. Mangini. 2000. Origins and development of Holocene coral reefs: A revisited model based on reef boreholes in the Seychelles, Indian Ocean. *Int J Earth Sci* 89, no. 2: 431–445. doi:10.1007/s005310000078.

Burt, J.A., Paparella, F., Al-Mansoori, N., Al-Mansoori, A., and H. Al-Jailani. 2019. Causes and consequences of the 2017 coral bleaching event in the southern Persian/Arabian Gulf. *Coral Reefs* 38, no. 4: 567–589. doi:10.1007/s00338-019-01767-y.

Clarke, K.R. 1993. Non-parametric multivariate analyses of changes in community structure. *Austral J Ecol* 18: 117–143.

DiBattista, J.D., Roberts, M.B., Bouwmeester, J., Bowen, B.W., Coker, D.J., Cortés, D.F., et al. 2016. A review of contemporary patterns of endemism for shallow water reef fauna in the Red Sea. *J Biogeogr* 43, no. 3: 423–439. doi:10.1111/jbi.12649.

Eakin, C.M., Sweatman, H.P.A., and R.E. Brainard. 2019. The 2014–2017 global-scale coral bleaching event: Insights and impacts. *Coral Reefs* 38, no. 4: 539–545. doi:10.1007/s00338-019-01844-2.

Elmak, A.M. 1988. *Zonation and community structure of the fringing reefs of Suakin (Sudanese Red Sea)*. M.Sc. Thesis, Department of Zoology, Faculty of Science, University of Khartoum, Khartoum, Sudan.

Fine, M., Gildor, H., and A. Genin. 2013. A coral reef refuge in the Red Sea. *Glob Change Biol* 19: 3640–3647. doi:10.1111/gcb.12356.

Furby, K.A., Bouwmeester, J., and M.L. Berumen. 2013. Susceptibility of central Red Sea corals during a major bleaching event. *Coral Reefs* 32, no. 2: 505–513. doi:10.1007/s00338-012-0998-5.

Gegner, H.M., Ziegler, M., Rädecker, N., Buitrago-López, C., Aranda, M., and C.R. Voolstra. 2017. High salinity conveys thermotolerance in the coral model Aiptasia. *Biol Open* 6, no. 12: 1943–1948. doi:10.1242/bio.028878.

Halpern, B.S., Frazier, M., Afflerbach, J., O'Hara, C., Katona, S., Stewart Lowndes, J.S., Jiang, N., Pacheco, E., Scarborough, C., and J. Polsenberg. 2017. Drivers and implications of change in global ocean health over the past five years. *PLoS One* 12, no. 7: e0178267. doi:10.1371/journal.pone.0178267.

Harrison, H.B., Álvarez-Noriega, M., Baird, A.H., Heron, S.F., MacDonald, C., and T.P. Hughes. 2019. Back-to-back coral bleaching events on isolated atolls in the Coral Sea. *Coral Reefs* 38, no. 4: 713–719. doi:10.1007/s00338-018-01749-6.

Hassan, M., Kotb, M.M.A., and A.A. Sofyani. 2002. Regional status of coral reefs in the Red Sea and the Gulf of Aden. In: *Status of coral reefs of the world*. Ed: C.R. Wilkinson, 45–52. Townsville: Australian Institute of Marine Science.

Head, S.M. 1980. *The ecology of corals in the Sudanese Red Sea*. Ph.D. Thesis, University of Cambridge, Cambridge, UK.

Hodgson, G., Kiene, W., Mihaly, J., Liebeler, J., Shuman, C., and L. Maun. 2004. Reef check instruction manual: A guide to reef check coral reef monitoring. *Reef Check*, Institute of the Environment, University of California at Los Angeles.

Hoegh-Guldberg, O., Pendleton, L., and A. Kaup. 2019. People and the changing nature of coral reefs. *Reg Studies Marine Sci* 30: 100699. doi:10.1016/j.rsma.2019.100699.

Hughes, T.P., Anderson, K.D., Connolly, S.R., Heron, S.F., Kerry, J.T., Lough, J.M., et al. 2018. Spatial and temporal patterns of mass bleaching of corals in the anthropocene. *Science* 359, no. 6371: 80–83. doi:10.1126/science.aan8048.

Hughes, T.P., Graham, N.A.J., Jackson, J.B.C., Mumby, P.J., and R.S. Steneck. 2010. Rising to the challenge of sustaining coral reef resilience. *Trends Ecol Evol* 25, no. 11: 633–642.

Hughes, T.P., Kerry, J.T., Álvarez-Noriega, M., Álvarez-Romero, J.G., Anderson, K.D., Baird, A.H., et al. 2017. Global warming and recurrent mass bleaching of corals. *Nature* 543, no. 7645: 373–377. doi:10.1038/nature21707.

Johnson, M.D., Scott, J.J., Leray, M., Lucey, N., Rodriguez Bravo, L.M., Wied, W.L., and A.H. Altieri. 2021. Rapid ecosystem-scale consequences of acute deoxygenation on a Caribbean coral reef. *Nat Comm* 12, no. 1: 4522. doi:10.1038/s41467-021-24777-3.

Kessel, S.T., Elamin, N.A., Yurkowski, D.J., Chekchak, T., Walter, R.P., Klaus, R., Hill, G., and N.E. Hussey. 2017. Conservation of reef manta rays (Manta alfredi) in a UNESCO World Heritage Site: Large-scale island development or sustainable tourism? *PLoS One* 12, no. 10: e0185419. doi:10.1371/journal.pone.0185419.

Klaus, R., Kemp, J., Samoilys, M., Anlauf, H., El Din, S., Abdalla, E.O., and T. Chekchak. 2008. *Ecological patterns and status of the reefs of Sudan*. 11th Int Coral Reef Symp, Ft. Lauderdale, Florida.

Lam, K., Shin, P.K.S., Bradbeer, R., Randall, D., Ku, K.K.K., Hodgson, P., and S.G. Cheung. 2006. A comparison of video and point intercept transect methods for monitoring subtropical coral communities. *J Exp Mar Biol Ecol* 333, no. 1: 115–128. doi:10.1016/j.jembe.2005.12.009.

Leujak, W., and R.F.G. Ormond. 2007. Comparative accuracy and efficiency of six coral community survey methods. *J Exp Mar Biol Ecol* 351, no. 1: 168–187. doi:10.1016/j.jembe.2007.06.028.

Mergner, H., and H. Schuhmacher. 1981. Quantitative analyse der korallenbesiedlung eines vorriffareals bei Aqaba (Rotes Meer). *Helgoländer Meeresunters* 34, no. 3: 337–354.

Mergner, H., and H. Schuhmacher. 1985. Quantitative analyse von korallengemeinschaften des Sanganeb-Atolls (mittleres Rotes Meer). I. Die besiedlungsstruktur hydrodynamisch unterschiedlich exponierter außen-und innenriffe. *Helgoländer Meeresunters* 39, no. 4: 375–417.

Mergner, H., Schuhmacher, H., and D.K. Kroll. 1992. Long-term changes in the coral community of a fore-reef area near Aqaba (Red Sea): 1976–1989. *Proc 7th Int Coral Reef Sym* 1: 104–113.

Monroe, A.A., Ziegler, M., Roik, A., Röthig, T., Hardenstine, R.S., Emms, M.A., Jensen, T., Voolstra, C.R., and M.L. Berumen. 2018. In situ observations of coral bleaching in the central Saudi Arabian Red Sea during the 2015/2016 global coral bleaching event. *PLoS One* 13, no. 4: e0195814. doi:10.1371/journal.pone.0195814.

Oksanen, J., Blanchet, F.G., Kindt, R., et al. 2022. *Vegan: Community Ecology Package*. R Package Version 2.6–4.

Olsen, E., Axelsen, B.E., Moland, E., Utne-Palm, A.C., Elamin, E.M., Mukhtar, M.A., Saleh, A.M., Elamin, S.M., Iragi, M.A., and S.G.F. Gumaa. 2021. Distribution and diversity of fish species along the Sudanese Red Sea coast based on three combined trap and gillnet surveys. *Fish Res* 242: 106032. doi:10.1016/j.fishres.2021.106032.

Ormond, R.F.G. 1976. The Red Sea. In: *Promotion of the establishment of marine park and reserves in the northern Indian Ocean including the Red Sea and Persian Gulf*. Proc Regional Meeting, Tehran, Iran, 6–10 March 1975, pp. 115–123.

Osman, E.O., Smith, D.J., Ziegler, M., Kürten, B., Conrad, C., El-Haddad, K.M., Voolstra, C.R., and D.J. Suggett. 2018. Thermal refugia against coral bleaching throughout the northern Red Sea. *Glob Change Biol* 24: e474–e484. doi:10.1111/gcb.13895.

PERSGA. 2010. *The status of coral reefs in the Red Sea and Gulf of Aden: 2009*. PERSGA, The Regional Organization for the Conservation of the Environment of the Red Sea & Gulf of Aden Technical Series. Jeddah, Saudi Arabia.

Rasul, N.M.A., Stewart, I.C.F., and Z.A. Nawab. 2015. Introduction to the Red Sea: Its origin, structure, and environment. In: *The Red Sea: The formation, morphology, oceanography and environment of a young ocean basin*. Eds: N.M.A. Rasul, and I.C.F. Stewart, 1–28. Berlin Heidelberg: Springer Earth System Sciences.

Reaka-Kudla, M.L. 1997. The global biodiversity of coral reefs: A comparison with rainforests. In: *Biodiversity II: Understanding and protecting our natural resources*. Eds: M.L. Reaka-Kudla, D.E. Wilson, and E.O. Wilson, 88–108. Washington, DC: Joseph Henry/National Academic Press.

Reinicke, G.B., Kroll, D.K., and H. Schuhmacher. 2003. Patterns and changes of reef-coral communities at the Sanganeb-Atoll (Sudan, central Red Sea): 1980 to 1991. *Facies* 49, no. 1: 271–297. doi:10.1007/s10347-003-0035-9.

Savary, R., Barshis, D.J., Voolstra, C.R., Cárdenas, A., Evensen, N.R., Banc-Prandi, G., Fine, M., and A. Meibom. 2021. Fast and pervasive transcriptomic resilience and acclimation of extremely heat-tolerant coral holobionts from the northern Red Sea. *Proc Natl Acad Sci* 118, no. 19: e2023298118. doi:10.1073/pnas.2023298118.

Schroeder, J.H., and D.H. Nasr. 1983. The fringing reefs of Port Sudan, Sudan: I. Morphology, sedimentology, zonation. *Esser Geogr Arb* 6: 29–44.

Spalding, M.D., Ravilious, C., and E.P. Green. 2001. *World atlas of coral reefs*. Berkeley, CA: University of California Press.

Tebbett, S.B., Connolly, S.R., and D.R. Bellwood. 2023. Benthic composition changes on coral reefs at global scales. *Nat Ecol Evol* 7, no. 1: 71–81. doi:10.1038/s41559-022-01937-2.

Teh, L.S.L., Teh, L.C.L., and U.R. Sumaila. 2013. A global estimate of the number of coral reef fishers. *PLoS One* 8, no. 6: e65397. doi:10.1371/journal.pone.0065397.

Veron, J.E.N., and M. Stafford-Smith. 2000. *Corals of the World*, vols. 1–3. Australian Institute of Marine Science. Townsville, Queensland, Australia: Odyssey Publishing.

Vine, P.J., and M.P. Vine. 1980. *Ecology of Sudanese coral reefs with particular reference to reef morphology and distribution of fishes*. Proc Symp Coast Mar Environ Red Sea and Gulf of Aden and Tropical Western Indian Ocean, University of Khartoum, Sudan, 2, pp. 87–140.

Appendix 3.1

Genera list of hard corals encountered across all study sites during surveys in 2016 and 2017

FAMILY	*GENUS*
Acroporidae	Acropora
	Anacropora
	Astreopora
	Montipora
Agariciidae	Leptoseris
	Pachyseris
	Pavona
Dendrophylliidae	Turbinaria
Faviidae	Cyphastrea
	Diploastrea
	Echinopora
	Erythrastrea
	Favia
	Favites
	Goniastrea
	Platygyra

FAMILY	*GENUS*
Fungiidae	Ctenactis
	Fungia
	Podobacia
Mussidae	Lobophyllia
	Symphyllia
Oculinidae	Galaxea
Pectiniidae	Echinophyllia
	Mycedium
	Oxypora
Pocilloporidae	Pocillopora
	Seriatopora
	Stylophora
Poritidae	Goniopora
	Porites
Siderastreidae	Coscinaraea
	Siderastrea
Trachyphylliidae	Trachyphyllia

Status of coral reefs along the Egyptian Red Sea coast

4

Ahmed Ghallab, Hussein N.M. Hussein,
Hashem Madkour, Alaa Osman and Aldoushy Mahdy

4.1 INTRODUCTION

The Egyptian coastline possesses a significant proportion and a considerable range of the coral reefs found in the Red Sea, with about 3800 km^2 of reef area (Spalding et al. 2001) distributed over a length of 1800 km (PERSGA 2010). The earliest studies on the coral reefs of the Red Sea were undertaken by Peter Forsskål as part of the Danish "Arabia Felix" expedition in 1761–1767 (Head 1987). Coral reefs extend along the length of the Red Sea to the tip of the Gulf of Aqaba. The basic form of coral reefs in the Red Sea is that of fringing reefs close to shore, with widths varying from a few metres to more than 1 km. The reefs are not continuous throughout the sea, but are separated by narrow channels known as Marsas or Sharms that formed from drowned river valleys connected to wadis (Head 1987). The Egyptian reefs extend in the north to the Gulfs of Suez and Aqaba and to Ras Hedarba in the south at the border of Sudan. They are not, however, continuous due to periodic flooding from wadis, which created gaps in the reef system.

The northern part of the Red Sea has the highest coral diversity and number of islands, while the south has the highest terrestrial biodiversity for the whole country (Hassan et al. 2002; Shaalan 2005). The Red Sea coral reefs are cited as the most diverse in the world (Loya 1972). The Red Sea is one of the most important repositories of marine biodiversity in the world and has unique marine habitats, including coral reefs, mangroves, and seagrass; the species endemism in the Red Sea is extremely high, particularly among some groups of reef fishes and reef-associated invertebrates (PERSGA 2010; Dawoud et al. 2016; Voolstra and Berumen 2019). The Red Sea is home to around 300 hermatypic coral species and 125 ahermatypic coral species; 70% of the stony coral species, including a variety of endemic species, are present in the Egyptian reefs (Kotb et al. 2008). These numbers are higher than those recorded for the Caribbean and are similar to those in the Indian Ocean (Hilmi et al. 2012). Several authors have studied the Red Sea reefs and their communities, e.g., Crossland (1938), Ammar (2004), and Al-Hammady (2011).

The coral reefs along the whole Egyptian coast have been reviewed by many authors who provided regional studies for the northern and southern Egyptian coasts. The coral distribution in some localities has been studied generally, referring to the community structure of coral reefs (Crossland 1938; Ammar and Nawar 1998; Ammar 2004; Al-Hammady 2011; Al-Hammady and Mahmoud 2013; Al-Hammady et al. 2015; Ismail et al. 2017), the ecology, and biology (Loya 1976; Kotb 1996; Kotb et al. 2001; Mohammed 2006; Dosoky et al. 2021). Generally, the Red Sea coral reef status at the Gulf of Aqaba is good, with an average live coral cover ranging from 20–50%, a positive ratio of live to dead coral cover, and high species diversity (Kotb et al. 2004). The average percentage of live coral cover for the Egyptian Red Sea is 45% at 5 m and 33% at 10 m (Hassan et al. 2002). The coral cover is significantly higher in the Red Sea than in the Gulf of Aqaba, ranging from 16–67% at 5 m depth, with an average of 45% in the Red Sea and 35% in the Gulf of Aqaba. The percentage of live cover varies depending on the geomorphological types of reefs in the Red Sea (Kotb et al. 2004).

On a global scale, the value of the total economic goods and services provided by coral reefs has been estimated to be US $375 billion per year, most of which is coming from recreation, sea defense services, and food production; this equates to an average value of around US$ 6,075 per hectare of coral reef per year (Edwards and Gomez 2007). According to a report on biodiversity conservation capacity building in Egypt (Egyptian Biodiversity CHM 2006), more than 8 million tourists visit Egypt annually, where coastal tourism is the largest subsector within the market. The Red Sea is one of the world's major tourist destinations, and reef-associated tourism is a major source of income for some Red Sea countries. For example, coral reef–related tourism contributes 3.5% to Egypt's Gross Domestic Product (GDP) (Hilmi et al. 2012; Mohamed et al. 2012). The Red Sea provides key resources for coastal populations such as food, shoreline protection and stabilization, and economic benefits from tourism (Ahmed 2010). In the last few decades, increased human settlement in coastal areas and the resultant increase in artisanal and commercial fishing activities to support local consumption, and to supply a growing international trade in coral reef species and products, have led to the depletion of many commercially valuable species (Bruckner et al. 2011).

DOI: 10.1201/9781003321392-5

Despite their obvious value, either inshore or offshore Red Sea reefs are subject to extensive anthropogenic damage (Al-Hammady 2011; Ammar 2011; Al-Hammady and Mahmoud 2013). The factors responsible for the observed declines in Red Sea offshore reefs include diving (Zakai and Chadwick-Furman 2002), fishing activities (Ault et al. 2009), and ship groundings (Lirman et al. 2010). Inshore reefs are impacted by a wide range of threats such as landfilling (Erftemeijer et al. 2012), sedimentation (Loya 1976; Rogers 1990), coastal development (Al-Hammady and Mahmoud 2013), swimming and snorkelling (Hannak et al. 2011), coral collection (Ginsburg et al. 2001), oil pollution (Sebens 1994; Al-Hammady 2011), and eutrophication (Koop et al. 2001). Coral bleaching as a result of global changes is another important factor that impacts both inshore and offshore reefs (Al-Hammady and Mahmoud 2013; Evensen et al. 2022). Furthermore, biotic factors may destroy inshore and offshore reefs, such as sea urchins *Diadema* spp. (Ammar 2011), the crown of thorns starfish *Acanthaster planci* (Ammar et al. 2007), macroalgal competition (Dustan 1999; El-Sadek et al. 2022), gastropods, *Coralliophila violacea* (Ghallab et al. 2021), and diseases (Richardson and Voss 2005; Al-Hammady and Mahmoud 2013). There is still a scarcity of data on the susceptibility of inshore and offshore reefs to decline and the extent of the impacts on Red Sea corals. The Egyptian coast of the Red Sea combines these complex issues. Tourism, along with destructive fishing methods, diving activities, and coastal development, pose the greatest threat (Cesar et al. 2003).

Coral reefs worldwide are subjected to extensive anthropogenic damage (Sebens 1994; Al-Hammady and Mahmoud 2013; Al-Hammady et al. 2015), and their existence is threatened by the economic activities they support (White et al. 2000). Although the rapidly expanding tourism industry is considered an extremely important economic activity, it has caused increasing pressure on the coral reefs of Egypt, and damage occurs from both direct and indirect impacts of tourism activities on coral communities (Barrania 2010). Two anthropogenic factors contributing to the coral reef decline are eutrophication (Koop et al. 2001) and damage from snorkelers and SCUBA divers (Zakai and Chadwick-Furman 2002). Among the direct impacts are trampling, coral breaking by divers or snorkelers, damage from recreational boat anchoring, and boat grounding. Riegl and Velimierov (1994) found that coral breakage was the most common damage, especially on highly frequented reefs.

Coral reef ecosystems have suffered an unprecedented loss of habitat-forming stony corals in recent decades. Although marine conservation has historically focused on passive habitat protection, demand for and interest in active habitat restoration have been growing in recent decades. The interaction of many factors affecting the distribution and coral bleaching includes sedimentation, overfishing, tourist activities, petroleum and phosphate production, and discharge of desalination plants into the marine environment (Mohammed and Mohamed 2005; Mohammed et al. 2009; Madkour 2013; Nasr et al. 2019; Boström-Einarsson et al. 2020).

The abundance, distribution, and percentage cover of the stony and soft corals have been studied by many authors in the Red Sea and the Indo-Pacific regions, including the distribution patterns along the Egyptian Red Sea coasts (Mohammed et al. 2010; Ismail et al. 2017; Dosoky et al. 2021). Coral cover in the Egyptian Red Sea had declined more than 30% in some places during recent decades, coupled with a significant increase of broken and damaged colonies (Jameson et al. 2007). Despite some southern Red Sea regions being more prone to bleaching, recent studies from the central Red Sea demonstrate that local corals can tolerate these high temperatures for brief periods of time (Voolstra et al. 2021; Evensen et al. 2022).

Reef flat areas typically have a live coral cover of 11–35%, while reef walls have a live coral cover of 12–85% and reef slopes of 2–62% (Abou Zaid 2000). On average, the percentage of stony coral cover remains stable from north to south, but soft coral cover slightly increases towards the south. The mean size of stony and soft corals increases towards the southern part of the Egyptian Red Sea (Kotb et al. 2001; Ismail et al. 2017). Seven common genera of soft corals have been recorded in the Egyptian Red Sea, including *Xenia*, *Heteroxenia*, *Sarcophyton*, *Lobophytum*, *Litophyton*, *Sinularia*, and *Dendronephthya* (Roushdy 1954; Vine 1986), although recently Ismail et al. (2017) recorded eleven soft coral genera in the Red Sea. Soft coral is among the most important reef communities that can produce natural products. Recently, soft corals are becoming more studied by pharmaceutical companies in their search for new medicines (Tammam et al. 2020; Abdelkarem et al. 2021).

Threats to coral reefs and the goods and services they provide are both natural and anthropogenic in origin. Natural disturbances have a high degree of natural variability due to large-scale episodic disturbances such as major storms, hurricanes, typhoons, and cyclones (Wilkinson 2008; Bruckner et al. 2011). Those disturbances can result in large-scale fluctuations in coral communities (Bruno and Selig 2007), which are part of the natural regime under which coral reefs have evolved (Moberg and Folke 1999). A recent assessment reported that the world has effectively lost 19% of its coral reefs, with 35% under threat in the next 10 to 40 years (Wilkinson 2008; ICRI 2012; Hoegh-Guldberg et al. 2017). Fortunately, 46% of the world's reefs are regarded as being relatively healthy and not under immediate threat of destruction (except from climate change). Declines in coral reef status vary around the world, reviewed by ICRI (2012), with a higher proportion of reefs effectively lost in the western Atlantic and the Indian Ocean than in the Red Sea, Australia, and the Pacific islands. The most serious declines have been recorded in Asia and the Arabian Gulf. Within the Asian Pacific region, the reefs of Southeast Asia are most seriously threatened, with 40% of reefs effectively lost, 45% under threat, and 15% at low threat. In contrast, more reefs in Australia and the Pacific islands are in better condition, with 2–8% effectively lost, 2–35% under threat, and 44–90% at low threat. Nonetheless, Bruno and Selig (2007) showed that the average coral cover on Indo-Pacific reefs has declined from approximately 50% to 22% in the last four decades, due to major storms and a variety of direct and indirect human impacts. The biological diversity of other animals and plants associated with coral reefs is high, and many endemic species make this region a globally significant repository of biodiversity. The reefs and associated seagrass beds also support endangered turtles and dugongs (Kotb et al. 2008).

Coral reefs in the Red Sea, like much of the rest of the world's reefs, are being degraded by a host of human and natural factors, such as wind-driven waves, tropical storms, tides, temperature, and salinity (Kotb et al. 2008; PERSGA 2010). Sedimentation, which may be enhanced by anthropogenic activities (Loya 1976; Rogers 1990), is also known to affect coral community structure and can damage coral colonies. Coral colonies affected by natural or anthropogenic stressors may suffer partial mortality, which has been shown to be a good indicator of reef condition (Ginsburg et al. 2001). Generally, eutrophication, increased sedimentation flowing from a disturbed terrestrial environment, diving, fishing activities, mining, and oil pollution are the main causes of reef destruction and decline (Sebens 1994). The coastal and marine resources of the Red Sea have contributed to the food, energy, industrial, and recreational needs of Egypt (Hilmi et al. 2012). However, environmental problems are found along the Red Sea coasts and Island due to recreation, tourism activities, landfilling, dredging, water pollution, solid waste disposal, phosphate pollution, and fishing practices, increased marine activities, including the number of fishing boats, as well as many petroleum pollution incidents. Since 1995, Law 102/1983 of the Ministry of State for Environment in Egypt (Nature Conservation Sector) has declared 22 islands in the Red Sea as protected areas, from Gifton Island off Hurghada (north) to Halayeb Island at the Egyptian border with Sudan. In 2006, the northern islands (Tawila, Ashrafi (Mokwarate), Ghanim, Small Gubal, North Umm El-Himmat, South Umm El-Himmat, North Geisum and South Geisum) were added to the Red Sea Protected Islands. Most of these northern islands are remote areas located near the entrance of the Gulf of Suez, and they have strategic importance for maritime transportation. In addition, these islands are surrounded by petroleum and fishing activities.

The majority of earlier studies on the substrate cover of the Egyptian Red Sea coast were dispersed, concentrated on the north (the Gulf of Aqaba and Ras Mohamed) or the south (Hurghada to Marsa Alam), and ignored the entire Red Sea and islands, particularly those in the north. The current study was created to fill this gap by examining the distribution and variety of living natural resources throughout the entire Red Sea of Egypt (the north, the middle, and the south). In light of the outcomes of the COP27 summit meeting, which was held in Sharm El-Sheikh, Egypt, in December 2022, the findings of this research may aid decision-makers in taking the necessary steps to protect coral reefs (both hard and soft) in the Red Sea.

4.2 MATERIALS AND METHODS

4.2.1 Study sites

This study was carried out at nineteen sites along the Egyptian Red Sea coasts (i.e., the northern and southern sections). Eight northern Red Sea islands that lie at the entrance of the Gulf of Suez were surveyed: Tawila, Ashrafi (also called Mokwarate), Ghanim, Small Gubal, North Umm El-Himmat, South Umm El-Himmat, North Geisum, and South Geisum Island (Figures 4.1 and 4.2). There were four sites offshore from Hurghada Island (El-Fanous Reef, Sabina Reef, Small Gifton Island, and Gota Abu-Rmada), and seven sites (Marsa Nekari, Elphinstone Reefs, Zabargad Island, Samadi Reef, El-Hamrawein, Al-Qusier Port, and Abu-Tartour) were chosen in the south (Figures 4.1 and 4.2). The data were collected during 2017–2019 and involved key marine habitat biota (flora and fauna), including substrate cover of stony corals, soft corals, dead corals, rubble, sand, algae, and sponges. The general characteristics and exact position of these sites are summarised as follows.

4.2.1.1 Tawila Island

This island lies at 27°35′15.24″N and 33°45′52″E, with a total area of 21.5 km^2. It lies at about 22 km offshore, with a sandy beach, surrounded by many shallow lagoons. It is used for sporting, tourism activities, and snorkeling. During the current study, the construction of hotels was starting on the island.

4.2.1.2 Ashrafi Island

This island lies at 27°45′57″N and 33°42′4.49″E, with a total area of 1.4 km^2 and is about 12 km offshore. Ashrafi Island consists of three longitudinal small islands located at the entrance of the Gubal Strait. It is of coral origin and is characterized by submerged coral reefs separated by narrow channels.

4.2.1.3 Ghanim Island

This island is located at 27°46′23″N and 33°35′51.7″E near the coast of Gabal El Zeit (Petroleum Company), with a total area of 4.6 km^2 and is about 3 km offshore. Ghanim Island is located near popular Marsa Ras El Bahar (a landing area for small fishing boats), and is surrounded by barrier submerged reefs that extend to Ashrafi Island.

4.2.1.4 Small Gubal Island

This island has a total area of 1.5 km^2 and is located at 27°41′23″N and 33°46′34.6″E, about 30 kilometres offshore at the southern entrance to the Gulf of Suez, and is distinguished by a sandy beach used for safari boat tourism, as well as a hawksbill turtle nesting site.

4.2.1.5 Northern umm El-Himmat Island

This island is located at 27°39′9.1″N and 33°38′19.45″E, and lies about 4.5 km from the beach at Ras Gemasha. It is surrounded by a barrier-submerged reef, and a petroleum platform is located about 3 km east of the island.

4.2.1.6 South umm El-Himmat Island

This island is located at 27°37′56.91″N and 33°40′29.91″E. It lies at the south of North Umm El-Himmat Island, and it is also surrounded by a barrier-submerged reef.

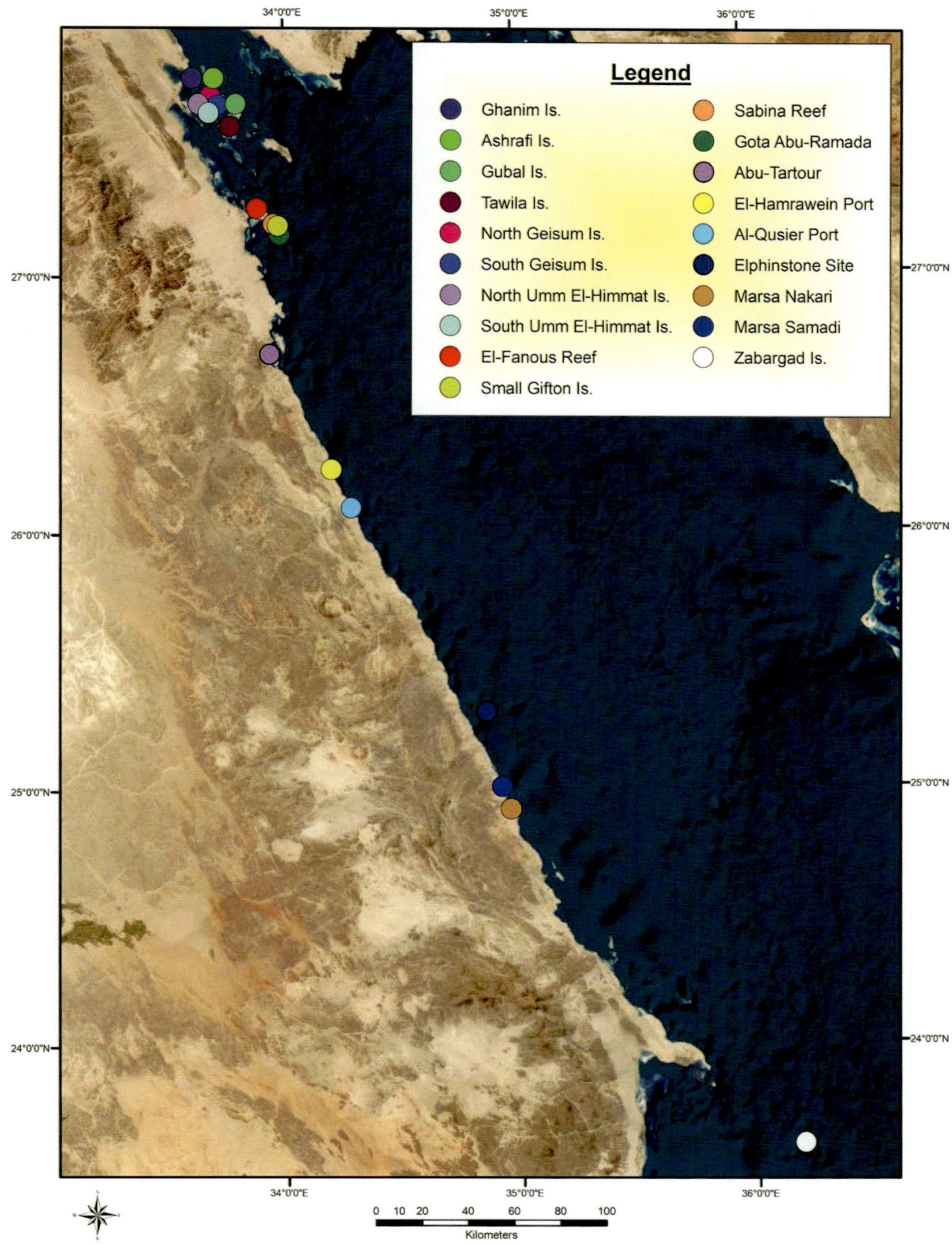

FIGURE 4.1 Location of the reef sites at the Egyptian Red Sea coast.

4.2.1.7 North Geisum Island

This island is located at 27°41′5.63″N and 33°41′26.78″E and is about 17 km from the beach. It is separated from South Geisum Island by a submerged back reef.

4.2.1.8 South Geisum Island

This island lies at 27°39′4.29″N and 33°42′33.25″E, about 5 km west of Gubal Island, with a total area of about 9.7 km^2. This island is characterized by a monospecific mangrove stand (*Avicennia marina*), and is surrounded by a submerged back reef and a few lagoons; it has an old landfill pier in the south that extends out for 200 m.

4.2.1.9 Gota Abu-Rmada

Gota is a local word meaning a small area arising from an offshore submerged reef. It is located at 27°8′24.24″N, 33°57′12.41″E, about 12 km from Hurghada City, with a semi-circular reef of 36 km^2 area, surrounded by a shallow flat sandy seabed, with a maximum depth of 20 m. The reef at this

FIGURE 4.2 Photographs of Red Sea northern islands.

site has two sides, one exposed on the northern side (the outer lagoon) and the sheltered southern reef side. Those sites are used daily by boats from Hurghada for diving and snorkeling, and they are under threat by overexploitation of these activities.

4.2.1.10 El-Fanous Reef

El-Fanous reef is named locally Shaab El-Fanous; Shaab means coral reef, and El-Fanous means the lighthouse found on top of the reef. This site is located at 27°15′52″N and 33°53′01″E, off the Hurghada City coast, and is 6 km from the new port of Hurghada Marina. Famous diving spots are situated around this reef's eastern and western corners. Many daily trips are undertaken to watch the Indo-Pacific bottlenose dolphins.

4.2.1.11 Sabina Reef (Shaab Sabina)

This site is located at 27°12′54.14″N, 33°57′9.24″E, in a channel between two islands, Big and Small Gifton. The Red Sea here is perfectly calm on the surface but has occasional strong currents and is a shallow dive site, with dive depths that range from 5 to 14 metres. The corner of the top reef divides the northern coral garden from the sheltered southern side; the southern part of this reef is mostly dead. At about 150 metres off the main reef is a picturesque shallow lagoon. The nutrient-rich water has made the north side of the reef wall sparkle with life, and the bottom around the erg is mainly sandy with some fire coral formations. Table and brain corals can be found at Sabina Reef.

4.2.1.12 Small Gifton

This site is located at 27°11′09″N, 33°58′53″E, at about 10 km from the Hurghada coast. It is a rocky island, surrounded with fringing reefs and a narrow reef flat. It has tourist recreational activities, anchorage for boats, and popular diving sites are found on the eastern and southern sides.

4.2.1.13 Abu-Tartour

Abu-Tartour is a harbour for mining transportation, located 6 km south of Safaga and 66 km from Hurghada City (the capital of the Red Sea governorate) at 26°41′13.58″N, 33°56′6.01″E.

4.2.1.14 El-Hamrawein Port

This is a harbour for shipping phosphates located 20 km north of Al-Qusier City at 26°14′33.65″N, 34°12′34.25″E. The tidal flat off Al-Hamrawein is very narrow and extends smoothly with a very gentle slope towards the sea.

4.2.1.15 Al-Qusier Port

This is at Al-Qusier City (impacted by eutrophication as a result of phosphate shipments), located at 26°5′59.18″N, 34°16′58.14″E. It is one of the oldest harbours on the Egyptian Red Sea coast, and the reefs have been impacted by phosphate shipments, sewage and anthropogenic activities from houses near the shore, and movement of fishing boats.

4.2.3.16 Samadi Reef

This site is located in the vicinity of the city of Marsa Alam at 24°59′15″N, 34°59′53″E. The reef is horseshoe-shaped, forming a shallow lagoon inside that ranges in depth from 6–7 m in the north, which is mainly sandy on the bottom except for patches of coral, to 12–15 m in the rest of the lagoon. The Red Sea Protectorates proposed a management plan with a simple subdivision into zones to simplify enforcement.

4.2.1.17 Elphinstone site

This offshore site is located 25 km north of Marsa Alam at 25°18′15.52″N, 34°51′46.57″E. It is a popular site in the southern Egyptian Red Sea for recreational diving activities, where oceanic whitetip sharks occur.

4.2.1.18 Marsa Nekari

This is a nearshore reef site located at 25 km south of Marsa Alam at 24°56′28.42″N, 34°59′0.17″E.

4.2.1.19 Zabargad Island

The Arabic name Zabargad refers to the mineral peridot. It lies at 23°35′27.31″N, 36°12′9.76″E. Currently the island is a natural protectorate, and it is considered to be one of the most important breeding sites of green turtles.

4.2.2 Field work

The data on coral reef assemblages were collected using SCUBA diving and snorkeling (free diving) during the period from 2017 to 2019, and were recorded using waterproof sheets at depths of 5 m. The benthic substrate cover survey was performed using the Line Intercept Transect (LIT) methods of English et al. (1997). Transects were laid along the depth contour from randomly chosen starting positions on the reef slopes. A 25-m-long tape transect was used to record the coral genera of the substrate cover (stony corals and soft corals). At the same time, data on dead corals, algae, sand, rubble, rocks, and sponges were recorded in order to evaluate the percentage cover of substrate every one-metre length along the LIT at each site. All colonies within each line transect were recorded, identified at generic level, counted, and photographed with a GoPro underwater camera. The different coral genera of the study areas were identified according to Sheppard and Sheppard (1991), Wallace (1999), Veron (2000), and Richards (2018).

4.2.3 Biotic and abiotic components surveyed

In the present study, biotic (living) and abiotic (nonliving) components were surveyed; abiotic components comprised sand, rubble, rock, and dead corals, while stony corals, soft corals, algae, and sponges represented the biotic components.

TABLE 4.1 Classification of coral reef genera (stony, soft, and fire corals) collected from study sites along the Egyptian Red Sea coast.

Phylum:					Porifera	(Sponges)
Phylum:					Cnidaria	
Class					Hydrozoa	
			Family		Milleporidae	
				Genus	*Millepora*	(Fire corals)
Class					Anthozoa	
	Subclass				Alcyonaria or Octocorallia	(Soft corals)
		Order			Alcyonacea	
			Family		Alcyoniidae	
				Genus	*Sarcophyton*	
					Sinularia	
					Lobophytum	
			Family		Xeniidae	
				Genus	*Xenia*	
		Suborder			Alcyoniina	
			Family		Nephtheidae	
				Genus	*Dendronephthya*	
	Subclass				Zoantharia or Hexacorallia	(Stony corals)
		Order			Scleractinia	
			Family		Acroporiidae	
				Genus	*Acropora*	
					Montipora	
			Family		Pocilloporidae	
				Genus	*Stylophora*	
					Pocillopora	
					Seriatopora	
			Family		Fungiidae	
				Genus	*Fungia*	
			Family		Poritidae	
				Genus	*Porites*	
			Family		Faviidae	
				Genus	*Favia*	
					Favites	
					Echinopora	
			Family		Merulinidae	
				Genus	*Platygyra*	
			Family		Oculinidae	
				Genus	*Galaxea*	

Sixteen living substrate taxa were surveyed (Table 4.1). Hydrocorals were the first taxa and belong to the genus *Millepora*, family Milleporidae, Class Hydrozoa, and Phylum Cnidaria. Ahermatypic corals (soft corals) were represented by four genera that comprised *Sarcophyton*, *Sinularia*, *Xenia*, and *Dendronephthya*, that belong to Subclass Alcyonaria (Octocorallia). The hermatypic corals (stony corals) comprised *Acropora*, *Montipora*, *Stylophora*, *Pocillopora*, *Seriatopora*, *Fungia*, *Porites*, *Favia* and *Platygyra*, which belong to Subclass Zoantharia (Hexacorallia). Both sponges and algae that were surveyed play a significant role in the marine environment (Table 4.1). The stony coral genera in the present study are classified according to life form into branching corals, comprising five genera, including *Acropora*, *Montipora*, *Stylophora*, *Pocillopora*, and *Seriatopora*, massive corals comprising three genera of *Porites*, *Platygyra*, and *Favia*, and solitary corals including only one genus *Fungia* (Table 4.1).

4.3 RESULTS

4.3.1 Distribution and percentage cover of living and nonliving components at northern islands

The percentage covers of both living and nonliving components are given in Table 4.2. Eight substrate cover categories (stony corals, soft corals, dead corals, rubble, algae, sand, rock, and sponges) in the eight northern studied islands were recorded and averaged 65.6 ± 27.0 %. The stony coral percentage covers in the northern islands were variable and ranged from the lowest percentage (41.3%) at Ashrafi Island to the highest (84%) at Ghanim Island (Table 4.3), with a mean cover of 60.1% ± 12.6. On the other hand, the soft coral percentage

TABLE 4.2 Substrate percentage cover (%) at the Egyptian Red Sea northern islands.

TYPE	*SUBSTRATE CATEGORY*	*TAWILA ISLAND*	*ASHRAFI ISLAND*	*GHANIM ISLAND*	*SMALL GUBAL*	*NORTH UMM EL-HIMMAT*	*SOUTH UMM EL-HIMMAT*	*NORTH GEISUM*	*SOUTH GEISUM*	*MEAN ± SD*	*TOTAL ±SD*
Living cover	Stony corals	52.0	41.3	84.0	64.5	65.0	64.0	61.1	48.5	60.1 ± 12.6	65.6 ± 27.0
	Soft corals	6.0	4.0	0.0	3.5	3.0	0.0	0.6	2.5	2.5 ± 2.1	
	Algae	5.0	9.3	0.0	0.0	4.0	0.0	0.0	1.5	2.5 ± 3.4	
	Sponges	0.3	0.0	0.0	0.0	0.0	0.0	0.0	4.5	0.6 ± 1.5	
Non-Living cover	Dead corals	30.7	36.0	10.7	31.5	26.0	34.4	34.3	39.0	30.3 ± 8.5	34.4 ± 13.4
	Rubble	3.3	4.0	0.0	0.5	1.0	0.0	1.1	2.0	1.5 ± 1.4	
	Sand	2.7	0.0	5.3	0.0	1.0	1.6	2.9	2.0	1.9 ± 1.7	
	Rocks	0.0	5.3	0.0	0.0	0.0	0.0	0.0	0.0	0.7 ± 1.8	

TABLE 4.3 Percentage cover (%) of stony coral genera at the Egyptian Red Sea northern islands.

STONY CORAL GENERA	*TAWILA ISLAND*	*ASHRAFI ISLAND*	*GHANIM ISLAND*	*SMALL GUBAL*	*NORTH UMM EL-HIMMAT*	*SOUTH UMM EL-HIMMAT*	*NORTH GEISUM*	*SOUTH GEISUM*	*MEAN ± SD*
Acropora	30.3	20.0	41.3	45.0	39.0	44.0	37.1	29.5	35.8 ± 8.3
Pocillopora	4.3	6.7	0.0	3.0	0.0	0.0	1.1	0.0	1.9 ± 2.5
Stylophora	10.7	12.0	38.7	5.0	23.0	17.6	19.4	14.5	17.6 ± 9.8
Porites	0.3	0.0	1.3	8.0	1.0	0.0	0.0	1.5	1.5 ± 2.6
Favia	0.7	1.3	0.0	2.5	0.0	0.8	0.0	1.5	0.9 ± 0.9
Millepora	1.0	0.0	0.0	0.5	0.0	0.0	0.0	0.0	0.2 ± 0.4
Montipora	2.0	0.0	0.0	0.0	0.0	1.6	0.6	1.0	0.7 ± 0.8
Seriatopora	1.7	1.3	2.7	0.0	1.0	0.0	2.9	0.5	1.3 ± 1.1
Platygyra	0.3	0.0	0.0	0.5	1.0	0.0	0.0	0.0	0.2 ± 0.4
Fungia	0.7	0.0	0.0	0.0	0.0	0.0	0.0	0.0	0.1 ± 0.2
Total	52	41.3	84.0	64.5	65.0	64.0	61.1	48.5	60.1 ± 12.6

TABLE 4.4 Percentage cover (%) of soft coral genera at the Egyptian Red Sea northern islands.

SOFT CORAL GENERA	*TAWILA ISLAND*	*ASHRAFI ISLAND*	*GHANIM ISLAND*	*SMALL GUBAL*	*NORTH UMM EL-HIMMAT*	*SOUTH UMM EL-HIMMAT*	*NORTH GEISUM*	*SOUTH GEISUM*	*MEAN ± SD*
Sarcophyton	1.0	0.0	0.0	0.5	0.0	0.0	0.6	1.0	0.4 ± 0.4
Sinularia	1.3	1.3	0.0	1.5	0.0	0.0	0.0	1.0	0.6 ± 0.7
Dendronephthya	0.0	2.7	0.0	1.5	3.0	0.0	0.0	0.5	1.0 ± 1.3
Xenia	3.7	0.0	0.0	0.0	0.0	0.0	0.0	0.0	0.5 ± 1.3
Total	6.0	4.0	0.0	3.5	3.0	0.0	0.6	2.5	2.5 ± 2.1

cover ranged from the lowest (0.6%) at North Geisum to the highest (6.0%) at Tawila Island (Table 4.4), with a mean of 2.5 ± 2.1% (Table 4.2). No soft corals were recorded at Ghanim and South Umm El-Himmat Islands (Table 4.4).

The nonliving component values varied but averaged 34.4 ± 13.4%. Dead coral cover ranged from 10.7% at Ghanim Island to 39% at South Geisum, with an average of 30.3 ± 8.5%. The cover of rubble had the lowest value of 0.5% at Small Gubal Island, the highest cover of 4.0% at Ashrafi Island, and was completely absent at Ghanim and South Umm El-Himmat Islands. The sand cover ranged from the lowest value of 1% at North Umm El-Himmat Island to a maximum of 5.3% at Ghanim Island, with a mean of 1.9 ± 1.7% (Table 4.2). Rocks were recorded only at Ashrafi Island, with a mean cover of 5.3%, and were absent at other sites.

The percentage cover and occurrence of other living components, such as algae and sponges, were variable. No algae were recorded at four islands (Ghanim, Small Gubal, South Umm El-Himmat, and North Geisum), while the highest algal cover recorded was 9.3% at Ashrafi Island, with a mean of

2.5%. Sponges were found on only two islands, with 0.3% and 4.5% coverage in Tawila and South Geisum Island, respectively (Table 4.2).

4.3.1.1 Percentage cover of stony coral genera

Nine genera of stony corals, in addition to one genus of hydrocorals, were represented in the northern Red Sea islands. These genera comprised the stony corals *Acropora*, *Pocillopora*, *Stylophora*, *Porites*, *Favia*, *Montipora*, *Seriatopora*, *Platygyra*, and *Fungia*, in addition to hydrocorals of the genus *Millepora* (Table 4.3). *Acropora* has the highest percentage cover of 45.0% at Small Gubal Island but declined to its lowest value of 20% at Ashrafi Island, with a mean cover of 35.8 ± 8.3 % (Table 4.3). With a mean of 17.6 ± 9.8%, the genus *Stylophora* came in second, with the highest value of 38.7% at Ghanim Island and the lowest (5.0%) at small Gubal Island. *Pocillopora* ranked third, with the highest value of 6.7% at Ashrafi Island and the lowest of 1.1% at North Geisum, for a mean of 1.9±2.5%, but was absent from four islands. The percentage covers of the remaining genera declined sharply and averaged 1.5, 0.9, 0.2, 0.7, 1.3, 0.2, and 0.1% for *Porites*, *Favia*, *Millepora*, *Montipora*, *Seriatopora*, *Platygyra*, and *Fungia*, respectively. The stony coral cover varies among the eight northern islands. The highest stony coral cover was 84% at Ghanim Island and the lowest 41.3% at Ashrafi Island, with a mean cover of 60.1 ± 12.6% (Table 4.3).

4.3.1.2 Percentage cover of soft coral genera

Four genera of soft corals were recorded at the northern Red Sea islands: *Sarcophyton*, *Sinularia*, *Dendronephthya*, and *Xenia* (Table 4.4). *Dendronephthya* had the highest percentage cover during this study, 3.0% at North Umm El-Himmat Island and a lowest value of 0.5% at South Geisum, with a mean cover of 1.0±1.3%. The genus *Xenia* was recorded only at Tawila Island, with a cover of 3.7% (Table 4.4). *Sinularia* recorded a highest cover of 1.5% at Small Gubal Island and a lowest value of 1.0% at South Geisum Island, with a mean of 0.6 ± 0.7%. The highest cover of *Sarcophyton* was 1% on Tawila and South Geisum Islands, and the lowest was 0.5% on Small Gubal Island, with a mean of 0.4±0.4% on the other four islands. Soft coral cover values varied among the eight northern islands. The highest averaged 6% on Tawila Island and dropped to 0.6% on South Geisum Island, with a mean of 2.5±2.1%. In the present study, no soft corals were recorded at Ghanim or South Umm El-Himmat Islands.

4.3.2 Distribution and percentage cover of living and nonliving components at Hurghada

The cover of living components at the Hurghada sites (Gota Abu-Rmada, El-Fanous Reef, Sabina Reef, and Small Gifton Island) averaged 66.87 ± 34.13%. The mean of stony corals was 54.8 ± 6.4%, and declined sharply to 2.7 ± 2.4% in soft corals, recorded 1.7 ± 0.66% for algae, and was 1.15 ± 0.17% for sponges (Table 4.5). On the other hand, the percentage cover of nonliving components was 33.5 ± 16.5% (Table 4.5), dominated by dead corals, which averaged 29.2 ± 3.03%, followed by 1.07 ± 0.63% for rubble, 0.075 ± 0.15% for sand, and 3.15 ± 0.95% for rock (Table 4.5).

4.3.2.1 Percentage cover of stony coral genera at Hurghada

The highest percentage cover of all stony and soft corals (Table 4.6) was 68.1% recorded in Sabina Reef, and the lowest cover was 54% in Small Gifton Island, while the cover of stony corals had moderate values of 62.5% at El-Fanous Reef and 60.5% at Gota Abu-Rmada. There are 10 genera of corals recorded in Gota Abu-Rmada, comprising 9 genera of stony corals and only one genus of fire coral, dominated by *Acropora* with a coverage of 18.5% compared with the lowest, *Seriatopora*, with only 0.5%. However, the number of coral genera declined to six in El-Fanous Reef, with five stony corals and one fire coral. *Acropora* was also the dominant species with a coverage of 43.5%, and the lowest was *Echinopora* with a coverage of 1%. At Sabina Reef, 9 genera were recorded, with 8 stony corals and one fire coral; *Acropora* was the dominant species with 37% coverage, while *Galaxea*, *Lobophyton*, and *Platygyra* each had 1% coverage (Table 4.6). At Small Gifton, 11 genera were recorded, 10 stony corals, and one fire coral; *Porites* was the dominant species with 14% coverage, while the lowest values were recorded for *Favia*, *Favites*, and *Galaxea*, with coverage of 0.5% for each.

TABLE 4.5 Substrate percentage cover (%) at Hurghada Island, Red Sea.

TYPE	*SUBSTRATE CATEGORY*	*GOTA ABU-RMADA*	*EL-FANOUS REEF*	*SABINA REEF*	*SMALL GIFTON IS.*	*MEAN ± SD*	*TOTAL ± SD*
Living cover	Stony corals	60.5	64	68.1	53.0	61.4±5.9	66.9 ± 27.6
	Soft corals	3.0	3.0	0.6	5.0	2.9±1.7	
	Algae	1.5	1.5	1.4	2.7	1.8±0.6	
	Sponges	0.0	1.2	0.9	1.3	0.9±0.6	
Non-Living cover	Dead corals	30.5	24.8	25.8	34.0	28.8±4.0	33.1 ± 12.7
	Rubble	1.2	1.2	0.2	1.7	1.1±0.6	
	Sand	0.0	0.0	0.0	0.3	0.1±0.1	
	Rocks	3.3	4.3	3.0	2.0	3.2±0.9	

TABLE 4.6 Percentage cover (%) of stony coral genera at Hurghada sites, Red Sea.

HARD CORAL GENERA	*GOTA ABU-RMADA*	*EL-FANOUS REEF*	*SABINA REEF*	*SMALL GIFTON IS.*	*MEAN ± SD*
Acropora	18.5	43.5	37	13	28 ± 14.6
Echinopora	1.5	1.0	0.0	1.0	0.88 ± 0.63
Stylophora	0.0	0.0	0.0	1.0	0.25±0.5
Favia	0.0	0.0	4.0	0.5	1.13 ± 1.93
Favites	2.5	0.0	1.1	0.5	1.04 ± 1.1
Galaxea	1.5	0.0	1.0	0.5	0.75±0.64
Millipora	12.5	7.0	7.0	12	9.625±3.04
Montipora	9.5	0.0	3.0	3.0	3.88 ± 4.00
Plategyra	1.0	0.0	1.0	0.5	0.63±0.47
Pocillopora	6.5	6.0	7.0	8.0	6.75±0.96
Porites	6.5	4.0	6.0	14	7.62±4.38
Seriatopora	0.5	1.0	0.0	0.0	0.38±0.48
Total	60.5	62.5	68.1	54	61.23±5.82

4.3.2.2 Percentage cover of soft coral genera at Hurghada

Five genera of soft corals were recorded at Hurghada Island: *Sarcophyton*, *Sinularia*, *Tubasteria*, *Xenia*, and *Lobophyton*, with total percentages of 3.0%, 1.3%, 1.6%, and 6.0% for the four sites, Gota Abu-Rmada, El-Fanous Sabina Reef, and Small Gifton Islands, respectively (Table 4.7). Out of them, only *Sinularia* and *Tubasteria* were recorded at most of the recorded sites, with decreasing average cover (1.0%, 1.0%, 0.5%, and 1.0%) for *Sinularia* and (2.0%, 0.3%, 0.1%, and 0.0 %) for *Tubasteria* (Table 4.7).

4.3.3 Coral reef status (living and nonliving components) at middle and southern sites

Seven sites were selected to be surveyed in the middle area, including Al-Qusier Port, El-Hamrawein Port, Abu-Tartoura, and the southern sites of Marsa Nekari, Elphinstone Reef, Zabargad Island, and Samadi Reef. The current survey revealed that the percentage cover of living components at those sites averaged 70.7 ± 11.5%, comprised of stony corals at 54.8 ± 6.3%, soft corals at 8.17 ± 3.37%, algae at 6.3 ± 1.4, and sponges at 1.6 ± 1.4% (Table 4.8). The percentage cover of non-living components was 29.3 ± 8.7, with dead corals accounting for 25.2 ± 5.1%, sand 1.7 ± 1.0%, rubble 1.5 ± 1.1%, and rocks 0.9 ± 0.7%.

At the southern sites, the percentage cover of living components was 60.5% at Marsa Nekari, an offshore site located 25 km south of Marsa Alam (Table 4.8); it increased to 74.5% at Elphinstone Reef, which is located 25 km north of Marsa Alam, to 5 m depth (Table 4.8), and increased slightly to 80.5% at Zabargad Island, which is located 70 km south of Marsa Alam and is characterised by the presence of dolphins and called "Dolphin House"; the total living cover averaged 73.5%. At the mid-sites, which included Al-Qusier Port located at Al-Qusier City that is impacted by eutrophication as a result of phosphate shipments, the percentage cover of living components of the substrate at a depth of 5 m averaged 65.5% (Table 4.8), while at El-Hamrawein Port the cover of live components averaged 70%, but declined slightly to 67% at Abu-Tartour, which lies 6 km south of Safaga and was impacted by phosphate mining.

4.3.3.1 Percentage cover of stony coral genera

Eleven genera of stony corals were recorded at the middle and southern sites (Table 4.9); 10 genera occurred at Marsa Nekari, with a total cover of 52%. *Acropora* was the dominant species, with 7.7% cover, while *Favia*, *Favites*, and *Galaxea* were the least common, each with a cover of 0.5%. At Elphinstone

TABLE 4.7 Percentage cover (%) of soft coral genera at Hurghada sites, Red Sea.

'	*GOTA ABU-RMADA*	*EL-FANOUS REEF*	*SABINA REEF*	*SMALL GIFTON IS.*	*MEAN ± SD*
Sinularia	1.0	1.0	0.5	1.0	0.87±0.25
Xenia	0.0	0.0	0.0	4.0	1.0±2.0
Sarcophyton	0.0	0.0	0.0	1.0	0.25±0.5
Tubasteria	2.0	0.3	0.1	0.0	0.6±0.94
Lobophyton	0.0	0.0	1.0	0.0	0.25 ± 0.5
Total	3.0	1.3	0.6	6.0	2.7±2.40

TABLE 4.8 Substrate percentage cover (%) at the middle and southern Egyptian Red Sea coast.

TYPE	*SUBSTRATE CATEGORY*	*MARSA NEKARI*	*ELPHINSTONE REEF*	*ZABARGAD IS*	*SAMADI REEF*	*AL-QUSIER PORT*	*EL-HAMRAWEIN SITE*	*ABU-TARTOUR*	*MEAN ± SD*	*TOTAL ± SD*
Living cover	Stony corals	52	56.5	60	64.5	43.5	52	55	54.8±6.4	70.7 ± 11.5
	Soft corals	3.5	11	9	7.5	10.5	11.5	3.5	8.1±3.3	
	Algae	5	7.5	7	0.5	11.5	5.5	7	6.3±1.4	
	Sponges	2	1	4.5	1	0	1	1.5	1.6±1.4	
Total living cover		62.5	76	80.5	73.5	65.5	70	67	70.7	
Non-Living cover	Dead corals	31.5	20	17	23.5	29	26.2	29	25.2±5.1	29.3 ± 8.7
	Rubble	1.5	0	1	1.5	3	2.8	1	1.5±1.0	
	Sand	3.5	2.5	1	1.5	0.5	1	2	1.7±1.0	
	Rocks	1	1.5	0.5	0	2	0	1	0.9±0.7	

TABLE 4.9 Percentage cover (%) of stony coral genera at middle and southern Egyptian Red Sea sites.

STONY CORAL GENERA	*MARSA NEKARI*	*ELPHINSTONE REEF*	*ZABARGAD IS*	*SAMADI REEF*	*AL-QUSIER PORT*	*EL-HAMRAWEIN SITE*	*ABU-TARTOUR SITE*	*MEAN ± SD*
Acropora	11	5.5	14.5	4	3.5	7	8.5	7.7 ± 3.8
Echinopora	2.5	3	2	1	1	2	0	1.6 ± 1.0
Stylophora	9	9	5	4.5	3	5	10.5	6.6±2.8
Pocillopora	7	2	3.5	2.5	4.5	6	7	4.6±2.0
Favia	0.5	5.5	5	0.5	1	2	3	2.5±2.0
Favites	0.5	0.5	2	0.5	1	1	2	1.1±0.6
Galaxea	0.5	0.5	2	0.5	1	3	3	1.5±1.1
Millipora	7.5	7	6	5	2	2	6	5.1±2.2
Montipora	3.5	3	4.5	2	0	0.5	1	2.1 ± 1.6
Platygyra	1.5	2	2	1.5	1	3	1	1.7±0.7
Porites	8.5	17.5	12.5	42.5	25	20	11	19.6±11.1
Serriatopora	0	1	1	0	0.5	0.5	2	0.7±0.7
Total	52.0	56.5	55	64	43.5	52	55	52.3±7.8

Reef, for the main substrate composition at 5 m, the stony coral cover was 56.5% (Table 4.9), and all 11 genera of stony corals were recorded, dominated by *Porites*, which has a coverage of 17.5%, while *Favites* and *Galaxea* had the lowest coverages, at 0.5% (Table 4.9). The average cover of stony corals on Zabargad Island was 55%, with *Acropora* and *Porites* dominating with 14.5% and 12.5%, respectively, and *Seriatopora* having the lowest cover of 1.0%. At Samadi Reef, there are 10 genera of stony corals, dominated by *Porites*, which averaged 42.5%, and the lowest genera were represented by *Favia*, *Favites*, and *Galaxea*, with only 0.5% (Table 4.9). At the middle sites, there are 10 genera of stony corals at Al-Qusier Port, with an average coverage of 43.5%, which represents the lowest coverage at the middle and southern sites. The branching corals were recorded with considerable values at 5 m depth; the types comprised *Acropora*, *Pocillopora*, and *Stylophora*, but *Porites* was the dominant species of massive corals recorded at 25%. However, at El-Hamrawein, all 11 genera were recorded, with a total coverage of 52%, dominated by *Porites* (20%), while *Montipora* and *Seriatopora* were the lowest with covers of 0.5%; in contrast, the branching corals of genera *Acropora*, *Stylophora*, and *Pocillopora* had covers of 7%, 5%, and 6%, respectively (Table 4.9). At Abu-Tartour, 11 genera of stony corals were recorded, with an average cover of 55%. The branching corals *Acropora*, *Stylophora*, and *Pocillopora* were dominant, with 8.5%, 10.5%, and 7%, respectively, while *Porites* had the highest cover of massive corals at 11%.

4.3.3.2 Percentage cover of soft coral genera

Six genera of soft corals, comprising *Sinularia*, *Sarcophyton*, *Heteroxenia*, *Xenia*, *Dendronephthya*, and *Lobophyton*, were recorded at the middle and southern Egyptian Red Sea islands (Table 4.10). The percentage cover of these genera was variable at those sites, represented by 3.5%, 11%, 9%, 7.5%, 10.5%, 11.5%, and 3.5% at Marsa Nekari, Elphinstone Reef, Zabargad Island, Samadi Reef, Al-Qusier Port, El-Hamrawein, and Abu-Tartour, respectively. *Sinularia* had the highest coverage of 4% at Elphinstone Reef and El-Hamrawein Port, but declined to 3.5% and 3% at Al-Qusier Port and Zabargad Island, respectively, and reached 1% at Marsa Nekari. Other soft coral coverage varied from 0.5% to 2% at different sites, with the exception of *Heteroxenia*, which increased to 3% at Al-Qusier Port.

TABLE 4.10 Percentage cover (%) of soft coral genera at middle and southern Egyptian Red Sea coast.

STONY CORAL GENERA	*MARSA NEKARI*	*ELPHINSTONE REEF*	*ZABARGAD IS*	*SAMADI REEF*	*AL-QUSIER PORT*	*EL-HAMRAWEIN PORT*	*ABU-TARTOUR*	*MEAN±SD*
Sinularia	1.0	4.0	3.0	2.0	3.5	4.0	0.5	2.6 ± 1.4
Sarcophyton	0.5	2.0	2.0	1.0	1.0	2.0	0.5	1.3±0.7
Heteroxenia	0.5	2.0	0.5	2.5	3.0	1.0	0.5	1.4±1.0
Xenia	0.5	1.0	0.5	1.0	1.5	2.0	0.5	1.0±0.6
Dendronephthya	0.5	1.0	1.0	1.0	1.0	1.0	0.5	0.9±0.2
Lobophyton	0.5	1.0	2.0	0.0	0.5	1.5	1.0	0.9 ± 0.6
Total	3.5	11	9.0	7.5	10.5	11.5	3.5	8.1±3.3

4.4 DISCUSSION

The main aim of this study was to investigate the status of reefs along the Egyptian Red Sea coast. Our findings indicate that the coral reefs in the southern area are generally in better condition than are those in the north. The mean percentage cover of living components (stony and soft corals, sponges, and algae) was 65.6 ± 27.0% at the northern sites and 66.9 ± 27.6% at Hurghada, while the middle and southern sites recorded 70.7±11.5%. The nonliving components (dead coral, sand, rocks, and rubble) had covers at the northern sites that averaged 34.4±13.4% and 33.1±12.7% at Hurghada, declining slightly to 29.3±8.7% in the south. These results agree with the findings of Hassan et al. (2002), who documented the average stony coral cover in the Red Sea of Egypt as 45% and 35% in the Gulf of Aqaba. Similar results were obtained by PERSGA (2010), which found that the status of reef cover in Egypt was as follows: the coral cover in north Hurghada at the reef flat was 30%, increased gradually to 65% at the reef edge, and 86% at the reef wall, but declined to 35% on the reef slope. At the same time, they found the cover at the reef flat averaged 18%, increased to 25% at the reef edge, 50% at the reef wall, and declined again to 20% on the reef slope in some locations from Hurghada to Safaga.

Mohamed et al. (2012) studied the coral cover in twelve sheltered and exposed localities along the Egyptian Red Sea coast, and measured coral cover at 42% in the north, 50% in the middle area, and 37% in the south. Hussein (2016) studied the coral reef cover at four sites, and found that the reef cover in Ras Gharib in the Gulf of Suez was 30.6%, increased to 32.3% at Small Gifton Island offshore from Hurghada, and reached 84.6% at Abu Ramada offshore from Hurghada, while it was 61% in Al-Qusier Port in the south. Al-Hammady et al. (2015) studied the cover of stony corals in three sites; in the north it was 31%, in the middle it increased to 35%, and in the south it reached 51.4%.

In the current study, the northern islands that are close together have the highest stony coral cover, with Ghanim Island having a high cover of 84% and no algal cover. At Ashrafi Island, a low cover of stony coral averaged 41.3%, but a high cover of algae was recorded at 9.3%. The algae are in competition with the corals, and the abundance of algae affects the growth and size of the coral reef (El-Sadek et al. 2022). The present results for the substrate components in the northern islands confirmed that an inverse relationship exists between the algal cover and the percentage cover of living coral reef. It was noticed that Ghanim Island has a high coral cover and zero cover of algae, whereas Ashrafi Island has the lowest living cover of coral reef and a high cover of algae. Hence, it can be assumed that a factor controlling the cover of coral reefs is the percentage cover of algae (El-Sadek et al. 2022). Ghanim Island's location near the shore of Gabal el Zait may protect it from activity such as illegal overfishing, given the substrate contains less than 10.7% dead coral and 0% rubble. The remaining northern islands may be affected by overfishing, petroleum, and tourism activities. Mohammed (2006), Ammar (2004), and Mohammed et al. (2009) showed that these activities have completely destroyed the benthic communities in the shallower areas.

In our results, there are four genera of soft corals recorded in the northern islands: *Sarcophyton*, *Xenia, Sinularia*, and *Dendronephthya*, with limited knowledge of the distribution and abundance of soft coral reefs in the Egyptian coast. Gohar (1940), Roushdy (1954), and Vine (1986) recorded eight genera of soft corals in the Egyptian Red Sea. The review of soft corals by Ismail et al. (2017) recorded an additional eleven genera at eight sites along the Red Sea and Gulf of Aqaba. The present analysis of the soft coral community structure on reefs in the northern islands shows that the cover ranges from a high of 6% at Tawila Island to a low of 0.6% at North Geisum. Hussein (2016) studied the percentage cover of soft coral reefs, finding 8.4% in Small Gifton Island and 2.7% in Abu Ramada, both offshore from Hurghada. Hassan et al. (2002) gave a similar result, estimating the average soft coral cover in the Egyptian Red Sea at 10%. These results are confirmed by our finding of the soft coral reef cover at nine sites in the northern Red Sea. Riegl and Velimirov (1994) gave the cover at Hurghada shores as Hurghada fringing reef (8.9%), Shaab Abu Rimathi (1.1%), Shaab Rur umm Qamar (8.5%), and Small Gubal Island 9.4%, while we found the cover of soft coral at Small Gubal to be 3.5%. Ismail et al. (2017) studied the distribution of soft corals along the Egyptian Red Sea coast and gave an average cover of soft corals for 5 genera (*Sinularia*, *Sarcophyton*, *Lobophyton*, *Xenia*, and *Nephthea*) at 0–5 m depth as 7%. In the current study, the average cover for four genera recorded in the northern islands at 0–5 m depth was 2.5%, with the dominant genera *Dendronephthya* (1%), followed by *Sinularia* (0.6%), *Xenia* (0.5%), and *Sarcophyton* (0.45).

The present study showed that the high percentage of dead coral in all sites may be due to stressors on the ecosystem, such as recreational activity, urban development, illegal fishing activity, petroleum operations in the northern sites, and sometime oil spills from ships. According to Northern Protection

Authority reports, many illegal small fishing boats approach the reefs surrounding the islands and use illegal fishing methods that rely on breaking the coral colonies to catch fish. However, the highest percentage of dead corals at a given site is a result of oil pollution, with many accidental petroleum leakages, as shown by the traces of heavy oil on the island beaches. This agrees with Rinkevich and Loya (1977), who reported that oil pollution is an important factor affecting coral reef distribution and reproduction. Al-Hammady and Mahmoud (2013) also documented oil pollution as the main threat responsible for increasing the cover of dead corals in the northern Red Sea. Salvat (1992) documented 20 categories of stress causing damage to coral reefs, including overfishing and oil pollution. In the northern islands, the cover of dead coral was highest in South Geisum (39%) and lowest in Ghanim Island (10.7%).

The current study shows that 10 genera of stony coral were recorded in the northern sites, with a dominance of breaching stony corals over massive corals where the genus *Acropora* has the highest cover, followed by *Stylophora* and *Pocillopora*. The dominance of the branching stony genus *Acropora* in the northern islands agrees with Ammar (2004) and Mohammed et al. (2010).

In general, the geographic position and geomorphology are other limiting factors for coral dominance besides the main factors (temperature, turbidity, landfilling, algal blooming, and overgrowth) that affect the distribution and growth. Mohammed (2006) and Mohammed et al. (2010) gave reviews on coral reef distribution in the Red Sea, and the comparison with these results indicates the dominance of branching *Acropora*, *Styllophora*, and *Porites* in all northern islands due to: (1) the surrounding reefs of the northern islands are sheltered areas; and (2) *Acropora* was fast growing under stress from anthropogenic factors. This agrees with Mohammed (2006), who found that *Stylophora*, *Acropora*, and *Porites* have highly competitive abilities and can tolerate the unfavourable conditions of the Gulf of Suez.

REFERENCES

Abdelkarem, F.M., Desoky, E.E.K., Nafady, A.M., Allam, A.E., Mahdy, A., Ashour, A., Mohamed, G.A., Miyamoto, T., and K. Shimizu. 2021. Two new polyhydroxylated steroids from Egyptian soft coral *Heteroxenia fuscescens* (Fam.; Xeniidae). *Nat Prod Res* 35: 236–243.

Abou Zaid, M. 2000. *Overview of the status of Red Sea coral reefs in Egypt*. PERSGA Technical Series No. 7, *Coral Reefs in the Red Sea and Gulf of Aden Surveys 1990 to 2000. Summary and Recommendation*. Unpublished report, 39 p.

Ahmed, B. 2010. *Cost of degradation of coral reefs and fisheries caused by tourism development, Egypt's Red Sea: A case study of Hurghada – Safaga Area*. Institute of National Planning, Cairo, Nasr City, Egypt. http://kenanaonline.com/users/drBarrania/posts/119025.

Al-Hammady, M.A.M. 2011. Patterns of bleaching and fertility in the two Red Sea corals *Stylophora pistillata* and *Acropora humilis* as biomonitors of environmental impacts. Ph.D. Thesis, Zoology Dept., Assiut University, Assiut, Egypt.

Al-Hammady, M.A.M., Fouda, F.M.A., Hussein, H.N.M., and A.A.M. El Sayed. 2015. Effect of anthropogenic activities on coral distribution at onshore and offshore reefs along the Egyptian coast, Red Sea. *Int J Environ MonitAnalysis* 3, no. 5: 1–9.

Al-Hammady, M.A.M., and M.A. Mahmoud. 2013. *The effect of expanding coastal urban, industrial centers, ports and tourism on the Red Sea coral reefs – Egypt*. Proc Internat Conf Environ Sci (ICES), pp. 21–22. https://www.researchgate.net/publication/261551476.

Ammar, M.S.A. 2004. Zonation of coral communities and environmental sensitivity offshore a resort site at Marsa Alam, Red Sea, Egypt. *Egypt J Zool* 42: 76–81.

Ammar, M.S.A. 2011. Coral diversity indices along the Gulf of Aqaba and Ras Mohammed, Red Sea, Egypt. *Biodiversitas* 12, no. 2: 92–98.

Ammar, M.S.A., Ghobachi, A.A., Omran, M.A., and A.M. Shabban. 2007. Status of coral reef affected by different impacts in some sites of the Red Sea. *Egypt J Aquat Res* 33: 224–237.

Ammar, M.S.A., and A.H. Nawar. 1998. *Quantitative study for the distribution of reef-building corals at Abu-Galawa, Hurghada, Red Sea*. International Conference Protection is a must. Euro-Arab Cooperation Center. Inter Sci Assoc, pp. 222–233.

Ault, J.S., Smith, S.G., and J.T. Tilmant. 2009. Are the coral reef finfish fisheries of south Florida sustainable? *Proc Internat Coral Reef Symp* 11: 989–993.

Barrania, A. 2010. *Cost of degradation of coral reefs and fisheries caused by tourism development, Egypt's Red Sea. A case study of Hurghada – Safaga area*. Inst Nat Plan, Cairo, Nasr City, Egypt.

Boström-Einarsson, L., Babcock, R.C., Bayraktarov, E., Ceccarelli, D., Cook, N., Ferse, S.C.A., Hancock, B., Harrison, P., Hein, M., Shaver, E., Smith, A.M., Suggett, D., Stewart-Sinclair, P.J., Vardi, T., and I.M. McLeod. 2020. Coral restoration – A systematic review of current methods, successes, failures and future directions. *PLoS One* 15, no. 1: e0226631. doi:10.1371/journal.pone.0226631.

Bruckner, T.A., Scheffler, R.M., Shen, G., Yoon, J., Chisholm, D., Morris, J., Fulton, B.D.,

Pozg, M.R.D., and S. Saxena. 2011. The mental health workforce gap in low- and middle-income countries: A needs-based approach. *Bull World Health Organ* 89: 184–194.

Bruno, J.F., and E.R. Selig. 2007. Regional decline of coral cover in the Indo-Pacific: Timing, extent and subregional comparisons. *PLoS One* 8: 1–8.

Cesar, H.J.S., Burke, L., and L. Pet-Soede. 2003. *The economics of worldwide coral reef degradation*. Cesar Environmental Economics Consulting, Arnhem, and WWF-Netherlands, Zeist, The Netherlands, 23 p.

Crossland, C. 1938. The coral reef at Ghardaqa, Red Sea. *Proc Zool Soc London Ser A* 108: 513–523.

Dawoud, W.A., Negm, A.S., Saleh, N.M., and M.F. Bady. 2016. Impact assessment of offshore pile driving noise on Red Sea marine mammals. *Int J Environ Sci Dev* 7, no. 1: 10–15.

Dosoky, Y.A., Ahmed, M., Madkour, F., and M.H. Hanafy. 2021. Coral bleaching occurrence along the Egyptian coast of the Red Sea during the summer heat stress period, 2020. *Egypt J Aquat Biol Fish* 25, no. 5: 17–37.

Dustan, P. 1999. Coral reefs under stress: Sources of mortality in the Florida Keys. *Nat Resource Forum* 23: 147–155.

Edwards, A.J., and E.D. Gomez. 2007. *Reef restoration concepts and guidelines: Making sensible management choices in the face of uncertainty*. Coral Reef Targeted Research & Capacity Building for Management Programme, St Lucia, Australia, iv + 38 p.

Egyptian Biodiversity CHM. 2006. *Biodiversity conservation capacity building in Egypt*. www.cbd.int/doc/pa/tools/Biodiversity%20conservation%20capacity%20planning%20in%20Egypt.pdf.

El-Sadek, A.M., El-Naggar, H.A., Hassan, A.M., El-Ganiny, A.A., and H.M.M. Khalaf-Allah. 2022. Benthic composition and the dynamics of coral-algal interactions in Abu Galum protected area, western coast of Aqaba Gulf. *Egypt J Aquat Biol Fish* 26, no. 4: 515–533.

English, S., Wilkinson, C., and V. Baker. 1997. *Survey manual for tropical marine resources*. Australian Institute of Marine Science, Townsville, pp. 34–49.

Erftemeijer, L.A., Riegl, B., Hoeksema, W.D., and A. Peter. 2012. Environmental impacts of dredging and other sediment disturbances on corals: A review. *Mar Pollut Bull* 64: 1737–1765.

Evensen, N.R., Voolstra, C.R., Fine, M., Perna, G., Buitrago-López, C., Cárdenas, A., Banc-Prandi, G., Rowe, K., and D. Barshis. 2022. Empirically derived thermal thresholds of four coral species along the Red Sea using a portable and standardized experimental approach. *Coral Reefs* 41: 239–252. doi:10.1007/s00338-022-02233-y.

Ghallab, A., Abou Zaid, M., Hanfy, M., Hassan, A., and A. Mahdy. 2021. Biological study, life cycle and fecundity of *Coralliophila violacea* (Coralliophilidae: Gastropoda: Mollusca) inhabiting the Red Sea coral reefs, Egypt. *Egypt J Aquat Biol Fish* 25, no. 3: 201–216.

Ginsburg, R.N., Gischler, E., and W.E. Kiene. 2001. Partial mortality of massive reef-building corals: An index of patch reef condition, Florida Reef Tract. *Bull Mar Sci* 69, no. 3: 1149–1173.

Gohar, H.A.F. 1940. Studies on Xeniidae of the Red Sea (their ecology, physiology, taxonomy and phylogeny). *Publ Mar Biol Station (University of Cairo, Egypt)* 2: 27–118.

Hannak, J.S., Kompatscher, S.M., and J. Herler. 2011. Snorkeling and trampling in shallow-water fringing reefs: Risk assessment and proposed management strategy. *J Environ Manage* 92: 2723–2733.

Hassan, M., Kotb, M.M.A., and A.A. Al-Sofyani. 2002. Status of coral reefs in the Red Sea-Gulf of Aden. In: *Status of coral reefs of the world: 2002*. Ed: C.R. Wilkinson, 45–52. GCRMN Report, Townsville: Australian Institute of Marine Science.

Head, S.M. 1987. Introduction. In: *Key environments: Red Sea*. Eds: A.J. Edwards, and S.M. Head, 1–21. Oxford: Pergamon Press.

Hilmi, N., Safa, A., Reynaud, S., and D. Allemand. 2012. Coral reefs and tourism in Egypt Red Sea. *Topics Middle Eastern Afr Econ* 14: 416–434.

Hoegh-Guldberg, O., Poloczanska, E.S., Skirving, W., and S. Dove. 2017. Coral reef ecosystems under climate change and ocean acidification. *Front Mar Sci* 4: 158. doi:10.3389/fmars.2017.00158.

Hussein, N.M.H. 2016. *The effect of some pollutants on coral reef reproduction and growth at Hurghada-Al-Qusier sector, Red Sea, Egypt*. Al-Azhar University, Faculty of Science, Zoology Department, 207 p.

ICRI. 2012. *International coral reef initiative*. Annual report, Status of coral reef around the world.

Ismail, H.A., Hanafy, M.H., Madkour, F.F., and M.I. Ahmed. 2017. Distribution of soft corals in the Egyptian coasts of the Red Sea and Gulf of Aqaba. *Int J Eng Sci Computing* 7, no. 9: 14944–14950.

Jameson, S.C., Ammar, M.S.A., Saadalla, E., Mostafa, H.M., and B. Riegl. 2007. A quantitative assessment of diving sites in the Egyptian Red Sea during a period of severe anchor damage: A baseline for restoration and management. *J Sustain Tour* 15, no. 3: 309–323.

Koop, K., Booth, D., Broadbent, A., Brodie, J., Bucher, D., Capone, D., Coll, J., Dennison, W., Erdmann, M., Harrison, P., Hoegh-Guldberg, O., Hutchings, P., Jones, G.B., Larkum, A.W.D., O'Neil, J., Steven, A., Tentori, E., Ward, S., Williamson, J., and D. Yellowlees. 2001. ENCORE: The effect of nutrient enrichment on coral reefs: Synthesis of results and conclusions. *Mar Pollut Bull* 42: 91–120.

Kotb, M.M.A. 1996. *Ecological and biological studies on the coral reefs at southern Sinai coasts, Red Sea, Egypt*. Ph.D. Thesis, Faculty of Science, Suez Canal Univ, 174 p.

Kotb, M.M.A., Abduaziz, M., Al-Agwan, Z., Al-Sheikh, K., Al-Yami, H., Banajah, A., Devantier, L., Eisinger, M., El Tayem, M., Hassan, M., Heiss, G., Howe, S., Kemb, J., Klaus, R., Krupp, F., Mohamed, N., Roupheal, T., Turner, J., and U. Zajonz. 2004. Status of coral reef in the Red Sea and Gulf of Aden. In: *Status of coral reefs of the world*. Ed: C. Wilkinson, 137–151. Global Coral Reef Monitoring Network and Australian Institute of Marine Science, Townsville, Queensland. https://www.researchgate.net/publication/230751435.

Kotb, M.M.A., Abou Zeid, M.M., and M.H. Hanafy. 2001. Overall evaluation of coral reef status along the Egyptian Red Sea coast. *Biol Mar Medit* 8, no. 1: 15–32.

Kotb, M.M.A., Hanafy, M.H., Rirache, H., Matsumara, S., Al-Sofyani, A.A., Ahmed, A.G., Bawazir, G., and F. Al-Horani. 2008. Status of coral reefs in the Red Sea and Gulf of Aden Region. In: *Status of coral reefs of the world: 2008*. Ed: C.E. Wilkinson, 67–78. Townsville, Australia: Global Coral Reef Monitoring Network and Reef and Rainforest Research Centre.

Lirman, D., Gracias, N., Gintert, B., Gleason, A.C.R., Deangelo, G., Dick, M., Martinez, E., and R.P. Reid. 2010. Damage and recovery assessment of vessel grounding injuries on coral reef habitats by use of georeferenced landscape video mosaics. *Limnol Oceanogr Methods* 8: 88–97.

Loya, Y. 1972. Community structure and species diversity of hermatypic corals at Eilat, Red Sea. *Mar Biol* 13: 100–123.

Loya, Y. 1976. Effects of water turbidity and sedimentation on the community structure of Puerto Rican corals. *Bull Mar Sci* 26, no. 4: 450–466.

Madkour, H.A. 2013. Impacts of human activities and natural inputs on heavy metal contents of many coral reef environments along the Egyptian Red Sea coast. *Arab J Geosci* 6: 1739–1752.

Moberg, F., and C. Folke. 1999. Ecological goods and services of coral reef ecosystems. *Ecol Econ* 29: 215–233.

Mohamed, A.R., Abdel-Hamid, A.M., and H.A. Abdel-Salam. 2012. *Status of coral reef health in the northern Red Sea, Egypt*. Proc. 12th Int Coral Reef Symp, Cairns, Australia, pp. 9–13.

Mohammed, T.A. 2006. Evaluation, distribution and the coral diversity in some coastal lagoons, Red Sea, Egypt. *Egypt J Aquat Res* 32, spec issue: 180–195.

Mohammed, T.A., Dar, M.A., and M.I. El-Saman. 2010. Distribution patterns of hard and soft corals along the Egyptian Red Sea Coast. *Egypt J Aquat Res* 36, no. 4: 543–555.

Mohammed, T.A., and M.A. Mohamed. 2005. Some ecological factors affecting coral reef assemblages off Ennadi, Red Sea, Egypt. *Egypt J Aquat Res* 31, no. 1: 133–145.

Mohammed, T.A., Shoukry, F.A., El-Komi, M.M., and M.A.H. Ezz El-Arab. 2009. Distribution and diversity of alcyonacean soft corals and scleractinia hard corals in the northern Red Sea, Egypt. *J Egypt German Soc Zool* 58(D): 67–83.

Nasr, H., Yousef, M., and H.A. Madkour. 2019. Impacts of discharge of desalination plants on marine environment at the southern part of the Egyptian Red Sea coast (case study). *Int J Ecotoxicol Ecobiol* 4, no. 3: 81–85.

PERSGA. 2010. *The status of coral reefs in the Red Sea and Gulf of Aden: 2009*. Regional Organization for the Conservation of the Environment of the Red Sea and Gulf of Aden. Technical Series no. 16. PERSGA, Jeddah, Saudi Arabia, 125 p.

Richards, Z. 2018. *Hard coral genus identification guide, version 2*. Perth: Western Australian Museum, 60 p.

Richardson, L.L., and J.D. Voss. 2005. Changes in a coral population on reefs of the northern Florida Keys following a coral disease epizootic. *Mar Ecol Prog Ser* 297: 147–156.

Riegl, B., and B. Velimirov. 1994. The structure of coral communities at Hurghada in the northern Red Sea. *Mar Ecol* 15, nos. 3–4: 213–231.

Rinkevich, B., and Y. Loya. 1977. *Harmful effects of chronic oil pollution on a Red Sea scleractinian coral population*. Third Int Coral Reef Symp, Miami, Florida, pp. 585–591.

Rogers, C. 1990. Responses of coral reefs and reef organisms to sedimentation. *Mar Ecol Prog Ser* 62: 185–202.

Roushdy, H.M. 1954. *Studies on Heteroxenia fuscescens (EHR.) in the Red Sea*. M.Sc. Thesis, University of Cairo, Egypt.

Salvat, B. 1992. Natural bleaching and mortality of scleractinian corals on Moorea Reefs (Society Archipelago) in 1991. *C R Acad. Sci Ser III – Sci de la Vie* 314: 105–111.

Sebens, K.P. 1994. Biodiversity of coral reefs: What are we losing and why? *Am Zool* 34: 115–133.

Shaalan, I.M. 2005. Sustainable tourism development in the Red Sea of Egypt: Threats and opportunities. *J Clean Prod* 13: 83–87.

Sheppard, C.R., and A.L. Sheppard. 1991. *Corals and coral communities of Arabia*. Natural History Museum, Basle, Switzerland, 419 p.

Spalding, M.D., Ravilious, C., and E.P. Green. 2001. *World atlas of coral reefs*. UNEP World Conservation Monitoring Centre, University of California Press, Berkeley, USA.

Tammam, M.A., Rárová, L., Kvasnicová, M., Gonzalez, G., Emam, A.M., Mahdy, A., Strnad, M., Ioannou, E., and V. Roussis. 2020. Bioactive steroids from the Red Sea soft coral *Sinularia polydactyla*. *Mar Drugs* 18, no. 12: 632. doi:10.3390/md18120632.

Veron, I. 2000. *Corals of the World*, 3 vols. Australian Institute of Marine Science, Townsville, Queensland, 1400 p. https://books.google.com.eg/books/about/Corals_of_the_World.html?id=Np3wAAAAMAAJ&redir_esc=y.

Vine, P. 1986. *Red Sea invertebrates*. London: Immel Publishing.

Voolstra, C.R., and M.L. Berumen. 2019. *Coral reefs of the Red Sea: Coral reefs of the world*. Cham, Switzerland: Springer International Publishing. ISBN 978-3-030-05800-5.

Voolstra, C.R., Valenzuela, J.J., Turkarslan, S., Cárdenas, A., Hume, B.C., Perna, G., Buitrago-López, C., Rowe, K., Orellana, M.V., Baliga, N.S., and S. Paranjape. 2021. Contrasting heat stress response patterns of coral holobionts across the Red Sea suggest distinct mechanisms of thermal tolerance. *Mol Ecol* 30, no. 18: 4466–4480.

Wallace, C.C. 1999. *Staghorn corals of the World: A revision of the coral genus Acropora*. CSIRO, Collingwood, Australia, 422 p. https://books.google.com.eg/books/about/Staghorn_Corals_of_the_World.html?id=Y5uVU4MfIKAC&redir_esc=y.

White, A.T., Vogt, H.P., and T. Arin. 2000. Philippine coral reefs under threat: The economic losses caused by reef destruction. *Mar Pollut Bull* 40, no. 7: 598–605.

Wilkinson, C. 2008. *Status of coral reefs of the World: 2008*. Global Coral Reef Monitoring Network and Reef and Rainforest Research Centre, Townsville, Australia, 296 p.

Zakai, D., and N.E. Chadwick-Furman. 2002. Impacts of intensive recreational diving on reef corals at Eilat, northern Red Sea. *Biol Conserv* 105: 179–187.

Thermal history of coral reefs along the Egyptian coast of the Red Sea

5

Mahmoud H. Hanafy, Muhammad Y.A. Dosoky and Amin R. Mohamed

5.1 INTRODUCTION

Over the past 30 years, there has been a remarkable increase in the severity and frequency of coral bleaching events, with attention being paid to these events globally since the early 1980s (Hughes et al. 2017, 2018). These bleaching events have led to a significant decline in global coral cover by 20%, with four pan-tropical coral bleaching events affecting hundreds of reefs (Lough et al. 2018; Hoegh-Guldberg et al. 2019). This decline in coral cover is strongly correlated with ocean warming, as a 1–3°C increase in SST may result in a 90% decline in global coral cover by 2100 (Bindoff et al. 2019; Cornwall et al. 2021). Moreover, according to the high-emission scenario of Representative Concentration Pathways (RCP8.5), severe annual bleaching is expected to occur before 2055, making it less likely for most corals to adapt to the +1.5°C rise in temperature (van Hooidonk et al. 2014, 2016).

The Red Sea is a home to a wide range of coral reefs that stretch across a north-south thermal gradient, which divides the entire region into four distinct thermal zones (Gajdzik et al. 2021a). Along the Egyptian coast, coral reefs are located in two thermal regions within the northern Red Sea (NRS). The first region, encompassing the Gulf of Aqaba and Gulf of Suez, is cooler and typically maintains thermal conditions below 28°C, while the other (the NRS basin) is warmer and characterized by higher temperatures ranging between 28°C and 31.25°C (Chaidez et al. 2017; Gajdzik et al. 2021a). In addition to the differences in thermal profiles, these two regions are situated within three different geographic and ecological areas (Gajdzik et al. 2021a, 2021b). Corals in the Gulf of Aqaba are well known for their resilience against coral bleaching; hence, they offer a potential thermal refuge for corals in the face of climate change (Fine et al. 2013; Osman et al. 2018). However, the north-south thermal gradient along the Egyptian Red Sea coast, which has not received much attention in previous research, provides a natural laboratory for studying how these coral populations respond to different thermal conditions (Banc-Prandi et al. 2022). Furthermore, the distinct thermal regimes present in this area may help corals adapt to future thermal stress events, thereby playing a crucial role in the conservation and restoration of coral reefs in the Red Sea.

Although the exact mechanisms behind this thermal resilience are still unclear, several factors have been proposed, including transcriptome resilience that promotes acclimatization (Fine et al. 2013; Grottoli et al. 2017; Savary et al. 2021; Banc-Prandi et al. 2022). However, one key factor that may mitigate heat stress in the NRS is the unique characteristic of water circulation, currents, and wind patterns in the region. The water circulation in the NRS might be primarily driven by the exchange of water masses between the Red Sea and the Gulfs of Suez and Aqaba (Sofianos and Johns 2015). This exchange occurs between the main basin and the two gulfs through the Straits of Gubal and Tiran (Biton and Gildor 2011; Eladawy et al. 2018). The NRS is also influenced by the Red Sea outflow, which is a deep and dense water mass flowing out from the Red Sea into the Gulf of Suez (Zhai et al. 2015; Yao and Hoteit 2018); this outflow contributes to the overall circulation patterns in the region. Although the Red Sea experiences a combination of surface and subsurface currents, the Red Sea coastal current is one of the main currents in the NRS; this current flows southward along the western coast of the Red Sea, driven primarily by prevailing winds and the pressure gradient between the northern and southern Red Sea (Zhai et al. 2015). Another important current in the region is the Sharm El-Sheikh Eddy, a mesoscale anticyclonic eddy that forms near Sharm El-Sheikh and can influence the circulation patterns (Raitsos et al. 2013). Additionally, localized coastal currents, influenced by local topography, can impact water movement along the shoreline (Yao et al. 2014).

It has been shown that corals in the NRS basin will not reach bleaching thresholds of 32oC before the end of the 21st century, and hence this region might host the last corals standing against bleaching (Eladawy et al. 2022). Despite the potential thermal refugium characteristics of the NRS basin (Fine

DOI: 10.1201/9781003321392-6

et al. 2013; Osman et al. 2018), recent studies have suggested that corals in this area on the Egyptian coast might not be immune to the effects of rising sea surface temperatures for long due to the rapidly rising incidence of high intensity marine heatwaves (Genevier et al. 2019). In this chapter, thermal conditions along the Egyptian coast between 1990 and 2020 were investigated to better understand the spatial and temporal patterns of the potential heat stress. We assessed the vulnerability of the coral reef environment in different sectors of the study area to thermal stress, in order to inform conservation efforts and support the sustainability of this important ecosystem.

5.2 METHODS

5.2.1 Study area

To investigate the heat stress pattern impacting coral reefs in the NRS basin along the Egyptian coast of the Red Sea, we selected a study area that extends for 627.15 km between 22.025°N and 27.675°N, and expands over 70.94±0.19 km from the shore (Figure 5.1). This approach ensures that the collected data is representative of the thermal climatology of the coral reef zone and includes most of the inshore and offshore reefs along the NRS basin on the Egyptian coast. To assess the spatial distribution of thermal stress and identify the most thermally vulnerable reefs from north to south, we divided the study area into five sectors (S1 to S5), with each geographic sector covering approximately one latitudinal degree.

5.2.2 Dataset

The spatiotemporal variation in the thermal conditions was inferred from NOAA's Coral Reef Watch Program (CRW) dataset. This dataset included daily 5 km (0.05°) high-resolution gridded records of sea surface temperature (SST v3.1) for the period from January 1, 1990, to December 31, 2020 (31 years). The CoralTemp satellite system was used to collect the data that represents the latest and most complete global satellite-derived SST dataset available for the period from 1985 to the present (Skirving et al. 2020). Time-series data were downloaded (https://coastwatch.pfeg.noaa.gov/erddap/griddap/; Dataset ID: NOAA_DHW), then trimming was applied to select the zone of our study (see Figure 5.1). Spatially, the final time series consisted of 1680 grid points (336±8.22 points/sector).

5.2.3 Methodology

The climatological mean SST over the 31-year period was calculated by averaging the appropriate annual monthly means of daily observed SSTs between 1990 and 2020 at each grid point

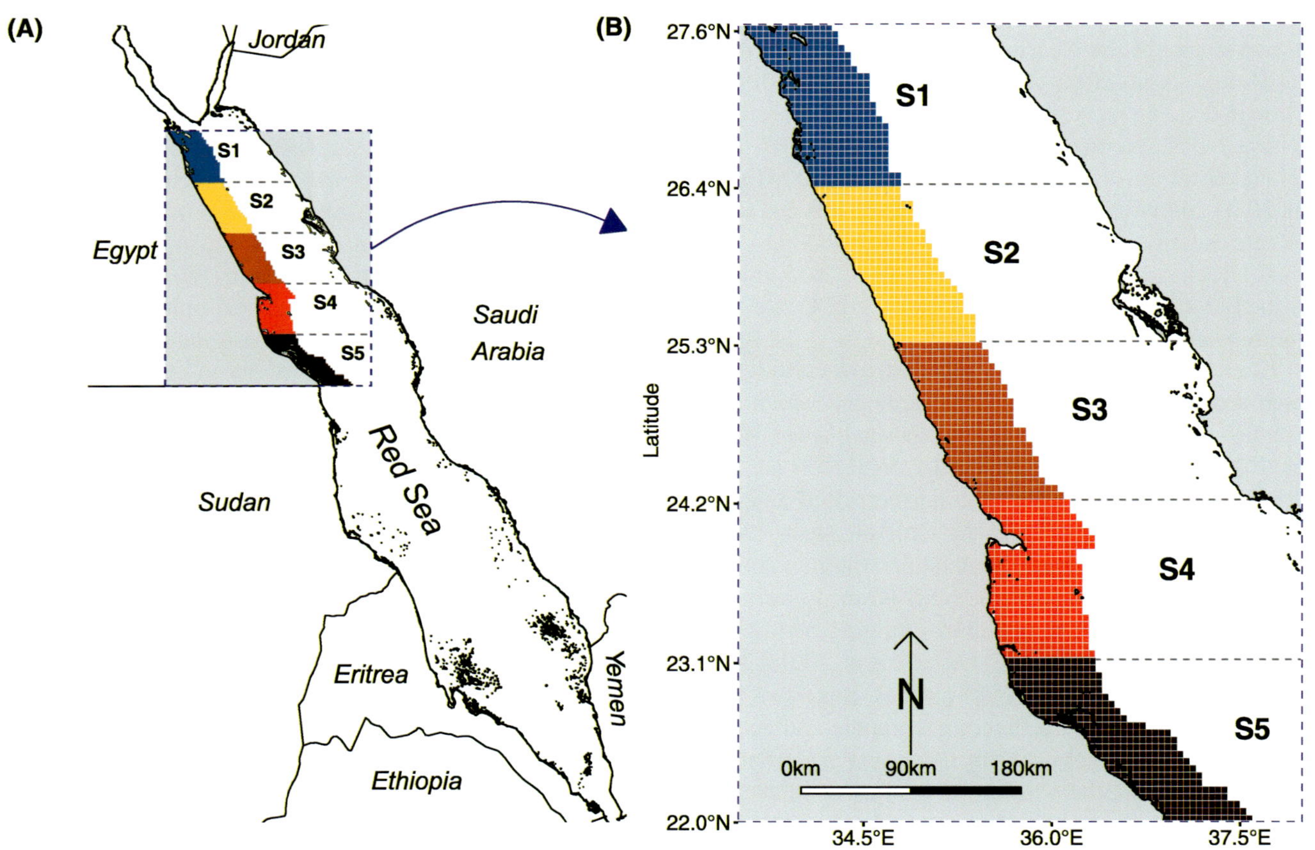

FIGURE 5.1 The study area located along the Egyptian coast of the Red Sea. The map in (A) provides an overview of the entire Red Sea, while the close-up in (B) highlights the specific area where most of the coastal reefs, both inshore and offshore, are located. This area is referred to as the coral reef zone and is represented by the coloured zone in the map. To facilitate analysis, we divided the coral reef zone into five geographic sectors, labelled S1 to S5 from north to south.

in the study area. To summarize the spatiotemporal variation of thermal conditions experienced by corals in each sector, we averaged the SSTs of the grid points and calculated the annual means for each month. To examine the intra-annual variation in SST, we averaged the standard deviations of the monthly mean SSTs.

To investigate how thermal conditions changed over the past 31 years in each sector, the normalized SST anomalies were used at each grid point. These anomalies were first normalized by dividing the monthly anomalies over the standard deviation. Then, the warming rate over the three decades (1990–2000, 2001–2010, and 2011–2020) was calculated using the normalized anomalies. The generalized linear regression model (GLM) was used to calculate the magnitude of spatial variation and to determine the warming rate in the study area.

To assess the heat stress patterns affecting coral reefs in the study area, we utilized three metrics: Maximum Sea Surface Temperature (SSTmax), Degree Heating Week (DHW), and Marine Heatwave (MHW). We first calculated the monthly mean SSTs for each grid point and then averaged the annual maximum SSTs to obtain the overall SSTmax for the study area. The SSTmax was used to identify months with high SSTs and determine the Maximum Monthly Mean (MMM) of SST over the 31-year period at each grid point. We then used the local MMMs to identify intensities and periods in which the daily SST exceeded the averaged MMM of the whole study area. The gridded MMMs were used to calculate the intensity of positive anomalies at each corresponding grid point.

However, the SSTmax alone may not necessarily indicate the presence of heat stress on coral reefs, particularly if exposure to high temperatures is brief or the maximum temperatures are within the usual reef climatology. Therefore, we used the DHW index to assess the magnitude of heat stress accumulation likely to affect corals in the study area. To calculate daily DHWs, we summed the positive anomalies (i.e., HotSpot, HS) of daily SSTs that exceed the MMM over a moving time window of 56 days and then divided the outcome by 7 (Equations 5.1 and 5.2). This approach was developed by Lachs et al. (2021) to refine the operational Degree Heating Week (DHW) algorithm used by NOAA to improve predictions of coral bleaching. The resulting optimized DHWs (hereafter, Opt DHWs) were used to describe the historical thermal stress accumulated in each grid point. We also used Opt DHWs to predict bleaching events according to NOAA's bleaching thresholds: 1) DHW of 0°C-week indicates no accumulated heat stress, 2) 0°C-week < DHWs < 4°C-week indicate a low level of heat stress and an unlikely occurrence of bleaching, 3) 4°C-week ≤ DHWs < 8°C-week reveal bleaching is more likely and can be expected, and 4) DHWs ≥ 8°C-week indicate that the heat stress can severely cause mass bleaching or even mass mortality events (Liu et al. 2018).

$$HS = \begin{cases} SST_{daily} - MMM, \ SST_{daily} > MMM \\ 0, \ SST_{daily} \leq MMM \end{cases} \tag{5.1}$$

$$\text{Opt DHW} = \frac{1}{7}\sum_{i=1}^{i=56}(HS_i, \ if \ HS_i > 0°\text{C}) \tag{5.2}$$

In addition, the number, duration, and intensity of MHWs were calculated using the definition of Hobday et al. (2016). Briefly, MHW was detected whenever SST anomalies exceeded the 90th percentile of the study area climatology for at least five consecutive days. We calculated the 90th percentile based on an 11-day shifting window of historical thermal conditions in the region. However, since this definition can include periods in which SSTs exceed the 90th percentile but still fall below the extreme conditions for the region (e.g., in winter months), we only considered SST exceedances above the long-term maximum 90th percentile as a detection threshold for summer MHWs, which are the most stressful heat conditions for coral reefs.

All data manipulations, calculations, and analyses were performed using R (R Core Team 2020), with the heatwaveR package (Schlegel and Smit 2018) used to detect and analyse MHWs.

5.3 RESULTS

5.3.1 Spatiotemporal variations in sea surface temperature across latitudinal gradient in the Red Sea

The average climatological SST from 1990 to 2020 was 26.22±2.61°C in the study area. Coral reefs in the study area experienced varying thermal conditions, ranging from 25.30±2.32°C in the north (S1) to 27.38±2.58°C in the south (S5) (Figure 5.2). Along a latitudinal gradient, the SST increased by 2.62±0.81°C from north to south, with the maximum annual temperature occurring in August across all sectors. The thermal conditions that exceeded the overall climatological SST of the study area by 1°C or more persisted for four months (July–October) in the northern sectors (S1 and S2), while similar or more extreme conditions lasted for 5–7 months (May–November) in the southern sectors. Coral reefs also experienced variations of >2°C per year, with the maximum SST variability occurring between February and August (approximately 7°C). The study revealed that while coral reefs in the northern sectors thrived in a monthly SST variation of 2.35±0.22°C/year and 2.44±0.21°C/year in S1 and S2, respectively, their counterparts in S3, S4, and S5 underwent higher intra-annual SST fluctuations (Table 5.1). The study also showed that although SST increased each decade in all sectors, the northern reefs in S1 experienced the highest long-term warming rate compared to the other sectors.

5.3.2 Increased annual sea surface temperature maxima in the southern sectors of the Red Sea

The study area experienced an annual SSTmax of 29.88±0.91°C. However, this threshold above which heat stress accumulates

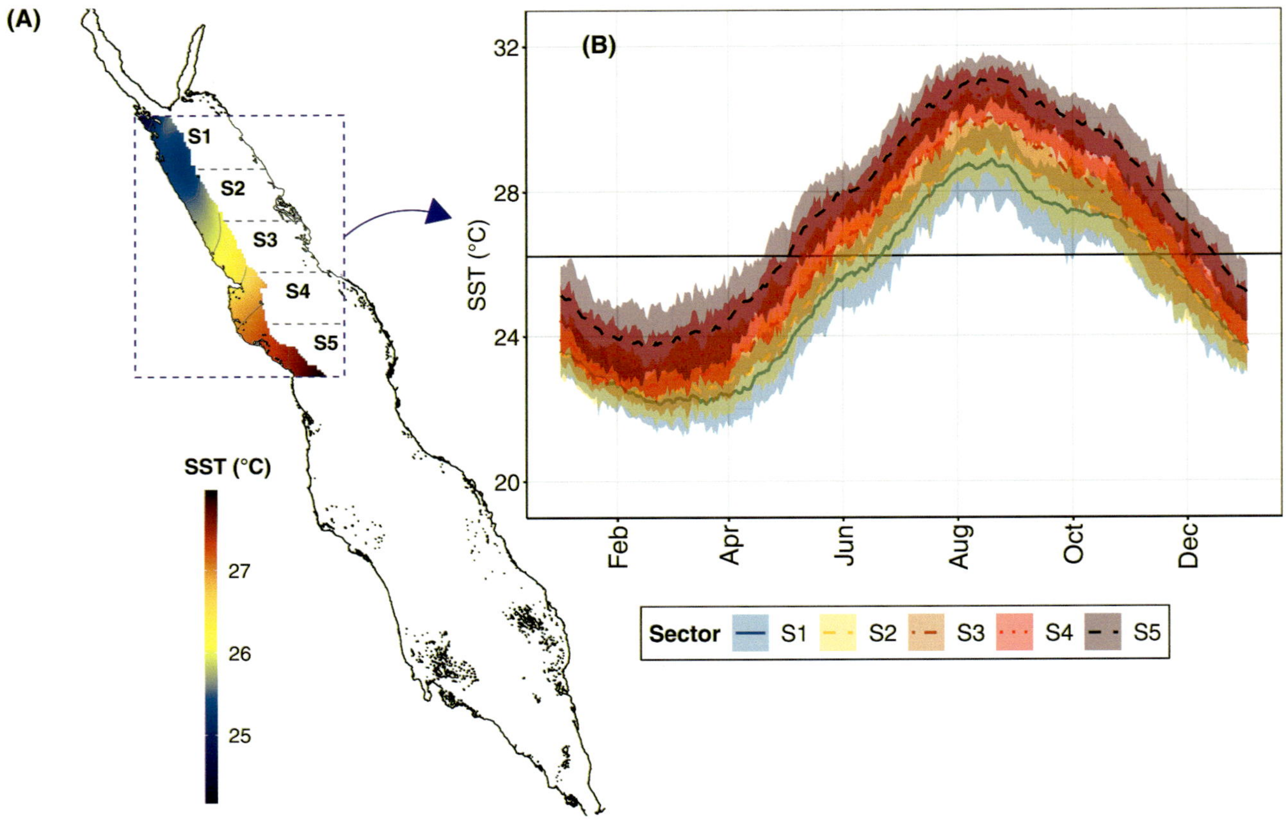

FIGURE 5.2 Spatiotemporal variation of sea surface temperature (SST) from 1990 to 2020. (A) The climatological mean SST along the study area. (B) Monthly SST variations for each sector are represented in relation to the mean SST of the entire study area (26.22°C), which is indicated by a horizontal solid black line.

TABLE 5.1 Summary of the SST conditions that prevailed in the study area from 1990 to 2020. The climatological mean SST (°C) is the long-term mean SST of all grid points in each sector (S1 to S5). Intra-annual variability reflects the extent of monthly changes in SST each year. The annual amplitude (°C) shows the difference between the maximum and minimum SST each year. The warming rate is the linear change of the normalized anomalies per decade (°C/decade).

SST CONDITIONS	*S1*	*S2*	*S3*	*S4*	*S5*
Climatological mean	25.30±2.32	25.55±2.39	26.10±2.53	26.78±2.62	27.38±2.58
Intra-annual variability	2.35±0.22	2.44±0.21	2.58±0.20	2.68±0.22	2.63±0.25
Annual amplitude	6.58±0.61	6.77±0.51	7.21±0.53	7.51±0.56	7.34±0.64
Warming rate	0.45±0.13	0.31±0.03	0.33±0.35	0.31±0.44	0.38±0.60

increased southward, following the pattern of SSTs. Corals in the southern sectors were therefore typically subjected to higher annual SSTmax than those in S1 and S2 (Figure 5.3). As a preliminary indicator of heat stress, maximum temperatures in the three southern sectors exceeded those in S1 and S2 by approximately 1.80±0.58°C. Notably, our results indicate that extreme thermal conditions have been increasing in all sectors over the last 10 years. Although the increase in annual SSTmax after 2010 in the southern sectors was not as great as that in the northern sectors, it raised the upper thermal threshold of the southern reef environment above the long-term MMM of the study area.

5.3.3 Northern Red Sea coral reefs have accumulated more heat stress compared to the south based on optimized DHWs

The study found that heat stress in the study area typically begins to accumulate above the long-term average (MMM) in late July, peaks in August, and subsides in September. As a result, corals are usually exposed to annual maximum thermal conditions that exceed the MMM of the study area

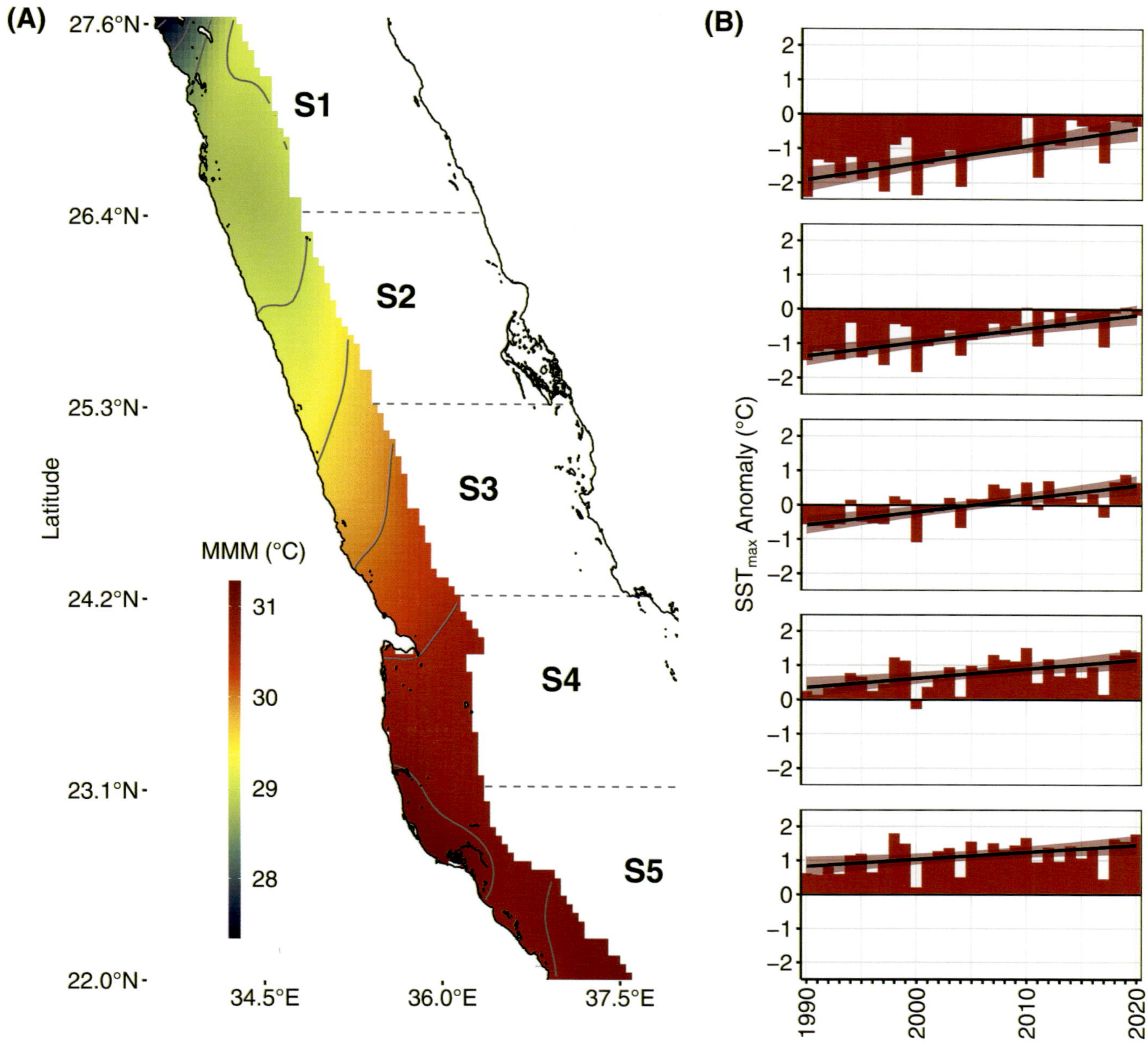

FIGURE 5.3 Spatiotemporal pattern of the maximum sea surface temperature (SSTmax) at the study area. (A) Mean of the annual SSTmax at each grid point. (B) Deviation of the annual SSTmax in each sector from the overall MMM of the study area. In all sectors, the SSTmax has increased with time (as shown by the linear trend line), and the number of positive anomalies has particularly increased in the three southern sectors in the last 10 years.

(29.88°C) for 27.14 ± 19.20 days between July and September (Figure 5.4). However, in 2011 and 2017, these conditions were limited to just three days and one day, respectively. Most of the extreme annual conditions occurred between 2010 and 2020, during which corals were subjected to 30.36 ± 0.30°C for 45.67 ± 11.95 days on average. Notably, in 2012, the reef environment experienced an unprecedented heat accumulation of 0.43 ± 0.26°C above the MMM for 60 days (July 12 to September 9). Similarly, in 2020, the heat stress persisted for 61 consecutive days (August 2 to September 30), but was more intense, with an increase of 0.60 ± 0.29°C above the MMM.

The yearly SST anomalies above the gridded MMMs were markedly high in the northern two sectors (0.40 ± 0.22°C in S1 and 0.38 ± 0.19°C in S2). In comparison, thermal stress anomalies exceeding the corresponding MMMs averaged 0.36 ± 0.17°C, 0.34 ± 0.16°C, and 0.31 ± 0.14°C in S3 to S5, respectively (Figure 5.5). The distribution of this stress varied along the study area, where most of the high DHWs were concentrated in the northern region (S1 and S2) between 2010 and 2020 (Figure 5.6). In S1, such heat stress of ≥4°C-week was detected in seven years and averaged 4.99 ± 0.61°C-week. Corals in S2 and S3 also experienced periods of exposure to 4.39 ± 0.42°C-week and 4.47 ± 0.36°C-week, respectively, during six and five years (Figure 5.7). In contrast, the detection of high levels of accumulated heat stress was much less common in the southernmost sectors, where corals in S4 were subjected to warming degrees of 4.66 ± 0.39°C-week in only two years, 2007 and 2020, while their counterparts in S5 were impacted

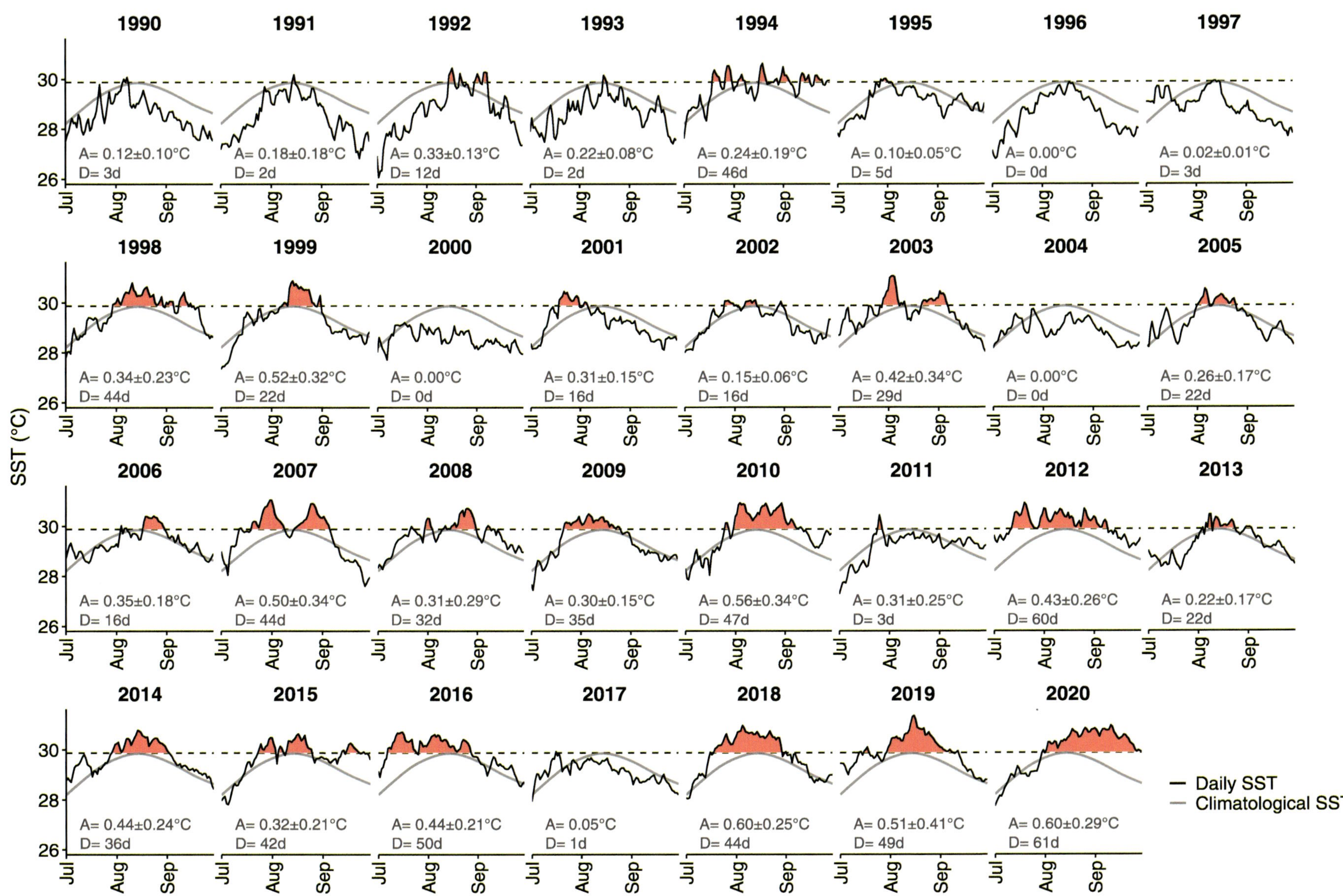

FIGURE 5.4 Time series of the summer thermal conditions prevailing in the study area between 1990 and 2020. The red shaded areas represent the periods during which the daily SST (black curve) exceeded the averaged MMM (horizontal dashed line). The grey curve summarizes the average 31 years daily SST. Positive SST anomalies (A) and the total number of days (D) during which the SST has exceeded the MMM are shown for each year.

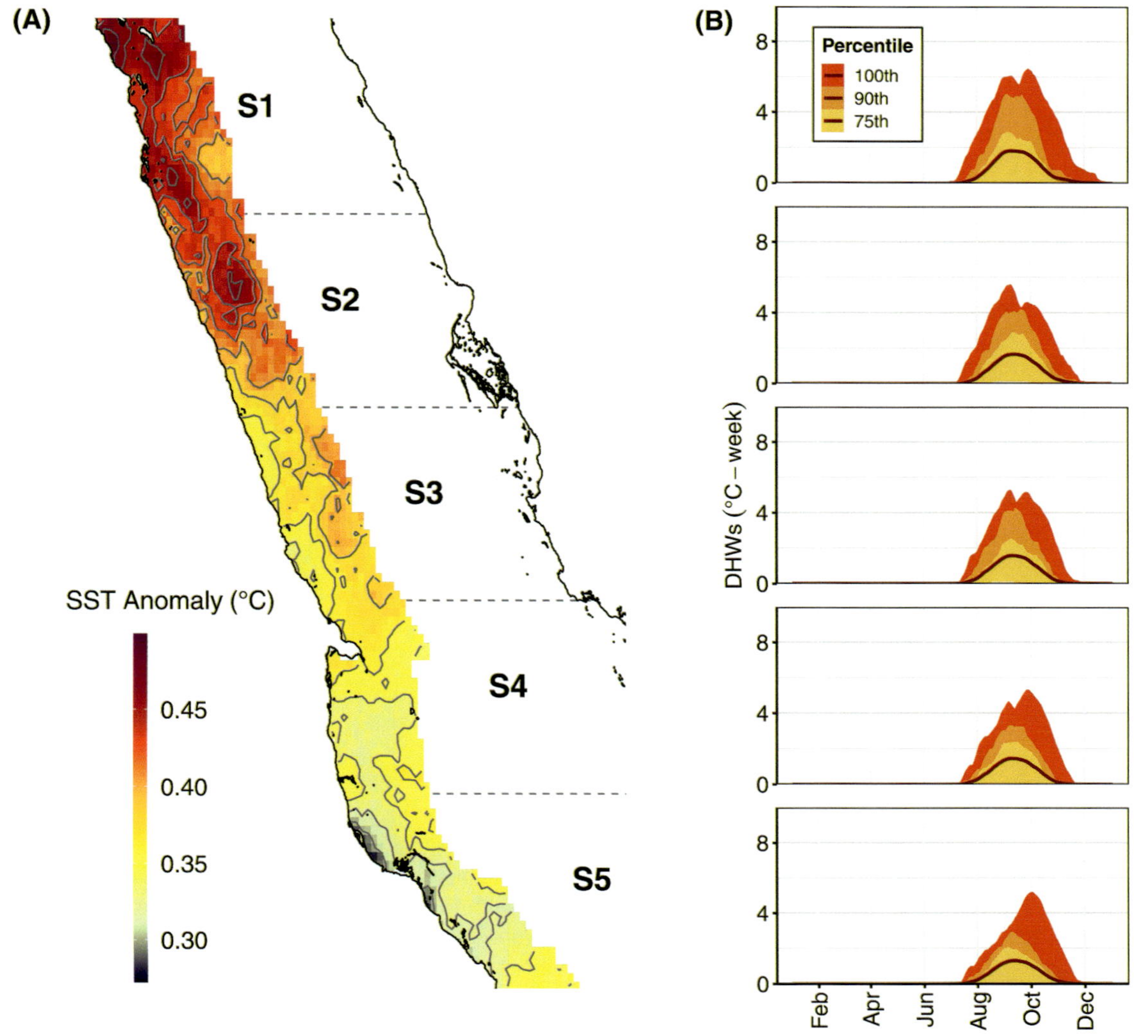

FIGURE 5.5 Anomalous SST and the corresponding accumulated heat stress in the studied sectors (S1–S5) along the Egyptian side of the Red Sea proper. (A) Pattern of the positive-only SST anomalies (above the MMM) in each grid point of the study area. (B) Daily time series of Opt DHWs in each sector. Percentiles were obtained from Opt DHWs over the 31 years.

by 4.77 ± 0.36°C-week in the summer of 2020. The anomalous SST values remained at low levels in both sectors during the summer season of the remaining years. It should be noted that the Opt DHWs showed the occurrence of thermal stress but to a lesser extent compared to operational DHWs of NOAA (Figure 5.8). The thermal stress events based on the Opt DHWs were generally less than 8°C-week during most of the studied period and less than 4°C-week in most grid points covering the study area. These levels, according to NOAA consensus, indicate that severe thermal stresses were not common on the Egyptian coast.

5.3.4 Increased frequency and intensity of marine heatwaves in the Red Sea over the past decade

Our investigation revealed 16 marine heatwave (MHW) summer events that occurred for 259 days over an 11-year period. During these events, sea surface temperatures exceeded the maximum 90th percentile threshold (which was determined to be 30.57°C) for at least five consecutive days (Figure 5.9). On average, the intensity of the detected summer heatwaves was 1.03 ± 0.18°C above the climatological SST, and was 0.25 ± 0.13°C above the 90th percentile threshold. However, the number, duration, and intensity of these events have increased in the last decade. Between 2010 and 2020, corals were exposed to 19.82 ± 24.13 days of 0.47 ± 0.35°C above the threshold during 11 MHW events. Of these events, the 2012 heatwave was the longest to date, lasting for 29 consecutive days (from June 25 to July 23) with a thermal stress intensity of 29.88 ± 0.59°C. This was twice as long as the 2010 event (Table 5.2). The corals encountered an even longer MHW in 2020, which lasted for an unprecedented 89 days (from 17 August to 13 November), with a mean intensity of 0.34°C above the 90th percentile threshold.

5.4 DISCUSSION

The current study provides a detailed assessment of the thermal stress conditions affecting coral reefs along the Egyptian coast of the Red Sea. Our analysis covers a zone that extends for

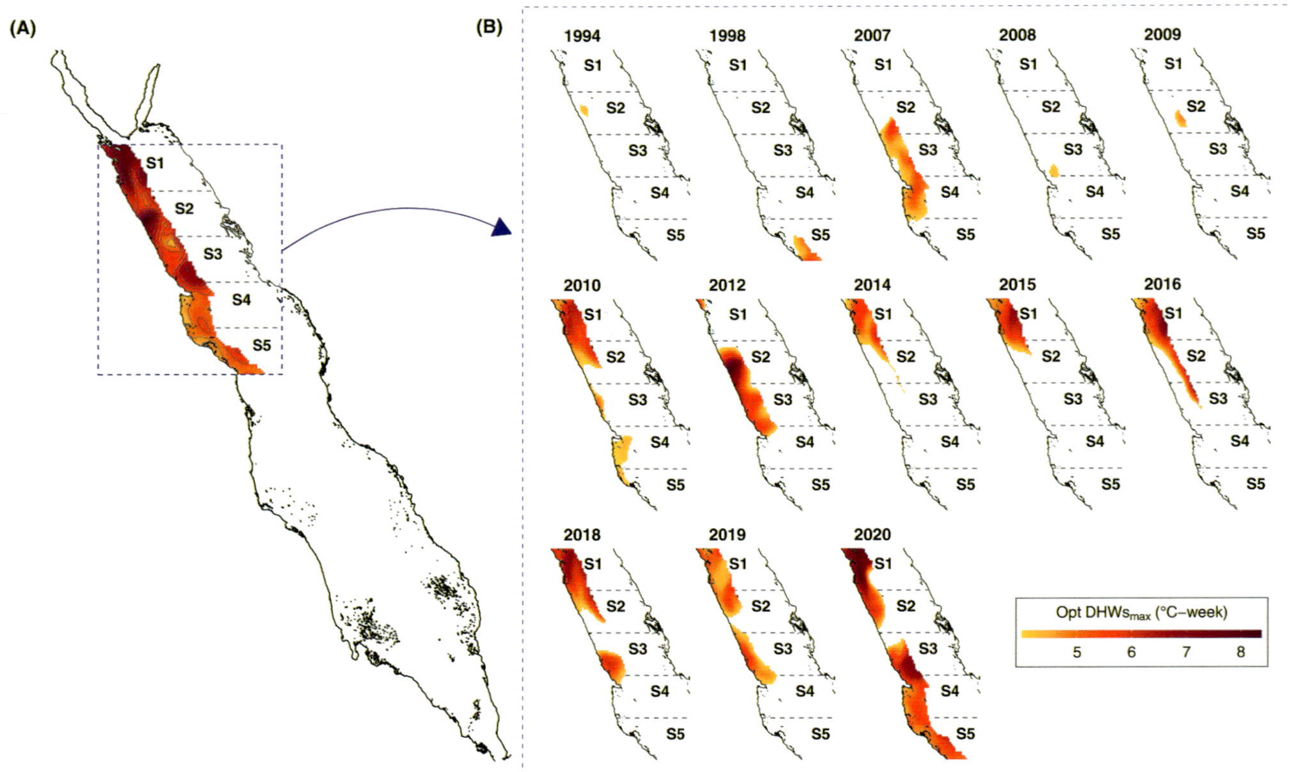

FIGURE 5.6 The accumulation of the heat stress in the Egyptian coral reef environment. (A) Maximum optimized DHWs (Opt DHWs$_{max}$) of the studied area. (B) Spatiotemporal distribution of the Opt DHWs$_{max}$ in the five studied sectors (S1–S5). Grid points with values lower than 4°C-week were removed from the maps and only high values (≥4°C-week) were retained to better represent the data and show the areas and years that were susceptible to high warming degrees. In cases where there was not at least one grid point equal to or greater than 4°C-week in one year, data for that year was ignored and completely removed from the map.

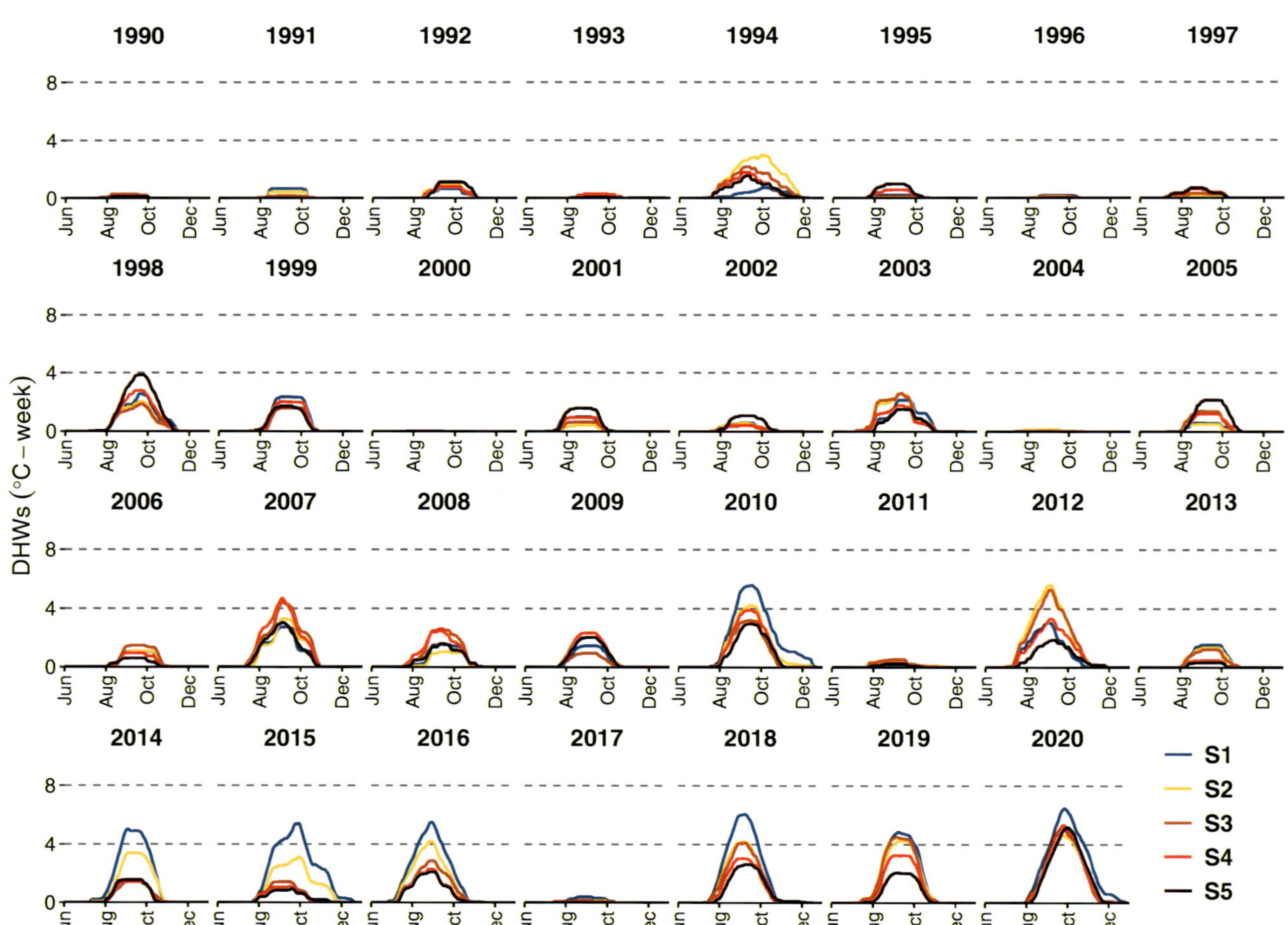

FIGURE 5.7 Daily Opt DHWs that were averaged from the gridded data of each sector. In the time series, only Opt DHWs between June and December are shown.

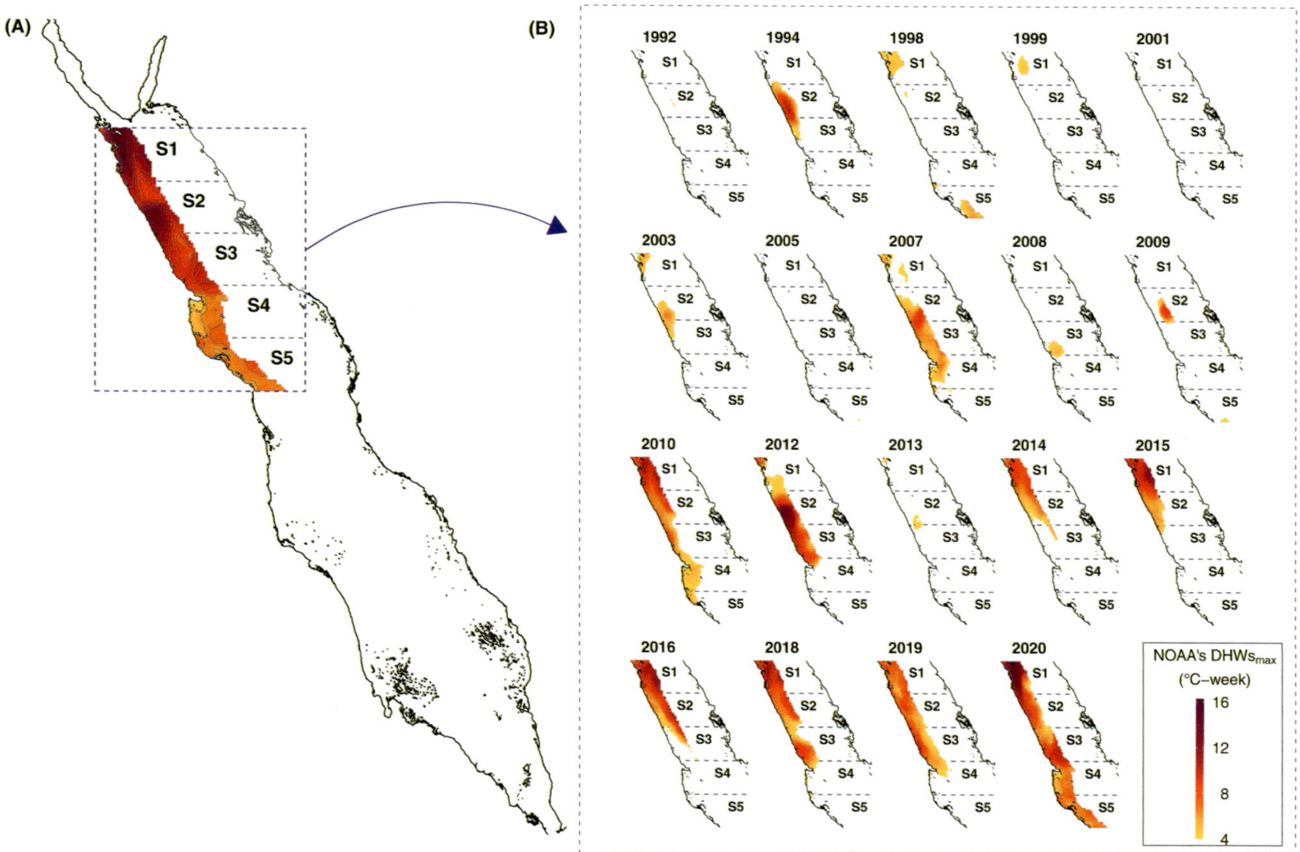

FIGURE 5.8 Operational NOAA's maximum DHWs on the Egyptian side of the Red Sea proper between 1990–2020. (A) Distribution of the maximum DHWs generated by NOAA. (B) Variation in the maximum DHWs over years. Only grid points and years of DHWs ≥4°C-week are shown. Data of NOAA's DHWs (Dataset ID: NOAA_DHW) were downloaded from https://coastwatch.pfeg.noaa.gov/erddap/griddap/.

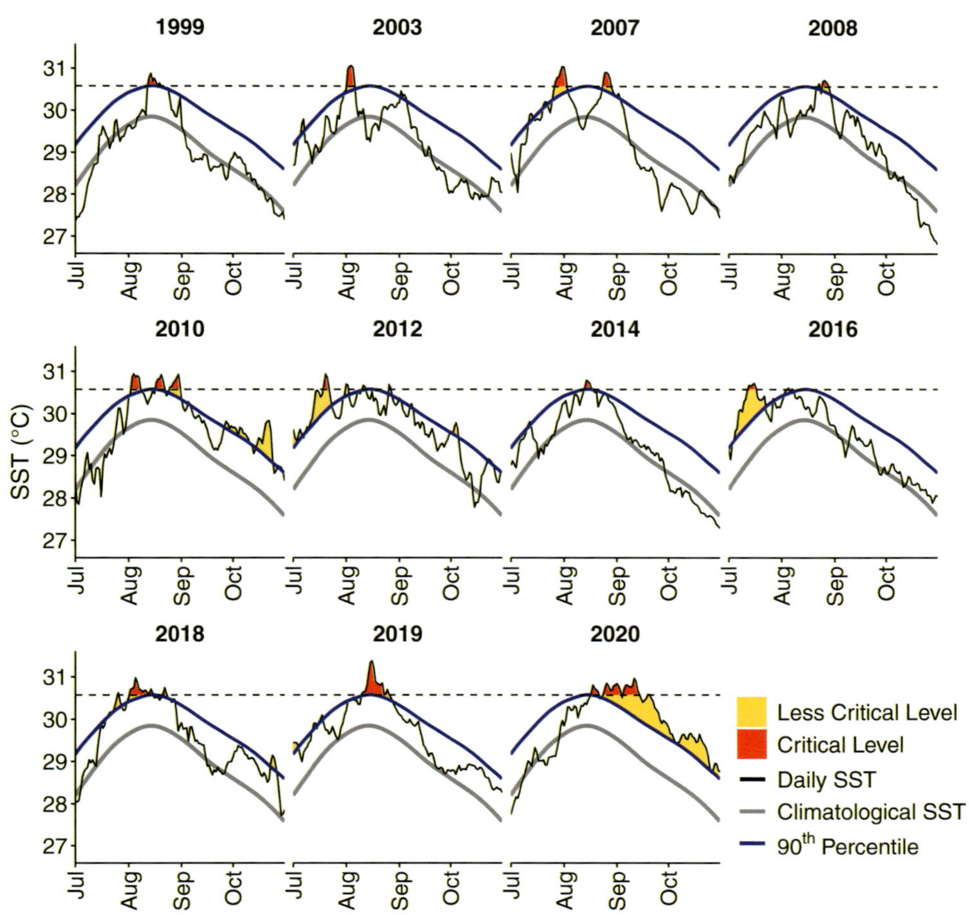

FIGURE 5.9 The summer MHWs detected at the Egyptian coast of the Red Sea. Whenever the daily SST (black solid curve) exceeds the 90th percentile threshold (blue curve), for at least five consecutive days, MHW is detected (the coloured area between the two curves). The width of this area reflects the duration, while its amplitude indicates the intensity of the MHW event. Nevertheless, the intensity of the MHW is considered critical (red area) if it exceeded the maximum 90th threshold (horizontal dashed line).

TABLE 5.2 Summary of the MHW events that impacted the study area during summertime. This summary includes dates (day/month), durations (number of days), and intensities (°C) of the detected MHW events. It also describes the rate (°C) of formation and decline of these events. Abbreviations are given below the table.

	DATE				INTENSITY RELATIVE TO THE CLIMATOLOGICAL MEAN SST				INTENSITY RELATIVE TO THE 90TH PERCENTILE THRESHOLD				ABSOLUTE INTENSITY				RATE	
YEAR	T_S	T_P	T_E	D	I_{MEAN}	I_{MAX}	I_{VAR}	I_{CUM}	I_{MEAN}	I_{MAX}	I_{VAR}	I_{CUM}	I_{MEAN}	I_{MAX}	I_{VAR}	I_{CUM}	R_{ONS}	R_{DEC}
1999	12/8	14/8	20/8	9	0.85	1.03	0.10	7.61	0.12	0.30	0.10	1.04	30.68	30.88	0.11	276.12	0.26	0.04
2003	31/7	3/8	6/8	7	1.12	1.35	0.26	7.81	0.37	0.61	0.27	2.61	30.81	31.07	0.28	215.70	0.18	0.21
2007	26/7	31/7	3/8	9	1.20	1.43	0.20	10.78	0.43	0.67	0.20	3.88	30.79	31.06	0.21	277.12	0.10	0.21
2007	24/8	26/8	3/9	11	0.98	1.23	0.17	10.76	0.20	0.46	0.18	2.21	30.59	30.92	0.25	336.51	0.16	0.06
2008	25/8	26/8	29/8	5	0.98	1.04	0.06	4.91	0.21	0.28	0.07	1.06	30.65	30.73	0.09	153.24	0.14	0.07
2010	2/8	4/8	8/8	7	1.04	1.21	0.18	7.26	0.31	0.48	0.18	2.14	30.78	30.94	0.18	215.46	0.18	0.11
2010	17/8	30/8	31/8	15	0.99	1.33	0.18	14.91	0.24	0.55	0.17	3.56	30.72	30.93	0.15	460.74	0.04	0.27
2012	25/6	20/7	23/7	29	1.24	1.70	0.25	35.95	0.33	0.88	0.27	9.62	29.88	30.93	0.59	866.46	0.03	0.25
2014	13/8	14/8	17/8	5	0.85	0.95	0.09	4.25	0.12	0.22	0.09	0.62	30.69	30.80	0.09	153.47	0.16	0.06
2016	5/7	13/7	22/7	18	1.43	1.80	0.26	25.67	0.55	0.92	0.25	9.87	30.32	30.72	0.33	545.77	0.09	0.10
2016	3/8	5/8	8/8	6	0.81	0.86	0.05	4.85	0.08	0.13	0.04	0.48	30.56	30.61	0.03	183.37	0.04	0.06
2018	19/7	26/7	27/7	9	0.92	1.15	0.11	8.32	0.12	0.37	0.12	1.08	30.28	30.63	0.20	272.51	0.05	0.22
2018	31/7	5/8	15/8	16	0.91	1.23	0.14	14.58	0.18	0.50	0.14	2.88	30.68	30.98	0.13	490.90	0.10	0.05
2018	21/8	22/8	25/8	5	0.84	0.97	0.08	4.19	0.09	0.22	0.09	0.43	30.59	30.75	0.10	152.95	0.14	0.07
2019	10/8	16/8	28/8	19	1.02	1.53	0.23	19.33	0.28	0.80	0.24	5.26	30.80	31.37	0.27	585.28	0.13	0.06
2020	17/8	11/9	13/11	89	1.24	1.76	0.23	110.20	0.34	0.90	0.23	29.87	29.82	30.96	0.88	2654.26	0.04	0.01

T_S: Starting date
T_P: Maximum peak date
T_E: Date of the end
D: Duration
I_{MEAN}: Mean intensity
I_{MAX}: Maximum intensity
I_{VAR}: Intensity variance
I_{CUM}: Cumulative intensity
R_{ONS}: Onset rate
R_{DEC}: Decline rate

70.94±0.19 km from the shore to ensure a comprehensive coverage of both inshore and offshore reefs. This approach allowed us to identify unusual thermal conditions and assess the potential vulnerability of the Egyptian reefs by describing the spatiotemporal distribution of heat stress from 1990 to 2020. Despite the limitations of satellite-derived SST data in nearshore zones, our results were highly consistent with the previous studies that used complete-basin datasets (Chaidez et al. 2017; Osman et al. 2018; Genevier et al. 2019; Mohamed et al. 2021).

Coral reefs along the Egyptian coast of the Red Sea are distributed along a latitudinal gradient, with thermal conditions increasing from north to south. This means that the corals found in the south of the Egyptian Red Sea coast have been exposed to a history of thermal environmental conditions that precede those in the north. Despite the contrasting coral reef environments between northern and southern regions, southern corals were not immune to thermal stress, exhibiting higher bleaching prevalence compared to their northern counterparts (Hanafy and Ismail 2012; Gajdzik et al. 2021a). However, through the current study, we highlight several points that still need further investigation: 1) the summer thermal conditions in the northern environment represent the lowest temperature levels in the study area, meaning that even if the northern sectors are exposed to unusual thermal conditions, they do not reach the stress limits to which southern corals are exposed; 2) despite the warming rate in the north (in S1) being higher than in the other sectors, this increase did not bring thermal conditions in this sector to the same thermal conditions as in the south, indicating that northern sectors are mostly geographically protected from the impact of climate changes; and 3) the southern corals may be thriving at extreme thresholds (upper limits) for coral tolerance to thermal stresses, meaning that a slight accumulation of positive anomalies of SST above the MMM may lead to more severe bleaching than their counterparts in the north, where the actual stress intensity is lower.

Determination of the bleaching thresholds is a complex process that depends on the amplitude and duration of exposure to thermal stress and on local conditions, even in corals of the same species (Savary et al. 2021). We used the Lachs et al. (2021) approach to calculate Opt DHWs for each grid point in the study area and compared this modified version of DHWs with the operational NOAA DHW product, on which many previous studies in the Red Sea have relied (Osman et al. 2018; DeCarlo et al. 2020; Hammerman et al. 2021). Although it appears that northern corals are exposed to Opt DHWs that may exceed 8°C-week (i.e., bleaching and mass mortality threshold according to NOAA consensus), temperatures representing this stress (Figure 5.3) are considered to be within the tolerance limits of Red Sea corals in general and northern corals in particular (Osman et al. 2018; Evensen et al. 2022). On the other hand, while it may appear that the northern region is more exposed to thermal stresses than the southern regions (Figure 5.5), in reality these stresses, even if unprecedented in that region, are false alarms and seem to have no effect on corals according to the bleaching history of the region (Eladawy et al. 2022).

Our results were largely consistent with the pattern of bleaching reported *in situ* in 2012 (Hanafy and Ismail 2012), as well as with the years in which the Egyptian coast was exposed to thermal stress, as revealed by remotely sensed data (Genevier et al. 2019; Shaltout 2019; Mohamed et al. 2021). Also, the predictions of the years in which summer thermal stresses occurred were consistent when using both MHW and Opt DHW stress indices. Moreover, our findings agreed with recent results from Eladawy et al. (2022), which showed somewhat high levels of DHWs (up to 20°C-week) in the north when using the 30°C-adjusted thermal threshold. However, the latter study showed that corals at the NRS can be far from bleaching events (DHWs < 5°C-week) when using the 32°C-adjusted thermal threshold. It is noteworthy that we used a baseline MMM that may represent the minimum limit for calculating thermal stress and may be more precautionary in predicting thermal stress (Lachs et al. 2021), especially with the presence of some coral species that have shown sensitivity to temperature increases in some areas of the Red Sea (Furby et al. 2013; Monroe et al. 2018).

Adaptation to the local environment is somehow underestimated in the Red Sea; indeed, latitudinal genetic differences in the Red Sea should not be ruled out. Despite the notion that corals in the Red Sea are genetically "the same" (Robitzch et al. 2015), based on nine microsatellite markers, recent work based on medium-scale genome-wide analyses (using RAD-Seq) suggests it is not always the case (Arrigoni et al. 2020; Sturm et al. 2020; van der Ven et al. 2021). Contrasting genetic differentiation patterns were observed in two different hard corals *Pocillopora verrucosa* and *Stylophora pistillata* populations across the Red Sea (Buitrago-López et al. 2023). It is expected that species with high genetic differentiation would accumulate variants with diverged degrees of allele frequencies across populations (van der Ven et al. 2021). Hence, local "genetic" adaptation might contribute to coral resilience in the Red Sea; however, the genetic signatures of local adaptation have not yet been established for the Egyptian coast.

Corals are metaorganisms as they live in close association with a myriad of microbes that are crucial for their survival (Rohwer et al. 2002; Rosenberg et al. 2007; Rosenberg and Zilber-Rosenberg 2018; Zilber-Rosenberg and Rosenberg 2021). Among these microbes, coral-associated bacteria are crucial for the resilience of corals against climate change (Bourne et al. 2016; Ziegler et al. 2019; Voolstra and Ziegler 2020; Meunier et al. 2021). Apart from a 16S ribosomal RNA (rRNA) gene amplicon sequencing study that showed site-specific relationships with mucus-associated bacteria along a latitudinal gradient in the Egyptian Red Sea (Osman et al. 2020), the roles that these coral-associated microbial assemblages play in bleaching resilience in general (and in particular in these coral populations) have not been explored. Studying the functional roles of the coral microbiomes using meta'omics approaches (shotgun metagenomes and metatranscriptomes) will unravel potential mechanisms of microbiome-mediated resilience. It is worth noting the importance of using an integrated coral-microbiota multiomics approach coupled with efficient reef monitoring when answering such questions (Mohamed et al. 2023).

While the water circulation, currents, and wind patterns in the NRS of Egypt create a dynamic marine environment that supports a diverse range of ecosystems, they also enable the mitigation process of heat stress due to vertical mixing of more deep, cool waters to the surface layer. Indeed, more efficient climate models are urgently needed to be implemented to accurately understand the local environments where these coral populations thrive. Finally, the thermal tolerance of corals can be attributed to environmental factors that shape the coral holobiont (both the host and its microbiome) at different molecular levels (Barshis et al. 2013; Dixon et al. 2015; Hughes et al. 2017; Voolstra et al. 2020, 2021; van Woesik et al. 2022); however, the exact molecular mechanisms underlying the thermal tolerance of corals in the NRS of Egypt have yet to be identified.

5.5 CONCLUSION

In this study, we highlighted the spatiotemporal variations in SST distributions and associated thermal stress in the coral reef environment along the Egyptian coast in the 1990–2020 period. Through the current study, coral reefs at the far north of the NRS basin have not mostly been exposed to thermal stresses that exceed actual bleaching thresholds. Therefore, we suggest that the high DHWs detected in the north (particularly in the last decade) do not pose a significant heat stress on corals. In contrast to this, in the southern reefs, although they appear to experience lower levels of thermal stress during summer months, it is feared that the level of SST causing this thermal stress may be at the upper limits of coral thermotolerance. We also demonstrated through our study that heatwaves have increased in intensity and frequency since 2010, resulting in coral thermal stress in the southern sectors that is evidenced by some bleaching events revealed from previous field surveys. Despite the increasing body of evidence suggesting corals in the northern Red Sea may offer a sanctuary against the perils of climate change ("thermal refugia"), assessment of coral health in this region in the light of the recent marine heatwaves would serve as a litmus test for this notion.

REFERENCES

Arrigoni, R., Berumen, M.L., Mariappan, K.G., et al. 2020. Towards a rigorous species delimitation framework for scleractinian corals based on rad sequencing: The case study of *leptastrea* from the indo-pacific. *Coral Reefs* 39: 1001–1025. doi:10.1007/s00338-020-01924-8.

Banc-Prandi, G., Evensen, N.R., Barshis, D.J., et al. 2022. Assessment of temperature optimum signatures of corals at both latitudinal extremes of the Red Sea. *Conserv Physiol* 10: coac002. doi:10.1093/conphys/coac002.

Barshis, D.J., Ladner, J.T., Oliver, T.A., et al. 2013. Genomic basis for coral resilience to climate change. *Proc Natl Acad Sci USA* 110: 1387–1392. doi:10.1073/pnas.1210224110.

Bindoff, N.L., Cheung, W.W., Kairo, J.G., et al. 2019. Changing ocean, marine ecosystems, and dependent communities. In: *The ocean and cryosphere in a changing climate: Special report of the intergovernmental panel on climate change.* Eds: H-O. Portner, D.C. Roberts, V. Masson-Delmotte, P. Zhai, M. Tignor, E. Poloczanska, et al., 447–587. Cambridge, UK and New York, NY: Cambridge University Press. doi:10.1017/9781009157964.007.

Biton, E., and H. Gildor. 2011. The general circulation of the Gulf of Aqaba (Gulf of Eilat) revisited: The interplay between the exchange flow through the Straits of Tiran and surface fluxes. *J Geophys Res* 116: 1–15. doi:10.1029/2010JC006860.

Bourne, D.G., Morrow, K.M., and N.S. Webster. 2016. Insights into the coral microbiome: Underpinning the health and resilience of reef ecosystems. *Ann Rev Microbiol* 70: 317–340. doi:10.1146/annurev-micro-102215-095440.

Buitrago-López, C., Cárdenas, A., Hume, B.C.C., et al. 2023. Disparate population and holobiont structure of pocilloporid corals across the Red Sea gradient demonstrate species-specific evolutionary trajectories. *Mol Ecol* 32: 2151–2173. doi:10.1111/mec.16871.

Chaidez, V., Dreano, D., Agusti, S., et al. 2017. Decadal trends in Red Sea maximum surface temperature. *Sci Rep* 7: 1–8. doi:10.1038/s41598-017-08146-z.

Cornwall, C.E., Comeau, S., Kornder, N.A., et al. 2021. Global declines in coral reef calcium carbonate production under ocean acidification and warming. *Proc Natl Acad Sci USA* 118: e2015265118. doi:10.1073/pnas.2015265118.

DeCarlo, T.M., Gajdzik, L., Ellis, J., et al. 2020. Nutrient-supplying ocean currents modulate coral bleaching susceptibility. *Sci Adv* 6: eabc5493. doi:10.1126/sciadv.abc5493.

Dixon, G.B., Davies, S.W., Aglyamova, G.V., et al. 2015. Genomic determinants of coral heat tolerance across latitudes. *Science* 348: 1460–1462. doi:10.1126/science.1261224.

Eladawy, A., Nakamura, T., Shaltout, M., et al. 2022. Appraisal of coral bleaching thresholds and thermal projections for the northern Red Sea refugia. *Front Mar Sci* 9: 938454. doi:10.3389/fmars.2022.938454.

Eladawy, A., Shaltout, M., Sousa, M.C., et al. 2018. Estimating the mean circulation and water exchange of the Gulf of Suez-Red Sea via a validated one-way atmospheric-hydrodynamic coupled model. *Earth Syst Environ* 2: 213–231. doi:10.1007/s41748-018-0049-7.

Evensen, N.R., Voolstra, C.R., Fine, M., et al. 2022. Empirically derived thermal thresholds of four coral species along the Red Sea using a portable and standardized experimental approach. *Coral Reefs* 41: 239–252. doi:10.1007/s00338-022-02233-y.

Fine, M., Gildor, H., and A. Genin. 2013. A coral reef refuge in the Red Sea. *Glob Change Biol* 19: 3640–3647. doi:10.1111/gcb.12356.

Furby, K.A., Bouwmeester, J., and M.L. Berumen. 2013. Susceptibility of central Red Sea corals during a major bleaching event. *Coral Reefs* 32: 505–513. doi:10.1007/s00338-012-0998-5.

Gajdzik, L., DeCarlo, T.M., Aylagas, E., et al. 2021a. A portfolio of climate-tailored approaches to advance the design of marine protected areas in the Red Sea. *Glob Change Biol* 27: 3956–3968. doi:10.1111/gcb.15719.

Gajdzik, L., Green, A.L., Cochran, J.E.M., et al. 2021b. Using species connectivity to achieve coordinated large-scale marine conservation efforts in the Red Sea. *Mar Pollut Bull* 166: 112244. doi:10.1016/j.marpolbul.2021.112244.

Genevier, L.G.C., Jamil, T., Raitsos, D.E., et al. 2019. Marine heatwaves reveal coral reef zones susceptible to bleaching in the Red Sea. *Glob Change Biol* 25, no. 7: 2338–2351. doi:10.1111/gcb.14652.

Grottoli, A.G., Tchernov, D., and G. Winters. 2017. Physiological and biogeochemical responses of super-corals to thermal stress from the northern Gulf of Aqaba, Red Sea. *Front Mar Sci* 4: 1–12. doi:10.3389/fmars.2017.00215.

Hammerman, N.M., Rodriguez-Ramirez, A., Staples, T.L., et al. 2021. Variable response of Red Sea coral communities to recent disturbance events along a latitudinal gradient. *Mar Biol* 168: 177. doi:10.1007/s00227-021-03984-y.

Hanafy, M.H., and M. Ismail. 2012. *Climate change and its effect on the Egyptian coast of the Red Sea: The first recorded coral mass bleaching event*. Report, Hurghada Environmental Protection and Conservation Association (HEPCA), Egypt, 30 p.

Hobday, A.J., Alexander, L.V., Perkins, S.E., et al. 2016. A hierarchical approach to defining marine heatwaves. *Prog Oceanogr* 141: 227–238. doi:10.1016/j.pocean.2015.12.014.

Hoegh-Guldberg, O., Jacob, D., Taylor, M., et al. 2019. The human imperative of stabilizing global climate change at 1.5°C. *Science* 365: eaaw6974. doi:10.1126/science.aaw6974.

Hughes, T.P., Anderson, K.D., Connolly, S.R., et al. 2018. Spatial and temporal patterns of mass bleaching of corals in the anthropocene. *Science* 359: 80–83. doi:10.1126/science.aan8048.

Hughes, T.P., Kerry, J.T., Álvarez-Noriega, M., et al. 2017. Global warming and recurrent mass bleaching of corals. *Nature* 543: 373–377. doi:10.1038/nature21707.

Lachs, L., Bythell, J.C., East, H.K., et al. 2021. Fine-tuning heat stress algorithms to optimise global predictions of mass coral bleaching. *Remote Sens* 13: 1–18. doi:10.3390/rs13142677.

Liu, G., Eakin, C.M., Chen, M., et al. 2018. Predicting heat stress to inform reef management: Noaa Coral Reef Watch's 4-month coral bleaching outlook. *Front Mar Sci* 5: 57. doi:10.3389/fmars.2018.00057.

Lough, J.M., Anderson, K.D., and T.P. Hughes. 2018. Increasing thermal stress for tropical coral reefs: 1871–2017. *Sci Rep* 8: 1–8. doi:10.1038/s41598-018-24530-9.

Meunier, V., Geissler, L., Bonnet, S., et al. 2021. Microbes support enhanced nitrogen requirements of coral holobionts in a high CO_2 environment. *Mol Ecol* 30: 5888–5899. doi:10.1111/mec.16163.

Mohamed, A.R., Ochsenkühn, M.A., Kazlak, A.M., et al. 2023. The coral microbiome: Towards an understanding of the molecular mechanisms of coral – Microbiota interactions. *FEMS Microbiol Rev* 47: 1–26. doi:10.1093/femsre/fuad005.

Mohamed, B., Nagy, H., and O. Ibrahim. 2021. Spatiotemporal variability and trends of marine heat waves in the Red Sea over 38 years. *J Mar Sci Eng* 9: 1–16. doi:10.3390/jmse9080842.

Monroe, A.A., Ziegler, M., Roik, A., et al. 2018. In situ observations of coral bleaching in the central Saudi Arabian Red Sea during the 2015/2016 global coral bleaching event. *PLoS One* 13: 1–13. doi:10.1371/journal.pone.0195814.

Osman, E.O., Smith, D.J., Ziegler, M., et al. 2018. Thermal refugia against coral bleaching throughout the northern Red Sea. *Glob Change Biol* 24: e474–e484. doi:10.1111/gcb.13895.

Osman, E.O., Suggett, D.J., Voolstra, C.R., et al. 2020. Coral microbiome composition along the northern Red Sea suggests high plasticity of bacterial and specificity of endosymbiotic dinoflagellate communities. *Microbiome* 8: 1–16. doi:10.1186/s40168-019-0776-5.

Raitsos, D.E., Pradhan, Y., Brewin, R.J.W., et al. 2013. Remote sensing the phytoplankton seasonal succession of the Red Sea. *PLoS One* 8: e64909. doi:10.1371/journal.pone.0064909.

R Core Team. 2020. *R: A language and environment for statistical computing*. R Foundation for Statistical Computing, Vienna, Austria. www.r-project.org/index.html.

Robitzch, V., Banguera-Hinestroza, E., Sawall, Y., et al. 2015. Absence of genetic differentiation in the coral *pocillopora verrucosa* along environmental gradients of the Saudi Arabian Red Sea. *Front Mar Sci* 2: 1–10. doi:10.3389/fmars.2015.00005.

Rohwer, F., Seguritan, V., Azam, F., and N. Knowlton. 2002. Diversity and distribution of coral-associated bacteria. *Mar Ecol Prog Ser* 243: 1–10. doi:10.3354/meps243001.

Rosenberg, E., Koren, O., Reshef, L., et al. 2007. The role of microorganisms in coral health, disease and evolution. *Nat Rev Microbiol* 5: 355–362. doi:10.1038/nrmicro1635.

Rosenberg, E., and I. Zilber-Rosenberg. 2018. The hologenome concept of evolution after 10 years. *Microbiome* 6: 1–14. doi:10.1186/s40168-018-0457-9.

Savary, R., Barshis, D.J., Voolstra, C.R., et al. 2021. Fast and pervasive transcriptomic resilience and acclimation of extremely heat-tolerant coral holobionts from the northern Red Sea. *Proc Natl Acad Sci USA* 118: e2023298118. doi:10.1073/pnas.2023298118.

Schlegel, R.W., and A.J. Smit. 2018. Heatwaver: A central algorithm for the detection of heatwaves and cold-spells. *J Open Source Softw* 3: 821. doi:10.21105/joss.00821.

Shaltout, M. 2019. Recent sea surface temperature trends and future scenarios for the Red Sea. *Oceanologia* 61: 484–504. doi:10.1016/j.oceano.2019.05.002.

Skirving, W., Marsh, B., De La Cour, J., et al. 2020. Coraltemp and the Coral Reef Watch coral bleaching heat stress product suite version 3.1. *Remote Sens* 12: 3856. doi:10.3390/rs12233856.

Sofianos, S., and W.E. Johns. 2015. Water mass formation, overturning circulation, and the exchange of the Red Sea with the adjacent basins. In: *The Red Sea: The formation, morphology, oceanography and environment of a young ocean basin*. Eds: N.M.A. Rasul, and I.C.F. Stewart, 343–353. Berlin, Heidelberg: Springer Earth System Sciences. doi:10.1007/978-3-662-45201-1_20.

Sturm, A.B., Eckert, R.J., Méndez, J.G., et al. 2020. Population genetic structure of the great star coral, *montastraea cavernosa*, across the Cuban archipelago with comparisons between microsatellite and snp markers. *Sci Rep* 10: 15432. doi:10.1038/s41598-020-72112-5.

van der Ven, R.M., Flot, J.-F., Buitrago-López, C., and M. Kochzius. 2021. Population genetics of the brooding coral *seriatopora hystrix* reveals patterns of strong genetic differentiation in the western Indian Ocean. *Heredity* 126: 351–365. doi:10.1038/s41437-020-00379-5.

van Hooidonk, R., Maynard, J., Tamelander, J., et al. 2016. Local-scale projections of coral reef futures and implications of the Paris agreement. *Sci Rep* 6: 1–8. doi:10.1038/srep39666.

van Hooidonk, R., Maynard, J.A., Manzello, D., and S. Planes. 2014. Opposite latitudinal gradients in projected ocean acidification and bleaching impacts on coral reefs. *Glob Change Biol* 20: 103–112. doi:10.1111/gcb.12394.

van Woesik, R., Shlesinger, T., Grottoli, A.G., et al. 2022. Coral-bleaching responses to climate change across biological scales. *Glob Change Biol* 28: 4229–4250. doi:10.1111/gcb.16192.

Voolstra, C.R., Buitrago-López, C., Perna, G., et al. 2020. Standardized short-term acute heat stress assays resolve historical differences in coral thermotolerance across microhabitat reef sites. *Glob Change Biol* 26: 4328–4343. doi:10.1111/gcb.15148.

Voolstra, C.R., Valenzuela, J., Turkarslan, S., et al. 2021. Contrasting heat stress response patterns of coral holobionts across the Red Sea suggest distinct mechanisms of thermal tolerance. *Mol Ecol* 30: 4466–4480. doi:10.1111/mec.16064.

Voolstra, C.R., and M. Ziegler. 2020. Adapting with microbial help: Microbiome flexibility facilitates rapid responses to environmental change. *BioEssays* 42: 1–9. doi:10.1002/bies.202000004.

Yao, F., and I. Hoteit. 2018. Rapid Red Sea deep water renewals caused by volcanic eruptions and the North Atlantic oscillation. *Sci Adv* 4: eaar5637. doi:10.1126/sciadv.aar5637.

Yao, F., Hoteit, I., Pratt, L.J., et al. 2014. Seasonal overturning circulation in the Red Sea: 1: Model validation and summer circulation. *J Geophys Res Oceans* 119: 2238–2262. doi:10.1002/2013JC009004.

Zhai, P., Bower, A.S., Smethie, W.M., and L.J. Pratt. 2015. Formation and spreading of Red Sea outflow water in the Red Sea. *J Geophys Res Oceans* 120: 6542–6563. doi:10.1002/2015JC010751.

Ziegler, M., Grupstra, C.G.B., Barreto, M.M., et al. 2019. Coral bacterial community structure responds to environmental change in a host-specific manner. *Nat Commun* 10: 3092. doi:10.1038/s41467-019-10969-5.

Zilber-Rosenberg, I., and E. Rosenberg. 2021. Microbial-driven genetic variation in holobionts. *FEMS Microbiol Rev* 45: 1–15. doi:10.1093/femsre/fuab022.

Coral reefs in the world's warmest sea

The paradoxical vulnerability of the world's most robust corals

6

Greta S. Aeby, Radhouane Ben-Hamadou, John A. Burt, Mauricio L. Padierna, Pedro A.G. Range, Bernhard Riegl, Mahsa Alidoost Salimi and Felipe De Oliveira Torquato

6.1 INTRODUCTION

Coral reefs are biologically rich ecosystems that support a multitude of marine species and provide critical ecosystem services to human populations, such as protecting coastlines and providing valuable sources of dietary protein and income from fisheries and tourism (Hicks et al. 2019). Despite their ecological and economical value, coral reefs are in serious decline (Pandolfi et al. 2003, 2011), as they are usually situated along coastlines and thus are subject to increasing interconnections of human populations and their associated impacts (Williams et al. 2019). Impacts such as overfishing, which removes critical herbivores, and land-based pollution with resultant sedimentation, excess nutrients, and other pollutants are cited as major local stressors contributing heavily to coral decline (Pandolfi et al. 2003; Hughes et al. 2010). Compounding these threats are the impacts of climate change with the resultant increases in sea surface temperatures, changing weather patterns, and ocean acidification (Pandolfi et al. 2011; Hughes et al. 2017).

Coral reefs in the Persian/Arabian Gulf (PAG) survive under extreme environmental conditions, with temperature and salinity ranges that exceed threshold limits of coral populations adapted to other tropical regions (Coles 2003; Sheppard et al. 1992). In recent decades, rapid economic and societal development has occurred in countries surrounding the PAG, primarily due to the discovery of vast oil and gas reserves (Burt et al. 2011; Sale et al. 2011; van Lavieren et al. 2011). This has led to a rise in the human populations along the PAG's coastal regions as well as subsequent urbanization (Edmonds et al. 2021). Hence, in addition to a naturally extreme environment, coral reefs in the PAG are threatened by impacts from eutrophication, industrial pollution, coastal re-claimation and dredging, as well as fishing damage and over-exploitation (Sheppard et al. 2010; Sheppard 2016; Burt 2014; Burt and Bartholomew 2019; Le Quesne et al. 2021). Even so, corals survive under these challenging environmental conditions off the coasts of Iran, Iraq, Kuwait, Saudi Arabia, Bahrain, Qatar, and the United Arab Emirates (Pohl et al. 2014; Edmonds et al. 2021; Bouwmeester et al. 2022). Approximately 55 to 60 coral species are recorded in the PAG, and these represent a stress-tolerant subset of corals found throughout the Indo-Pacific (Riegl 1999; Purkis and Riegl 2012; Coles and Riegl 2013; Alidoost Salimi et al. 2018). In general, coral communities are comprised of poritids, merulininds, acroporids, and siderastreids (Foster et al. 2013; Alidoost Salimi et al. 2018). However, coral reefs in the PAG have a lower three-dimensional structure compared to other regions and instead appear as "coral carpets" (Purkis and Riegl 2012) (Figure 6.1). Despite their ecological and economical importance, the coral reefs of the PAG are severely understudied but are of interest to scientists studying the impacts of global climate change. Corals in the PAG already exist in a thermal environment that is equal to, or even worse than, what is predicted to occur throughout the tropical oceans (Hoegh-Guldberg et al. 2018). In this chapter, we offer current information about coral reefs within the PAG, including an overview of adaptations corals have developed to survive in the challenging PAG, examine patterns of connectivity within the PAG, report on coral bleaching and disease events, and discuss how PAG coral reefs may be changing through time.

DOI: 10.1201/9781003321392-7

FIGURE 6.1 Example of a low-relief coral reef offshore of Qatar.

6.2 OVERVIEW OF THE PAG CORAL REEFS AND ADAPTATIONS TO THE EXTREME ENVIRONMENT

Coral reefs in the PAG are uniquely challenged by a combination of extreme natural environmental conditions and increasing local anthropogenic activities. The PAG sits astride the subtropical high-pressure zone where global atmospheric circulation patterns result in an environment characterized by extreme aridity, limited cloud cover with concomitant high solar insolation, and highly variable air temperatures. Added to this is the PAG's unusually shallow depth (mean < 30 m) and its semi-enclosed nature that constricts exchange with the Indian Ocean through the narrow (42 km) Strait of Hormuz (Vaughan et al. 2019). Together, these factors result in a marine system that is characterized by extreme and highly variable temperatures (range: <12°C to >36°C annually), hyper-salinity (typically 40–42 PSU, to >50 PSU in lagoons), recurrent summer hypoxia, and high turbidity (Coles 2003; de Verneil et al. 2021; Lachkar et al. 2022). These stressors are particularly acute in the southern basin of the PAG between Bahrain and Dubai, where shallow depths (typically <20 m) exacerbate these conditions, making this the most environmentally variable and extreme portion of the PAG. As an example, coral reefs in the southern PAG occur at the highest temperatures known to be experienced by corals globally, where summer sea temperatures often exceed 36°C and persist at >34°C for over a month (Riegl et al. 2012a; Burt et al. 2019; Paparella et al. 2019). These conditions represent an extreme stressor for corals, with consequent responses from the community to the genetic levels. Coral diversity drops markedly from the Strait of Hormuz at the entrance of the PAG, with diversity decreasing along an environmental gradient towards the southwest (Sheppard and Sheppard 1991; Bauman et al. 2013; Bento et al. 2016). As a result, overall coral species richness in the PAG is an order of magnitude lower than that in the Indo-Pacific (Coles 2003; Bauman et al. 2013; Feary et al. 2012), with the lowest species diversity of corals as well as reef fishes occurring in the shallow and most environmentally extreme southern basin along the coast of Abu Dhabi (Feary et al. 2010; Burt et al. 2011; Riegl et al. 2012a; Hoey et al. 2016; Bouwmeester et al. 2022). Coral assemblages in the southern basin are heavily dominated by stress-tolerant taxa such as poritids and merulinids (Sheppard and Sheppard 1991; Riegl 1999; Burt et al. 2011). More environmentally sensitive groups, such as acroporids and pocilloporids, are rare to non-existent in the southern PAG, where they had historically been more common before recurrent bleaching events (Burt et al. 2008; Riegl et al. 2018). However, they become progressively more common members of the assemblage on reefs along the coast of Iran, Kuwait, and the offshore islands of the central PAG, where deeper waters buffer extremes in temperatures and salinity (Shokri et al. 2000; Hamza Alhazeem et al. 2017; Bejarano et al. 2022). The 'winnowing' effect of the gradient in extreme environmental conditions across the PAG results in coral assemblages that, relative to corals in other tropical regions, are extremely robust and resilient to thermal stress, particularly in the southern basin (Bauman et al. 2013; Riegl et al. 2017), with the extreme thermal tolerance of PAG corals first recognized as early as the 1960s (Kinsman 1964). Corals in Abu Dhabi and Qatar are known to have the highest bleaching thresholds globally (Riegl et al. 2011), with bleaching onset only beginning at temperatures that persist beyond 34.5°C for consecutive days and lethal impacts not observed until temperatures exceed 35.5°C (Shuail et al. 2016; Burt et al. 2019). As a result, reefs of the southern PAG have become a hotbed for research on how corals cope with and respond to extreme temperatures (Burt 2013; Vaughan and Burt 2016), giving global relevance to the regional reefs as threats from climate change loom across the tropics (Burt et al. 2020).

The Gulf is a young sea, yet corals colonized and rapidly adapted to the extreme environmental conditions. A recent population genomics study focused on the pan-tropical and locally abundant brain coral, *Platygyra daedalea*, and explored the demographic history of PAG populations, showing that this species colonized the PAG as it flooded in the wake of the Holocene glacial transgression (Smith et al. 2022). As the PAG flooded, coral species migrated from the Indian Ocean into the deep channel along the Iranian coast near the Strait of Hormuz between 12,000 and 10,000 YBP (years before present), before splitting into a northern population (e.g., near modern Kuwait) and a central/southern population (e.g., offshore from modern Qatar/UAE) 8,000 YBP as they continued to expand into the growing proto-PAG. In the southern PAG, new subpopulations subsequently formed in a stepwise manner away from the pioneer populations, after the establishment of the PAG's modern coastlines 6,000 years ago. Although this is extremely recent from an evolutionary perspective, the strong selective environment and limited genetic exchange with the Indian Ocean led to rapid adaptation in PAG populations that allowed them to cope with extreme conditions (Smith et al. 2022).

Recent studies have shed light on the mechanisms underpinning the remarkable thermal tolerance of PAG corals. Several studies have now demonstrated that PAG coral populations are genetically distinct from those in the Indian Ocean (Howells et al. 2016; Smith et al. 2017a; Alidoost Salimi et al. 2021; Torquato et al. 2022), with an over-representation of genes associated with adaptations to the local environmental extremes (Smith et al. 2022). Additionally, studies of non-symbiotic larval survival and performance at elevated temperatures demonstrate that this thermal tolerance is inherent in the host (Howells et al. 2016), rather than being contingent on a relationship with thermally tolerant symbionts. These genetic differences in corals are echoed in divergent gene expression across populations, with PAG corals showing enhanced expression of genes related to core cellular processes related to components of coral stress modulation, further underscoring their adaptive tolerance to the PAG environment (Kirk et al. 2018). Several studies have now also shown that the superior thermal tolerance of PAG corals is heritable, with crossbreeding experiments showing that crossbred offspring, sired by PAG corals, have significantly enhanced survival and performance at elevated temperatures compared with purebred offspring from populations outside the PAG (Kirk et al. 2018; Howells et al. 2021).

The unique suite of genetic adaptations suggests that PAG corals represent a valuable genetic resource for human-assisted approaches to reef conservation in other regions. Although earlier researchers have suggested that PAG corals were a resource worth propagating for translocation to other regions, such a process would be logistically challenging due to the need to keep corals alive during transport and the limited numbers that could be transferred (Coles and Riegl 2013). Recent advances in husbandry and crossbreeding offer great promise for transferring genetic resources through sperm cryopreservation, like that long used in the horse breeding industry, allowing for rapid 'assisted evolution' of thermal tolerance of coral populations in other parts of the world (van Oppen et al. 2015, 2017; Anthony et al. 2017). However, coral cryopreservation is still an area of active research, as some cryopreservation chemicals (e.g., DMSO) impair sperm survival, and such challenges must be addressed before crossbreeding can be successfully applied to transfer heat tolerance traits across ocean basins (Howells et al. 2022).

In addition to genotypic adaptation, recent studies have also demonstrated phenotypic adaptation to environmental stress in PAG corals through epigenetic modifications, whereby adaptive changes in stress tolerance occur within an organism's lifetime as opposed to the many generations needed for genomic changes, thereby offering an opportunity for much more rapid responsiveness to environmental change (Liew et al. 2020; Smith et al. 2022). Liew et al. (2020) showed that methylation patterns that control gene expression in *P. daedalea* were transmitted from adults to sperm and into larval offspring via genome-wide inheritance, with hypermethylation patterns in adults and sperm suggesting intergenerational acclimation to local temperature and salinity extremes. These authors also demonstrated that under thermal stress, genotype-independent adjustments of methylation levels in stress-related genes were strongly correlated with offspring survival rates (Liew et al. 2020). Further research by Smith et al. (2022) explored methylation patterns in a stress-related gene (XRN1) in corals from the PAG and the more environmentally benign Sea of Oman. They identified numerous outlier positions in XRN1 where methylation was more than four-fold higher in PAG individuals, with these positions associated with enhanced larval thermal tolerance. Subsequent sequencing of a subset of these outlier sites showed that intact methylation sites occurred in 87% of PAG corals compared with only 8% of corals from the Sea of Oman, indicating that selection is acting to preserve methylation in this genomic region in the PAG population. Such heritable epigenetic changes provide a further mechanism through which the superior thermal tolerance of PAG corals is maintained, with the capacity for much more rapid acclimation to environmental change than is possible from genetic means alone.

While the coral host clearly exhibits inherent characteristics that support its capacity to cope with extreme temperatures, their symbiotic dinoflagellate algal partners also play an important role for PAG corals. Early work in the PAG region had indicated that the coral-associated zooxanthellae assemblages were unusual, with seven common coral species from the southern PAG shown to be heavily dominated by a generalist group, then termed *Symbiodinium* clade C3, rather than groups more typically associated with stress tolerance (e.g., Dla, C15) in other regions (Hume et al. 2013). Improvements in sequencing techniques in subsequent years showed that PAG corals were, in fact, dominated by a novel symbiont species new to science, *Symbiodinium thermophilum* (later renamed to *Cladocopium thermophilum* during revision of Symbiodinacaea nomenclature), that was genetically distant from all C3 types occurring outside the PAG (Hume et al. 2015, 2018; LaJeunesse et al. 2018). Further work showed *C. thermophilum* to be the prevalent symbiont in all coral species examined, including across highly variable changes in temperature across seasons as well as through bleaching events during extreme heat stress (Smith et al. 2017b; Hume et al. 2018; Howells et al. 2020a).

With the ubiquity of *C. thermophilum* among taxonomically divergent coral species in the southern PAG, it has been suggested that the symbiont also plays a role in thermal tolerance in addition to adaptations of the host (Hume et al. 2015; Howells et al. 2016; Smith et al. 2017a; Howells et al. 2020a). Analyses of genetic diversity of coral symbionts across the PAG region showed *C. thermophilum* to be a member of a highly diverse but cryptically distributed ancient group of thermally tolerant symbionts that experienced selection after colonizing the PAG following the Holocene transgression, resulting in their present-day dominance in corals across the thermally extreme southern PAG (Hume et al. 2016). Although further research is needed to discern the relative importance of host- versus symbiont-specific adaptations as well as host-symbiont interactions, recent developments in approaches such as CRISPR-Cas9-based genome editing tools are opening new avenues to gain insights into these processes (Smith et al. 2022).

6.3 CORAL REEF CONNECTIVITY IN THE PAG

Corals are sessile organisms inhabiting discrete patches separated from each other by a few metres to thousands of kilometres of unsuitable habitats (e.g., deep sea and sandy bottoms).

Similar to many marine organisms, corals utilize a pelagic larval dispersal phase that connects coral populations across habitats or regions. Connectivity among populations is an important demographic process underlying the maintenance of genetic diversity and persistence of populations and species through time (Hughes and Tanner 2000; Foster et al. 2012). For coral reefs, connectivity between reefs is critical to ensure their resilience against disturbances, especially in the face of changing environmental conditions associated with global climate change (McCook et al. 2009; Frys et al. 2020).

The study of connectivity in the PAG region is particularly interesting due to the geographic features and oceanographic events. The PAG is a semi-enclosed body of water, located on the corner of the vast Indo-Pacific Ocean, between 24°N–30°N latitude and 48°E–57°E longitude. The PAG is connected to the Western Indian Ocean (WIO) via the Strait of Hormuz, which can act as a physical barrier by impeding or decreasing the movement of larvae between the semi-enclosed basin and the open sea (Torquato and Møller 2022). Other oceanographic barriers isolate not only the PAG but also the entire ROPME Sea Area (RSA), which includes the eight nations bordering the PAG and the Sea of Oman (https://gcrmn.net/regional-organization-for-the-protection-of-the-marine-environment-region/). For example, the upwelling off Somalia and off Oman creates semipermeable barriers for larval movement between the PAG and the WIO, and between east and west sides of the PAG, respectively (DiBattista et al. 2016; Priest et al. 2016; Torquato and Møller 2022).

Although understanding both the demographic dynamics and evolutionary processes in the PAG is urgently needed, the study of connectivity involving corals and coral-associated organisms are still incipient in the region. The PAG has been subject to reef habitat loss due to recurrent bleaching events and widespread anthropogenic coastal development (Burt 2014). Hence, the persistence of some coral species in the PAG depends on larval transport from source habitats to degraded areas (Riegl et al. 2017). Furthermore, the role played by the PAG in a metapopulation context is fundamental to providing insights to both evolutionary processes and conservation strategies that will cope with near-future climate change (Torquato et al. 2022).

Investigations on connectivity carried out in the PAG have used biophysical modelling and population genetics, which have sought to describe both the connectivity within the PAG and between the PAG and its adjacent Sea of Oman. In the following sections, we describe the most important findings obtained from these studies.

6.3.1 Biophysical modelling

In brief, the biophysical modelling approach uses information on currents and environmental conditions to simulate both deterministic and/or stochastic displacements of large numbers of virtual particles (i.e., larvae) through space and time (Paris et al. 2013). In the PAG, biophysical modelling methods were mainly applied for assessing the effective network of natural reserves and for describing metapopulation dynamics among the reef habitats located in the central and southern regions of the PAG (e.g., Cavalcante et al. 2020; Bento et al. 2021). In addition, the same approach was used to measure the strength of the Strait of Hormuz as a barrier for the connectivity between the PAG and its adjacent Sea of Oman (e.g., Torquato and Møller 2022). Overall, the investigations on connectivity using biophysical modelling in the PAG draw attention to both high levels of local retention and low connections among nearby reefs.

Bento et al. (2021) used biophysical modelling to investigate the recruitment from Sir Bu Nair Island (SBN), a natural reserve in the United Arab Emirates (UAE), to degraded areas. Specifically, the authors targeted the bleaching-sensitive species *Acropora downingi* to test whether SBN acts as a source, exporting larvae to vulnerable sink reefs. The study showed that although SBN served as a source habitat, exporting larvae to nearby reefs, the reserve exhibited mostly local larval retention. Moreover, the model showed a low probability of the SBN receiving larvae from neighbouring reefs.

In the central region of the PAG, Cavalcante et al. (2020) described the connectivity among several coral reefs. The model showed that, depending on specific combinations of physical and environmental conditions, coral larvae could be transported for large distances within the PAG. However, the connectivity matrix displayed high levels of larval retention to natal reefs with only low contributions to surrounding reefs. Similar results indicating restricted connectivity, and high levels of larval retention were found in the southern PAG (Cavalcante et al. 2016; Riegl et al. 2017).

Studying the connectivity between the PAG and the adjacent Sea of Oman, Torquato and Møller (2022) studied the processes underlying bio- and phylo-geographic patterns observed in the northeastern PAG. The authors found that the biological traits of coral larvae affect the connectivity between both seas by interrupting, reducing, or causing asymmetry of the larval movements through the Strait of Hormuz. For example, vertical migration and lower mortality rates increased the larval exportation from the PAG to the Sea of Oman. On the other hand, vertical migration constrained the movement in the opposite direction.

Improvements from future studies are needed but require reliable mapping and habitat characterization. In addition, species-specific models need biological information that are not yet available, for most species of the PAG, such as the vertical distribution of eggs and larvae, spawning times, pelagic larval duration, etc.

6.3.2 Population genetics

The exchange of only a few individuals, per generation, either among habitats within the PAG or between the PAG and the adjacent Sea of Oman, is enough to maintain genetic homogeneity among the populations (Slatkin 1993). As such, high connectivity in the region can be inferred from genetic homogeneity (i.e., panmixia), and vice versa. On the other end of the spectrum (i.e., a scenario of complete isolation), species integrity is not maintained due to the accumulation of genetic differences between the two groups. Alternatively, if the gene flow is limited but not completely interrupted over many generations, a spatial genetic structure can be observed.

Population genetics studies in the PAG have shown both panmixia and genetic structure between populations, depending on the genetic markers used and the target species.

Although population genetics studies can use several types of genetic markers, only two of them – namely, mitochondrial DNA (mtDNA) and single nucleotide polymorphisms (SNPs) – have been used to investigate evolutionary connectivity in the PAG. Using SNPs, the genetic structure of some coral species has been identified not only within the PAG, but also between the PAG and the Sea of Oman. Smith et al. (2022) targeted the brain coral, *Platygyra daedalea*, as the study system and found a genetic discontinuity between both seas (PAG and Sea of Oman), indicating a decrease in gene flow likely boosted by natural selection. Furthermore, the authors showed that the offshore reefs in the PAG and those in the vicinity of the Strait of Hormuz were more admixed with colonies from the Sea of Oman than with the PAG inshore reefs, indicating more connectivity among these habitats. Studying the table coral, *Acropora downingi*, between the PAG and the Sea of Oman, Torquato et al. (2022) found a constraint in the gene flow between the PAG and the Sea of Oman. Given that the PAG is just 15,000 years old, the authors suggested that the rapid differentiation process between the two populations might be due to a combination of restricted connectivity and either bottleneck events or recent foundation effects.

Coral-associated fish species have also been the subject of population genetic studies in the PAG region, using both mtDNA and SNPs, and the results for this group showed no genetic structure both within the PAG and between the PAG and the adjacent Sea of Oman. For example, *Cephalopholis hemistiktos* (Priest et al. 2016), *Nemipterus japonicus* (Farivar et al. 2017), *Pomacanthus maculosus* (Torquato et al. 2019), and *Istiblennius pox* (Mehraban et al. 2020) showed no restrictions in connectivity either within the PAG or through the Strait of Hormuz. Yet, sea urchins exhibited a pattern like corals with limited gene flow through the strait (Ketchum et al. 2020).

According to the hypotheses depicted by Torquato and Møller (2022), the constraint of gene flow in the region might not only be due to the physical-biological interactions between the larvae and the oceanographic circulation, but also a combination of this with natural selection or genetic drift. The next step in the study of molecular connectivity in the PAG region requires additional sampling sites throughout the region to provide a finer resolution for the role the PAG may play in the persistence of both present-day and future coral reef metapopulations.

6.4 CORAL DISEASE IN THE PAG

Coral disease is a significant factor impacting coral reefs with localized disease outbreaks occurring worldwide (Harvell et al. 2007; Heron et al. 2010). Coral disease has severely altered coral reefs in the Caribbean (Gladfelter 1982; Porter et al. 2001; Sutherland et al. 2004), which is currently suffering from the single most damaging disease outbreak in recent history (stony coral tissue loss disease). Stony coral tissue loss disease emerged on the Florida Reef Tract in 2014 (Precht et al. 2016; Walton et al. 2018; Muller et al. 2020) and has spread to coral reefs throughout the neighbouring Caribbean regions (Alvarez-Filip et al. 2019; Estrada-Saldívar et al. 2020; Brandt et al. 2021; Heres et al. 2021; Forrester et al. 2022). Disease is a natural component of all populations, but disease outbreaks indicate a shift in the normal relationship among host, pathogen, and the environment. Outbreaks of coral disease have increased through time (Ward and Lafferty 2004) and have been linked to anthropogenic impacts such as dredging activities (Pollock et al. 2014), terrestrial runoff (Haapkylä et al. 2011; Aeby et al. 2016), overfishing (Raymundo et al. 2009), plastic pollution (Lamb et al. 2018), and increased ocean temperatures (Bruno et al. 2007; Maynard et al. 2011).

Coral reefs face different threats depending on their geographic region. The PAG is a challenging environment for corals with extremes in temperatures and salinities as well as a multitude of anthropogenic factors. Coral disease has not been well-studied in this region, but there have been reports from Iran, Qatar, and the United Arab Emirates (Figure 6.2). Of the available studies, six types of diseases are reported to be affecting 16 coral genera. Diseases include three tissue loss diseases (white syndromes, yellow-banded tissue loss disease, black band disease), growth anomalies, focal bleaching, and dark spot disease; these are summarized in the following sections (Figure 6.3). PAG coral diseases are similar to what has been found on coral reefs in other regions, apart from yellow-banded tissue loss, which has only been reported from the PAG (Korrubel and Riegl 1998; Riegl et al. 2012b; Aeby et al. 2020) and the Gulf of Kachchh, India (Bharath et al. 2023). Disease in the PAG is commonly reported on *Acropora*, *Dipsastraea*, *Platygyra*, and *Porites* (Table 6.1). In other regions, disease is frequently reported in *Acropora* (Patterson et al. 2002; Williams and Miller 2005; Aeby 2005; Aeby et al. 2011a; Williams et al. 2011; Roff et al. 2011; Gignoux-Wolfsohn et al. 2012) and *Porites* (Raymundo et al. 2005; Aeby et al. 2011b; Haapkylä et al. 2007; Thinesh et al. 2011; Séré et al. 2012) but is not common in the other two coral genera (*Dipsastraea* and *Platygyra*).

Corals are simple animals, and disease manifests as tissue loss, growth anomalies, or discolouration (Work and Aeby 2006). In the following sections, disease descriptions are organized according to the type of lesion: tissue loss (white syndrome, black band disease, yellow-banded tissue loss), growth anomalies (excluding invertebrate galls), and discolourations (focal bleaching, dark spot).

6.4.1 White syndrome (WS)

White syndrome refers to disease in corals resulting in progressive tissue loss but with no pathogens grossly evident on lesions in the field (Bourne et al. 2015). The white syndromes represent a lesion type that is known to have different underlying etiologies, ecologies, and differences in host specificity, even though the gross lesions (i.e., lesion determined underwater) may look similar (Work et al. 2012; Aeby et al. 2015a). The underlying disease agents reported for white syndromes include pathogenic bacteria (Ben-Haim and Rosenberg 2002; Sussman et al. 2008; Ushijima et al. 2012, 2014), chimeric parasites (Work et al. 2011), programmed cell death (Ainsworth et al. 2007), as well as a variety of other organisms (Work et al. 2012). White syndromes are some of the most virulent of the coral diseases, resulting in substantial colony mortality (Bourne et al. 2015). Coral diseases that fit the description of

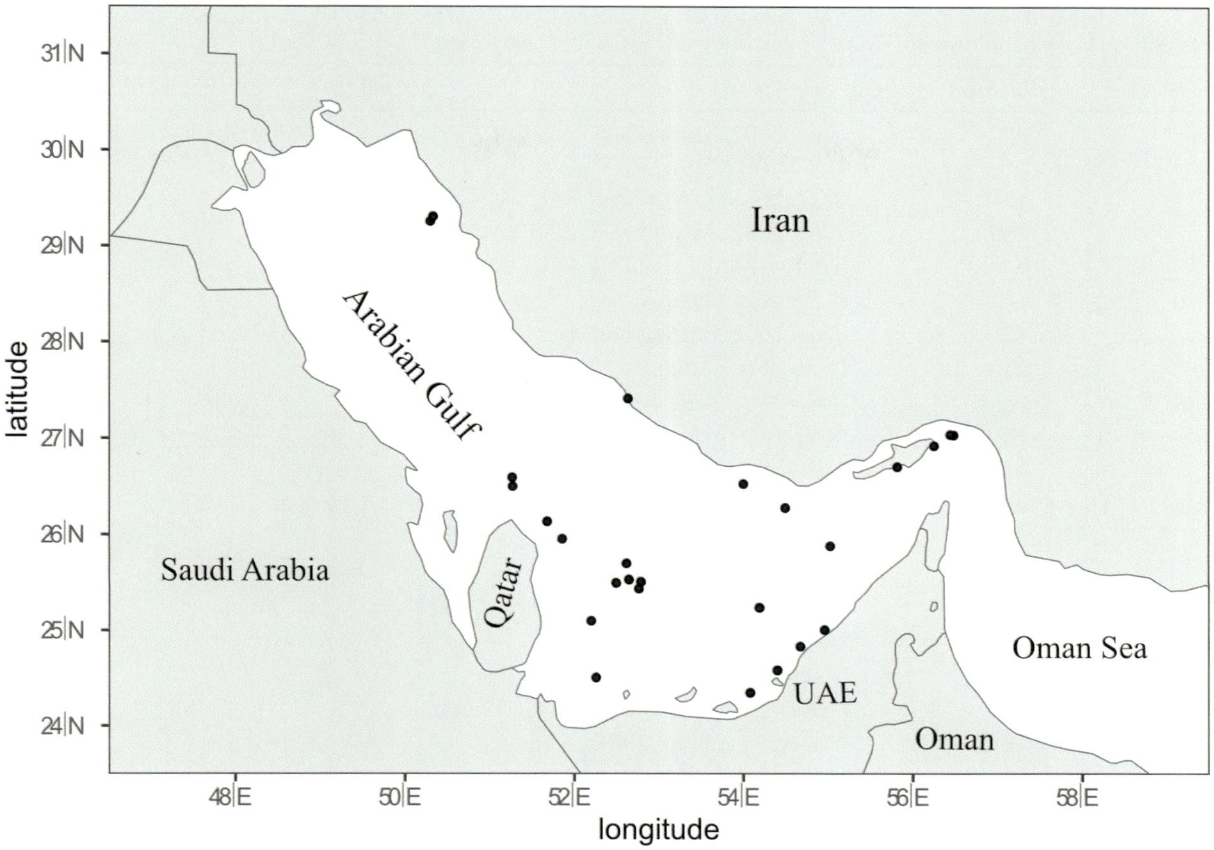

FIGURE 6.2 Map of Arabian Gulf showing sites (black dots) surveyed for coral disease.

FIGURE 6.3 Representative photos of types of diseases found in the PAG. A and B) black band disease; C and D) focal bleaching; E) dark spot; F-H) growth anomalies; I-K) white syndromes; and L-N) yellow banded tissue loss.

TABLE 6.1 The types of diseases reported for affected coral genera within the PAG. WS = white syndrome, YBTL = yellow-banded tissue loss, BBD = black band disease, GA = growth anomaly, FBL = focal bleaching, DS = dark spot.

CORAL GENUS	*DISEASE*	*SOURCES*
Acropora	WS	(Riegl 2002; Kavousi and Rezai 2011; Howells et al. 2020b; Aeby et al. 2020; Hazraty-Kari et al. 2021)
	YBTL	(Korrubel and Riegl 1998; Riegl 2002)
	BBD	(Riegl 2002; Aeby et al. 2020; Hazraty-Kari et al. 2021)
	GA	(Hazraty-Kari et al. 2021)
Coscinarea	WS	(Aeby et al. unpub.)
	YBTL	(Riegl 2002; Hazraty-Kari et al. 2021)
	BBD	(Aeby et al. unpub.)
Cyphastrea	WS	(Aeby et al. 2020; Aeby et al. unpub.)
	YBTL	(Korrubel and Riegl 1998)
	GA	(Aeby et al. unpub.)
Dipsastraea	WS	(Alidoost Salimi et al. 2017; Aeby et al. 2020; Aeby et al. unpub.)
	YBTL	(Fatemi and Shokri 2001; Riegl 2002)
	BBD	(Aeby et al. 2020)
	GA	(Kavousi and Rezai 2011; Aeby et al. 2020; Aeby et al. unpub.)
	FBL	(Aeby et al. unpub.)
Echinopora	BBD	(Aeby et al. 2020)
Favites	WS	(Howells et al. 2020b; Aeby et al. unpub.)
	YBTL	(Kavousi and Rezai 2011)
	GA	(Aeby et al. unpub.)
Goniopora	WS	(Aeby et al. 2020; Aeby et al. unpub.)
	FBL	(Aeby et al. 2020)
Pavona	WS	(Aeby et al. unpub.)
	BBD	(Aeby et al. 2020)
Platygyra	WS	(Kavousi and Rezai 2011; Howells et al. 2020b; Aeby et al. 2020; Aeby et al. unpub.)
	YBTL	(Fatemi and Shokri 2001; Riegl 2002)
	BBD	(Aeby et al. 2020)
	GA	(Kavousi and Rezai 2011; Aeby et al. 2020; Hazraty-Kari et al. 2021; Aeby et al. unpub.)
Plesiastrea	WS	(Aeby et al. unpub.)
Porites	WS	(Alidoost Salimi et al. 2017; Howells et al. 2020b; Aeby et al. 2020, unpub.)
	YBTL	(Korrubel and Riegl 1998; Fatemi and Shokri 2001; Riegl 2002; Alidoost Salimi et al. 2017; Aeby et al. 2020; Hazraty-Kari et al. 2021)
	GA	(Aeby et al. 2020; Hazraty-Kari et al. 2021; Aeby et al. unpub.)
	FBL	(Aeby et al. 2020; Hazraty-Kari et al. 2021; Aeby et al. unpub.)
Psammocora	WS	(Aeby et al. unpub.)
Siderastrea	WS	(Howells et al. 2020b)
Stylophora	WS	(Howells et al. 2020b)
Turbinaria	WS	(Riegl 2002; Aeby et al. unpub.)
	YBTL	(Korrubel and Riegl 1998; Riegl 2002; Aeby et al. unpub.)
	GA	(Aeby et al. unpub.)
	FBL	(Aeby et al. unpub.)
	DS	(Aeby et al. unpub.)

WS (white band disease, tissue loss diseases (chronic, subacute or acute)) were found in all three regions studied (UAE, Qatar, Iran) affecting 14 coral genera (Riegl 2002; Kavousi and Rezai 2011; Riegl et al. 2012b; Alidoost Salimi et al. 2017; Aeby et al. 2020; Howells et al. 2020b; Hazraty-Kari et al. 2021; Aeby et al., unpub. data). Of the studies that report disease prevalence (proportion of corals affected), WS prevalence ranging from 0.2%–2.5% occurred on coral reefs along the UAE (Aeby et al. 2020), two coral species with WS had a prevalence of ~1% on Kish Island, Iran (Alidoost Salimi et al. 2017), and a recent baseline study from Qatar reports WS prevalence from 0.13%–8.3% (Aeby et al., unpub. data). WS appears to be endemic in the PAG, having been first observed in 1996 (Riegl 2002); however, outbreaks of white syndrome have also occurred. Riegl et al. (2012b) noted an outbreak of WS on multiple coral genera following the 2010 bleaching event in the UAE. Howells et al. (2020b) documented a multi-year outbreak along the coast of the UAE between 2014 and 2016, affecting

10 different coral genera. Authors report that *Acropora* were the most susceptible to WS, with reports of up to 42% disease prevalence among surveyed colonies. They also report that WS prevalence was high in other affected coral genera including *Platygyra* (prevalence of up to 33%), *Porites* (prevalence = 33%), and *Stylophora* (prevalence = 9%). The study states that outbreaks occurred annually, beginning in summer, with WS prevalence increasing exponentially with cumulative heat stress and outbreaks stopped once cooler temperatures returned, with disease prevalence remaining at 0.05% in the other months of the years. White syndrome outbreaks have occurred on reefs in many other regions of the world, and the field patterns vary. Similar to the PAG, Willis et al. (2004) and Bruno et al. (2007) document WS outbreaks across the Great Barrier Reef (GBR) affecting multiple coral genera with clear links to heat stress. In contrast, WS outbreaks occurred in American Samoa and the northwestern Hawaiian Islands, but only affected *Acropora* spp., and there were no links to heat stress (Aeby et al. 2011a).

6.4.2 Black band disease (BBD)

Black band disease is caused by a microbial consortium, visually dominated by filamentous cyanobacteria, which creates the characteristic black band (Richardson 2004). Other constituents of the BBD lesion include sulfide-oxidizing bacteria (*Beggiatoa* spp.), sulfate-reducing bacteria that include members of the *Desulfovibrio* genus, and numerous heterotrophic bacteria (Cooney et al. 2002; Frias-Lopez et al. 2002; Barneah et al. 2007). The sulfate-reducing bacteria are responsible for the highly concentrated sulfide and anoxic conditions underneath the BBD mat that is lethal to coral tissue (Richardson 2004). Similar to other regions in the world, BBD affects multiple coral genera in the PAG, with *Acropora* the most common coral genus reported with the disease (Riegl 2002; Aeby et al. 2020; Hazraty-Kari et al. 2021). As in other regions, BBD infections on PAG reefs are seasonal, with increased prevalence in warmer months (Riegl 2002), and a BBD outbreak was noted following the 2010 bleaching event in the region (Riegl et al. 2012b). Although BBD is found on reefs worldwide, the specific bacterial species associated with this polymicrobial disease that have been genetically identified varies across regions (Frias-Lopez et al. 2002; Barneah et al. 2007; Miller and Richardson 2011; Aeby et al. 2015b; Hadaidi et al. 2018). It has been suggested that the specific microbial communities found in BBD may be primarily derived from the local reef environment (Miller and Richardson 2011), and future studies are needed to examine this in the PAG.

6.4.3 Yellow-banded tissue loss disease (YBTL)

Yellow-banded tissue loss disease (previously reported as Arabian yellow band) is a progressive tissue loss disease that leaves a distinct yellow band in the denuded skeleton adjacent to the lesion, which distinguishes this lesion from the standard 'white syndrome' classification. It was first described from PAG by Korrubel and Riegl (1998) affecting multiple coral species on reefs near Jebel Ali in Dubai. There have been limited studies on the ecology of the disease, but as with white syndrome, the etiology is unknown, it appears to occur year-round (no seasonality), and it can progress rapidly (up to 2 cm per week) (Riegl et al. 2012b). Aeby et al. (2020) conducted baseline surveys on reefs in the UAE and reported yellow-banded tissue loss (YBTL) at four out of eight survey sites in the PAG, one out of five sites in the Strait of Hormuz, and no cases of the disease at four sites in the Oman Sea. Authors report YBTL was only on *Porites*, and prevalence ranged from 0.08% to 1.14% of colonies within transects. Alidoost Salimi et al. (2017) found *Porites* colonies with similar disease lesions at Kish Island, Iran, identified it as *Porites* peeling tissue loss disease, and reported an average prevalence of 7.25%. Recent surveys in Qatar report YBTL in *Turbinaria* at one out of ten survey sites (Aeby et al., unpub. data). Corals with the distinct lesions identified as YBTL were thought to be found exclusively in the PAG (Riegl et al. 2012b; Aeby et al. 2020), but similar disease signs have recently been reported from the Gulf of Kachchh, India (Bharath et al. 2023).

6.4.4 Growth anomalies (GA)

Growth anomalies appear on colonies as distinct, protuberant masses that affect a variety of coral genera from many regions of the world (Work et al. 2016). The etiology of GAs in corals are unknown, but they are associated with reduced colony growth (Bak 1983; Stimson 2011), partial colony mortality (Peters et al. 1986; Irikawa et al. 2011; Work et al. 2008a), reduced lipid storage (Yamashiro et al. 2000; Palmer and Baird 2018), and decreased reproduction (Yamashiro et al. 2000; Work et al. 2008a; Irikawa et al. 2011; Palmer and Baird 2018). Growth anomalies occur in all three regions in the PAG affecting eight coral genera, but *Porites* and *Platygyra* are the coral genera commonly reported with GAs. Growth anomalies in *Porites* are widespread across the Indo-Pacific (Aeby et al. 2011b), but there are limited reports of growth anomalies in *Platygyra*, with only one report from the Great Barrier Reef (Loya et al. 1984) and one from New Caledonia (Aeby et al. 2015a). *Platygyra* spp. in the PAG appear to be more prone to this disease compared to other regions.

6.4.5 Focal bleaching (FB)

Focal bleaching appears as discrete, focal bleached patches on colonies that have significantly fewer endosymbiotic dinoflagellates, but the coral tissue remains intact (Roff et al. 2008; Lawrence et al. 2015). The lesions are thought to be due to a viral infection of the endosymbiotic dinoflagellates (Lawrence et al. 2015; Kenkel et al. 2020) but little else is known about this disease. Focal bleaching has primarily been observed on massive poritids (Roff et al. 2008; Lawrence et al. 2015; Aeby et al. 2021) but has been reported in other coral genera (Aeby et al. 2015a; Work et al. 2015). In the PAG, focal bleaching is found in all three regions, most frequently on *Porites* spp., but is also reported from *Goniopora* spp. and *Dipsastrea* spp. (Aeby et al. 2020; Hazraty-Kari et al. 2021; Aeby et al., unpub. data). In Qatar, *Turbinaria* spp. had focal bleaching, which appeared as discrete pie-shaped wedges of bleached tissue (Aeby et al., unpub. data).

6.4.6 Dark spot disease (DSD)

Dark spot disease presents as focal or multi-focal spots of dark discolouration on coral colonies. DSD has been reported from coral reefs across the wider Caribbean affecting 16 coral species (Work and Weil 2016). Corals with similar disease signs are also reported from the Indo-Pacific (Work et al. 2008b, 2014, 2015) and the Red Sea (Aeby et al. 2021), where the disease is called endolithic hypermycosis, based on histopathology that showed the lesions to be associated with endolithic fungal infections (Work et al. 2008b, 2014, 2015; Aeby et al. 2021). In the Caribbean, DSD can progress to tissue loss, resulting in partial to total colony mortality (Work and Weil 2016), or disease signs may disappear without any evident damage and reappear on different areas of the colony (Borger 2005; Gochfeld et al. 2006). No ecological studies have been conducted on DSD from the Indo-Pacific. Dark spot has only been reported in *Turbinaria* from reefs in Qatar (Aeby et al., unpub. data) but histology is required to determine whether endolithic fungal infections are underlying these lesions.

The types and levels of coral disease within the PAG appear to be similar to what has been found in some regions of the Indo-Pacific. However, disease surveys are lacking in many areas of the PAG, and so the health status of PAG corals is still unclear. Baseline disease surveys are a critical first step in understanding coral disease, giving scientists the capacity to identify and respond to changes in disease levels through time. Research is critically needed to fill in the knowledge gaps on disease distribution, prevalence, and pathogenesis for the reefs of the PAG.

6.5 CORAL BLEACHING

Coral reefs were historically widespread in the PAG, with fringing reefs extending along the coastlines of all the bordering countries (Kinsman 1964; Shinn 1976; Sheppard and Wells 1988). Despite the resilience of these PAG coral populations, severe bleaching leading to massive coral mortality (>80% loss) was reported in the 1970s, 1996, 1998, and 2017. Milder bleaching events with limited coral mortality (0–20% loss) occurred in 2002, 2010, 2011, and 2012 (Sheppard and Loughland 2002; Riegl and Purkis 2015; Shuail et al. 2016; Riegl et al. 2018; Burt et al. 2019). The increasing frequency of these coral bleaching events caused an overall loss of 40% of the living coral cover between 1996 and 2019, at a rate of almost 20% per decade (Burt et al. 2021).

The primary cause for recent coral bleaching and mortality events in the PAG, as in other regions of the world, is the increasing frequency and duration of marine heatwaves, driven by global emissions of greenhouse gases (Burt et al. 2019; Paparella et al. 2019), and the rate of warming in the Gulf in recent decades has been twice the global average due to its unique geography and bathymetry (Lachkar et al. 2022). The precise mechanisms determining the periodicity and duration of these heatwaves remain, however, poorly understood. Some authors suggest a connection to the El Nino Southern Oscillation and the Indian Ocean Zonal Mode on a five-year cycle (Riegl 2003), but the increasing trend of bleaching events during the last decades has not followed this pattern. The severe bleaching recorded in 1996, 2010, and 2017 occurred during La Niña not El Niño years, suggesting that other factors are involved in inducing thermal anomalies (Purkis and Riegl 2005). Winds in the PAG are primarily from the NNW and are highly seasonal, reaching 40–50 km/hr and gusts of up to 100 km/hr during *shamal* (north in Arabic) events, which can occur both in winter and in summer and can significantly reduce seawater temperatures. Accordingly, Paparella et al. (2019) proposed that the occurrence of *shamal* wind during the summer mitigates the incidence and severity of coral bleaching. They observed that, in years with reported bleaching events, long periods of slow wind speeds (<4 m/s) caused seawater temperatures to persist above the coral bleaching threshold for more than 5 weeks, whereas occasional stronger winds during intervening years produced sufficient wind-driven evaporative cooling, which provided temporary relief from thermal stress and allowed corals to persist. These extreme temperature excursions have recently been associated with the occurrence of hours-long hypoxic and anoxic events on the PAG reefs at night, which likely represent a previously unrecognized additional mechanism driving bleaching (de Verneil et al. 2021).

The severity of coral bleaching and mortality is linked to the cumulative number of days of exposure to high temperatures, rather than the maximum temperature itself. Observations from multiple bleaching events across several reefs in the southern PAG have shown that, on average, corals will bleach if exposed to temperatures above 35°C for more than 3 weeks or above 34°C for more than 8–9 weeks (Riegl et al. 2011, 2012c). Significant bleaching-induced mortality only occurs when temperatures exceed 35.5°C for extended periods (Riegl et al. 2011, 2012c).

The bleaching thresholds in the PAG are the highest recorded anywhere in the world and, on average, PAG corals can tolerate approximately 5°C more heat than corals on the Great Barrier Reef and in the Caribbean (Riegl et al. 2011). This indicates that coral physiology can adapt to rising temperatures, providing hope that at least some coral species will have the ability to adapt to a warming world (Burt et al. 2014, 2020).

Despite their strong heat tolerance, relative to other regions, there is substantial variability among PAG coral species, in terms of their susceptibility to bleaching. There are currently 66 scleractinian coral species recorded for the PAG (DiBattista et al. 2016), although only 63 of the 66 species are listed in the IUCN Red List of Threatened Species, and information on susceptibility to bleaching is only provided for 33 species; two-thirds of those species are classified as highly susceptible (22 species, Figure 6.4). All seven species of *Acropora* are considered as highly susceptible to bleaching, but only two species are listed as vulnerable to extinction (*A. horrida* and *A. nasuta*). This conservation status clearly needs to be revised, as populations of all species of *Acropora* are collapsing in the PAG, including the historically dominant reef-builder *A. downingi*, which is now threatened with regional extinction (Riegl et al. 2018). In Qatar, annual bleaching events of variable severity have been observed since 2019, including in deeper offshore reefs, which were previously unaffected by bleaching (Fanning et al. 2021; Bouwmeester et al. 2022). These recurrent disturbances have completely eradicated *A. downingi* from several reefs, including Umm Al-Arshan (Figure 6.5), where some of the largest populations of reproductive size colonies previously occurred (Burt et al. 2016). The consistent loss of adult colonies is now causing a decrease in the overall reproductive output of the species,

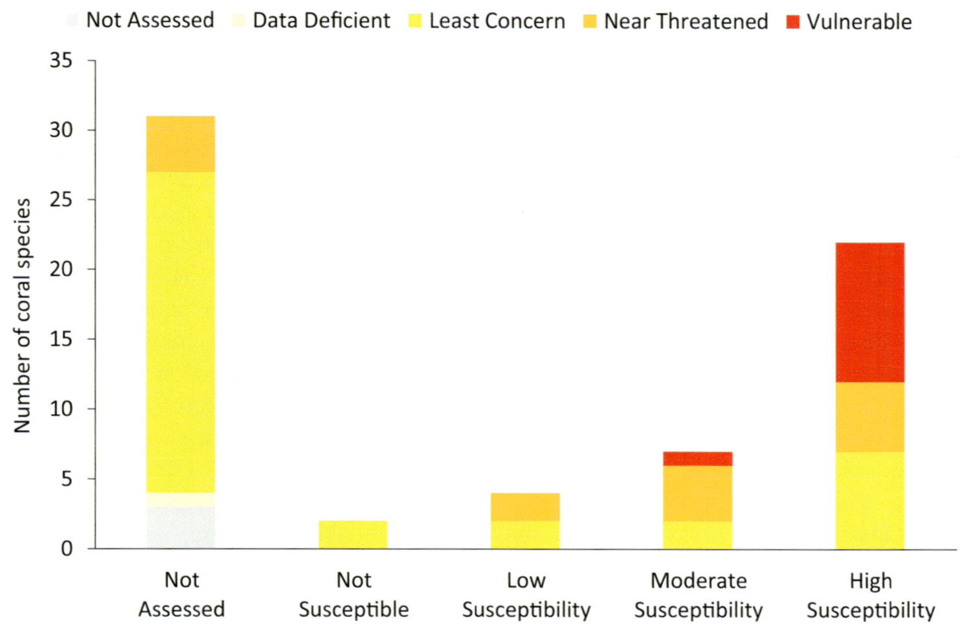

FIGURE 6.4 Classification of the 66 scleractinian coral species recorded for the PAG according to their conservation status in the IUCN Red List of Threatened Species (colour code in the legend) and susceptibility to bleaching (categories in the x-axis).

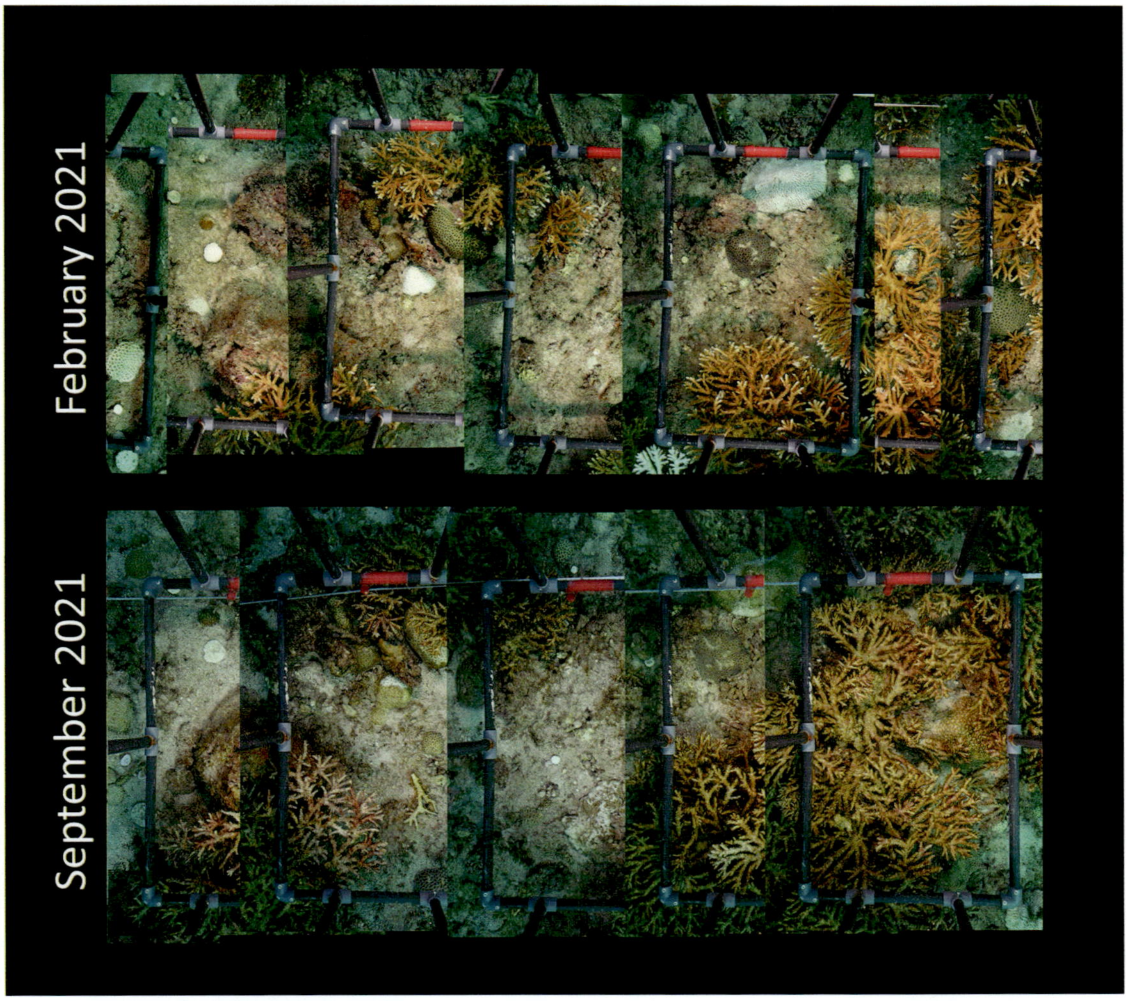

FIGURE 6.5 Photomosaic from Umm Al-Arshan reef in Qatar depicting the same colonies of *Acropora downingi* in February (healthy) and September 2021 (recently dead).

hindering its ability to recover on these reefs. In contrast, only two out of 14 species of the family Merulinidae are classified as susceptible to bleaching (*Cyphastrea serailia* and *Favites acuticollis*), and none is classified as threatened with extinction. These inter-specific differences in susceptibility to bleaching are clearly reflected in the changes to the community structure of coral reefs in the PAG, in response to increasingly frequent bleaching disturbances, whereby the previously dominant *Acropora* species are consistently being replaced by merulinids and poritids.

Bleaching severity and recovery are highly variable among reefs in the PAG, with some reefs experiencing relatively minor losses and rapid recovery, whereas others are completely devastated by a single bleaching event. The recovery of live coral cover after a bleaching disturbance can occur through asexual regeneration of surviving colonies and by recruitment of larvae (Baker et al. 2008). Within the PAG, settlement studies have shown that larval colonization rates are extremely low and highly biased towards merulinids and poritids on the degraded reefs of the southern PAG (Bento et al. 2017; Burt and Bauman 2019), with settlement of acroporids only found on a limited number of offshore sites where coral stands remain intact (Bento et al. 2021). Riegl et al. (2012c) found a west-to-east gradient in the impact of the 2010 bleaching in the PAG, with the quickest recovery in the eastern sites and the longest lasting event in the western sites. Burt et al. (2019) reported that bleaching and paling of coral colonies persisted for up to six months after the peak of the 2017 bleaching event (September), despite temperatures remaining above the bleaching thresholds for only a few weeks. We observed a similar pattern in several offshore reefs in Qatar during the 2017, 2019, 2020, and 2021 bleaching events (pers. obs.). The long extent of these bleaching periods suggests that PAG corals have strong physiological plasticity, whereby perhaps they can compensate the reduction of autotrophic energy inputs from their symbionts by heterotrophic suspension feeding as found in other coral species (Grotelli et al. 2006).

The Middle East region is warming almost two times faster than the global average, at a rate of 0.45°C and 0.59°C per decade, for air and seawater temperatures, respectively (Al Senafi 2022; Zittis et al. 2022). Climate projections indicate more severe and prolonged heatwaves during the summer, and these trends are expected to continue for the foreseeable future, unless strong measures are taken to reduce greenhouse gas emissions (Zittis et al. 2022). This raises the question of whether the extreme bleaching thresholds seen in PAG corals are due to selection caused from exposure to recurring heat events. Recent studies suggest that bleaching can indeed increase thermal tolerance of surviving coral populations, by promoting the selection and intergenerational inheritance of heat-resistance traits of corals and their symbionts (Howells et al. 2016, 2021; Kirk et al. 2018). The potential application of these heat-evolved populations to increase the resilience of coral reefs to future bleaching events, through assisted evolution, is a promising avenue for the development of new restoration technologies.

6.6 CORAL POPULATION DYNAMICS

Coral community dynamics in the southeastern PAG have changed in response to variations in disturbance patterns. Disturbance frequency and severity were inferred to have been similar throughout much of the Holocene until the early 21st century (Burt et al. 2011; Purkis and Riegl 2012). Disturbances due to unusual heat or cold events occurred every ~15–20 years, during which coral mortality reached up to 50%, mostly affecting the genus *Acropora* (Riegl 1999; Riegl and Purkis 2009, 2012). From 1996–2010 moderate disturbances recurred every five years, again preferentially killing *Acropora*, with ~10% overall coral mortality. However, since 2010 the frequency of high temperature anomalies has increased, causing annual or biennial severe bleaching and subsequent mortality with no intermittent regeneration possible due to the limited time between successive events. Coral mortality of ~20–50% with >90% *Acropora* mortality was observed in 2010–12 and 2015–17 (Riegl 1999; Burt et al. 2008, 2011; Riegl and Purkis 2009, 2012, 2015). This has led to a significant decrease in coral cover across the PAG within the space of only two decades. Drastic declines across some coral taxa have been documented in many regions (Riegl et al. 2017; Burt et al. 2019; Pratchett et al. 2020) and throughout the coastal waters of most neighbouring countries: Bahrain, Qatar, UAE, Saudi Arabia, and Kuwait (Sheppard et al. 2010; Burt 2013). Underlying the changes in coral cover are changes in population structure and diversity. Changes in coral populations deserve close and urgent attention, as a greater understanding of these demographic changes can provide the information necessary to generate projections and test hypotheses on the proximal causes of coral decline (Edmunds and Riegl 2020).

One of the most conspicuous changes in the coral composition of the reefs of the PAG has been the loss of corals of the genus *Acropora* that were once abundant on coastal reefs and offshore banks. For example, *Acropora* were found around the mainland and on many shoals in Qatar (Shinn 1976; Emara et al. 1985) and on the ocean-facing side of the PAG of Salwah (Sheppard and Salm 1988), as well as the northern tip of the Qatar Peninsula (around Ras Rakan) (Kinsman 1964; Shinn 1976). Today, these areas harbour only occasional, increasingly sparse, fringes of merulinids and poritids (Burt et al. 2016; Purkis et al. 2017). On offshore banks around the United Arab Emirates (Fasht Adhm, Abu Kafan, Halat Dalma), *Acropora*-dominated coral cover declined from ~60% in 1985 to zero by 2007 with very few corals, if any, remaining at these sites (Sheppard, Riegl and Purkis, unpubl. data).

However, it is not just *Acropora* that has been affected. In Abu Dhabi, many offshore islands and shoals as well as patch reefs leading to lagoons had dense coral fringes (Evans et al. 1973; Purser and Seibold 1973), but these have mostly disappeared (Purkis and Riegl 2012; Bauman et al. 2013). Even *Porites harrisoni*, a very resilient coral that was formerly dense in Abu Dhabi's western regions, is now severely depleted. Densely coral-covered nearshore reefs once existed in the western UAE (Sila to Al Dabiyah), but few corals remain on the mostly dead frameworks (Burt et al. 2019). Coral patches off Abu Dhabi (Shinn 1973) and Dubai (Riegl 1999; Purkis et al. 2006) were lost to heat and landfill from coastal construction.

The densest remaining coral populations in the southeastern PAG, which still harbour some of the now rare *Acropora* populations, exist at the offshore islands Halul (Qatar) and Sir Abu Nuair (Sharjah, UAE) (Bento et al. 2021). Relatively dense coral populations remain in coastal Abu Dhabi at Saadiyat, Ras Ghanada (Burt et al. 2011), and Umm al Quwain (Bauman et al. 2013; Bento et al. 2016). Apparently healthy coral populations exist offshore from Iran, particularly around Hengam Island (Kabiri et al. 2013; Vajed Samiei et al. 2016). A few offshore sites around Qatar also harbour relatively high coral cover, with Fasht East Halul and Binzayan exceeding 30% coral cover and more than 20 coral species (Bouwmeester et al. 2022). However, Qatar's nearshore sites, such as Fasht al Dibal, Fasht al Hurabi, and Fasht al Udayd, have suffered similar fates to the sites around Abu Dhabi, with low species richness and a continued decline in coral cover and diversity due to prolonged bleaching events (Bouwmeester et al. 2022). Umm al Arshan off the northern coast of Qatar had several large colonies of *A. downingi* but succumbed to bleaching in 2021 (pers. obs.). Human-made disturbances (dredging, filling, thermal and chemical pollution) have damaged many areas (Sale et al. 2011; Sheppard et al. 2010), but temperature-induced bleaching and diseases seem to be the major causes of decline (Riegl and Purkis 2009, 2015; Burt et al. 2011; Purkis and Riegl 2012; Bauman et al. 2013).

These phases of disturbance-spacing are reflected in coral communities due to species-specific differences in mortality. With the decrease in coral cover across the PAG, a progression in dominance from *Acropora* to *Porites* to merulinids has also been observed (Riegl and Purkis 2015). Using different disturbance severity and frequency scenarios, Riegl et al. (2017) used size-class specific survival, growth, and estimated fertility and connectivity to project future populations of the five most common reef-coral species: *Acropora downingi*, *Porites harrisoni*, *Dipsastraea pallida*, *Platygyra lamellina*, and *Cyphastrea microphthalma*. *A. downingi* and *P. harrisoni* were predicted to be most affected, where growth into the largest size class ceased under the severe disturbance scenarios. Furthermore, retention within size-classes, which indicated an absence of growth, became much more common and shrinkage strongly increased. In the other species, no size-classes were lost, but growth transitions also decreased, and shrinkages increased with the effect of stronger contributions to the smaller classes. This has strong implications for the survival of local populations as fecundity depends on coral size, so a shift to smaller, less fecund corals can lead to an overall loss of fertility in the population. Lower fecundity means that populations become increasingly dependent on connectivity to other populations, which is also compromised as population sizes decrease and become patchier. Species that are less affected by whole colony mortality showed a lower connectivity requirement than those where whole colony mortality was common. However, there are limitations for this connectivity as self-seeding appears to be a strong component of these populations (Cavalcante et al. 2020).

In conclusion, the increased frequency and severity of disturbances affecting coral reefs in the PAG has led to the extirpation of local populations of corals at nearshore sites. The population structure in the remaining sites has shifted towards smaller, less fecund, and more vulnerable colonies of corals, which are less connected. Unless these disturbances can be mitigated, the outlook for these populations remains bleak, with some pockets of small corals perhaps surviving long enough to hopefully rebound when, and if, conditions improve. A regionally integrated approach to future research within the PAG is essential. There is a growing need for large-scale research to guide management of reefs and their stressors, as these operate at much larger scales than the national borders within which most research currently occurs.

REFERENCES

Aeby, G.S. 2005. Outbreak of coral disease in the Northwestern Hawaiian Islands. *Coral Reefs* 24, no. 3: 481.

Aeby, G.S., Bourne, D.G., Wilson, B., and T.M. Work. 2011a. Coral diversity and the severity of disease outbreaks: A cross-regional comparison of *Acropora* white syndrome in a species-rich region (American Samoa) with a species-poor region (Northwestern Hawaiian Islands). *J Mar Biol* 2011: 490198.

Aeby, G.S., Callahan, S., Cox, E.F., et al. 2016. Emerging coral diseases in Kāne'ohe Bay, O'ahu, Hawai'i (USA): Two major disease outbreaks of acute *Montipora* white syndrome. *Dis Aquat Org* 119, no. 3: 189–198. doi:10.3354/DAO02996.

Aeby, G.S., Howells, E., Work, T., et al. 2020. Localized outbreaks of coral disease on Arabian reefs are linked to extreme temperatures and environmental stressors. *Coral Reefs* 39, no. 3: 829–846. doi:10.1007/s00338-020-01928-4.

Aeby, G., Tribollet, A., Lasne, G., et al. 2015a. Assessing threats from coral and crustose coralline algae disease on the reefs of New Caledonia. *Mar Freshw Res* 67, no. 4: 455–465. doi:10.1071/MF14151.

Aeby, G.S., Shore, A., Jensen, T., Ziegler, M., Work, T., and C.R. Voolstra. 2021. A comparative baseline of coral disease in three regions along the Saudi Arabian coast of the central Red Sea. *PLoS One* 16, no. 7: e0246854. doi:10.1371/JOURNAL.PONE.0246854.

Aeby, G.S., Williams, G.J., Franklin, E.C., et al. 2011b. Growth anomalies on the coral genera *Acropora* and *Porites* are strongly associated with host density and human population size across the Indo-Pacific. *PLoS One* 6, no. 2: e16887. doi:10.1371/JOURNAL.PONE.0016887.

Aeby, G.S., Work, T.M., Runyon, C.M., Shore-Maggio, A., Ushijima, B., Videau, P., Beurmann, S., and S.M. Callahan. 2015b. First record of black band disease in the Hawaiian archipelago: Response, outbreak status, virulence, and a method of treatment. *PLoS One* 10, no. 3: e0120853. doi:10.1371/journal.pone.0120853.

Ainsworth, T.D., Kvennefors, E.C., Blackall, L.L., Fine, M., and O. Hoegh-Guldberg. 2007. Disease and cell death in white syndrome of Acroporid corals on the Great Barrier Reef. *Mar Biol* 151: 19–29. doi:10.1007/s00227-006-0449-3.

Alidoost Salimi, P., Mostafavi, P., Chen, C.A., Fatemi, S., and M. Pichon. 2018. The Scleractinia (Cnidaria: Anthozoa) of Abu-Musa and Sirri Islands, Persian Gulf. *Zool Stud* 57: e56. doi:10.6620/ZS.2018.57-56. PMID: 31966296; PMCID: PMC6409733.

Alidoost Salimi, P., Mostafavi, P., Chen, C.A., Pichon, M., and M. Alidoost Salimi. 2021. Molecular phylogeny of some coral species from the Persian Gulf. *Mol Biol Rep* 48: 2993–2999. doi:10.1007/s11033-021-06251-x.

Alidoost Salimi, M., Mostafavi, P., Fatemi, S.M.R., and G.S. Aeby. 2017. Health status of corals surrounding Kish Island, Persian Gulf. *Dis Aquat Org* 124, no. 1: 77–84. doi:10.3354/DAO03105.

Al Senafi, F. 2022. Atmosphere-ocean coupled variability in the Arabian/Persian Gulf. *Front Mar Sci* 9: 809355.

Alvarez-Filip, L., Estrada-Saldívar, N., Pérez-Cervantes, E., Molina-Hernández, A., and F.J. González-Barrios. 2019. A rapid spread of the stony coral tissue loss disease outbreak in the Mexican Caribbean. *Peer J* 11: e8069. doi:10.7717/PEERJ.8069/SUPP-3.

Anthony, K., Bay, L.K., Costanza, R., et al. 2017. New interventions are needed to save coral reefs. *Nat Ecol Evol* 1, no. 10: 420–1422. doi:10.1038/s41559-017-0313-5.

Bak, R.P.M. 1983. Neoplasia, regeneration and growth in the reef-building coral *Acropora palmata*. *Mar Biol* 77, no. 3: 221–227. doi:10.1007/BF00395810.

Baker, A.C., Glynn, P.W., and B. Riegl. 2008. Climate change and coral reef bleaching: An ecological assessment of long-term impacts, recovery trends and future outlook. *Estuar Coast Shelf Sci* 80: 435–471. doi:10.1016/j.ecss.2008.09.003.

Barneah, O., Ben-Dov, E., Kramarsky-Winter, E., and A. Kushmaro. 2007. Characterization of black band disease in Red Sea stony corals. *Environ Microbiol* 9, no. 8: 1995–2006. doi:10.1111/J.1462-2920.2007.01315.X.

Bauman, A.G., Feary, D.A., Heron, S.F., Pratchett, M.S., and J. Burt. 2013. Multiple environmental factors influence the spatial distribution and structure of reef communities in the northeastern Arabian Peninsula. *Mar Pollut Bull* 72, no. 2: 302–312. doi:10.1016/J.MARPOLBUL.2012.10.013.

Bejarano, I., Orenes-Salazar, V., Bento, R., García-Charton, J.A., and D. Mateos-Molina. 2022. Coral reefs at Sir Bu Nair Island: An offshore refuge of *Acropora* in the southern Arabian Gulf. *Mar Pollut Bull* 178: 113570. doi:10.1016/J.MARPOLBUL.2022.113570.

Ben-Haim, Y., and E. Rosenberg. 2002. A novel *Vibrio* sp. pathogen of the coral *Pocillopora damicornis*. *Mar Biol* 141, no. 1: 47–55. doi:10.1007/s00227-002-0797-6.

Bento, R., Cavalcante, G., Mateos-Molina, D., Riegl, B., and I. Bejarano. 2021. Recruitment and larval connectivity of a remnant *Acropora* community in the Arabian Gulf, United Arab Emirates. *Coral Reefs* 40, no. 6: 1889–1898. doi:10.1007/S00338-021-02187-7.

Bento, R., Feary, D.A., Hoey, A.S., and J.A. Burt. 2017. Settlement patterns of corals and other benthos on reefs with divergent environments and disturbances histories around the northeastern Arabian Peninsula. *Front Mar Sci* 4: 1–12. doi:10.3389/fmars.2017.00305.

Bento, R., Hoey, A.S., Bauman, A.G., Feary, D.A., and J.A. Burt. 2016. The implications of recurrent disturbances within the world's hottest coral reef. *Mar Pollut Bull* 105, no. 2: 466–472. doi:10.1016/J.MARPOLBUL.2015.10.006.

Bharath, M., Chandran, R., Aeby, G., et al. 2023. First report of yellow-banded tissue loss disease on coral reefs outside the Arabian/Persian Gulf. *Dis Aquat Org* 153: 1–8. doi:10.3354/dao03713.

Borger, J.L. 2005. Dark spot syndrome: A scleractinian coral disease or a general stress response? *Coral Reefs* 24, no. 1: 139–144. doi:10.1007/s00338-004-0434-6.

Bourne, D.G., Ainsworth, T.D., Pollock, F.J., and B.L. Willis. 2015. Towards a better understanding of white syndromes and their causes on Indo-Pacific coral reefs. *Coral Reefs* 34, no. 1: 233–242. doi:10.1007/s00338-014-1239-x.

Bouwmeester, J., Ben-Hamadou, R., Range, P., Al Jamali, F., and J.A. Burt. 2022. Spatial patterns of reef fishes and corals in the thermally extreme waters of Qatar. *Front Mar Sci* 9: 989841. doi:10.3389/FMARS.2022.989841.

Brandt, M.E., Ennis, R.S., Meiling, S.S., et al. 2021. The emergence and initial impact of Stony Coral Tissue Loss Disease (SCTLD) in the United States Virgin Islands. *Front Mar Sci* 8: 715329. doi:10.3389/FMARS.2021.715329/FULL.

Bruno, J.F., Selig, E.R., Casey, K.S., et al. 2007. Thermal stress and coral cover as drivers of coral disease outbreaks. *PLoS Biol* 5, no. 6: e124. doi:10.1371/journal.pbio.0050124.

Burt, J.A. 2013. The growth of coral reef science in the Gulf: A historical perspective. *Mar Pollut Bull* 72, no. 2: 289–301. doi:10.1016/J.MARPOLBUL.2013.05.016.

Burt, J.A. 2014. The environmental costs of coastal urbanization in the Arabian Gulf. *City* 18, no. 6: 760–770.

Burt, J.A., Al-Harthi, S., and A. Al-Cibahy. 2011. Long-term impacts of coral bleaching events on the world's warmest reefs. *Mar Environ Res* 72, no. 4: 225–229. doi:10.1016/J.MARENVRES.2011.08.005.

Burt, J.A., and A. Bartholomew. 2019. Towards more sustainable coastal development in the Arabian Gulf: Opportunities for ecological engineering in an urbanized seascape. *Mar Pollut Bull* 142: 93–102. doi:10.1016/j.marpolbul.2019.03.024.

Burt, J.A., Bartholomew, A., and P. Usseglio. 2008. Recovery of corals a decade after a bleaching event in Dubai, United Arab Emirates. *Mar Biol* 154, no. 1: 27–36. doi:10.1007/S00227-007-0892-9.

Burt, J.A., and A.G. Bauman. 2019. Suppressed coral settlement following mass bleaching in the southern Persian/Arabian Gulf. *Aquat Ecosyst Health Manag* 23, no. 2: 1–9. doi:10.1080/14634988.2019.1676024.

Burt, J.A., Camp, E.F., Enochs, I.C., et al. 2020. Insights from extreme coral reefs in a changing world. *Coral Reefs* 39, no. 3: 495–507. doi:10.1007/s00338-020-01966-y.

Burt, J.A., Feary, D., Bauman, A., Usseglio, P., Cavalcante, G., and P. Sale. 2011. Biogeographic patterns of reef fish community structure in the northeastern Arabia Peninsula. *ICES J Mar Sci* 68: 1875–1883. doi:10.1016/j.marpolbul.2013.05.016.

Burt, J.A., Smith, E.G., Warren, C., and J. Dupont. 2016. An assessment of Qatar's coral communities in a regional context. *Mar Pollut Bull* 105, no. 2: 473–479. doi:10.1016/J.MARPOLBUL.2015.09.025.

Burt, J.A., Paparella, F., Al-Mansoori, N., Al-Mansoori, A., and H. Al-Jailani. 2019. Causes and consequences of the 2017 coral bleaching event in the southern Persian/Arabian Gulf. *Coral Reefs* 38, no. 4: 567–589.

Burt, J.A., Range, P., Claerboudt, M., et al. 2021. *Chapter 4: Status and trends of coral reefs of the ROPME Sea Area*. Status of Coral Reefs of the World: 2020. International Coral Reef Initiative (ICRI), Global Coral Reef Monitoring Network (GCRMN), Australian Institute of Marine Science (AIMS), United Nations Environment Programme (UNEP), 12 p.

Burt, J.A., van Lavieren, H., and D. Feary. 2014. Persian Gulf reefs: An important asset for climate science in urgent need of protection. *Ocean Challenge* 20: 49–56.

Cavalcante, G.H., Feary, D.A., and J.A. Burt. 2016. The influence of extreme winds on coastal oceanography and its implications for coral population connectivity in the southern Arabian Gulf. *Mar Pollut Bull* 105, no. 2: 489–497. doi:10.1016/J.MARPOLBUL.2015.10.031.

Cavalcante, G., Vieira, F., Mortensen, J., et al. 2020. Biophysical model of coral population connectivity in the Arabian/Persian Gulf. *Adv Mar Biol* 87, no. 1: 193–221. doi:10.1016/BS.AMB.2020.07.001.

Coles, S.L. 2003. Coral species diversity and environmental factors in the Arabian Gulf and the Gulf of Oman: A comparison to the Indo-Pacific region. *Atoll Res Bull* 507: 1–19.

Coles, S.L., and B.M. Riegl. 2013. Thermal tolerances of reef corals in the Gulf: A review of the potential for increasing coral survival and adaptation to climate change through assisted translocation. *Mar Pollut Bull* 72, no. 2: 323–332. doi:10.1016/J.MARPOLBUL.2012.09.006.

Cooney, R.P., Pantos, O., le Tissier, M.D.A., Barer, M.R., O'Donnell, A.G., and J.C. Bythell. 2002. Characterization of the bacterial consortium associated with black band disease in coral using molecular microbiological techniques. *Environ Microbiol* 4, no. 7: 401–413. doi:10.1046/J.1462-2920.2002.00308.X.

de Verneil, A., Burt, J.A., Mitchell, M., and F. Paparella. 2021. Summer oxygen dynamics on a southern Arabian Gulf coral reef. *Front Mar Sci* 8: 1676. doi:10.3389/FMARS.2021.781428.

DiBattista, J.D., Roberts, M.B., Bouwmeester, J., et al. 2016. A review of contemporary patterns of endemism for shallow water reef fauna in the Red Sea. *J Biogeogr* 43: 423–439. doi:10.1111/jbi.12649.

Edmonds, N.J., Al-Zaidan, A.S., Al-Sabah, A.A., et al. 2021. Kuwait's marine biodiversity: Qualitative assessment of indicator habitats and species. *Mar Pollut Bull* 163: 111915.

Edmunds, P.J., and B. Riegl. 2020. Urgent need for coral demography in a world where corals are disappearing. *Mar Ecol Prog Ser* 635: 233–242. doi:10.3354/MEPS13205.

Emara, H.I., El-Samra, M.I., El-Debb, K.Z., and I.F. Ahmed. 1985. A Preliminary study of the chemical characteristics of the coral reef areas in Qatari Waters (Gulf Area). *Proc 5th Internat Coral Reef Symp* 6: 13–16.

Estrada-Saldívar, N., Molina-Hernández, A., Pérez-Cervantes, E., et al. 2020. Reef-scale impacts of the stony coral tissue loss disease outbreak. *Coral Reefs* 39: 861–866. doi:10.1007/s00338-020-01949-z.

Evans, G., Murray, J.W., Biggs, H.E.J., Bate, R., and P.R. Bush. 1973. The oceanography, ecology, sedimentology and geomorphology of parts of the Trucial Coast barrier island complex, Persian Gulf. In: *The Persian Gulf.* Ed: B.H. Purser, 233–277. Springer. doi:10.1007/978-3-642-65545-6_14.

Fanning, L.M., Al-Naimi, M.N., Range, P., et al. 2021. Applying the ecosystem services – EBM framework to sustainably manage Qatar's coral reefs and seagrass beds. *Ocean Coast Manag* 205: 105566. doi:10.1016/j.ocecoaman.2021.105566.

Farivar, S., Jalil-Piran, Z., Zarei, F., and H. Hosseinzadeh Sahafi. 2017. Intraspecific phylogeography of the Japanese threadfin bream, *Nemipterus japonicus* (Perciformes: Nemipteridae), from the Persian Gulf and Indo-West Pacific: A preliminary study based on mitochondrial DNA sequence. *Iran J Fish Sci* 16, no. 2: 587–604. http://jifro.ir/article-1-2723-en.html.

Fatemi, S., and M. Shokri. 2001. Iranian coral reefs status with particular reference to Kish Island. *Proc Indian Ocean Regional Workshop* 1–3.

Feary, D.A., Burt, J.A., Bauman, A.G., Usseglio, P., Sale, P.F., and G.H. Cavalcante. 2010. Fish communities on the world's warmest reefs: What can they tell us about the effects of climate change in the future? *J Fish Biol* 77, no. 8: 1931–1947.

Feary, D.A., Burt, J.A., Cavalcante, G.H., and A.G. Bauman. 2012. Extreme physical factors and the structure of Gulf fish and reef communities. In: *Coral reefs of the Gulf: Adaptation to climatic extremes.* Eds: B.M. Riegl, and S.J. Purkis, 163–170. Netherlands: Springer. doi:10.1007/978-94-007-3008-3_9.

Forrester, G.E., Arton, L., Horton, A., Nickles, K., and L.M. Forrester. 2022. Antibiotic treatment ameliorates the impact of Stony Coral Tissue Loss Disease (SCTLD) on coral communities. *Front Mar Sci* 9: 859740. doi:10.3389/FMARS.2022.859740/FULL.

Foster, K.A., Foster, G., and S. Al-Harthi. 2013. Coral Assemblages in the southeastern Arabian Gulf (Qatar and Abu Dhabi, UAE): Various stages of *Acropora* recovery a decade after recurrent elevated temperature anomalies. *Open J Marine Sci* 3: 28–39.

Foster, N.L., Paris, C.B., Kool, J.T., et al. 2012. Connectivity of Caribbean coral populations: Complementary insights from empirical and modelled gene flow. *Mol Ecol* 21, no. 5: 1143–1157.

Frias-Lopez, J., Zerkle, A.L., Bonheyo, G.T., and B.W. Fouke. 2002. Partitioning of bacterial communities between seawater and healthy, black band diseased, and dead coral surfaces. *Appl Environ Microbiol* 68, no. 5: 2214–2228. doi:10.1128/AEM.68.5.2214-2228.2002.

Frys, C., Saint-Amand, A., Le Hénaff, M., et al. 2020. Fine-scale coral connectivity pathways in the Florida Reef Tract: Implications for conservation and restoration. *Front Mar Sci* 7: 312.

Gignoux-Wolfsohn, S.A., Marks, C.J., and S.V. Vollmer. 2012. White band disease transmission in the threatened coral, *Acropora cervicornis*. *Sci Rep* 2, no. 1: 1–3.

Gladfelter, W.B. 1982. White-band disease in *Acropora palmata*: Implications for the structure and growth of shallow reefs. *Bull Mar Sci* 32, no. 2: 639–643.

Gochfeld, D.J., Olson, J.B., and M. Slattery. 2006. Colony versus population variation in susceptibility and resistance to dark spot syndrome in the Caribbean coral *Siderastrea siderea*. *Dis Aquat Org* 69, no. 1: 53–65. doi:10.3354/DAO069053.

Grotelli, A., Rodrigues, L., and J. Palardy. 2006. Heterotrophic plasticity and resilience in bleached corals. *Nature* 440: 1186–1189.

Haapkylä, J., Seymour, A.S., Trebilco, J., and D. Smith. 2007. Coral disease prevalence and coral health in the Wakatobi Marine Park, south-east Sulawesi, Indonesia. *J Mar Biol Assoc UK* 87, no. 2: 403–414.

Haapkylä, J., Unsworth, R.K.F., Flavell, M., Bourne, D.G., Schaffelke, B., and B.L. Willis. 2011. Seasonal rainfall and runoff promote coral disease on an inshore reef. *PLoS One* 6, no. 2: e16893. doi:10.1371/JOURNAL.PONE.0016893.

Hadaidi, G., Ziegler, M., Shore-Maggio, A., Jensen, T., Aeby, G., and C.R. Voolstra. 2018. Ecological and molecular characterization of a coral black band disease outbreak in the Red Sea during a bleaching event. *Peer J* 6, no. 7: e5169. doi:10.7717/peerj.5169.

Hamza Alhazeem, S., Burt, J.A., Hasan Alsaffar, A., Chen, W., Abdulla Al-Kandari, M., and M. Abdulla Al. 2017. Long-term coral community stability in a disturbed marginal reef in Kuwait. *J Water Resour Ocean Sci* 6, no. 6: 85–89. doi:10.11648/J.WROS.20170606.12.

Harvell, D., Jordán-Dahlgren, E., Merkel, S., et al. 2007. Coral disease, environmental drivers, and the balance between coral and microbial associates. *Oceanography* 20: 172–195.

Hazraty-Kari, S., Tavakoli-Kolour, P., Das, R.R., et al. 2021. Baseline assessment of coral diseases in an environmentally extreme environment of the northern Persian Gulf. *Mar Pollut Bull* 171: 112707. doi:10.1016/J.MARPOLBUL.2021.112707.

Heres, M.M., Farmer, B.H., Elmer, F., and H. Hertler. 2021. Ecological consequences of stony coral tissue loss disease in the Turks and Caicos Islands. *Coral Reefs* 40, no. 2: 609–624. doi:10.1007/S00338-021-02071-4.

Heron, S.F., Willis, B.L., Skirving, W.J., Eakin, C.M., and C.A. Page. 2010. Summer hot snaps and winter conditions: Modelling white syndrome outbreaks on Great Barrier Reef corals. *PLoS One* 5, no. 8: 12210. doi:10.1371/journal.pone.0012210.

Hicks, C.C., Cohen, P.J., Graham, N.A.J., et al. 2019. Harnessing global fisheries to tackle micronutrient deficiencies. *Nature* 574, no. 7776: 95–98. doi:10.1038/s41586-019-1592-6.

Hoegh-Guldberg, O., Jacob, D., Taylor, M., et al. 2018. Impacts of 1.5° C global warming on natural and human systems. In: *Global warming of 1.5°C: An IPCC special report on the impacts of global warming of 1.5°C above pre-industrial levels and related global greenhouse gas emission pathways, in the context of strengthening the global response to the threat of climate change, sustainable development, and efforts to eradicate poverty.* Eds: V. Masson-Delmotte, P. Zhai, H.-O. Pörtner, et al., 175–312. Cambridge, UK and New York, NY: Cambridge University Press. doi:10.1017/9781009157940.005.

Hoey, A.S., Feary, D.A., Burt, J.A., Vaughan, G., Pratchett, M.S., and M.L. Berumen. 2016. Regional variation in the structure and function of parrotfishes on Arabian reefs. *Mar Pollut Bull* 105, no. 2: 524–531.

Howells, E.J., Abrego, D., Liew, Y.J., Burt, J.A., Meyer, E., and M. Aranda. 2021. Enhancing the heat tolerance of reef-building corals to future warming. *Sci Adv* 7: eabg6070. https://coralreefwatch.noaa.gov/product/5km/.

Howells, E.J., Abrego, D., Meyer, E., Kirk, N.L., and J.A. Burt. 2016. Host adaptation and unexpected symbiont partners enable reef-building corals to tolerate extreme temperatures. *Glob Change Biol* 22, no. 8: 2702–2714. doi:10.1111/GCB.13250.

Howells, E.J., Bauman, A.G., Vaughan, G.O., Hume, B.C.C., Voolstra, C.R., and J.A. Burt. 2020a. Corals in the hottest reefs in the world exhibit symbiont fidelity not flexibility. *Mol Ecol* 29, no. 5: 899–911. doi:10.1111/MEC.15372.

Howells, E.J., Hagedorn, M., van Oppen, M.J.H., and J.A. Burt. 2022. Challenges of sperm cryopreservation in transferring heat adaptation of corals across ocean basins. *Peer J* 10: e13395. doi:10.7717/PEERJ.13395/SUPP-2.

Howells, E.J., Vaughan, G.O., Work, T.M., Burt, J.A., and D. Abrego. 2020b. Annual outbreaks of coral disease coincide with extreme seasonal warming. *Coral Reefs* 39, no. 3: 771–781. doi:10.1007/S00338-020-01946-2/FULLTEXT.HTML.

Hughes, T.P., Barnes, M.L., Bellwood, D.R., et al. 2017. Coral reefs in the anthropocene. *Nature* 546, no. 7656: 82–90. doi:10.1038/nature22901.

Hughes, T.P., Graham, N.A.J., Jackson, J.B.C., Mumby, P.J., and R.S. Steneck. 2010. Rising to the challenge of sustaining coral reef resilience. *Trends Ecol Evol* 25, no. 11: 633–642. doi:10.1016/J.TREE.2010.07.011.

Hughes, T.P., and J.E. Tanner. 2000. Recruitment failure, life histories, and long-term decline of Caribbean corals. *Ecology* 81, no. 8: 2250–2263.

Hume, B.C.C., Angelo, C.D., Smith, E.G., Stevens, J.R., Burt, J.A., and J. Wiedenmann. 2018. Validation of the binary designation symbiodinium thermophilum (dinophyceae). *J Phycol* 54, no. 5: 762–764. doi:10.1111/jpy.12764.

Hume, B.C.C., D'Angelo, C., Burt, J., Baker, A.C., Riegl, B., and J. Wiedenmann. 2013. Corals from the Persian/Arabian Gulf as models for thermotolerant reef-builders: Prevalence of clade C3 Symbiodinium, host fluorescence and ex situ temperature tolerance. *Mar Pollut Bull* 72, no. 2: 313–322. doi:10.1016/J.MARPOLBUL.2012.11.032.

Hume, B.C.C., D'Angelo, C., Smith, E.G., Stevens, J.R., Burt, J., and J. Wiedenmann. 2015. *Symbiodinium thermophilum sp. nov.*, a thermotolerant symbiotic alga prevalent in corals of the world's hottest sea, the Persian/Arabian Gulf. *Nature Sci Rep* 5, no. 1: 8562. doi:10.1038/srep08562.

Hume, B.C.C., Voolstra, C.R., Arif, C., et al. 2016. Ancestral genetic diversity associated with the rapid spread of stress-tolerant coral symbionts in response to Holocene climate change. *Proc Nat Acad Sci* 113: 4416–4421. doi:10.1073/pnas.1601910113.

Irikawa, A., Casareto, B.E., Suzuki, Y., Agostini, S., Hidaka, M., and R. van Woesik. 2011. Growth anomalies on *Acropora cytherea* corals. *Mar Pollut Bull* 62, no. 8: 1702–1707. doi:10.1016/J.MARPOLBUL.2011.05.033.

Kabiri, K., Pradhan, B., Samimi-Namin, K., and M. Moradi. 2013. Detecting coral bleaching, using QuickBird multi-temporal data: A feasibility study at Kish Island, the Persian Gulf. *Estuar Coast Shelf Sci* 117: 273–281.

Kavousi, J., and H. Rezai. 2011. Coral lesions around some Iranian Islands in the Persian Gulf. *Galaxea J Coral Reef Studies* 13, no. 1: 5–6. doi:10.3755/GALAXEA.13.5.

Kenkel, C.D., Mocellin, V.J.L., and L.K. Bay. 2020. Global gene expression patterns in *Porites* white patch syndrome: Disentangling symbiont loss from the thermal stress response in reef-building coral. *Mol Ecol* 29: 3907–3920.

Ketchum, R.N., Smith, E.G., DeBiasse, M.B., et al. 2020. Population genomic analyses of the sea urchin *Echinometra* sp. EZ across an extreme environmental gradient. *Genome Biol Evolution* 12, no. 10: 1819–1829. doi:10.1093/GBE/EVAA150.

Kinsman, D.J.J. 1964. Reef coral tolerance of high temperatures and salinities. *Nature* 202: 1280–1282.

Kirk, N.L., Howells, E.J., Abrego, D., Burt, J.A., and E. Meyer. 2018. Genomic and transcriptomic signals of thermal tolerance in heat-tolerant corals (*Platygyra daedalea*) of the Arabian/Persian Gulf. *Mol Ecol* 27, no. 24: 5180–5194. doi:10.1111/MEC.14934.

Korrubel, J.L., and B. Riegl. 1998. A new coral disease from the southern Arabian Gulf. *Coral Reefs* 17: 22.

Lachkar, Z., Mehari, M., Lévy, M., Paparella, F., and J.A. Burt. 2022. Recent expansion and intensification of hypoxia in the Arabian Gulf and its drivers. *Front Mar Sci* 9: 1616. doi:10.3389/FMARS.2022.891378.

LaJeunesse, T.C., Parkinson, J.E., Gabrielson, P.W., et al. 2018. Systematic revision of symbiodiniaceae highlights the antiquity and diversity of coral endosymbionts. *Curr Biol* 28, no. 16: 2570–2580.e2576. doi:10.1016/J.CUB.2018.07.008.

Lamb, J.B., Willis, B.L., Fiorenza, E.A., et al. 2018. Plastic waste associated with disease on coral reefs. *Science* 359, no. 6374: 460–462.

Lawrence, S.A., Davy, J.E., Wilson, W.H., Hoegh-Guldberg, O., and S.K. Davy. 2015. *Porites* white patch syndrome: Associated viruses and disease physiology. *Coral Reefs* 34, no. 1: 249–257. doi:10.1007/S00338-014-1218-2.

Le Quesne, W.J.F., Fernand, L., Ali, T.S., Andres, O., et al. 2021. Is the development of desalination compatible with sustainable development of the Arabian Gulf? *Mar Pollut Bull* 173: 112940.

Liew, Y.J., Howells, E.J., Wang, X., Michell, C.T., Burt, J.A., Idaghdour, Y., and M. Aranda. 2020. Intergenerational epigenetic inheritance in reef-building corals. *Nat Clim Change* 10, no. 3: 254–259. doi:10.1038/S41558-019-0687-2.

Loya, Y., Bull, G., and M. Pichon. 1984. Tumor formations in scleractinian corals. *Helgolander Meeresunters* 37: 99–112. doi:10.1007/BF01989297s.

Maynard, J.A., Anthony, K.R.N., Harvell, C.D., et al. 2011. Predicting outbreaks of a climate-driven coral disease in the Great Barrier Reef. *Coral Reefs* 30, no. 2: 485–495. doi:10.1007/s00338-010-0708-0.

McCook, L.J., Almany, G.R., Berumen, M.L., et al. 2009. Management under uncertainty: Guide-lines for incorporating connectivity into the protection of coral reefs. *Coral Reefs* 28: 353–366.

Mehraban, H., Esmaeili, H.R., Zarei, F., Ebrahimi, M., and A. Gholamhosseini. 2020. Genetic diversification, population structure, and geophylogeny of the Scarface rockskipper *Istiblennius pox* (Teleostei: Blenniidae) in the Persian Gulf and Oman Sea. *Mar Biodiv* 50, no. 2: 1–12. doi:10.1007/S12526-020-01045-Z.

Miller, A.W., and L.L. Richardson. 2011. A meta-analysis of 16S rRNA gene clone libraries from the polymicrobial black band disease of corals. *FEMS Microbiol Ecol* 75, no. 2: 231–241. doi:10.1111/J.1574-6941.2010.00991.X.

Muller, E.M., Sartor, C., Alcaraz, N., and R. van Woesik. 2020. Spatial epidemiology of the stony coral tissue loss disease in Florida. *Front Mar Sci* 7: 163.

Palmer, C.V., and A.H. Baird. 2018. Coral tumor-like growth anomalies induce an immune response and reduce fecundity. *Dis Aquat Org* 130, no. 1: 77–81. doi:10.3354/DAO03258.

Pandolfi, J.M., Bradbury, R.H., Sala, E., et al. 2003. Global trajectories of the long-term decline of coral reef ecosystems. *Science* 301, no. 5635: 955–958.

Pandolfi, J.M., Connolly, S.R., Marshall, D.J., and A.L. Cohen. 2011. Projecting coral reef futures under global warming and ocean acidification. *Science* 333, no. 6041: 418–422. doi:10.1126/SCIENCE.1204794.

Paparella, F., Xu, C., Vaughan, G.O., and J.A. Burt. 2019. Coral bleaching in the Persian/Arabian Gulf is modulated by summer winds. *Front Mar Sci* 6: 1–15. doi:10.3389/FMARS.2019.00205/BIBTEX.

Paris, C.B., Helgers, J., Van Sebille, E., and A. Srinivasan. 2013. Connectivity modeling system: A probabilistic modeling tool for the multi-scale tracking of biotic and abiotic variability in the ocean. *Environ Model Softw* 42: 47–54.

Patterson, K., Porter, J.W., Ritchie, K.B., et al. 2002. The etiology of white pox, a lethal disease of the Caribbean elkhorn coral, *Acropora palmata*. *Proc Nat Acad Sci* 99, no. 13: 8725–8730. doi:10.1073pnas.092260099.

Peters, E.C., Halas, J.C., and H.B. McCarty. 1986. Calicoblastic neoplasms in *Acropora palmata*, with a review of reports on anomalies of growth and form in corals. *J Nati Cancer Institute* 76, no. 5: 895–912. doi:10.1093/JNCI/76.5.895.

Pohl, T., Al-Muqdadi, S.W., Ali, M.H., Fawzi, N.A.M., Ehrlich, H., and B. Merkel. 2014. Discovery of a living coral reef in the coastal waters of Iraq. *Sci Rep* 4: 4250. doi:10.1038/SREP04250.

Pollock, F.J., Lamb, J.B., Field, S.N., et al. 2014. Sediment and turbidity associated with offshore dredging increase coral disease prevalence on nearby reefs. *PLoS One* 9, no. 7: e102498. doi:10.1371/JOURNAL.PONE.0102498.

Porter, J.W., Dustan, P., Jaap, W.C., et al. 2001. Patterns of spread of coral disease in the Florida Keys. In: *The ecology and etiology of newly emerging marine diseases*. Ed: J.W. Porter, 1–24. Netherlands: Springer. doi:10.1007/978-94-017-3284-0_1.

Pratchett, M.S., McWilliam, M.J., and B. Riegl. 2020. Contrasting shifts in coral assemblages with increasing disturbances. *Coral Reefs* 39, no. 3: 783–793. doi:10.1007/S00338-020-01936-4.

Precht, W.F., Gintert, B.E., Robbart, M.L., Fura, R., and R. van Woesik. 2016. Unprecedented disease-related coral mortality in southeastern Florida. *Sci Rep* 6: 31374. doi:10.1038/srep31374.

Priest, M.A., Dibattista, J.D., McIlwain, J.L., Taylor, B.M., Hussey, N.E., and M.L. Berumen. 2016. A bridge too far: Dispersal barriers and cryptic speciation in an Arabian Peninsula grouper (*Cephalopholis hemistiktos*). *J Biogeogr* 43, no. 4: 820–832. doi:10.1111/JBI.12681.

Purkis, S.J., and B.M. Riegl. 2005. Spatial and temporal dynamics of Arabian Gulf coral assemblages quantified from remote-sensing and in situ monitoring data. *Mar Ecol Prog Ser* 287: 99–113. doi:10.3354/meps287099.

Purkis, S.J., and B.M. Riegl. 2012. Geomorphology and reef building in the SE Gulf. In: *Coral reefs of the Gulf: Adaptation to climatic extremes*. Eds: B.M. Riegl, and S.J. Purkis, 33–50. Netherlands: Springer. doi:10.1007/978-94-007-3008-3_3.

Purkis, S.J., Riegl, B.M., and R. Dodge. 2006. Fractal patterns of coral communities: Evidence from remote sensing (Arabian Gulf, Dubai, U.A.E.). Proc 10th International Coral Reef Symposium, Okinawa, Japan.

Purkis, S.J., Rivers, J., Strohmenger, C.J., et al. 2017. Complex interplay between depositional and petrophysical environments in Holocene tidal carbonates (Al Ruwais, Qatar). *Sedimentology* 64, no. 6: 1646–1675. doi:10.1111/SED.12368.

Purser, B.H., and E. Seibold. 1973. The principal environmental factors influencing Holocene sedimentation and diagenesis in the Persian Gulf. In: *The Persian Gulf*. Ed: B.H. Purser, 1–9. Springer. doi:10.1007/978-3-642-65545-6_1.

Raymundo, L.J., Halford, A.R., Maypa, A.P., and A.M. Kerr. 2009. Functionally diverse reef-fish communities ameliorate coral disease. *Proc Nat Acad Sci US* 106, no. 40: 17067–17070. doi:10.1073/PNAS.0900365106.

Raymundo, L.J., Rosell, K.B., Reboton, C.T., and L. Kaczmarsky. 2005. Coral diseases on Philippine reefs: Genus *Porites* is a dominant host. *Dis Aquat Org* 64, no. 3: 181–191.

Richardson, L.L. 2004. Black band disease. In: *Coral health and disease*. Eds: E. Rosenberg, and Y. Loya, 325–336. Berlin, Heidelberg: Springer. doi:10.1007/978-3-662-06414-6_18.

Riegl, B. 1999. Corals in a non-reef setting in the southern Arabian Gulf (Dubai, UAE): Fauna and community structure in response to recurring mass mortality. *Coral Reefs* 18, no. 1: 63–73. doi:10.1007/s003380050156.

Riegl, B. 2002. Effects of the 1996 and 1998 positive sea-surface temperature anomalies on corals, coral diseases and fish in the Arabian Gulf (Dubai, UAE). *Mar Biol* 140, no. 1: 29–40. doi:10.1007/S002270100676/METRICS.

Riegl, B. 2003. Climate change and coral reefs: Different effects in two high-latitude areas (Arabian Gulf, South Africa). *Coral Reefs* 22: 433–446.

Riegl, B.M., Benzoni, F., Samimi-Namin, K., and C. Sheppard. 2012a. Environmental constraints for reef building in the Gulf. In: *Coral reefs of the Gulf: Adaptation to climatic extremes*. Eds: B.M. Riegl, and S.J. Purkis, 187–224. Netherlands: Springer.

Riegl, B.M., Bruckner, A.W., Samimi-Namin, K., and S.J. Purkis. 2012b. Diseases, Harmful Algae Blooms (HABs) and their effects on Gulf coral populations and communities. In: *Coral reefs of the Gulf: Adaptation to climatic extremes*. Eds: B.M. Riegl, and S.J. Purkis, 107–125. Netherlands: Springer.

Riegl, B.M., Cavalcante, G., Bauman, A.G., Feary, D.A., Steiner, S., and S. Purkis. 2017. Demographic mechanisms of reef coral species winnowing from communities under increased environmental stress. *Front Mar Sci* 4, no. 6: 344. doi:10.3389/FMARS.2017.00344/BIBTEX.

Riegl, B.M., Johnston, M., Purkis, S., et al. 2018. Population collapse dynamics in *Acropora downingi*, an Arabian/Persian Gulf ecosystem-engineering coral, linked to rising temperature. *Glob Change Biol* 24, no. 6: 2447–2462. doi:10.1111/GCB.14114.

Riegl, B.M., and S.J. Purkis. 2009. Model of coral population response to accelerated bleaching and mass mortality in a changing climate. *Ecol Modell* 220, no. 2: 192–208. doi:10.1016/J.ECOLMODEL.2008.09.022.

Riegl, B.M., and S.J. Purkis. 2012. Coral reefs of the Gulf: Adaptation to climatic extremes in the world's hottest sea. In: *Coral reefs of the Gulf: Adaptation to climatic extremes*. Eds: B.M. Riegl, and S.J. Purkis, 1–4. Netherlands: Springer. doi:10.1007/978-94-007-3008-3_1.

Riegl, B.M., and S.J. Purkis. 2015. Coral population dynamics across consecutive mass mortality events. *Glob Change Biol* 21, no. 11: 3995–4005. doi:10.1111/GCB.13014.

Riegl, B.M., Purkis, S.J., Al-Cibahy, A.S., Abdel-Moati, M.A., and O. Hoegh-Guldberg. 2011. Present limits to heat-adaptability in corals and population-level responses to climate extremes. *PLoS One* 6, no. 9: e24802. doi:10.1371/JOURNAL.PONE.0024802.

Riegl, B.M., Purkis, S.J., Al-Cibahy, A.S., et al. 2012c. Coral bleaching and mortality thresholds in the SE Gulf: Highest in the world. In: *Coral reefs of the Gulf: Adaptation to climatic extremes*. Eds: B.M. Riegl, and S.J. Purkis, 95–105. Netherlands: Springer.

Roff, G., Kvennefors, E.C.E., Fine, M., et al. 2011. The ecology of 'Acroporid white syndrome', a coral disease from the southern Great Barrier Reef. *PLoS One* 6, no. 12: e26829. doi:10.1371/journal.pone.0026829.

Roff, G., Ulstrup, K.E., Fine, M., Ralph, P.J., and O. Hoegh-Guldberg. 2008. Spatial heterogeneity of photosynthetic activity within diseased corals from the Great Barrier Reef. *J Phycol* 44, no. 2: 526–538. doi:10.1111/J.1529-8817.2008.00480.X.

Sale, P.F., Feary, D.A., Burt, J.A., et al. 2011. The growing need for sustainable ecological management of marine communities of the Persian Gulf. *Ambio* 40: 4–17. doi:10.1007/s13280-010-0092-6.

Séré, M.G., Schleyer, M.H., Quod, J.P., and P. Chabanet. 2012. *Porites* white patch syndrome: An unreported coral disease on Western Indian Ocean reefs. *Coral Reefs* 31, no. 3: 739–739.

Sheppard, C.R.C. 2016. Coral reefs in the Gulf are mostly dead now, but can we do anything about it? *Mar Pollut Bull* 105: 593–598.

Sheppard, C.R.C., Al-Husiani, M., Al-Jamali, F., et al. 2010. The Gulf: A young sea in decline. *Mar Pollut Bull* 60, no. 1: 13–38. doi:10.1016/J.MARPOLBUL.2009.10.017.

Sheppard, C.R.C., and R. Loughland. 2002. Coral mortality and recovery in response to increasing temperature in the southern Arabian Gulf. *Aquat Ecosyst Health Manag* 5: 395–402. doi:10.1080/14634980290002020.

Sheppard, C.R.C., Price, A., and C. Roberts. 1992. *Marine ecology of the Arabian region: Patterns and processes in extreme tropical environments*. London, UK: Academic Press.

Sheppard, C.R.C., and R.V. Salm. 1988. Reef and coral communities of Oman, with a description of a new coral species (Order Scleractinia, genus *Acanthastrea*). *J Nat Hist* 22, no. 1: 263–279. doi:10.1080/00222938800770201.

Sheppard, C.R.C., and A. Sheppard. 1991. Corals and coral communities of Arabia. *Fauna of Arabia* 12: 3–170.

Sheppard, C.R.C., and S. Wells. 1988. *Coral reefs of the world: Volume 2: Indian ocean, Red Sea and Gulf*. Nairobi, Kenya and Gland, Switzerland: IUCN, UNEP.

Shinn, E.A. 1973. Recent intertidal and nearshore carbonate sedimentation around Rock Highs, E Qatar, Persian Gulf. In: *The Persian Gulf*. Ed: B.H. Purser, 193–198. Springer. doi:10.1007/978-3-642-65545-6_11.

Shinn, E.A. 1976. Coral reef recovery in Florida and the Persian Gulf. *Environ Geol* 1, no. 4: 241–254.

Shokri, M., Haeri-Ardakani, O., and P. Abdollahi. 2000. Coral reef resources of the Islamic Republic of Iran. *Reef Encounter* 27: 26–29.

Shuail, D., Wiedenmann, J., D'Angelo, C., et al. 2016. Local bleaching thresholds established by remote sensing techniques vary among reefs with deviating bleaching patterns during the 2012 event in the Arabian/Persian Gulf. *Mar Pollut Bull* 105, no. 2: 654–659. doi:10.1016/J.MARPOLBUL.2016.03.001.

Slatkin, M. 1993. Isolation by distance in equilibrium and non-equilibrium populations. *Evolution* 47, no. 1: 264–279. doi:10.1111/J.1558-5646.1993.TB01215.X.

Smith, E.G., Hazzouri, K.M., Choi, J.Y., et al. 2022. Signatures of selection underpinning rapid coral adaptation to the world's warmest reefs. *Sci Adv* 8, no. 2: 7287.

Smith, E.G., Hume, B.C.C., Delaney, P., Wiedenmann, J., and J.A. Burt. 2017a. Genetic structure of coral-Symbiodinium symbioses on the world's warmest reefs. *PLoS One* 12, no. 6: e0180169. doi:10.1371/JOURNAL.PONE.0180169.

Smith, E.G., Vaughan, G.O., Ketchum, R.N., McParland, D., and J.A. Burt. 2017b. Symbiont community stability through severe coral bleaching in a thermally extreme lagoon. *Sci Rep* 7, no. 1: 1–9. doi:10.1038/s41598-017-01569-8.

Stimson, J. 2011. Ecological characterization of coral growth anomalies on *Porites compressa* in Hawai'i. *Coral Reefs* 30, no. 1: 133–142. doi:10.1007/S00338-010-0672-8.

Sussman, M., Willis, B.L., Victor, S., and D.G. Bourne. 2008. Coral pathogens identified for White Syndrome (WS) epizootics in the Indo-Pacific. *PLoS One* 3, no. 6: e2393. doi:10.1371/JOURNAL.PONE.0002393.

Sutherland, K., Porter, J., and C. Torres. 2004. Disease and immunity in Caribbean and Indo-Pacific zooxanthellate corals. *Mar Ecol Prog Ser* 266: 273–302. doi:10.3354/meps266273.

Thinesh, T., Mathews, G., and J.K.P. Edward. 2011. Coral disease prevalence in the Palk Bay, Southeastern India – With special emphasis to black band. *Indian J Geo-Mar Sci* 40, no. 6: 813–820.

Torquato, F., Bouwmeester, J., Range, P., et al. 2022. Population genetic structure of a major reef-building coral species *Acropora downingi* in northeastern Arabian Peninsula. *Coral Reefs* 41, no. 3: 743–752. doi:10.1007/S00338-021-02158-Y/FIGURES/3.

Torquato, F., and P.R. Møller. 2022. Physical-biological interactions underlying the connectivity patterns of coral-dependent fishes around the Arabian Peninsula. *J Biogeogr* 49, no. 3: 483–496. doi:10.1111/JBI.14318.

Torquato, F., Range, P., Ben-Hamadou, R., et al. 2019. Consequences of marine barriers for genetic diversity of the coral-specialist yellowbar angelfish from the Northwestern Indian Ocean. *Ecol Evol* 9, no. 19: 11215–11226. doi:10.1002/ECE3.5622.

Ushijima, B., Smith, A., Aeby, G.S., and S.M. Callahan. 2012. *Vibrio owensii* induces the tissue loss disease *Montipora* white syndrome in the Hawaiian reef coral *Montipora capitata*. *PLoS One* 7, no. 10: e46717. doi:10.1371/JOURNAL.PONE.0046717.

Ushijima, B., Videau, P., Burger, A.H., et al. 2014. *Vibrio coralliilyticus* strain OCN008 is an etiological agent of acute *Montipora* white syndrome. *Appl Environ Microbiol* 80, no. 7: 2102–2109. doi:10.1128/AEM.03463-13.

Vajed Samiei, J., Saleh, A., Shirvani, A., Sheijooni Fumani, N., Hashtroudi, M., and M.S. Pratchett. 2016. Variation in calcification rate of *Acropora downingi* relative to seasonal changes in environmental conditions in the northeastern Persian Gulf. *Coral Reefs* 35, no. 4: 1371–1382. doi:10.1007/S00338-016-1464-6.

van Lavieren, H., Burt, J., Feary, D., et al. 2011. *Managing the growing impacts of development on fragile coastal and marine systems: Lessons from the Gulf.* A Policy Report, United Nations University – Institute for Water, Environment, and Health. Hamilton, ON, Canada.

Van Oppen, M.J.H., Gates, R.D., Blackall, L.L., et al. 2017. Shifting paradigms in restoration of the world's coral reefs. *Glob Change Biol* 23, no. 9: 3437–3448. doi:10.1111/gcb.13647.

Van Oppen, M.J.H., Oliver, J.K., Putnam, H.M., and R.D. Gates. 2015. Building coral reef resilience through assisted evolution. *Proc Nat Acad Sci* 112: 2307–2313. doi:10.1073/pnas.1422301112.

Vaughan, G.O., Al-Mansoori, N., and J. Burt. 2019. The Arabian Gulf. In: *World Seas: An environmental evaluation, second edition.* Ed: C. Sheppard, 1–23. Amsterdam, NL: Elsevier Science. doi:10.1016/B978-0-08-100853-9.00001-4.

Vaughan, G.O., and J.A. Burt. 2016. The changing dynamics of coral reef science in Arabia. *Mar Pollut Bull* 105, no. 2: 441–458. doi:10.1016/J.MARPOLBUL.2015.10.052.

Walton, C.J., Hayes, N.K., and D.S. Gilliam. 2018. Impacts of a regional, multi-year, multi-species coral disease outbreak in southeast Florida. *Front Mar Sci* 5: 323. doi:10.3389/fmars.2018.00323.

Ward, J.R., and K.D. Lafferty. 2004. The elusive baseline of marine disease: Are diseases in ocean ecosystems increasing? *PLoS Biol* 2, no. 4: e120. doi:10.1371/journal.pbio.0020120.

Williams, D.E., and M.W. Miller. 2005. Coral disease outbreak: Pattern, prevalence and transmission in *Acropora cervicornis*. *Mar Ecol Prog Ser* 301: 119–128.

Williams, G.J., Graham, N.A.J., Jouffray, J.B., et al. 2019. Coral reef ecology in the anthropocene. *Funct Ecol* 33, no. 6: 1014–1022. doi:10.1111/1365-2435.13290.

Williams, G.J., Knapp, I.S., Work, T.M., and E.J. Conklin. 2011. Outbreak of *Acropora* white syndrome following a mild bleaching event at Palmyra Atoll, Northern Line Islands, Central Pacific. *Coral Reefs* 30: 621.

Willis, B.L., Page, C.A., and E.A. Dinsdale. 2004. Coral disease on the Great Barrier Reef. In: *Coral health and disease*. Eds: E. Rosenberg, and Y. Loya, 69–104. Berlin, Heidelberg: Springer.

Work, T.M., and G.S. Aeby. 2006. Systematically describing gross lesions in corals. *Dis Aquat Org* 70: 155–160. doi:10.3354/dao070155.

Work, T.M., Aeby, G.S., and S.L. Coles. 2008a. Distribution and morphology of growth anomalies in *Acropora* from across the Indo-Pacific. *Dis Aquat Org* 78, no. 3: 255–264.

Work, T.M., Aeby, G.S., and K.A. Hughen. 2015. Gross and microscopic lesions in corals from Micronesia. *Vet Pathol* 53, no. 1: 153–162. doi:10.1177/0300985815571669.

Work, T.M., Aeby, G.S., Lasne, G., and A. Tribollet. 2014. Gross and microscopic pathology of hard and soft corals in New Caledonia. *J Invertebr Pathol* 120: 50–58. doi:10.1016/j.jip.2014.05.007.

Work, T.M., Aeby, G.S., Stanton, F.G., and D. Fenner. 2008b. Overgrowth of fungi (endolithic hypermycosis) associated with multifocal to diffuse distinct amorphous dark discoloration of corals in the Indo-Pacific. *Coral Reefs* 27: 663.

Work, T.M., Forsman, Z.H., Szabó, Z., Lewis, T.D., Aeby, G.S., and R.J. Toonen. 2011. Inter-specific coral chimerism: Genetically distinct multicellular structures associated with tissue loss in *Montipora capitata*. *PLoS One* 6, no. 7: e22869. doi:10.1371/journal.pone.0022869.

Work, T.M., Kaczmarsky, L., and E.C. Peters. 2016. Skeletal growth anomalies in corals. In: *Diseases of coral*. Eds: C.M. Woodley, C.A. Downs, A.W. Bruckner, J.W. Porter, and S.B. Galloway, 291–299. Hoboken, NJ: John Wiley & Sons.

Work, T.M., Russell, R., and G.S. Aeby. 2012. Tissue loss (white syndrome) in the coral *Montipora capitata* is a dynamic disease with multiple host responses and potential causes. *Proc Biol Sci* 279, no. 1746: 4334–4341. doi:10.1098/rspb.2012.1827.

Work, T.M., and E. Weil. 2016. Dark-spots disease. In: *Diseases of coral*. Eds: C.M. Woodley, C.A. Downs, A.W. Bruckner, J.W. Porter, and S.B. Galloway, 354–360. Hoboken, NJ: John Wiley & Sons.

Yamashiro, H., Yamamoto, M., and R. van Woesik. 2000. Tumor formation on the coral *Montipora informis*. *Dis Aquat Org* 41: 211–217. doi:10.3354/dao041211.

Zittis, G., Almazroui, M., Alpert, P., et al. 2022. Climate change and weather extremes in the Eastern Mediterranean and Middle East. *Rev Geophys* 60: e2021RG000762. doi:10.1029/2021RG000762.

Coral reefs in the pearl of the Gulf – Bahrain

7

Reem AlMealla and Leanne Hepburn

7.1 INTRODUCTION

Bahrain is an archipelago of 33 natural low-lying islands (Figure 7.1), with coral reefs that were historically known to be among the most extensive in the southern basin of the Arabian Gulf (Riegl 2003; Purkis and Riegl 2005; Kavousi et al. 2014). However, in the last four decades, reefs have undergone a significant decline due to intense bleaching events (Uwate and Shams 1999) and other anthropogenic impacts. Large-scale coastal development in pursuit of socio-economic growth has added 12.8% to the Kingdom's total land area between 1961 (690 km^2) and 2020 (785 km^2) (World Bank 2020).

Similar to other reefs in the Arabian Gulf, coral reefs in Bahrain are subjected to extreme environmental conditions, with seasonal temperatures ranging between 16–35°C (Burt et al. 2011; AlMealla 2020), coupled with high salinity levels of about 45 psu in the eastern waters and 55.1 psu in western waters (Al-Wedaei et al. 2011). Reefs are found in shallow waters at depths of 4 m to 18 m depending on their location. To date, there is only one scientific peer-reviewed paper from over a decade ago (Burt et al. 2013) published specifically on the reefs of Bahrain. Although there are a few published reports and chapters that have reported on the Kingdom's reefs, these remain fairly limited, they lack new local knowledge, and are now outdated. Therefore, through this chapter we aim to provide an overview of coral reefs in Bahrain that combine both local knowledge and scientific published information, thus providing new insights and generating a better understanding of the reefs in Bahrain.

7.2 DISTRIBUTION OF CORAL REEFS IN BAHRAIN

Coral reefs occupy a total area of 850 km^2, which is larger than the terrestrial lands of Bahrain (Burt et al. 2013). Reefs are distributed mainly in the north and east of Bahrain's territorial waters (Figure 7.1), with Fasht Al Adhm being the largest (170 km^2) followed by Fasht Al Jarim (139.7 km^2), and the smallest being Khor Fasht (18.6 km^2). In Bahrain, the word *fasht* is a local geological terminology that is used to refer to and/or describe a shallow area (<10 m) with a hard substrate formed by a mixture of bedrock and limestone structures such as dead coral reefs and fossils. *Fashts* are usually submerged in water, which could in certain areas be exposed to air during low tide, and are commonly known to be dominated by coral reef habitats. However, *fashts* could also host and be connected to other marine habitats such as seagrass beds, pearl oyster beds, and sandy beds within their vicinity (Alkhuzai et al. 2009; Sheppard et al. 2010; Burt et al. 2013).

The northern waters of Bahrain also host a variety of habitats, with the majority of areas being dominated by pearl oyster beds; however, coral habitats have been reported in this region (Arfa et al. 2012). In 2012, the Northern Hayrat area, which comprises three of the largest pearl oyster beds in the Kingdom – namely, Hayr Shtayyah, Hayr Bu Amamah and Hayr Bul Thamah – was inscribed as a UNESCO World Heritage Site (Figure 7.1). The site also embodies a coral reef area locally referred to as Najwat Bul Thamah (also known as Reef Bul Thamah). The local term *Hayr* (plural *Hayrat*) is used to refer to a dominant oyster bed habitat occurring in deep waters (7 to >25 m). On the eastern side of the mainland, in addition to the Fasht Al Adhm area, both the areas surrounding Umm Jaileed island (which is a sandy island) and Fasht Tglailib were historically known to also host coral reef habitats. Individual coral colonies were observed and reported previously by Burt et al. (2013) in the surrounding area of Umm Jaileed island, where the habitat is flat, shallow (similar to a *fasht*) and consists of a hard substrate providing an ideal environment for coral establishment. In an effort to ensure consistency in the documentation of coral information in Bahrain, in this manuscript, this area will be referred to as Umm Jaileed.

The nature of the topography of the reefs in Bahrain is influenced by its locality; nearshore reefs such as those located in Fasht Al Adhm are usually shallow in nature with a gently sloping reef that goes to a maximum depth of 10 m. The reefs are of minimal relief due to the sites being relatively homogenous and lacking extensive vertical structural complexity; this also applies to reefs in the southeastern waters such as Umm Jaleed and Fasht Tagilib (Figure 7.2). Offshore reefs are also present within the vicinity of areas such as Najwat Bul Thamah; *Najwa(t)* is the local term that refers to a coral reef fisheries area which encompasses Reef Bul Thanah. It is important to note that Ministerial Decision no. 2 of 2017 regarding

DOI: 10.1201/9781003321392-8

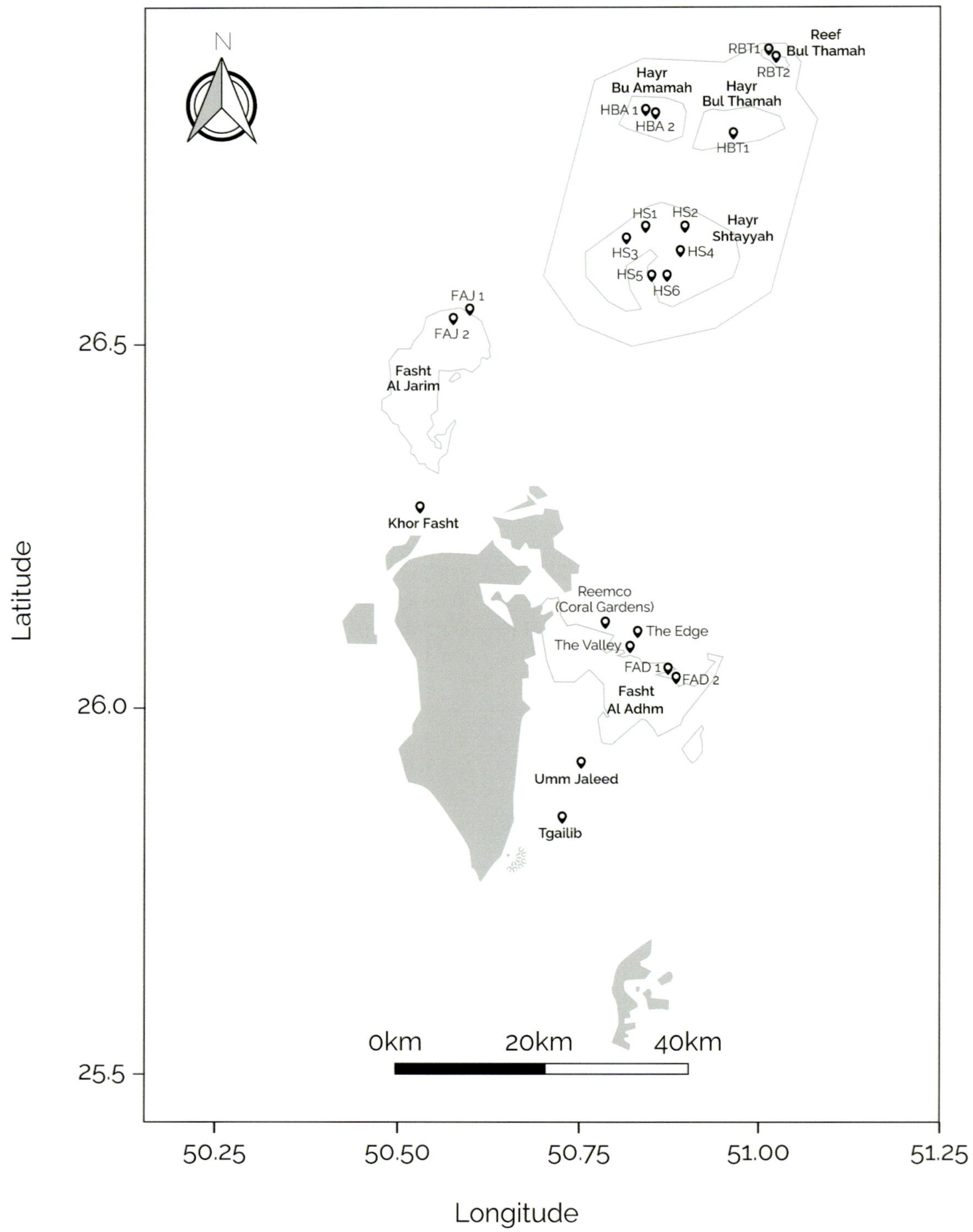

FIGURE 7.1 Distribution of coral reefs in the Kingdom of Bahrain (Shapefiles were created using Google Earth and base map was created using RStudio).

designating Najwat Bul Thamah as a marine protected area defines a *Najwat* as an area that consists of pointed rock formations (similar to a sea mount structure) that are not exposed to air regardless of tide movement and host a rich marine biodiversity. The official definition differs slightly from the way the term is used on a local basis, as local fishermen use the term to refer to areas of fisheries due to the coral reef presence. *Najwas* are relatively homogenous, with a gentle slope starting at 10 m and continuing to 12 m, where Pleistocene bedrock continues to a depth of 18 m, after which the substrate becomes more patchy with sand and bedrock formations as it slopes off into the deep (>50 m). The northern Hayrat area hosts very productive habitats for important fisheries as they are situated in clear and deeper waters (10–25 m) that experience strong currents. Patch reefs are also present in the Northern Hayrat area; these reefs are usually small in size, and dominated by coral colonies that are isolated in nature. The patch reefs in this area are integrated throughout the pearl oyster beds. These reefs are formed on flat rocky beds that lack structural complexity and are relatively homogenous (Figure 7.2). Local divers in Hayr Shtayyah have observed that the patch reefs found here host the most abundant hermatypic coral in the northern Hayrat area.

FIGURE 7.2 (a) Site overview of Reemco (coral gardens), Fasht Al Adhm (July 2022); (b) Site overview of Najwat (Reef) Bul Thamah, Northern Hayrat (July 2022); (c) Patch reef, Hayr Shtayyah (Site: HS4) (October 2022).

Hayr Bu Amamah and Hayr Bul Thamah host more soft corals (Octocorallia) than hard corals (Scleractinia).

In terms of distance from the main island, Najwat (Reef) Bul Thamah, Hayr Bul Thamah, Hayr Bu Amamah and Hayr Shtayyah are located approximately 75 km (Alkhuzai et al. 2009), 58 km, 57 km and 30 km from the mainland, respectively (Arfa et al. 2012). Both Khor Fasht and Fasht Al Jarim are located around 20 km from the mainland (Burt et al. 2013), while Fasht Al Adhm is less than 2 km away.

7.3 DIVERSITY OF CORAL REEFS IN BAHRAIN

7.3.1 Coral species richness

In the 1980s, a total of 30 hermatypic coral species were recorded, with *Acropora* spp. dominating the shallow areas (<5m) while poritids and faviids dominated the deeper areas (<20 m) (Barratt and Ormond 1985; Sheppard 1988; Vousden 1988, 1995; Burt et al. 2013). A total of 32 coral hermatypic species were reported as present in the Kingdom of Bahrain's report to the Convention on Biological Diversity in 2006 (Al-Kuwari 2006). However, only 23 species were documented in scientific surveys conducted by Burt et al. (2013). Furthermore, the first scientific documentation of coral patch reef presence in the northern Hayrat region was done by Arfa et al. (2012), which recorded a total of 32 coral species (27 hard corals and 5 soft corals) in Hayr Shtayyah while Hayr Bu Amamah, Hayr Bul Thamah and the Buffer zone had 18 (13 hard corals and 5 soft corals), 19 (14 hard corals and 5 soft corals), and 16 species (12 hard corals and 4 soft corals), respectively.

Based on recent studies (AlMealla 2020; AlMealla et al. 2024) and ongoing surveys conducted by AlMealla (2017–2022), 22 species of hermatypic coral were recorded; however, not all species correspond to the government list and that outlined by Arfa et al. (2012) and Burt et al. (2013) (Table 7.1; Figure 7.3). Compared to the original list (Al-Kuwari 2006), additional species such as *Dipsastraea favus* and *Fungia sp.* were observed by Burt et al. (2013) and AlMealla (2020), while Arfa et al. (2012) additionally recorded the presence of species such as *Heterocyathus aequicostatus* and *Goniopora sp.* (Table 7.1). Lastly, it is noted that the deeper waters of the

TABLE 7.1 Familial classification and presence of coral species observed in the Kingdom of Bahrain.

	FAMILY	*SPECIES (AL-KUWARI 2006)*	*ARFA ET AL. (2012)*	*BURT ET AL. (2013)*	*OBSERVED BY ALMEALLA (2017–2022)*
1	Acroporidae	*Acropora muricata (formerly A. formosa)*			
		Acropora sp.	Y		
					Montipora sp.
2	Agariciidae	*Pavona cactus*		Y	Y
3	Caryophylliidae	*Paracyathus sp.*			
			Heterocyathus aequicostatus		
			Polycyathus sp.		
4	Coscinaraeidae	*Anomastrea irregularis*	Y		Y
		Coscinaraea monile	Y		Y
			Coscinaraea columna		
			Coscinaraea sp.		
5	Dendrophylliidae	*Duncanopsammia peltata (formerly Turbinaria peltata)*	Y		Y
			Turbinaria reniformis		
6	Diploastraeidae		*Diploastrea sp.*		
7	Faviidae	*Dipsastraea speciosa (formerly Favia speciosa)*	Y	Y	Y
		Dipsastraea pallida (formerly Favia pallida		Y	Y
			Dipsastraea sp.	*Dipsastraea favus (formerly F. cf. favus*	Y
8	Fungiidae		Y		*Fungia sp.*
			Fungia cycloites		
9	Leptastreidae	*Leptastrea transversa*		Y	
		Leptastrea purpurea		Y	
					Leptastrea pruinosa
10	Merulinidae	*Cyphastrea serailia*	Y	Y	Y
		Cyphastrea microphthalma	Y	Y	Y
		Favites pentagona		Y	Y
		Favites chinesis		Y	Y
		Favites abdita		Y	Y
		Platygyra crosslandi	Y (*Platygyra sp.*)	Y	
		Platygyra sinensis		Y	
		Platygyra daedalea		Y	Y
		Platygyra lamellina		Y	
11	Mussidae	*Acanthastrea echinata*		Y	
			Acanthastrea maxima		
12	Plesiastreidae	*Plesiastrea versipora*	Y	Y	Y
13	Pocilloporidae	*Madracis kirbyi*		Y	
		Stylophora pistillata	Y (*Stylophora sp.*)		
14	Poritidae		*Goniopora sp.*		
		Porites lobata	Y (*Porites sp.*)	Y	Y
		Porites lutea		Y	Y
		Porites harrisoni		Y	
		Porites nodifera			
		Porites compressa			Y
15	Psammocoridae	*Psammocora contigua*	Y (*Psammocora sp.*)	Y	Y (*Psammocora sp.*)
16	Rhizangiidae	*Siderastrea savignyana*	Y (*Siderastrea sp.*)	Y	Y

FIGURE 7.3 Coral colonies recorded in Fasht Al Adhm (July 2022) with: (a) *Porites lutea*, (b) *Leptastrea pruinosa*, (c) *Dipsastraea favus*, and (d) *Cyphastrea microphthalma*.

northern region of Bahrain harbour greater coral species richness than the eastern region (Arfa et al. 2012; Burt et al. 2013; observation by the author, 2022).

7.3.2 Reef-associated benthic organisms

Similar to other reefs in the Arabian Gulf, Bahraini reefs are of significant and important ecological value despite hosting a lower (i.e., 10%) species diversity compared to regions such as the Indo-Pacific (Coles 2003). Low species diversity could be attributed to the region's geographic isolation and its extreme and stressful environmental conditions defining it as the hottest sea on the planet, which in turn, limits the diversity of species it hosts (Coles 2003; Coles and Riegl 2013; Hume et al. 2013). The reefs in Bahrain are connected to other ecosystems such as seagrass beds, mangroves and pearl oyster beds, all of which support each other. For example, mangroves and seagrass beds are well known worldwide to act as nurseries for many important reef fish and other organisms. On the other hand, coral reefs play an important role in supporting the life cycle of pearl oysters through: (1) the provision of hard substrate space for reproduction, (2) provision of nutrients required by oysters, and (3) hosting fish and other organisms that influence oyster health. In return, oyster beds, which in general are associated with epi and macrofauna, and enhanced biodiversity (Lown et al. 2021) also provide additional biomass and food to fish, thereby potentially promoting fish abundance (Cartier and Carpenter 2014). The reefs in Bahrain host diverse communities of fish,

FIGURE 7.4 Calcareous algae (Halimeda sp.) and Bryozoa (*Celleporaria pigmentaria*) recorded on the Najwat (Reef) Bul Thamah in Bahrain.

crustaceans, mollusks and echinoderms as well as migratory megafauna such as whale sharks (*Rhincodon typus*), Arabian carpet sharks (*Chiloscyllium arabicum*) and other elasmobranchs, including various species of rays such as eagle rays (Myliobatidae).

The presence of benthic algae on coral reefs is a common occurrence worldwide and is exacerbated by a combination of stressors, including overfishing of herbivores, excess nutrient, and organic matter input, etc. (Diaz-Pulido et al. 2009; Barott et al. 2012; Eich et al. 2019; Figures 7.4, 7.5 and 7.6). Although high cover of algae on a reef is deemed negative, in some cases, algae can have positive effects on reefs; for example, crustose coralline algae (CCA) promote coral recruitment (Ritson-Williams et al. 2009) and suppresses macroalgae recruitment (Vermeij et al. 2011; Eich et al. 2019; Figure 7.5). In addition, some algae epiphytes are important for reef-associated fish such as damselfish (Pomacentridae), butterflyfish (Chaetodontidae), and parrotfish (Scaridae).

Furthermore, Bahraini reefs host several species of sponges (Porifera), which are essential to reef function and dynamics; they are an important food source for reef-associated fish, e.g., angelfish (Pomacanthidae) and other organisms, e.g., brittle stars (Ophiothricidae) and shrimps (*Penaeus semisulcatus*), as they provide food as well as shelter (Figure 7.7). Conversely, excessive bioeroding sponge and sea urchin activity can have negative effects on the reef, and it is therefore important to monitor community composition and shifts happening on reefs, which can contribute towards further structural erosion (Perry and Hepburn 2008). It is important to highlight that no studies on the sponges of Bahrain have yet been published.

To date, quantitative and qualitative information regarding reef-associated species in Bahrain remains deficient, with the exception of commercially important fisheries (Ali and Abahussain 2013), making it difficult to understand and predict how reef systems in Bahrain will cope in the future with synergistic climate impacts.

7.4 STATUS AND TRENDS OF BAHRAINI CORAL REEFS

Historically, coral reefs in Bahrain were said to be among the most extensive and dense in the south of the Arabian Gulf (Riegl 2003; Purkis and Riegl 2005). However, decades of being subjected to intense anthropogenic activities including recurring bleaching events have caused a significant decline in reef health. Prior to 1995, coral cover was reported to be dense (50–75%) in shallow waters, both in areas of higher current flow and more saline areas to the southeast (Vousden 1995; Alkhuzai et al. 2009). Branching *Acropora* was said to be widespread and dominated the reef communities in shallow areas, whilst dense *Porites* and Faviidae colonies dominated the deeper waters (Vousden 1995; Alkhuzai et al. 2009; Burt et al. 2013). However, the mass bleaching events in 1996 and 1998, where temperatures reached 37.7°C and 38°C, respectively, have caused the reefs in the Kingdom to suffer an estimated loss of >90% of coral cover (Rezai et al. 2004; Al-Kuwari 2006; AlMealla 2020). In between these two bleaching events, Uwate et al. (1997) recorded a coral cover of 43.1% in the nearshore Fasht Al Adhm site; however, following the 1998 episode, coral cover declined to 0% in Fasht Al Adhm (Uwate et al. 2000), thereby making Bahrain the worst Gulf country affected by these bleaching events (Wilkinson 1998; Uwate and Shams 1999; Lough 2000).

Live coral cover increased from 0% in 2000 to 9.8% on reefs in Fasht Al Adhm over the next eleven years, indicating promising reef recovery and new coral establishments (Figure 7.8). This increasing trend continued for the next six years up to 2017 when coral cover reached 13.6% (Burt et al. 2013). Unfortunately, following the 2017 bleaching event (Burt et al. 2019), coral cover decreased to 1.4% in some sites, e.g., the Valley in Fasht Al Adhm (AlMealla 2020; AlMealla et al. 2024). To date, there is no published data on the bleaching rates on Bahraini reefs. Nevertheless, as previous temporal

FIGURE 7.5 Selected species of red algae recorded on reefs in the Northern Hayrat region in Bahrain with: (a) Coralline red algae, *Asparagopsis taxiformis* (Order: Ceramiales)| July 2022, Najwat (Reef) Bul Thamah; (b) Coralline red algae, *Negoniolithon mamillosum*, July 2022, Najwat (Reef) Bul Thamah; (c) Coralline red algae, *Lithothamnion* sp., July 2022, Najwat (Reef) Bul Thamah; and (d) Red algae, Rhodophyta sp., June 2022, Hayr Bu Amamah.

data indicates reef ability to recover, the recovery witnessed between the years 2000 and 2017 provides a positive outlook, despite encountering multiple bleaching events in previous years (e.g., 1998, 2010; AlMealla 2020; AlMealla et al. 2024). Ongoing surveys indicate that coral cover is recovering in areas such as Fasht Al Adhm and Fasht Al Jarim. This is supported by observations of stabilisation (i.e., binding) of rubble substrate with crustose coralline algae on these reefs (Figure 7.9; Gómez-Lemos et al. 2018; AlMealla et al. 2024; Jorissen et al. 2021; Cornwall et al. 2023). In addition, sites such as Reemco (Coral Gardens) and the Edge were noted to host a healthy amount of coral cover of about 15% (observation by the author and unpublished data, 2023).

Contrasting with this, the offshore reef in Najwat (Reef) Bul Thamah, which is a deeper site (>10 m), has maintained a stable coral cover (11.9–12.5%) following the 1995 and 1997 bleaching events (Uwate et al. 2000). In subsequent years, it proved that it was able to recover well, with coral cover reaching approximately 50% in 2008 accompanied by the presence of *Acropora* colonies (Alkhuzai et al. 2009). In 2012, coral cover decreased significantly (16.3%; Burt et al. 2013); however, it then remained relatively stable to 2018 (14.8%; AlMealla et al. 2024).

To date, no studies have been conducted to gather data related to coral cover in the northern Hayrat region, including Hayr Bul Thamah, Hayr Bu Amamah and Hayr Shtayyah, with the exception of Najwat (Reef) Bul Thamah, hence leaving a large area unexplored. The only surveys done in this region were specific to studies of the pearl oyster beds by Arfa et al. (2012), who made notes of the coral colonies encountered.

Closer to the north of the mainland, historical data collected in 1991 indicate that Fasht Al Jarim used to host dense assemblages of coral from 10 species (5 families) including *Acropora* (Fadlallah et al. 1993; Alkhuzai et al. 2009; Burt et al. 2013). However, surveys conducted in 2011 by Burt et al. (2013) report that coral cover was very low (0.2%) with only four species recorded, mainly comprised of faviids and siderastreids with no *Acropora* observed. Ongoing recent surveys indicate that the percentage of coral cover in this area has not increased; however, evidence of a high percentage of crustose coralline algae (CCA) is noted (observation by the author, 2022). No historic data is known regarding Khor Fasht;

FIGURE 7.6 Selected species of brown algae (Ochrophyta) recorded on the reef at Khor Fasht in Bahrain with: (a) Funnelweed algae, *Padina* sp., January 2023, Khor Fasht; (b) Tropical seaweed, *Sargassum decurrens*, January 2023, Khor Fasht; (c) Brown algae, *Hormophysa cuneiformis*, January 2023, Khor Fasht; and (d) Unidentified brown algae, January 2023, Khor Fasht.

however, Burt et al. (2013) reported that the coral cover in this area is similar (<1%) with turf algae dominating the site (~50%).

In the early 1980s, the southeastern reefs of Fasht Tgailib and Umm Jaleed were known for their more saline waters, which harboured a more dominant poritid and faviid community, with coral cover reported between 12–70% (Sheppard 1988) and 55% (Vousden 1995; Burt et al. 2013). This dominance in coral taxa continued in 2011 but with low coral cover (~2%), with evidence of the reefs shifting towards algae dominance (40.8%) recorded (Burt et al. 2013).

Despite reports of the decline in coral cover, it is worth noting that in 2017, CCA, an important carbonate producer, was observed to be extremely high in percentage cover (42.2%) in Fasht Al Adhm (Figure 7.9; AlMealla et al. 2024). CCA contributes extensively towards the cementation and stability of reefs as it plays an important role in binding the reef together through colonising reef substrate, which promotes accretion and enhances larval recruitment of various organisms (Rasser and Riegl 2002; Mallela and Perry 2007; Perry and Hepburn 2008; Langer 2008; Fujita et al. 2009). Similar trends were observed on Fasht Al Jarim during the recent surveys in June 2022, where high coverage of CCA was observed to colonise the reefs, therefore preparing the substrate for new recruits and indicating opportunities for recovery (Figure 7.9; observation by the author, 2022).

7.5 THREATS AND FUTURE CONSERVATION EFFORTS

In the last four decades, coral reefs in Bahrain have suffered severe impacts due to various environmental and anthropogenic factors that have resulted in their degradation and destruction (Uwate et al. 2000; Naser 2012; Burt et al. 2013), some of which we detail in the following sections.

7.5.1 Coastal development

Between 1960 and 2020, Bahrain has increased its land area by 13.8% as part of its Vision 2030 in order to cater for a growing

FIGURE 7.7 Selected species of sponges (Porifera) observed in Fasht Al Adhm with: (a) Bioeroding sponge, *Cliona orientalis*, January 2023; (b) Bioeroding sponge, *Cliothosa hancocki*, January 2023; (c) *Chondrilla nucula*, January 2023; (d) Possibly *Didemnum perlucidum*, January 2023; (e) *Halicona* sp., January 2023; and (f) *Desmapsamma anchorata*, August 2020.

Year	1997	2000	2011	2017	2018
Coral Cover	43.1%	0%	9.8%	13.6%	1.4%
	Uwate et al. 1997	Uwate et al. 2000	Burt et al. 2013	AlMealla 2020	

FIGURE 7.8 Timeline of change in coral cover between 1997 and 2018 in Fasht Al Adhm area.

FIGURE 7.9 Evidence of reef and rubble binding by colonization of crustose coralline algae (CCA) (a) The Valley, Fasht Al Adhm (May, 2017); (b) Fasht Al Jarim (June, 2022).

population and the need for urban expansion while pursuing socio-economic development (World Bank 2020). In the coming half century, coastal development and activities of dredging and reclamation remain top priorities for the Kingdom. Unfortunately, dredging and reclamation works have multiple negative impacts on coral reefs and their recovery rates. It is well documented that pollution arising from urban and coastal developments have resulted in altering coral reef ecosystems worldwide, threatening their functionality (Lafferty et al. 2004; Browne et al. 2012a; Fan et al. 2013). In addition, reef accretion and coral growth rates are negatively impacted from anthropogenic pollution caused by sedimentation and eutrophication (Edinger et al. 2000; Mallela and Perry 2007; Browne et al. 2012b). Impacts of high sedimentation include declining calcification rates (Mallela 2013), increase in coral mortality due to smothering and burial (Loya 1976; Nugues and Roberts 2003), increase in coral disease (Bruno et al. 2003), reduced larval settlement (Fabricius et al. 2005) and reduced light availability necessary for photosynthesis (Rogers 1990; Wolanski and De'ath 2005; Browne et al. 2012a). Sedimentation has also been noted to negatively impact secondary reef builders, especially encrusters, which are vital contributors and supporters of the net positive reef carbonate budget (Fabricius and De'ath 2001; Mallela and Perry 2007). To date, there are no studies that record and quantify the impacts of dredging and reclamation on reefs in Bahrain. Of concern is that Bahrain is currently planning on building five new islands as part of its ambitious $30 billion plan for economic recovery and as part of its efforts to shift from an oil-dependent economy to one that is built on private business, tourism and manufacturing (National Geographic 2022). The project intends to double the size of Bahrain's land mass; however, one of the proposed islands is to be built on the largest reefs in the Arabian Gulf, Fasht Al Adhm, which not only is of significant ecological value but is also a source of livelihood, including external markets, and protein to many fishing communities. Similarly, the second island will be built on Fasht Al Jarim, while the third island may impact the reef areas in Khor Fasht. The remaining two islands will destroy other nearshore marine habitats such as seagrass beds and pearl oyster beds amongst others, all of which act as a nursery ground for juvenile fish and other marine organisms. The fact that there is very limited data or published literature on the reefs of Bahrain is concerning, particularly given the planned developments due to take place on Bahraini reefs. Moreover, the consequences of such developments will be severe and long lasting or permanent, and hence it is vital for environmental regulators to manage the areas and projects properly through sound planning and consultation, identifying and designating areas with live coral for protection. Finally, it is pertinent to carefully consider coastal features, hydrodynamics and offshore bathymetric conditions in addition to the impact on fisheries supported by the ecosystems and the fishing communities that rely on them for their livelihood.

7.5.2 Elevated sea temperatures and ocean acidification

Ocean warming and ocean acidification, both climate impacts, are globally recognised as a key threat to marine ecosystems including coral reefs as they alter ocean temperature and chemistry (Pandolfi et al. 2011; IPCC 2015). The impacts of ocean acidification include decreasing the concentration of carbonate ions, which means that reef-building corals will find it challenging to produce and build their skeletons (Gattuso et al. 1998; Kleypas et al. 1999; Orr et al. 2005; Anthony et al. 2008). In addition, when carbonate concentrations fall too low, already formed CaCO3 structures will start to dissolve (Orr et al. 2005). This in turn will impact coral growth and the accretion rate of reefs, reducing their ability to keep pace with rising sea levels (Perry et al. 2018).

In the last two decades, reefs in the Arabian Gulf including Bahrain have been exposed to severe temperature anomalies (~2℃ above the average) at a rapidly recurring rate (Burt et al. 2012; Riegl 2002; Sheppard and Loughland 2002). Alarmingly, but not unexpectedly, the world is currently experiencing a global coral bleaching event in 2024. This is the fourth global event on record and the second in the last 10 years (NOAA 2024). To date, there are no data regarding the bleaching extent in Bahrain; however, the impacts of the bleaching events have been witnessed and in turn have contributed to the decline in coral cover (see Section 4).

7.5.3 Pressure on fisheries

Fishing in Bahrain has extensive cultural and historical significance, as well as supporting the livelihood of many fishing communities (Ali and Abahussain 2013); the sector contributes to food security, which is especially important given that Bahrain imports 90% of its food (Efron et al. 2018).

Coastal development and intensive land reclamation projects have resulted in the destruction and degradation of important habitats for fish, including seagrass beds and coral reefs (Zainal et al. 1993) in addition to mangroves (Abido et al. 2011), which have severely impacted the maintenance and availability of fish stocks (Zainal et al. 2012). This, coupled with intensive fishing, has slowed the recovery of fishery production. In 2020, the total fisheries production in Bahrain was reported to be 14,420 metric tons, which is lower than the production in 2012, when the highest productivity was recorded (27,092.72 metric tons; World Bank 2023). Unfortunately, enforcement of environmental regulations remains relatively weak (Naser 2012; Zainal et al. 2008), with illegal fishing occurring in marine protected areas, thus impacting the recovery of fisheries and habitats (Al-Cibahy et al. 2012). This is especially noted in Najwat Bul Thamah, where extensive fishing damage is evident with repeated reports of illegal use of nets such as *gargoor*, which are local handmade fishing traps and gill nets (Alkhuzai et al. 2009; Uwate et al. 2000).

7.5.4 Desalination

The negative impact of desalination plants on the marine environment is well-documented worldwide (Mabrook 1994; Miri and Chouikhi 2005; Dupavillon and Gillanders 2009; Roberts et al. 2010). The output of the desalination process includes brine water discharged into the marine environment, which carries chemical pollutants that impact the chemical properties of seawater and sediment quality (Altayaran and Madany 1992; Hoepner and Lattemann 2002). Based on a study by Macpherson et al. (2006), which assessed the impacts of the Al Dur desalination plant (pre- and post-establishment), located in the southeast of Bahrain, on macro benthic assemblages, the results revealed negative changes in the benthic species richness and a reduction in biodiversity and abundance of microbenthic assemblages such as polychaeta and amphipod organisms, all of which is largely attributed to the brine discharge. In addition, an extreme elevation in temperatures (> 38°C) and hypersaline waters (> 55‰) were recorded by Al-Osaimi et al. (2019) at locations near the discharge outlet in waters of depths > 3 m during both seasons, including the high and low tide cycles, with exceptional levels noted in summer. The temperature of the brine discharge is one of the major concerns, as several fish kill cases, along with the disappearance of marine benthic species such as echinoderms, have been previously reported in the vicinity of desalination plants (Khordagui 2002; Fernandez-Torquemada et al. 2005). The extremely high temperature impacts marine fauna and flora through reducing their growth rate, impacting their osmoregulatory mechanisms, causing shrinkage of body cells, and contributing to the malfunction of the endocrine system (Al-Osaimi et al. 2019).

It is important to note that the area surrounding Al Dur is dominated by seagrass beds, which are a vital and sensitive habitat that plays an important ecological role as a feeding and spawning ground for fish, in addition to being a nursery for a broad array of marine megafaua including dugongs, dolphins and green turtles (Al-Osaimi et al. 2019). In the south-eastern side of Bahrain, it is thought that expanding industrial and desalination works have likely impacted the reefs of Umm Jaileed and Fasht Tghailib, impacting their recovery (Zainal et al. 2008).

7.5.5 Conservation efforts

The economic value of the ecosystem services provided by the marine environment in Bahrain (including coral reefs) was estimated to be US$1.8 billion per year (Sheppard 2007; Al Zayani and Loughland 2009). In an effort to conserve and protect marine areas of high biodiversity, in 2017 the government of Bahrain declared the Northern Hayrat region as a marine protected area (MPA). The area is currently protected based on Law no. 2 of 2017 regarding the declaration of Najwat Bul Thamah as an MPA and Law no. 3 of 2017 regarding the declaration of Hayr Bul Thamah, Hayr Shtayyah and Hayr Bu Amamah as MPAs. Furthermore, Article 74 (b), paragraph 3 of Law no. 7 of 2022 regarding the environment entails that it is prohibited to cause harm to any species. This includes a clause specific to the prohibition of cutting, damaging or destroying coral reefs, shells, snails and other endangered species.

In addition, to help reduce pressures on fisheries, the Ras Hayan Mariculture Centre (i.e., the National Aquaculture Centre) has been working at enhancing its capacity and developing its hatchery area. The centre is currently producing juveniles of various species such as Sobaity seabream (*Sparidentex hasta*) and the very popular with locals, brown-spotted grouper (*Epinephelus coioides*). In addition, a partnership has been established between the Directorate of Agriculture and Marine Resources and the Taiwan Authority for International Cooperation to farm new species. The partnership includes training Bahraini citizens through equipping them with aquaculture techniques and supporting local enterprises to establish aquaculture practices. Moreover, additional national initiatives such as encouraging the private sector to invest in aquaculture, programs to train citizens in responsible fish farming, and allocating sites for local fish farming have been set up to develop aquaculture. These initiatives represent some of the efforts put in place to help reduce fishing pressures on natural habitats, thereby providing space and time for their recovery.

In an effort to create new habitats for fisheries as a replacement for those that were either lost and/or impacted by coastal development, the Kingdom has resorted to deploying artificial reefs (Erftemeijer et al. 2004; Barber 2006). Artificial reefs take various forms with differing degrees of success, they can be expensive to construct, and they may only develop low-diversity fouling communities (Burt et al. 2009, 2012).

7.6 CONCLUSION

In conclusion, coral reefs in Bahrain remain understudied and more research is required in order to better advise decision makers as to the best course of action in terms of conservation and also to propose effective restoration initiatives. Finally, it is crucial to continue generating data through setting up a long-term monitoring program that will enable reef scientists to detect changes in the reefs and be better able to protect the structural integrity and abundant natural resources of the Kingdom.

REFERENCES

Abido, M., Abahussain, A., and H. Abdel Munsif. 2011. Status and composition of mangrove plant community in Tubli Bay of Bahrain during the years 2005 and 2010. *Arab Gulf J Sci Res* 29: 100–111.

Al-Cibahy, A.S., Al-Khalifa, K., Böer, B., and K. Samimi-Namin. 2012. Conservation of marine ecosystems with a special view to coral reefs in the Gulf. In: *Coral reefs of the Gulf: Adaptation to climatic extremes*. Eds: B.M. Riegl, and R.E. Dodge, 337–348. Dordrecht, Netherlands: Springer.

Ali, T.S., and A.A. Abahussain. 2013. Status of commerical fisheries in the Kingdom of Bahrain. *Basrah J Agric Sci* 26, no. 1: 220–238.

Al-Kuwari, Z. 2006. *Bahrain first national report to the convention on biological diversity*. Manama, Bahrain. www.sce.gov.bh/Media/Downloads/reports/pdf/report-1.pdf; downloaded 20 May 2022.

Alkhuzai, J., Sheppard, C., Abdulqader, E.A.A., Alkhuzai, S.A., and R.A. Loughland. 2009. Subtidal habitats. In: *Marine atlas of Bahrain*. Eds: R.A. Loughland, and A.M. Zainal, 115–167. Kingdom of Bahrain: Center for Studies and Research.

AlMealla, R.K. 2020. *Reef carbonate budgets & symbiodiniaceae diversity in the Arabian Gulf, Western Indian Ocean & Central Indo-Pacific*. Ph.D. Thesis, University of Essex. https://repository.essex.ac.uk/28940/; downloaded 15 January 2023.

AlMealla, R.K., Edullantes, B., and Hepburn, L.J. 2024. Bleaching threatens positive carbonate budgets on Bahraini reefs. *Marine Biology* 171: 39. https://doi.org/10.1007/s00227-023-04351-9

Al-Osaimi, A., Ali, T.S., Al-Zubari, W., and H. Naser. 2019. Effect of brine discharge from Al-Dur RO desalination plant on the infauna species composition in the east coast of Bahrain. *Management Studies* 7, no. 6: 609–623.

Altayaran, A., and I. Madany. 1992. Impact of a desalination plant on the physical and chemical properties of seawater: Bahrain. *Water Res* 26, no. 4: 435–441.

Al-Wedaei, K., Naser, H., Al-Sayed, H., and A. Khamis. 2011. Assemblages of macro-fauna associated with two seagrass beds in Kingdom of Bahrain: Implications for conservation. *J Assoc Arab Univ Basic Appl Sci* 10, no. 1: 1–7. doi:10.1016/j.jaubas.2011.06.004.

Al Zayani, A.K., and R.A. Loughland. 2009. Conservation of Bahrain's marine enviornment. In: *Marine atlas of bahrain*. Eds: R. Loughland, and Z. Zainal, 371–383. Manama, Bahrain: Miracle Publishing.

Anthony, K.R.N., Kline, D.I., Diaz-Pulido, G., Dove, S., Hoegh-Guldberg, O., and D.M. Karl. 2008. Ocean acidification causes bleaching and productivity loss in coral reef builders. *PNAS* 105: 17442–17446.

Arfa, S.B., AlMealla, R., Edwards, D., and E. Cruz. 2012. *Monitoring programme for the oyster bed sites*. Directorate of Human Resources & Finance, Ministry of Culture, Kingdom of Bahrain.

Barber, T. 2006. *Designed artificial reef program for Bahrain*, 1–17. Greenville, NC: Reef Ball Development Group Inc.

Barott, K.L., Williams, G.J., Vermeij, M.J.A., Harris, J., Smith, J.E., Rohwer, F.L., and S.A. Sandin. 2012. Natural history of coral-algae competition across a gradient of human activity in the Line Islands. *Mar Ecol Prog Ser* 460: 1–12. doi:10.3354/meps09874.

Barratt, L., and R. Ormond. 1985. Fasht Al Adhm environmental studies phase 1 report, technical volume. In: *A Report to the Ministry of Housing, State of Bahrain*. Ed: W.S. Atkins. Manama, Bahrain: Partners Overseas.

Browne, N.K., Smithers, S.G., and C.T. Perry. 2012a. Coral reefs of the turbid inner-shelf of the Great Barrier Reef, Australia: An environmental and geomorphic perspective on their occurrence, composition and growth. *Earth-Sci Rev* 115: 1–20.

Browne, N.K., Smithers, S.G., Perry, C.T., and P.V. Ridd. 2012b. A field-based technique for measuring sediment flux on coral reefs: Application to turbid reefs on the Great Barrier Reef. *J Coast Res* 28: 1247–1262.

Bruno, J.F., Petes, L.E., Harvell, C.D., and A. Hettinger. 2003. Nutrient enrichment can increase the severity of coral diseases. *Ecol Lett* 6: 1056–1061.

Burt, J., Al-Harthi, S., and A. Al-Cibahy. 2011. Long-term impacts of coral bleaching events on the world's warmest reefs. *Mar Environ Res* 72, no. 4: 225–229.

Burt, J.A., Al-Khalifa, K., Khalaf, E., AlShuwaikh, B., and A. Abdulwahab. 2013. The continuing decline of coral reefs in Bahrain. *Mar Pollut Bull* 72: 357–363. doi:10.1016/j.marpolbul.2012.08.022.

Burt, J., Bartholomew, A., and D. Feary. 2012. Man-made structures as artificial reefs in the Gulf. In: *Coral reefs of the Gulf*. Eds: B.M. Riegl, and S.J. Purkis, 171–186. Dordrecht, The Netherlands: Springer. doi:10.1007/978-94-007-3008-3.

Burt, J., Bartholomew, A., Usseglio, P., Bauman, A., and P.F. Sale. 2009. Are artificial reefs surrogates of natural habitats for corals and fish in Dubai, United Arab Emirates? *Coral Reefs* 28: 663–675.

Burt, J.A., Paparella, F., Al-Mansoori, N., Al-Mansoori, A., and H. Al-Jailani. 2019. Causes and consequences of the 2017 coral bleaching event in the southern Persian/Arabian Gulf. *Coral Reefs* 38: 567–589. doi:10.1007/s00338-019-01767-y.

Cartier, L.E., and K.E. Carpenter. 2014. The influence of pearl oyster farming on reef fish abundance and diversity in Ahe, French Polynesia. *Mar Pollut Bull* 78: 43–50. doi:10.1016/j.marpolbul.2013.11.027.

Coles, S.L. 2003. Coral species diversity and environmental factors in the Arabian Gulf and the Gulf of Oman: A comparison to the Indo-Pacific region. *Atoll Res Bull* 507: 1–19.

Coles, S.L., and B.M. Riegl. 2013. Thermal tolerances of reef corals in the Gulf: A review of the potential for increasing coral survival and adaptation to climate change through assisted translocation. *Mar Pollut Bull* 72: 323–332.

Cornwall, C.E., Carlot, J., Branson, O., et al. 2023. Crustose coralline algae can contribute more than corals to coral reef carbonate production. *Commun Earth Environ* 4: 105. doi:10.1038/s43247-023-00766-w.

Diaz-Pulido, G., McCook, L.J., Dove, S., Berkelmans, R., Roff, G., Kline, D.I., Weeks, S., Evans, R.D., Williamson, D.H., and O. Hoegh-Guldberg. 2009. Doom and boom on a resilient reef: Climate change, algal overgrowth and coral recovery. *PLoS One* 4, no. 4: e5239. doi:10.1371/journal.pone.0005239.

Dupavillon, J.L., and B.M. Gillanders. 2009. Impacts of seawater desalination on the giant Australian cuttlefish *Sepia apama* in the upper Spencer Gulf, South Australia. *Mar Environ Res* 67, no. 4–5: 207–218. doi:10.1016/j.marenvres.2009.02.002. Epub 2009 Mar 6. PMID: 19332355.

Edinger, E.N., Limmon, G.V., Jompa, J., Widjatmoko, W., Heikoopà, J.M., and M.J. Riskà. 2000. Normal coral growth rates on dying reefs: Are coral growth rates good indicators of reef health? *Mar Pollut Bull* 40: 404–425.

Efron, S., Fromm, C., Gelfed, B., Nartaraj, S., and C. Sova. 2018. *Food security in the Gulf Cooperation Council*. Rand Corporation. www.rand.org/content/dam/rand/pubs/external_publications/EP60000/EP67748/RAND_EP67748.pdf; downloaded 29 April 2023.

Eich, A., Ford, A.K., Nugues, M.M., McAndrews, R.S., Wild, C., and S.C.A. Ferse. 2019. Positive association between epiphytes and competitiveness of the brown algal genus *Lobophora* against corals. *Peer J* 7: e6380. doi:10.7717/peerj.6380.

Erftemeijer, P.L.A., DeGraaff, R., and G. Boot. 2004. Site selection for artificial reefs in Bahrain (Arabian Gulf) based on GIS technology and hydrodynamic modeling. *J Marine Sci Environ* 2: 29–38.

Erftemeijer, P.L.A., Riegl, B., Hoeksema, B.W., and P.A. Todd. 2012. Environmental impacts of dredging and other sediment disturbances on corals: A review. *Mar Pollut Bull* 64, no. 9: 1737–1765.

Fabricius, K., and G. De'ath. 2001. Environmental factors associated with the spatial distribution of crustose coralline algae on the Great Barrier Reef. *Coral Reefs* 19: 303–309.

Fabricius, K., De'ath, G., McCook, L., Turak, E., and D.M. Williams. 2005. Changes in algal, coral and fish assemblages along water quality gradients on the inshore Great Barrier Reef. *Mar Pollut Bull* 51: 384–398.

Fadlallah, Y.H., Eakin, M., Allen, K., Rahim, R., Reaka-Kudla, M., and S. Earle. 1993. *Reef coral distribution and reproduction, community structure, and reef health (Qatar, Bahrain, Saudi Arabia, Kuwait): Results of the RV Mt. Mitchell Cruise*. Proc Scientific Workshop on the Results of the Mt. Mitchell Cruise in the ROPME Sea Area, January, Kuwait, p. 1–26.

Fan, L., Liu, M., Simister, R., Webster, N.S., and T. Thomas. 2013. Marine microbial symbiosis heats up: The phylogenetic and functional response of a sponge holobiont to thermal stress. *ISME J* 7: 991–1002.

Fernandez-Torquemada, Y., Sanchez-Lizaso, J., and J. Gonzalez-Correa. 2005. Preliminary results of the monitoring of the brine discharge produced by the SWRO desalination plant of Alicante. *Desalination* 182, nos. 1–3: 395–402.

Fujita, K., Osawa, Y., Kayanne, H., Ide, Y., and H. Yamano. 2009. Distribution and sediment production of large benthic foraminifers on reef flats of the Majuro Atoll, Marshall Islands. *Coral Reefs* 28: 29–45.

Gattuso, J.P., Frankignoulle, M., Bourge, I., Romaine, S., and R.W. Buddemeier. 1998. Effect of calcium carbonate saturation of seawater on coral calcification. *Glob Planet Change* 18: 37–46.

Gómez-Lemos, L.A., Doropoulos, C., Bayraktarov, E., and G. Diaz-Pulido. 2018. Coralline algal metabolites induce settlement and mediate the inductive effect of epiphytic microbes on coral larvae. *Sci Rep* 8, no. 1: 17557. doi:10.1038/s41598-018-35206-9.

Hoepner, T., and S. Lattemann. 2002. Chemical impacts from seawater desalination plants – A case study of the northern Red Sea. *Desalination* 152: 133–140.

Hume, B., D'Angelo, C., Burt, J., Baker, A.C., Riegl, B., and J. Wiedenmann. 2013. Corals from the Persian/Arabian Gulf as models for thermotolerant reef-builders: Prevalence of clade C3 *Symbiodinium*, host fluorescence and ex situ temperature tolerance. *Mar Pollut Bull* 72: 313–322.

IPCC. 2015. *Climate change 2014: Synthesis report: Contribution of working groups I, II and III to the fifth assessment report of the intergovernmental panel on climate change*. IPCC, Geneva, Switzerland, 151 p.

Jorissen, H., Galand, P.E., Bonnard, I., et al. 2021. Coral larval settlement preferences linked to crustose coralline algae with distinct chemical and microbial signatures. *Sci Rep* 11: 14610. doi:10.1038/s41598-021-94096-6.

Kavousi, J., Tavakoli-Kolour, P., Mohammadizadeh, M., Bahrami, A., and A. Barkhordari. 2014. Mass coral bleaching in the northern Persian Gulf, 2012. *Sci Mar* 78: 397–404.

Khordagui, H. 2002. Power and desalination plant. In: *The Gulf ecosystem: Health and sustainability*. Eds: Khan, et al., 173–191. Leiden: Backhuys Publishers.

Kleypas, J.A., Buddemeier, R.W., Archer, D., Gattuso, J.P., Langdon, C., and B.N. Opdyke. 1999. Geochemical consequences of increased atmospheric carbon dioxide on coral reefs. *Science* 284: 118–120.

Lafferty, K.D., Porter, J.W., and S.E. Ford. 2004. Are diseases increasing in the ocean? *Ann Rev Ecol Evol Syst* 35: 31–54.

Langer, M.R. 2008. Assessing the contribution of foraminiferan protists to global ocean carbonate production. *J Eukaryot Microbiol* 55: 163–169.

Lough, J.M. 2000. 1997–1998: Unprecedented thermal stress to coral reefs? *Geophys Res Lett* 27: 3901–3904. doi:10.1029/2000GL011715.

Lown, A.E., Hepburn, L.J., Heywood, J.L., and T.C. Cameron. 2021. European native oysters and associated species richness in the presence of non-native species in a southern North Sea estuary complex. *Conserv Sci Practice* 3:e361. doi:10.1111/csp2.361

Loya, Y. 1976. Effects of water turbidity and sedimentation. *Bull Mar Sci* 26: 450–466.

Mabrook, B. 1994. Environmental impact of waste brine disposal of desalination plants, Red Sea, Egypt. *Desalination* 97: 453–465.

Macpherson, E., Raventos, N., and A. Garcia-Rubies. 2006. Effects of brine discharge from desalination plant on macrobenthic communities in the NW Mediterranean. *Mar Environ Res* 62, no. 1: 1–14.

Mallela, J. 2013. Calcification by reef-building sclerobionts. *PLoS One* 8: e60010.

Mallela, J., and A.C.T. Perry. 2007. Calcium carbonate budgets for two coral reefs affected by different terrestrial runoff regimes, Rio Bueno, Jamaica. *Coral Reefs* 26: 129–145.

Miri, R., and A. Chouikhi. 2005. Ecotoxicological marine impacts from seawater desalination plants. *Desalination* 182: 403–410.

Naser, H.A. 2012. Evaluation of the environmental impact assessment system in Bahrain. *J Environ Prot* 3: 233–239.

National Geographic. 2022. Bahrain expansion into the sea threatens fisheries. *Environment, national geographic*. www.nationalgeographic.com/environment/article/bahrains-expansion-into-the-sea-threatens-fisheries; downloaded 31 January 2023.

NOAA. 2024. NOAA confirms 4th global coral bleaching event. Last accessed, 21 April 2024 via https://www.noaa.gov/news-release/noaa-confirms-4th-global-coral-bleaching-event

Nugues, M.M., and C.M. Roberts. 2003. Coral mortality and interaction with algae in relation to sedimentation. *Coral Reefs* 22: 507–516.

Orr, J.C., Fabry, V.J., Aumont, O., Bopp, L., Doney, S.C., Feely, R.A., Gnanadesikan, A., Gruber, N., Ishida, A., Joos, F., Key, R.M., Lindsay, K., Maier-Reimer, E., Matear, R., Monfray, P., Mouchet, A., Najjar, R.G., Plattner, G.K., Rodgers, K.B., Sabine, C.L., Sarmiento, J.L., Schlitzer, R., Slater, R.D., Totterdell, I.J., Weirig, M.F., Yamanaka, Y., and A. Yool. 2005. Anthropogenic ocean acidification over the twenty-first century and its impact on calcifying organisms. *Nature* 437: 681–686.

Pandolfi, J.M., Connolly, S.R., Marshall, D.J., and A.L. Cohen. 2011. Acidification projecting coral reef futures under global warming and ocean projecting coral reef futures under global warming and ocean acidification. *Science* 333: 418–422.

Perry, C.T., Alvarez-Filip, L., Graham, N.A.J., Mumby, P.J., Wilson, S.K., Kench, P.S., Manzello, D.P., Morgan, K.M., Slangen, A.B.A., Thomson, D.P., Januchowski-Hartley, F., Smithers, S.G., Steneck, R.S., Carlton, R., Edinger, E.N., Enochs, I.C., Estrada-Saldivar, N., Haywood, M.D.E., Kolodziej, G., Murphy, G.N., Perez-Cervantes, E., Suchley, A., Valentino, L., Boenish, R., Wilson, M., and C.

Macdonald. 2018. Loss of coral reef growth capacity to track future increases in sea level. *Nature* 558: 396–400.

Perry, C.T., and L.J. Hepburn. 2008. Syn-depositional alteration of coral reef framework through bioerosion, encrustation and cementation: Taphonomic signatures of reef accretion and reef depositional events. *Earth-Sci Rev* 86: 106–144.

Purkis, S.J., and B. Riegl. 2005. Spatial and temporal dynamics of Arabian Gulf coral assemblages quantified from remote-sensing and in situ monitoring data. *Mar Ecol Prog Ser* 287: 99–113.

Rasser, M.W., and A.B. Riegl. 2002. Holocene coral reef rubble and its binding agents. *Coral Reefs* 21: 57–72.

Rezai, H., Wilson, S., Claereboudt, M., and B. Riegl. 2004. Coral reef status in the ROPME sea area: Arabian/Persian Gulf, Gulf of Oman and Arabian Sea. In: *Status of coral reefs of the World*. Ed: C. Wilkinson, 155–169. Townsville: Australian Institute of Marine Science.

Riegl, B. 2002. Effects of the 1996 and 1998 positive sea-surface temperature anomalies on corals, coral diseases and fish in the Arabian Gulf (Dubai, UAE). *Mar Biol* 140: 29–40.

Riegl, B. 2003. Climate change and coral reefs: Different effects in two high-latitude areas (Arabian Gulf, South Africa). *Coral Reefs* 22: 433–446.

Ritson-Williams, R., Arnold, S.N., Fogarty, N.D., Steneck, R.S., Vermeij, M.J.A., and V.J. Paul. 2009. New perspectives on ecological mechanisms affecting coral recruitment on reefs. *Smithson Contrib Mar Sci* 38: 437–457.

Roberts, D.A., Johnston, E.L., and N.A. Knott. 2010. Impacts of desalination plant discharges on the marine environment: A critical review of published studies. *Water Res* 44, no. 18: 5117–5128. doi:10.1016/j.watres.2010.04.036.

Rogers, C.S. 1990. Responses of coral reefs and reef organisms to sedimentation. *Mar Ecol Prog Ser* 62: 185–202.

Sheppard, C.R. 1988. *Similar trends, different causes: Responses of corals to stressed environments in Arabian Seas*. Proc 6th International Coral Reef Symp, Townsville, Australia, 3, pp. 297–302.

Sheppard, C.R. 2007. *The status of coral reef in Fasht Al Adhm, Bahrain*. Bahrain: GEOMATEC.

Sheppard, C., Al-Husiani, M., Al-Jamali, F., Al-Yamani, F., Baldwin, R., Bishop, J., Benzoni, F., Dutrieux, E., Dulvy, N.K., Durvasula, S.R.V., Jones, D.A., Loughland, R., Medio, D., Nithyanandan, M., Pilling, G.M., Polikarpov, I., Price, A.R.G., Purkis, S., Riegl, B., Saburova, M., Namin, K.S., Taylor, O., Wilson, S., and K. Zainal. 2010. The Gulf: A young sea in decline. *Mar Pollut Bull* 60: 13–38.

Sheppard, C., and R. Loughland. 2002. Coral mortality and recovery in response to increasing temperature in the southern Arabian Gulf. *Aquat Ecosyst Health Manag* 5: 395–402.

Uwate, K.R., Almeshkhas, J., and H. Almaghani. 1997. *Field observations of coral bleaching in Bahrain in 1996*. Directorate of Fisheries, Ministry of Works and Agriculture, State of Bahrain, 13 p.

Uwate, K.R., Bader, H., Torabi, M., Bailey, S., and P. Keenan. 2000. *Reef check surveys in Bahrain, Arabian Gulf*. Directorate of Fisheries and Marine Resources, Manama, Bahrain.

Uwate, K.R., and A.J. Shams. 1999. *Summary of recent coral and coral reef related activities in Bahrain*. Regional Meeting on the Assessment of the Health of Coral Reefs. Manama, Bahrain, pp. 1–6.

Vermeij, M.J.A., Dailer, M.L., and C.M. Smith. 2011. Crustose coralline algae can suppress macroalgal growth and recruitment on Hawaiian coral reefs. *Mar Ecol Prog Ser* 422: 1–7. doi:10.3354/meps08964.

Vousden, D.H. 1995. *Bahrain marine habitats and some environmental effects on seagrass beds: A study of the marine habitats of Bahrain with particular reference to the effects of water temperature, depth and salinity on seagrass biomass and distribution*. Ph.D. Thesis, University of Wales, Cardiff, Wales, 261 pp.

Vousden, D.H. 1988. *The Bahrain marine habitat survey: A study of the marine environment of Bahrain using remote sensing as a rapid assessment methodology*. Proc ROPME Workshop on Coastal Area Development, p. 3–34. Regional Seas Reports and Studies No. 90/ROPME. Publication No. GC-5/006.

Wilkinson, C. 1998. The 1997–1998 mass bleaching event around the world. In: *Status of coral reefs of the world: 1998*. Ed: C.R. Wilkinson, 15–38. Townsville: Australian Institue of Marine Sciences.

Wolanski, E., and G. De'ath. 2005. Predicting the impact of present and future human land-use on the Great Barrier Reef. *Estuar Coast Shelf Sci* 64: 504–508.

World Bank. 2020. *Data: Land area (sq km)*. Bahrain. https://data.worldbank.org/indicator/AG.LND.TOTL.K2?locations=BH; downloaded 31 January 2023.

World Bank. 2023. *Total fisheries production (metric tons) Bahrain*. Data, Food and Agriculture Organisation. https://data.worldbank.org/indicator/ER.FSH.PROD.MT?locations=BH; downloaded 15 January 2023.

Zainal, A., Dalby, D., and I. Robinson. 1993. Monitoring marine ecological changes on the east coast of Bahrain with Landsat TM. *Photogramm Eng Remote Sens* 59: 415–421.

Zainal, K., Al-Madany, I.M., Al-Sayyed, H., Khamis, A., Al Shuhaby, S., Al Hisaby, A., Elhoussiny, W., and E. Khalaf. 2012. The cumulative impacts of reclamation and dredging on the marine ecology and land-use in the Kingdom of Bahrain. *Mar Pollut Bull* 64: 1452–1458.

Zainal, K., Al-Sayed, H., and I. Al-Madany. 2008. Coastal pollution in Bahrain and its management. In: *Protecting the Gulf's marine ecosystems from pollution*. Eds: A.H. Abuzinada, H.-J. Barth, F. Krupp, B. Böer, and T.Z. Abdessalaam, 147–162. Basel: Birkhäuser.

PART 2

Fish

Mesopelagic fishes in the Red Sea

8

Stein Kaartvedt, Svenja Christiansen and Anders Røstad

8.1 THE MESOPELAGIC ZONE, FISHES AND DIEL VERTICAL MIGRATION

The mesopelagic, or the ocean twilight zone, is defined as the depth interval from 200 to 1000 m. There, the light intensity is too low for photosynthesis, but fishes with specially adapted eyes can utilize the weak remaining light (Warrant and Locket 2004). Below 1000 m depth sunlight generally does not penetrate. Kaartvedt et al. (2019a) related the twilight zone to light levels rather than a certain depth interval. In their analysis, the light intensities defining the twilight zone would range from 10^{-1} to 10^{-9} mol quanta m^{-2} s^{-1}. The highest intensity corresponds to empirical findings on the upper light exposure of pearlsides (*Maurolicus* spp.), which have unique eyes adapted to higher light intensities than the deeper-living lanternfishes (de Busserolles et al. 2017). The lowest intensity of this twilight habitat corresponds to the visual threshold of lanternfishes (Myctophidae) (Denton 1990). By this definition, the daytime twilight habitat will vary relative to water clarity, extending deeper than 1000 m in very clear oceanic water, while it may even be mostly located above 200 m in very murky coastal waters (Kaartvedt et al. 2019a).

Mesopelagic fishes prevail in all oceans, with an estimated global abundance of 1–10 billion tonnes (Gjøsaeter and Kawaguchi 1980; Irigoien et al. 2014). Abundance, vertical distribution and diel vertical migrations are commonly studied using echosounders as mesopelagic fishes form acoustic scattering layers (SLs) (e.g., Irigoien et al. 2014; Klevjer et al. 2016; Proud et al. 2017). Such layers may contain a high diversity of fishes, but also invertebrate taxa (e.g., Olivar et al. 2017). Mesopelagic SLs often consist of several sub-layers, which may – or may not – reflect different species and size distributions (Ariza et al. 2016; Christiansen et al. 2021; Giske et al. 1990). The SLs' daytime vertical distribution varies throughout the world's oceans, related to factors such as oxygen and light levels (Aksnes et al. 2017; Bianchi et al. 2013; Klevjer et al. 2016).

On a global scale, organisms forming about 50% of the mesopelagic backscatter appear to carry out diel vertical migration (DVM; Klevjer et al. 2016), migrations explained (as for other organisms) by the trade-off between access to food in shallow waters and predator avoidance at depth (e.g., Pearre 2003). The degree of DVM varies geographically. Some SLs remain at depth throughout the diel cycle, with persistent deep layers repeatedly ascribed to *Cyclothone* spp. (e.g., Olivar et al. 2012; Peña et al. 2020). In some locations, entire SLs ascend at night and descend at day (Klevjer et al. 2016). The degree of DVM may also vary seasonally. At high latitudes, the nocturnal ascent typically fades during winter (Dypvik et al. 2012; Staby et al. 2011). At this time, there is reduced access to food in upper waters and seasonally migrating prey (copepods) appear at depth. Both these circumstances would diminish the reward of ascent, affecting the trade-off between foraging and predator avoidance.

8.2 THE UNIQUE RED SEA ENVIRONMENT – THE WORLD'S WARMEST TWILIGHT ZONE

The (northern) Red Sea is an oligotrophic, clear water ocean, with a vertically extended mesopelagic zone in terms of light intensities (Klevjer et al. 2012; Røstad et al. 2016). The Red Sea is especially warm, and therefore of particular interest relative to coral reefs and climate change (Voolstra et al. 2021). However, the most unique temperature characteristic of this ocean is the warm deep waters, being about 21.5°C down to the bottom > 2000 m. This makes the Red Sea by far the warmest deep-sea environment in the world, with implied high respiration rates and requirements for food. The Red Sea has particularly low concentrations of zooplankton in deep waters (Dypvik and Kaartvedt 2013; Weikert 1982; Wishner 1980). Moreover, there is an oxygen minimum zone with lowest values of 0.5 ml O_2 l^{-1} at depths of ~350–450 m (Klevjer et al. 2012). In sum, mesopelagic fishes of the Red Sea must cope with particularly warm waters, particularly low concentrations of prey at depth as well as possible oxygen limitations. There are synergistic effects, as the warm waters would enhance both digestion of prey and utilization of oxygen (Seibel et al. 2016).

DOI: 10.1201/9781003321392-10

8.3 THE RED SEA MESOPELAGIC FISH FAUNA

In contrast to other tropical waters, the Red Sea mesopelagic fauna contains few fish species. Johnson and Feltes (1984) only referred to eight species, a very small fraction of the ~300 deep-water pelagic fish species estimated to exist in the Indian Ocean. Correspondingly, the Red Sea mesopelagic fish diversity is much lower than in the adjacent Gulf of Aden (Dalpadado and Gjøsaeter 1987, and references therein). In describing a new species of *Vinciguerria* (*mabahiss*) from the Red Sea, Johnson and Feltes (1984) chose the title "*Vinciguerria* (Salmoniformes: Photichthyidae) from the Red Sea and Gulf of Aqaba, with comments on the depauperacy of the Red Sea mesopelagic fish fauna," i.e., underlining the poor species richness of the ocean. Yet, this ocean houses both *Maurolicus* sp. and lanternfishes, i.e., fishes that span the full vertical domain of the twilight zone (cf., the definition of the twilight zone relative to light and species distribution; Kaartvedt et al. 2019a).

The environment affects growth rates and maximum size of the fishes. Generally, growth rate increases, while size at maturity and maximum size decreases with increasing temperatures (reviewed by Ohlberger 2013). However, the oligotrophic and extremely warm Red Sea environment might limit growth rates, and Aldanondo et al. (2023) estimated lower somatic growth rates than reported for other mesopelagic fishes at similar tropical and subtropical latitudes. Dalpadado and Gjøsæter (1987) found size at first maturity of Red Sea individuals of the mesopelagic fish *Maurolicus muelleri*, now classified as *M. mucronatus* (de Busserolles et al. 2017) at ~20 mm, which is less than recorded for *Maurolicus* in colder waters (e.g., Gjøsæter 1981). Trawl catches of the common *Benthosema pterotum* might at first suggest they mature at a larger size in the Red Sea than individuals from other oceans, but this may well relate to methods (Dalpadado and Gjøsaeter 1987).

There is some seasonality both in terms of temperature and production in the Red Sea, with a yearly winter bloom (Gittings et al. 2021). Isari et al. (2017) addressed the presence of fish larvae in the Red Sea, among which *Vinciguerria* sp. and *Benthosema* spp. were the most common mesopelagic species. The larvae of *Vinciguerria* sp. (likely the endemic *V. mabahiss*) displayed a gradual increase in abundances from January to April, with lower abundance during the rest of the year. Larvae of *Benthosema* were recorded throughout most of the year, except for a few months in mid-summer (July and August).

8.3.1 Red Sea mesopelagic scattering layers

While low in diversity, mesopelagic SLs are prevalent throughout the Red Sea wherever the bottom topography (depth) permits (Klevjer et al. 2012). A big uncertainty worldwide is the contribution from targets other than fish to the backscatter, with particular focus on gas-bearing siphonophores. Gas-filled chambers – like the swim bladder of fishes – may represent more than 90% of the backscatter (Proud et al. 2019). However, for the Red Sea it appears that mesopelagic SLs indeed are primarily composed of fishes (Kaartvedt et al. 2019b).

The mesopelagic SLs in the Red Sea span from ~200–800 m, roughly distributed in three main layers, though the layer structure may vary somewhat (Figures 8.1 and 8.2). The shallowest layer appears centred at ~200 m during mid-day; a second layer at ~500 m, while the deepest spans down to 600–800 m (Dypvik and Kaartvedt 2013; Kaartvedt et al. 2019b; Klevjer et al. 2012). Since the vertical distributions are related to light, there is some variation related to changing weather and water clarity (Kaartvedt et al. 2017; Røstad et al. 2016). The literature and our own unpublished results suggest that *Maurolicus mucronatus* forms the shallow-most layer (previously called *M. muelleri*; Dalpadado and Gjøsaeter 1987). *Vinciguerria mabahiss* is a main component in the central layer, with the possible inclusion of some *Benthosema pterotum* (Dypvik and Kaartvedt 2013). *B. pterotum* prevails in the deepest layer (Dypvik and Kaartvedt 2013). Other species may contribute to the backscatter, though to what extent is unknown.

8.4 THE DVM BEHAVIOUR IN THE RED SEA

The organisms of the upper layer ascribed to *Maurolicus* respond to changing light throughout the day, with the deepest distribution mid-day. There normally are some faint traces of dusk and dawn ascents, but this layer never moves all the way to the surface at night (Kaartvedt et al. 2017, 2019b; Klevjer et al. 2012). This pattern represents an interesting deviation from the dusk and dawn migrations characterizing *M. muelleri* in Norwegian fjords (Christiansen et al. 2021; Staby et al. 2011). Throughout much of the year, that species migrates to the surface of the fjord, foraging in surface waters for short periods at dusk and dawn (~40 ±20 min), with a "midnight sinking" in between. Such behaviour is ascribed to its special eye, adapted for vision at twilight (also the case for *M. mucronatus*), but not in darkness (Christiansen et al. 2019; de Busserolles et al. 2017). Kaartvedt et al. (2022) suggested that the discrepancy between the sites can be related to the duration of the surface twilight periods. During the equinoxes, twilight periods last for about double the time at 61°N (Norwegian fjord) than at 22°N (Red Sea study site). The rapid shifts between day and night may prevent *M. mucronatus* from exploiting shallow Red Sea food sources in twilight. If correct, the "antipredation windows" (c.f., Clark and Levy 1988) in near-surface waters at dusk and dawn – i.e., periods with some light for foraging, yet still relatively safe from predators – are less available niches for mesopelagic fishes at low latitudes. This would suggest other/additional strategies for foraging, which remain to be established (Kaartvedt et al. 2022).

In contrast, basically the whole populations of the fishes forming the two deeper Red Sea layers ascend towards upper waters at night, regardless of time of year (Figure 8.2). Additional examples from other times and parts of the Red Sea are given by Dypvik and Kaartvedt (2013), Kaartvedt et al. (2022) and Klevjer et al. (2012). The species that appear segregated at daytime may mix at night. Yet, the layer ascribed to *Vinciguerria* occasionally spends a very short time in the uppermost metres. Kaartvedt et al. (2022) reported on very

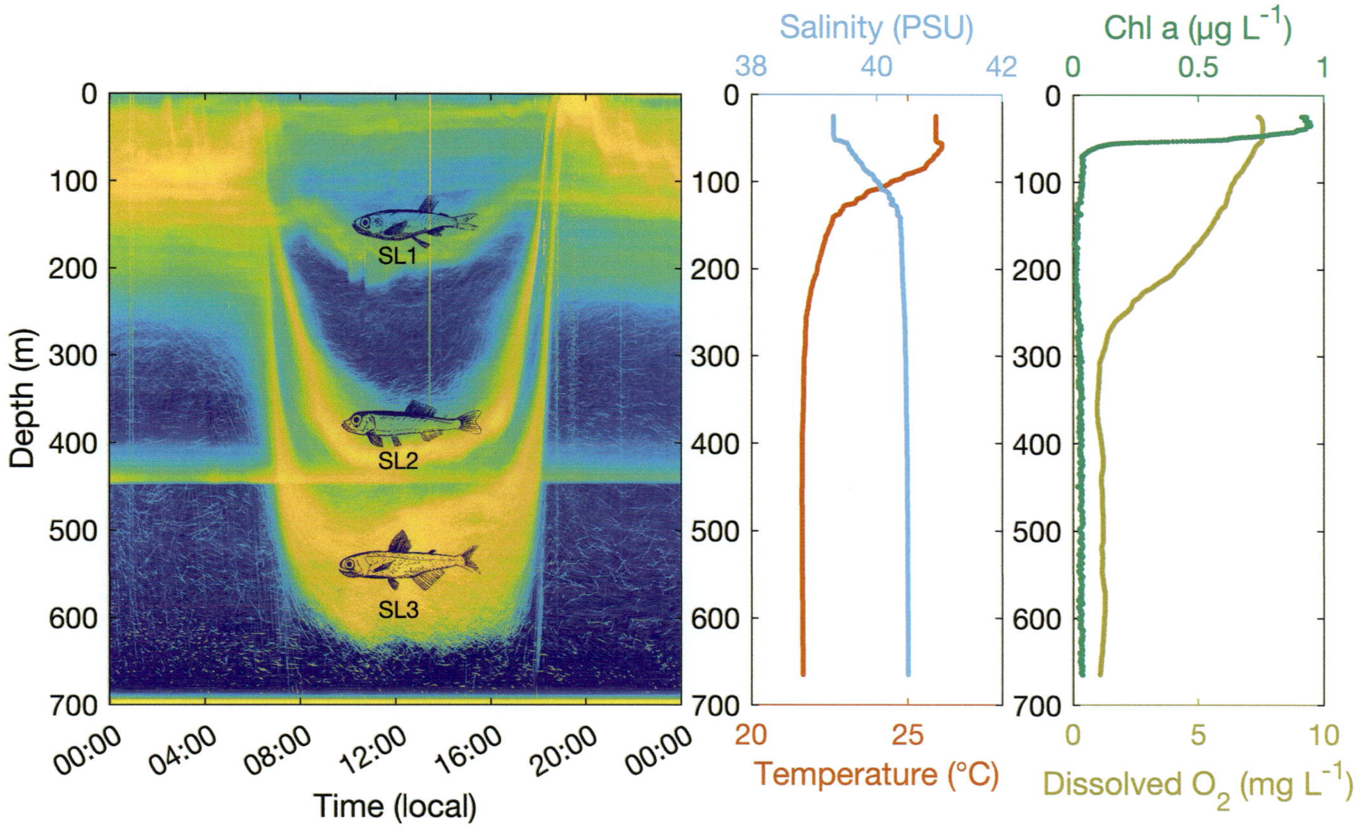

FIGURE 8.1 Diel echogram and environmental data (January 2014) collected in a ~700-m-deep basin located near King Abdullah University of Science and Technology (22.5°N, 39.03°E). The acoustic data are tentatively dominated by *Maurolicus mucronatus* (SL1), *Vinciguerria mabahiss* (SL 2) and *Benthosema pterotum* (SL3). The acoustic data were obtained by an upward-looking, bottom-mounted Simrad EK 60 38 kHz echosounder, methods further described by Kaartvedt et al. (2017). The horizontal line at ~450 m and vertical spikes in the backscatter originating at the surface are noise. The sampling location and collection of environmental data are described by Wiebe et al. (2016). The oxygen profile in this enclosed basin slightly deviates from the open Red Sea, where oxygen contents tend to increase by depth below a minimum zone at ~350–450 m (see Klevjer et al. 2012; Kaartvedt et al. 2020).

rapid and short-lasting darts to surface waters in the afternoon, the fish occasionally spending only about 10 minutes at the surface, before rapidly descending. *Vinciguerria* superficially resembles *Maurolicus* spp. but is adapted to lower light levels. Kaartvedt et al. (2022) speculated that the combination of rapid switches between day and night, and the later arrival of dark-adapted predators to upper waters, might explain the striking behaviour at dusk.

Although flexible and peculiar patterns occur at dusk and dawn, otherwise invariant migrations to the upper layer every night may relate to the combination of low access to prey in deep waters and the unusually warm waters. Dypvik and Kaartvedt (2013) found that *B. pterotum* was foraging on zooplankton in upper waters throughout the night, while rapidly digesting the preceding nocturnal meal in the warm mesopelagic daytime habitat. The constrained epipelagic distribution of zooplankton and the unusually warm waters of the Red Sea likely forced the whole population to ascend and feed in epipelagic waters every night, as the prey eaten at night becomes fully digested at mesopelagic depths during daytime. In this sense, the extreme Red Sea helps researchers to understand migrations in other parts of the world's oceans with other vertical profiles of prey and temperature, and where migrations differ.

The twilight zone organisms experience a different set of light conditions than only related to day and night. Their "light comfort zones" (c.f., Røstad et al. 2016) move vertically throughout the diel cycle, but also within day and night. In the Red Sea (as elsewhere), the fishes' daytime distributions adjust relative to weather, with associated changes in light that may vary by one to two orders of magnitude (Kaartvedt et al. 2017). Nocturnal light levels may vary even more, with surface light spanning five orders of magnitude from a bright, full moon to an overcast night (de Busserolles et al. 2017; Ryer and Olla 1999). Constraints of upward migrations by mesopelagic scattering layers at full moon have been repeatedly documented worldwide (Benoit-Bird et al. 2009; Prihartato et al. 2016; Wang et al. 2014), yet there appears to have been little analysis of how varying nocturnal light may affect fishes geographically. The Red Sea is characterized by clear water, and it stretches through a region with generally clear skies, suggesting on average higher nocturnal light levels than in many other oceanic regions. Some fishes may be constrained by too high light (like at full moon), whereas others may be constrained by too low light in dark nights, depending on their eye sensitivities. The role of nocturnal light in the success of mesopelagic fishes remains to be established.

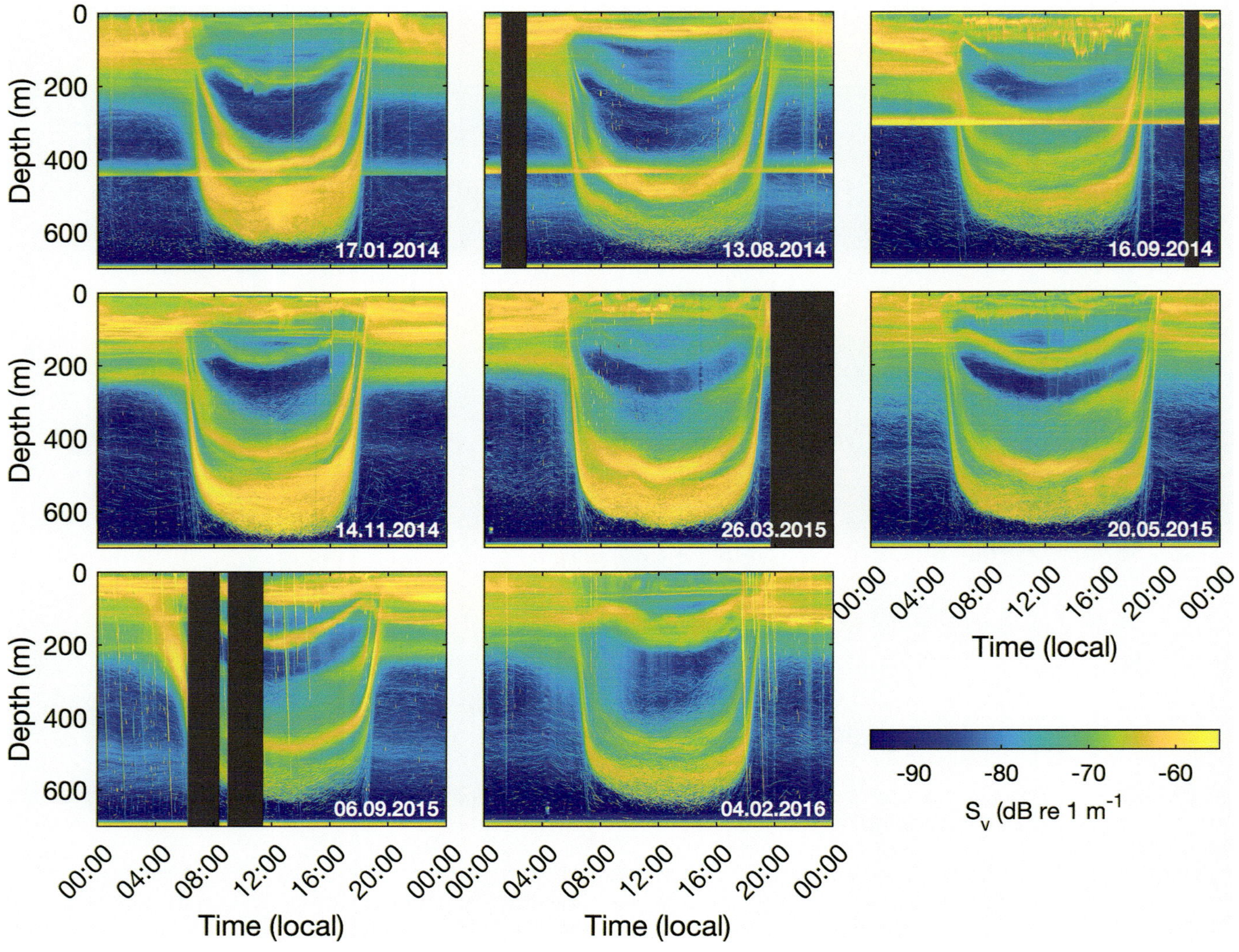

FIGURE 8.2 Diel echograms collected during different times of the year and in different years. Occasional horizontal lines and vertical spikes in the backscatter originating at the surface are noise. Methods and location are the same as for Figure 8.1.

8.4.1 Daytime activity

Reports on torpid behaviour (Barham 1971; Kaartvedt et al. 2009) have led to an apparent notion that vertically migrating mesopelagic fishes are relatively inactive at day (Belcher et al. 2020; Boswell 2020; Catul et al. 2011; Torres et al. 1979). However, this does not apply to the Red Sea, where vertically migrating mesopelagic fishes are actively swimming in deep water throughout the day (Sobradillo et al. 2022). Swimming at mesopelagic depth affects respiration (Torres et al. 1979). Currently, there is much interest in the possible contribution from vertically migrating mesopelagic fishes for the active carbon pump. In current models, estimates of fish mediated vertical carbon export are particularly sensitive to respiration (Davison et al. 2013; Saba et al. 2021), and hence their activity levels in the mesopelagic zone.

The fishes that migrate vertically every day from the upper, productive to mesopelagic layers contribute to the vertical flux of organic matter, fuelling microbial production and respiration in the twilight zone. Estimates from the Red Sea suggest that daytime excretion by the migrating fishes provides an important source of labile dissolved organic matter, generating a hotspot for heterotrophic prokaryotes in the twilight zone (Calleja et al. 2018; Morán et al. 2022). Current modelling efforts aim at assessing the role of mesopelagic fishes in the Red Sea ecosystem.

8.5 MESOPELAGIC FISHES AS PREDATORS AND PREY (ROLE IN THE FOOD CHAIN)

Worldwide, there is increasing interest in the role of mesopelagic fishes in the food chain, as high abundances suggest their importance as predators and prey, including for charismatic macrofauna (Catul et al. 2011; Irigoien et al. 2014; St. John et al. 2016). The fishes forming the acoustic scattering layers are small zooplankton feeders. Movements affect interaction with prey as well as predators (Gerritsen and Strickler 1977; O'Brien et al. 1990). In the Red Sea, the low zooplankton concentration in deep water as well as stomach contents suggest little foraging during daytime (Dypvik and Kaartvedt 2013). Scant available data suggest that foraging of the fishes of the upper mesopelagic layer peaks in the morning and late afternoon (Dalpadado and Gjøsaeter 1987). Fishes from deeper waters forage during their nocturnal migrations to upper waters. Copepods are the main prey for the mesopelagic fishes (Dalpadado and Gjøsaeter 1987; Dypvik and Kaartvedt 2013), though with an ontogenetic shift towards larger prey items such as euphausiids, amphipods and fishes (Dalpadado and Gjøsaeter 1987).

Little is known about the role of mesopelagic fishes as prey in the Red Sea food chain, and any evidence so far is circumstantial. The use of satellite telemetry assessing diving behaviour and vertical habitat use of sharks (including whale sharks) and manta rays suggests their link between epi- and mesopelagic layers in the Red Sea (Berumen et al. 2014; Braun et al. 2014; Spaet et al. 2017). Kaartvedt et al. (2019b) observed a shark voraciously foraging on accumulations of mesopelagic fish caused by artificial light in near-bottom waters, but due to the unnatural conditions, this only suggests potential predators being present. Long-term acoustic studies in near-bottom waters (at 550–800 m) indeed suggest little interaction with predators at depth (Sobradillo et al. 2022; Kaartvedt and Røstad, unpublished). This contrasts with results from studies in a Norwegian fjord, where acoustic observations of interactions between mesopelagic fish and their potential predators are repeatedly recorded (Christiansen et al. 2021; Godø et al. 2014; Kaartvedt et al. 2009, 2012). Squids may be central predators in upper waters at night (Kaartvedt et al. 2020, 2022), but again their potential interaction with the mesopelagic fish remains to be established.

Apart from behavioural observations and stomach data (and other data on foraging), assessing the roles of mesopelagic fishes as predators and prey requires abundance estimates of the fishes. The highest abundance estimates of mesopelagic fishes globally are based on the use of echosounders (Davison et al. 2015; Gjøsaeter and Kawaguchi 1980; Irigoien et al. 2014), which provide an easy way for large-scale mapping. The use of echosounders in assessing the abundance of mesopelagic fishes is normally associated with high levels of uncertainties, due to unknown acoustic properties and target identities, including the relative contribution of fishes to the backscatter (Irigoien et al. 2014; Proud et al. 2019). Yet, assessing the abundance of mesopelagic fishes in the Red Sea by acoustic mapping should be an achievable task. The Red Sea scattering layers are largely formed by fishes (Kaartvedt et al. 2019b), the diversity is low (e.g., Dalpadado and Gjøsaeter 1987; Johnson and Feltes 1984), the acoustic backscatter of individual fish (target strength) can be assessed in situ (Sobradillo et al. 2022), and the backscatter of the small mesopelagic fishes is little affected by behaviour (own unpublished results).

8.6 MESOPELAGIC FISHES AND CLIMATE CHANGE

In various settings it has been argued that the unusually warm and rapidly warming waters of the Red Sea (Chaidez et al. 2017) make this ocean a suitable laboratory for studying the effects of climate change (Kleinhaus et al. 2020). That notion cannot be applied similarly for the mesopelagic fishes, as the temperatures of the deep waters of the Red Sea markedly exceed those of any realistic temperature change in the world's twilight zone. Such high temperatures in deep water are made possible by the presence of a shallow sill (137 m) at the opening of the Red Sea to the Indian Ocean (e.g., Zarubin et al. 2017), largely confining the deep waters in this warm region from the cold waters of the global deep ocean. Nonetheless, the Red Sea may teach us at least two lessons. It suggests a direction towards lower diversity among mesopelagic fishes. There is no comparable environment worldwide. The Mediterranean is another warm (yet much cooler than the Red Sea) deep ocean. The mesopelagic fish community of the Mediterranean is characterized by relatively low diversity compared to other oceanic regions of the world (Olivar et al. 2012). High diversity among mesopelagic fishes may otherwise occur in tropical waters, including oligotrophic regions (Olivar et al. 2017). The Red Sea also teaches us that at least some species of mesopelagic fish may thrive under extreme, warm conditions, far above what will occur in the future World Ocean. It moreover offers the possibility to address interactions between low oxygen and high temperatures. These are pertinent questions as the oceans' oxygen minimum zones are expanding and minimum oxygen concentrations declining (Deutsch et al. 2011; Seibel et al. 2016; Stramma et al. 2011).

The low oxygen zones are key predictors for the depth of mesopelagic scattering layer migration depth on a global scale (Bianchi et al. 2013; Klevjer et al. 2016), although alternative explanations to hypoxia *per se*, such as associated reduction of light penetration, may apply (Aksnes et al. 2017). Mesopelagic fishes appear to have a high tolerance for hypoxia (Childress and Seibel 1998; Olivar et al. 2017; Seibel 2011), and some species are very abundant in the hypoxic Arabian Sea (Gjøsaeter and Kawaguchi 1980). Klevjer et al. (2012) found no evidence that low oxygen contents constrained the vertical distribution of the Red Sea mesopelagic SLs (see also Figure 8.1). The Red Sea mesopelagic acoustic layers appear above, within and below a mesopelagic oxygen minimum zone (OMZ; Kaartvedt et al. 2020; Klevjer et al. 2012). Nevertheless, the hypoxic waters can affect interactions between predators and prey. OMZs can provide refuges for predators during daytime (Wishner et al. 2018), but can also force the fishes into shallower waters, making them more vulnerable to visual predators (Klevjer et al. 2016; Koslow et al. 2011; Netburn and Koslow 2015).

The mesopelagic fishes of the Red Sea inhabit an intermediate oxygen minimum zone, but one in which the temperature at depth is much higher than elsewhere (Seibel et al. 2016). High temperature will elevate oxygen demand, and Seibel et al. (2016) argued that the combination of high temperature and moderate hypoxia is more physiologically challenging than much lower oxygen levels at low temperature. A primary mechanism facilitating low oxygen tolerance is an ability to dramatically reduce energy expenditure while in low oxygen waters. Accordingly, Red Sea euphausiids occupying the oxygen minimum zone reduced aerobic metabolism by more than 50% during exposure to hypoxia (Seibel et al. 2016). Yet, the mesopelagic fishes of the deepest scattering layer in the Red Sea appeared more active than reports on torpid individuals from elsewhere suggest (Barham 1971; Sobradillo et al. 2022). This calls for further understanding of how the mesopelagic fishes of the Red Sea accommodate to their exceptional environment.

ACKNOWLEDGEMENT

The fieldwork providing data for Figures 8.1 and 8.2 was funded by King Abdullah University of Science and Technology (KAUST), Saudi Arabia. Stein Kaartvedt and Svenja Christiansen were funded by the SUMMER project within the EU Horizon 2020 research and innovation programme (grant agreement number 817806; Sustainable Use of Mesopelagic Resources) during the preparation of the manuscript.

REFERENCES

Aksnes, D.L., Røstad, A., Kaartvedt, S., Martinez, U., Duarte, C.M., and X. Irigoien. 2017. Light penetration structures the deep acoustic scattering layers in the global ocean. *Sci Adv* 3, no. 5: e1602468. doi:10.1126/sciadv.1602468.

Aldanondo, N., Kaartvedt, S., and X. Irigoien. 2023. Growth patterns of two Red Sea mesopelagic fishes. *Mar Biol* 170: Article 8. doi:10.1007/s00227-022-04144-6.

Ariza, A., Landeira, J.M., Escánez, A., Wienerroither, R., Aguilar de Soto, N., Røstad, A., Kaartvedt, S., and S. Hernández-León. 2016. Vertical distribution, composition and migratory patterns of acoustic scattering layers in the Canary Islands. *J Mar Syst* 157: 82–91. doi:10.1016/j.jmarsys.2016.01.004.

Barham, E.G. 1971. Deep-sea fishes: Lethargy and vertical orientation. In: *Proceedings of international symposium on biological sound scattering in the ocean*. Ed: G.B. Farquhar, 100–118. Washington, DC: Maury Center for Ocean Science.

Belcher, A., Cook, K., Bondyale-Juez, D., Stowasser, G., Fielding, S., Saunders, R.A., Mayor, D.J., and G.A. Tarling. 2020. Respiration of mesopelagic fish: A comparison of respiratory Electron Transport System (ETS) measurements and allometrically calculated rates in the Southern Ocean and Benguela Current. *ICES J Mar Sci* 77, no. 5: 1672–1684. doi:10.1093/icesjms/fsaa031.

Benoit-Bird, K.J., Au, W.W.L., and D.W. Wisdoma. 2009. Nocturnal light and lunar cycle effects on diel migration of micronekton. *Limnol Oceanogr* 54, no. 5: 1789–1800. doi:10.4319/lo.2009.54.5.1789.

Berumen, M.L., Braun, C.D., Cochran, J.E.M., Skomal, G.B., and S.R. Thorrold. 2014. Movement patterns of juvenile whale sharks tagged at an aggregation site in the Red Sea. *PLoS One* 9, no. 7: e103536. doi:10.1371/journal.pone.0103536.

Bianchi, D., Galbraith, E.D., Carozza, D.A., Mislan, K.A.S., and C.A. Stock. 2013. Intensification of open-ocean oxygen depletion by vertically migrating animals. *Nature Geosci* 6, no. 7: 545–548. doi:10.1038/ngeo1837.

Boswell, K.M. 2020. Oceanographic structure and light levels drive patterns of sound scattering layers in a low-latitude oceanic system. *Front Mar Sci* 7: 51. doi:10.3389/fmars.2020.00051.

Braun, C.D., Skomal, G.B., Thorrold, S.R., and M.L. Berumen. 2014. Diving behavior of the reef manta ray links coral reefs with adjacent deep pelagic habitats. *PLoS One* 9, no. 2: e88170. doi:10.1371/journal.pone.0088170.

Calleja, M.L., Ansari, M.I., Røstad, A., Silva, L., Kaartvedt, S., Irigoien, X., and X.A. Morán. 2018. The Mesopelagic scattering layer: A hotspot for heterotrophic prokaryotes in the Red Sea twilight zone. *Front Mar Sci* 5: 259. doi:10.3389/fmars.2018.00259.

Catul, V., Gauns, M., and P.K. Karuppasamy. 2011. A review on mesopelagic fishes belonging to family Myctophidae. *Rev Fish Biol Fish* 21, no. 3: 339–354. doi:10.1007/s11160-010-9176-4.

Chaidez, V., Dreano, D., Agusti, S., Duarte, C.M., and I. Hoteit. 2017. Decadal trends in Red Sea maximum surface temperature. *Sci Rep* 7, no. 1: 8144. doi:10.1038/s41598-017-08146-z.

Childress, J.J., and B.A. Seibel. 1998. Life at stable low oxygen levels: Adaptations of animals to oceanic oxygen minimum layers. *J Exp Biol* 201: 1223–1232.

Christiansen, S., Klevjer, T.A., Røstad, A., Aksnes, D.L., and S. Kaartvedt. 2021. Flexible behaviour in a mesopelagic fish (*Maurolicus muelleri*). *ICES J Mar Sci* 78, no. 5: 1623–1635. doi:10.1093/icesjms/fsab075.

Christiansen, S., Titelman, J., and S. Kaartvedt. 2019. Nighttime swimming behavior of a mesopelagic fish. *Front Mar Sci* 6, no. 787: 1–12. doi:10.3389/fmars.2019.00787.

Clark, C.W., and D.A. Levy. 1988. Diel vertical migrations by juvenile sockeye salmon and the antipredation window. *Am Nat* 131, no. 2: 271–290. doi:10.1086/284789.

Dalpadado, P., and J. Gjøsaeter. 1987. Observations on mesopelagic fish from the Red Sea. *Mar Biol* 96: 173–183.

Davison, P.C., Checkley, D.M., Koslow, J.A., and J. Barlow. 2013. Carbon export mediated by mesopelagic fishes in the northeast Pacific Ocean. *Prog Oceanogr* 116: 14–30. doi:10.1016/j.pocean.2013.05.013.

Davison, P.C., Koslow, J.A., and R.J. Kloser. 2015. Acoustic biomass estimation of mesopelagic fish: Backscattering from individuals, populations, and communities. *ICES J Mar Sci* 72, no. 5: 1413–1424. doi:10.1093/icesjms/fsv023.

de Busserolles, F., Cortesi, F., Helvik, J.V., Davies, W.I.L., Templin, R.M., Sullivan, R.K.P., Michell, C.T., Mountford, J.K., Collin, S.P., Irigoien, X., Kaartvedt, S., and J. Marshall. 2017. Pushing the limits of photoreception in twilight conditions: The rod-like cone retina of the deep-sea pearlsides. *Sci Adv* 3, no. 11: eaao4709. doi:10.1126/sciadv.aao4709.

Denton, E.J. 1990. Light and vision at depths greater than 200 metres. In: *Light and life in the sea*. Eds: P.J. Herring, A.K. Campbell, M. Whitfield, and L. Maddock, 127–148. Cambridge: Cambridge University Press.

Deutsch, C., Brix, H., Ito, T., Frenzel, H., and L. Thompson. 2011. Climate-forced variability of ocean hypoxia. *Science* 333, no. 6040: 336–339. doi:10.1126/science.1202422.

Dypvik, E., and S. Kaartvedt. 2013. Vertical migration and diel feeding periodicity of the skinnycheek lanternfish (*Benthosema pterotum*) in the Red Sea. *Deep Sea Res Part I: Oceanogr Res Pap* 72: 9–16. doi:10.1016/j.dsr.2012.10.012.

Dypvik, E., Røstad, A., and S. Kaartvedt. 2012. Seasonal variations in vertical migration of glacier lanternfish, *Benthosema glaciale*. *Mar Biol* 159, no. 8: 1673–1683. doi:10.1007/s00227-012-1953-2.

Gerritsen, J., and J.R. Strickler. 1977. Encounter probabilities and community structure in zooplankton: A mathematical model. *J Fish Res Board Can* 34, no. 1: 73–82. doi:10.1139/f77-008.

Giske, J., Aksnes, D.L., Baliño, B.M., Kaartvedt, S., Lie, U., Nordeide, J.T., Salvanes, A.G.V., Wakili, S.M., and A. Aadnesen. 1990. Vertical distribution and trophic interactions of zooplankton and fish in Masfjorden, Norway. *Sarsia* 75, no. 1: 65–81. doi:10.1080/00364827.1990.10413442.

Gittings, J.A., Raitsos, D., Brewin, R.J., and I. Hoteit. 2021. Links between phenology of large phytoplankton and fisheries in the northern and central Red Sea. *Remote Sens* 13, no. 2: 231.

Gjøsæter, J. 1981. Life history and ecology of *Maurolicus muelleri* (Gonostomatidae) in Norwegian waters. *Serie Havundersokelser–Fiskeridirektoratets Skrifter* 17, no. 3: 109–131.

Gjøsaeter, J., and K. Kawaguchi. 1980. *A review of the world resources of mesopelagic fish*, vol. 193. Rome, Italy: Food and Agriculture Organization of the United Nations.

Godø, O.R., Handegard, N.O., Browman, H.I., Macaulay, G.J., Kaartvedt, S., Giske, J., Ona, E., Huse, G., and E. Johnsen. 2014. Marine Ecosystem Acoustics (MEA): Quantifying processes in the sea at the spatio-temporal scales on which they occur. *ICES J Mar Sci* 71, no. 8: 2357–2369. doi:10.1093/icesjms/fsu116.

Irigoien, X., Klevjer, T.A., Røstad, A., Martinez, U., Boyra, G., Acuña, J.L., Bode, A., Echevarria, F., Gonzalez-Gordillo, J.I., Hernandez-Leon, S., Agusti, S., Aksnes, D.L., Duarte, C.M., and S. Kaartvedt. 2014. Large mesopelagic fishes biomass and trophic efficiency in the open ocean. *Nat Comm* 5, no. 3271: 1–10. doi:10.1038/ncomms4271.

Isari, S., Pearman, J.K., Casas, L., Michell, C.T., Curdia, J., Berumen, M.L., and X. Irigoien. 2017. Exploring the larval fish community of the central Red Sea with an integrated morphological and molecular approach. *PLoS One* 12, no. 8: e0182503. doi:10.1371/journal.pone.0182503.

Johnson, R.K., and R.M. Feltes. 1984. *A new species of Vinciguerria (Salmoniformes, Photichthyidae) from the Red Sea and Gulf of Aqaba, with comments on the depauperacy of the Red Sea mesopelagic fish fauna.* Chicago, IL: Field Museum of Natural History.

Kaartvedt, S., Christiansen, S., Røstad, A., and D. Aksnes. 2022. Mesopelagic fishes in a hurry at low latitudes. *Mar Ecol Prog Ser* 694: 149–156. doi:10.3354/meps14114.

Kaartvedt, S., Langbehn, T.J., and D.L. Aksnes. 2019a. Enlightening the ocean's twilight zone. *ICES J Mar Sci* 76, no. 4: 803–812. doi:10.1093/icesjms/fsz010.

Kaartvedt, S., Røstad, A., and D.L. Aksnesa. 2017. Changing weather causes behavioral responses in the lower mesopelagic. *Mar Ecol Prog Ser* 574: 259–263. doi:10.3354/meps12185.

Kaartvedt, S., Røstad, A., Christiansen, S., and T.A. Klevjer. 2020. Diel vertical migration and individual behavior of nekton beyond the ocean's twilight zone. *Deep Sea Res Part I: Oceanogr Res Pap* 160: 103280. doi:10.1016/j.dsr.2020.103280.

Kaartvedt, S., Røstad, A., Klevjer, T., and A. Staby. 2009. Use of bottom-mounted echo sounders in exploring behavior of mesopelagic fishes. *Mar Ecol Prog Ser* 395: 109–118. doi:10.3354/meps08174.

Kaartvedt, S., Røstad, A., Opdal, A.F., and D.L. Aksnes. 2019b. Herding mesopelagic fish by light. *Mar Ecol Prog Ser* 625: 225–231. doi:10.3354/meps13079.

Kaartvedt, S., Staby, A., and D. Aksnes. 2012. Efficient trawl avoidance by mesopelagic fishes causes large underestimation of their biomass. *Mar Ecol Prog Ser* 456: 1–6. doi:10.3354/meps09785.

Kleinhaus, K., Al-Sawalmih, A., Barshis, D.J., Genin, A., Grace, L.N., Hoegh-Guldberg, O., Loya, Y., Meibom, A., Osman, E.O., Ruch, J.-D., Shaked, Y., Voolstra, C.R., Zvuloni, A., and M. Fine. 2020. Science, diplomacy, and the Red Sea's unique coral reef: It's time for action. *Front Mar Sci* 7: 90. doi:10.3389/fmars.2020.00090.

Klevjer, T.A., Irigoien, X., Røstad, A., Fraile-Nuez, E., Benítez-Barrios, V.M., and S. Kaartvedt. 2016. Large scale patterns in vertical distribution and behaviour of mesopelagic scattering layers. *Sci Rep* 6, no. 1: 19873. doi:10.1038/srep19873.

Klevjer, T.A., Torres, D.J., and S. Kaartvedt. 2012. Distribution and diel vertical movements of mesopelagic scattering layers in the Red Sea. *Mar Biol* 159, no. 8: 1833–1841. doi:10.1007/s00227-012-1973-y.

Koslow, J.A., Goericke, R., Lara-Lopez, A., and W. Watson. 2011. Impact of declining intermediate-water oxygen on deepwater fishes in the California Current. *Mar Ecol Prog Ser* 436: 207–218.

Morán, X.A.G., García, F.C., Røstad, A., Silva, L., Al-Otaibi, N., Irigoien, X., and M.L. Calleja. 2022. Diel dynamics of dissolved organic matter and heterotrophic prokaryotes reveal enhanced growth at the ocean's mesopelagic fish layer during daytime. *Sci Total Environ* 804: 150098. doi:10.1016/j.scitotenv.2021.150098.

Netburn, A.N., and J.A. Koslow. 2015. Dissolved oxygen as a constraint on daytime deep scattering layer depth in the southern California current ecosystem. *Deep Sea Res Part I: Oceanogr Res Pap* 104: 149–158. doi:10.1016/j.dsr.2015.06.006.

O'Brien, W.J., Browman, H.I., and B.I. Evans. 1990. Search strategies of foraging animals. *Am Sci* 78, no. 2: 152–160.

Ohlberger, J. 2013. Climate warming and ectotherm body size – From individual physiology to community ecology. *Funct Ecol* 27, no. 4: 991–1001. doi:10.1111/1365-2435.12098.

Olivar, M.P., Bernal, A., Molí, B., Peña, M., Balbín, R., Castellón, A., Miquel, J., and E. Massutí. 2012. Vertical distribution, diversity and assemblages of mesopelagic fishes in the western Mediterranean. *Deep Sea Res Part I: Oceanogr Res Pap* 62: 53–69. doi:10.1016/j.dsr.2011.12.014.

Olivar, M.P., Hulley, P.A., Castellón, A., Emelianov, M., López, C., Tuset, V.M., Contreras, T., and B. Molí. 2017. Mesopelagic fishes across the tropical and equatorial Atlantic: Biogeographical and vertical patterns. *Prog Oceanogr* 151: 116–137. doi:10.1016/j.pocean.2016.12.001.

Pearre, S. 2003. Eat and run? The hunger/satiation hypothesis in vertical migration: History, evidence and consequences. *Biol Rev Cambridge Phil Soc* 78, no. 1: 1–79.

Peña, M., Cabrera-Gámez, J., and A.C. Domínguez-Brito. 2020. Multi-frequency and light-avoiding characteristics of deep acoustic layers in the North Atlantic. *Mar Environ Res* 154: 104842. doi:10.1016/j.marenvres.2019.104842.

Prihartato, P., Irigoien, X., Genton, M., and S. Kaartvedt. 2016. Global effects of moon phase on nocturnal acoustic scattering layers. *Mar Ecol Prog Ser* 544: 65–75. doi:10.3354/meps11612.

Proud, R., Cox, M.J., and A.S. Brierley. 2017. Biogeography of the global ocean's mesopelagic zone. *Curr Biol* 27, no. 1: 113–119. doi:10.1016/j.cub.2016.11.003.

Proud, R., Handegard, N.O., Kloser, R.J., Cox, M.J., and A.S. Brierley. 2019. From siphonophores to deep scattering layers: Uncertainty ranges for the estimation of global mesopelagic fish biomass. *ICES J Mar Sci* 76, no. 3: 718–733. doi:10.1093/icesjms/fsy037.

Røstad, A., Kaartvedt, S., and D.L. Aksnes. 2016. Light comfort zones of mesopelagic acoustic scattering layers in two contrasting optical environments. *Deep Sea Res Part I: Oceanogr Res Pap* 113: 1–6. doi:10.1016/j.dsr.2016.02.020.

Ryer, C., and B. Olla. 1999. Light-induced changes in the prey consumption and behavior of two juvenile planktivorous fish. *Mar Ecol Prog Ser* 181: 41–51. doi:10.3354/meps181041.

Saba, G.K., Burd, A.B., Dunne, J.P., Hernández-León, S., Martin, A.H., Rose, K.A., Salisbury, J., Steinberg, D.K., Trueman, C.N., Wilson, R.W., and S.E. Wilson. 2021. Toward a better understanding of fish-based contribution to ocean carbon flux. *Limnol Oceanogr* 66, no. 5: 1639–1664. doi:10.1002/lno.11709.

Seibel, B.A. 2011. Critical oxygen levels and metabolic suppression in oceanic oxygen minimum zones. *J Exp Biol* 214, no. 2: 326–336. doi:10.1242/jeb.049171.

Seibel, B.A., Schneider, J.L., Kaartvedt, S., Wishner, K.F., and K.L. Daly. 2016. Hypoxia tolerance and metabolic suppression in oxygen minimum zone Euphausiids: Implications for ocean deoxygenation and biogeochemical cycles. *Integr Comp Biol* 56, no. 4: 510–523. doi:10.1093/icb/icw091.

Sobradillo, B., Christiansen, S., Røstad, A., and S. Kaartvedt. 2022. Individual daytime swimming of mesopelagic fishes in the world's warmest twilight zone. *Deep Sea Res Part I: Oceanogr Res Pap* 190: 103897. doi:10.1016/j.dsr.2022.103897.

Spaet, J.L.Y., Lam, C.H., Braun, C.D., and M.L. Berumen. 2017. Extensive use of mesopelagic waters by a Scalloped hammerhead shark (*Sphyrna lewini*) in the Red Sea. *Anim Biotelemetry* 5, no. 1: 20. doi:10.1186/s40317-017-0135-x.

Staby, A., Røstad, A., and S. Kaartvedt. 2011. Long-term acoustical observations of the mesopelagic fish *Maurolicus muelleri* reveal novel and varied vertical migration patterns. *Mar Ecol Prog Ser* 441: 241–255. doi:10.3354/meps09363.

St. John, M.A., Borja, A., Chust, G., Heath, M., Grigorov, I., Mariani, P., Martin, A.P., and R.S. Santos. 2016. A dark hole in our understanding of marine ecosystems and their services: Perspectives from the mesopelagic community. *Front Mar Sci* 3: 31. doi:10.3389/fmars.2016.00031.

Stramma, L., Prince, E.D., Schmidtko, S., Luo, J., Hoolihan, J.P., Visbeck, M., Wallace, D.W.R., Brandt, P., and A. Körtzinger. 2011. Expansion of oxygen minimum zones may reduce available habitat for tropical pelagic fishes. *Nat Clim Change* 2, no. 1: 33–37.

Torres, J.J., Belman, B.W., and J.J. Childress. 1979. Oxygen consumption rates of midwater fishes as a function of depth of occurrence. *Deep Sea Res* 26A: 185–197.

Voolstra, C.R., Valenzuela, J.J., Turkarslan, S., Cárdenas, A., Hume, B.C.C., Perna, G., Buitrago-López, C., Rowe, K., Orellana, M.V., Baliga, N.S., Paranjape, S., Banc-Prandi, G., Bellworthy, J., Fine, M., Frias-Torres, S., and D.J. Barshis. 2021. Contrasting heat stress response patterns of coral holobionts across the Red Sea suggest distinct mechanisms of thermal tolerance. *Mol Ecol* 30, no. 18: 4466–4480. doi:10.1111/mec.16064.

Wang, Z., DiMarco, S.F., Ingle, S., Belabbassi, L., and L.H. Al-Kharusi. 2014. Seasonal and annual variability of vertically migrating scattering layers in the northern Arabian Sea. *Deep Sea Res Part I: Oceanogr Res Pap* 90: 152–165. doi:10.1016/j.dsr.2014.05.008.

Warrant, E.J., and N.A. Locket. 2004. Vision in the deep sea. *Biol Rev* 79, no. 3: 671–712. doi:10.1017/S1464793103006420.

Weikert, H. 1982. The vertical distribution of zooplankton in relation to habitat zones in the area of the Atlantis II Deep, central Red Sea. *Mar Ecol Prog Ser* 8: 129–143. doi:10.3354/meps008129.

Wiebe, P.H., Bucklin, A., Kaartvedt, S., Røstad, A., and L. Blanco-Bercial. 2016. Vertical distribution and migration of euphausiid species in the Red Sea. *J Plankton Res* 38: 888–903. doi:10.1093/plankt/fbw038.

Wishner, K.F. 1980. The biomass of the deep-sea benthopelagic plankton. *Deep Sea Res Part A: Oceanogr Res Pap* 27, nos. 3–4: 203–216. doi:10.1016/0198-0149(80)90012-6.

Wishner, K.F., Seibel, B.A., Roman, C., Deutsch, C., Outram, D., Shaw, C.T., Birk, M.A., Mislan, K.A.S., Adams, T.J., Moore, D., and S. Riley. 2018. Ocean deoxygenation and zooplankton: Very small oxygen differences matter. *Sci Adv* 4, no. 12: eaau5180. doi:10.1126/sciadv.aau5180.

Zarubin, M., Lindemann, Y., and A. Genin. 2017. The dispersion-confinement mechanism: Phytoplankton dynamics and the spring bloom in a deeply-mixing subtropical sea. *Prog Oceanogr* 155: 13–27. doi:10.1016/j.pocean.2017.05.005.

Fish larvae distribution in the western Arabian Gulf

9

Spatiotemporal patterns in relation to water column characteristics

Thadickal V. Joydas, Madhusoodhanan Rakhesh, Javed A. Hussain, Mohamed Asharaf, Karuppasamy P. Manikandan, Priya B. Das, Muaadh A. Alnuwairah and Abdulaziz Al-Suwailem

9.1 INTRODUCTION

Larval fish studies are essential to fisheries science since they provide crucial information on spawning periodicity, breeding season, breeding grounds, and the annual recruitment pattern of fish in a particular area. The survival of larval fish depends upon many physicochemical and biological factors in the sea. Their roles in regulating fishery stocks are paramount in extreme marine environments such as the Arabian Gulf (hereafter the Gulf). The Gulf is a unique marginal sea of the northwestern Indian Ocean, where the oceanographic conditions are subjected to environmental extremes emanating from both anthropogenic and natural stressors. The hypersaline waters resulting from high evaporation over precipitation, low fresh water supply, high brine rejection from coastal desalination activities, and extreme seasonal temperature fluctuations exert tremendous pressure on resident communities and biogeochemical processes (Al-Yamani 2021). In the past two decades, the drastic environmental changes from non-climatic and climatic stressors have threatened the pelagic food web in the Gulf. They have compromised the flow of matter and energy through diverse microbial pathways, thus disturbing marine biodiversity, productivity, biogeochemical cycling, and fishery stocks. Connecting plankton diversity and microbial food web structure with fish larval recruitment and its resilience to environmental perturbations will be critical to understanding anthropogenic and climate-change-related ecosystem alterations.

Previous studies on fish larvae in the region covered aspects such as composition, abundance, and distribution (Ahlstrom 1966; Ali-Khan 1976, 1985, 1989, 1994; Ali-Khan and Hempel 1974; Ali-Khan and Waqar 1981, 1985; Nellen 1973; Aftab and Ali-Khan 1992; Ali-Khan and Aftab 1993; Ali-Khan et al. 2001). Nellen (1973) conducted pioneering work on the fish larvae from the northern and southern regions of the Eastern Gulf. Houde et al. (1986) conducted a comprehensive study on the fish larvae of Kuwaiti waters, offshore waters of Saudi Arabia, and the southern part of the Gulf. They identified fish larvae belonging to 53 families, of which 41 were identified down to genera and 24 down to species levels. A study conducted in Saudi Arabian waters during 1984–1985 (KFUPM/RI 1986) reported ichthyoplankton from the northern areas of Saudi Arabian territorial waters in the Gulf. Nevertheless, their findings were restricted to six commercially important families from the coastal waters. Soon after the 1991 oil spill, McCain and Hassan (1993) worked on the effect of oil slicks and sheen on the abundance of fish eggs and larvae. They found a drastic decline in the abundance of fish eggs and larvae in the oil-impacted areas of Abu Ali in the Saudi Arabian waters of the Gulf. The Regional Organization for the Protection of the Marine Environment (ROPME) of the Gulf conducted a Gulf-wide oceanographic expedition in the winter of 2006 (ROPME 2012). Fish larvae were an integral part of their study. Apart from these studies, there were no concerted attempts to study the abundance, diversity, and distribution of ichthyoplankton from the western Gulf.

Hence, a knowledge gap exists in basic information on larval fish populations and their environmental preferences from

DOI: 10.1201/9781003321392-11

many parts of the Gulf. Therefore, there is an urgent need to make any unpublished data on ichthyoplankton from the Gulf available to the public. Understanding the density, diversity, and spatiotemporal aspects of larval fish populations will help marine managers formulate effective management practices for the sustainable harvesting of fishery resources in the northern Gulf, where fishery stocks have already been declining for the past two decades.

The present study thus contributes to achieving the United Nations Sustainable Development Goal (SDG) 14: Conserve and sustainably use oceans, seas, and marine resources for sustainable development for the period 2021–2030 (International Decade of Ocean Science for Sustainable Development) with the strapline "The Ocean We Need for the Future We Want." Also, the present study from the Exclusive Economic Zone (EEZ) of the Kingdom of Saudi Arabia (KSA) in the Gulf, being of high national and international importance, will lead to collaborative international research ventures with neighbouring Gulf countries (contributing to UN's SDG-17: Partnership for the goals – Targets 17.6 (Knowledge sharing and cooperation for access to science, technology, and innovation), 17.7 (Encourage effective partnership), and 17.8 (Enhance availability of reliable data)) in addressing numerous challenges that the Gulf ecosystem is currently facing.

Here, we elucidate the spatiotemporal fluctuations in the abundance, composition, diversity, and distribution of planktonic fish larvae and eggs in the Saudi Arabian waters of the Gulf between November 2002 and August 2003. Despite the archival nature of the data set, the information obtained offers reliable baseline information on which the subsequent changes in the fish larval community of the northwestern Gulf could be investigated.

9.2 MATERIALS AND METHODS

This study is based on extensive sampling conducted in 2002 and 2003 under the *Oceanographic Survey in Support of the Marine and Coastal Damage Assessment* project intended to assess the long-term impact of the 1991 oil spill on the marine resources and environment of the Kingdom of Saudi Arabia. Five cruises were conducted using *R/V Al-Gosaibi*, during fall 2002, winter 2003, spring 2003, early summer 2003, and peak summer 2003. High-resolution plankton and environmental sampling were carried out during each cruise along five transects. Transects A (coastal) and B (offshore) were placed parallel to the coast, while transects C (northern), D (central), and E (southern) were located perpendicular to the coast connecting the coastal and offshore transects (Figure 9.1 and Table 9.1). While processing the data, coastal and offshore stations overlaying transects C, D, and E were included under their respective transects.

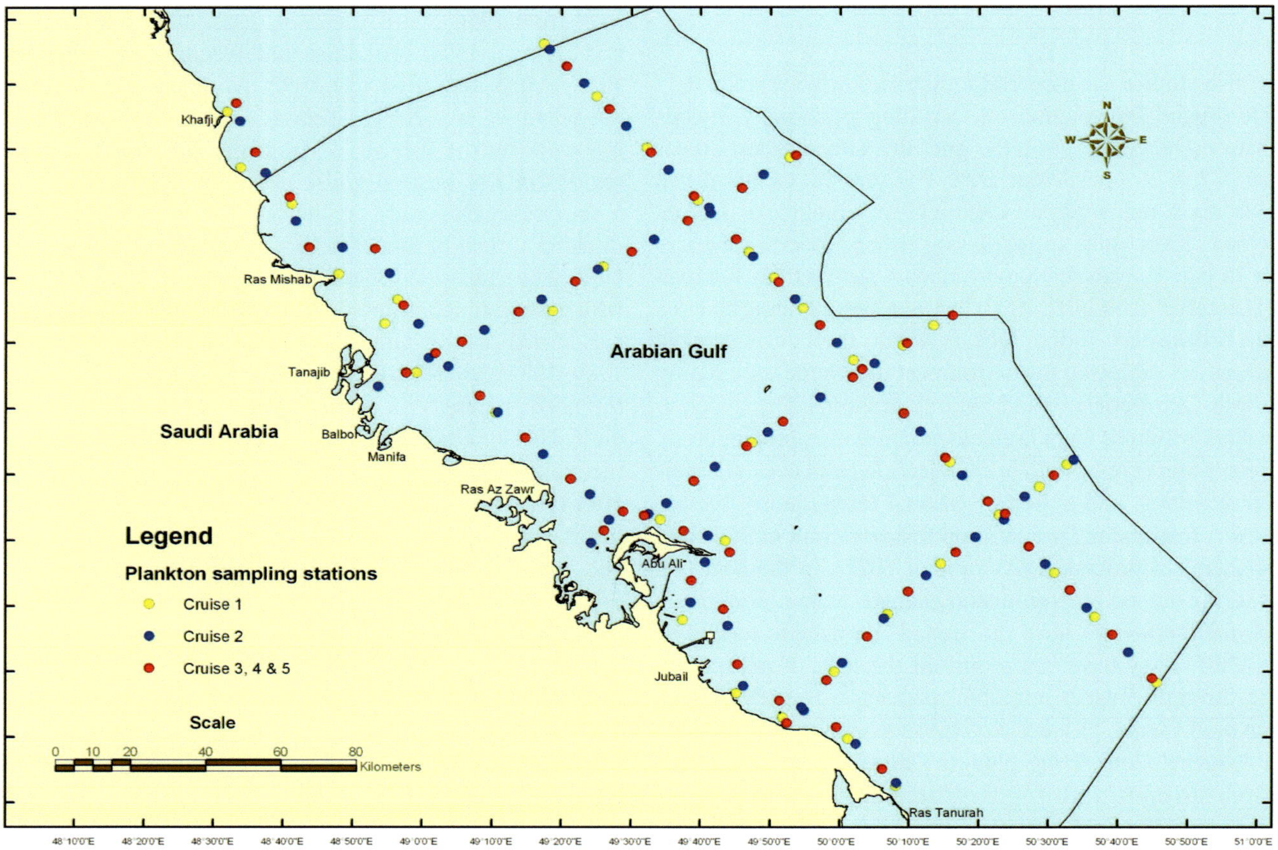

FIGURE 9.1 Sampling locations for ichthyoplankton in the Saudi Arabian territorial waters of the western Gulf.

TABLE 9.1 Station locations for physicochemical variables and larval fish samples in the Saudi Arabian territorial waters of the Gulf.

TRANSECTS	*ST. NO.*	*DEPTH (M)*	*LATITUDE (ºN)*	*LONGITUDE (ºE)*
Transect A	A010	13.4	28.45	48.56
	A100	11.5	28.32	48.60
	A200	2.6	28.21	48.68
	A300	6.2	28.08	48.74
	A400	3.9	28.06	48.90
	A500	3.8	27.93	48.96
	A600	5.6	27.81	49.03
	A700	9.4	27.70	49.14
	A800	6.2	27.59	49.25
	A900	9.6	27.49	49.36
	A1000	5.6	27.41	49.48
	A1100	7.7	27.36	49.63
	A1200	9.8	27.30	49.74
	A1300	6.9	27.23	49.65
	A1400	31.4	27.16	49.72
	A1500	8.5	27.02	49.76
	A1600	9.4	26.93	49.86
	A1700	6.8	26.86	50.00
	A1800	7.7	26.75	50.10
Transect B	B1900	37.1	28.55	49.35
	B2000	45.2	28.43	49.45
	B2100	36.1	28.32	49.55
	B2200	38.1	28.21	49.65
	B2300	47	28.10	49.75
	B2400	46	27.99	49.85
	B2500	44.2	27.88	49.95
	B2600	48	27.77	50.05
	B2700	53.2	27.65	50.16
	B2800	54	27.54	50.25
	B2900	53.6	27.43	50.36
	B3000	61.8	27.32	50.45
	B3100	61.1	27.21	50.55
	B3200	46.1	27.10	50.65
	B3300	49	26.98	50.75
Transect C	C3320	5.2	27.67	48.91
	C3400	8.6	27.76	48.96
	C3500	8.9	27.84	49.10
	C3600	18.2	27.91	49.23
	C3700	32	27.99	49.37
	C3800	44.2	28.07	49.50
	C3900	43.2	28.15	49.64
	C4000	47.3	28.23	49.77
	C4100	50.2	28.32	49.90
Transect D	D4200	8.2	27.36	49.44
	D4300	7.3	27.40	49.53
	D4400	25.4	27.48	49.65
	D4500	43.6	27.57	49.78
	D4600	52.9	27.66	49.90
	D4700	50.8	27.75	50.03
	D4800	52.8	27.83	50.16
	D4890	57.3	27.91	50.27

(Continued)

TABLE 9.1 (Continued)

TRANSECTS	*ST. NO.*	*DEPTH (M)*	*LATITUDE (ºN)*	*LONGITUDE (ºE)*
Transect E	E4900	5.2	26.87	49.88
	E5000	24.2	26.98	49.97
	E5100	43.2	27.09	50.07
	E5200	53.8	27.20	50.17
	E5300	54.5	27.30	50.28
	E5400	52.5	27.40	50.40
	E5500	62.5	27.50	50.51

Note:

- Transect A is a coastal transect stretching from the Kuwait border to the Dammam port area. This transect followed the 5 m bathymetry line along the coast, which is about 291 km long.
- Transect B is an offshore transect between Karan Island and Arabiyah Island. It extends from the Kuwait border to the Bahrain border. This transect does not follow a particular depth contour and is about 230 km long.
- Transect C is the northern transect running perpendicular to the shoreline in a northeast-southwest direction off the Tanajib area and is 127 km long.
- Transect D is the central transect running perpendicular to the shoreline in a northeast-southwest direction off the Abu Ali area and is 120 km long.
- Transect D is the southern transect running perpendicular to the shoreline in a northeast-southwest direction between Ras Tanurah and Jubail and is 103 km long.

9.2.1 Environmental sampling

A Conductivity-Temperature-Depth (CTD; Sea-Bird SBE-25 SEALOGGER) profiler was used to obtain each station's vertical salinity, temperature, and depth profiles. Additionally, auxiliary sensors attached to the CTD collected information on the vertical distribution of pH, dissolved oxygen (DO), and Chlorophyll *a* (Chl *a*). The accuracy of DO measurements was verified with a modified Winkler titration method. Chl *a* was estimated using a CTD-mounted fluorometer (SCUFA-Self-Contained Underwater Fluorescence Apparatus, Turner Design). The minimum detection limit was 0.02 mg Chl *a*/m^3. The Chl *a* values were corrected for turbidity. All environmental data were subjected to quality checks to remove spikes and bin averaging. Although the SBE-25 sensors were factory calibrated, the quality of the CTD sensors was maintained by periodic laboratory calibrations. In total, 555 vertical profiles were collected per cruise, with 2775 environmental profiles processed during the study.

9.2.2 Ichthyoplankton sampling and sample processing

A total of 272 samples (fall 2002, n = 45; winter 2003, n = 56; spring 2003, n = 57; early summer, n = 58; peak summer, n = 56) were collected for ichthyoplankton. The tow duration of the vertical hauls of 330 µm mesh net for ichthyoplankton (mouth area: 0.19625 m^2; net length: 2 m) varied depending upon the depth sampled. The nets were towed at a speed of 1.0–1.5 knots. Most of the sampling was undertaken during the day. A digital flow meter (General Oceanics) attached to the mouth of each net determined the volume of seawater filtered. The initial and final flow meter readings were noted at each station while sampling. The organisms that adhered to the nets were washed down with copious seawater as soon as the samples reached onboard. The ichthyoplankton samples were preserved in 5% buffered (with borax) formaldehyde.

The preservative was diluted before the fish larvae were sorted under zoom stereo-microscopes. The samples were again preserved in 5% buffered formaldehyde for taxonomic enumeration. Subsequently, meristematic counts, body proportions, and other morpho-taxonomical characters were used to identify the larvae up to the lowest taxonomic level possible using available taxonomic literature for fish larvae (Moser et al. 1984; Moser 1996; Jeyaseelan 1998; Leis and Carson-Ewart 2000; D'Ancona and Cavinato 1965; Hayashi and Tadokoro 1962; Hoda 1978; Houde 1977, 1981; Houde and Schekter 1983; Houde and Fore 1973; Houde et al. 1979; Leak 1981; Sreekumari 1976), and their abundance (nos/100 m^3) estimated taking account of the water filtered through the net. Based on the larval abundance, the fish families were classified as Principal (>50 larvae/family), Common (20–50 larvae/family), and Rare (<20 larvae/family). The larval specimens not identified down to the family level (due to damage and other technical difficulties) were included under the list of unclassified fish larvae. A mark of interrogation (?) was used to denote such specimens in the tables.

9.3 RESULTS

9.3.1 Environmental scenario – vertical patterns

9.3.1.1 Water temperature

The maximum water temperature in the Saudi Arabian territorial waters of the Gulf was 34.9°C during summer, and the minimum was 15.2°C during winter. The vertical temperature distribution revealed perceptible seasonal changes in the north-south direction along the coastal waters, with the northern part being colder than the central and southern parts. Such a north-south trend, however, was not evident in the offshore areas. Comparatively cold waters occupied the coastal areas during late fall, winter, and spring, which increased towards the offshore areas, whereas a reverse trend was observed in summer. The water temperature during spring was very similar to the late fall with moderate temperature ranges except in offshore waters, where the lowest water temperature was reported during spring, especially in the northern part. The entire water column heated up at the onset of summer. In summer, the water temperature attained the seasonal peak (34.9°C). Although the coastal waters showed vertically well-mixed waters in all the seasons, there were perceptible north-south differences in vertical thermal structure in the offshore waters (Figure 9.2). The water temperature showed a large seasonal cycle along all the zonal as well as the meridional transects. The amplitude of the seasonal cycle in surface temperature was largest (19.35°C) along the coastal waters from winter to summer (15.59–34.94°C), which was 5°C greater than that of the offshore waters (20.74–35.36°C). The perpendicular transects (C, D, and E) also showed a large seasonal cycle, the amplitude of which varied from 13°C to 19°C. The largest seasonal cycle was near the western coastal area of the northern waters, while the least occurred near the eastern offshore area of the southern waters.

9.3.1.2 Salinity

The vertical salinity distribution in the coastal areas showed comparatively low-salinity waters in the south and high-salinity waters in the central area. Salinity decreased towards offshore waters. The maximum salinity (>43.5) was observed along the coastal bays of Abu-Ali and Manifa. Unlike water temperature, the salinity distribution did not show a seasonal disparity in the coastal waters. The seasonal peak was observed in the northern area during spring (salinity: 41.2). Due to the coastal intrusion of low-salinity surface offshore waters (salinity <39) at the onset of summer that reached the seasonal peak in summer, a very weak halocline was formed in the central and southern offshore waters, which intensified to 20–25 m thickness in summer. Conversely, offshore advection of high-salinity coastal waters along the bottom started in late fall and continued until spring (Figure 9.3).

9.3.1.3 Dissolved oxygen

The vertical distributions of DO showed evidence of an oxygen minimum layer in the bottom waters of the southern offshore region. During winter, the northern coastal region had higher DO concentrations (6.25 mg.l^{-1}) that decreased towards the southeast. During spring, the DO levels increased from northwest to southeast in the surface waters, whereas the bottom waters had higher DO in the northeastern region. The oxygen levels decreased considerably from the onset of summer to summer. During summer, the oxygen distribution patterns differed between coastal (1.84–6.30; ca. 4.75 mg.l^{-1}) and offshore waters (2.33–7.52; ca. 4.28 mg.l^{-1}). Although surface waters increased from the coast to the offshore, bottom waters displayed a reverse trend (Figure 9.4).

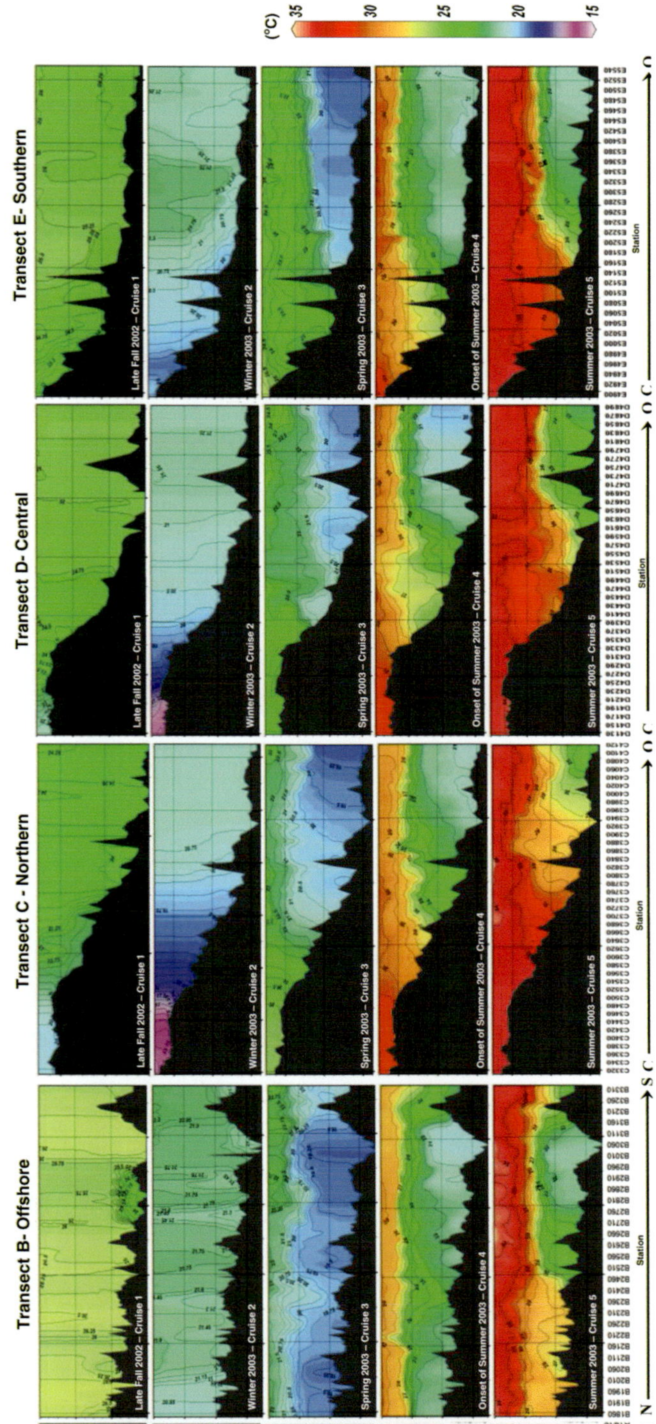

FIGURE 9.2 Seasonal vertical distribution of water temperature along the coastal, offshore, northern, central, and southern oceanographic transects in the Saudi Arabian territorial waters of the western Arabian Gulf. N → S: North to South; C → O: Coastal to Offshore.

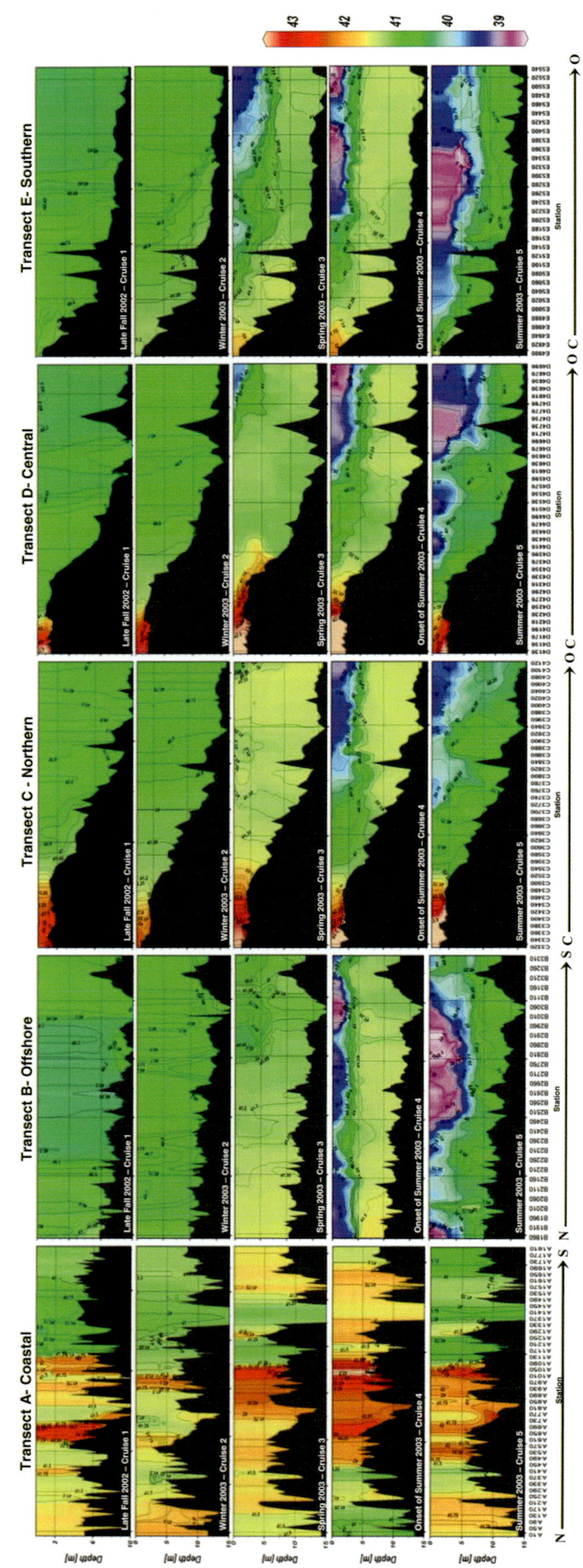

FIGURE 9.3 Seasonal vertical salinity distribution along the coastal, offshore, northern, central, and southern oceanographic transects in the Saudi Arabian territorial waters of the western Arabian Gulf. N → S: North to South; C → O: Coastal to Offshore.

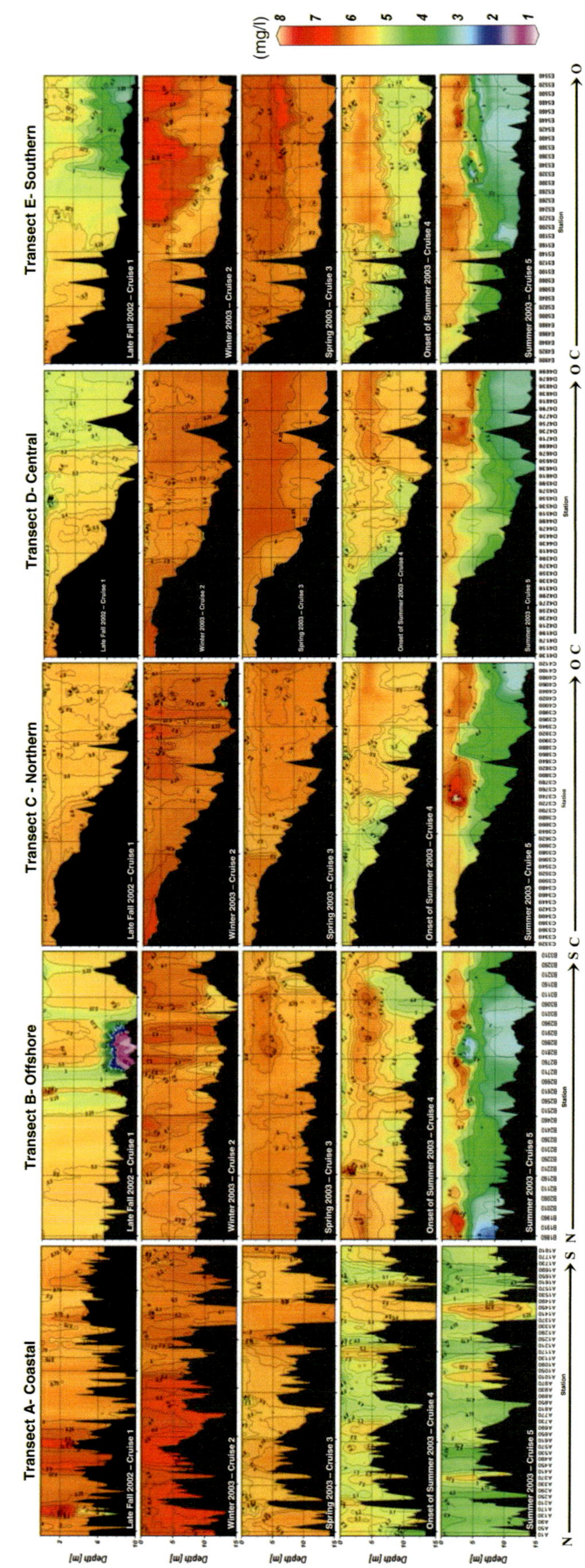

FIGURE 9.4 Seasonal vertical dissolved oxygen distribution along the coastal, offshore, northern, central, and southern oceanographic transects in the Saudi Arabian territorial waters of the western Arabian Gulf. N → S: North to South; C → O: Coastal to Offshore.

9.3.1.4 *Chlorophyll* a

The vertical distribution of Chl *a* along the five transects showed the lowest minimum (0.23 mg/m^3), maximum (1.93 mg/m^3), and mean (0.93 mg/m^3) Chl *a* levels in spring. The Chl *a* levels gradually increased in summer through late fall and peaked (mean: 1.75 mg/m^3) in winter. Albeit a similar trend between the coastal and offshore areas, Chl *a* was higher in the coastal waters than the offshore waters during spring and the onset of summer. Associated with the thermocline, the subsurface Chl *a* maximum was observed in offshore waters during spring through summer at 25–30 m (Figure 9.5). The coastal subsurface waters off Manifa to Jubail showed relatively high Chl *a* concentration in all seasons.

9.3.2 Fish larvae distribution

Fish larvae belonging to 62 families, of which 47 were identified down to genera and 11 down to species levels were recorded. The mean relative abundance (in percentage) of larval fish belonging to 22 families of commercial importance revealed the predominance of Gobiidae, Engraulidae, and Clupeidae in Saudi Arabian waters (Table 9.2). Although the commercially important family Scombridae showed high relative abundance in offshore waters, larvae of Mullidae and Callionymidae exhibited high abundance in central and southern waters, respectively. Among the commercially important families, Haemulidae, Sparidae, and Carangidae were higher in the northern waters, while Nemipteridae and Sphyraenidae were higher in the central and southern waters, respectively.

9.3.2.1 *Seasonal distribution of fish larvae according to the PCR criterion (Principal-Common-Rare)*

The larval fish specimens obtained in the fall (1380 specimens) were classified under 28 families, 21 genera, seven species, and four types, while 26 specimens remained unidentified. Engraulidae (57.6%) and Clupeidae (29.6%) were the most abundant families. Bregmacerotidae (2.6%), a commonly occurring family in fall, was represented by three species: *Bregmaceros arabicus*, *B. nectabanus*, and *B. rarisquamosus*. Leiognathidae (1.3%) was the most dominant among the 24 rare families reported in this season (Table 9.3).

In winter, 410 larval fish specimens belonging to 29 families, 25 genera, and eight species were identified, while 65 specimens remained unclassified. Engraulidae (42.1%) was the most abundant family during this season and was represented by two genera, *Stolephorus* and *Thryssa*. Among the commonly occurring families, Haemulidae (10.1%), represented by a single genus, *Pomadasys*, was the most dominant family. Under the rare category, 26 fish families were recorded, among which Mugilidae (4.0%), Bregmacerotidae (3.8%), and Synodontidae (3.2%) were the predominant ones (Table 9.4).

Across the seasons, larval fish maximal abundance and diversity were observed during spring. Out of 2114 larval specimens obtained, 1971 larvae were identified and grouped under 49 families, 30 genera, seven species, and 27 types, while 143 specimens remained unidentified. Clupeidae (38.7%) and Gobiidae (17.1%) were the most abundant among principal

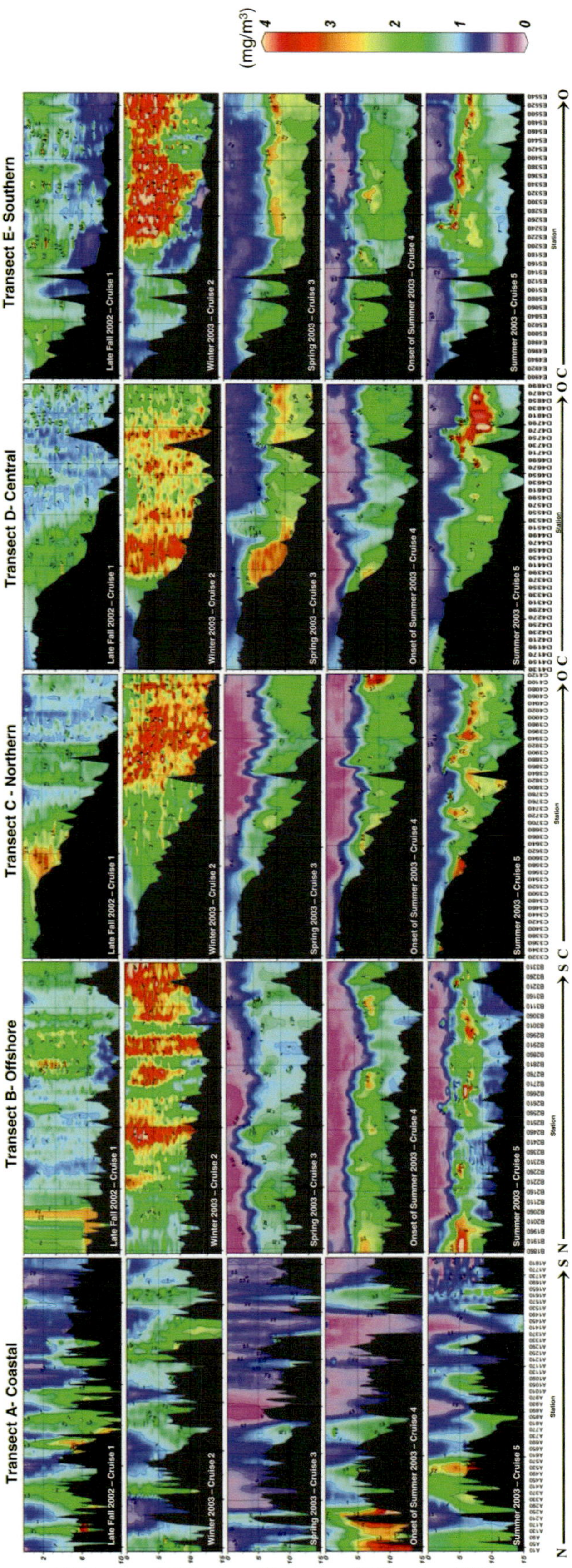

FIGURE 9.5 Seasonal vertical distribution of Chlorophyll *a* (Chl *a*) along the coastal, offshore, northern, central, and southern oceanographic transects in the Saudi Arabian territorial waters of the western Arabian Gulf. N → S: North to South; C → O: Coastal to Offshore.

TABLE 9.2 Mean relative abundance (%) of the most dominant and commercially important larval fish families in Saudi Arabian waters. The contribution of the most dominant families (≥5%) in each transect/location is given in bold italicized numbers.

		MEAN RELATIVE ABUNDANCE (%)				
S. NO.	FAMILY	COASTAL	OFFSHORE	NORTHERN	CENTRAL	SOUTHERN
1	Blenniidae	0.49	0.30	1.16	0.99	0.56
2	Bothidae	0.00	2.32	1.10	1.29	1.36
3	Bregmacerotidae	0.18	2.00	1.26	***5.57***	2.21
4	Callionymidae	3.81	0.87	0.52	1.46	***4.94***
5	Carangidae	1.44	1.99	3.62	2.22	0.97
6	Clupeidae	***22.65***	***12.09***	***26.96***	***23.34***	***30.13***
7	Cynoglossidae	2.91	2.50	0.37	0.54	1.59
8	Engraulidae	***25.75***	***44.38***	***26.64***	***28.66***	***33.37***
9	Gobiidae	***29.53***	***5.32***	***10.64***	***11.17***	***4.63***
10	Haemulidae	2.10	3.59	***6.82***	0.43	1.31
11	Labridae	0.64	0.26	0.00	0.08	0.43
12	Lutjanidae	0.00	0.10	0.00	0.00	0.03
13	Mullidae	3.57	1.62	0.89	***7.07***	1.18
14	Nemipteridae	1.81	0.72	0.70	2.72	0.22
15	Sciaenidae	0.32	0.38	0.76	0.24	1.83
16	Scombridae	0.19	***10.67***	1.43	0.18	0.64
17	Serranidae	0.00	0.99	0.96	0.79	0.05
18	Sillaginidae	1.04	0.01	0.77	1.46	0.16
19	Sparidae	0.12	0.47	4.01	3.07	0.28
20	Sphyraenidae	0.00	1.47	0.23	0.70	3.39
21	Synodontidae	0.12	2.58	1.38	2.71	1.74
22	Trichiuridae	0.07	0.34	0.08	0.00	0.15

families, while Blenniidae (2.0%) and Callionymidae (1.9%) were dominant among the common families. Many families were classified as rare (36) during this season. Trichonotidae formed the most abundant family (0.8%) (Table 9.5).

During the onset of summer, 1366 larval fish specimens were classified under 30 families, 29 genera, seven species, and 16 types, while 113 specimens remained unidentified. The most abundant principal family was Clupeidae (30.6%). Cynoglossidae (2.3%), Mullidae (2.2%), and Carangidae (1.3%) were abundant among the families classified as common. Bothidae (1.2%) and Leiognathidae (1.0%) were the most dominant among the 23 rare families (Table 9.6).

In the peak of summer, 729 larval specimens were identified from 25 families, 19 genera, four species, and 14 types, while 49 specimens remained unidentified. Engraulidae (38.5%) and Clupeidae (26.8%) were the most abundant among the principal families. Sphyraenidae (5.0%) was the most abundant among the commonly occurring larval groups, and Carangidae (2.4%) and Bothidae (1.9%) were among the 20 rare fish families (Table 9.7). Representative illustrations of fish larvae sampled during the present study are given in Figure 9.6.

9.3.2.2 Spatiotemporal distribution of fish eggs and fish larvae

The mean density of fish eggs varied from 297 eggs/100 m^3 (central waters) to 472 eggs/100 m^3 (northern waters) with clear seasonal distribution patterns (Table 9.8). Although a high abundance of fish eggs was reported from offshore (990 eggs/100 m^3) and southern (1020 eggs/100 m^3) waters during late fall, northern waters reported the maximal abundance during spring (1002 eggs/100 m^3) and summer (865 eggs/100 m^3). Among all seasons, the highest fish egg density was reported in spring (Range and mean: 395–1002, 575) and the lowest in winter (Range and mean: 33–301, 159) (Figures 9.7A and 9.8A). One-way ANOVA analysis showed homogeneous spatial distribution of fish larvae (F = 2.0218, P = >0.05), except in the central (F = 4.2837, P = <0.01) and offshore waters (F = 2.8192, P = <0.05).

Most fish larvae collected during this study were newly hatched with attached yolk, unpigmented eyes, and signs of dorsal and anal fin thickenings. Although the average larval fish abundance did not support a coastal (305 larvae/100 m^3) – offshore (328 larvae/100 m^3) trend, a north-south trend was apparent with higher average density in the southern waters (329 larvae/100 m^3) than the northern waters (213 larvae/100 m^3) (Table 9.8).

Seasonally, the highest average larval fish density for any season was reported during spring in the coastal and southern waters (862–864 larvae/100 m^3). In late fall, the offshore waters (547 larvae/100 m^3), mainly the central and northern areas, exhibited the highest larval density. The central waters in the offshore area (575 larvae/100 m^3) and coastal areas of the northern waters (270 larvae/100 m^3) showed the highest densities at the onset of summer and summer, respectively. The seasonal larval fish densities varied from 37–862 (av. 305 larvae/100 m^3) in the coastal waters, 56–547 (av. 328 larvae/100 m^3) in the

TABLE 9.3 Principal, Common, and Rare fish families collected during fall 2002.

	S. NO.	*FAMILY*	*GENERA/SPECIES*	*PERCENTAGE*
Principal	1	Engraulidae	*Stolephorus* sp.	53.56
			Thryssa sp.	4.05
	2	Clupeidae	*Sardinella* sp. 1	23.83
			Sardinella sp. 2	5.76
			Total	**87.2**
Common	1	Bregmacerotidae	*Bregmaceros arabicus*	0.21
			Bregmaceros nectabanus	0.92
			Bregmaceros rarisquamosus	1.56
	2	Synodontidae	*Saurida* sp.	1.92
			Total	**4.61**
Rare	1	Antennariidae		0.07
	2	Apogonidae	*Apogon* sp.	0.14
	3	Blenniidae	*Omobranchus* sp.	0.07
	4	Bothidae	*Engyprosopon grandisquma*?	1.00
	5	Callionymidae	*Callionymus* sp.	0.36
	6	Carangidae	*Gnathanodon speciousus*	0.07
			Scomberoides sp.	0.07
	7	Cepolidae		0.07
	8	Chirocentridae		0.07
	9	Cynoglossidae	*Cynoglossus* sp.	0.71
	10	Gobiidae Type 1		0.57
		Gobiidae Type 2		0.14
		Gobiidae Type 3		0.14
		Gobiidae Type 4		0.07
	11	Haemulidae	*Pomadasys* sp.	0.07
	12	Leiognathidae	*Leiognathus* sp.	1.28
	13	Mullidae	*Upeneus tragula*	0.07
	14	Nemipteridae		0.50
	15	Pinguipedidae	*Parapercis* sp.	0.07
	16	Platycephalidae	*Platycephalus* sp.	0.07
	17	Sciaenidae		0.07
	18	Scorpaenidae		0.07
	19	Soleidae	*Solea elongatis*	0.07
	20	Sparidae		0.07
	21	Tetraodontidae		0.07
	22	Trichiuridae	*Lapturacanthus*?	0.14
	23	Trichonotidae		0.14
	24	Triglidae	*Lepidotrigla* sp.	0.07
			Total	**6.31**

offshore areas, 80–344 (av. 213 larvae/100 m^3) in the northern waters, 93–575 (av. 269 larvae/100 m^3) in the central waters, and 112–864 (av. 329 larvae/100 m) in the southern waters (Figures 9.7B and 9.8B).

Results of ANOVA indicated significant seasonal heterogeneity in the larval fish distribution in the coastal and southern waters (F = 10.1296, P = <0.001); the distribution was homogenous in all other areas (P = >0.05).

9.3.2.3 Fish larval diversity

The ichthyoplankton diversity was assessed using three indices: the Shannon-Wiener index (H') for measuring species diversity, the Margalef Index (d') for taxa richness, and Pielou's index for evenness (J'). The average species richness index ranged from 3.3 (nearshore) to 7.6 (southern) on a spatial scale and from 4.06 (summer) to 7.55 (spring) seasonally; species diversity from 1.0 (nearshore) to 1.9 (offshore) spatially and from 1.3 (summer) to 1.98 (spring) seasonally; and evenness index from 0.78 (nearshore) to 0.85 (central) spatially and from 0.74 (late fall) to 0.89 (winter) seasonally. The richness and evenness indices showed an increasing trend from Manifa to Abu Ali and a decreasing trend towards the south. Comparatively, higher species diversity (H': 1.94) was observed in offshore waters than in coastal waters (1.05). In the northern waters, the highest diversity index of 2.017 was recorded off Tanajib. The

TABLE 9.4. Principal, Common, and Rare fish families collected during winter 2003.

	S. NO	*FAMILY*	*GENERA/SPECIES*	*PERCENTAGE*
Principal	1	Engraulidae	*Stolephorus* sp. 1	16.63
			Stolephorus sp. 2	20.63
			Thryssa?	4.84
			Total	**42.11**
Common	1	Callionymidae	*Callionymus* sp.	4.21
	2	Haemulidae	*Pomadasys* sp.	10.11
			Total	**14.32**
Rare	1	Bothidae	*Engyprosopon grandisquma*?	0.42
			Bothid Type 2	0.21
	2	Bregmacerotidae	*Bregmaceros nectabanus*	1.26
			Bregmaceros rarisquamosus	2.53
	3	Carangidae	*Carangoides*	0.63
	4	Clupeidae	*Etrumeus teres*	1.26
			Dussumieria sp.	0.21
	5	Cynoglossidae	*Cynoglossus* sp.	1.89
	6	Ephippidae	*Platax*?	0.63
	7	Gobiidae	Gobid Type 1	0.63
			Gobid Type 4	0.21
	8	Kyphosidae	*Kyphosus* sp.	0.21
	9	Mugilidae	*Mugil* sp.	4.00
	10	Mullidae	*Upeneus tragula*	1.26
	11	Ophidiidae	*Ophidion* sp.	0.84
	12	Paralichthyidae	*Pseudorhombus elevatus*	0.21
	13	Platycephalidae	*Platycephalus* sp.	0.21
	14	Polynemidae?		0.21
	15	Pomacanthidae		0.42
	16	Sciaenidae		2.74
	17	Scombridae		0.21
	18	Serranidae		1.05
	19	Sillaginidae	*Sillago* sp.	0.42
	20	Soleidae	*Solea elongatis*	0.21
	21	Sparidae	*Acanthopagarus* sp.	1.26
			Argyrops?	0.21
	22	Synodontidae	*Saurida* sp.	2.95
			Saurida undosquamis	0.21
	23	Trichiuridae	*Trichiurus* sp.	0.42
	24	Trichonotidae		0.84
	25	Triglidae	*Lepidotrigla* sp.	1.89
	26	Tripterygidae		0.21
			Total	**29.86**

richness index (*d′*) in the central waters showed an increasing trend towards the offshore area, especially off Abu Ali (1.132), while in the southern waters, the richness index showed high values in the offshore areas. In winter, high diversity was observed in the central waters (3.0957), followed by southern waters (2.951). In spring, relatively high values of the richness index were recorded from offshore (*d′*: 2.176), northern (*d′*: 2.636), and southern waters (*d′*: 2.794). During the onset of summer, the *d′* values of northern, central, and southern waters showed an increasing trend towards offshore waters. The fish larval diversity was higher in the offshore region compared to the coastal waters. In summer, the species richness and evenness were significantly higher in the northern parts of coastal waters. On the other hand, offshore waters showed an increasing trend towards the southern parts, with high values recorded north of Abu Ali (2.506). Although the highest species diversity was recorded from stations off Ras Tanurah (3.602), there was no regular trend for richness index in these regions. Relatively lesser richness was observed in central (0.934) and southern waters (0.899). The Evenness index indicated a homogeneous distribution of larval fish species in winter, spring, and the onset of summer.

TABLE 9.5 Principal, Common, and Rare fish families collected during spring 2003.

	S. NO.	*FAMILY*	*GENERA/SPECIES*	*PERCENTAGE*
Principal	1	Carangidae	Carangids	0.09
			Caranx sp.	0.09
			Decapterus sp.	1.37
			Trachurus sp.	1.37
	2	Clupeidae	*Dussumieria* sp.	0.19
			Etrumeus teres	3.03
			Herklosichthys sp.	19.87
			Sardinella sp.	11.64
			Clupid type 4	1.42
			Clupid type 5	0.33
			Clupid type 6	2.27
	3	Engraulidae	*Stolephorus* sp. 1	10.97
			Stolephorus sp. 2	0.47
	4	Gobiidae	Gobid (13 different type)	17.12
	5	Synodontidae	*Saurida* sp.	2.41
			Total	**72.64**
Common	1	Blenniidae	*Blenniids*	1.42
			Omobranchus sp.	0.28
			Parablennius sp.	0.33
	2	Bregmacerotidae	*Bregmaceros arabicus*	0.14
			Bregmaceros nectabanus	0.61
			Bregmaceros rarisquamosus	0.19
	3	Callionymidae	*Callionymus* sp.1	1.84
			Callionymus sp.2	0.14
	4	Drepanidae	Drepanids	1.14
	5	Labridae	Labrid type 1	0.66
			Labrid type 2	0.28
	6	Leiognathidae	*Leiognathus* sp.	1.28
	7	Monocanthidae	Monocanthid	0.95
	8	Mullidae	*Upeneus tragula*	1.84
			Total	**11.10**
Rare	1	Balistidae?	Balistid	0.05
	2	Bothidae	*Engyprosopon grandisquma*?	0.33
			Bothid Type 2	0.05
			Bothid Type 3	0.09
	3	Cynoglossidae	*Cynoglossus* sp.	0.66
	4	Gerridae	Gerrids	0.47
	5	Gobisocidae	Gobisocids	0.28
	6	Glaucostomatidae	Glaucostomatid	0.05
	7	Haemulidae	*Pomadasys* sp.	0.8
	8	Kyphosidae	*Kyphosus* sp.	0.14
	9	Lutjanidae	*Lutjanid* sp.	0.05
	10	Monodontidae	Monodont	0.05
	11	Mugilidae	*Mugil* sp.	0.24
	12	Nemipteridae	Nemipterid	0.71
	13	Nomeidae	Nomids	0.57
	14	Paralichthyidae	*Pseudorhombus elevatus*	0.09
	15	Pempheridae?	Pempherids?	0.09
	16	Platycephalidae	*Platycephalus* sp.	0.28
	17	Polynemidae	Polynemid?	0.05
	18	Pomacanthidae	Pomacanthid	0.05
	19	Pomacentridae	Pomacentrid	0.19
	20	Priacanthidae	*Priacanthus* sp.	0.05
	21	Rachycentridae	*Rachycentron* sp.	0.05

(Continued)

TABLE 9.5 (Continued)

S. NO.	FAMILY	GENERA/SPECIES	PERCENTAGE
22	Scaridae	Scarid	0.24
23	Schindleridae?	Schindlerids?	0.05
24	Sciaenidae	Sciaenids	0.38
25	Scorpaenidae	Scorpaenids	0.05
26	Scombridae	Scombrids	0.05
27	Serranidae	Serranids	0.47
28	Sillaginidae	*Sillago* sp.	0.61
29	Sparidae	Sparids	0.43
30	Sphyraenidae	*Sphyraena* Type 1	0.14
		Sphyraena Type 2	0.05
31	Syngnathidae	*Syngnathus* sp.	0.05
32	Tetraodontidae	Tetraodontid type 1	0.05
		Tetraodontid type 2	0.05
		Tetraodontid type 3	0.09
33	Terapontidae	Terapontids	0.19
34	Trichiuridae	*Trichiurus* sp.	0.24
35	Trichonotidae	Trichonotid	0.85
36	Triglidae	*Lepidotrigla* sp.	0.09
		Total	**9.47**

9.3.2.4 Fish larvae-environmental interactions

Spearman rank correlation-based (ρ) multivariate analysis of biotic-environmental interactions (BIO-ENV analysis) revealed the relationship between ichthyoplankton and environmental variables. The importance of environmental combinations varied between seasons. The results showed that among a set of environmental variables used, water temperature, salinity, dissolved oxygen, and Chl *a* were the most crucial background variables explaining the abundance and spatiotemporal distribution patterns of fish larvae in Saudi Arabian territorial waters. Although the combined effect of all these variables showed a maximum influence in late fall (ρ: 0.623), the synergistic seasonal changes in water temperature and salinity influenced the larval fish distribution in winter (ρ: 0.379) and summer (ρ: 0.321), where the combination has benefited from the marker environmental variable characterizing each season (i.e., salinity in winter, ρ: 0.376, and water temperature in summer, ρ: 0.295). Other variables, such as TPH and pH, did not significantly impact the larval abundance and distribution. The correlation values of the five best environmental combinations influencing fish larvae distribution in Saudi Arabian territorial waters are shown in Table 9.9.

9.4 DISCUSSION

9.4.1 Environmental scenario

Based on the Master Oceanographic Observations Data Set (MOODS), Swift and Bower (2003) observed north-south temperature gradients with colder temperatures in the northern end of the Gulf than in the south. The present study revealed a seasonal cycle in water temperature, which was the coldest in winter and the warmest in summer. The winter-to-summer contrast was about 18°C (16.5–34.5°C). Yoshida et al. (1998) attributed the lowest temperature in January to winter cooling. Although this seasonal cycle was expected to make the water denser and uniformly well-mixed during winter and light and stratified in summer, the very shallow nature (~10 m) of the coastal areas makes it easy for the vertical transfer of heat to occur. The surface and bottom mixing can stir the water column over its full depth. Although the offshore area resembled the coastal waters during late fall and winter with vertically well-mixed waters, a contrasting thermal structure was found from spring to summer. The thermal structure developed a weak stratification with the warming in spring and formed a weak thermocline. The seasonal thermocline became more pronounced and well-defined as the heating increased in summer. All three perpendicular transects followed a similar seasonal cycle, with coastal waters remaining vertically well-mixed but with strong horizontal gradients during late fall and winter. Summer zonal temperature distribution showed a weak two-layer structure, with an upper isothermal layer, the thickness of which increased towards the south.

While the mixing of offshore waters in late fall and winter could be attributed to evaporation and winter cooling effects, which increased the density of surface waters and initiated convection, coastal advection of low-salinity offshore waters during spring reduced the salinity in the upper layers. Reynolds (1993) suggested a density-driven circulation in the Gulf, wherein the surface current from the northwestern Indian Ocean, having relatively low salinity and high temperature, intrudes through the Strait of Hormuz and flows along the Iranian coast to the area offshore of Qatar, where it forms a cyclonic gyre. The present observation on the intrusion of low-salinity water from the offshore area, especially

TABLE 9.6 Principal, Common, and Rare fish families collected during the onset of summer 2003.

	S. NO.	*FAMILY*	*GENERA/SPECIES*	*PERCENTAGE*
Principal	1	Clupeidae	Clupeid	0.47
			Dussumieria sp.	0.34
			Etrumeus teres	0.07
			Herklosichthys?	26.03
			Sardinella sp.	3.65
			Clupid type 6	0.07
	2	Scombridae	Scombrid	0.95
			Auxis sp.	0.14
			Scomber sp.	21.97
	3	Engraulidae	Engraulids	0.34
			Stolephorus sp. 1	6.49
			Stolephorus sp. 2	11.63
	4	Gobiidae	Gobiid	5.68
			Total	**77.83**
Common	1	Cynoglossidae	*Cynoglossus* sp.	2.37
	2	Mullidae	*Upeneus tragula*	2.23
	3	Carangidae	*Carangoides* sp.	0.34
			Caranx sp.	0.41
			Decapterus sp.	0.14
			Scomberoides sp.	0.07
			Trachurus sp.	0.41
			Total	**5.97**
Rare	1	Blenniidae	Blenniids	0.14
	2	Bothidae	*Engyprosopon grandisquma*?	1.15
			Bothid	0.07
	3	Bregmacerotidae	*Bregmaceros* sp.	0.07
			Bregmaceros nectabanus	0.41
			Bregmaceros rarisquamosus	0.2
	4	Callionymidae	*Callionymus* sp.1	0.2
	5	Centriscidae		0.07
	6	Chaetodontidae	*Chaetodon* sp.	0.27
	7	Ephippidae	*Platax* sp.	0.07
	8	Haemulidae	*Pomadasys* sp.	0.14
	9	Leiognathidae	*Leiognathus* sp.	1.01
	10	Monocanthidae	*Paramonocanthus choirocephalus*	0.2
	11	Nemipteridae		0.47
	12	Nomeidae		0.47
			Psenes indica	0.47
	13	Ophichthidae	*Ophichthus* sp.	0.14
	14	Pomacentridae		0.14
	15	Priacanthidae	*Priacanthus* sp.	0.07
	16	Sciaenidae		0.07
	17	Scorpaenidae		0.07
	18	Serranidae		0.14
			Epinephelus sp.	0.07
	19	Sillaginidae	*Sillago* sp.	0.14
	20	Sparidae		0.07
	21	Sphyraenidae	*Sphyraena* Type 1	0.68
			Sphyraena Type 2	0.07
			Sphyraena Type 3	0.2
	22	Synodontidae	*Saurida* sp.	0.68
			Saurida undosquamis	0.07
	23	Trichonotidae		0.61
			Total	**8.63**

TABLE 9.7 Principal, Common, and Rare fish families collected during summer 2003.

	S. NO.	FAMILY	GENERA/SPECIES	PERCENTAGE
Principal	1	Engraulidae	Engraulids	4.37
			Stolephorus sp. 1	13.88
			Stolephorus sp. 2	20.18
			Thryssa?	0.13
	2	Clupeidae	Clupeids	16.07
			Sardinella sp.	0.13
			Clupid type 6	10.67
	3	Gobiidae	Gobiid (7 different types)	8.23
			Total	**73.66**
Common	1	Sphyraenidae	*Sphyraena* Type 1	3.85
			Sphyraena Type 3	1.15
	2	Nemipteridae		2.57
			Total	**7.57**
Rare	1	Blenniidae	Blenniids	0.26
	2	Bothidae	Bothids	0.51
			Engyprosopon grandisquma?	1.41
	3	Bregmacerotidae	*Bregmaceros nectabanus*	0.64
			Bregmaceros rarisquamosus	1.03
	4	Callionymidae	*Callionymus* sp.	1.03
	5	Carangidae	Carangids	0.39
			Caranx sp.	0.13
			Decapterus sp.	0.39
			Trachurus sp.	1.54
	6	Cepolidae		0.39
	7	Cynoglossidae	*Cynoglossus* sp.	0.13
	8	Drepanidae		0.39
	9	Labridae	Labrids (3 types)	0.39
	10	Leiognathidae	*Leiognathus* sp.	0.13
	11	Leptobramidae?	Laptobramids?	0.26
	12	Lutjanidae	*Lutjanid* sp.	0.13
	13	Mullidae	*Upeneus tragula*	0.39
	14	Sciaenidae		0.13
	15	Scombridae	Scombrids	1.16
			Auxis sp.	0.13
	16	Serranidae	*Epinephelus* sp.	0.26
	17	Sillaginidae		0.26
	18	Sparidae	Sparid	0.64
			Pagellus?	0.13
	19	Synodontidae	*Saurida* sp.	0.13
	20	Tetraodontidae		0.13
			Total	**12.51**

in the central and southern Saudi Arabian waters, supports this explanation. In the present study, the coastal advection of offshore waters intensified with time and peaked in summer, when the thickness of the low-salinity layer deepened and extended almost up to the coast. The high-salinity coastal waters started flowing towards the offshore area along the bottom from late fall to spring, which was most evident in the northern and central waters. Yoshida et al. (1998), based on hydrographic data collected onboard *RT/V Umitaka-Maru*, made similar observations of high-salinity bottom waters in the Gulf. They attributed this to intense evaporation and winter cooling. Yamamoto et al. (2022) report that seasonality controls the advection of high-salinity bottom waters from Kuwait Bay to the Gulf, where the outward-flowing high-salinity Gulf Deep Water (GDW) originates in the northern Gulf. Thus, the most prominent seasonally influenced physical processes such as winter convection and mixing, the spring to summer coastal advection of low-salinity offshore waters, and the fall to spring offshore advection of high-salinity bottom waters stabilize and destabilize the water column in the western Gulf.

TABLE 9.8 The mean spatiotemporal abundance of fish eggs and larvae in the Saudi Arabian waters of the Gulf.

	CRUISE NUMBER	SEASONS	NUMBER OF SAMPLES	TRANSECTS/REGIONS (MEAN ABUNDANCE/100 M³)					OVERALL (ABUNDANCE/100 M³)
				COASTAL	OFFSHORE	NORTHERN	CENTRAL	SOUTHERN	MIN-MAX; MEAN (MEDIAN)
Fish eggs	**1**	Late Fall	45	161	990	168	128	1020	128–1020; 493 (168)
	2	Winter	56	242	33	136	301	82	33–301; 159 (136)
	3	Spring	57	546	414	1002	395	517	395–1002; 575 (517)
	4	Onset of summer	58	648	471	190	191	469	190–648; 394 (469)
	5	Summer	56	516	390	865	469	196	196–865; 487 (469)
	Annual mean			**423**	**460**	**472**	**297**	**457**	**297–472; 422 (457)**
Fish larvae	**1**	Late Fall	45	37	547	344	107	144	37–547; 236 (144)
	2	Winter	56	287	56	80	93	112	56–287; 126 (93)
	3	Spring	57	862	461	260	328	864	260–864; 555 (461)
	4	Onset of summer	58	75	442	113	575	380	75–575; 317 (380)
	5	Summer	56	263	135	270	240	147	135–270; 211 (240)
	Annual mean			**305**	**328**	**213**	**269**	**329**	**213–329; 289 (305)**

Several factors such as temperature, salinity, wind speed and direction, photosynthesis, and respiration of organisms affect the DO concentration in the marine environment. Temperature is vital in regulating DO concentrations (Cordoba and Vargas 1996). The water temperature was lower during late fall and winter and higher during spring, the onset of summer, and summer near the coast. The DO levels were higher near the coast during winter and late fall and lower during spring, the onset of summer, and summer. The coastal waters were colder than offshore during winter and late fall, then warmer during spring and summer. This is due to the rapid response of the coastal water temperatures to the seasonal air temperature changes. Low water temperature in the northern end of the coastal transect in winter increased the DO concentrations. Stratification, both thermal and haline, which develops during summer due to increased solar radiation and advection of low-salinity waters from the eastern offshore region towards the coast, also leads to a stratified DO distribution, with high values in the upper layers and low values in the near-bottom layer. Wind speed and wave action are the other factors that influence the DO levels. The shallow nature of the Gulf and the strong winds result in the Gulf waters being highly oxygenated in winter. The lowest dissolved oxygen levels of 3.96 mg/l observed previously in the Gulf were reported by Brewer and Dyrssen (1985), while during the present study, the minimum DO levels reported were about 0.81 mg/l over a comparatively smaller area. Shriadah and Al-Ghais (1999) observed an annual DO variation from 3.63 mg/l to 9.02 mg/l along the UAE coast, with a mean value of 5.98 mg/l. In Kuwaiti waters, the DO levels ranged from 5.92 to 9.69, with a mean of 7.46 mg/l. In the present study, the mean DO levels were comparable between the northern (5.62 mg/l), central (5.54 mg/l), and southern (5.44 mg/l) waters, with a slight decreasing trend from north to south. The mean DO in the study area across all seasons was 5.53 mg/l. As in a KFUPM/RI (1988a) report, most DO values recorded in Saudi Arabian waters are between 5.7 mg/l and 7.15 mg/l.

9.4.2 Fish eggs and larvae

The present study identified 62 larval fish families, 47 genera, and 11 species in Saudi Arabian territorial waters in the Gulf. The highest number (49) of larval fish families was recorded during spring and the lowest (25) during summer. Twelve larval fish families (Bothidae, Bregmacerotidae, Callionymidae, Carangidae, Clupeidae, Cynoglossidae, Engraulidae, Gobiidae, Mullidae, Sciaenidae, Sparidae, and Synodontidae) were frequently recorded throughout the seasons. Houde et al. (1986) reported nine families (Engraulidae, Clupeidae, Gobiidae, Sparidae, Callionymidae, Bregmacerotidae, Cynoglossidae, Soleidae, and Bothidae) from Kuwaiti waters. The codlet (Bregmacerotidae) are small pelagic fish inhabiting the neritic and oceanic waters of tropical and subtropical seas. Houde et al. (1986) identified all three species of *Bregmaceros* (*Bregmaceros arabicus*, *B. nectabanus*, and *B. rarisquamosus*) from the Northwest Arabian Gulf (NWAG) off Kuwait in July (Ahmed and Hussain 2000). Also, the present study confirmed few changes in the family composition of ichthyoplankton in Saudi Arabian waters since 1980.

In an earlier study (1979–1980) from Kuwaiti waters, Houde et al. (1986) reported the highest ichthyoplankton density (2283/100 m^3) during the onset of summer (June 1980) and the lowest (490/100 m^3) during winter (January 1980). The mean fish larval density for other seasons remained 711/100 m^3 in late fall, 793/100 m^3 in spring, and 865/100 m^3 in summer. The present study (2002–2003) in the western Gulf, south of Kuwaiti waters, reported the lowest larval fish density in winter (January, 308/100 m^3) and the highest (1147/100 m^3) in spring (March–April). Similar to seasonal patterns reported by Houde et al. (1986) in Kuwaiti waters, the mean ichthyoplankton density in Saudi Arabian waters did not differ significantly between other seasons (762/100 m^3 in the late fall, 729/100 m^3 during the onset of summer, and 699/100 m^3 in summer). Earlier studies (KFUPM/RI 1988b) in Saudi Arabian waters of the Gulf indicate high larval fish densities in nearshore

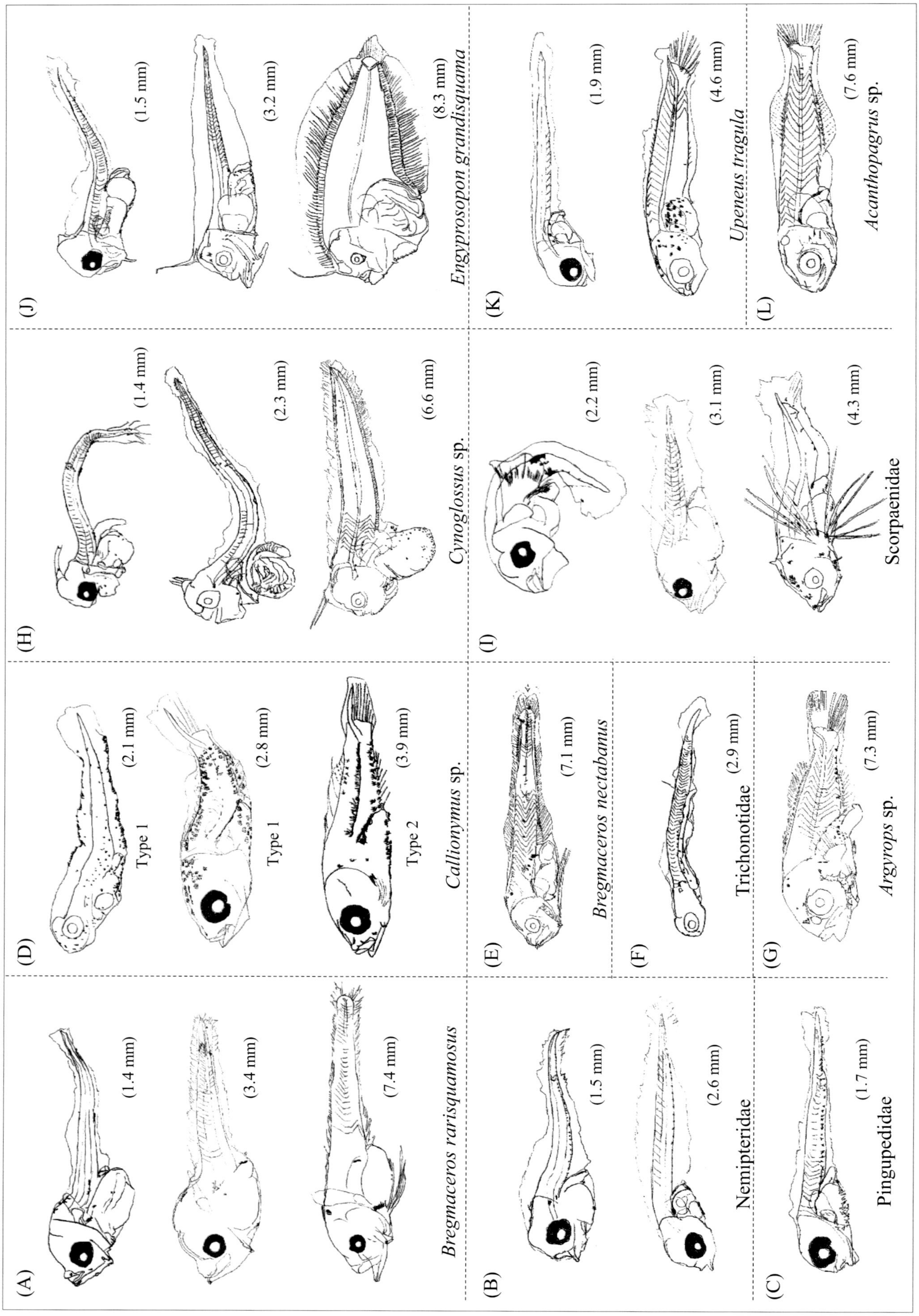

FIGURE 9.6 Representative illustrations of fish larvae from the Saudi Arabian waters of the Arabian Gulf in 2002–2003.

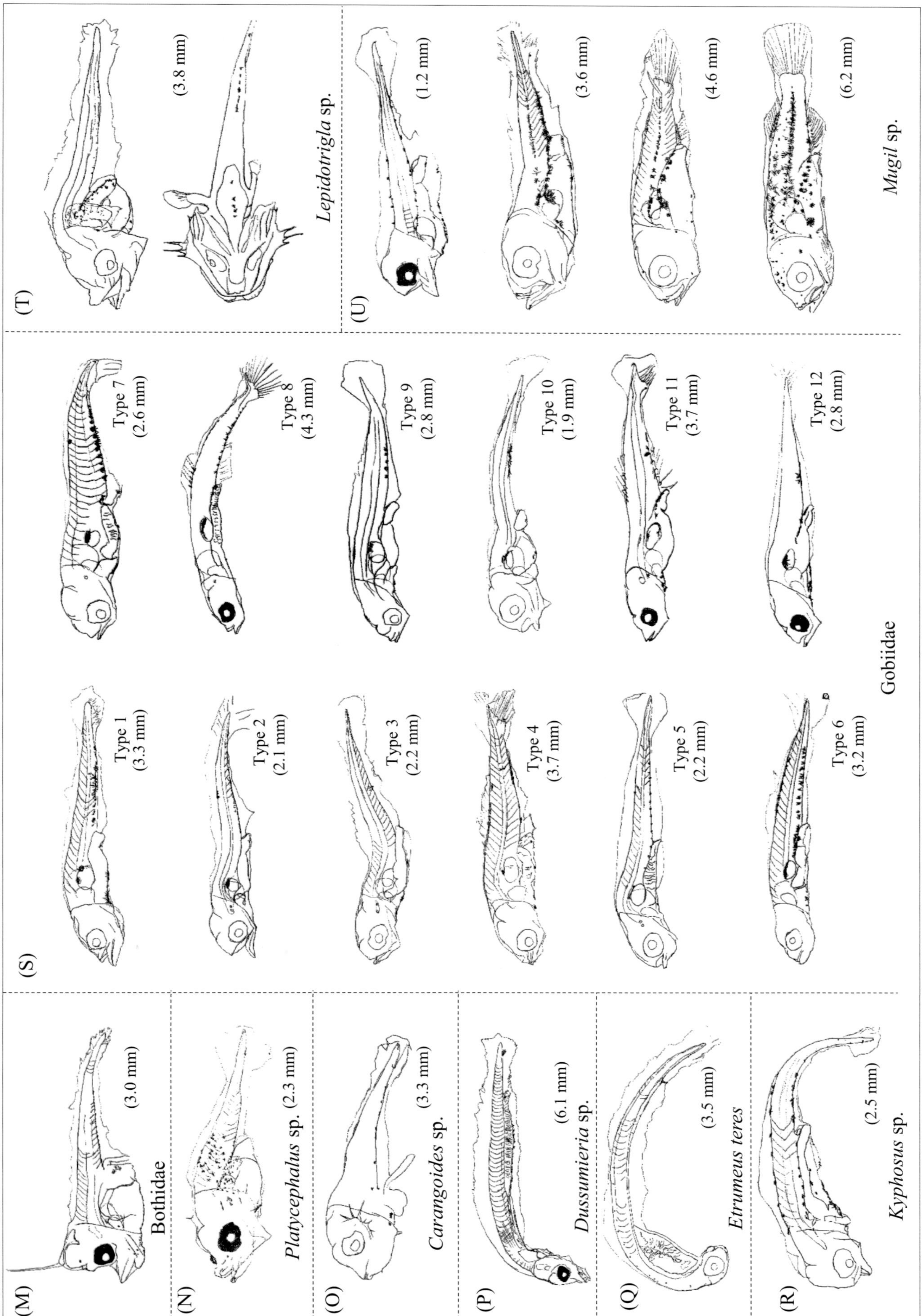

FIGURE 9.6 (Continued)

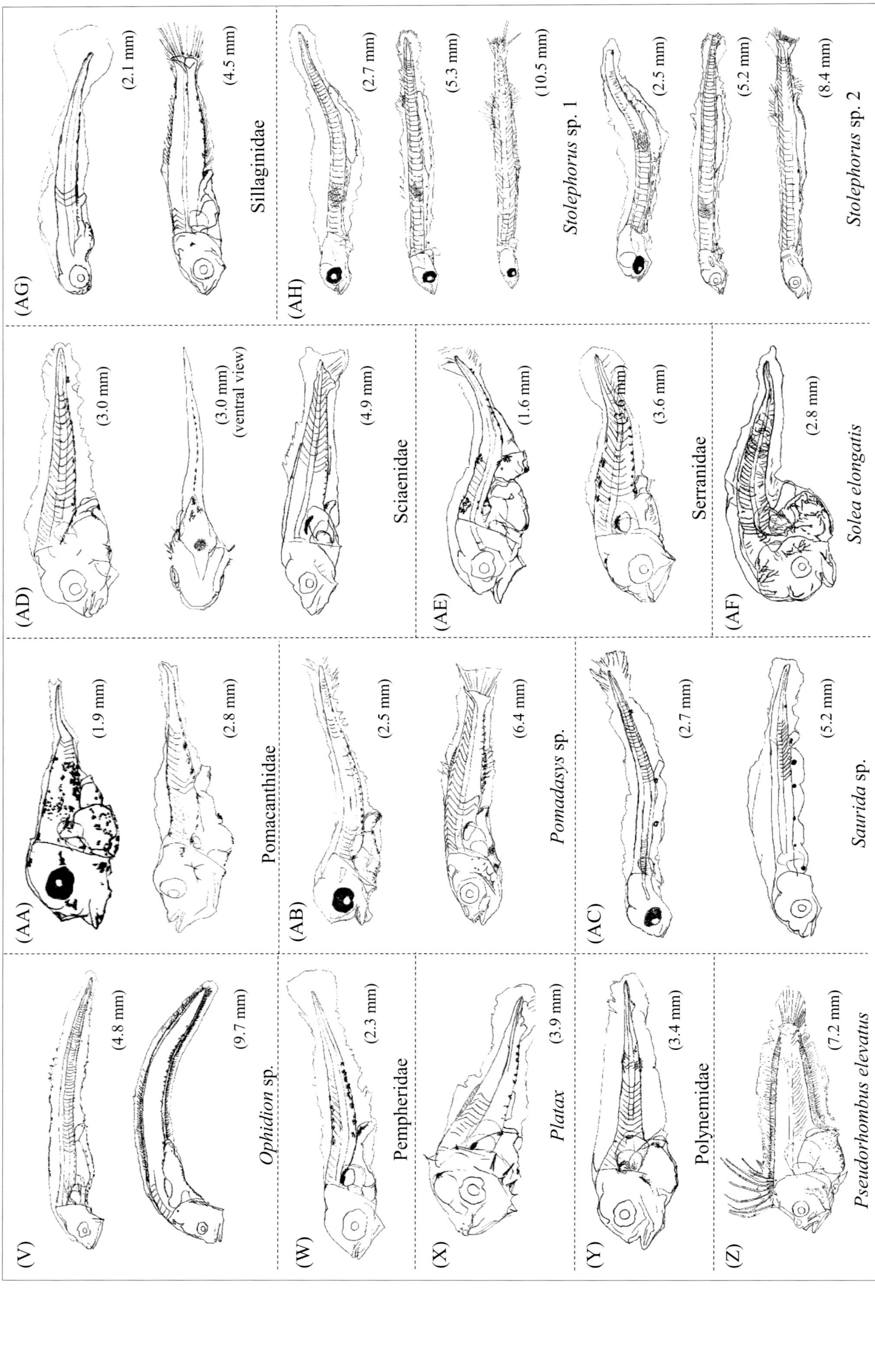
(V)
(4.8 mm)
(9.7 mm)
Ophidion sp.
(W)
(2.3 mm)
Pempheridae
(X)
(3.9 mm)
Platax
(Y)
(3.4 mm)
Polynemidae
(Z)
(7.2 mm)
Pseudorhombus elevatus
(AA)
(1.9 mm)
(2.8 mm)
Pomacanthidae
(AB)
(2.5 mm)
(6.4 mm)
Pomadasys sp.
(AC)
(2.7 mm)
(5.2 mm)
Saurida sp.
(AD)
(3.0 mm)
(3.0 mm) (ventral view)
(4.9 mm)
Sciaenidae
(AE)
(1.6 mm)
(3.6 mm)
Serranidae
(AF)
(2.8 mm)
Solea elongatis
(AG)
(2.1 mm)
(4.5 mm)
Sillaginidae
(AH)
(2.7 mm)
(5.3 mm)
(10.5 mm)
Stolephorus sp. 1
(2.5 mm)
(5.2 mm)
(8.4 mm)
Stolephorus sp. 2

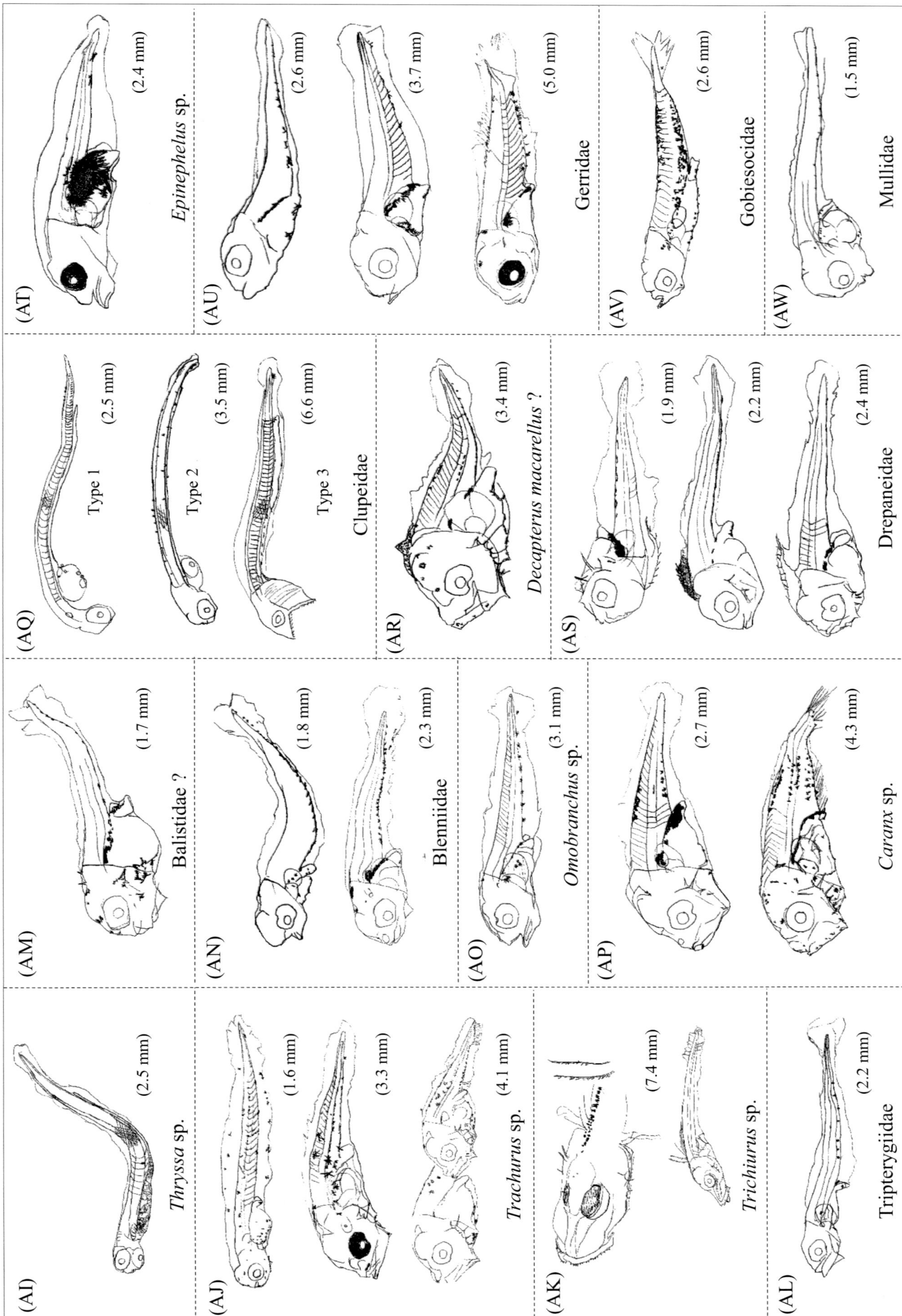

FIGURE 9.6 (Continued)

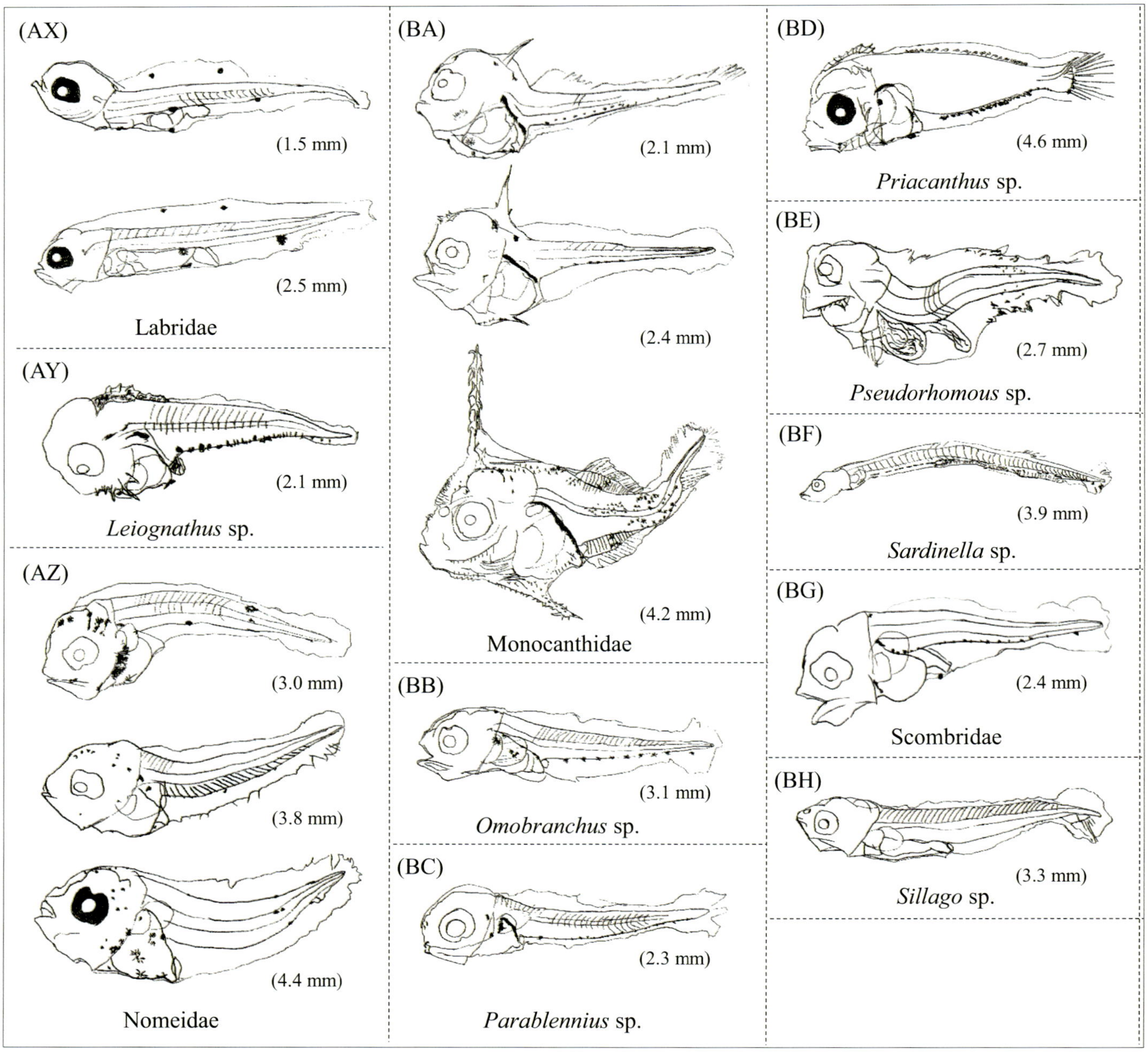

FIGURE 9.6 (Continued)

waters. Along the shore, the lowest densities were recorded in the Saudi Arabian part of the Gulf of Salwa, a narrow, shallow embayment separating the peninsula of Qatar from Saudi Arabia, where harsh environmental conditions prevail (salinity up to 63 and water temperature up to 40°C during summer (Joydas et al. 2015). In this study, although the mean fish larval density in the coastal (314/100 m^3) and offshore (325/100 m^3) waters remained comparable, ichthyoplankton exhibited a clear north-south (latitudinal) gradient in their abundance (southern waters, 341/100 m^3; central waters, 288/100 m^3; northern waters, 201/100 m^3). Interestingly, fish eggs had their seasonal peak (spring) in northern waters and fish larvae in southern waters.

The number of principal fish families and the mean percentage abundance of fish larvae varied spatially within seasons. Engraulidae and Clupeidae were the most abundant families in all seasons, while Scombridae was among the dominant families during the onset of summer only. Along the coastal waters, Gobiidae, Engraulidae, and Clupeidae were dominant. In offshore waters, the predominant families were Engraulidae, Clupeidae, and Scombridae. Although Clupeidae, Engraulidae, and Gobiidae were dominant in northern waters, Engraulidae, Clupeidae, and Gobiidae were the most abundant families in the central waters. Engraulidae, Clupeidae, and Callionymidae were dominant in southern waters, followed by Gobiidae, Sphyraenidae, Bregmacerotidae, Sciaenidae, Synodontidae, Cynoglossidae, Bothidae, Haemulidae, and Mullidae in their order of abundance.

Nellen (1973) reported fish larvae belonging to 107 families, 120 genera, and 65 species (mesh size: 500 µ) in the Arabian Sea and eastern Gulf. Gobiidae and Apogonidae were predominant in the southern part of the eastern Gulf, constituting 84% and 65%, respectively. In the northern part of the eastern Gulf, Gobiidae, Clupeidae, and Pomadasyidae were dominant. Similarly, Houde et al. (1986) reported fish larvae belonging to 53 families, 41 genera, 24 species, and 84 specific

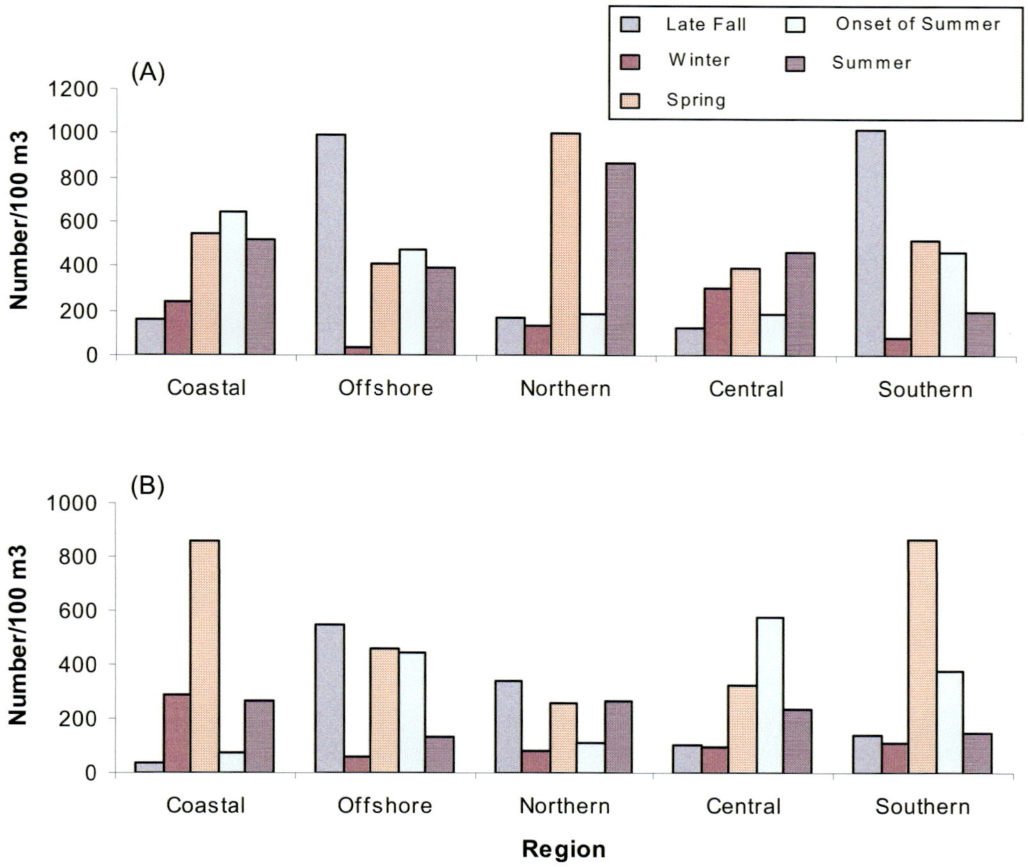

FIGURE 9.7 Seasonal variation in abundance of (A) fish eggs, (B) fish larvae.

types (by using a net with 333 μ mesh size) in Kuwaiti waters, offshore areas of Saudi Arabia, and in the southern part of the Gulf in 1979–80. Engraulidae, Gobiidae, and Clupeidae were the most dominant families in Kuwaiti waters. An earlier study from Kuwaiti waters also indicated the dominance of Clupeidae from April to October (Grabe et al. 1992).

In the present study, Engraulidae, Clupeidae, Gobiidae, Haemulidae, Sparidae, and Carangidae were the dominant families in Saudi Arabian waters of the western Gulf. Thus, the type and abundance, and seasonal peaks in fish larvae observed in the present study are similar to the observations by Hussain and Ahmed (1999) from Iraqi waters, Houde et al. (1986) and Grabe et al. (1992) from Kuwaiti waters. The spring-early summer abundance peaks of clupeid larvae were also reported from northern Iranian coastal waters, coinciding with the regional spawning season (Van Zaling et al. 1993).

Fish larvae collected during 1984–1985 from the coastal waters off Manifa and bay areas of Tanajib in the northern Gulf (KFUPM/RI 1986) contained only six families, with Carangidae, Gobiidae, and Serranidae as abundant families during summer and Clupeidae and Mugilidae during spring. Sparidae showed peaks during July and December but also high abundance during March. The present study also reported all those families, again indicating that the species composition has not changed much, even locally. Clupeidae and Gobiidae are the most dominant families of fish larvae (collected with mesh size 505 μ), even in the eastern Gulf of Mexico (Houde et al. 1979).

In a study in Kuwaiti waters in October 1991, Al-Yamani et al. (1992) reported mean fish egg and larval densities of 100 and 300/100 m^3, respectively. In comparison, McCain and Hassan (1993) reported a range of 0–567 eggs and 0–130 larvae/100 m^3 in Saudi Arabian waters of the Gulf for the same month. In yet another study (using a 64-micron mesh net) from Kuwaiti waters in April–May 1992, Al-Yamani et al. (1993) reported fish egg densities varying between 500 and 10,600 eggs/100 m^3 (mean: 2,368 eggs/100 m^3) and fish larval density varying between 350 and 5,200 larvae/100 m^3 (mean: 1,546 larvae/100 m^3). In the present study, densities (from 330-micron mesh net) of fish eggs varied between 31 and 3,056 eggs/100 m^3 (mean: 550 eggs/100 m^3), and that of fish larvae varied between 10 and 3,056 larvae/100 m^3 (mean: 597 larvae/100 m^3) during April–May 2003. The higher abundances reported by Al-Yamani et al. (1993) might be due to the smaller mesh size used for larval fish collection.

While comparing the present findings with earlier reports in 1980 (Houde et al. 1986; fish eggs: 652/100 m^3, fish larvae: 375/100 m^3) and in 1991 (McCain and Hassan 1993; fish eggs: 207/100 m^3, fish larvae: 14/100 m^3), it could be inferred that after a severe reduction in the ichthyofaunal abundance following the 1991 oil spill, the larval fish population had recuperated by 2002–2003 (present study; fish eggs: 431/100 m^3, fish larvae: 298/100 m^3). During the 1991 oil spill, the northern part of the Saudi Arabian waters was significantly impacted. However, in 2002–2003, the mean abundance of fish eggs and larvae was highest in the northern and southern waters, respectively.

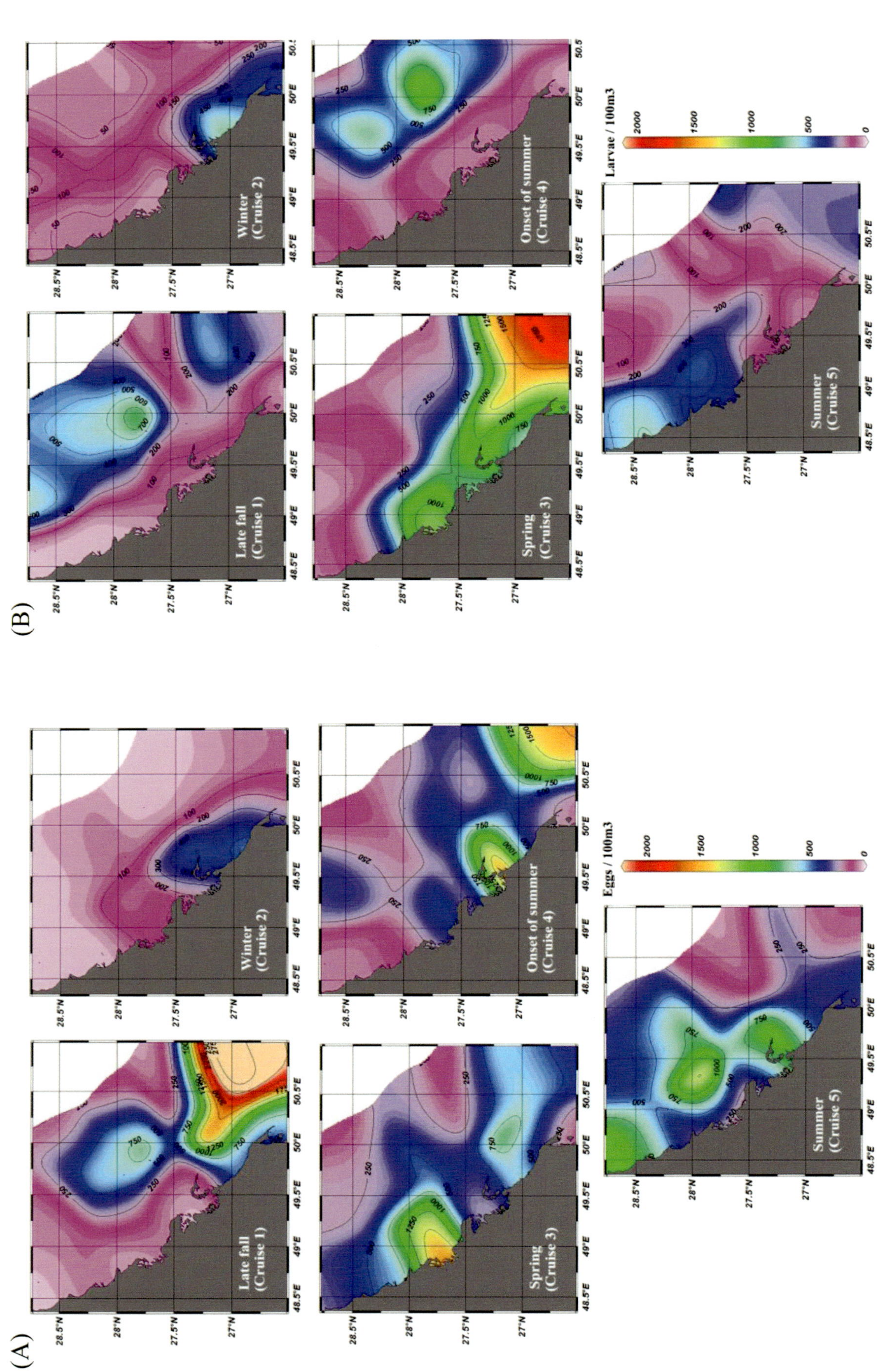

FIGURE 9.8 Spatial and seasonal distribution patterns of (A) fish eggs, (B) fish larvae in the Saudi Arabian territorial waters of the Gulf.

TABLE 9.9 BIOENV analysis showing the five best environmental combinations that influenced the seasonal fish larval distribution in the Saudi Arabian territorial waters of the Arabian Gulf.

SEASONS/NO. OF VARIABLES	*CORRELATION*	*ENVIRONMENTAL VARIABLES*
Late Fall		
1	0.648	Salinity, Oxygen
2	0.643	Temperature, Oxygen
3	0.629	Temperature, Salinity, Oxygen, TPH
4	0.627	Salinity, Oxygen, TPH
5	0.623	Temperature, Salinity, Oxygen, Chlorophyll
Winter		
1	0.379	Temperature, Salinity
2	0.376	Salinity
3	0.351	Temperature, Salinity, Oxygen
4	0.35	Temperature, Salinity, pH
5	0.35	Temperature, Salinity, Chlorophyll
Spring		
1	0.329	Temperature
2	0.299	Temperature, Salinity
3	0.266	Temperature, TPH
4	0.265	Temperature, Oxygen
5	0.258	Temperature, Salinity, Oxygen
Onset of Summer		
1	0.384	Temperature, Salinity, pH
2	0.371	Temperature, pH
3	0.366	Temperature, Salinity
4	0.362	Temperature, Salinity, pH, Chlorophyll
5	0.36	Temperature, Salinity, pH, TPH
Summer		
1	0.321	Temperature, Salinity
2	0.295	Temperature
3	0.268	Salinity
4	0.262	Temperature, Salinity, TPH
5	0.256	Temperature, Salinity, pH

ROPME organized a large-scale oceanographic expedition to study the physical, chemical, and biological oceanographic settings in the Gulf and Gulf of Oman in the winter of 2006 (ROPME 2012). Ichthyoplankton was an essential part of this expedition. The study revealed a decreasing gradient in fish egg density from coastal to offshore waters. The fish spawning areas differed from nursery and larval feeding areas; therefore, areas of high larval concentration did not correspond to areas of high fish egg density. The abundance of fish eggs in the Gulf was slightly higher than in the Gulf of Oman, especially at the northern end. However, maximal larval abundance was encountered in the southern end of the Gulf near the Strait of Hormuz. The winter 2006 mean fish larval density at stations corresponding to the 2002–2003 study (present study) shows comparable levels in the central sector but higher larval densities in the northern and southern waters in 2006. However, this comparison must be taken warily, as the studies used different sampling methodologies (such as vertical hauls from near bottom to surface in 2002–2003 and surface hauls in winter 2006). Also, the winter 2006 study failed to establish any relationships between ichthyoplankton and physicochemical and biological variables in the Gulf, as zooplankton data and seasonal environmental variability, the crucial factors affecting fish larvae, were not included in their analysis. Similar to the present observation, the winter 2006 study reported the most diverse fish larvae northwest of Bahrain, closer to the southern Saudi Arabian waters. In winter 2006, the Shannon-Wiener diversity for fish larvae in Saudi Arabian waters (H': 0–1.69) was significantly lower than the winter fish larval H' reported from the same area in 2002–2003.

9.4.3 Environmental regulation of fish larval populations

The larval stages of fish species in the Gulf are influenced by various factors that arise from its unique oceanographic setting, including the enormous seasonal changes in water temperature, high salinity, density-driven circulation patterns, and biological interactions. In this study, the biological-environmental interaction analysis (BIOENV) suggested that among the background variables used to explore environmental regulation of fish larvae, the cumulative effect of water temperature, salinity, dissolved oxygen, and Chl *a* influenced larval fish distribution in Saudi Arabian waters. An earlier study in Tarut Bay (KFUPM/RI 1990) reported temperature as the primary factor influencing reproductive activity and fish recruitment in Saudi Arabian waters. The seawater temperature always influences the seasonal structure of fish larvae in the Gulf (Rabbaniha et al. 2015). Studies in the northern Iranian coastal waters revealed a positive correlation between the abundance of pelagic-spawned fish larvae (e.g., Clupeidae, Terapontidae) and increased water temperature. The demersal recruited larvae (e.g., Gobiidae), on the other hand, were supported by higher Chl *a* levels in winter (Rabbaniha et al. 2015). Studies in Kuwaiti waters suggest that the spring-summer conditions in coastal waters of the northern Gulf favour a high density of nanoplankton (even frequent occurrence of HABs, harmful algal blooms) and microzooplankton (Ahmed et al. 2022; Saburova et al. 2022; Yamamoto et al. 2022) that offer the perfect prey-size spectrum for the survival of the pelagic spawned fish larvae. Conversely, the higher winter density of diatoms in the coastal waters ensures the benthic food supply for the demersal fish and their larvae.

As phytoplankton are the base of the marine food web and provide an important food source for micro- and mesozooplankton that form critical food sources for different developmental stages of fish larvae, areas with higher Chl *a* levels may be more attractive to spawners and fish larvae looking for food. Earlier studies on ichthyoplankton in the northern Gulf revealed that zooplankton availability and larval fish abundance have a strong positive relationship (Hussain and Ahmed 1999; Grabe et al. 1992). A recent study proved that zooplankton communities of the Arabian Gulf and the Gulf of Oman are shaped mainly by the combined effect of temperature, salinity, and dissolved oxygen, especially during summer and spring (Abedi et al. 2023). As suggested by the BIOENV analysis in this study, along with water temperature and salinity, the spatial and seasonal variations in Chl *a* levels (and subsequent changes in zooplankton communities) influence larval fish recruitment. Therefore, the highest abundance of fish larvae recorded during the spring-summer months suggests that a temperature rise and supply of optimal prey size are crucial in supporting larval fish recruitment, dominated by pelagic spawners, in the northern Gulf.

9.4.4 Ocean circulation and fish larval dispersal in Saudi Arabian waters

In the present study, based on the thermohaline structure and density distribution, it was found that low-salinity water of offshore southern origin intruded into the central and southern areas during spring. As a low-salinity tongue, this was most prominent in the upper layers and seen advected into the western region. The advection continued and peaked in summer. The strong inflow in summer extends almost to the head of the Gulf (Reynolds 2002). Combining the present observation with the previous studies (Johns et al. 1999; Reynolds 1993), it could be stated that the basic features of the circulation of the Gulf can be divided into two regimes – a northern regime and a southern regime. The northern regime is dominated by wind forcing (wind stress over the sea surface). The wind along the axis of the Gulf drives a southerly flow along the Arabian and Iranian coasts (Reynolds 1993). Along the Arabian coast, the wind-driven response is a downwelling coastal current that skirts the coast.

On the other hand, along the Iranian coast, the wind-driven response is an upwelling coastal current. The southerly coastal current along the coast of Kuwait and Saudi Arabia is modulated to some extent by the freshwater input from the Shatt Al-Arab River system at the Gulf's head. This freshwater inflow forms a strong coastal current approximately 20 km wide along the Kuwaiti and Saudi Arabian coasts (Reynolds 2002). The Acoustic Doppler Current Profiler (ADCP)-derived current vectors from the present study also corroborated the southerly flow. The centre of the northern Gulf appears to be pretty stagnant (Reynolds 1993). The warmer, low-salinity Indian Ocean surface water that intrudes the Gulf along the Iranian coast forms a cyclonic gyre off Qatar and creates a waterfront north of Qatar. This density-driven waterfront in the offshore waters north of Qatar moves further northwards and coastal as the season progresses through winter to summer, with its maximum effect in Saudi Arabian waters in summer.

Thus, the intrusion of low-salinity Indian Ocean waters and the formation of a waterfront near the southern Saudi Arabian waters seem to support diverse plankton stock that sustains a high density and diversity of fish larval populations in the southern parts of Saudi Arabian waters. Supporting evidence is available from a recent study in the central Gulf. Liu et al. (2022) report consistently higher mesozooplankton abundance, biomass, and diversity in the northeastern offshore locations off Qatar that periodically encounter nutrient and species-rich but low-salinity water entering the Gulf from the northwestern Indian Ocean.

9.5 FUTURE DIRECTIONS

9.5.1 DNA sequencing overcomes problems related to species-level taxonomy of fish larvae

Because of the small size, complex external morphology, anatomy, and ontogenetic variability, the morphological characteristics used to identify ichthyoplankton are few and challenging, hampering their taxonomic descriptions and often limiting their identifications to higher taxonomic levels. Many studies have shown that DNA sequencing offers better insights into the genetic diversity of fish larvae in marine ecosystems. A combination of molecular and morphology-based taxonomy approaches could yield invaluable information on the spatial and seasonal distribution of larval fish communities in the WAG. Nearly 75% of the known adult ichthyofauna from the Gulf have been barcoded, and sequences are available on Barcode Of Life Data Systems (BOLD Systems) (e.g., Clupeidae (11/14), Terapontidae (3/4), Gobiidae (15/20) (sequenced/known from the area)). Still, species-level descriptions of their larval forms are unknown mainly due to the inherent difficulties in identifying various developmental stages (especially the yolk-sac stage, pre-flexion, flexion, and post-flexion stages) of the same species.

To begin with, the focus should be on the larval stages of the principal and common families reported from the northwestern Gulf (Saudi Arabian and Kuwaiti waters). Several other selection criteria, such as abundant year-round, commercially important, pelagic, and demersal species, could also be used. The species-level identification can be made using DNA barcoding by matching the DNA sequences of the collected larval specimens to those of adults in the database. The Mitochondrial Cytochrome c Oxidase subunit I (COI) gene, owing to its low intraspecific but high interspecific genetic variation, is considered an efficient species-specific marine metazoan molecular marker, and it is typically adequate for discrimination of fish species (Ward et al. 2009). The DNA sequences of larval fish specimens could be aligned with available genetic sequences of adult and larval fish from NCBI and BOLD systems. Morphological characteristics such as melanophore patterns (larval pigment) of the genetically confirmed species could be used for subsequent morphological identification of larval stages, which is more applied and cost-effective. Such an approach could provide invaluable information on the larval fish populations of the Gulf, as well as their geographic distribution and migration patterns, which can be used to inform conservation and management efforts.

9.5.2 Micro- and mesozooplankton and fish larval recruitment

Microzooplankton, the heterotrophic organisms that feed on phytoplankton, nanoflagellates, and bacteria in the water column, are vital in transferring energy and nutrients to higher trophic levels, including fish larvae, especially during the early stages of development when they are still feeding on small plankton. Studies from the North Sea have highlighted the importance of microzooplankton for larval herring feeding ecology (de Figueiredo et al. 2007; Peck et al. 2012). The consumption of microzooplankton by higher trophic levels, such as copepods, indirectly benefits the post-larval stages of fish by providing a more stable and abundant food source. Thus, both microzooplankton and copepods are essential for the survival and growth of larval and post-larval stages of fish, but their relative importance varies depending on the specific fish species and the seasonal characteristics of the ecosystem.

Copepod community composition influences larval fish survival. Although the relative importance of cyclopoid and calanoid copepods on fish larvae dynamics in the Gulf is not well understood, a recent multi-year, multi-disciplinary study from a fish spawning and nursery ground in Kuwaiti waters (by the Kuwait Institute for Scientific Research, Kuwait, manuscript under processing) suggests that the larval fish peaks are adjusted to seasonal peaks in cyclopoid copepods in the northern Gulf. Cyclopoids (e.g., *Oithona* spp.) and calanoid copepods (e.g., *Parvocalanus crassirostris*, *Acartia* spp.) that are essential food sources for many fish larvae are present in high abundance in the northern Gulf. Cyclopoid copepods are smaller and more abundant than calanoid copepods and may be a more important food source for the initial developmental stages of fish larvae that cannot yet feed on larger prey. On the other hand, Calanoid copepods are larger, and because of their trophic position, they are more nutritious than cyclopoid copepods (that channel the primary production through grazing on microzooplankton) and may be a more important food source for larger or more mature fish larvae and juveniles. Seasonal environmental factors such as temperature and salinity regulate the relative abundance and spatiotemporal dynamics of cyclopoid (in summer and fall) and calanoid (in winter and spring) copepods in the coastal waters of the northern Gulf. This could affect the availability of these copepod groups as a food source for fish larvae in different parts of the Gulf. Therefore, information on zooplankton dynamics (micro- and mesozooplankton) in spawning areas and the size-structured plankton composition would significantly improve our understanding of the potential trophic relationships between fish larvae and zooplankton in the northern Gulf.

9.6 CONCLUSIONS

Covering a major part of the Saudi Arabian waters of the Gulf and data from various seasons, this study provides valuable baseline data on fish eggs and larvae of the region. Fish larvae belonging to 62 families (47 down to genera and 11 down to species levels) were recorded. Similar to the reports from Kuwaiti waters, Gobiidae, Engraulidae, and Clupeidae were the dominant families. The higher numbers (49) of families were reported in spring and lower numbers (25) in summer. Spatially, species richness and diversity showed increasing gradients towards offshore and the south, and seasonally, spring showed higher values compared to other seasons. Ichthyoplankton abundance was higher in the northern waters and lower in the southern waters. Similarly, fish larvae exhibited significant seasonal heterogeneity in the coastal and southern waters. The available environmental data suggested that the seasonal changes in water temperature, salinity, DO, Chl *a*, and circulation patterns were responsible for the observed fish larval distribution patterns in Saudi Arabian waters of the Gulf. As zooplankton (micro- and mesozooplankton) forms the most critical prey source for fish larvae, future studies must include the zooplankton standing stock (preferably in terms of dry mass or carbon equivalents for copepods) as an important explanatory variable. More concerted studies employing DNA sequencing for taxonomic resolution, experimental approaches, and field observations are needed to document the larval fish diversity and understand the importance of zooplankton composition on fish larval survival and how this may vary depending on the specific fish species and environmental conditions in the Gulf. Therefore, the present data and the inferences derived on fish larval standing stock, community structure, and the spatiotemporal aspects shall assist marine managers in formulating effective management practices for the sustainable harvesting of fishery resources in the northern Gulf in line with the UN's SDG-14.

REFERENCES

Abedi, E., Seyfabadi, J., Saleh, A., and A. Sari. 2023. Mesozooplankton communities related to water masses in the Persian Gulf and the Gulf of Oman. *Mar Pollut Bull* 188: 114571.

Aftab, J., and J. Ali-Khan. 1992. Composition and distribution of fish larvae in EEZ of Pakistan during January and February, 1977. *Proc Pakistan Congress Zool* 12: 501–514.

Ahlstrom, E.H. 1966. *Distribution and abundance of sardine and anchovy larvae in the California current region off California and Baja California, 1951–1964: A summary*. US Department of the Interior, Fish and Wildlife Service, No. 534, 71 p.

Ahmed, A., Al-Said, T., Madhusoodhanan, R., Naqvi, S.W.A., Sarkar, A., Fernandes, L., Thuslim, F., Al-Zakri, W., and F. Al-Yamani. 2022. Environmental impact of a series of flash flood events on a hypersaline subtropical system in the northwestern Arabian Gulf. *Mar Pollut Bull* 175: 113394.

Ahmed, S.M., and N.A. Hussain. 2000. Abundance and distribution of eggs and larvae of Clupeiformes in the north western Arabian Gulf. *Basrah Sci* 18: 159–164.

Ali-Khan, J. 1976. Distribution and abundance of fish larvae off the coast of West Pakistan. *Mar Biol* 37: 305–324.

Ali-Khan, J. 1985. Occurrence and distribution of fish larvae in Gulf of Aden during October to November 1966 and February to March 1967. *Indian J Fish* 32: 198–214.

Ali-Khan, J. 1989. Kinds and abundance of fish larvae in the Arabian Sea during a SW monsoon period. *Meeresforsch* 32: 218–240.

Ali-Khan, J. 1994. Day and night differences in the larval fish catch from the Gulf of Aden. *Meeresforschung* 28, no. 1: 10–18.

Ali-Khan, J., and J. Aftab. 1993. Relative abundance of fish larvae in Pakistani waters based on "Dr. Fridtjof Nansen" cruises 1 and 2, 1997. *Marine Res* 2: 1–9.

Ali-Khan, J., and G. Hempel. 1974. Relation of fish larvae and zooplankton biomass in the Gulf of Aden. *Mar Biol* 28: 311–316.

Ali-Khan, J., and T. Waqar. 1981. Abundance of fish larvae in Manora Channel during 1968–1970. *Biologia* 27, no. 2: 128–142.

Ali-Khan, J., and T. Waqar. 1985. On the occurrence of fish larvae in the Manora Channel, Karachi. *Indian J Fish* 32: 90–100.

Ali-Khan, S., Aftab, J., and J. Ali-Khan. 2001. Morphology of some ichthyoplankton from the backwaters of Sandspit. *Pakistan J Mar Biol* 7: 329–342.

Al-Yamani, F.Y. 2021. *Fathoming the northwestern Arabian Gulf: Oceanography and marine biology*. Kuwait: Kuwait Institute for Scientific Research. ISBN 978-99966-37-40-7.413 p.

Al-Yamani, F.Y., Al-Rifaie, K., and W. Ismail. 1993. Post-spill zooplankton distribution in the NW Gulf. *Mar Pollut Bull* 27: 239–243.

Al-Yamani, F.Y., Ismail, W., and K. Al-Rifaie. 1992. *Post-war assessment of zooplankton resources in the Kuwait waters*. Proc 1992 Middle East Environ Conf, April, World Trade Centre, Dubai, pp. 7–9.

Brewer, P.G., and D. Dyrssen. 1985. Chemical oceanography of the Persian Gulf. *Progress Oceanog* 14: 41–55.

Cordoba, R., and A.J. Vargas. 1996. Temperature, salinity, oxygen, and nutrient profiles at a 200 m deep station in Golfo Dulce, Pacific coast of Costa Rica. *Rev Biol Trop* 44: 233–236.

D'Ancona, U., and G. Cavinato. 1965. The fishes of the family Bregmacerotidae. *Dana Report* 64: 1–92.

De Figueiredo, G.M., Nash, R.D., and D.J. Montagnes. 2007. Do protozoa contribute significantly to the diet of larval fish in the Irish Sea? *J Mar Biol Assoc UK* 87, no. 4: 843–850.

Grabe, S.A., Lees, D.C., and H.P. Allaire. 1992. Macrozooplankton studies in Kuwait Bay (Arabian Gulf). I: Distribution and composition of the ichthyoplankton. *J Plankton Res* 14, no. 4: 607–623.

Hayashi, S., and A. Tadokoro. 1962. Occurrence of the Taiwan-Ainoko, Stolephorus zollingeri (Bleeker). *Bull Japan Soc Sci Fish* 28, no. 1: 26–29.

Hoda, S.M.S. 1978. On some post-larval fishes from Iraqi waters. *Biologia* 24, no. 2: 223–235.

Houde, E.D. 1977. Food concentration and stocking density effects on survival and growth of laboratory-reared larvae of bay anchovy *Anchoa mitchilli* and lined sole *Achirus lineatus*. *Mar Biol* 43: 333–341.

Houde, E.D. 1981. Distribution and abundance of four types of codlet (Pisces: Bregmacerotidae) larvae from the eastern Gulf of Mexico. *Biol Oceanog* 1, no. 1: 81–104.

Houde, E.D., Almatar, A.H., Leak, J.C., and C.E. Down. 1986. *Ichthyoplankton abundance and diversity in the western Gulf*. Kuwait Bull Marine Sci, KISR, Kuwait. No. 8, 60 p.

Houde, E.D., and P.L. Fore. 1973. *Guide to identify of eggs and larvae of some Gulf of Mexico clupeid fishes*. Florida Department of Natural Resources, Marine Research Laboratory, Leafl. Ser. IV, Pt 1 (23), 14 p.

Houde, E.D., Leak, J.C., Dowd, C.E., Berkeley, S.A., and W.J. Richards. 1979. *Ichthyoplankton abundance and diversity in the eastern Gulf of Mexico*. Rep to Bur Land Mgt, Contract AA550-CT7–28. NTIS PB-299839. xxxii plus 546 p.

Houde, E.D., and R.C. Schekter. 1983. Oxygen uptake and comparative energetics among eggs and larvae of three subtropical marine fishes. *Mar Biol* 72, no. 3: 283–293.

Hussain, N.A., and S.M. Ahmed. 1999. Influence of hydrographic conditions on the interaction between ichthyoplankton and macrozooplankton at Khor Al-Zubair lagoon, Iraq, Arabian Gulf. *Qatar Univ Sci J* 18: 247–259.

Jeyaseelan, M.J.P. 1998. *Manual of fish eggs and larvae from Asian mangrove waters*. Paris, France: UNESCO, 193 p.

Johns, W.E., Jacob, G.A., Kindle, J.C., Murray, S.B., and M. Carron. 1999. *Arabian marginal seas and gulfs*. University of Miami RSMAS Technical Report 2000–01, Report of a Workshop held at Stennis Space Centre, 11–13 May, Mississippi.

Joydas, T.V., Qurban, M.A., Manikandan, K.P., Ashraf, T.T.M., Ali, S.M., Al-Abdulkader, K., Qasem, A., and P.K. Krishnakumar. 2015. Status of macrobenthic communities in the hypersaline waters of the Gulf of Salwa, Arabian Gulf. *J Sea Res* 99: 34–46.

KFUPM/RI. 1986. *Final report – Northern area biological study, 1984/85*. Prepared for the Arabian American Oil Company by the Water Resources and Environment Division, Research Institute, King Fahd University of Petroleum and Minerals, Saudi Arabia, Report Project No. 24059, pp. 1–134.

KFUPM/RI. 1988a. *Marine environmental effects of the Abu Ali-Berri causeway extension*. Prepared for the Arabian American Oil Company by Water Resources and Environment Division, Research Institute, King Fahd University of Petroleum and Minerals, Dhahran, Saudi Arabia. Report Project No. 24108.

KFUPM/RI. 1988b. *Aramco sustaining research project – Environmental studies: Third annual report 1986/87. Vol. II, Oceanographic investigation*. Prepared for the Arabian American Oil Company by Water Resources and Environment Division, Research Institute, King Fahd University of Petroleum and Minerals, Dhahran, Saudi Arabia. Report Project No. 24079.

KFUPM/RI. 1990. *Aramco sustaining research project-environmental studies, vol. III, Tarut bay biotopes investigation – Plankton and disheries*. Prepared for Saudi Aramco by the Water Resources and Environmental Division, Research Institute, King Fahd University of Petroleum and Minerals, Dhahran, Saudi Arabia. Report Project No. 24079.

Leak, J.C. 1981. Distribution and abundance of carangid fish larvae in the eastern Gulf of Mexico, 1971 – 1974. *Biol Oceanog* 1, no. 1: 1–28.

Leis, J.M., and B.M. Carson-Ewart (Eds.). 2000. *The larvae of Indo-Pacific coastal fishes: An identification guide to marine fish larvae*, vol. 2. Brill.

Liu, H., El-Din, N.N., Rowe, G., Al-Ansi, M., Wei, C.L., Soliman, Y., Nunnally, C., Quigg, A., Al-Ansari, I.S., Al-Maslamani, I., and M.A. Abdel-Moati. 2022. Characteristics and renewal of zooplankton communities under extreme environmental stresses in the oligotrophic hypersaline Arabian Gulf. *Prog Oceanogr* 201: 102643.

McCain, J.C., and M. Hassan. 1993. Assessment of ichthyoplankton in Saudi Arabian Gulf waters before and after the 1991 Gulf oil spill. *Arab J Sci Eng* 18, no. 2: 269–291.

Moser, H.G. (Ed.). 1996. *The early stages of fishes in the California: Current region*. Lawrence, KS: Allen Press.

Moser, H.G., Ahlstrom, E.H., and J.R. Paxton. 1984. Myctophidae: Development. *Ontogeny Systematics Fishes* 1: 218–239.

Nellen, W. 1973. Kinds and abundance of fish larvae in the Arabian Sea and Persian Gulf. In: *The biology of the Indian Ocean*. Ed: B. Zeitzschel, 415–430. Berlin: Springer-Verlag.

Peck, M.A., Kanstinger, P., Holste, L., and M. Martin. 2012. Thermal windows supporting survival of the earliest life stages of Baltic herring (Clupea harengus). *ICES J Mar Sci* 69, no. 4: 529–536.

Rabbaniha, M., Molinero, J.C., Lopez-Lopez, L., Javidpour, J., Primo, A.L., Owfi, F., and U. Sommer. 2015. Habitat association of larval fish assemblages in the northern Persian Gulf. *Mar Pollut Bull* 97, nos. 1–2: 105–110.

Reynolds, R.M. 1993. Physical oceanography of the Gulf, Strait of Hormuz, and the Gulf of Oman – Results from the Mt Mitchell expedition. *Mar Pollut Bull* 27: 35–59.

Reynolds, R.M. 2002. Oceanography. In: *The Gulf ecosystem, health and sustainability*. Eds: N.Y. Khan, M. Munavar, and A.R.G. Price, 53–64. Leiden, The Netherlands: Backuys Publishers.

ROPME. 2012. *ROPME oceanographic cruise-Winter 2006*. Icthyoplankton in the ROPME Sea Area. Technical Report No. 8, 102 p.

Saburova, M., Al-Kandari, M., Polikarpov, I., Akbar, A., Hussain, S., Rahmeh, R., Al-Zakri, W., and F. Al-Yamani. 2022. Alien toxic dinoflagellate Heterocapsa circularisquama from the western pacific in Kuwait, NW Indian ocean. *Deep Sea Res Part II: Top Stud Oceanogr* 196: 105027. doi:10.1016/j.dsr2.2022.105027.

Shriadah, M., and S.M. Al-Ghais. 1999. Environmental characteristics of the United Arab Emirates waters along the Arabian Gulf: Hydrographical survey and nutrient salts. *Indian J Mar Sci* 28, no. 3: 225–232.

Sreekumari, A. 1976. *Development and distribution of the larvae of the white bait, Stolephorus zoolingeri Bleeker (Engraulidate: Pisces) along the west cost of India*. Proc Symp Warm Waters Zooplankton, Special publication, UNESCO, NIO, Goa.

Swift, S.A., and A.S. Bower. 2003. Formation and circulation of dense water in the Persian/Arabian Gulf. *J Geophys Res Oceans* 108, no. C1: 4–1. doi:10.1029/2002JC001360.

Van Zaling, N.P., Owfi, F., Ghasemi, S., Khorshidian, K., and N. Niamaimandi. 1993. *Resources of small pelagic in Iranian waters, a review*. Rome: Food and Agriculture Organisation.

Ward, R.D., Hanner, R., and P.D. Hebert. 2009. The campaign to DNA barcode all fishes, FISH-BOL. *J Fish Biol* 74, no. 2: 329–356.

Yamamoto, T., Madhusoodhanan, R., Al-Said, T., Ahmed, A., Fernandes, L., Nithyanandan, M., Thuslim, F., Alghunaim, A., Al-Zekri, W., Naqvi, S.W.A., and F. Al-Yamani. 2022. Dynamic hydrographic and water-quality variations in the northwestern Arabian Gulf, a sinking zone of reverse estuarine circulation. *Mar Pollut Bull* 179: 113714.

Yoshida, J., Matsuyama, M., Senjyu, T., Ishimaru, T., Morinaga, T., Arakawa, H., Kamatani, A., Maeda, M., Otsuki, A., Hashimoto, S., and I. Kasuga. 1998. *Hydrography in the RSA during the RT/V Umitaka-Maru cruises: Offshore Environment of the ROPME Sea Area after the war-related oil spill*, 1–22. Tokyo: Terra Scientific Publishing Company.

Photo-identification of the oceanic whitetip shark (*Carcharhinus longimanus*) at Elphinstone Reef, Egypt, Red Sea

10

Ahmed M. Shawky and Alessandro De Maddalena

10.1 INTRODUCTION

The Red Sea is a well-known area for the presence of sharks (Randall 1986). A total of 49 species of sharks occurs in the area encompassing the Red Sea and the Gulf of Aden (Bonfil and Abdallah 2004). Human impacts negatively affect shark abundance and behaviour, but there are relatively few studies data regarding the diversity and ecological significance of sharks in the Red Sea (Shawky and De Maddalena 2013; Braun et al. 2014; Spaet et al. 2016). The program "Shark ecology and its sustainable use in the Southern Red Sea, Egypt" was carried out in 2008 by the authors to assess the presence of sharks and to study shark–human interactions (fishing and diving activities) at different diving sites in the areas of Marsa Alam (Elphinstone Reef) and the far southern reefs and islands, southern Egyptian Red Sea, by means of underwater observations and data collection (Shawky and De Maddalena 2013).

The oceanic whitetip shark *Carcharhinus longimanus* (Poey 1861) belongs to the order Carcharhiniformes and to the family Carcharhinidae (Compagno 1984). *C. longimanus* is considered to be one of the most important shark species supporting the Red Sea biodiversity. It is a pelagic species that has a cosmopolitan distribution, having been recorded in the Atlantic, Pacific and Indian Oceans, and tends to stay on continental and insular shelves, slopes and oceanic basins, at depths ranging from the surface to at least 152 m (Compagno 1984; Matsunaga and Nakano 1999; Camhi et al. 2008; Young and Carlson 2020). This species is mainly found in waters with temperatures over 20°C (Flmnh 2016), but it can also occur between 15°C and 28°C (Compagno 1984). The oceanic whitetip shark is active and usually slow, but able to swim at high speed up to 4.6 m/second (Papastamatiou et al. 2018). Equally nocturnal and diurnal, it tends to be solitary but can be found in large to enormous groups around a food source. This shark may approach divers closely, usually without showing any aggressive behaviour, and divers should use caution (Shawky and De Maddalena 2013). It can segregate by sex and size. It can occasionally ingest inedible items. It feeds on schooling fishes by simply taking bites from the tightly massed prey and possibly swimming through the school with its mouth wide open and ingesting prey that inadvertently swims into its jaws (Madigan et al. 2015; Andrzejaczek et al. 2018). Possibly their white fin apex serves to bait prey into the strike zone (seen from a distance its white patches may appear like a school of small fishes and when the prey approaches, the shark quickly accelerates and attacks) (Myrberg 1987). *Carcharhinus longimanus* is a migratory shark and a top predator found in tropical open water worldwide and has significant importance in marine ecosystems (Compagno 1984; Camhi et al. 2008; McCauley et al. 2013; Camargo et al. 2016).

C. longimanus has a very distinctive morphology that makes it unmistakable. The first dorsal fin is very large and tall, originating over the pectoral fin inner margin. The second dorsal fin is small, compared to the first. The anal fin is larger than the second dorsal and with a long apex, and its posterior margin is deeply concave. Pectoral fins are very long and wide, and their apex slightly rounded. The caudal fin has a relatively long lower lobe and very long upper lobe with a medium-sized terminal lobe. A low interdorsal ridge is present or absent. The snout is dorso-ventrally depressed and relatively long, rounded in a dorso-ventral view. The mouth is wide and parabolic in ventral view. Labial folds are short, the eyes are relatively small, the nostrils are small, and there are no spiracles. There are five pairs of short gill slits, the fourth and fifth

DOI: 10.1201/9781003321392-12

located over the pectoral fin base. Dorsal surfaces are bronze-grey and ventral surfaces white. The first dorsal, pectoral, pelvic and caudal fin upper and lower apex have a conspicuous irregular white patch; the second dorsal and anal fins have a conspicuous irregular black patch at the apex, more extended on the anal fin, and additional small dark spots on the second dorsal fin apex. The boundary separating dorsal from ventral colouration is sharp and indented. The pectoral fin ventral surface is partially white and partially covered by the same dark indented colouration of the dorsal surface, particularly at the inner margin, free rear tip and posterior margin. The upper teeth have one cusp, are large, triangular, and with strongly serrated edges; the lower teeth are smaller, with one cusp, narrow, pointed, with very finely serrated edges in the upper part, and erect to oblique. The dental formula is 14 to 15–1 to 2–14 to 15/13 to 15–1–13 to 15.

The maximum size is 395 cm. The size at birth is 60–70 cm, and at maturity is 175–198 cm (male) and 180–200 cm (female), at 4–7 years of age (Lessa et al. 1999; Bonfil et al. 2008). The embryonic development is placental viviparous, and the gestation is 9–12 months with a litter size of 1–15 young. (Backus et al. 1956; Seki et al. 1998). It has a recorded range of litter sizes from 1–15 pups (Backus et al. 1956; Stevens 1984; Bonfil et al. 2008), while the Red Sea population has recorded litter sizes of 10–15 pups (Gohar and Mazhar 1964). The maximum age is at least 18 years in the central Pacific (D'Alberto et al. 2017). The oceanic whitetip shark is both a predator of bony fishes, stingrays, cephalopods, sea turtles, cetaceans, gastropods, and crustaceans, and a scavenger that consumes 1 to 1.5 kg of prey per day (Madigan et al. 2015).

The Red List of the International Union for Conservation of Nature (IUCN) currently lists *C. longimanus* as "Critically Endangered" (Rigby et al. 2019). It was listed in Appendix II of the Convention on International Trade in Endangered Species (CITES) in 2014 (Camargo et al. 2016). This species used to be extremely abundant around the world but fisheries, mostly because of by-catch, greatly reduced its populations everywhere (Bigelow and Schroeder 1948; Cousteau and Cousteau 1970; Garrick 1982; Castro 1983; Compagno 1984; Randall 1986; Myrberg 1987; Last and Stevens 2009; De Maddalena et al. 2016; De Maddalena 2017; Rigby et al. 2019). In the southern Egyptian Red Sea, the oceanic whitetip shark was recorded with more specimens observed for each shark species (Shawky and De Maddalena 2013). A 2012 Pacific-wide stock assessment for oceanic whitetip sharks showed that the populations are presently overfished and stocks have decreased to levels potentially below sustainable levels (Rice and Harley 2012). For the past 50 years, this species has had a global decline due to overfishing and by-catch of tuna and swordfish and the shark fin trade (Hazin et al. 1990; Baum et al. 2003; Baum and Myers 2004; Dulvy et al. 2014; Camargo et al. 2016). These data support the importance and the urgency of preserving the populations of *C. longimanus* in areas where they are still recorded on a regular basis.

Photo-identification (photo-ID) is a low-cost technique that can provide an alternative tool to identify individual animals, based on photos of body markings, without physically capturing the animal (Hammond et al. 1990; Karczmarski et al. 2005; Jean et al. 2010; Bicknell et al. 2016; Dodd et al. 2018). Underwater photography, using common photo-ID techniques, is an essential tool to determine the temporal and spatial migratory patterns, and to monitor the occurrence, distribution, sex ratio and the behaviour over time and space without any physical impacts (Strong et al. 1992; Reisser et al. 2008; Dodd et al. 2018). This method has been successfully used in studies on marine mammals, such as cetaceans (Shawky et al. 2015), dugongs (Anderson 1995; Shawky et al. 2017), sea turtles (Dunbar et al. 2021), whale sharks (Holmberg et al. 2008; Cochran et al. 2016; McCoy et al. 2018), basking sharks (Gore et al. 2016), white sharks (Domeier and Nasby-Lucas 2007; Anderson et al. 2011), and flapper skates (Dodd et al. 2018).

In this chapter, we present new findings of this assessment, specifically on the photo-ID of oceanic whitetip sharks occurring at Elphinstone Reef, Egypt. The goals include the following:

1. Identifying the different individual oceanic whitetip sharks at Elphinstone Reef and monitoring visiting patterns monthly through the creation of an identification catalogue.
2. Examining the sex ratios of visiting sharks and their re-occurrence patterns monthly.
3. Assessing the normal or abnormal behaviour of identified oceanic whitetip sharks toward divers.

10.2 MATERIALS AND METHODS

10.2.1 Study sites

A photo-ID project of oceanic whitetip sharks has been carried out to monitor the sharks inhabiting the area. The project was conducted over a total of 24 months at Elphinstone Reef, in the Marsa Alam region, from 28 June 2016 to 5 January 2019, between 6 a.m–3 p.m. The study site is a submerged reef with a maximum length and width of 500 m and 90 m, respectively. Its approximate location is 25.308°N and 34.860°E, 6.5 nautical miles (10 km) off the Egyptian coast in the Red Sea (Figure 10.1). All dives to observe the sharks were performed using SCUBA starting from the reef wall, passing over the south plateau and drifted more than 500 metres using our surface marker buoy (SMB) to allow the skipper to follow our bubbles for safety. Even if most observations of sharks were done while diving, sometimes after finishing the dive and getting out of the water into the speedboat, an oceanic whitetip shark ascended to the surface, and it was possible to film it from the boat by inserting the camera into the water on a 1.25-metre stick.

The operations were conducted aboard speedboats of the Red Sea Diving Safari and Blue Ocean diving centres; only on one day a boat of the Orca Diving Centre was used. No chum or bait was used to attract the sharks to the boat. The sharks were instead located by the experience of the diving team, who know where and when these predators can be found in the study area.

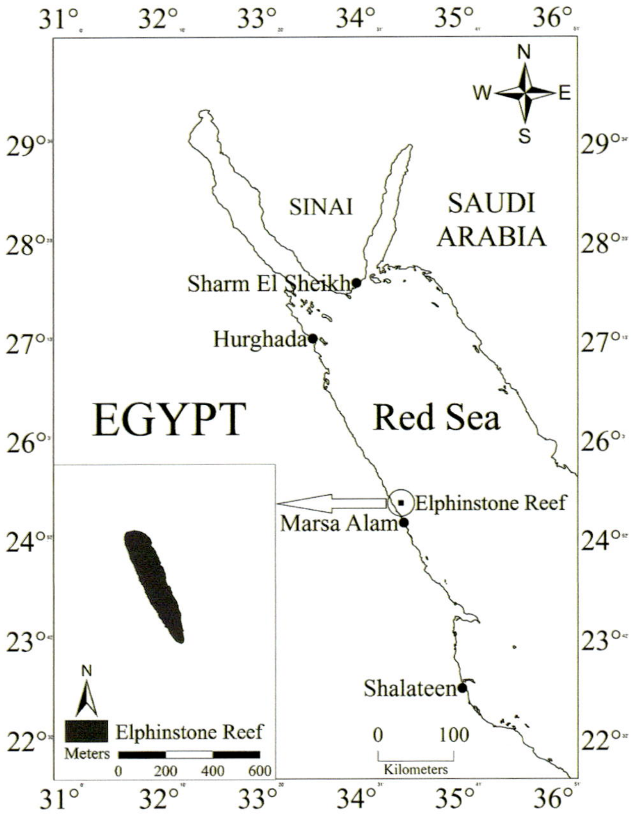

FIGURE 10.1 Map of the Egyptian Red Sea showing the location of Elphinstone Reef; the detail map indicated by arrow.

FIGURE 10.2 The first author taking photos of an oceanic whitetip shark using HD camera GoPro Hero 4 at Elphinstone Reef. Photo credit: Lorella Nimis.

10.2.2 Methodology

Sharks can be identified using each individual shark's unique shape and colouration of the dorsal fin. This technique is named finprinting and is totally non-invasive because the sharks are identified by photographs or video frames (Peschak and Scholl 2006). An attempt to apply finprinting in this study of the Red Sea oceanic whitetip sharks has been initiated. For every specimen encountered, an effort was made to collect a clear photo of both left and right surfaces of the first dorsal fin. Sharks were photographed or filmed using an HD GoPro Hero 4 Silver Edition with Red Filter (set in 4K mode) for subsequent analyses of their morphology and colouration, and whenever possible, photo-identification of the specimens (Figure 10.2). Although most of the photos and videos were taken by the first author, some images and footage were also acquired from volunteers. Sharks can be reliably identified using each individual shark's unique trailing-edge notch pattern, the shape of the tip and pigmentation patches of the dorsal fin, lower lobe of the caudal fin and pelvic fin for sex determination (Figure 10.3). All observed sharks were listed and assigned a five-digit identifier. The first two letters of the species name and the third letter for the site name were recorded, followed by the sequential number of the individual sighting. For example, the code of the *Carcharhinus longimanus* number 01 at Elphinstone Reef would be CLE01. All photos were compiled for every specimen encountered, with a specific effort to collect a clear photo of both left and right surfaces of the first dorsal fin. Gender was established by documenting the presence of claspers for males or absence for females (D'Alberto et al. 2017).

FIGURE 10.3 Main parts in the body of oceanic whitetip sharks that were used in photo-ID: (a) the shape of the tip and pigmentation patches of the dorsal fin, (b) lower lobe of the caudal fin, and (c) the sex of the shark. Photo credit: Ahmed M. Shawky.

10.3 RESULTS

10.3.1 Data collection

A total of 78 shark encounters were recorded during 148 dives. The details of the data collection are summarized in Table 10.1.

10.3.2 Photo-ID

Following this study, it has been possible to identify a total of 105 oceanic whitetip sharks at Elphinstone Reef (8 males and 97 females). Examining all the data obtained on the form for recorded encounters, a photo-ID catalogue of all the oceanic whitetip sharks observed during the study was prepared. For each specimen, the catalogue features the identification number, code, photos of the dorsal fin (both the right and the left sides), sex, and number of times the shark has been observed during the study period. A sample of the photo-ID catalogue is shown in Figure 10.4. For a few specimens, it was possible to take the photo of only one side of the first dorsal fin because the shark did not turn to the opposite side or the diver was unable to move to the other side because of the strong current. The identification of the individual sharks was made easier not only by the differences in the posterior margin of the dorsal fin and the shape of the fin apex, but also by the wide white patch at the apex, which is always strongly irregular both in extension and shape. Moreover, some individuals featured a deformed first dorsal fin, strongly bent or with deep cuts or incisions. In addition to the features of the first dorsal fin, those of the lower lobe of the caudal fin and the pelvic fins were also used.

For each encounter, from one to a maximum of six sharks were observed. The identification of the 105 individuals was distributed as follows over the years of study: 57 sharks were identified in 2016, 22 in 2017, 25 in 2018, and one in 2019. The number of newly identified (capture) oceanic whitetip sharks and those previously identified (re-capture) per month were recorded (Figure 10.5). Identified sharks were re-sighted up to seven times. The degree of residency for 105 individuals (i.e., the frequency of re-sighting) was recorded. Most sharks were encountered only once (64 individuals, or 61%), 22 individuals were sighted more than once (21%), 12 individuals (11%) were sighted three times, four individuals (4%) were sighted four times, five individuals (2%) were sighted five times, while only one individual (1%) was sighted seven times (Figure 10.6).

Concerning the sex of the observed sharks, of the 105 individuals, eight were males and 97 were females. This corresponds to 7.6% of males and 92.4% of females, or a male to female ratio of 1: 13.125. The sex distribution differed

TABLE 10.1 Collected data for oceanic whitetip sharks at Elphinstone Reef, including new identified shark individuals (capture) and those previously identified (re-capture).

YEAR	*MONTHS*	*NO. OF SIGHTING*	*DAYS*	*DIVES*	*DIVE HOURS*	*CAPTURE*	*RE-CAPTURE*
2016	June	2	1	2	2	3	0
	July	2	2	4	3	3	0
	August	8	12	20	17	22	0
	September	4	11	20	16	10	11
	October	4	9	18	15	11	8
	November	5	10	16	14	8	12
	December	3	1	2	2	0	3
2017	January	1	1	2	2	1	1
	June	2	1	2	2	2	2
	July	5	3	4	8	5	5
	August	4	2	4	7	3	6
	September	4	3	6	7	2	0
	October	1	1	2	2	2	6
	November	6	5	7	8	3	0
	December	6	4	4	3	4	1
2018	January	2	1	1	2	1	0
	May	4	2	8	10	5	0
	June	2	1	4	5	7	1
	July	2	1	4	4	2	1
	August	1	1	2	2	0	3
	September	1	1	2	2	4	4
	November	6	4	10	12	6	5
	December	2	1	2	2	0	5
2019	January	1	1	2	2	1	0
Total	24	78	79	148	149	105	74

ID	Right side	Left side	Re-sighting	
			Days	Dates
CLE011	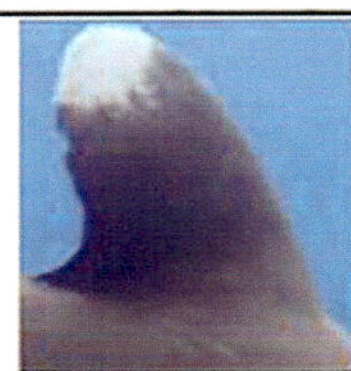		326	07/11/2016 29/09/2017
CLE012			378	18/08/2016 31/08/2017
CLE019	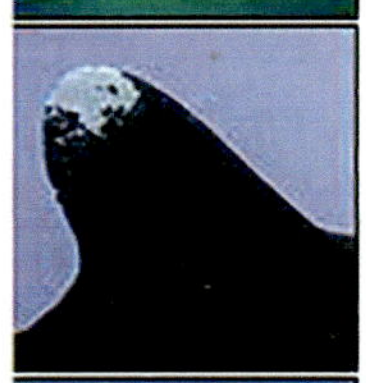		834	08/09/2016 21/12/2018
CLE021	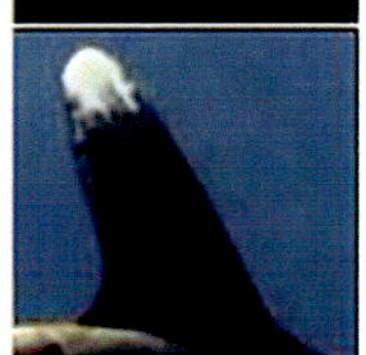	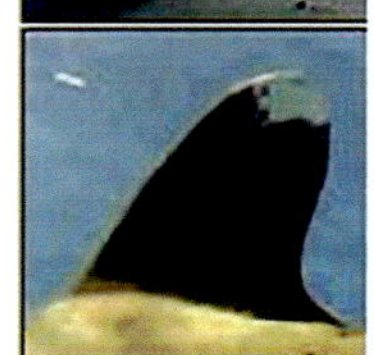	274	02/11/2016 03/08/2017
CLE031			364	30/12/2016 29/12/2017
CLE036			291 162	15/10/2016 02/08/2017 11/01/2018
CLE045			736	21/11/2016 27/11/2018
CLE067		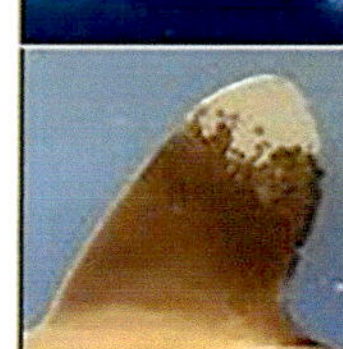	377	07/11/2017 19/06/2018

FIGURE 10.4 Long-term re-sighting for eight female shark individuals identified at Elphinstone Reef.

significantly within months (one-way ANOVA: $df = 6$, $F = 4.4$, $P = 0.029$). The highest number of individuals was recorded in August 2016 (20 females and 2 males, N = 22). It appears that female oceanic whitetip sharks arrived at Elphinstone a month earlier than males. The earliest female sighting was at the beginning of this study in June 2016, whereas the earliest male sighting was in August (Figure 10.7). The rate of male sightings increased as the season progressed, peaking in November (five males). Female sightings decreased in September and peaked again slightly in November. In 2017, female sharks repeated the same scenario of the previous year with the earliest sighting, followed by males later in September, then increased again with a high peak in November. In 2018, the females start increasing in May and then June, followed by males in August, while females were re-sighted more in November.

The discovery curve shows a steady occurrence of new shark individuals at the study site, increasing in August to November 2016. In 2017, it increased obviously in July and

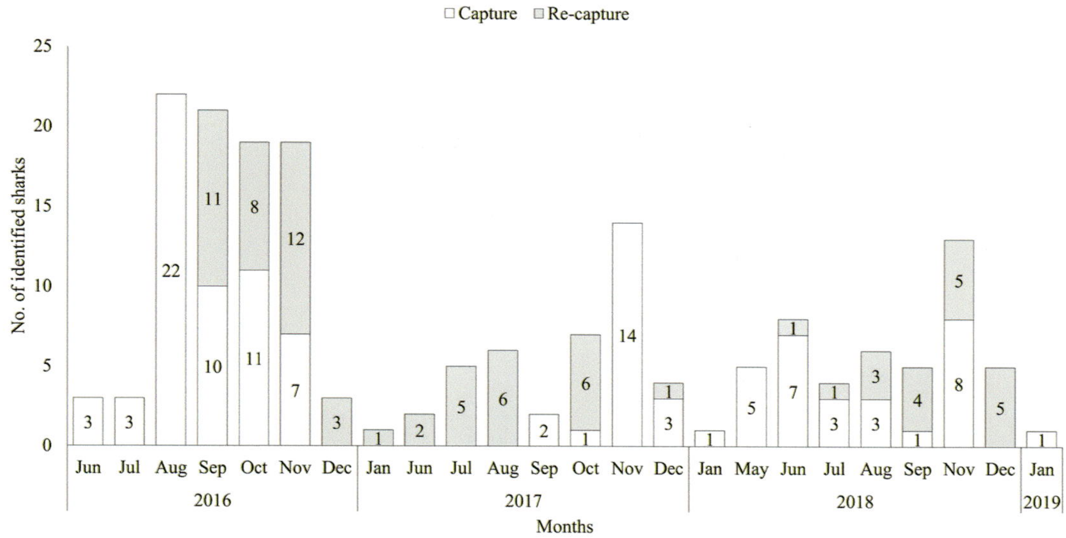

FIGURE 10.5 Number of newly identified (capture) oceanic whitetip sharks and those previously identified (re-capture) per month.

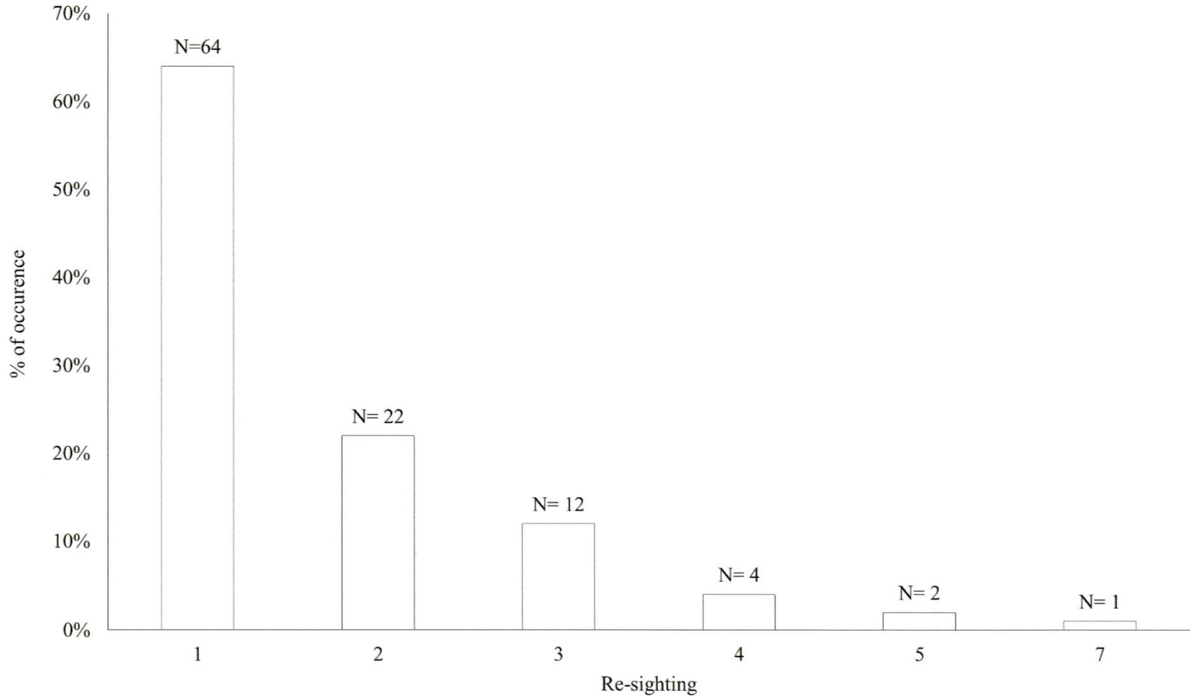

FIGURE 10.6 Frequency of the re-sighting patterns for the identified sharks.

December, while in 2018 it started increasing again in May, June and November (Figures 10.8 and 10.9). A total of eight females were re-sighted in the long-term annually, with a range of 162–834 days (376 ± 199, Mean±SD) (Table 10.2). A female shark with code CLE011 was re-sighted pregnant on 29 September 2017.

10.3.3 Other records

Unusual shark behaviour with divers was recorded for two identified sharks. The first individual, with the code CLE41, approached a diver continuously from different directions, sometimes rolling itself to push its cranial end to the divers. After photo-ID analysis for this shark, it was observed that a fishing hook was entangled in the subject's mouth. The second individual, with a code of CLE37, approached the diver's fins on the surface at the end of the dive (Figure 10.10). The female shark with code CLE011 was recorded with a fishing hook in its mouth and swimming fast among divers. During the recording of this incident, a support diver emitted a continuous sound signal with a diving accessory, to attempt to lure the shark away from the diver. This appeared to cause the shark to change its direction, in response to the sound. However, the shark returned its attention to the diver one more time and appeared to be approaching to attack the diver at the surface, but again changed direction when more sound signals were emitted by the support diver.

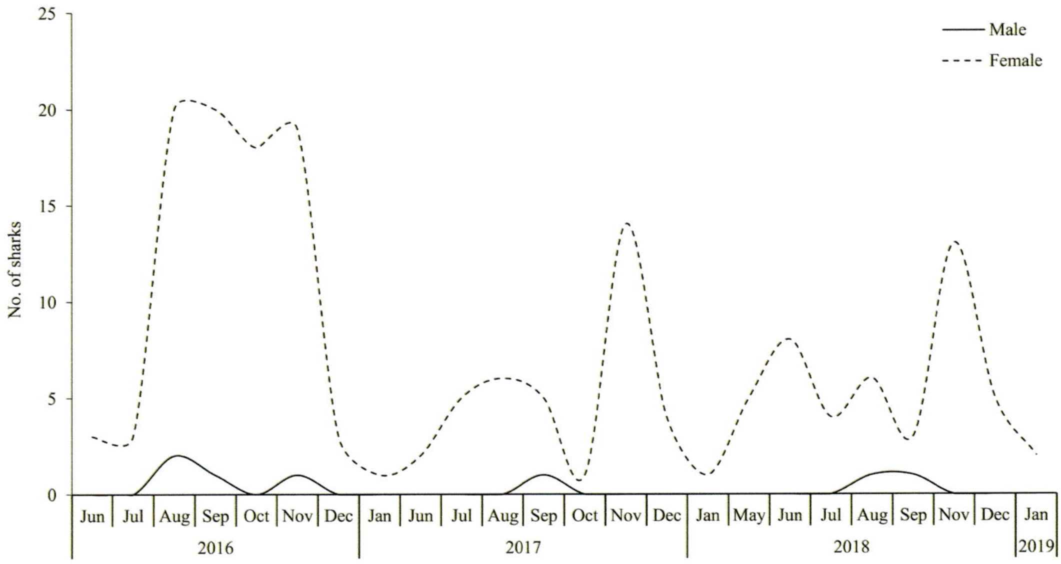

FIGURE 10.7 Number of male and female oceanic whitetip sharks per month.

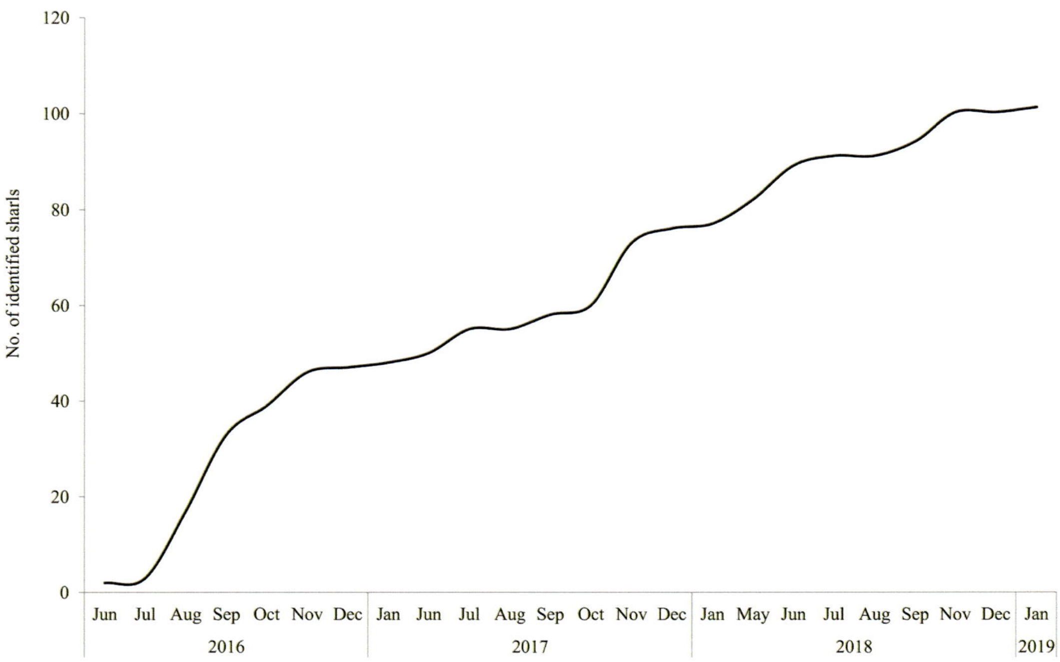

FIGURE 10.8 Rate of discovery for oceanic whitetip sharks at Elphinstone Reef between June 2016 and January 2019.

Oceanic whitetip sharks were accompanied by pilotfish, *Naucrates ductor* (Linnaeus 1758), in 13 cases, and in one case an individual was observed with a big group of 65 pilotfish, while another one was observed with 29 pilotfish. In 10 cases the presence of shark suckers, *Remora remora* (Linnaeus 1758), was also observed, swimming close to the shark or attached to its skin with their suction disk. No predatory behaviour was observed during this study, but in one case an oceanic whitetip shark was observed following a large bottlenose dolphin (Acevedo-Gutiérrez 2002).

During this study, the presence of a high number of boats and recreational divers were recorded constantly at Elphinstone Reef, with a maximum of 27 boats and approximately 330 divers concentrated at a single moment in close proximity to the observation site.

10.4 DISCUSSION

A total of 105 individuals of *C. longimanus* were clearly identified using photo-identification, which confirms that it is a useful tool for a variety of subjects, including population composition, abundance estimates, residency, movement, demography, and social behaviour (Speed et al. 2007; Marshall and Pierce 2012). The particular shape and colouration of the oceanic whitetip's first dorsal fin has proven to be a perfect key to the identification of individuals. The presence of natural identification marks on *C. longimanus* in fact eliminates the need for physical marker tags, and it is also a more permanent means of identification. The association of a large dorsal fin with a very

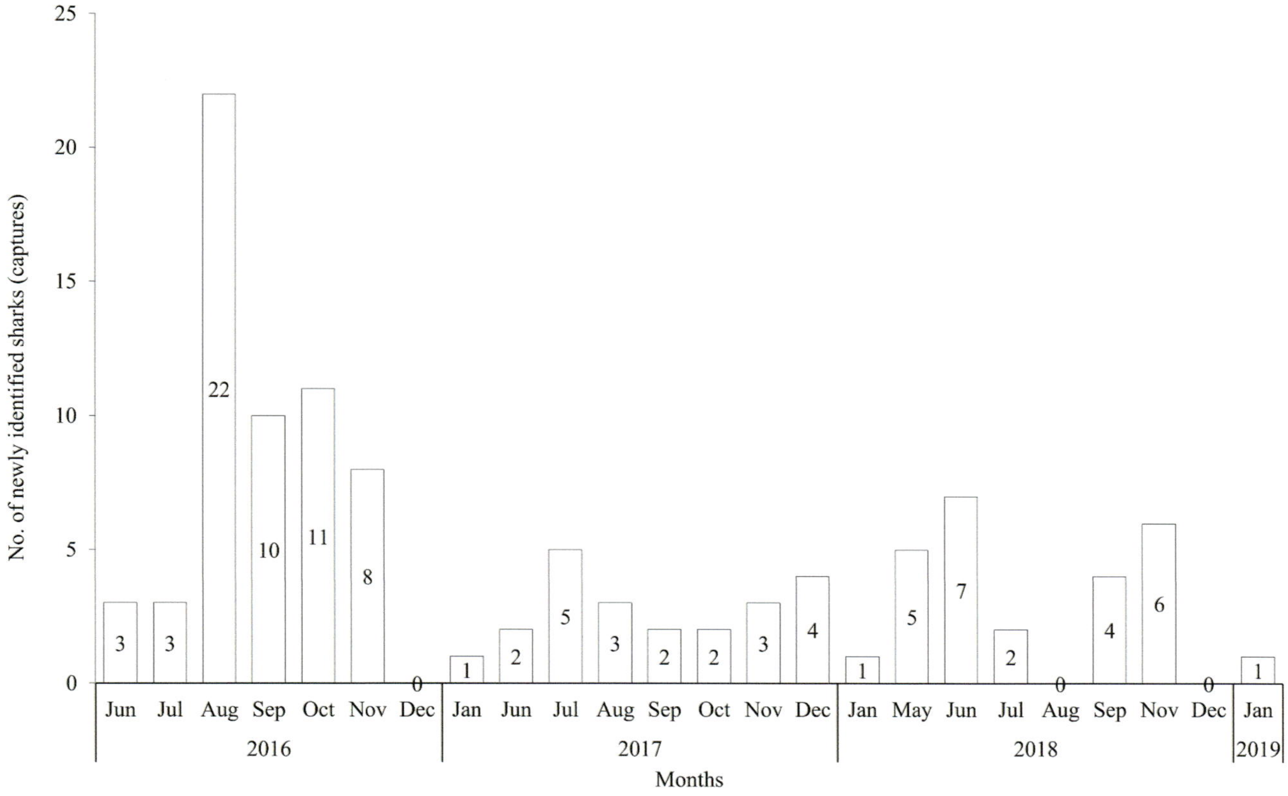

FIGURE 10.9 Number of oceanic whitetip sharks identified per month.

evident shape and a brightly coloured pattern at its apex makes it almost impossible to mistake the identification of individuals. In some other species of sharks, a smaller dorsal fin with a lesser degree of intraspecific variation in morphology and a uniform colouration make photo-identification less immediate and less likely to be used with success on a large percentage of the individuals encountered. However, even for *C. longimanus*, it is important that the study we started is continued and conducted on a regular basis to check the possibility that some of the features observed in the shape and colouration of the dorsal fin may vary with the passing of time. As Marshall and Pierce (2012) have pointed out referring to sharks in general, not only scars, wounds, nicks, and scratches may transform over time, but even prominent spots may change during the animal's growth. The degree of change in the shape and colour of the oceanic whitetip's first dorsal fin over time needs to be investigated.

10.4.1 Philopatry

Photo-identification projects in other parts of the world have shown philopatry or site preference in sharks (Howey-Jordan et al. 2013). Philopatry has been defined as the return of individuals to their birthplace, or another adopted locality, often for reasons linked to their diet. For example, individual great white sharks, *Carcharodon carcharias* (Linnaeus 1758), have been reported to revisit the waters of specific sites in South Africa, South Australia, New Zealand, Mexico, California, New England, and Italy (De Maddalena 2014). Howey-Jordan et al. (2013) observed philopatry in *C. longimanus* in the Bahamas waters, and suggested that this behaviour may be due to the local abundance of large teleost prey such as billfishes, tuna, and dolphinfish. Recently, it was recorded for putative mating activity on mature females (Talwar et al. 2023). The data collected in our study shows philopatry in oceanic whitetip sharks at Elphinstone Reef. This is likely to be due to the presence of many suitable prey. However, we should also consider the idea that the food provided by some dive operators in the area may play an additional reason for these large sharks to return to the area or remain around it. Similarly to what has been reported by Madigan et al. (2015) for the oceanic whitetip sharks showing philopatry to the Bahamas waters, at Elphinstone Reef the sharks observed were mostly female.

In this study, eight females returned to Elphinstone reef in the next years. One of the tagged oceanic whitetip sharks returned back to the same study site after 127 days after travelling 500 km (Tolotti et al. 2015). White et al. (2017) found that more than 60% of their studied sharks remained within their study site for the majority of the study's duration. Gender distribution was also similar to that reported by Howey-Jordan et al. (2013) and Gallagher et al. (2014) for male and female oceanic whitetip sharks. Observations of more individual female sharks in August is one of the most important results of this study, and this time may be due to mating and the aggregation of dominant mature females (Talwar et al. 2023). Sharks that live in shallow coastal habitats have migratory cycles and repeat their movements between different stationary sites (Speed et al. 2010; Chapman et al. 2015). These movements are due to global climate change (Dulvy et al. 2008, 2021; Kibria et al. 2017; Crear et al. 2020; Bettcher et al. 2023) and are part of their life cycle along the coast, offshore, and over the

TABLE 10.2 Residency pattern of the 105 oceanic white tip sharks identified at Elphinstone Reef between 28 June 2016 and 5 January 2019.

	2016							2017								2018								2019	RE-	
CODE	JUN	JUL	AUG	SEP	OCT	NOV	DEC	JAN	JUN	JUL	AUG	SEP	OCT	NOV	DEC	JAN	MAY	JUN	JUL	AUG	SEP	NOV	DEC	JAN	SIGHTING	SEX
1	■																								1	F
2	■																								1	F
3	■																								1	F
4		■																							1	F
5		■																							1	F
6		■																							1	F
7			■	■																					2	F
8			■																						1	F
9			■																						1	F
10			■																						1	F
11			■		■	■						■													4	F
12			■								■														2	F
13			■								■														3	F
14			■																						1	F
15			■	■																					2	F
16			■																						1	F
17			■																						1	M
18			■		■																				2	F
19			■	■																					2	F
20			■	■																					2	F
21			■	■		■					■														4	F
22			■	■																					3	M
23			■	■								■													3	F
24			■	■																					2	F
25			■	■																					2	F
26			■	■	■																				3	F
27			■																						1	F
28			■	■																					1	F
29				■	■																				3	F
30				■	■																				2	F
31				■	■	■	■	■															■		6	F
32				■																					1	F
33				■	■	■																			3	F
34				■																					2	F
35				■																					1	F
36				■	■	■					■										■	■	■	■	7	F
37					■																				1	F
38				■																					3	F
38				■																					3	F
40					■																				1	F
41					■	■																			2	F
42					■	■																			2	F
43					■	■																			2	F
44					■																				1	F
45					■																	■			2	F
46					■	■																			2	M
47					■																				1	M
48					■		■																		3	F
49					■	■		■																	3	F
50						■			■																3	F
51						■			■																2	F
52						■				■															1	F
53						■				■															2	F
54						■				■	■	■													5	F
55						■				■	■	■													4	F
56						■				■															1	F

TABLE 10.2 (Continued)

	2016							*2017*								*2018*								*2019*	*RE-*	
CODE	*JUN*	*JUL*	*AUG*	*SEP*	*OCT*	*NOV*	*DEC*	*JAN*	*JUN*	*JUL*	*AUG*	*SEP*	*OCT*	*NOV*	*DEC*	*JAN*	*MAY*	*JUN*	*JUL*	*AUG*	*SEP*	*NOV*	*DEC*	*JAN*	*SIGHTING*	*SEX*
57												■													1	F
58												■													1	F
59													■												1	F
60														■											1	F
61														■											1	F
62														■											1	F
63														■											1	F
64														■											1	F
65														■											1	F
66														■											1	F
67														■											1	F
68														■				■							2	F
69														■											1	F
70														■											1	F
71														■	■										1	F
72														■											2	F
73														■											1	F
74															■										1	F
75															■										1	F
76															■										1	F
77																■						■			2	F
78																	■					■			2	F
79																	■								1	F
80																	■								1	F
81																	■								1	F
82																	■								1	F
83																		■							1	F
84																		■							1	F
85																		■							1	F
86																		■							1	F
86																		■							1	F
88																		■							1	F
89																		■	■	■					3	F
90																			■	■		■	■		4	F
91																			■						1	F
92																			■	■					1	F
93																				■	■				1	F
94																				■	■				1	M
95																				■	■				1	F
96																						■			1	F
97																						■			1	F
98																						■			1	F
99																						■			1	F
100																							■		1	F
101																						■			1	F
102																						■			1	F
103																						■			1	F
104																						■	■		2	F
105																								■	1	F

FIGURE 10.10 Shark of code CEL41 approaches the diver at the surface. Photo credit: Tarek Noaman.

continental shelf. These individuals migrate long distances and show residency, seasonally (Bigelow et al. 1999; Block et al. 2011; Howey-Jordan et al. 2013; Weigmann 2017).

10.4.2 Shark attacks and monitoring behaviour

Recently, photo-identification has become a helpful tool in identifying individual sharks responsible for shark attacks in Egypt. Levine et al. (2014) reported that the use of underwater photographs of *C. longimanus* showed that a single individual was responsible for three shark attacks in Sharm El Sheikh, Egypt, at the end of 2010. This individual was recognized with a distinctive crescent-shape feature in the upper lobe of the caudal fin. On 5 August 2017, a shark attack by a small female oceanic whitetip shark was documented, 60 km north of Marsa Alam city. The shark attacked a 20-year-old Austrian girl while she was following dolphins in the open water, but she survived the attack. A scientific team from the Nature Conservation Sector of the Ministry of the Environment in Egypt took quick action and caught the individual shark after it was identified, using photo-ID. On 5 August 2018, another attack happened 20 km north of Marsa Alam city. A 41-year-old Czech tourist was attacked by a large tiger shark (pers. obs.). Recently, on 9 June 2023, a 23-year-old Russian man was killed in an attack by a tiger shark while snorkelling off the coast of the Egyptian Red Sea resort of Hurghada, and the shark was caught after using photo-identification to recognise a permanent notch to the dorsal fin while swimming near the surface. From these observations and the results of this study, it may be suggested that mating activities start in May and increase in the next months, when the males were first recorded, with an increase in sightings of new females. This information may explain the shark attacks in the Egyptian Red Sea, during this time of the year.

Photo-identification in this study has also been shown to be extremely useful for monitoring the behaviour of specific individuals showing abnormally aggressive behaviour (Gallagher et al. 2014). It is well known that the threat that sharks represent to humans has been greatly exaggerated in the past and that attacks on humans by any species of shark are extremely rare events. However, we must also take into account the fact that *C. longimanus* is one of the most dangerous species of sharks (Levine et al. 2014). Therefore, the large number of people diving in close proximity to a number of oceanic whitetip sharks at Elphinstone Reef, with basically a lack of regulation in these interactions, creates a particularly dangerous situation where accidents with this species are much more likely to occur (Shawky and De Maddalena 2013). Having the chance to specifically monitor potentially dangerous individuals of *C. longimanus* is without doubt a valuable tool that should be used by local authorities on a daily basis.

10.5 IMPLICATIONS FOR CONSERVATION

The photo-ID technique is an essential monitoring method for the oceanic whitetip shark. This study reinforces the importance of Elphinstone Reef as a hotspot for the presence and observation of several species of sharks, especially the oceanic whitetip shark (Shawky and De Maddalena 2013). This research team hopes that this study will provide additional support to convince the authorities of the Red Sea Protectorates to declare Elphinstone Reef and the area of Marsa Alam a marine park. As this type of action has already been taken for other sites in the area, this study should support a similar action for Elphinstone Reef and the area of Marsa Alam. Specifically, this study concluded that during August, many individual sharks visited the area of Marsa Alam, especially around the study site, and special regulations are strongly recommended to avoid the occurrence of shark attacks. A code of conduct for those frequenting this area needs to be established, and the number of divers and boats frequenting Elphinstone Reef should be regulated. This program of photo-identification for *C. longimanus*, begun with this study, should be continued including other sites in the southern Red Sea, such as the Brothers Islands, Daedalus Reef, Zabargad Island, and Ras Mohamed National Park in southern Sinai, and Sharm El Sheikh. The divers frequenting these areas could be involved in the collection of a continuous stream of data to properly monitor the presence of the species at these sites. The total number of adult and subadult oceanic whitetip sharks that visit Elphinstone Reef is still unknown. Therefore, further study using laser photogrammetry will ensure continued observation of shark aggregations at Elphinstone Reef for mating or breeding, so significant measures can be taken to avoid shark attacks along the Egyptian Red Sea coast.

Photo-identification is a useful tool in the study of a variety of subjects, including population composition, abundance estimates, residency and movement, demography, and social behaviour. Even if this study is only a first effort to identify the sharks frequenting the area, and the results are therefore still limited, this may represent the start of a long-term study that will necessarily produce a more detailed insight into the ethology of this species, in view of possible future collaborative work to be conducted on a global scale by using automated recognition programmes. Laser photogrammetry techniques are highly recommended to estimate the sexual maturity of individual sharks by measuring the total lengths of the sharks. It is an effective method to examine the sexual maturity of sharks, by measuring the body size of sharks underwater, like manta rays (Deakos 2010), white sharks (Leurs et al. 2015), whale sharks (Rohner et al. 2015), and catsharks (Rogers et al. 2017). Measurements of the body size, or morphometrics, is a standard methodology for studying size class segregation in a population (Cubbage and Calambokidis 1987), physical and sexual maturity (Waters and Whitehead 1990), animal growth (Clark 2001), and shark life history (Deakos 2010). Furthermore, it is very important to have more information about shark maturity and sex composition to be used in species conservation and management implementation (Rohner et al. 2015).

ACKNOWLEDGEMENT

Very special thanks to Dr. Khaled Fahmy, former Minister of Environment, for his guidance in implementing a monitoring program for sharks, and Dr. Gamal Gomaa and Dr. Ahmed Ghalab from the Nature Conservation Sector for logistical support. Thanks also to the diving centres of Red Sea Diving Safari, Blue Ocean and Orca. Special thanks to my dive buddy Mr. Tarek Noaman, and to Dr. Eric Glenn Haenni who kindly edited the English text of this work. Particular thanks from Alessandro De Maddalena go to his wife Alessandra and his son Antonio.

CONTRIBUTIONS

S.A.M. – Editing the manuscript and final format, the fieldwork and data analysis.

D.A. – Writing the first draft of the manuscript.

REFERENCES

Acevedo-Gutiérrez, A. 2002. Interactions between marine predators: Dolphin food intake is related to number of sharks. *Mar Ecol Prog Ser* 240: 267–271.

Anderson, P.K. 1995. Scarring and photo identification of dugongs (*Dugong dugon*) in Shark Bay, Western Australia. *Aquat Mamm* 21: 205–211.

Anderson, S.D., Chapple, T.K., Jorgensen, S.J., Klimley, A.P., and B.A. Block. 2011. Long term individual identification and site fidelity of white sharks, *Carcharodon carcharias*, off California using dorsal fins. *Mar Biol* 158: 1233–1237.

Andrzejaczek, S., Gleiss, A.C., Pattiaratchi, C.B., and M.G. Meekan. 2018. First insights into the fine-scale movements of the sandbar shark, *Carcharhinus plumbeus*. *Front Mar Sci* 5: 483.

Backus, R.H., Springer, S., and E.L. Arnold. 1956. A contribution to the natural history of the white-tip shark, *Pterolamiops longimanus* (Poey). *Deep-Sea Res* 3: 178–188.

Baum, J.K., and R.A. Myers. 2004. Shifting baselines and the decline of pelagic sharks in the Gulf of Mexico. *Ecol Lett* 7: 135–145.

Baum, J.K., Myers, R.A., Kehler, D.G., Worm, B., Harley, S.J., and P.A. Doherty. 2003. Collapse and conservation of shark populations in the Northwest Atlantic. *Science* 299: 389–392.

Bettcher, V.B., Franco, A.C.S., and L.N. dos Santos. 2023. Habitat-use of the vulnerable Atlantic Nurse Shark: A review. *Peer J* 11: e15540. doi:10.7717/peerj.15540.

Bicknell, A.W.J., Godley, B.J., Sheehan, E.V., Votier, S.C., and M.J. Witt. 2016. Camera technology for monitoring marine biodiversity and human impact. *Front Ecol Environ* 14: 424–432.

Bigelow, H.B., and W.C. Schroeder. 1948. Sharks. In: *Fishes of the Western North Atlantic. Part one: Lancelets, ciclostomes, sharks*. Eds: J. Tee-Van, C.M. Breder, S.F. Hildebrand, A.E. Parr, and W.C. Schroeder, 53–576. New Haven, CT: Memoir Sears Foundation for Marine Research, Yale University.

Bigelow, K.A., Boggs, C.H., and X. He. 1999. Environmental effects on swordfish and blue shark catch rates in the US North Pacific longline fishery. *Fish Oceanogr* 8: 178–198.

Block, B.A., Jonsen, I.D., Jorgensen, S.J., Winship, A.J., et al. 2011. Tracking apex marine predator movements in a dynamic ocean. *Nature* 475: 86–90.

Bonfil, R., and M. Abdallah. 2004. *Field identification guide to the sharks and rays of the Red Sea and Gulf of Aden*. FAO Species Identification Guide for Fishery Purposes. Rome: FAO, 71 p.

Bonfil, R., Clarke, S., and H. Nakano. 2008. The biology and ecology of the oceanic whitetip shark, *Carcharhinus longimanus*. In: *Sharks of the open ocean: Biology, fisheries and conservation*. Eds: M.D. Camhi, E.K. Pikitch, and E.A. Babcock, 128–139. Oxford: Blackwell Science.

Braun, C.D., Skomal, G.B., Thorrold, S.R., and M.L. Berumen. 2014. Diving behavior of the reef manta ray links coral reefs with adjacent deep pelagic habitats. *PLoS One* 9: e88170.

Camargo, S.M., Coelho, R., Chapman, D., Howey-Jordan, L., Brooks, E.J., Fernando, D., et al. 2016. Structure and genetic variability of the Oceanic Whitetip Shark, *Carcharhinus longimanus*, determined using mitochondrial DNA. *PLoS One* 11, no. 5: e0155623. doi:10.1371/journal.pone.0155623.

Camhi, M.D., Lauck, E., Pikitch, E.K., and E.A. Babcock. 2008. A global overview of commercial fisheries for open ocean sharks. In: *Sharks of the open ocean: Biology, fisheries and conservation*. Eds: M.D. Camhi, E.K. Pikitch, and E.A. Babcock. Oxford, UK: Blackwell Publishing.

Castro, J. 1983. *The sharks of North American waters*. College Station, TX: Texas A&M University Press, 180 p.

Chapman, D.D., Feldheim, K.A., Papastamatiou, Y.P., and R.E. Hueter. 2015. There and back again: A review of residency and return migrations in sharks, with implications for population structure and management. *Annu Rev Mar Sci* 7: 547–570.

Clark, T.B. 2001. *Population structure of Manta birostris (Chondrichthyes: Mobulidae) from the Pacific and Atlantic Oceans*. M.S. Thesis, Texas A&M University, Galveston, TX.

Cochran, J.E.M., Hardenstine, R.S., Braun, C.D., Skomal, G.B., Thorrold, S.R., Xu, K., Genton, M.G., and M.L. Berumen. 2016. Population structure of a whale shark *Rhincodon typus* aggregation in the Red Sea. *J Fish Biol* 89: 1570–1582.

Compagno, L.J.V. 1984. FAO species catalogue. Vol. 4: Sharks of the World: An annotated and illustrated catalogue of shark species known to date. Parts 2. *FAO Fish Synop* 125: 251–655.

Cousteau, J.P., and P. Cousteau. 1970. *The shark: Splendid savage of the sea*. London: Cassell.

Crear, D.P., Latour, R.J., Friedrichs, M.A., St-Laurent, P., and K.C. Weng. 2020. Sensitivity of a shark nursery habitat to a changing climate. *Mar Ecol Prog Ser* 652: 123–136. doi:10.3354/meps13483.

Cubbage, J., and J. Calambokidis. 1987. Size-class segregation of bowhead whales discerned through aerial stereo photogrammetry. *Mar Mamm Sci* 3: 179–185.

D'Alberto, B.M., Chin, A., Smart, J.J., Baje, L., White, W.T., and C.A. Simpfendorfer. 2017. Age, growth and maturity of oceanic whitetip shark (*Carcharhinus longimanus*) from Papua New Guinea. *Mar Freshw Res* 68: 1118–1129.

Deakos, M.H. 2010. Paired-laser photogrammetry as a simple and accurate system for measuring the body size of free-ranging manta rays *Manta alfredi*. *Aquat Biol* 10: 1–10.

De Maddalena, A. 2014. *Biologia dello squalo bianco*. Milano: Editoriale Magenes, 190 p.

De Maddalena, A. 2017. *Sharks of the Maldives*. Milano: Editoriale Magenes, 216 p.

De Maddalena, A., Bansch, H., and W. Heim. 2016. *Sharks of the Mediterranean*. Jefferson: McFarland, 236 p.

Dodd, J., Thorburn, J., Milway, V.A., Campbell, R., and D.M. Bailey. 2018. Evaluating the potential of photo-identification as a monitoring tool for flapper skate (*Dipturus intermedius*) Steven Benjamins. *Aquat Conserv Mar Freshw Ecosyst* 28, no. 6: 1360–1373. doi:10.3354/ab00258.

Domeier, M.L., and N. Nasby-Lucas. 2007. Annual re-sightings of photographically identified white sharks (*Carcharodon carcharias*) at an eastern Pacific aggregation site (Guadalupe Island, Mexico). *Mar Biol* 150: 970–984.

Dulvy, N.K., Baum, J.K., Clarke, S., Compagno, L.J.V., et al. 2008. You can swim but you can't hide: The global status and conservation of oceanic pelagic sharks and rays. *Aquat Conserv* 18: 459–482.

Dulvy, N.K., Fowler, S.L., Musick, J.A., Cavanagh, R.D., Kyne, P.M., Harrison, L.R., Carlson, J.K., Davidson, L.N., Fordham, S.V., and M.P. Francis. 2014. Extinction risk and conservation of the world's sharks and rays. *eLife* 3: e00590. doi:10.7554/ELIFE.00590.

Dulvy, N.K., Pacoureau, N., Rigby, C.L., Pollom, R.A., Jabado, R.W., Ebert, D.A., Finucci, B., Pollock, C.M., Cheok, J., Derrick, D.H., Herman, K.B., Sherman, C.S., VanderWright, W.J., Lawson, J.M., Walls, R.H., Carlson, J.K., Charvet, P., Bineesh, K.K., Fernando, D., Ralph, G.M., Matsushiba, J.H., HiltonTaylor, C., Fordham, S.V., and C.A. Simpfendorfer. 2021. Overfishing drives over one-third of all sharks and rays toward a global extinction crisis. *Curr Biol* 31, no. 21: 1–15. doi:10.1016/j.cub.2021.08.062.

Dunbar, S.G., Anger, E.C., Parham, J.R., Kingen, C., Wright, M.K., Hayes, C.T., et al. 2021. HotSpotter: Using a computer-driven photo-id application to identify sea turtles. *J Exp Mar Biol Ecol* 535: 151490. doi:10.1016/j.jembe.2020.151490.

Flmnh. 2016. *Carcharhinus longimanus*. www.flmnh.ufl.edu/fish/discover/species-profiles/carcharhinus-longimanus; downloaded 31 January 2016.

Gallagher, A.J., Romeiro, J., Canabal, D., Canabal, V., and N. Hammerschlag. 2014. Novel social behaviors in a threatened apex marine predator, the oceanic whitetip shark *Carcharhinus longimanus*. *Ethol Ecol Evol* 26, no. 4: 413–417. doi:10.1080/03949370.2014.896831.

Garrick, J.A.F. 1982. *Sharks of the genus Carcharhinus*. NOAA Technical Report NMFS Circular 445, pp. 1–194.

Gohar, H.A.F., and F.M. Mazhar. 1964. The elasmobranchs of the north-western Red Sea: Publications of the marine biological station. *Al-Ghardaqa (Red Sea)* 13: 1–144.

Gore, M.A., Frey, P.H., Ormond, R.F., Allan, H., and G. Gilkes. 2016. Use of photo-identification and mark – recapture methodology to assess basking shark (*Cetorhinus maximus*) populations. *PLoS One* 11: e0150160.

Hammond, P.S., Mizroch, S.A., and G.P. Donovan. 1990. *Individual recognition of cetaceans: Use of photo-identification and other techniques to estimate population parameters*. Report of the International Whaling Commission, Cambridge. Special Issue 12, p. 440.

Hazin, F.H.V., Couto, A.A., Kihara, K., Otsuka, K., and M. Ishino. 1990. Distribution and abundance of pelagic sharks in the south-western equatorial Atlantic. *J Tokyo Univ Fish* 77, no. 1: 51–64.

Holmberg, J., Norman, B., and A. Arzoumanian. 2008. Robust, comparable population metrics through collaborative photo-monitoring of whale sharks *Rhincodon typus*. *Ecol Appl* 18, no. 1: 222–233.

Howey-Jordan, L.A., Brooks, E.J., Abercrombie, D.L., Jordan, L.K.B., Brooks, A., William, S., Gospodarczyk, E., and D.D. Chapman. 2013. Complex movements, philopatry and expanded depth range of a severely threatened pelagic shark, the Oceanic Whitetip (*Carcharhinus longimanus*) in the western North Atlantic. *PLoS One* 8, no. 2: e56588. doi:10.1371/journal.pone.0056588.

Jean, C., Ciccione, S., Talma, E., Ballorain, K., and J. Bourjea. 2010. *Photoidentification method for green and hawksbill turtles – First results from Reunion*. International Sea Turtle Soc (Hg.): Indian Ocean Turtle Newsletter S, 8–13.

Karczmarski, L., Wursig, B., Gailey, G., Larson, W.K., and C. Vanderlip. 2005. Spinner dolphins in a remote Hawaiian atoll: Social grouping and population structure. *Behav Ecol* 16, no. 4: 675–685.

Kibria, G., Haroon, A.K.Y., and D. Nugegoda. 2017. *Climate change and its effects on global shark fisheries*. doi:10.13140/RG.2.2.15363.81441.

Last, P.R., and J.D. Stevens. 2009. *Sharks and rays of Australia*. Australia: CSIRO, 656 p.

Lessa, R., Santana, F.M., and R. Paglerani. 1999. Age, growth and stock structure of the oceanic whitetip shark, *Carcharhinus longimanus*, from the southwestern equatorial Atlantic. *Fish Res* 42: 21–30.

Leurs, G., O'Connell, C.P., Andreotti, S., Rutzen, M., and H.V. Noordegraaf. 2015. Risks and advantages of using surface laser photogrammetry on free-ranging marine organisms: A case study on white sharks *Carcharodon carcharias*. *J Fish Biol* 86: 1713–1728. doi:10.1111/jfb.12678.

Levine, M., Collier, R.S., Ritter, E., Fouda, M., and V. Canabal. 2014. Shark cognition and a human mediated driver of a spate of shark attacks. *Open J Anim Sci* 4: 263–269.

Linnaeus, C. 1758. *Systema Naturae per regna tria naturae, secundum classes, ordines, genera, species, cum characteribus, differentiis, synonymis, locis*. Vol. 1. Regnum animale. Stockholm: Laurentius Salvius, 824 p.

Madigan, D.J., Brooks, E.J., Bond, M.E., Gelsleichter, J., Howey, L.A., Abercrombie, D.L., Brooks, A., and D.D. Chapman. 2015. Diet shift and site-fidelity of oceanic whitetip sharks *Carcharhinus longimanus* along the Great Bahama Bank. *Mar Ecol Prog Ser* 529: 185–197.

Marshall, A.D., and S.J. Pierce. 2012. The use and abuse of photographic identification in sharks and rays. *J Fish Biol* 80: 1361–1379.

Matsunaga, H., and H. Nakano. 1999. Species composition and CPUE of pelagic sharks caught by Japanese longline research and training vessels in the Pacific Ocean. *Fisheries Sci* 65: 16–22.

McCauley, D.J., Power, E.A., Bird, D.W., Mcinturff, A., Dunbar, R.B., Durham, W.H., Micheli, F., and H.S. Young. 2013. Conservation at the edges of the world. *Biol Conserv* 165: 139–145.

McCoy, E., Burce, R., David, D., Aca, E.Q., Hardy, J., Labaja, J., Snow, S.J., Ponzo, A., and G. Araujo. 2018. Long-term photo-identification reveals the population dynamics and strong site fidelity of adult whale sharks to the coastal waters of Donsol, Philippines. *Front Mar Sci* 5: 271. doi:10.3389/fmars.2018.00271.

Myrberg, A. 1987. Shark behaviour. In: *Sharks*. Ed: J.D. Stevens, 84–92. Hong Kong: Intercontinental Publishing Corp Ltd, 240 p.

Papastamatiou, Y.P., Iosilevskii, G., Leos-Barajas, V., Brooks, E.J., Howey, L.A., Demian, D., Chapman, D.D., and Y.Y. Watanabe. 2018. Optimal swimming strategies and behavioral plasticity of oceanic whitetip sharks. *Sci Rep* 8, no. 1: 551. doi:10.1038/s41598-017-18608-z.

Peschak, T.P., and M.C. Scholl. 2006. *South Africa's great white shark*. Cape Town: Struik, 98 p.

Poey, F. 1861. *Memorias sobre la historia natural de la Isla de Cuba*. Vol. 2. Havana: Viuda de Barcina, 442 p.

Randall, J.E. 1986. *Sharks of Arabia*. London: IMMEL Publishing, 148 p.

Reisser, J., Proietti, M., Kinas, P., and I. Sazima. 2008. Photographic identification of sea turtles: Method description and validation, with an estimation of tag loss. *Endang Species Res* 5, no. 1: 73–82.

Rice, J., and S. Harley. 2012. *Stock assessment of oceanic whitetip sharks in the western and central Pacific Ocean*. WCPFC-SC8–2012/SA-WP-06 Rev 1. Report to the Western and Central Pacific Fisheries Commission Scientific Committee. Eighth Regular Session, 7–15 August, Busan, Korea.

Rigby, C.L., Barreto, R., Carlson, J., Fernando, D., Fordham, S., Francis, M.P., Herman, K., Jabado, R.W., Liu, K.M., Marshall, A., Pacoureau, N., Romanov, E., Sherley, R.B., and H. Winker. 2019. *Carcharhinus longimanus*. IUCN Red List of Threatened Species 2019: T39374A2911619.

Rogers, T.D., Cambiè, C., and M.J. Kaiser. 2017. Determination of size, sex and maturity stage of free swimming catsharks using laser photogrammetry. *Mar Biol* 164: 213. doi:10.1007/s00227-017-3241-7.

Rohner, C.A., Richardson, A.J., Prebble, C.E.M., Marshall, A.D., Bennett, M.B., Weeks, S.J., Cliff, G., Wintner, S.P., and S.J. Pierce. 2015. Laser photogrammetry improves size and demographic estimates for whale sharks. *Peer J* 3: e886. doi:10.7717/peerj.886.

Seki, T., Taniuchi, T., Nakano, H., and M. Shimizu. 1998. Age, growth and reproduction of the oceanic whitetip shark from the Pacific Ocean. *Fisheries Sci* 64: 14–20.

Shawky, M.A., Alwany, A.M., Zakaria, S., and G.S. El-Etreby. 2015. Estimation of the abundance of the spinner dolphin *Stenella longirostris* using photo identification technique in Samadai Reef, Red Sea, Egypt. *Catrina: Int J Environ Sci* 283: 1–26. doi:10.12816/0010696.

Shawky, A.M., and A. De Maddalena. 2013. Human impact on the presence of sharks at diving sites of the Southern Red Sea, Egypt. *Boll Mus St Nat Venezia* 64: 51–62.

Shawky, A.M., Sallam, W.S., Alwany, M.A., Zakarea, S., Mohamed, D.A., and D.A. Mohamed. 2017. Photo identification of Dugongs in Marsa Alam and Wadi El Gemal National Park, western Egyptian coast of the Red Sea. *Al Azhar Bull Sci* 28, no. 2: 1–10.

Spaet, J.L., Nanninga, G.B., and M.L. Berumen. 2016. Ongoing decline of shark populations in the Eastern Red Sea. *Biol Conserv* 201: 20–28.

Speed, C.W., Field, I.C., Meekan, M.G., and C.J.A. Bradshaw. 2010. Complexities of coastal shark movements and their implications for management. *Mar Ecol Prog Ser* 408: 275–293.

Speed, C.W., Meekan, M.G., and J.A. Bradshaw. 2007. Spot the match – wildlife photo identification using information theory. *Front Zool* 4: 1–11.

Stevens, J.D. 1984. Biological observations on sharks caught by sport fishermen off New South Wales. *Aust J Mar Freshw Res* 35: 573–590.

Strong, W.R., Murphy, R.C., Bruce, B.D., and D.R. Nelson. 1992. Movements and associated observations of bait-attracted white sharks, *Carcharodon carcharias*: A preliminary report. *Aust J Mar Freshw Res* 43: 13–20.

Talwar, B.S., Bond, M.E., Williams, S., Brooks, E.J., Chapman, D.D., Howey, L.A., Knotek, R., and J. Gelsleichter. 2023. Reproductive timing and putative mating behavior of the oceanic whitetip shark *Carcharhinus longimanus* in the eastern Bahamas. *Endang Species Res* 50: 181–194. doi:10.3354/esr01231.

Tolotti, M.T., Bach, P., Hazin, F., Travassos, P., and L. Dagorn. 2015. Vulnerability of the oceanic whitetip shark to pelagic longline fisheries. *PLoS One* 10: e0141396. doi:10.1371/journal.pone.0141396.

Waters, S., and H. Whitehead. 1990. Population and growth parameters of Galapagos sperm whales estimated from length distributions. *Report of the International Whaling Commission* 40: 225–235.

Weigmann, S. 2017. Reply to Borsa (2017): Comment on 'Annotated checklist of the living sharks, batoids and chimaeras (Chondrichthyes) of the world, with a focus on biogeographical diversity by Weigmann (2016)'. *J Fish Biol* 90: 1176–1181. doi:10.1111/jfb.13234.

White, T.D., Carlisle, A.B., Kroodsma, D.A., Block, B.A., Casagrandi, R., De Leo, G.A., Gatto, M., Micheli, F., and D.J. McCauley. 2017. Assessing the effectiveness of a large marine protected area for reef shark conservation. *Biol Conserv* 207: 64–71.

Young, C.N., and J.K. Carlson. 2020. The biology and conservation status of the oceanic whitetip shark (*Carcharhinus longimanus*) and future directions for recovery. *Rev Fish Biol Fish* 30, no. 2: 293–312.

PART 3

Invertebrates

Pearl oysters in Bahrain *Economic, heritage, and biological aspects*

11

Layla J. Hazeem, Mohamed Yusuf and Hashim Al-Sayed

11.1 INTRODUCTION

Pearling was the main fishing industry of the Arabian Gulf and the backbone of the economy of Bahrain and the region until the 1930s. This industry, which helped the livelihood of most local people, faced a sudden downward trend before the middle of the twentieth century when the rise of oil began. With the introduction of the oil industry, divers found more rewarding onshore jobs, which encouraged them to end their traditional occupations. Although fishing for pearls was common in most Arabian Gulf countries, the pearls found in the pearl oysters fished in Bahrain waters were superior in quality compared to those harvested elsewhere. Bahrain was the center of pearl trading in the Arabian Gulf, and all pearls from the neighbouring countries were brought here for trade.

The pearl banks are situated in shallow waters, mainly off the Arabian shore of the Gulf. The 1877–1878 Arabian Gulf administration report by the Government of India noted that the pearl banks in Bahrain were highly productive (Crouzet 2019). In the period between the early 1890s and early 1900s, the total value of pearls exported from the Gulf tripled. The approximate average yearly export values of pearls and mother of pearl (shells) were 561,353 and 269,788 Pounds Sterling, respectively, in the whole Gulf region (Lorimer 1915) for the period 1873 to 1905. Approximately 35,000 tons was the annual catch for the entire Arabian Gulf, an estimate calculated from literature reports of catch rate, number of boats, and the number of fishermen (Lorimer 1915).

The growth of the pearl industry was particularly remarkable in Bahrain, where Lorimer's Gazetteer reported that the value of pearl exports rose by nearly 600% between 1873 and 1906, when Indian merchants joined others from Paris, London, and New York in Bahrain, all competing to secure the finest pearls (Belgrave 1934; Al-Rumaihi 1973; Onley 2014; Peterson 2016; BACA 2020).

Pearl oysters belong to the group of bivalves that form the most important class of mollusks, either in the number of species or in commercial importance. They are very common for consumption and for producing pearls. From an ecological point of view, bivalves occupy all marine ecological niches, from the coast to great depths. They intervene at lower trophic levels in marine ecosystems by transferring energy through links in the food chain (Gosling 1992). Moreover, pearl oysters play an ecological role as biofilters, help to maintain functioning of the ecosystem, and are an important biological indicator for monitoring environmental pollution.

The Supreme Council for the Environment (SCE) issued several decisions and laws to protect the environment and monitor the status of the marine environment in the Kingdom of Bahrain on a regular basis. Recently, the SCE has focused on protecting the marine environment (SCE 2015) and has implemented the national biodiversity strategy, which was developed to improve livelihoods, to protect the country's ecosystem, and to guarantee food security in a country with less than 3% arable land. The strategy intends to re-establish marine resources and fish stocks to secure sustainable limits besides protecting agricultural lands from municipal development, which has become an increasing problem for the rapidly expanding country. In 2016, the UN Environment operated closely with the Bahrain government to certify that their National Biodiversity Strategy and Action Plan is associated with the Sustainable Development Goals (SDGs) and other biodiversity-correlated conventions, comprising desertification and climate change, Ramsar on wetlands, and the Convention on International Trade in Endangered Species of Wild Fauna and Flora (SCE 2016; UNEP 2023). Furthermore, the Bahrain Institute for Pearl and Gemstones (DANAT) has launched several training courses and research proposals that focus on pearls and gemstones, which play a vital role in ensuring, protecting, and enhancing public trust in pearls, gemstones, and jewellery (DANAT 2019). In addition, there is the prominent interest of the Bahrain Authority for Culture and Antiquities (BACA) in shedding light on Bahrain's ancient history in the pearling industry through launching of the Pearl Path project (BACA 2020).

DOI: 10.1201/9781003321392-14

11.2 HISTORY AND ECONOMY OF THE PEARL OYSTER IN THE ARABIAN GULF AND THE KINGDOM OF BAHRAIN

Since about 1050 CE, when oysters were collected for food and pearls, the Gulf's oyster beds have been subjected to anthropogenic influences (Carter 2005), such as heavy fishing for pearls between the late 1700s to early 1900s (Bowen 1951). The magnitude of the fishing activity was enormous; in 1878 Captain Durand of the East India Trading Company estimated that over 4000 pearling boats were working in the Arabian Gulf oyster banks (Carter 2005; Hightower 2013). However, a pearl dealer named Justamond highlighted worries about over-exploitation of the oyster beds as early as 1770, describing the beds as "overfished and depleted" (Carter 2005). A century later, in 1878, Durand reaffirmed this warning and stated that the yield of pearls and oysters had reduced significantly. The decline in the pearls resulted in a near doubling of prices, which led to an increase in exploitation (Hightower 2013).

In 1905, Mr. James Hornell, a Madras Fisheries Bureau officer, recommended a scientific study of the beds to determine the true status of the standing stocks, but this investigation was not completed due to the outbreak of the First World War (Carter 2005). The pearling economy reached its peak in 1911–1912, after which a series of catastrophes including wars, price crashes, the arrival of cheap cultivated pearls, the US Wall Street crash, and its impact on the market for luxury goods, all led to the decline in the 1930s and ultimately the total collapse of the industry by 1950 (BACA 2020). The development of cultured pearl farming in Japan and the prosperity brought about by the discovery and exploitation of oil, led to a catastrophic collapse in the pearl industry and the regional pearl economy (Burdett 1995; Mohammed and Yassien 2003). The industry lingered, much-reduced, throughout the twentieth century, which brought to an end over 7000 years of pearling in the region (Al-Rumaihi 1973; Crouzet 2019).

The Kingdom of Bahrain in particular was famous for its pearl fisheries and even now is regarded as a valuable center for the natural pearl trade, as it is among the few countries where fishing for natural pearls is permitted today (Alatawi et al. 2019). For centuries, the oyster beds located off the northern shores of Bahrain were recognized for the quality of pearls, the density of the pearl oysters, and the abundance of pearls during the collecting seasons. In his *Natural History*, Pliny the Elder described the island of Tylos (Bahrain) as: "It is famous for the vast number of its pearls" (Bostock and Riley 1855). It established the largest market for pearls in the Gulf by the nineteenth century (Bishara et al. 2016). The oyster beds in the north of Bahrain were the focal point of a natural pearl fishery that dominated the Arabian Gulf from at least the third century BC until the early twentieth century. At the peak years of the pearling economy, the high demand for pearls starting in the nineteenth century produced a distinguished product economy in Bahrain. Pearls exports contributed three-quarters of Bahrain's total exports in 1877, and the major traders were from Bombay, Persia, and Turkey. Europe emerged as a major direct market for Bahrain's pearl exports following the turn of the century. According to BACA, by 1904–1905 an estimated 97.3% of the Gulf's turnover in pearls was traded through Bahrain. Bahrain's annual pearl fishery was a community-wide endeavour, from pearling merchants, divers, and dhow captains to boat builders, timber merchants, and general goods suppliers. Nearly every profession found in Bahrain existed to serve the pearling economy (Belgrave 1934; Carter 2005; BACA 2020). Fuccaro (2009) noted that between 1929 and 1931, Bahrain's merchants lost two-thirds of their capital as a result of the market collapse. In 1932, the value of Bahrain's pearl exports to the United States alone dropped to almost 5% of what it had been three years earlier. An average of 238 boats sailed out for the pearl dive every year between 1935 and 1945, compared to the 500-plus boats that went out before 1930. The emergence of new economic sectors dramatically reduced incentives to remain in the pearling trades. Although the discovery of oil did not instantly do away with the pearling industry, the emergence of well-paid wage work in the oil fields, coupled with the steady decline in the market value of pearls during the 1930s, made diving for pearls less attractive than it had been a decade earlier. For those looking for economic stability in hard times, there was no comparison between the confidence of a monthly wage and the instability of continually falling advance payments of the pearling industry. During the 1953 season, only 11 boats (Table 11.1) sailed to the banks (oyster beds) from Bahrain; since much money could be earned onshore, few men were willing to leave their jobs to dive during the summer months (Al-Rumaihi 1973).

In the 1955 diving season, pearl prices continued to be the same, but the amounts that were earned by divers and pullers were considerably better than in the previous seasons; the average earnings of a diver were Rs. 1,008 (Rs., or Gulf rupee) and of a puller Rs. 672 for the diving season of four months and ten days. Pearl prices were between Rs. 100 to 120 per Chow. Table 11.2 shows the estimated value of the revenue from pearls in Bahrain between 1833 and 1954 (Belgrave 1934; Cummins 1957; Carter 2005; Peterson 2016; Crouzet 2019).

The main pearl-diving season (*il-ghos il-od*, 'main diving') ran from mid-May until roughly the autumnal equinox in late September. In early April, before the beginning of the main season, there were three to four weeks of 'pre-season' diving (*il-khanchiyyah* or *il-ghos il-bard*, 'cold diving'), and afterward in October, a similarly short period of 'late season' diving (*or-raddah*, 'the return' or *Mujannah*) running from October to March. During these two periods, small boats work close inshore and do not venture far from the island

TABLE 11.1 Number of boats and men working in the diving industry in Bahrain between 1835 and 1954.

YEARS	*BOATS*	*MEN*
1835	1,500	–
1889	400	11,200 (approx.)
1896	900	–
1926–1931	500	20,000
1935	360	11,550
1940	191	7,500
1945	121	5,100
1952	20	563
1954	11 to 12	500

TABLE 11.2 Estimated value of pearl revenue in Bahrain between 1833 and 1954.

YEARS	*REVENUES*
1833	£200,000 to £240,000
1896	£303,941
1900	£500,000
1901	£400,000
1913	£1,850,000
1922	£2,000,000
1925	£220,000
1926	£192,000
1926–1931	£1,500,000
1946	£62,000
1949	£54,300
1954	£27,400

(Al-Rumaihi 1973; Onley 2014). In Bahrain, the precise times for the start and end of the season were not fixed, up until the pearling reforms of the mid-1920s, but from that point on, the government issued a yearly public statement specifying the dates of the different stages of the season (Bishara et al. 2016).

The choice of pearl beds for fishing (*Fasht*) depends on the captain's (*Al-Nogtha*) experience; however, divers who had considerable knowledge of the pearl beds could contribute to the selection process. The first day of the main season was marked by the boats leaving port en masse with flags flying, and the people singing and beating drums.

11.3 BIOLOGICAL ASPECTS OF THE PEARL OYSTER *PINCTADA RADIATA*: GROWTH PATTERNS AND DISTRIBUTION

Seven *Pinctada* species may exist in the waters of Bahrain (Khamdan 2004). These are *Pinctada imbricata* (this includes *P. radiata* and *p. fucata*), *Pinctada margaritifera*, *Pinctada maculata*, *Pinctada sugillata*, *Pinctada nigra*, *Pinctada anomioides*, and *Pinctada sp.* However, it is usually agreed that no more than three species occur in Bahrain (*Pinctada imbricata* (the most commonly found of all species), *Pinctada margaritifera*, which has been noted to be the least found of all species, and *Pinctada nigra* (Arfa et al. 2012). Arfa et al. (2012) recommended that further research should be conducted to identify the different species present in the Kingdom of Bahrain. The local name of the pearl oyster in the Kingdom of Bahrain is *Mahhara* (محارة). Figure 11.1 shows the ventral and dorsal sides of *Pinctada radiata*.

The pearl oyster, *Pinctada radiata* (Leach 1814), has an extensive distribution between tropical belts and is truly marine. It is commonly distributed throughout the tropical regions but is not recorded on the west coast of Africa. Additionally, *Pinctada radiata* can also be found in the Red Sea and the Mediterranean Sea off Tunisia (Al-Sayed 2009). It thrives in shallow and relatively clear waters, generally in zones of coral reefs, at depths from 5 to 30 m (Tlig-Zouari et al. 2009).

In Bahrain, pearl oyster beds are found in most surrounding areas. However, the majority of gem-quality pearls can be obtained in waters located 50 km north to northeast of Bahrain, at Hayrs (*Hayr* means "oyster bed"), such as Hayr Shutayah, Hayr Bulthama, and Hayr Bu Amamah (Alatawi et al. 2019). Figure 11.2 shows a recent map for the distribution of oyster beds around Bahrain.

Knowledge of the reproductive biology of *Pinctada radiata* is crucial in understanding the population dynamics of the stock. Studies on the spawning period of *Pinctada radiata* have been published since the 1950s in Australia (Tranter 1958) and Japan (Derbali et al. 2009). The gonadal development, maturation, and spawning have also been well documented (Behzadi et al. 1997). In Bahrain, information on the reproductive cycle of *Pinctada radiata* is relatively scarce, despite the few studies focused on population structure (e.g., Al-Sayed et al. 1993), on spat settlement and growth (e.g., Al-Sayed 1996; Khamdan 1998), and periodicity and reproduction (e.g., Al-Sayed 1996).

In the Kingdom of Bahrain, few scientists have evaluated the biological aspects of oyster beds. One of these studies was accomplished by Nayar and Al-Rumaidh (1993a, 1993b) from the Bahrain Centre for Studies and Research (BCSR), which lasted for four years. The density and size structure of the oyster populations were assessed, and some essential biological aspects such as reproduction, yield, and the size composition of pearls were examined. They also assessed a total of 15 sites around Bahrain, including Hayr Shutayyah, Hayr Bu-l-Thamah, Hayr Bu Am'amah, Hayr Al Waadi, Hayr Arjila, Hayr Abu Haqul, Hayr Shiqita, Hayr Al Jaal, Hayr Al Kharab, Hayr Mayyana, Hayr Bin Zayaan, Hayr Bu Sawr, Hayr Al Sayah, Fasht Al Jarim, and Fasht Al Adham. A total of 1,145,420 pearl oysters were collected between 1986 and 1989, and pearls were found in approximately 54,338 individuals (4.7%). The authors found that the average number of pearls that were larger than 0.2 cm corresponded to 1 in 99 oysters, and 1 in 459 for those larger than 0.3 cm. The density of pearl oysters was found to be high in the above-mentioned sites, and the average density of pearl oysters in the main beds (Hayr Shutayyah, Hayr Bu-l-Thamah, Hayr Bu Am'amah) was > 4 oysters per m^2. Moreover, the oyster size was found to vary from 0.5 cm (small spat 2–3 weeks old) to 9 cm across all study sites. The seasonality and spawning behavior were examined, and they found that small-sized spats of 0.5–1.5 cm were present in large numbers during July–August and November–December. They showed that the development of gonads occurred during the months March–April, and they mature and spawn during July. Most oysters were observed to be in a spent condition, or some of the follicles were found with few residual eggs or empty. The same cycle was noticed during October–November, yet a larger number of oysters were observed in the developing stage with follicles nearly full of developing eggs. Oysters collected in the months January–March were at developed stages of maturity and ready to spawn again in April–May. The constant environmental conditions, in particular temperature and salinity, explained the occurrence of spawning year-round with a peak period between July and December.

The age of some bivalves can be defined through the annual rings. Growth lines can be detected using acetate-peels of shell sections of oysters. The growth lines are formed

FIGURE 11.1 The ventral and dorsal sides of *Pinctada radiata*.

annually and can be reliably used to determine age (Richardson et al. 1993). In the pearl oysters of Bahrain, many rings can be formed in rapid succession, which makes it difficult to determine the age of the oysters. Growth rates are controlled by food availability and the quality of water, which suggest that environmental factors are very crucial for their development (Martínez-Fernández et al. 2006). Most research conducted on pearl oysters showed that their growth is confined to the periods when the water temperatures are high (Nayar and Al-Rumaidh 1993a). In addition, oysters of 40–45 mm have grown to 60–70 mm in 12 months, and therefore it is believed that oysters of this size could be approximately one year and eight months old. When these oysters reach the size of 70–75 mm, it could be said that they are around two years and two months old. This demonstrated the slightly faster growth rate of oysters in Bahrain compared to other areas (Nayar and Al-Rumaidh 1993a).

There were many attempts to collect young spats and enable them to grow into adults to nourish the pearl fishery resource. Spat collection using different setting materials that are locally available and low cost (e.g., empty oyster shells, nylon rope, coir rope type, steel wire, plastic baskets, and wooden frames) was assessed by Al-Sayed et al. (1997). The maximum spat settlement was recorded in empty oyster shells and in the mid-depth (0.5–1.5m) region. The highest settlement was recorded in Fasht Al-Jarim (north of Bahrain). Additionally, it was found that after spat settlement in summer (July–August 1989), the spat showed a higher rate of growth (0.204–0.248 mm/day) until the following winter (December–January), when spat growth reduced at a gradual rate due to the drop in water temperature from 27–33°C to 17–18°C. The mean size of 57.69 mm was achieved after eighteen months of settlement, which agrees with the findings of Nayar and Al-Rumaidh (1993b). In nature, pearl oyster settlement appears in a wide-ranging substrate. In addition to that, the settlement of pearl oyster larvae is gregarious and affected by adult preferences; consequently, the highest attachment densities are documented on the live and dead shells of its species (Gosling 2003). The settlement on macrophyte, in particular *Halodule uninervis*, might occur in large numbers, and the life cycle of pearl oysters is directly

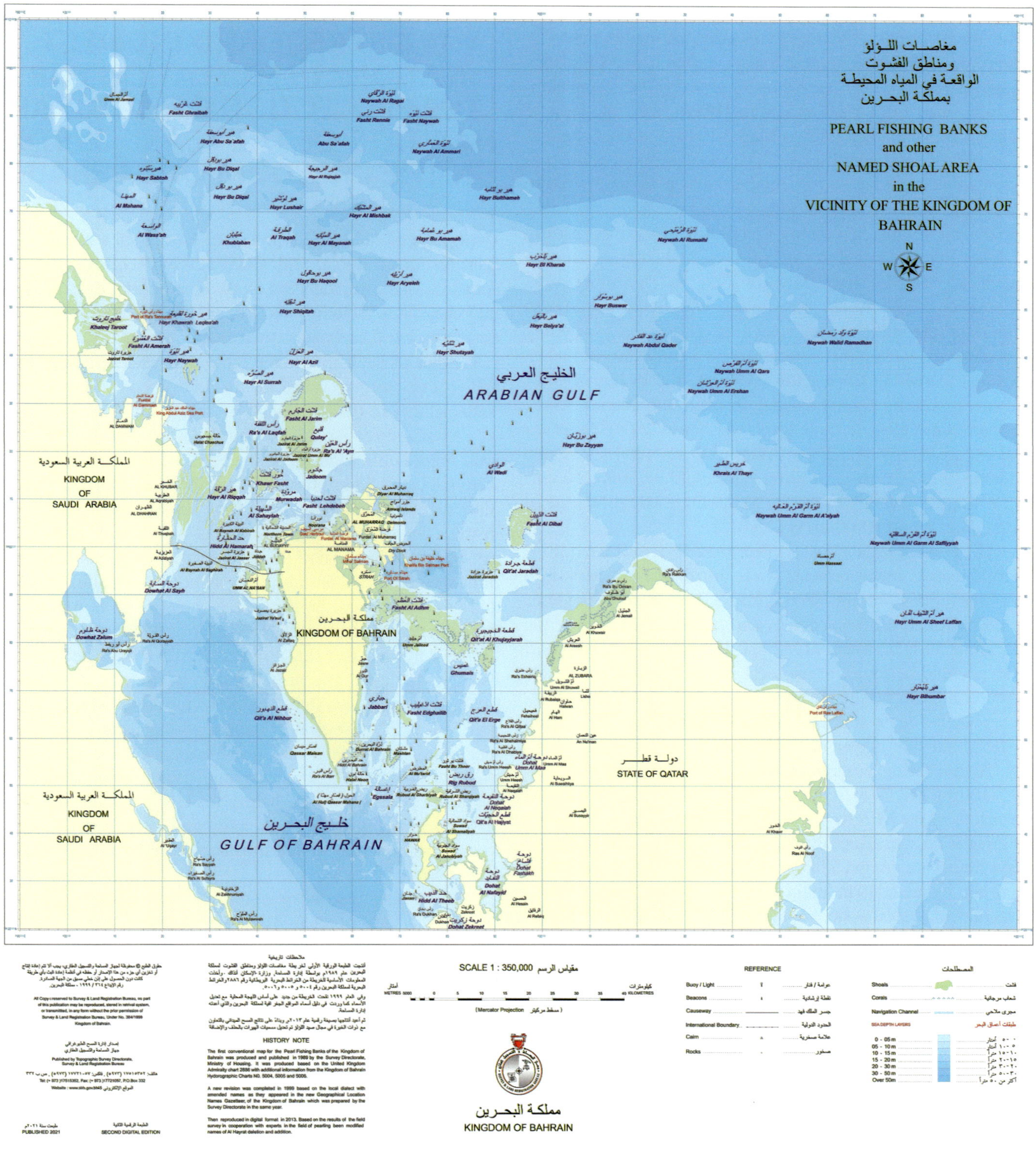

FIGURE 11.2 Pearl fishing banks and other named shoal areas in the vicinity of the Kingdom of Bahrain (Source: Survey and Land Registration Bureau).

associated with the existence and death of the seagrass (Basson et al. 1977). Khamdan (1988) found that settlement can also be achieved on algae such as *Hormophysa sp.* and *Sargassum sp.* Furthermore, oyster spat displays a better ability of alternate attachment and detachment on seagrass than do the older individuals (Basson et al. 1977). Seasonal changes in spat growth were also recorded, and rapid growth coincided with warm temperatures and declined during winter months (Khamdan 1988; Al-Sayed et al. 1997).

The population structure and the growth pattern of *Pinctadata radiata* during 1992 in the Ras Hayan area on the east coast of Bahrain and the Barbar area in the north were described by Al-Sayed et al. (1997). The results revealed that the larger and relatively older oysters were found in the Ras Hayan area, which suggested good growth conditions prevailing in that area. The maximum and minimum sizes of pearl oysters were 110 mm and 35 mm, respectively, and the 50–70 mm group was the most numerous and noticeable. The monthly mean size measurements showed an increase in size only every three months, after which there was a decline in oyster growth. The reproductive biology, maturation, duration, and frequency of spawning were investigated by Al-Sayed (1996) using samples collected from the Ras Hayan area on the east coast of Bahrain from January 1989 to March 1990. Seasonal changes in gamete count were observed, and the maximum count was recorded during May and December. Spawning was found to occur over an extended period, with a maximum spawning rate during August, September, and October. A similar pattern of spawning was found in samples collected from Al-Jadoum (north of Bahrain) from January 1992 to December 1993 (Khamdan 1998). Water temperature was found to have an important role in the start of spawning. The peak of spawning was observed during the summer months (July–August) when seawater temperature is usually at its uppermost value of 34°C. However, spawning might occur in other months while the temperature is close to these levels (Al-Sayed 2009). *Pinctada radiata* are highly fecund and Alansari et al. (1994) estimated an average of 0.95 to 1.7 million eggs could be produced during a single spawning event, of which there could be several throughout a year (Ibrahim et al. 2018).

Differences in pearl oyster *Pinctada radiata* (Leach 1814) morphometric measurements were attributed to variations in the salinity level around Bahrain. Large-sized individuals could be found on the northern and eastern sides, where the salinity level is between 40–42 ppt. In contrast, smaller pearl oysters could be found in the west due to high salinity levels that reach up to 60 ppt. However, the phylogenetic investigation revealed no noticeable variations between oysters from different sites (Al-Sayed et al. 1997).

A monitoring program for the oyster bed sites was conducted from March to September 2012 by Environment Arabia on behalf of the Ministry of Culture. The study area consisted of three main sites: Hayr Bu-l-Thamah, Hayr Bu Am'amah, and Hayr Shtayyah, around 58 km, 57 km, and 30 km, respectively, from the mainland. The main objectives of that study were to determine the pearl oyster populations and size structure across the selected sites, and to produce a map showing the density of pearl oysters. The results indicated that Hayr Shutayyah has the highest density of pearl oysters compared to the other sites. Oysters were not detected in areas characterized by "sand" habitats, whereas "coral, rock and sand" habitats retained "abundant" oyster densities (Arfa et al. 2012). The results confirmed that pearl oysters prefer to inhabit hard substrates. The results also showed that the minimum size was less than 2 cm in more than half of the sampled stations, which revealed that juvenile oysters had recently occurred at many stations during the sampling period. The maximum size of pearl oysters was more than 7 cm, which could be approximately 2 years old according to Nayar and Al-Rumaidh (1993a). The largest collected oyster was 8.9 cm from Hayr Bu-l-Thama in June 2012. Spat settlement was found to occur between May and September 2012 during the mid-to-late spring and the summer months. This agrees with the earlier findings of Nayar and Al-Rumaidh (1993a).

There is a lack of information about the present status of the pearl oyster *Pinctada radiata* in the Kingdom of Bahrain. In Qatar, there was a dramatic decline in the standing stock of oysters, with an estimated reduction rate of 580:1 between 2002 and 2016 (Ibrahim et al. 2018). Fishing seemed to be a substantial contributor to this damage.

11.4 DIRECT AND INDIRECT NATIONAL LEGISLATION FOR THE PROTECTION OF NATURAL HABITATS OF THE PEARL OYSTER IN THE KINGDOM OF BAHRAIN

Dredging and land-filling processes have been carried out in Bahrain for economic and social development over the past decades. This ultimately led to the destruction of many natural ecosystems including the oyster beds, as well as seagrass meadows, coral reefs, mangroves, and fisheries. Thus, for the protection of such important ecological systems, strict laws and regulations are crucial for the recovery and to ensure the sustainability of these systems.

As stated earlier, Alatawi et al. (2019) demonstrated that most gem-quality pearls have been traditionally found in waters located 50 km north and northeast of Bahrain at Hayr Shutayah, Hayr Bulthama, and Hayr Bu Amamah. The shoals in these areas contain an average of about 4–30 oysters per m^2. For the protection of these resources, the government has implemented strict laws and regulations on mollusk fishing and has recognized the three areas mentioned as part of the Pearling Path project and UNESCO World Heritage Sites.

In addition, Pearling Protected Areas have been established to revitalize the pearl industry as well as the national heritage associated with it. The protected zones were designed with the help of several members of the local community, including pearl divers and pearl traders. There have been suggestions to establish diving and fishing training centers. Bahrain's SCE enlisted the help of the UN Environment's West Asia office in preserving and maintaining the oyster habitats, as well as developing policies, strategies, and national development plans.

There have also been initiatives to safeguard the three oyster beds described in Bahrain's northern region. These three oyster beds, as well as their maritime buffer zone, are protected at the national level by Decree No. 2 of 1995 on Wildlife Protection, Legislative Decree No. 21 of 1996 on Environmental Protection, and Decree No. 20 of 2002 on Fishing and Exploitation of Marine Resources. In 2011, a legislative decree was enacted that conferred and designated the marine sites and buffer zone with the designation of 'National Marine Protected Areas.' Bahrain launched a national plan in 2017 to revitalize the pearling industry. Several studies conducted on the seafloor concluded that insufficient pearl oyster harvesting in previous years had resulted in overcrowding and consequently loss of many oysters from Bahrain's oyster beds (e.g., Al-Sayed 2009; Naser 2016).

Moreover, the government has taken up measures to regulate pearl diving, protect Bahrain's oyster beds, strengthen Bahrain as an international center for natural pearls, as well as transform the national pearls and gemstones laboratory, DANAT, into a global center for pearl examination and assessment. In 2017, the SCE issued two orders targeted at increasing site protection and regulating fishing and other maritime activities. Najwat Bul Thama, the three oyster beds, and a marine buffer zone around the oyster beds are among the orders. As previously stated, all of these sites have been classified

TABLE 11.3 Summary of the national legislation for the protection of natural habitats of pearl oysters in the Kingdom of Bahrain.

LEGISLATION	*YEAR*
Decree-Law No. 10: Regarding the control of pearls and precious stones	**1990**
Decree-Law No. 2: The three oyster beds, as well as their maritime buffer zone, are protected at the national level.	**1995**
Decree-Law No. 21: Environmental Protection	**1996**
Decree-Law No. 20: Fishing and Exploitation of Marine Resources	**2002**
A legislative decree was enacted that conferred and designated the marine sites and buffer zone with the designation of "National Marine Protected Areas."	**2011**
Bahrain launched a national plan to revitalize its pearl industry.	**2017**
Establishment of the national pearls and gemstones laboratory, Bahrain Institute for Pearls and Gemstones, DANAT.	**2017**
The Supreme Council for the Environment issued two orders targeted at increasing site protection and regulating fishing and other maritime activities.	**2017**
The Ministry of Works, Municipalities, and Urban Planning issued an order to regulate the ways of getting a pearl diving license and established specific norms and requirements that diving centres must follow to operate pearl-diving trips to preserve a sustainable oyster population.	**2017**

as UNESCO World Heritage Sites. The Ministry of Works, Municipalities, and Urban Planning issued an order regulating the ways of getting a pearl diving license and establishing specific norms and requirements that diving centers must follow to operate pearl-diving trips to preserve a sustainable oyster population. Pearls collected in Bahrain waters should be tested, analysed, and verified by DANAT to ensure that they are genuine saltwater pearls. The legislations are summarized in Table 11.3.

11.5 CONCLUSION AND RECOMMENDATIONS

Research on the current status of pearl oyster beds in Bahrain is scarce. The oyster beds of Bahrain have faced centuries of overutilization and have been subjected to a substantial rise in environmental stresses and anthropogenic disturbances during the last two decades (Naser 2016). Thus, it is crucial to assess the existing environmental condition of the historic oyster beds and to establish a long-term ecological monitoring program. The following are some recommendations that might enhance the conservation of pearl oyster beds and maintain their sustainability:

- Establishing a standard monitoring program of oyster beds once every two years (including the biological, ecological, social, and economic status of the pearl oyster).
- Investigating the reproduction of pearl oysters.
- Conducting more research to determine the different species (if any) present in the waters of Bahrain.
- Enhancing public awareness about their role in the sustainability of oyster beds.
- Taking actions such as protection of productive shallow offshore habitats and enforcing restriction on fishing to control the decline.
- Re-seeding of selected areas could develop the recovery of damaged oyster beds.
- Implementing of existing legislation that will lead to the protection and sustainability of marine resources.

Pearling activities have the possibility of being fully restored. They surely have a promising role in securing the future of this sector, as well as maintaining the country's culture and heritage. The revival of pearl trading and the pearling industry can be beneficial for pearl divers, traders, merchants, and modern boat suppliers through creating a wide variety of specialized job opportunities. It is important to highlight that during the designing of marine management plans, priority should be given to conserving habitat-forming species, such as oysters.

ACKNOWLEDGMENTS

The authors would like to thank Mr. Ahmed Fathi Ebrahim Al Thawadi and Mr. Isa Ali Abdulla Isa from the Survey and Land Registration Bureau for supporting the work with the map that shows the pearl oyster beds in the waters of the Kingdom of Bahrain. We would like to thank DANAT (Bahrain Institute for Pearls and Gemstones) for their continuous support to the Ph.D. student Mohamed Yusuf.

REFERENCES

Alansari, A., Shams, A., and A. Aldaggal. 1994. *Trial production of pearl oyster of Pinctada radiata trochophore larvae for use as larval food*. Ministry of Commerce and Agriculture, Directorate of Fisheries, National Mariculture Center, Bahrain.

Alatawi, A., Karampelas, S., Abdulla, H., Alsheari, H., Abdulla, M., and Z. Taresh. 2019. Diving for natural pearls off Bahrain: A different type of gemmological sample collection. *J Gemmol* 36, no. 8: 702–704. doi:10.15506/JoG.2019.36.8.702.

Al-Rumaihi, M. 1973. *Social and political change in Bahrain since the First World War*. Ph.D. Thesis, Durham University. http://etheses.dur.ac.uk/7942/.

Al-Sayed, H. 1996. Periodicity of reproduction in pearl oyster *Pinctada radiata* in Bahrain. *J Egypt German Soc Zool* 21(D): 121–133.

Al-Sayed, H. 2009. Review of spawning studies of some commercially important marine organisms in the coastal waters of Bahrain. *J Assoc Arab Univ Basic Appl Sci* 8: 75–92.

Al-Sayed, H., Al-Rumaidh, M., and N. Nayar. 1997. Spat settlement and growth of yearling of the pearl oyster *Pinctada radiata* in Bahrain water. *Arab Gulf J Sci Res* 15, no. 2: 467–480.

Al-Sayed, H., Al-Rumaihi, E., and M. Al-Rumaidhm. 1993. *Some morphometric measurements and population structure of the pearl oyster Pinctada radiata at two different coasts in Bahrain (1992)*. Bahrain Center for Studies and Scientific Research, Kingdom of Bahrain.

Arfa, S., AlMealla, R., Edwards, D., and E. Cruz. 2012. *Monitoring program for the oyster bed sites*. Final report, Environment Arabia, Bahrain, 136 p.

BACA. 2020. *History of pearling: Pearling path*. Retrieved from Bahrain Authority for Culture and Antiquities. https://pearlingpath.bh/en/history-of-pearling/.

Basson, P., Burchard, J., Hardy, J., and A. Price. 1977. *Biotopes of the Western Arabian Gulf.* Dhahran: Saudi ARAMCO, Department of Loss Prevention and Environmental Affairs.

Behzadi, S., Parivar, K., and P. Roustaian. 1997. Gonadal cycle of pearl oyster, *Pinctada fucata* (Gould) in the northeast Persian Gulf, Iran. *Oceanogr Lit Rev* 12, no. 44: 1543.

Belgrave, C. 1934. Pearl diving in Bahrain. *J Roy Centr Asian Soc* 21, no. 3: 450–452. doi:10.1080/03068373408725323.

Bishara, F., Haykel, B., Hertog, S., Holes, C., and J. Onley. 2016. *The traditional economy of the Gulf.* London: Bloomsbury Academic. http://hdl.handle.net/10871/20537.

Bostock, J., and H. Riley. 1855. *Pliny the Elder: The natural history*. London: Taylor and Francis. *Perseus Tufts* 6, no. 32: 28.

Bowen, R. 1951. The pearl fisheries of the Persian Gulf. *Middle East J* 5, no. 2: 161–180. www.jstor.org/stable/4322269.

Burdett, A. 1995. *Records of the Persian Gulf pearl fisheries 1857–1962*. Cambridge: Cambridge University Press.

Carter, R. 2005. The history and prehistory of pearling in the Persian Gulf. *J Econ Soc Hist Orient* 48, no. 2: 139–209. doi:10.1163/1568520054127149.

Crouzet, G. 2019. The British Empire in India, the Gulf pearl and the making of the Middle East. *Middle Eas Stud* 55, no. 6: 864–878. doi:10.1080/00263206.2019.1588121.

Cummins, J.W. 1957. *Report of an inquiry into the working of the system of government in Bahrain and the structure of the Bahrain civil service*. Foreign Office Records. British Residency, Bahrain. http://www.agda.ae/en/catalogue/tna/fo/371/126897/n/319.

DANAT. 2019. *Proceedings of the 1st pearl symposium: Conference proceedings report*. www.danat.bh/wp-content/uploads/2019/11/PEARL-SYMPOSIUM-V12.pdf.

Derbali, A., Jarboui, O., Ghorbel, M., and K. Dhieb. 2009. Reproductive biology of the pearl oyster, *Pinctada radiata* (Mollusca: Pteriidae), in Northern Kerkennah Island (Gulf of Gabes). *Cah Biol Mar* 50, no. 3: 215–222.

Fuccaro, N. 2009. *Histories of city and state in the Persian Gulf: Manama since 1800*. New York: Cambridge University Press.

Gosling, E. 1992. *The mussel Mytilus: Ecology, physiology, genetics, and culture*, 87–169. Developments in Aquaculture and Fisheries Sci 25. Amsterdam: Elsevier. www.researchgate.net/publication/255981293_Population_and_communtiy_ecology_of_Mytilus.

Gosling, E. 2003. Reproduction, settlement, and recruitment. In: *Bivalve molluscs: Biology, ecology, and culture*, 131–168. Oxford: Fishing News Books, Blackwell Publishing.

Hightower, V. 2013. Pearls and the Southern Persian/Arabian Gulf: A lesson in sustainability. *Environ Hist* 18, no. 1: 44–59. doi:10.1093/envhis/ems113.

Ibrahim, A., David, S., Bruno, G., Mark, C., Mohammed, A., and L. Lewis. 2018. Decline in oyster populations in traditional fishing grounds; is habitat damage by static fishing gear a contributory factor in ecosystem degradation? *J Sea Res* 140: 40–51. doi:10.1016/j.seares.2018.07.006.

Khamdan, S. 1988. *The Bahrain pearl oyster: Their genetics and systematics*. M.Sc. Thesis, University College of North Wales, UK, 132 p.

Khamdan, S. 1998. Occurrence of Bucephalus sp. trematode in the gonad of the pearl oyster, *Pinctada radiata. Environment Intl* 24, no. 1: 117–120. doi:10.1016/S0160-4120(97)00128-1.

Khamdan, S. 2004. *The Gulf encyclopaedia for pearls and oysters*. Manama, Bahrain: Bahrain Centre for Studies and Research.

Lorimer, J. 1915. *Gazetteer of the Persian Gulf, appendix C: The pearl and mother-of-pearl fisheries of the Persian Gulf*, vol. 2, 2220–2293. Calcutta, India: Superintendent Government Printing.

Martínez-Fernández, E., Acosta-Salmón, H., and P. Southgate. 2006. The nutritional value of seven species of tropical microalgae for black-lip pearl oyster (*Pinctada margaritifera, L.*). *Aquaculture* 257, no. 1–4: 491–503. doi:10.1016/j.aquaculture.2006.03.022.

Mohammed, S., and M. Yassien. 2003. Population parameters of the pearl oyster *Pinctada radiata* (Leach) in Qatari Waters, Arabian Gulf. *Turk J Zool* 27, no. 4: 339–343. https://journals.tubitak.gov.tr/zoology/issues/zoo-03-27-4/zoo-27-4-10-0207-7.pdf.

Naser, H. 2016. Management of marine protected zones – case study of Bahrain, Arabian Gulf. In: *Applied studies of coastal and marine environments*. Ed: M. Marghany, 323–350. INTECH Publishing. doi:10.5772/62132.

Nayar, K.N., and M.J. Al-Rumaidh. 1993a. *Studies on the pearl oysters of Bahrain waters*. Kingdom of Bahrain: Bahrain Centre for Studies and Research.

Nayar, K.N., and M.J. Al-Rumaidh. 1993b. The rate of growth of spat and yearlings of the pearl oyster *Pinctada radiata* (Leach) of Bahrain waters. *J Mar Biol Assoc India* 35, nos. 1–2: 1–13.

Onley, J. 2014. Indian communities in the Persian Gulf, c. 1500–1947. In: *The Persian Gulf in modern times: People, ports, and history*. Ed: L.G. Potter, 231–266. New York: Palgrave Macmillan. doi:10.1057/9781137485779_10.

Peterson, J. 2016. *The emergence of the Gulf States: Studies in modern history*. London: Bloomsbury Academic Press. https://lccn.loc.gov/2015047965.

Richardson, C.A., Collis, S.A., Ekaratne, K., Dare, P., and D. Key. 1993. The age determination and growth rate of the European flat oyster, Ostrea edulis, in British waters determined from acetate peels of Umbo growth lines. *ICES J Mar Sci* 50: 493–500.

SCE. 2015. *A protected areas strategy for the Kingdom of Bahrain*. The National Biodiversity Strategy and Action Plan (NBSAP) of the Kingdom of Bahrain.

SCE. 2016. *The national biodiversity strategy and action plan*. Kingdom of Bahrain 2016–2021. The Supreme Council for Environment of the Kingdom of Bahrain for the UN. www.cbd.int/doc/world/bh/bh-nbsap-v2-en.pdf.

Tlig-Zouari, S., Rabahoui, I., Irathani, I., and O.K. Ben Hassine. 2009. Distribution, habitat and population densities of the invasive species Pincatada radiata (Mollusca: Bivalvia) along the northern and eastern coasts of Tunisia. *Cah Biol Mar* 50: 131–142.

Tranter, D. 1958. Reproduction in Australian pearl oysters (Lamellibranchia). IV: *Pinctada margaritifera* (Linnaeus). *Mar Freshw Res* 9, no. 4: 509–525. doi:10.1071/MF9580509.

UNEP. 2023. www.unep.org/news-and-stories/story/bahrain-shows-way-protect-biodiversity-its-region.

An annotated checklist of heterobranch molluscs of the Arabian Gulf and their zoogeographic affinities with adjacent seas

12

Manickam Nithyanandan, Nathalie Yonow, Manal Al-Kandari, Rakhesh Madhusoodhanan, Gopikrishna Mantha and Fathima Thuslim

12.1 INTRODUCTION

Molluscs are essential marine ecosystem components and are economically important as food, jewellery, decorative products, and natural compounds. In the phylum Mollusca, the subclass Heterobranchia includes diverse marine, freshwater, brackish water, and terrestrial invertebrates, which occur in various shapes, sizes, and sometimes in flamboyant colours. The marine species are found in a wide range of habitats, such as sand, mud, rocks, coral reefs, mangroves, saltmarshes, seagrass meadows, and the deep sea. Until recently, heterobranch diversity and distribution in the Arabian/Persian Gulf (APG) region have received little research attention. Recent studies on heterobranch molluscs of the APG have partially revealed their regional diversity and distribution (Smythe 1982; Glayzer et al. 1984; Jones 1986; Malaquias and Reid 2008; George 2012; Al-Yamani et al. 2012; Al-Rifaie et al. 2012; Nithyanandan 2012; Yonow 2012; Rezai et al. 2016; Amini-Yekta et al. 2016, 2019; Amouei and Fatemi 2021; Amini-Yekta and Dekker 2021; Nithyanandan et al. 2021). Knowledge gaps exist in understanding the regional diversity and distribution patterns of heterobranch molluscs and their zoogeographic affinities with adjacent seas, although George (2012) began listing the known non-coral macro-invertebrates in the southeastern corner of the APG, including the Opisthobranchia, as they were then known. In this study, we review the records of heterobranchs from the APG to prepare an annotated and updated checklist and to establish a baseline species list for future research and conservation efforts. Furthermore, we examined the beta diversity patterns to understand endemicity and zoogeographic affinities with the neighbouring seas.

The APG is a shallow and young sea, approximately 1000 km long, with depths of less than 90 m at the Strait of Hormuz and an average depth of 36 m (Figure 12.1) (UK Admiralty Guides 1982). The northeast shore is mountainous with deeper waters and terrigenous sediments and the Arabian side is a flatter desert comprising carbonate and sandy sediments (UK Admiralty Guides 1982). Due to high temperatures and constant winds, sea temperatures are higher than 30°C, and salinities are some of the highest in the world, averaging 38–45‰ in the north and central APG and 40–50‰ in the southern APG; the average tidal range is approximately 2 m with maxima of 4 m (UK Admiralty Guides 1982). Despite these extreme environmental conditions, diverse marine habitats including sandy beaches, mud flats, rocky shores, coral reefs, and seagrass habitats support unique marine biodiversity (Basson et al. 1977). Heterobranchs in this sea are commonly encountered animals, with a clear division of assemblages between sandy and coral-reef habitats (see 'Habitat' in the Checklist, section 3.2).

In this study, we used three components of species diversity: alpha, beta, and gamma diversity, as proposed by Whittaker (1960). Alpha diversity refers to species diversity (number of species) within each area. Beta diversity refers to the change in species composition from one area to another and thus measures how different the species compositions are between any two sample areas. Gamma diversity refers to the overall diversity in the region, considering the alpha and beta diversity of the individual seas within the zoogeographic region of interest (Whittaker 1960). These concepts have been

DOI: 10.1201/9781003321392-15

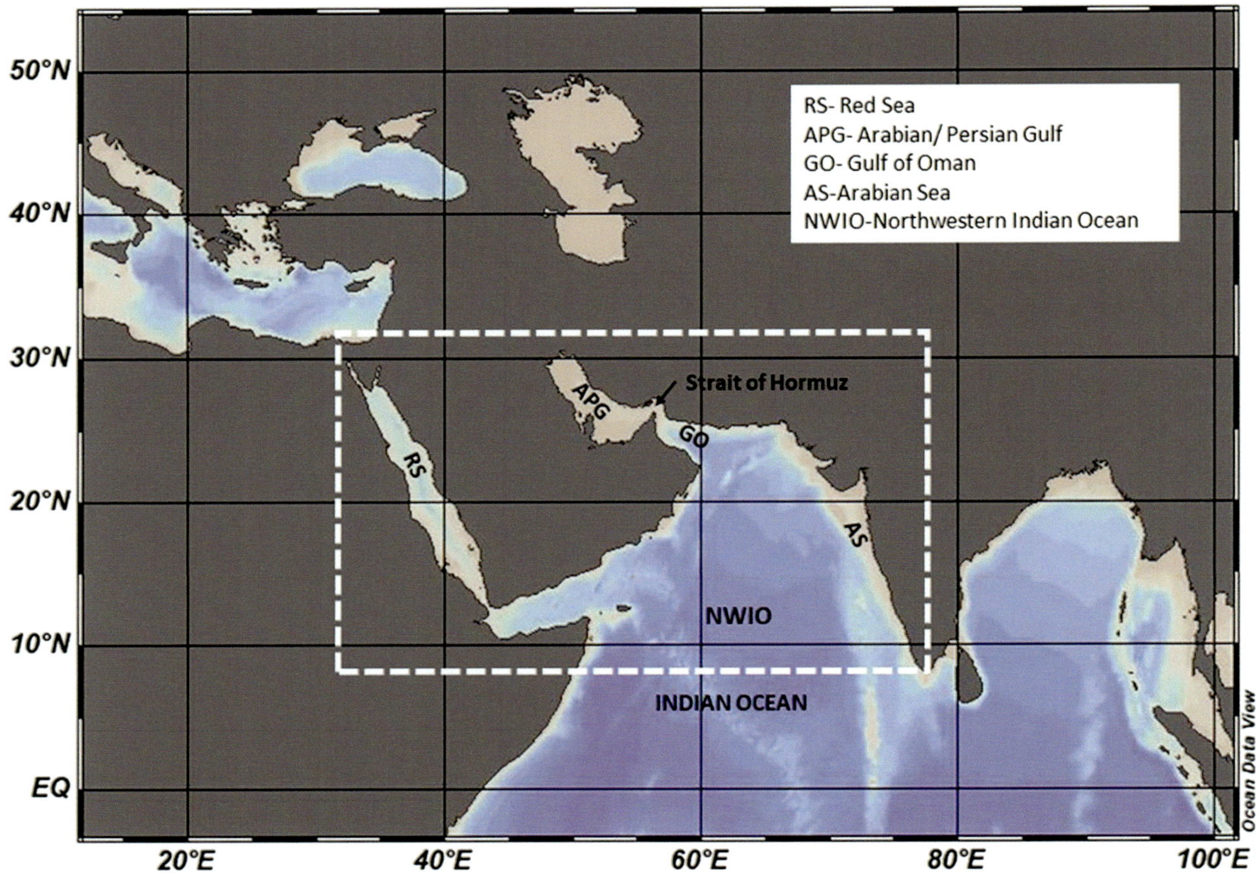

FIGURE 12.1 Map showing the Arabian Gulf and its adjacent marginal seas included in this study. Map courtesy of Ocean Data View 5.6.3 (Schlitzer 2021; http://odv.awi.de).

widely adopted and play a central role in biodiversity and ecological studies of terrestrial and marine ecosystems. In this study, we used alpha, beta, and gamma diversity measures for the first time to describe the sea slug diversity in the northwestern Indian Ocean zoogeographic region encompassing the APG, the Red Sea, the Gulf of Oman, and the western Arabian Sea. Emphasis was given to beta diversity to explore how such measures could improve biodiversity conservation efforts. Understanding the drivers of beta diversity in marine communities is vital for conservation and management efforts, as it provides insights into the processes that determine the composition of communities and the factors that influence species distributions. This study also provides the first faunal list of marine heterobranchs from the APG since it is important to establish regional inventories as baselines against which to monitor future change (e.g., Nimbs and Smith 2016; Nimbs et al. 2020).

12.2 MATERIALS AND METHODS

The published records (1982–2021) of heterobranchs of the APG were critically reviewed in order to prepare an updated checklist (Smythe 1982; Glayzer et al. 1984; Jones 1986; Malaquias and Reid 2008; Nithyanandan 2012; Yonow 2012; Al-Yamani et al. 2012; Rezai et al. 2016; Amini-Yekta et al. 2019; Fatemi and Attaran-Fariman 2015; Fatemi et al. 2020; Amouei and Fatemi 2021; Amini-Yekta and Dekker 2021; Nithyanandan et al. 2021). Bouchet et al. (2017) and the World Register of Marine Species (WoRMS 2024) were followed for classification and nomenclature. We categorized the heterobranch species based on the seas of Arabia: the Red Sea (RS), the Arabian/Persian Gulf (APG), the Gulf of Oman (GO), and the Arabian Sea (AS), to examine the zoogeographic similarities of species shared between the seas, endemicity, and ubiquitous species in the northwestern Indian Ocean (NWIO) region by constructing Venn diagrams using the free online software, Bioinformatics and Evolutionary Genomics (https://bioinformatics.psb.ugent.be/webtools/Venn; Ghent, Belgium).

Beta diversity (β) may reflect two different phenomena, the spatial turnover of species and nestedness of assemblages or areas, which result from two antithetical processes, species replacement and species loss. We explored the Nudipleura β diversity in the NWIO region using two dissimilarity indices: the Sørensen dissimilarity index and the Simpson dissimilarity index. The Sørensen dissimilarity index (β_{sor}) incorporates actual spatial turnover and differences in species richness (Koleff et al. 2003) and accounts for all aspects of β diversity. In contrast, the Simpson dissimilarity index (β_{sim}) describes and accounts for spatial turnover without the influence of richness gradients. Since β_{sor} and β_{sim} are equal without nestedness, their difference measures the nestedness component (β_{nes}) of β diversity (Baselga 2010). According to Patterson and Atmar (1986), nestedness is defined as a "pattern where the species present at species-poor sites form proper subsets of the species in species – richer sites."

The formulae used are as follows:

Total beta diversity: Sørensen (β_{sor}) $= (b + c)/(2a + b + c)$

Spatial turnover component of beta diversity: Simpson (β_{sim}) $= \min(b, c)/(a + \min(b, c))$

Nestedness component of beta diversity (β_{nes}, i.e., species loss due to dispersal Limitation):

$$(\beta_{nes}) = \beta_{sor} - \beta_{sim}$$

where a = species common in both sites/areas, b = species in area 1 only, c = species in area 2 only, and min (b, c) = the minimum of b and c.

In the checklist, all new records for the APG are given in bold typeface, either based on specimens examined and identified by the second author or from photographic records. In the case of photographic records, we examined citizen science data collected by recreational divers, which is considered an essential source of record for the marine sea slugs (Smith and Davis 2019): Facebook® (www.facebook.com/uaebranchers), iNaturalist® (www.inaturalist.org/), and the Mediterranean slug site (www.medslugs.de/). Records from the United Arab Emirates (UAE) by Gosliner et al. (2008, 2015, 2018), wherein location (east or west coast) was not mentioned, are marked with the symbol †. Species validated after this manuscript's completion are indicated with the symbol # and excluded from the data analysis.

12.3 RESULTS

The annotated checklist presented as follows includes 140 species in 77 genera, 36 families, eight orders, and two superorders (Euthyneura and Nudipleura): 61 of these species (in bold) are new records for the APG. Where an unidentified species has a number, it refers to one of the three editions of the identification guide by Gosliner et al. (2008, 2015, 2018) listed in the distribution of that species. Unfortunately, the species numbers changed with different editions, and precise localities of individual photographs were omitted in later editions.

12.3.1 Zoogeographic similarities

Of the 759 species of heterobranchs reported from the NWIO, 554 were sea slugs (Nudipleura) (Table 12.1, Figure 12.2). Along the APG-GO-AS oceanographic continuum, maximal alpha

TABLE 12.1 Heterobranchs of the Arabian Peninsula and adjacent seas.

ORDER/FAMILY	*GENUS*	*SPECIES*	*APG*	*RS*	*GO*	*AS*
ACTEONIMORPHA						
ACTEONIDAE	*Acteon*	*Acteon* sp.		+		+
	Pupa	*affinis* (A. Adams, 1855)	+	+		
		cf. *solidula* (Linnaeus, 1758)	+	+		+
		tessellata (Reeve, 1842)		+		
		Pupa sp.		+		+
APLUSTRIDAE (= HYDATINIDAE)	*Hydatina*	*albocincta* (van der Hoeven, 1839)				+
		physis (Linnaeus, 1758)		+	+	+
		zonata (Lightfoot, 1786)		+	+	+
	Bullina	*Bullina* sp. a		+		
		Bullina sp. b		+		
		Bullina sp. c		+		
	Micromelo	*scriptus* (Garrett, 1857)	+		+	
		guamensis (Quoy & Gaimard, 1825)			+	
RINGICULIMORPHA						
RINGICULIDAE	*Ringicula*	*acuta* R. A. Philippi, 1849		+		
		encarpoferens de Folin, 1867				+
		minuta H. Adams, 1872		+		
CEPHALASPIDEA						
ALACUPPIDAE	*Roxania*	*lithensis* (Sturany, 1903)		+		
BULLIDAE	*Bulla*	*arabica* Malaquias & Reid, 2008	+	+		+
	Aliculastrum	*cylindricum* (Helbling, 1779)	+	+		
	Atys	*ehrenbergi* (Issel, 1869)		+		
		naucum (Linnaeus, 1758)		+		
		pellyi (Smith, 1872)	+			

(Continued)

TABLE 12.1 (Continued)

ORDER/FAMILY	*GENUS*	*SPECIES*	*APG*	*RS*	*GO*	*AS*
		semistriatus Pease, 1860		+		+
		Atys sp. a	+		+	
		Atys sp. b	+			
		Atys sp. c	+			
	Bakawan	*rotundata* (A. Adams, 1850)	+		+	+
	Diniatys	*dentifer* (A. Adams, 1850)	+	+		
		dubius (Schepman, 1913)		+		
	Haloa	*aptei* (Bharate et al., 2018)				+
		constricta (A. Adams, 1850)				+
		cf. *pemphis* (Philippi, 1847)	+	+		+
	Haminoea	*natalensis* (Krauss, 1848)				+
		glabra (A. Adams, 1850)				+
		Haminoea sp.	+			+
	Lamprohaminoea	*ovalis* (Pease, 1868)		+	+	+
	Liloa	*mongii* (Audouin, 1826)		+		
		porcellana (A. Gould, 1859)		+		
		Liloa sp.			+	
	Phanerophthalmus	*albocollaris* Heller & Thompson, 1983		+	+	
		olivaceus (Ehrenberg, 1828)		+		+
		Phanerophthalmus sp.			+	
	Smaragdinella	*calyculata* (Broderip & Sowerby I, 1829)		+		+
		sieboldi A. Adams, 1864		+		+
RETUSIDAE	*Pyrunculus*	*fourierrii* (Audouin, 1826)		+		
	Retusa	*desgenettii* (Audouin, 1826)		+		
		tarutana Smythe, 1979		+		
		Retusa sp.	+			
TORNATINIDAE (= SCPAHANDRIDAE)	*Acteocina*	*fusiformis* (A. Adams, 1850)		+		
		inconspicua (H. Adams, 1852)	+	+		
		involuta (G. Nevill & H. Nevill, 1871)	+	+		
		mucronata (Philippi, 1849)		+		
		simplex (A. Adams, 1850)		+		
	Tornatina	*persiana* Smith, 1872	+			
		Tornatina sp.	+			
CYLICHNIDAE	*Cylichna*	*collyra* Melvill, 1906	+		+	
		cylindracea (Pennant, 1777)	+			
		villersii (Audouin, 1826)		+		
		Cylichna sp.	+			
MNESTIIDAE	*Mnestia*	*girardi* (Audouin, 1826)		+		
AGLAJIDAE	*Biuve*	*fulvipunctata* (Baba, 1938)	+	+	+	
	Chelidonura	*electra* Rudman, 1970				+
		flavolobata Heller & Thompson, 1983		+		
		hirundinina (Quoy & Gaimard, 1833)				
		livida Yonow, 1994	+	+	+	+
		pallida Risbec, 1951				+
		punctata Eliot, 1903			+	+
		cf. *punctata* Eliot, 1903				+
	Mariaglaja	*inornata* (Baba, 1949)		+		
		tsurugensis (Baba & Abe, 1959)		+		
	Odontoglaja	*mosaica* Gosliner, 2011		+		
	Philinopsis	*ctenophoraphaga* Gosliner, 2011		+		
		speciosa Pease, 1860	+	+	+	+
		Philinopsis sp.		+		
	Tubulophilinopsis	*gardineri* (Eliot, 1903)				+
		lineolata (H. Adams & A. Adams, 1854)			+	

TABLE 12.1 (Continued)

ORDER/FAMILY	*GENUS*	*SPECIES*	*APG*	*RS*	*GO*	*AS*
		pilsbryi (Eliot, 1900)		+		
		reticulata (Eliot, 1903)		+		
	Noalda	*Noalda* sp.		+		
COLPODASPIDIDAE	*Colpodaspis*	*thompsoni* Brown, 1979		+		+
	Spinoaglaja	cf. *orientalis* (Baba, 1949)		+		
	'Spinophallus'	*falciphallus* (Gosliner, 2011)		+		
GASTROPTERIDAE	*Gastropteron*	*bicornutum* Baba & Tokioka, 1965			+	
	Sagaminopteron	*nigropunctatum* Carlson & Hoff, 1973		+		
		cf. *ornatum* Tokioka & Baba, 1964		+		
		psychedelicum Carlson & Hoff, 1974		+		+
	Siphopteron	cf. *makisig* Ong & Gosliner, 2017	+			
		tigrinum Gosliner, 1989				+
		Siphopteron sp.		+		
PHILINIDAE	*Philine*	*kurodai* Habe, 1946				+
		Philine sp. a	+			
		Philine sp. b				+
		Philine sp. c		+		
		Philine sp. d		+		
APLYSIIDA						
AKERIDAE	*Akera*	*soluta* (Gmelin, 1791)		+		
APLYSIIDAE	*Aplysia*	*argus* Rüppell & Leuckart, 1830		+		+
		cornigera Sowerby, 1869	+	+		+
		ghanimii Golestani et al., 2019		+	+	+
		juliana Quoy & Gaimard, 1832			+	+
		nigrocincta von Martens, 1880		+		+
		oculifera A. Adams & Reeve, 1850		+	+	+
		Aplysia sp.	+			
	Dolabella	*auricularia* (Lightfoot, 1786)		+	+	+
	Dolabrifera	*dolabrifera* (Rang, 1828)		+		+
	Phyllaplysia	*lafonti* (Fischer, 1870)			+	
		Phyllaplysia sp.		+	+	+
	Bursatella	*leachii* Blainville, 1817	+	+	+	+
	Notarchus	*indicus* Schweigger, 1820		+		
	Petalifera	*petalifera* (Rang, 1828)		+		
		ramosa Baba, 1959		+		+
	Stylocheilus	*longicauda* (Quoy & Gaimard, 1825)	+	+	+	
		striatus (Quoy & Gaimard, 1832)	+	+	+	+
	Syphonota	*geographica* (A. Adams & Reeve, 1850)		+		+
SACOGLOSSA						
JULIIDAE	*Berthelinia*	*ganapati* Sarma, 1975				+
		limax (Kawaguti & Baba, 1959)				+
		waltairensis Sarma, 1975				+
	Julia	*exquisita* Gould, 1862		+		
		zebra Kawaguchi, 1981		+		
OXYNOIDAE	*Lobiger*	*viridis* Pease, 1863		+		+
		Lobiger sp.			+	
	Oxynoe	*jordani* Krug, Berriman & Valdés, 2018		+		
		viridis (Pease, 1861)			+	+
VOLVATELLIDAE	*Volvatella*	*vigourouxi* (Montrouzier in Souverbie, 1861)				+
	Ascobulla	*Ascobulla* sp.		+		
COSTASIELLIDAE	*Costasiella*	*kuroshimae* Ichikawa, 1993		+		+
		cf. *kuroshimae* Ichikawa, 1993	+	+	+	
		cf. *usagi* Ichikawa, 1993		+		
		Costasiella sp. a			+	
		Costasiella sp. b		+		

(Continued)

TABLE 12.1 (Continued)

ORDER/FAMILY	*GENUS*	*SPECIES*	*APG*	*RS*	*GO*	*AS*
		Costasiella sp. c		+		
		Costasiella sp. d		+		
	Panderevela	*Panderevela* sp.		+		
HERMAEIDAE	*Caliphylla*	*viridis* (Deshayes, 1857)				+
	Cyerce	cf. *elegans* Bergh, 1870		+		
		elegans Bergh, 1870				+
		nigra Bergh, 1871				+
		Cyerce sp.		+		
	Polybranchia	*burni* Medrano, Krug, Gosliner, Biju Kumar & Á. Valdés, 2018		+		
		jensenae Medrano, Krug, Gosliner, Biju Kumar & Á. Valdés, 2018		+		
		orientalis (Kelaart, 1858)				+
		rubicunda (Bergh, 1871)		+		
		samanthae Medrano, Krug, Gosliner, Biju Kumar & Á. Valdés, 2018		+		
PLAKOBRANCHIDAE (= ELYSIIDAE)	*Elysia*	*bengalensis* Swennen, 2011				+
		bangtawaensis Swennen, 1998				+
		chilkensis Eliot, 1916				+
		expansa (O'Donoghue, 1924)				+
		cf. *hirasei* Baba, 1955	+			+
		grandifolia-marginata complex	+	+	+	+
		mercieri (Pruvot-Fol, 1930)	+	+	+	
		obtusa Baba, 1938			+	
		pusilla (Bergh, 1871)		+	+	+
		rufescens (Pease, 1871)				+
		thompsoni Jensen, 1993				+
		cf. *thompsoni* Jensen, 1993		+		
		tomentosa Jensen, 1997 complex	+	+		+
		trilobata Heller & Thompson, 1983		+		
		Elysia sp. 13 (2015)			+	
		Elysia sp. 34 (2015)		+		
		Elysia sp. 35 (2015)			+	
		Elysia sp. a	+		+	
		Elysia sp. b			+	
		Elysia sp. c			+	
		Elysia sp. d		+		
		Elysia sp. e		+		
		Elysia sp. f		+		
		Elysia sp. g		+		
		Elysia sp. h		+		
	Plakobranchus	*ocellatus* van Hasselt, 1824	+	+	+	+
		cf. *ocellatus* van Hasselt, 1824		+		
		Plakobranchus sp.		+		
	Thuridilla	cf. *albopustulosa* Gosliner, 1995		+		
		carlsoni Gosliner, 1995				+
		decorata (Heller & Thompson, 1983)		+	+	
		gracilis (Risbec, 1828)				+
		livida (Baba, 1955)				+
		cf. *indopacifica* Gosliner, 1995		+	+	
		vataae (Risbec, 1928)				+
		Thuridilla sp. 3 (2015)			+	
		Thuridilla sp.			+	

TABLE 12.1 (Continued)

ORDER/FAMILY	*GENUS*	*SPECIES*	*APG*	*RS*	*GO*	*AS*
LIMAPONTIDAE (= STILIGERIDAE)	*Ercolania*	*annelyleorum* Wägele, Stemmer, Burghardt & Händeler, 2010	+			
		cf. *coerulea* Trinchese, 1892	+			
		coerulea Trinchese, 1892	+			
		gopalai (K.V. Rao, 1937)				+
		irregularis (Eliot, 1905)				+
		pica (Annandale & Prashad, 1922)				+
		raorum (Ev. Marcus & Er. Marcus, 1970)				+
		varians (Eliot, 1905)				+
	Stiliger	*ornatus* Ehrenberg, 1828		+		
	Placida	*cremoniana* (Trinchese, 1892)		+		
		daguilarensis Jensen, 1990	+			+
SYSTELLOMMATOPHORA						
ONCHIDIDAE	*Onchidium*	*durum* Labbé, 1934		+		
	Paromoionchis	*penangensis* Dayrat & Goulding, 2019				+
	Paraperonia	*gondwanae* Labbé, 1934		+		
		madagascariensis (Labbé, 1934)		+		
	Peronia	*anomala* Labbé, 1934		+		
		persiae Maniei, Espeland, Mohavedi & Wägele, 2020	+			
		madagascariensis (Labbé, 1934)			+	
		verruculata (Cuvier, 1830)	+	+	+	+
	Peronina	*tenera* (Stoliczka, 1869)				+
	Platevindex	*tigrinus* (Stoliczka, 1869)				+
PLEUROBRANCHIDA						
PLEUROBRANCHIDAE	*Berthella*	*tupala* Er. Marcus, 1957				+
		Berthella sp.		+		
	Berthellina	*citrina* (Rüppell & Leuckart, 1828)		+		+
		delicata (Pease, 1861)		+		+
		minor Bergh, 1905				+
		edwardsii (Vayssière, 1897)				+
	Pleurehedra	*stellata* (Risso, 1826)		+		+
	Tomoberthella	*martensi* (Pilsbry, 1896)	+	+		
	Pleurobranchaea	*brockii* Bergh, 1897			+	+
	Pleurobranchus	*albiguttatus* (Bergh, 1905)		+		+
		forskalii Rüppell & Leuckart, 1828		+	+	+
		grandis Pease, 1868		+		
		cf. *grandis* Pease, 1868		+		
		mamillatus Quoy & Gaimard, 1832	+			+
		peronii Cuvier, 1804			+	
	Euselenops	*luniceps* (Cuvier, 1816)		+	+	+
UMBRACULIDA						
UMBRACULIDAE	*Umbraculum*	*umbraculum* (Lightfoot, 1786)			+	
NUDIBRANCHIA – DORIDINA 1						
AEGIRIDAE	*Aegires*	*citrinus* Pruvot-Fol, 1930		+	+	
		villosus Farran, 1905	+	+	+	
	Notodoris	*gardineri* Eliot, 1906		+	+	+
		minor Eliot, 1904			+	
CALYCIDORIDIDAE	*Diaphorodoris*	*mitsuii* (Baba, 1938)			+	
		Diaphorodoris sp.			+	
GONIODORIDIDAE	*Ceratodoris*	*pilosa* (Bouchet & Ortea, 1983)				+
		cf. *stellata* (Rudman, 2004)			+	
		virginiae (Gosliner, 2004)			+	
	Goniodoridella	*savignyi* Pruvot-Fol, 1933		+	+	

(Continued)

TABLE 12.1 (Continued)

ORDER/FAMILY	*GENUS*	*SPECIES*	*APG*	*RS*	*GO*	*AS*
	Murphydoris	*adusta* Paz-Sedano, Smirnoff, Candás, Gosliner & Pola, 2022			+	
	Okenia	*echinata* Baba, 1949				+
		rhinorma Rudman, 2007		+		
		Okenia sp. a			+	+
		Okenia sp. b				+
		Okenia sp. c			+	
	Pelagella	*castanea* (Alder & Hancock, 1845)		+		
		joubini (Risbec, 1928)		+		+
		kolabana (Winckworth, 1946)				+
		modesta (Alder & Hancock, 1864)				+
		Goniodoris sp.			+	
	Bermudella	*brunneomaculata* (Gosliner, 2004)			+	
		pellucida (Burn, 1967)	?		+	+
	Trapania	*aurata* Rudman, 1987			+	
		euryeia Gosliner & Fahey, 2008		+		+
		japonica (Baba, 1935)	+		+	
		lemnalioides Smirnoff, Donohoo & Gosliner, 2022		+		
		palmula Gosliner & Fahey, 2008		+		
		toddi Rudman, 1987		+	+	
		Trapania sp.			+	
GYMNODORIDIDAE	*Gymnodoris*	*alba* (Bergh, 1877)	+	+	+	
		ceylonica (Kelaart, 1858)		+	+	+
		citrina (Bergh, 1877)		+	+	+
		impudica (Rüppell & Leuckart, 1830)	+	+	+	
		inornata (Bergh, 1880)		+	+	
		cf. *nigricolor* Baba, 1960		+		
		okinawae Baba, 1936		+		+
		striata (Eliot, 1908)		+		+
		Gymnodoris sp. a	+			
		Gymnodoris sp. b	+		+	
		Gymnodoris sp. c			+	
		Gymnodoris sp. d		+		
		Gymnodoris sp. e		+		
		Gymnodoris sp. f		+		
		Gymnodoris sp. g		+		
		Gymnodoris sp. h		+		
		Gymnodoris sp. i		+		
		Gymnodoris sp. j				+
POLYCERIDAE	*Polycera*	*abei* (Baba, 1960)			+	+
		melanosticta Miller, 1996		+	+	
		risbeci Odhner, 1941		+	+	
		Polycera sp.	+			
	Tambja	*affinis* (Eliot, 1904)		+	+	
	Martadoris	*limaciformis* (Eliot, 1908)		+		
	Thecacera	*pacifica* (Bergh, 1884)		+	+	
		pennigera (Montagu, 1813)			+	+
		picta Baba, 1972			+	
		Thecacera sp.			+	
	Plocamopherus	*ceylonicus* (Kelaart, 1858)			+	+
		indicus Bergh, 1890		+	+	
		maculapodium Vallès & Gosliner, 2006			+	
		margaretae Vallès & Gosliner, 2006			+	
		ocellatus Rüppell & Leuckart, 1828	+	+	+	

TABLE 12.1 (Continued)

ORDER/FAMILY	*GENUS*	*SPECIES*	*APG*	*RS*	*GO*	*AS*
		tilesii Bergh, 1877		+	+	
		cf. *lemur* Vallès & Gosliner, 2006	+			
	Nembrotha	*aurea* Pola, Cervera & Gosliner, 2008			+	
		kubaryana Bergh, 1877				+
		megalocera Yonow, 1990		+		
	Kalinga	*ornata* Alder & Hancock, 1864				+
	Kaloplocamus	*peludo* Vallès & Gosliner, 2006			+	
		ramosus (Cantraine, 1935)		+	+	
		Kaloplocamus sp.			+	
	Limacia	*langavi* Toms, Pola, Heyden & Gosliner, 2021			+	
	Tyrannodoris	*luteolineata* (Baba, 1936)				+
HEXABRANCHIDAE	*Hexabranchus*	*sanguineus* (Rüppell & Leuckart, 1830)		+	+	+
		giganteus Tibiriçá, Pola & Cervera, 2023		+		
DORIDINA 2						
ACTINOCYCLIDAE	*Actinocyclus*	*sordida* (Rüppell & Leuckart, 1830)		+		
		verrucosus Ehrenberg, 1831		+		+
	Hallaxa	cf. *albopunctata* Gosliner & Johnson, 1994				+
		indecora (Bergh, 1905)		+		+
		michaeli Gosliner & Johnson, 1994				+
		Hallaxa sp.				+
DISCODORIDIDAE	*Asteronotus*	*cespitosus* (van Hasselt, 1824)	+	+	+	+
		denticulata (Eliot, 1908)		+		+
		hepaticus (Abraham, 1877)	+			
		markaensis Donohoo & Gosliner, 2020		+		
		mimeticus Gosliner & Valdés, 2002		+		
		namuro Donohoo & Gosliner, 2020		+		
		raripilosus (Abraham, 1877)				+
		spongicolus Gosliner & Valdés, 2002		+		
	Alloiodoris	*Alloiodoris* sp.		+		
	Atagema	*intecta* (Kelaart, 1858)			+	+
		ornata (Ehrenberg, 1831)		+		
		osseosa (Kelaart, 1859)				+
		rugosa Pruvot-Fol, 1951				+
		spongiosa (Kelaart, 1858)		+	+	+
		tristis (Alder & Hancock, 1864)				+
		Atagema sp. a		+		
		Atagema sp. b		+		
		Atagema sp. c		+		
	Carminodoris	*estrelyado* (Gosliner & Behrens, 1998)		+	+	
		flammea (Fahey & Gosliner, 2003)		+		+
		grandiflora (Pease, 1860)		+		+
		nodulosa (Angas, 1864)			+	
		pustulata Abraham, 1877	+	+		+
		cf. *armata* Baba, 1993		+		
	Discodoris	*boholiensis* Bergh, 1877				+
		cebuensis Bergh, 1877		+		
		coerulescens Bergh, 1888		+		
		cf. *coerulescens* Bergh, 1888		+		
		Discodoris sp.	+			
	Doriopsis	*granulata* (Ehrenberg, 1831)		+		
	Paradoris	*erythraeensis* (Vayssière, 1912)	+	+		
		hypocrita Yonow, 2020		+		
		Paradoris sp. a			+	

(Continued)

TABLE 12.1 (Continued)

ORDER/FAMILY	GENUS	SPECIES	APG	RS	GO	AS
		Paradoris sp. b		+		
		Paradoris sp. c		+		
	Peltodoris	*murrea* (Abraham, 1877)	+	+	+	+
		rubra (Bergh, 1905)		+		
	Sebadoris	*fragilis* (Alder & Hancock, 1864)	+	+	+	+
		nubilosa (Pease, 1871)		+	+	
		Sebadoris sp.			+	
	Taringa	*sublutea* (Abraham, 1877)				+
	Tayuva	*lilacina* (Gould, 1852)			+	+
	Thordisa	*sanguinea* Baba (1955)				+
		villosa (Alder & Hancock, 1864)				+
	Avaldesia	*tahala* Chan & Gosliner, 2007		+		
	Halgerda	*biqiea* Donohoo & Gosliner, 2023		+		
		paulayi Donohoo & Gosliner, 2023		+		
		punctata Farran, 1902				+
		tessellata (Bergh, 1880)				+
		cf. *willeyi* Eliot, 1904			+	
	Jorunna	*funebris* (Kelaart, 1858)	+	+	+	+
		labialis Eliot, 1908		+		
		parva (Baba, 1938)		+		
		cf. *parva* (Baba, 1938)		+		
		ramicola Miller, 1996			+	
		rubescens (Bergh, 1876)				+
		Jorunna sp. a	+	+		
		Jorunna sp. b			+	
		Jorunna sp. c		+		
		Jorunna sp. d		+		
	Platydoris	*cruenta* (Quoy & Gaimard, 1832)				+
		ellioti (Alder & Hancock, 1864)				+
		formosa (Alder & Hancock, 1864)				+
		inframaculata (Abraham, 1877)		+		
		pulchra Eliot, 1904		+		
		scabra (Cuvier, 1804)		+		
		striata (Kelaart. 1858)		+		
		townsendi Eliot, 1905	+			
	Rostanga	*arbutus* (Angas, 1864)				+
		lutescens (Bergh, 1905)		+		
		muscula (Abraham, 1877)				+
		Rostanga sp.			+	+
	Sclerodoris	*apiculata* (Alder & Hancock, 1864)		+		+
		rugosa (Vayssière, 1912)		+		
		cf. *rugosa* (Vayssière, 1912)		+		
		tuberculata Eliot, 1904		+		+
		Sclerodoris sp. a		+		
		Sclerodoris sp. b		+		
		Sclerodoris sp. c		+		
	Spongiodoris	*Spongiodoris* sp.		+		
DORIDIDAE	*Artachaea*	*clavata* Eliot, 1907		+		
		intermedia Eales, 1938		+		
		verrucosa Eliot, 1908		+		
		Artachaea sp.				+
	Archidoris	*vayssierea* O'Donoghue, 1929		+		
	Doriopsis	*granulosa* Pease, 1860		+		+
		pecten (Collingwood, 1881)			+	
	Doris	*immonda* Risbec, 1928		+	+	+

TABLE 12.1 (Continued)

ORDER/FAMILY	*GENUS*	*SPECIES*	*APG*	*RS*	*GO*	*AS*
		Doris sp. a	+			
		Doris sp. b	+			
		Doris sp. c			+	
		Doris sp. d		+		
		Doris sp. e		+		
		Doris sp. f		+		
		Doris sp. g		+		
CADLINELLIDAE	*Cadlinella*	*ornatissima* (Risbec, 1928)		+	+	
		subornatissima Baba, 1996		+		
		Cadlinella sp.		+	+	
CADLINIDAE	*Aldisa*	cf. *zavorensis* Tibiriçá, Pola & Cervera, 2017		+		
		Aldisa sp.		+		
CHROMODORIDIDAE	*Ardeadoris*	*angustolutea* (Rudman, 1990)				+
		egretta Rudman, 1984		+	+	
		electra Rudman, 1990		+		
		symmetrica (Rudman, 1990)		+	+	
		undaurum (Rudman, 1985)			+	
	Ceratosoma	*tenue* Abraham, 1876		+		
		trilobatum (J.E. Gray, 1850)		+	+	
	Chromodoris	*africana* Eliot, 1904		+		
		aspersa (A. Gould, 1852)				+
		baqe Bonomo & Gosliner, 2020		+		
		elisabethina Bergh, 1877				+
		hamiltoni Rudman, 1977			+	
		mandapamensis (Valdés, Mollo & Ortea, 1999)				+
		quadricolor (Rüppell & Leuckart, 1830)		+		
		strigata Rudman, 1987		+		
		tenuilinearis Farran, 1905			+	
		Chromodoris sp.			+	
	Diversidoris	*aurantionodulosa* Rudman, 1987		+	+	
		flava (Eliot, 1904)		+		
	Doriprismatica	*atromarginata* (Cuvier, 1804)		+		+
		kyanomarginata Yonow, 2018	+	+		
		plumbea (Pagenstecher, 1877)		+	+	
		rossi S. B. Matsuda & Gosliner, 2018		+		
	Glossodoris	*andersonae* Matsuda & Gosliner, 2018		+		
		cincta (Bergh, 1888) complex	+	+	+	+
		hikuerensis (Pruvot-Fol, 1954)		+		
		kahlbrocki Yonow, 2018		+		
		pallida (Rüppell & Leuckart, 1830)	+	+	+	+
		poecila Winckworth, 1946				+
		rufomarginata (Bergh, 1890)		+		+
		semeion (Winckworth, 1946)				+
		Glossodoris sp.	+			
	Goniobranchus	*alius* (Rudman, 1987)			+	+
		cf. *albonares* (Rudman, 1990)	+		+	
		albopustulosus (Pease, 1860)			+	
		annulatus (Eliot, 1904)	+	+	+	+
		aureopurpureus (Collingwood, 1881)			+	
		bombayanus (Winckworth, 1946)	+			+
		cavae (Eliot, 1904)			+	+
		cazae (Gosliner & Behrens, 2004)	+		+	
		charlottae (Schrödl, 1999)		+		
		conchyliatus (Yonow, 1984)				+

(Continued)

TABLE 12.1 (Continued)

ORDER/FAMILY	*GENUS*	*SPECIES*	*APG*	*RS*	*GO*	*AS*
		decorus (Pease, 1860)			+	+
		fidelis (Kelaart, 1858)		+		+
		geminus (Rudman, 1987)		+		+
		glenei (Kelaart, 1858)				+
		hintuanensis (Gosliner & Behrens, 1998)		+		
		kuniei (Pruvot-Fol, 1930)			+	
		obsoletus (Rüppell & Leuckart, 1830)	+	+	+	
		pseudodecorus Yonow, 2018	+	+	+	
		rufomaculatus (Pease, 1871)				+
		setoensis (Baba, 1938)			+	+
		tennentanus (Kelaart, 1859)		+		
		tinctorius (Rüppell & Leuckart, 1830)		+	+	
		trimarginatus (Winckworth, 1946)				+
		tumuliferus (Collingwood, 1881)				+
		verrieri (Crosse, 1875)		+		+
		cf. *verrieri* (Crosse, 1875)		+	+	
		Goniobranchus sp. 23 (2018)	+			
		Goniobranchus sp. 24 (2018)			+	
		Goniobranchus sp. 33 (2015)			+	
		Goniobranchus sp. a	+			
		Goniobranchus sp. b	+			
	Hypselodoris	*alburtuqali* Gosliner & Johnson, 2018		+		
		capensis (Barnard, 1927)			+	+
		carnea (Bergh 1889)	+		+	+
		dollfusi (Pruvot-Fol, 1933)	+	+	+	
		fucata Gosliner & Johnson, 1999			+	
		ghardaqana (Gohar & Aboul-Ela, 1957)		+	+	+
		infucata (Rüppell & Leuckart, 1830)	+	+		+
		kanga Rudman, 1977	+		+	+
		maculosa (Pease, 1871)	+		+	+
		maridadilus Rudman, 1977		+	+	+
		nigrostriata (Eliot, 1904)	+	+	+	+
		pulchella (Rüppell & Leuckart, 1830)	+	+	+	+
		regina Ev. Marcus & Er. Marcus, 1970			+	
		whitei A. Adams & Reeve, 1850	+		+	
		yarae Gosliner & Johnson, 2018		+		
		Hypselodoris sp. 5 (2015)			+	
		Hypselodoris sp. 12 (2015)			+	
		Hypselodoris sp.	+			
	Mexichromis	*katalexis* Yonow, 2001	+	+	+	
		lemniscata (Quoy & Gaimard, 1832)		+	+	
		mariei (Crosse, 1872)			+	+
		multituberculata (Baba, 1953)			+	
		pusilla (Bergh, 1874)		+		+
	Thorunna	*africana* Rudman, 1984		+		
		florens (Baba, 1949)			+	
		horologia Rudman, 1984			+	
	Verconia	*alboannulata* Rudman, 1986		+		
		laboutei (Rudman, 1986)			+	
		norba (Er. Marcus & Ev. Marcus, 1970)		+	+	
		simplex (Pease, 1871)		+		
		sudanica Rudman, 1985		+		
		Verconia sp.		+	+	
	Miamira	*magnifica* Eliot, 1910		+		
	Orodoris	*miamirana* Bergh, 1875		+		

TABLE 12.1 (Continued)

ORDER/FAMILY	*GENUS*	*SPECIES*	*APG*	*RS*	*GO*	*AS*
DORIDINA 3						
DENDRODORIDIDAE	*Dendrodoris*	*atromaculata* (Alder & Hancock, 1864)				+
		albopurpurea Burn, 1957		+		
		coronata Kay & Young, 1969		+	+	
		cuprea (Ehrenberg, 1831)		+		
		elizabethina (Kelaart, 1859)			+	
		elongata Baba, 1936		+		
		fumata (Rüppell & Leuckart, 1830)	+	+	+	+
		goani Rao & Kumary, 1973				+
		immaculata (Audouin, 1826)		+		
		jousseaumei (Vayssière, 1912)		+		
		krusensternii (J. E. Gray, 1850)		+	+	+
		leptopus Ehrenberg, 1831		+		
		lugubris Ehrenberg, 1831		+		
		nigra (Stimpson, 1855)	+	+	+	+
		nigropunctata (Vayssière, 1912)		+		
		tigrina (Audouin, 1826)		+		
		tuberculosa (Quoy & Gaimard, 1832)		+		+
		Dendrodoris sp. a	+			
		Dendrodoris sp. b	+			
	Doriopsilla	*miniata* (Alder & Hancock, 1866)				+
		cf. *miniata* (Alder & Hancock, 1866)	+			+
		nigrocera Yonow, 2012	+			+
		Doriopsilla sp. a	+			+
		Doriopsilla sp. b	+			
		Doriopsilla sp. c		+		
PHYLLIDIIDAE	*Phyllidia*	*alyta* Yonow, 1996				+
		coelestis Bergh, 1905				+
		elegans Bergh, 1869			+	
		marindica (Yonow & Hayward, 1991)				+
		multifaria Yonow, 1986		+		
		multituberculata R.C. Boettger, 1918			+	
		ocellata Cuvier, 1804			+	+
		picta Pruvot-Fol, 1957			+	+
		rueppelii (Bergh, 1869)	+	+	+	
		schupporum Fahrner & Schrödl, 2000	+	+	+	+
		sudanensis (Heller & Thompson, 1983)		+		
		undula Yonow, 1986		+		
		varicosa Lamarck, 1801		+	+	+
		Phyllidia sp. a	+		+	
		Phyllidia sp. b		+		
	Phyllidiella	*amphitrite* Yonow, 2020		+		
		melanocera (Yonow, 1986)		+		
		pustulosa (Cuvier, 1804) complex		+	+	+
		rosans (Bergh, 1873)				+
		striata (Bergh, 1889)				+
		zeylanica (Kelaart, 1858)		+		+
	Phyllidiopsis	*cardinalis* Bergh, 1876		+	+	
		dautzenbergi (Vayssière, 1912)	+	+		
		gemmata (Pruvot-Fol, 1957)				+
		krempfi Pruvot-Fol, 1957				+
		monacha Yonow, 1986		+		
		phiphiensis Brunckhorst, 1993				+
		sinaiensis (Yonow, 1988)		+		

(*Continued*)

TABLE 12.1 (Continued)

ORDER/FAMILY	*GENUS*	*SPECIES*	*APG*	*RS*	*GO*	*AS*
		xishaensis (Lin, 1983)				+
		Phyllidiopsis sp. a		+		
		Phyllidiopsis sp. b		+		
CLADOBRANCHIA 1						
BORNELLIDAE	*Bornella*	*stellifera* (Adams & Reeve in Adams, 1848)		+		+
		cf. *johnsonorum* Pola, Rudman & Gosliner, 2009		+		
HANCOCKIIDAE	*Hancockia*	*papillata* (O'Donoghue, 1932)				+
DENDRONOTIDAE	*Pseudobornella*	*orientalis* Baba, 1932	?		?	
DOTIDAE	*Doto*	*kabretiana* (O' Donoghue, 1929)		+		
		cf. *kya* Er. Marcus, 1961	+	+		+
		Doto sp. 7 (2015)	+			
		Doto sp. a	+			
		Doto sp. b	+			
		Doto sp. c				+
	Kabeiro	*orcha* (Yonow, 2000)		+		
		rubroreticulata Shipman & Gosliner, 2015	+	+	+	
		Kabeiro sp.	+			
LOMANOTIDAE	*Lomanotus*	*vermiformis* Eliot, 1908	+	+	+	+
		Lomanotus sp. a			+	+
		Lomanotus sp. b		+		
		Lomanotus sp. c		+		
		Lomanotus sp. d		+		
SCYLLAEIDAE	*Scyllaea*	*fulva* Quoy & Gaimard, 1824		+		+
	Crosslandia	*viridis* Eliot, 1902		+		
	Notobryon	*wardi* Odhner, 1936		+		
TETHYDIDAE	*Melibe*	*bucephala* Bergh, 1902		+		
		engeli Risbec, 1937		+	+	
		japonica Eliot, 1913				+
		megaceras Gosliner, 1987	+		+	
		viridis (Kelaart, 1858)		+	+	+
		Melibe sp.			+	
TRITONIDAE	*Marianina*	*rosea* (Pruvot-Fol, 1930)				+
	Marionia	*dakini* (O'Donoghue, 1924)			+	
		glama (Rüppell & Leuckart, 1828)		+		
		pambanensis O'Donoghue, 1932				+
		rubra (Rüppell & Leuckart, 1828)		+		
		Marionia sp. a			+	
		Marionia sp. b		+		
	Marioniopsis	*cyanobranchiata* (Rüppell & Leuckart, 1828)		+		
		levis (Eliot, 1904)		+		
		viridescens (Eliot, 1904)	+	+	+	+
	Tritonia	*bollandi* Smith & Gosliner, 2003		+	+	
		Tritonia sp.		+	+	
	Tritoniopsis	*elegans* (Audouin, 1826)		+	+	
		cf. *frydis* Er. Marcus & Ev. Marcus, 1970		+		
		graviera Vayssiére, 1912		+		
		Tritoniopsis sp.		+		
CLADOBRANCHIA 2						
ARMINIDAE	*Armina*	*babai* (Tchang, 1934)				+
		cinerea (Farran, 1905)				+
		gracilis (Bergh, 1874)				+
		magna Baba, 1955		+	+	
		punctilopsis G.-Y. Lin, 1992				+
		semperi (Bergh, 1886)		+	+	

TABLE 12.1 (Continued)

ORDER/FAMILY	*GENUS*	*SPECIES*	*APG*	*RS*	*GO*	*AS*
		Armina sp. 4 (2008)			+	
		Armina cf. sp. 6 (2008)			+	
		Armina sp. 7 (2008)	?		?	
		Armina sp. 8 (2008)	?		?	
		Armina sp. 10 (2015)			+	
		Armina sp. 18 (2015)			+	
		Armina sp. a		+		
		Armina sp. b			+	
	Pleurophyllidiella	*paucidentata* O'Donoghue, 1932				+
	Dermatobranchus	*albus* (Eliot, 1904)		+	+	
		caeruleomaculatus Gosliner & Fahey, 2011		+	+	
		dendronephthyphagus Gosliner & Fahey, 2011		+	+	
		cf. *fortunatus* (Bergh, 1888)		+		+
		glabrus (Eliot, 1908)		+		
		gonatophorus van Hasselt, 1824		+		
		cf. *leoni* Gosliner & Fahey, 2011		+		
		ornatus (Bergh, 1874)		+	+	
		piperoides Gosliner & Fahey, 2011			+	+
		pustulosus van Hasselt, 1824		+		
		semistriatus Baba, 1949			+	+
		striatus van Hasselt, 1824		+		
		tuberculatus Gosliner & Fahey, 2011		+	+	
		Dermatobranchus sp. a		+	+	+
		Dermatobranchus sp. b		+		
		Dermatobranchus sp. c		+		
		Dermatobranchus sp. d		+		
		Dermatobranchus sp. e		+		
		Dermatobranchus sp. f		+		
MADRELLIDAE	*Madrella*	*ferruginosa* Alder & Hancock, 1864		+		
		gloriosa Baba, 1949		+		
JANOLIDAE	*Janolus*	*flavoannulatus* Pola & Gosliner, 2019		+	+	
		mirabilis Baba & Abe, 1970		+		
		toyamensis Baba & Abe, 1970			+	+
		savinkini Martynov & Korshunova, 2012		+	+	
		Janolus sp. a		+		
		Janolus sp. b		+		
CLADOBRANCHIA 3						
AEOLIDIIDAE	*Aeolidia*	*habessinica* Ehrenberg, 1828		+		
	Aeolidiopsis	*palythoae* (Gosliner, 1985)				+
		cf. *ransoni* Pruvot-Fol, 1956		+		
	Anteaeolidiella	*fijensis* Carmona at al., 2014				+
		indica (Bergh 1888)		+		+
		poshitra Carmona et al., 2014				+
		Anteaeolidiella sp.				+
	Baeolidia	*moebii* Bergh, 1888		+	+	+
		salaamica (Rudman, 1982)				+
	Bulbaeolidia	*alba* (Risbec, 1928)	+	+	+	
	Cerberilla	*ambonensis* Bergh, 1905			+	
		annulata (Quoy & Gaimard, 1832)		+		+
		Cerberilla sp. 5 (2008)	?		?	
		Cerberilla sp. 8 (2015)			+	
		Cerberilla sp.		+	+	
	Limenandra	*barnosii* Carmona, Pola, Cervera & Gosliner, 2014		+	+	

(Continued)

TABLE 12.1 (Continued)

ORDER/FAMILY	*GENUS*	*SPECIES*	*APG*	*RS*	*GO*	*AS*
		confusa Carmona, Pola, Cervera & Gosliner, 2014		+		
		fusiformis (Baba, 1949)	+	+	+	
		nodosa Haefelfinger & Stamm, 1958		+		
		rosanae Carmona, Pola, Cervera & Gosliner, 2014	+	+		
		Limenandra sp.	+			
FACELINIDAE	*Cratena*	cf. *affinis* (Baba, 1949)		+		
		capensis Barnard, 1927			+	
		lineata (Eliot, 1905)	+		+	+
		poshitraensis Bharate, Padula, Apte & Shimpi, 2020				+
		pawarshindeorum Bharate, Padula, Apte & Shimpi, 2020				+
		Cratena sp.	+		+	+
	Spurilla	*braziliana* MacFarland, 1909		+		
	Caloria	*indica* (Bergh, 1896)	+	+	+	
		Caloria sp.		+		
	Facelina	*lineata* Eliot, 1905	+		+	+
		Facelina sp. a		+		
		Facelina sp. b		+		
	Favorinus	*japonicus* Baba, 1949		+		
		mirabilis Baba, 1955		+	+	
		tsuruganus Baba & Abe, 1964		+		
		Favorinus sp.	+			
	Herviella	*albida* Baba, 1966	+	+		+
		affinis Baba, 1960				+
		yatsui (Baba, 1930)				+
	Moridilla	*brockii* Bergh, 1888				+
	Myja	*longicornis* Bergh, 1896			+	
		cf. *longicornis* Bergh, 1896	+			
		Myja sp.			+	
	Noumeaella	*isa* Ev. Marcus & Er. Marcus, 1970				+
	Phidiana	*anulifera* (Baba, 1949)		+		+
		bourailli (Risbec, 1928)		+	+	
		militaris (Alder & Hancock, 1864)	+		+	+
		semidecora (Pease, 1860)				+
	Pruvotfolia	*rhodopos* (Yonow, 2000)		+		
	Pteraeolidia	*semperi* (Bergh, 1870) complex		+	+	+
		Pteraeolidia sp. a		+		
		Pteraeolidia sp. b		+		
	Sakuraeolis	*gujaratica* Rudman, 1980	?		?	+
		kirembosa Rudman, 1980			+	
		Sakuraeolis sp.			+	
FLABELLINIDAE	*Coryphella*	*amabilis* (Hirano & Kuzirian, 1991)	+			
	Coryphellina	*rubrolineata* O'Donoghue, 1929	+	+	+	+
		Coryphellina sp.			+	
	Samla	*bicolor* (Kelaart, 1858)	+	+	+	+
		bilas (Gosliner & Willan, 1991)		+		
		Samla sp.		+		
	Flabellina	*Flabellina* sp.		+	+	+
GLAUCIDAE	*Glaucus*	*atlanticus* Forster, 1777				+
	Glaucilla	*marginata* Reinhardt &Bergh, 1868		+		
MYRRHINIDAE	*Godiva*	*quadricolor* (Barnard, 1927)			+	
		rachelae Rudman, 1980			+	

TABLE 12.1 (Continued)

ORDER/FAMILY	*GENUS*	*SPECIES*	*APG*	*RS*	*GO*	*AS*
	Phyllodesmium	*colemani* Rudman, 1991		+		
		crypticum Rudman, 1981		+		
		fastuosum Ehrenberg, 1831		+		
		hyalinum Ehrenberg, 1831		+		
		koehleri Burghardt, Schrödl & Wägele, 2008		+		
		magnum Rudman, 1991	+	+		
		poindimiei (Risbec, 1928)	+		+	+
		serratum (Baba, 1949)			+	+
		Phyllodesmium sp.				+
EUBRANCHIDAE	*Eubranchus*	*echizenicus* Baba, 1975				+
		mandapamensis (K.P. Rao, 1968)	+			
		cf. *mandapamensis* (K.P. Rao, 1968)			+	+
		mimeticus Baba, 1975				+
		ocellatus (Alder & Hancock, 1864)		+	+	
		rubropunctatus Edmunds, 1969		+	+	+
		virginalis (Baba, 1949)				+
		Eubranchus sp. 19 (2015)	+			
		Eubranchus sp. a	+			
		Eubranchus sp. b	+			
		Eubranchus sp. c		+		
		Eubranchus sp. d		+		
	Leostyletus	*misakiensis* (Baba, 1960)	+			
TRINCHESIIDAE	*Catriona*	*susa* Ev. Marcus & Er. Marcus, 1960		+		
		cf. *columbiana* (O'Donoghue, 1922)				+
	Phestilla	*goniophaga* Hu et al., 2020		+		
		lugubris (Bergh, 1870)	+	+		+
		melanobrachia Bergh, 1874		+	+	+
		minor Rudman, 1981		+		+
	Tenellia	*Tenellia* sp. 2 (2018)			+	
		Tenellia sp. 14 (2018)			+	
		Tenellia sp.			+	
	Trinchesia	*albocrusta* (MacFarland, 1966)	+			
		ornata (Baba, 1937)			+	+
		sibogae (Bergh, 1905)		+	+	
		yamasui (Hamatani, 1993)	+	+	+	+
		Trinchesia sp. a	+			
		Trinchesia sp. b		+		
		Trinchesia sp. c		+		
		Trinchesia sp. d		+		
		Trinchesia sp. e		+		
		Trinchesia sp. f		+		
		Trinchesia sp. g		+		
		Trinchesia sp. h		+		
		Trinchesia sp. i		+		
		Trinchesia sp. j		+		
		Trinchesia sp. k		+		
UNIDENTIIDAE	*Unidentia*	*angelvaldesi* Millen & Hermosillo, 2012				+
EMBLETONIIDAE	*Embletonia*	*gracilis* Risbec, 1928		+		+
PLEUROLIDIIDAE	*Pleurolidia*	*juliae* Burn, 1966		+		
	Protaeolidiella	*atra* Baba, 1995		+		

Abbreviations of regions: APG – Arabian/Persian Gulf; RS – Red Sea; GO – Gulf of Oman; AS – Arabian Sea; ? – location unknown.

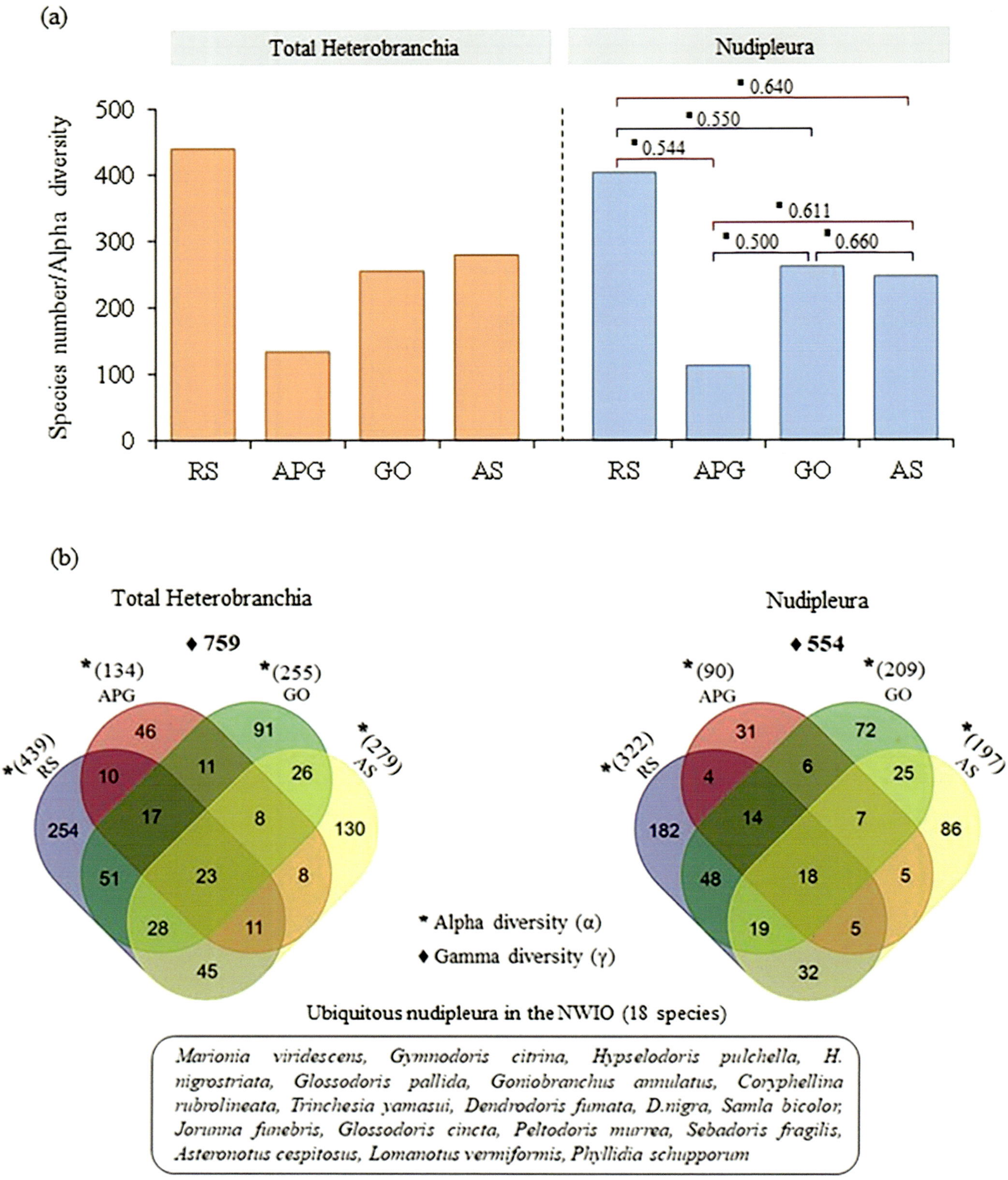

FIGURE 12.2 (a) Alpha diversity for each sea. The total number of heterobranchs (759 species) and nudipleurans (554 species) recorded from the NWIO region (i.e., gamma diversity) are provided. (b) Venn diagrams illustrating the overlapping numbers of the total heterobranchs and nudipleuran species between the four seas: Arabian Persian Gulf (APG), Red Sea (RS), Gulf of Oman (GO), and Arabian Sea (AS).

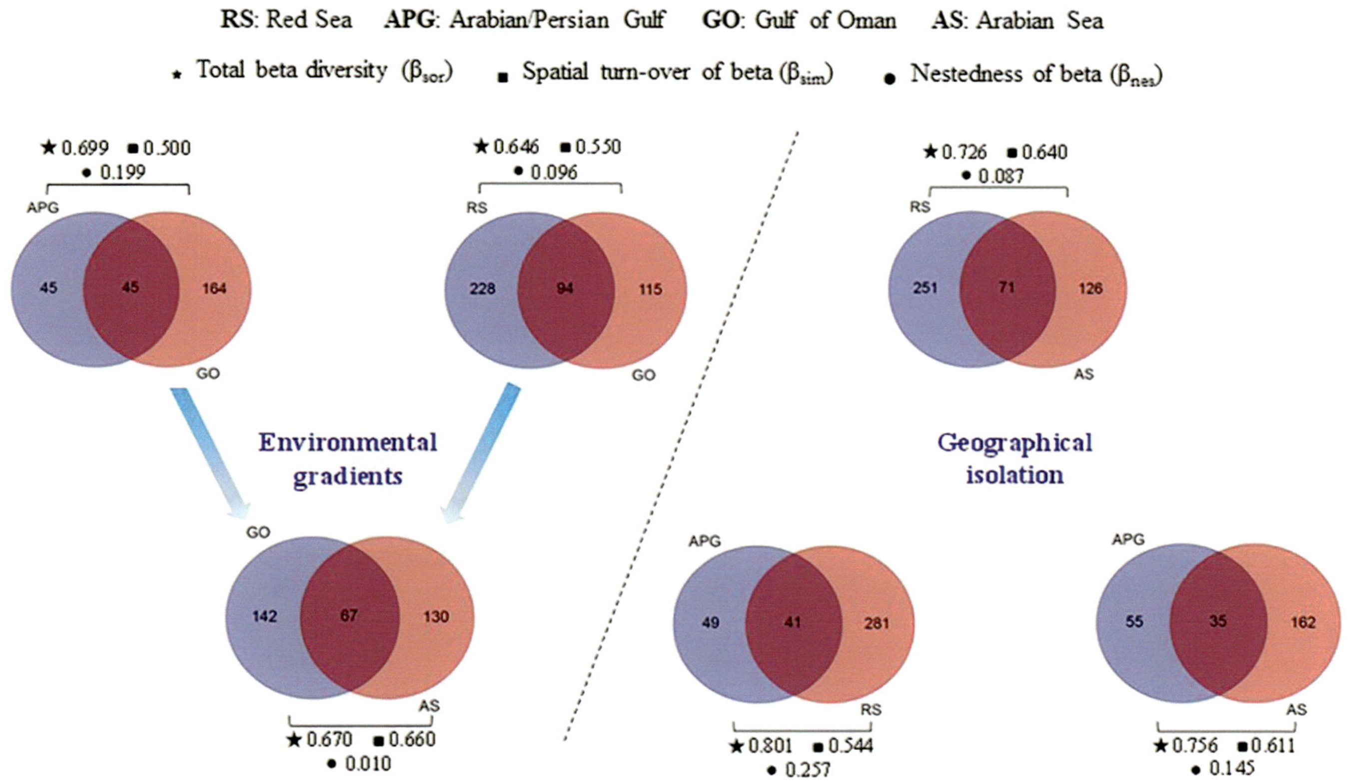

FIGURE 12.3 Patterns of nudipleuran beta diversity and its components between the marginal seas of the NWIO.

diversity of Nudipleura was recorded in the GO (209 species), followed by the AS (197) and APG (90 species). Along the RS-GO-AS continuum, maximal alpha diversity was recorded in the RS (322 species), followed by the GO (209 species) and AS (197 species). Thus, the RS had the highest nudipleuran diversity, and the APG had the lowest. The Venn diagram clearly shows the endemic taxa and the taxa shared between the seas (Figure 12.2). For example, the RS and APG share 41 species, of which only four species were shared only between the RS and the APG; 14 species were collectively present in the RS, APG, and GO; and five species commonly occurred in the APG, RS, and AS, while only 18 species were common across all four seas.

Using the Sørensen dissimilarity index and the Simpson dissimilarity index, we disentangled the nestedness and turnover components of nudipleuran beta diversity in the NWIO. Total beta diversity (β_{sor}) along the APG-GO-AS continuum ranged from 0.670 (GO vs. AS) to 0.699 (APG vs. GO). Along the RS-GO-AS continuum, the beta diversity varied from 0.646 (RS vs. GO) to 0.670 (GO vs. AS). High spatial turnover (or species replacement) of nudipleurans was primarily responsible for high beta diversity values. Although minimal, the role of nestedness in the beta diversity was higher between GO and APG, followed by RS and GO, and GO and AS. Despite the vast differences in the (reported) nudipleuran alpha diversity (and endemicity) between the APG (90 (39) species) and the RS (322 (188) species), the beta diversity (β_{sor}: 0.646–0.699) and its spatial turnover component (β_{sim}: 0.500–0.550) remained comparable between the APG-GO and RS-GO continua.

Among the geographically isolated areas, spatial replacement of species (β_{sim}: 0.544 (APG vs. RS), 0.611 (APG vs. AS), 0.640 (RS vs. AS)) was mainly responsible for the observed beta diversity patterns (β_{sor}: 0.801 (APG vs. RS), 0.756 (APG vs. AS), 0.726 (RS vs. AS)). Among the nudipleuran species reported from the APG, 46% were present in the RS and 39% in the AS. Similarly, 37% of species reported from the AS were also present in the RS. Consequently, the nestedness was the highest between the RS and APG (β_{nes}: 0.257), followed by the APG and AS (β_{nes}: 0.145) and RS and AS (β_{nes}: 0.087) (Figure 12.3).

12.3.2 Checklist of species recorded in the APG with notes on habitat and distribution

the species in **bold** indicate new records.

Superorder EUTHYNEURA
Order ACTEONIMORPHA
Family ACTEONIDAE

1. *Pupa affinis* (A. Adams, 1855)
   Habitat: Sandy substrates
   Distribution: Kuwait (Al-Yamani et al. 2012: pl. 141, p. 166), Iran, APG (Kohan et al. 2012)
2. ***Pupa* cf. *solidula*** (Linnaeus, 1758)
   Habitat: Sandy substrates
   Distribution: Abu Dhabi, UAE, APG (this study)

Family APLUSTRIDAE

3. *Micromelo scriptus* (Garrett, 1857)
   Habitat: Rocky substrata and coral reefs
   Distribution: Indo-Pacific (Feliciano et al. 2021); UAE, APG (Dipper and Woodward 1989: p. 10; George 2012 as *M. undata*; this study)

Order CEPHALASPIDEA
Family BULLIDAE

4. *Bulla arabica* (Malaquias and Reid, 2008)
   Habitat: Intertidal hard substrates, seagrass beds
   Distribution: Red Sea, Aden, APG, Gulf of Oman, and Arabian Sea (Malaquias and Reid 2008; Al-Yamani et al. 2012: pl. 154a, b, p. 177 as *Bulla ampulla*; George 2012; Kohan et al. 2012; Al-Kandari et al. 2020; Amini-Yekta and Dekker 2021)

Family HAMINOEIDAE

5. *Aliculastrum cylindricum* (Helbling, 1779)
   Habitat: Muddy substrates
   Distribution: Iran, Qatar, APG (Kohan et al. 2012; Al-Khayat et al. 2021; as *Atys cylindricus*)
6. *Atys pellyi* (Smith, 1872)
   Habitat: Muddy substrates
   Distribution: Kuwait, Iran, APG (Al-Yamani et al. 2012: pl. 152a, p. 175; Amini-Yekta et al. 2014)
7. *Atys* sp. a
   Habitat: Muddy substrates
   Distribution: Kuwait, APG (Al-Yamani et al. 2012: pl. 152b, p. 175)
8. *Atys* sp. b
   Habitat: Muddy substrates
   Distribution: Kuwait, APG (Al-Yamani et al. 2012)
9. *Atys* sp. c
   Habitat: Muddy substrates
   Distribution: Qatar, APG (Al-Khayat et al. 2021)
10. *Bakawan rotundata* (A. Adams, 1850)
   Habitat: Intertidal rocky substrates and coral reefs
   Distribution: Tropical Indo-West Pacific (Oskars and Malaquias 2019); Kuwait, APG (Al-Yamani et al. 2012: pl. 153, p. 176 as *Haminoea vitrea*)
11. ***Diniatys dentifer*** (A. Adams, 1850)
   Habitat: Shallow reef areas
   Distribution: (Gosliner et al. 2018: p. 352), APG (this study)
12. ***Haloa* cf. *pemphis*** (R. A. Philippi, 1847) (Figure 12.4a)
   Habitat: Muddy substrates
   Distribution: UAE, APG (this study)
13. *Haminoea* sp.
   Habitat: Coral reefs and rocky habitats
   Distribution: Farsi Island, Iran, APG (Rezai et al. 2016)

Family RETUSIDAE

14. *Retusa* sp.
   Habitat: Muddy substrates
   Distribution: Kuwait and Qatar, APG (Al-Yamani et al. 2012: pl. 156a, b, p. 179; Al-Khayat et al. 2021)

Family TORNATINIDAE

15. *Acteocina inconspicua* (H. Adams, 1872)
   Habitat: Muddy substrates
   Distribution: Kuwait, APG (Al-Yamani et al. 2012: pl. 157a, p. 180 as *Tornatina inconspicua*)
16. *Acteocina involuta* (G. Nevill & H. Nevill, 1871)
   Habitat: Sandy Habitats
   Distribution: Iran, APG (Kohan et al. 2012)
17. *Tornatina persiana* (E.A. Smith, 1872)
   Habitat: Muddy substrates
   Distribution: Kuwait, APG (Al-Yamani et al. 2012: pl. 157b, p. 180)
18. *Tornatina* sp.
   Habitat: Muddy substrates
   Distribution: Kuwait, APG (Al-Yamani et al. 2012: pl. 157c, p. 180)

Family CYLICHNIDAE

19. *Cylichna collyra* (Melvill, 1906)
   Habitat: Muddy substrates
   Distribution: Kuwait, APG (Al-Yamani et al. 2012: pl. 155a, p. 178)
20. *Cylichna cylindracea* (Pennant, 1777)
   Habitat: Muddy substrates
   Distribution: Kuwait, APG (Al-Yamani et al. 2012: pl. 155b, p. 178)
21. *Cylichna* sp.
   Habitat: Muddy substrates
   Distribution: Kuwait, APG (Al-Yamani et al. 2012)

Family AGLAJIDAE

22. ***Biuve fulvipunctata*** (Baba, 1938) (Figure 12.4b)
   Habitat: Sandy-rocky habitats
   Distribution: Indo-Pacific (Gosliner et al. 2008); Saudi Arabia, APG (this study)
23. *Chelidonura livida* (Yonow 1994)
   Habitat: Sandy substrates
   Distribution: Red Sea (Yonow 1994, 2008); Western Indian Ocean (Gosliner et al. 2015); SE APG (George 2012); Kuwait, APG (Nithyanandan et al. 2021: figure 12.2, p. 94)
24. ***Philinopsis speciosa*** (Pease, 1860)
   Habitat: Sandy habitats
   Distribution: Indo-Pacific and Eastern Pacific (Gosliner et al. 2018); APG (this study)

Family GASTROPTERIDAE

25. ***Siphopteron* cf. *makisig*** (Ong & Gosliner, 2017) (Figure 12.4c)
   Habitat: Coral reefs
   Distribution: Indo-Pacific (Gosliner et al. 2018); Qatar, APG (this study)

Family PHILINIDAE

26. ***Philine* sp.**
   Habitat: Coral reefs, among rubble
   Distribution: Dubai, APG (this study)

FIGURE 12.4 Rare and new records of heterobranch sea slugs from the Arabian/Persian Gulf: (a) *Haloa* cf. *pemphis* (photograph by Stewart Clarke, Abu Dhabi, UAE); (b) *Biuve fulvipunctata* (photograph by Baleegh Ahmed Abukaboos, Saudi Arabia); (c) *Siphopteron* cf. *makisig* (photograph by Dany, Qatar); (d) *Costasiella* cf. *kuroshimae* (photograph by Stewart Clarke, Abu Dhabi, UAE); (e) *Elysia* cf. *hirasei* (photograph by Baleegh Ahmed Abukaboos, Abu Ali and Tarut islands, Saudi Arabia); (f) *Elysia mercieri* (photograph by Baleegh Ahmed Abukaboos, Saudi Arabia); (g) *Ercolania annelyleorum* (photograph by Baleegh Ahmed Abukaboos, Saudi Arabia); (h) *Ercolania caerulea* (photograph by Baleegh Ahmed Abukaboos, Saudi Arabia); (i) *Aegires villosus* (photograph by Michael Rall, Dubai); (j) *Gymnodoris alba* (photograph by Michael Rall, Dubai); (k) *Gymnodoris* sp. a. (photograph by Baleegh Ahmed Abukaboos, Abu Ali and Tarut islands, Saudi Arabia).

Order APLYSIIDA
Family APLYSIIDAE

27. *Aplysia cornigera* G.B (Sowerby II, 1869)
    Habitat: Soft substrates
    Distribution: Kuwait, Iran, and SE APG (Glayzer et. 1984; Dipper and Woodward 1989: p. 19; Rezai et al. 2016)
28. ***Aplysia* sp.**
    Habitat: Soft substrates
    Distribution: Kuwait and Iran, APG (this study)
29. *Bursatella leachii* (Blainville, 1817)
    Habitat: Soft substrates
    Distribution: Indo-Pacific (Gosliner et al. 2015); Iran, APG (Rezai et al. 2016)
30. *Stylocheilus longicauda* (Quoy & Gaimard, 1825)
    Habitat: Soft substrates
    Distribution: Circumtropical (Yonow 2008); Iran, APG (Amouei and Fatemi 2021)
31. *Stylocheilus striatus* (Quoy & Gaimard, 1832)
    Habitat: Soft substrates

Distribution: Circumtropical (Yonow 2008), Iran, APG (Rezai et al. 2016)

Family COSTASIELLIDAE

32. ***Costasiella* cf. *kuroshimae*** (Ichikawa, 1993) (Figure 12.4d)
    Habitat: Sand patches between coral reefs
    Distribution: Indo-West Pacific (Gosliner et al. 2015); Red Sea (Yonow, Chapter 13, this volume); Abu Dhabi, APG (this study)

Order SACOGLOSSA
Family PLAKOBRANCHIDAE

33. ***Elysia* cf. *hirasei*** (Baba, 1955) (Figure 12.4e)
    Habitat: Green algal beds
    Distribution: West coast of India (Apte and Desai 2017); Abu Ali and Tarut Islands, Saudi Arabia, APG (this study)
34. ***Elysia mercieri* (Pruvot-Fol, 1930)** (Figure 12.4f)
    Habitat: Shallow reef areas
    Distribution: Indian and Pacific oceans (Gosliner et al. 2018); Saudi Arabia, APG (this study)
35. *Elysia grandifolia-marginata* complex
    Habitat: Green algal beds
    Distribution: Circumtropical (Gosliner et al. 2008); Red Sea (Yonow 2008); UAE, Iran, and SE APG (Dipper and Woodward 1989: p. 32; George 2012; Amini-Yekta et al. 2016; this study)
36. *Elysia* cf. *tomentosa* (K. R. Jensen, 1997)
    Habitat: Mostly found on *Caulerpa* sp. beds
    Distribution: Indo-Pacific (Gosliner et al. 2015); Red Sea (Yonow 2008); Iran, APG (Oladi et al. 2018)
37. *Elysia* sp.
    Habitat: Green algal beds
    Distribution: Iran, APG (Rezai et al. 2016)
38. *Plakobranchus ocellatus* (van Hasselt, 1824)
    Habitat: Sandy habitats
    Distribution: Indo-Pacific (Gosliner et al. 2015); Red Sea (Yonow 2008); Iran, APG (Rezai et al. 2016)

Family LIMAPONTIIDAE

39. ***Ercolania annelyleorum*** (Wägele et al., 2010) (Figure 12.4g)
    Habitat: Algal beds
    Distribution: Lizard Island, Australia, and Solomon Islands (Wägele et al. 2010); Saudi Arabia, APG (this study)
40. ***Ercolania caerulea*** (Trinchese 1892) (Figure 12.4h)
    Habitat: Algal beds
    Distribution: Tanzania, Hong Kong, and Philippines (Gosliner et al. 2018); Saudi Arabia, APG (this study)

41 ***Ercolania* cf. *caerulea*** (Trinchese, 1892)
    Habitat: Algal beds
    Distribution: Saudi Arabia, APG (this study)

42 *Placida daguilarensis* (Jensen, 1990)
    Habitat: Found among green algae, *Bryopsis plumosa* (Al-Yamani et al. 2012)
    Distribution: Kuwait, APG (Al-Yamani et al. 2012: pl. 166, p. 188)

Order SYSTELLOMMETAPHORA
Family ONCHIDIIDAE

43. *Peronia persiae* (Maniei et al., 2020)
    Habitat: Hard substrates
    Distribution: Persian Gulf, APG (Maniei et al. 2020)
44. *Peronia verruculata* (Cuvier, 1830)
    Habitat: Hard substrates
    Distribution: Indo-Pacific (Dayrat et al. 2020); Kuwait, APG (Jones 1986; Al-Yamani et al. 2012: pl. 180, p. 199 as *Peronia peronii*; Al-Kandari et al. 2020).

Super Order NUDIPLEURA
Order PLEUROBRANCHIDA
Family PLEUROBRANCHAEIDAE

45. ***Pleurobranchus mamillatus*** (Quoy & Gaimard, 1832)
    Habitat: Coral reefs
    Distribution: Indo-West Pacific (Gosliner et al. 2015); Dubai, APG (this study)

Order DORIDINA
Family AEGIRIDAE

46. ***Aegires villosus*** (Farran, 1905) (Figure 12.4i)
    Habitat: Coral reefs
    Distribution: Indo-West Pacific (Gosliner et al. 2018); APG (this study)

Family GONIODORIDIDAE

47. ***Trapania japonica*** (Baba, 1935)
    Habitat: Coral rubble
    Distribution: Indo-Pacific (Gosliner et al. 2018); APG (this study)

Family GYMNODORIDIDAE

48. ***Gymnodoris alba*** (Bergh, 1877) (Figure 12.4j)
    Habitat: Hard substrates
    Distribution: Indo-Pacific (Gosliner et al. 2015); Red Sea (Yonow 2008); Qatar, APG (this study)
49. ***Gymnodoris* sp. a** (Figure 12.4k)
    Habitat: Hard substrates
    Distribution: Qatar, APG (this study)
50. ***Gymnodoris* sp. b** (Figure 12.5a)
    Habitat: Hard substrates
    Distribution: Indo-Pacific (Gosliner et al. 2015); Red Sea (Yonow 2008); Saudi Arabia, APG (this study)
51. *Gymnodoris impudica* (Rüppell & Leuckart, 1830)
    Habitat: Hard substrates
    Distribution: Indo-Pacific (Gosliner et al. 2015 as *G. rubropapulosa*); Red Sea (Yonow 2008); Iran, APG (Rezai et al. 2016 as *G. rubropapulosa*)

FIGURE 12.5 Rare and new records of heterobranch sea slugs from the Arabian/Persian Gulf: (a) *Gymnodoris* sp. b. (photograph by Angelo Pia, Qatar); (b) *Asteronotus cespitosus* (photograph by Keith P. Wilson, Dubai, UAE); (c) *Carminodoris pustulata* (photograph by Jay R-Oliveras, Qatar, UAE); (d) *Peltodoris murrea* (photograph by Baleegh Ahmed Abukaboos, Saudi Arabia); (e) *Jorunna funebris* (photograph by Dragan Petkovich, Abu Dhabi, UAE); (f) *Jorunna* sp. (photograph by Kirstin N. Binoth, Dubai, UAE); (g) *Doris* sp. a. (photograph by Valeriy Skryabin, Kuwait); (h) *Doriprismatica kyanomarginata* (photograph by Michael Rall, Dubai, UAE); (i) *Goniobranchus pseudodecorus* (photograph by Stewart Clarke, Abu Dhabi, UAE); (j) *Goniobranchus* sp. 23 (photograph by Stewart Clarke, Abu Dhabi, UAE); (k) *Hypselodoris kanga* (photograph by Stewart Clarke, Abu Dhabi, UAE); (l) *Hypselodoris nigrostriata* (photograph by Stewart Clarke, Abu Dhabi, UAE).

Family POLYCERIDAE

52. *Plocamopherus ocellatus* (Rüppell & Leuckart, 1828)
    Habitat: Coral reefs
    Distribution: Red Sea (Yonow 2008; Gosliner et al. 2015); Kuwait and Iran, APG (Nithyanandan 2012; Rezai et al. 2016; Hoeksema and Yonow 2021); Lessepsian migrant
53. ***Plocamopherus* cf. *lemur*** (Vallès & Gosliner, 2006)
    Habitat: Coral reefs
    Distribution: Qatar, APG (this study)
54. ***Polycera* sp.**
    Habitat: Sandy areas.
    Distribution: Qatar, APG (this study)

Family DISCODORIDIDAE

55. *Asteronotus cespitosus* (van Hasselt, 1824) (Figure 12.5b)
    Habitat: Coral reefs
    Distribution: Indo-Pacific (Gosliner et al. 2015); Red Sea (Yonow 2008); Dubai and Iran, APG (George 2012; Rezai et al. 2016; this study)
56. ***Carminodoris pustulata*** (Abraham, 1877) Figure 12.5c)
    Habitat: Coral reefs
    Distribution: Western Pacific Ocean (Gosliner et al. 2015); Dubai, APG (this study)
57. *Discodoris* sp.
    Habitat: Hard substrates
    Distribution: Indo-Pacific (Gosliner et al. 2015); Kuwait, APG (Al-Kandari et al. 2020)
58. *Paradoris erythraeensis* (Vayssière, 1912)#
    Habitat: Coral reefs
    Distribution: Indian and Western Pacific oceans (Gosliner et al. 2018); Red Sea (Yonow 2008); Abu Dhabi, APG (George 2012); Saudi Arabia, APG (this study).
59. ***Peltodoris murrea*** (Abraham, 1877) (Figure 12.5d)
    Habitat: Intertidal rocky areas
    Distribution: Indian and Western Pacific oceans (Gosliner et al. 2018); Red Sea (Yonow 2008); Saudi Arabia, APG (this study)
60. *Jorunna funebris* (Kelaart, 1858) (Figure 12.5e)
    Habitat: Coral reefs
    Distribution: Western Indian Ocean (Gosliner et al. 2015); Red Sea (Yonow 2008); Iran, APG (Rezai et al. 2016)
61. ***Jorunna* sp.** (Figure 12.5f)
    Habitat: Coral reefs
    Distribution: Dubai, APG (this study)
62. *Platydoris townsendi* (Eliot, 1905)
    Habitat: Hard substrates, preferentially coral reefs
    Distribution: Iran, APG (Dorgan et al. 2002)
63. ***Platydoris* sp.**#
    Habitat: Coral rubble and sandy substrates
    Distribution: Saudi Arabia, APG (this study)
64. *Sebadoris fragilis* (Alder & Hancock, 1864)
    Habitat: Coral reefs
    Distribution: Indian Ocean (Gosliner et al. 2015); Red Sea (Yonow 2008); Kuwait, APG (Jones 1986)

Family DORIDIDAE

65. ***Doris* sp. a** (Figure 12.5g)
    Habitat: Rocky substrates
    Distribution: Kuwait, APG (this study)
66. ***Doris* sp. b**
    Habitat: Rocky substrates
    Distribution: APG (this study)

Family CHROMODORIDIDAE

67. *Doriprismatica kyanomarginata* (Yonow, 2018) (Figure 12.5h)
    Habitat: Coral reefs
    Distribution: Red Sea (Yonow 2018); Jubail, Saudi Arabia, and Dubai, APG (this study); UAE and Dubai, APG (Dipper and Woodward 1989: p. 58)
68. *Glossodoris cincta* (Bergh, 1888) complex
    Habitat: Coral reefs
    Distribution: Indo-Pacific (Gosliner et al. 2015); Red Sea (Yonow 2008); Qatar and UAE, APG (George 2012; Yonow 2012)
69. *Glossodoris pallida* (Rüppell & Leuckart, 1830)
    Habitat: Coral reefs
    Distribution: Indo-Pacific (Gosliner et al. 2015); Red Sea (Yonow 2008); Kuwait and SE APG (Dipper and Woodward 1989: p. 58; Nithyanandan 2012: figure 12.4, p. 4 as *G.*cf. *pallida*)
70. *Glossodoris* sp.
    Habitat: Coral reefs
    Distribution: Iran, APG (Rezai et al. 2016)
71. ***Goniobranchus* cf. *albonares*** (Rudman, 1990)
    Habitat: Coral reefs
    Distribution: Western Pacific Ocean (Gosliner et al. 2018) Iran, APG (this study)
72. *Goniobranchus annulatus* (Eliot, 1904)
    Habitat: Coral reefs
    Distribution: Indo-West Pacific (Gosliner et al. 2015); Red Sea (Yonow 2008); Kuwait, Iran, and SE APG (Dipper and Woodward 1989: p. 59; Nithyanandan 2012; George 2012; Yonow 2012; Rezai et al. 2016)
73. *Goniobranchus bombayanus* (Winckworth, 1946)
    Habitat: Coral reefs
    Distribution: Gulf of Kutch and Mandapam, India (Valdés et al. 1999); Kuwait, APG (Nithyanandan et al. 2021: figure 12.3, p. 95)
74. *Goniobranchus cazae* (Gosliner & Behrens, 2004)
    Habitat: Coral reefs
    Distribution: Gulf of Oman (Gosliner et al. 2015); UAE and Qatar, APG (George 2012; Kardousha et al. 2016; this study)
75. *Goniobranchus obsoletus* (Rüppell & Leuckart, 1830)
    Habitat: Coral reefs
    Distribution: Red Sea and APG (Yonow 2008; Gosliner et al. 2008; Nithyanandan 2012: figure 12.3a, b, p. 4; George 2012; Rezai et al. 2016)
76. ***Goniobranchus pseudodecorus*** (Yonow, 2018) (Figure 12.5i)
    Habitat: Coral reefs
    Distribution: Red Sea (Yonow 2018); Abu Dhabi, APG (this study)
77. ***Goniobranchus* sp. 23 (Gosliner et al. 2018)** (Figure 12.5j)
    Habitat: Coral reefs
    Distribution: Abu Dhabi, UAE, APG (this study)
78. ***Goniobranchus* sp. a**
    Habitat: Hard substrates
    Distribution: Only known from Kuwait (Nithyanandan et al. 2021: figure 12.4, p. 96), Qatar and Dubai, APG (this study)

79. *Goniobranchus* sp. b
Habitat: Hard substrates
Distribution: Kuwait, APG (Nithyanandan et al. 2021: figure 12.5, p. 97)
80. ***Hypselodoris carnea*** (Bergh, 1889)
Habitat: Intertidal to coral reef habitats
Distribution: Western Indian Ocean (Gosliner et al. 2018); Dubai, APG (this study)
81. *Hypselodoris dollfusi* (Pruvot-Fol, 1933)
Habitat: Coral reefs
Distribution: Indian Ocean (Gosliner et al. 2015); Red Sea (Yonow 2018); Kuwait and UAE, APG (Glayzer et al. 1984; Dipper and Woodward 1989: p. 59); Oman (Gosliner and Behrens 2000).
82. *Hypselodoris infucata* (Rüppell & Leuckart, 1830)
Habitat: Coral reefs
Distribution: Indian Ocean (Gosliner et al. 2015); Red Sea (Yonow 2008); Kuwait, Qatar, and UAE, APG (George 2012; Nithyanandan et al. 2021: figure 12.6a, b, p. 98; this study)
83. ***Hypselodoris kanga*** (Rudman, 1977) (Figure 12.5k)
Habitat: Coral reefs
Distribution: Indian Ocean (Gosliner et al. 2015); Dubai, APG (this study)
84. *Hypselodoris maculosa* (Pease, 1871)
Habitat: Coral reefs
Distribution: Indo-Pacific (Gosliner et al. 2015); Kuwait (Glayzer et al. 1984), Dubai, APG (this study)
85. *Hypselodoris nigrostriata* (Eliot, 1904) (Figure 12.5l)
Habitat: Coral reefs
Distribution: Indo-Pacific (Gosliner et al. 2015); Red Sea and Oman (Yonow 2008); Dubai, APG (George 2012; this study)
86. *Hypselodoris pulchella* (Rüppell & Leuckart, 1830)
Habitat: Coral reefs
Distribution: Indian Ocean (Gosliner et al. 2015); Red Sea (Yonow 2008); Iran and UAE, APG (Dipper and Woodward 1989: p. 59; Rezai et al. 2016)
87. *Hypselodoris whitei* (A. Adams & Reeve, 1850)
Habitat: Coral reefs
Distribution: Indo-Pacific (Gosliner et al. 2008); Iran, APG (Rezai et al. 2016)
88. *Hypselodoris* sp.
Habitat: Hard substrates
Distribution: Kuwait, APG (Nithyanandan et al. 2021: figure 12.7, p. 100)
89. ***Mexichromis katalexis*** (Yonow, 2001) (Figure 12.6a)
Habitat: Coral reefs
Distribution: Indo-Pacific (Yonow 2001); Red Sea (Yonow 2008); APG (this study)

Family DENDRODORIDIDAE

90. *Dendrodoris fumata* (Rüppell & Leuckart, 1830)
Habitat: Hard substrates
Distribution: Indo-Pacific (Gosliner et al. 2015); Red Sea (Yonow 2008); Kuwait, Iran, and Dubai, APG (Jones 1986; Rezai et al. 2016; this study)
91. *Dendrodoris nigra* (Stimpson, 1855)
Habitat: Hard substrates
Distribution: Indo-Pacific (Gosliner et al. 2015); Red Sea (Yonow 2008); Kuwait, APG (Jones 1986)
92. ***Dendrodoris* sp. a** (Figure 12.6b)
Habitat: Hard substrates
Distribution: Abu Dhabi (UAE), APG (this study)
93. ***Dendrodoris* sp. b**
Habitat: Hard substrates
Distribution: Saudi Arabia, APG (this study)
94. *Doriopsilla* cf. *miniata* (Alder & Hancock, 1866)
Habitat: Hard substrates
Distribution: Indo-Pacific (Gosliner et al. 2015); Kuwait, APG (Nithyanandan 2012: figure 12.7, p. 5)
95. *Doriopsilla nigrocera* (Yonow, 2012)
Habitat: Hard substrates
Distribution: Saudi Arabia, APG (Yonow 2012)
96. ***Doriopsilla* sp. a**
Habitat: Hard substrates
Distribution: Iran, APG (Rezai et al. 2016)
97. ***Doriopsilla* sp. b** (Figure 12.6c)
Habitat: Hard substrates
Distribution: Qatar, APG (this study)

Family PHYLLIDIIDAE

98. *Phyllidia rueppelii* (Bergh, 1869)
Habitat: Coral reefs
Distribution: Red Sea (Yonow 2008; APG (Rezai et al. 2016 as *Fryeria rüeppelii* [sic])
99. *Phyllidia schupporum* (Fahrner and Schrödl, 2000)
Habitat: Coral reefs
Distribution: Red Sea (Gosliner et al. 2015; Yonow 2020); Kuwait, APG (Nithyanandan 2012: figure 12.6, p. 5 as *Phyllidia (Fryeria) rueppelii*)
100. *Phyllidia* sp.
Habitat: Coral reefs
Distribution: Iran, APG (Rezai et al. 2016)
101. ***Phyllidiopsis dautzenbergi*** (Vayssière, 1912) (Figure 12.6d)
Habitat: Coral reefs
Distribution: Red Sea (Yonow 2008); Saudi Arabia, APG (this study)

Order CLADOBRANCHIA

Family DENDRONOTIDAE

102. *Pseudobornella orientalis* (Baba, 1932)[†]
Habitat: Sandy substrata
Distribution: UAE, APG (Gosliner et al. 2008)

Family DOTIDAE

103. *Doto* cf. *kya* Er. (Marcus, 1961)
Habitat: Hard substrata

FIGURE 12.6 Rare and new records of heterobranch sea slugs from the Arabian/Persian Gulf: (a) *Mexichromis katalexis* (photograph by Dragan Petkovich, Abu Dhabi UAE); (b) *Dendrodoris* sp. a (photograph by Stewart Clarke, Abu Dhabi, UAE); (c) *Doriopsilla* sp. b (photograph by Angelo Piya, Qatar, UAE); (d) *Phyllidiopsis dautzenbergi* (photograph by Baleegh Ahmed Abukaboos, Saudi Arabia); (e) *Doto* sp. 7 (photograph by Dany, Qatar); (f) *Kabeiro rubroreticulata* (photograph by Dragan Petkovich, Abu Dhabi, UAE); (g) *Lomanotus vermiformis* (photograph by Stewart Clarke, Abu Dhabi, UAE); (h) *Bulbaeolidia alba* (photograph by Baleegh Ahmed Abukaboos, Saudi Arabia); (i) *Limenandra rosanae* (photograph by Gordon Smith, Dubai, UAE); (j) *Limenandra* sp. (photograph by Kirstin N. Binoth, Dubai, UAE); (k) *Cratena* sp. (photograph by Stewart Clarke, Abu Dhabi, UAE); (l) *Facelina lineata* (photograph by Kirstin N. Binoth, Dubai, UAE).

Distribution: East Pacific (Er. Marcus 1961); Red Sea (Yonow, Chapter 13, this volume); Kuwait and Qatar, APG (Al-Yamani et al. 2012: pl. 162, p. 184; this study)

104. ***Doto* sp. 7** (Figure 12.6e)
Habitat: Coral rubble
Distribution: Seychelles (Gosliner et al. 2015); Qatar, APG (this study)

105. ***Doto* sp. a**
Habitat: Hard substrata
Distribution: Dubai, APG (this study)

106. ***Doto* sp. b**
Habitat: Hard substrata
Distribution: Qatar, APG (this study)

107. ***Kabeiro rubroreticulata*** (Shipman & Gosliner, 2015) (Figure 12.6f)
Habitat: Seagrass beds
Distribution: Philippines (Gosliner et al. 2015); Dubai, APG (this study)

108. ***Kabeiro* sp.**
Habitat: Mostly seen on hydroids
Distribution: Qatar, APG (this study)

Family LOMANOTIDAE

109. ***Lomanotus vermi formis*** (Eliot, 1908) (Figure 12.6g)
   Habitat: Coral reefs
   Distribution: Circumtropical (Gosliner et al. 2015); Dubai, APG (this study)

Family TRITONIDAE

110. *Marioniopsis viridescens* (Eliot, 1904)
   Habitat: Coral reefs
   Distribution: Red Sea (Yonow 2008 as *Marioniopsis*); Iran and Dubai, APG (Rezai et al. 2016 as *Marioniopsis viridescens*; this study)

Family ARMINIDAE

111. *Armina* sp. 7†
   Habitat: Gravelly substrata
   Distribution: UAE (Gosliner et al. 2008)
112. *Armina* sp. 8†
   Habitat: Sandy substrata
   Distribution: UAE (Gosliner et al. 2008)

Family AEOLIDIIDAE

113. *Bulbaeolidia alba* (Risbec, 1928) (Figure 12.6h)
   Habitat: Rocky substrata and coral reefs
   Distribution: Indo-Pacific (Gosliner et al. 2008); Red Sea (Yonow 2008 as *Aeolidiella*); Kish Island, Iran and Saudi Arabia, APG (Rezai et al. 2016 as *Aeolidiella? alba*; this study)
114. *Cerberilla* sp. 5†
   Habitat: Coral reefs
   Distribution: UAE (Gosliner et al. 2008)
115. ***Limenandra fusiformis*** (Baba, 1949)
   Habitat: Coral reefs
   Distribution: Indian and western Pacific oceans (Gosliner et al. 2015); Red Sea (Yonow 2008); Qatar, APG (this study)
116. ***Limenandra rosanae*** (Carmona et al., 2014) (Figure 12.6i)
   Habitat: Coral reefs
   Distribution: Philippines and Hawaii Islands (Gosliner et al. 2015); Red Sea (Yonow 2008 as *Limenandra* sp.); Qatar, APG (this study)
117. ***Limenandra* sp.** (Figure 12.6j)
   Habitat: Sandy areas
   Distribution: Jumeriah beach, Dubai, APG (this study)

Family FACELINIDAE

118. *Caloria indica* (Bergh, 1896)
   Habitat: Coral reefs
   Distribution: Indo-Pacific (Gosliner et al. 2015); Red Sea (Yonow 2008); Kuwait, APG (Nithyanandan et al. 2021)
119. ***Cratena lineata*** (Eliot, 1905)
   Habitat: Coral reefs
   Distribution: Indo-West Pacific (Gosliner et al. 2015); Qatar and Dubai, APG (this study)
120. ***Cratena* sp.** (Figure 12.6k)
   Habitat: Coral reefs
   Distribution: Qatar, APG (this study)
121. ***Facelina lineata*** (Eliot, 1905) (Figure 12.6l)
   Habitat: Coral reefs
   Distribution: India and Tanzania (Gosliner et al. 2018); Red Sea (Yonow 2015: fig. 42); Qatar, APG (this study)
122. ***Favorinus* sp.** (Figure 12.7a)
   Habitat: Coral reefs
   Distribution: Qatar, APG (this study)
123. ***Herviella albida*** (Baba, 1966) (Figure 12.7b)
   Habitat: Coral reefs
   Distribution: Indo-Pacific (Gosliner et al. 2015); Qatar, APG (this study)
124. ***Myja* cf. *longicornis*** (Bergh, 1896) (Figure 12.7c)
   Habitat: Coral reefs
   Distribution: Dubai, APG (this study)
125. *Phidiana militaris* (Alder & Hancock, 1864)
   Habitat: Coral rubble
   Distribution: Indo-West Pacific (Gosliner et al. 2015); Iran and UAE, APG (Rezai et al. 2016; George 2012)
126. *Sakuraeolis gujaratica* (Rudman, 1980)†
   Habitat: Coral reefs
   Distribution: Indian Ocean (Gosliner et al. 2015); APG (Gosliner et al. 2008)

Family FLABELLINIDAE

127. *Coryphella amabilis* (Hirano & Kuzirian, 1991)
   Habitat: Hard substrates, coral reefs
   Distribution: Kuwait, APG (Al-Yamani et al. 2012: pl. 164, p. 186 as *Flabellina amabilis*)
128. *Coryphellina rubrolineata* (O'Donoghue, 1929) (Figure 12.7d)
   Habitat: Coral reefs
   Distribution: Red Sea and Gulf of Aden (Ekimova et al. 2022); Qatar and UAE, APG (Dipper and Woodward 1989: p. 60; this study)
129. *Samla bicolor* (Kelaart, 1858)
   Habitat: Rocky intertidal areas, coral reefs
   Distribution: Indo-Pacific (Gosliner et al. 2015); Kuwait, APG (Nithyanandan 2012: figure 12.8, p. 5)

Family MYRRHINIDAE

130. *Phyllodesmium magnum* (Rudman, 1991)#
   Habitat: Sand
   Distribution: Indo-Pacific (Gosliner et al. 2018); UAE, APG (George 2012; this study)
131. ***Phyllodesmium poindimiei*** (Risbec, 1928) (Figure 12.7e)
   Habitat: Coral reefs
   Distribution: Indo-Pacific (Gosliner et al. 2015); Qatar and Dubai, APG (this study)

Family EUBRANCHIDAE

132. ***Eubranchus mandapamensis*** (Rao, 1968) (Figure 12.7f)
   Habitat: Coral reefs, on hydroids
   Distribution: Indo-Pacific (Gosliner et al. 2015); Qatar, APG (this study)
133. ***Eubranchus* sp. 19**
   Habitat: On hydroid *Sertularia* sp. and concrete walls (Gosliner et al. 2015)
   Distribution: Indo-Pacific (Gosliner et al. 2015); Qatar, APG (this study)

FIGURE 12.7 Rare and new records of heterobranch sea slugs from the Arabian/Persian Gulf: (a) *Favorinus* sp. (photograph by Dem Panaligan); (b) *Herviella albida* (photograph by Baleegh Ahmed Abukaboos, Saudi Arabia); (c) *Myja* cf. *longicornis* (photograph by Kirstin N. Binoth, Dubai, UAE); (d) *Coryphellina rubrolineata* (photograph by Gordon Smith, Dubai, UAE); (e) *Phyllodesmium poindimiei* (photography by Dem Panligan, Qatar, UAE); (f) *Eubranchus mandapamensis* (photograph by Henrik Rasmussen, Qatar, UAE); (g) *Eubranchus* sp. a (photograph by Henrik Rasmussen, Qatar, UAE); (h) *Eubranchus* sp. b (Jay R-Oliveras, Qatar); (i) *Trinchesia* sp. a (photograph by Kirstin N. Binoth, Dubai, UAE).

134. ***Eubranchus* sp. a** (Figure 12.7g)
Habitat: Hard substrates
Distribution: Dubai (UAE), APG (this study)
135. ***Eubranchus* sp. b** (Figure 12.7h)
Habitat: Hard substrates
Distribution: Qatar, APG (this study)
136. *Leostyletus misakiensis* (Baba, 1960)
Habitat: soft-substrate habitats and harbours feeding on the hydroid *Obelia*
Distribution: Northern Pacific (Japan and California, USA: https://invasions.si.edu/nemesis/calnemo/species_summary/78612; www.seaslugforum.net/showall/eubrmisa); Kuwait, APG (Al-Yamani et al. 2012: pl. 163a, b, p. 185 as *Eubranchus miskaiensis* [sic])

Family TRINCHESIIDAE
137. *Phestilla lugubris* (Bergh, 1870)
Habitat: Coral reefs
Distribution: Indo-Pacific (Gosliner et al. 2015); Red Sea (Yonow 2008); Iran and SE APG (George 2012; Rezai et al. 2016)
138. *Trinchesia albocrusta* (MacFarland, 1966)
Habitat: Hard substrates
Distribution: Kuwait, APG (Al-Yamani et al. 2012: pl. 165, p. 187, as *Cuthona albocrusta*)
139. *Trinchesia yamasui* (Hamatani, 1993)
Habitat: Coral reefs
Distribution: Indo-Pacific (Gosliner et al. 2015), Saudi Arabia, Iran, APG (Rezai et al. 2016; this study)
140. ***Trinchesia* sp. a** (Figure 12.7i)
Habitat: Coral reefs
Distribution: Ras Al Khaimah (UAE), APG (this study)

12.4 DISCUSSION

The compilation of nudipleuran diversity from the marginal seas of the NWIO revealed maximum species diversity in the RS with a minimum in the APG. Compared to the GO and the AS, the extreme environmental conditions facilitated high nudipleuran endemism in the RS and the APG, thereby contributing to high spatial turnover of nudipleuran diversity along the oceanographic continua between the RS and the AS, and the APG and the AS. The role of dispersal-limiting processes, such as lack of direct connectivity between water bodies, differences in physicochemical and biological conditions, depth contours and flow dynamics, habitat/substratum availability, etc., contributing to beta diversity should be the highest among the areas with maximum geographic discontinuity. However, from the highest and lowest nestedness values of beta diversity (between RS and APG: $\beta_{nes} = 0.257$, between RS and AS: $\beta_{nes} = 0.087$), it is clear that, although geographic discontinuity is the highest between the RS and the APG, the extent of environmental similarity between these seas (as both contain some of the world's hottest and saltiest seawater) plays a crucial role in nudipleuran species endemism. Along the oceanic continuum, environmental gradients can play a significant role in shaping the patterns of biodiversity and can result in both nested and turnover beta diversity. The relative importance of these two processes varies depending on the specific system being studied. In the case of geographically isolated areas, we propose that the extent of environmental similarity and dissimilarity regulates the magnitude of nestedness and turnover, respectively. The geographical and environmental similarity between the Mediterranean-type basins, where the biotic communities are exposed to some of the world's hottest and saltiest seawater, seems to be one of the principal factors responsible for the high nestedness in nudipleuran diversity between the RS and the APG.

Nudipleura beta diversity reflects two antithetical phenomena, spatial turnover and nestedness. Spatial turnover occurs when some species are replaced by others due to environmental or spatial constraints (Qian et al. 2005). Nestedness occurs when sites with a lower number of species are subsets of sites with a higher number (Ulrich and Gotelli 2007), reflecting a non-random process of species loss. Although nestedness reduces the conservation effort aimed at protecting and maximizing species diversity by decreasing the number of habitats to be protected, spatial turnover increases the effort and resources spent on it. Also, mistaking nestedness for turnover may cause overestimations of global biodiversity (Baselga 2010). Therefore, it is essential to disentangle these components of beta diversity to understand the factors or processes contributing to changes in species diversity/richness across the systems and to draw effective biodiversity conservation strategies.

The high endemism of the Nudipleura in the RS resulting from its unique environmental settings (Yonow 2008) is the primary reason for the Red Sea's highest nudipleuran spatial turnover and nestedness compared with the other areas explored in this study. Although the RS and the APG are Mediterranean-type systems, vast differences in their water depths result in different hydrographic profiles. As the differences in species composition are often associated with differences in environmental conditions (Al-Said et al. 2017; Al-Yamani et al. 2019), the differences in water depth (maximum depth: APG ≤ 90 m, RS ~ 3000 m), wave exposure, or habitat complexity (abundance of coral habitats and submerged vegetation in the RS) also favour high nudipleuran species diversity in the RS (Yonow 2008). The high evaporation rate, low precipitation, and riverine influx contribute to high salinity and low nutrient availability in the APG (Al-Yamani 2021) and the RS (Maillard and Soliman 1986). In the RS, the abundant growth and periodic blooming of the diazotrophic cyanobacteria *Trichodesmium erythraeum*, which accounts for approximately 60–80% of marine biological N_2 fixation in the world's oceans (Boatman et al. 2017), helps to a great extent in meeting the biological nitrogen requirements of the system.

The Red Sea and the Arabian/Persian Gulf have different geological histories (Yonow 2008; Sheppard et al. 2010), shaping the physical structure of their coastlines and influencing the distribution of habitats and species. In general, the high Nudipleura beta diversity between the seas has likely resulted from a combination of physicochemical (including differences in water temperature, salinity, and nutrient levels), biological (including the presence of different species of coral and other benthic organisms, which is essential given the strong host-specificity for many heterobranch species), and historical factors that shape the patterns of biodiversity in these ecosystems.

12.5 CONCLUSIONS

The nudipleuran diversity of the APG, RS, GO, and AS varies among these seas; in general, the Red Sea has one of the highest levels of biodiversity in the world, including a high diversity of the Nudipleura. This is partly due to the Red Sea's warm, low-turbidity, saline waters that support a unique and diverse community of coral and other benthic organisms. In comparison, the APG has a lower diversity of nudipleurans due in part to the unique oceanographic settings (high salinity, a wide range of temperatures and shallow water bodies, etc.) in this region. Higher endemicity of the APG fauna (e.g., echinoderms) has been established in previous studies (Price 1982). However, with the predicted extinction risk for biodiversity in the region (Wabnitz et al. 2018), habitat conservation efforts for coral reefs and seagrass beds are crucial to prevent species loss. The Gulf of Oman and the western Arabian Sea have higher nudipleuran diversities than the APG, but this high diversity is still substantially lower than that of the RS. Numerous species in these regions are reported from the RS and APG, further supporting the 'native range' terminology used in Hoeksema and Yonow (2021) as applied for *Plocamopherus ocellatus*. The native range indicates that a species is indigenous to the area and here, some species previously considered endemic to the Red Sea occur all around the Arabian Peninsula. These species have been indicated in the chapter on sea slugs of the Red Sea (Yonow, Chapter 13, this volume).

It is important to note that these comparisons are based on the diversity trends revealed by the present compilation and that the documented Nudipleura diversity of any given area is likely to be strongly dependent on the number of studies available from each area; at present, the APG is the most understudied water body for nudipleuran diversity. Therefore, concerted studies from little-explored areas like the APG could improve the sea slug gamma diversity in the NWIO and provide more detailed information on the consistency of beta diversity patterns between its marginal seas. As understanding the drivers of beta diversity is vital for conserving and managing marine communities, partitioning the beta diversity into its components, such as nestedness and turnover, provides insights into the processes determining the composition of communities and the factors that influence species distributions.

ACKNOWLEDGMENTS

We thank the Kuwait Institute for Scientific Research for the facilities provided to the authors MN, MAK, RM, GM, and FT. We extend our thanks to the following individuals who came forward to share their beautiful photographs to validate the heterobranch records from the APG and GO: Stewart Clarke, Dragan Petkovich, Kirstin N. Binoth, Baleegh Ahmed Abukaboos, Gordon Smith, Jay R-Oliveras, Henrik Rasmussen, Keith P. Wilson, Michael Rall, Michal Rysniak, Dany, Angelo Pia, Dem Panaligan, Sven Kahlbrock, and Carole Harris, members of UAE Branchers, NudiBase and Nudibranch Central Facebook® pages, and iNaturalist®. We also extend our sincere thanks to: Dr. Friedhelm Krupp, Senckenberg Museum, Germany, and David George, Natural History Museum, UK, for sending material to NY for identification; Kuwait Petroleum Company (KPC) for the research grant (FM075C) awarded to MAK; Dr. Valeriy Skryabin in field collections and photography; and also to the reviewers for their critical comments that helped improve the manuscript.

REFERENCES

Al-Kandari, M., Oliver, P.G., Chen, W., Skryabin, V., Raghu, M., Yousif, A., Al-Jazzaf, S., Taqi, A., and A. Al-Hamad. 2020. Diversity and distribution of the intertidal Mollusca of the State of Kuwait, Arabian Gulf. *Reg Stud Mar Sci* 33: 100905. doi:10.1016/j.rsma.2019.100905.

Al-Khayat, J.A., Vethamony, P., and M. Nanajkar. 2021. Molluscan diversity influenced by mangrove habitats in the Khors of Qatar. *Wetlands* 41, no. 4: 45. doi:10.1007/s13157-021-01441-6.

Al-Rifaie, K., Al-Yamani, F., Lennox, A., Boltachova, N., Revkov, N., Grintsov, V., Kolesnikova, E., and V. Murina. 2012. Macrozoobenthos community structure during four seasons in Kuwait Bay. *Intl J Oceans Oceanogr* 6, no. 1: 45–67.

Al-Said, T., Al-Ghunaim, A., Subba Rao, D.V., Al-Yamani, F., Al-Rifaie, K., and A. Al-Baz. 2017. Salinity-driven decadal changes in phytoplankton community in the NW Arabian Gulf of Kuwait. *Environ Monit Assess* 189, no. 6: 268. doi:10.1007/s10661-017-5969-4.

Al-Yamani, F.Y. 2021. *Fathoming the northwestern Arabian Gulf: Oceanography and marine biology*. Kuwait: Kuwait Institute for Scientific Research, 413 p.

Al-Yamani, F.Y., Madhusoodhanan, R., Skryabin, V., and T. Al-Said. 2019. The response of microzooplankton (tintinnid) community to salinity related environmental changes in a hypersaline marine system in the northwestern Arabian Gulf. *Deep Sea Res Part II: Top Stud Oceanogr* 166: 151–170. doi:10.1016/j.dsr2.2019.02.005.

Al-Yamani, F.Y., Skryabin, V., Botlachova, N., Revkov, N., Makarov, M., Grinstov, V., and E. Kolesnikova. 2012. *Illustrated atlas on the zoobenthos of Kuwait*. Kuwait: Kuwait Institute for Scientific Research, 383 p.

Amini-Yekta, F., and H. Dekker. 2021. An updated checklist of marine gastropods of the Persian Gulf and Gulf of Oman. *Zootaxa* 4957, no. 1: 1–71. doi:10.11646/zootaxa.4957.1.1.

Amini-Yekta, F., Asgari, M., and S. Izadi. 2016. *Molluscs, crustaceans and echinoderms of the intertidal area of the Qeshm Island (with illustrated list of other invertebrates of this region)*. Persian Gulf Biotechnology Park, Qeshm Island, 134 p. [in Persian].

Amini-Yekta, F., Jalili, M., Pourjomeh, F., Hakim Elahi, M., and H. Rezaei. 2014. Distribution of Molluscs in the Eastern Persian Gulf, PG-GOOS cruise. *J Persian Gulf* 5, no. 17: 37–48.

Amini-Yekta, F., Shokri, M.R., Maghsoudlou, A., and H. Rajabi-Maham. 2019. Intertidal gastropod assemblages shaped by key environmental variables across the northern Persian Gulf and the Gulf of Oman. *Mar Ecol* 40, no. 3: e12545. doi:10.1111/maec.12545.

Amouei, M., and Y. Fatemi. 2021. On the occurrence *of Stylocheilus longicauda* (Quoy & Gaimard, 1825) Heterobranchia, Aplysiidae in the Persian Gulf. *Iranian J Animal Biosystem* 17, no. 1: 101–102.

Apte, D., and D. Desai. 2017. *Field guide to the sea slugs of India*. Mumbai, India: Bombay Natural History Society, 456 p.

Baselga, A. 2010. Partitioning the turnover and nestedness components of beta diversity. *Global Ecol Biogeogr* 19: 134–143. doi:10.1111/j.1466-8238.2009.00490.x.

Basson, P.W., Burchard, J.E., Hardy, J.T., and A.R.G. Price. 1977. *Biotopes of the Western Arabian Gulf: Marine life and environments of Saudi Arabia*. Dhahran: Aramco, 184 p.

Boatman, T.G., Lawson, T., and R.J. Geider. 2017. A key marine diazotroph in a changing ocean: The interacting effects of temperature, CO_2 and light on the growth of *Trichodesmium erythraeum* IMs101. *PLoS One* 12, no. 1: e0168796. doi:10.1371/journal.pone.0168796.

Bouchet, P., Rocroi, J.P., Hausdorf, B., et al. 2017. Revised classification, nomenclator and typification of gastropod and monoplacophoran families. *Malacologia* 61, nos. 1–2: 1–526. doi:10.4002/040.061.0201.

Dayrat, B., Goulding, T.C., Apte, D., et al. 2020. Systematic revision of the genus *Peronia* Fleming, 1822 (Gastropoda, Euthyneura, Pulmonata, Onchidiidae). *ZooKeys* no. 972: 1–224. doi:10.3897/zookeys.972.52853.

Dipper, F., and T. Woodward. 1989. *The living seas: Marine life of the Southern Gulf*. Dubai: Motivate Publishing, Emirates Printing Press, 95 p.

Dorgan, K.M., Valdés, Á., and T.M. Gosliner. 2002. Phylogenetic systematics of the genus *Platydoris* (Mollusca, Nudibranchia, Doridoidea) with descriptions of six new species. *Zoologica Scripta* 31, no. 3: 271–319.

Ekimova, I., Deart, Y., Antokhina, T., Mikhlina, A., and D. Schepetov. 2022. Stripes matter: Integrative systematics of *Coryphellina rubrolineata* species complex (Gastropoda: Nudibranchia) from Vietnam. *Diversity* 14: 294. doi:10.3390/d14040294.

Fatemi, Y., and G. Attaran-Fariman. 2015. Checklist of the opisthobranchs (Heterobranchia: Gastropoda) along the Iranian coasts of the Gulf of Oman. *J Biodivers Environ Sci* 6, no. 3: 1–7.

Fatemi, Y., Taherizadeh, M., and A. Shadadi. 2020. First record of *Thuridilla indopacifica* Gosliner, 1995 (Sacoglossa: Plakobranchidae) from the Gulf of Oman, Iran. *Int J Aq Biol* 8, no. 6: 421–423.

Feliciano, K., Malaquias, M.A.E., Stout, C., Brenzinger, B., Gosliner, T.M., and A. Valdés. 2021. Molecular and morphological analysis reveal pseudocryptic diversity in *Micromelo undatus* (Bruguière, 1792) (Gastropoda: Heterobranchia: Aplustridae). *System Biodiv* 19, no. 7: 834–858. doi:10.1080/14772000.2021.1939458.

George, J.D. 2012. Reef-associated macroinvertebrates of the SE Gulf. In: *Coral reefs of the Gulf: Adaptation to climatic extremes, coral reefs of the World 3*. Eds: B.M. Riegl, and S.J. Purkis, 253–308. Berlin, Heidelberg: Springer Science+Business Media.

Glayzer, B.A., Glayzer, D.T., and K.R. Smythe. 1984. The marine molluscs of Kuwait Arab. Gulf. *J Conchol London* 31, no. 5: 311–330.

Gosliner, T.M., and D.W. Behrens. 2000. Two new species of Chromodorididae (Mollusca: Nudibranchia) from the tropical Indo-Pacific, with a redescription of *Hypselodoris dollfusi* (Pruvot-Fol, 1933). *Proc California Acad Sci Series 4* 52, no. 10: 111–124.

Gosliner, T.M., Behrens, D.W., and A. Valdés. 2008. *Indo-Pacific nudibranchs and sea slugs: A field guide to the world's most diverse fauna*. Sea Challengers Natural History Books and the California Academy of Sciences, California, 426 p. doi:10.1080/14772000.2021.1939458.

Gosliner, T.M., Valdés, A., and D.W. Behrens. 2015. *Nudibranch and sea slug identification, Indo-Pacific*. Jacksonville, FL: New World Publications, 408 p.

Gosliner, T.M., Valdés, A., and D.W. Behrens. 2018. *Nudibranch and sea slug identification, Indo-Pacific*. Jacksonville, FL: New World Publications, 451 p.

Hoeksema, B.W., and Y. Yonow. 2021. Rarity in the native range of Lessepsian migrant *Plocamopherus ocellatus* (Nudibranchia): Fact or artifact? *Ecology* 102, no. 11: e03481. doi:10.1002/ecy.3481.

Jones, D.A. 1986. *A field guide to the seashores of Kuwait and Arabian Gulf*. University of Kuwait, distributed by Blandford Press, Dorset, 192 p.

Kardousha, M.M., Al-Muftah, A., and J.A. Al-Khayat. 2016. Exploring Sheraoh Island at South-Eastern Qatar: First distributional records of some inland and offshore biota with annotated checklist. *J Marine Sci Res Dev* 6: 191. doi:10.4172/2155-9910.1000191.

Kohan, A., Zahra, B., and M. Shokri. 2012. The gastropod along the Bushehr province intertidal zone of the Persian Gulf. *J Persian Gulf (Mar Sci)* 3, no. 9: 33–42.

Koleff, P., Gaston, K.J., and J.K. Lennon. 2003. Measuring beta diversity for presence-absence data. *J Anim Ecol* 72: 367–382. doi:10.1046/j.1365-2656.2003.00710.x.

Maillard, C., and G. Soliman. 1986. Hydrography of the Red Sea and exchanges with the Indian Ocean in summer. *Oceanologica Acta* 9, no. 3: 249–269.

Malaquias, M.A.E., and D.G. Reid. 2008. Systematic revision of the living species of Bullidae (Mollusca: Gastropoda: Cephalaspidea), with a molecular phylogenetic analysis. *Zool J Linn Soc* 153, no. 3: 453–543. doi:10.1111/j.1096-3642.2008.00369.x.

Maniei, F., Espeland, M., Mohavedi, M., and H. Wägele. 2020. Description of a new *Peronia* species (Gastropoda: Eupulmonata: Onchidiidae) from Iran, Persian Gulf. *Zootaxa* 4758, no. 3: 501–531. doi:10.11646/zootaxa.4758.3.5.

Nimbs, M.J., and S.D.A. Smith. 2016. An illustrated inventory of the sea slugs of New South Wales, Australia (Gastropoda: Heterobranchia). *Proc Royal Soc Victoria* 128, no. 2: 44–113. doi:10.1071/RS16011.

Nimbs, M.J., Hutton, I., Davis, T.R., Larkin, M.F., and S.D.A. Smith. 2020. The heterobranch sea slugs of Lord Howe Island, NSW, Australia (Mollusca: Gastropoda). *Proc Royal Soc Victoria* 132, no. 1: 12–41. doi:10.1071/RS20002.

Nithyanandan, M. 2012. New and rare nudibranch records from Kuwait, Arabian Gulf (Mollusca: Opisthobranchia). *Mar Biodiv Rec* 5: e115. doi:10.1017/S1755267212000954.

Nithyanandan, M., Al-Kandari, M., and G. Mantha. 2021. New records of nudibranchs and a cephalaspid from Kuwait, northwestern Arabian Gulf (Mollusca: Heterobranchia). *ZooKeys* no. 1048: 91–107. doi:10.3897/zookeys.1048.66250.

Oladi, M., Nasrolahi, A., Ahmadzadeh, F., Alizadeh, O., and P.J. Krug. 2018. Molecular confirmation of the occurrence of *Elysia* cf. *tomentosa* (Mollusca: Heterobranchia) in the Persian Gulf. *Zool Middle East* 64, no. 2: 124–130. doi:10.1080/09397140.2018.1444571.

Oskars, T., and M. Malaquias. 2019. A molecular phylogeny of the Indo-West Pacific species of *Haloa sensu lato* gastropods (Cephalaspidea: Haminoeidae): Tethyan vicariance, generic diversity, and ecological specialization. *Molecul Phylogen Evol* 139: 106557. doi:10.1016/j.ympev.2019.106557.

Patterson, B.D., and W. Atmar. 1986. Nested subsets and the structure of mammalian faunas and archipelagos. In: *Island biogeography of mammals*. Eds: L.R. Heaney, and B.D. Patterson. *Biol J Linn Soc* 28, nos. 1 and 2: 65–82.

Price, A.R.G. 1982. Echinoderms of Saudi Arabia, comparison between Echinoderm faunas of Arabian Gulf, SE Arabia, Red Sea and Gulf of Aqaba and Suez. *Fauna Saudi Arabia* 4: 3–21.

Qian, H., Ricklefs, R.E., and P.S. White. 2005. Beta diversity of angiosperms in temperate floras of eastern Asia and eastern North America. *Ecol Lett* 8: 15–22. doi:10.1111/j.1461-0248.2004.00682.x.

Rezai, H., Mohtarami, S.A., Dehghani, H., Tavakoli, P., Bargahi, H.R., and K. Kabiri. 2016. Nudibranchs from the Northern Persian Gulf. *J Persian Gulf (Marine Sci)* 7, no. 24: 1–8.

Schlitzer, R. 2021. *Ocean data view 5.6.3*. http://odv.awi.de; downloaded 1 May 2023.

Sheppard, C., Al-Husiani, M., Al-Jamali, F., Al-Yamani, F., Baldwin, R., Bishop, J., Benzoni, F., Dutrieux, E., Dulvy, N., Durvasula, S.R., Jones, D.A., Loughland, R., Medio, D., Nithyanandan, M., Pilling, G., Polikarpov, I., Price, A.R.G., Purkis, S., Riegl, B., Saburova, M., Namin, K.S., Taylor, O., Wilson, S., and K. Zainal. 2010. The Gulf: A young sea in decline. *Mar Pollut Bull* 60, no. 1: 3–38. doi:10.1016/j.marpolbul.2009.10.017.

Smith, S.D.A., and T.R. Davis. 2019. Slugging it out for science: Volunteers provide valuable data on the diversity and distribution of heterobranch sea slugs. *Molluscan Res* 39: 214–223. doi:10.1080/13235818.2019.1594600.

Smythe, K. 1982. Seashells of the Arabian Gulf. In: *The natural history of the Arabian Gulf*. Eds: J. Cloudsley-Thompson, and A.C. Cloudsley-Thompson. London: Allen & Unwin, 123 p.

UK Admiralty Guides. 1982. *Port approach gthes: The Persian Gulf and its approaches*, 1.95–1.122. Somerset: UK Hydrographic Office.

Ulrich, W., and N.J. Gotelli. 2007. Null model analysis of species nestedness patterns. *Ecology* 88: 1824–1831. doi:10.1890/06-1208.1.

Valdés, Á., Mollo, E., and J. Ortea. 1999. Two new species of *Chromodoris* (Mollusca, Nudibranchia, Chromodorididae) from southern India, with a redescription of *Chromodoris trimarginata* (Winckworth, 1946). *Proc California Acad Sci* 51, no. 13: 461–472.

Wabnitz, C.C.C., Lam, V.W.Y., Reygondeau, G., et al. 2018. Climate change impacts on marine biodiversity, fisheries and society in the Arabian Gulf. *PLoS One* 13, no. 5: e0194537. doi:10.1371/journal.pone.0194537.

Wägele, H., Stemmer, K., Burghardt, I., and K. Händeler. 2010. Two new sacoglossan sea slug species (Opisthobranchia, Gastropoda): *Ercolania annelyleorum* sp. nov. (Limapontioidea) and *Elysia asbecki* sp. nov. (Plakobranchoidea), with notes on anatomy, histology and biology. *Zootaxa* 2676, no. 1: 1–28. doi:10.11646/zootaxa.2676.1.1.

Whittaker, R.H. 1960. Vegetation of the Siskiyou Mountains, Oregon and California. *Ecol Monogr* 30, no. 3: 279–338. doi:10.2307/1943563.

WoRMS Editorial Board. 2024. *World Register of Marine Species*. VLIZ. https://www.marinespecies.org.

Yonow, N. 1994. A new species and new record of *Chelidonura* from the Red Sea (Cephalaspidea: Aglajidae). *J Conchology* 35, no. 2: 141–147.

Yonow, N. 2001. Results of the rumphius biohistorical expedition to Ambon (1990). Part 11: Doridacea of the families Chromodorididae and Hexabranchidae (Mollusca, Gastropoda, Opisthobranchia, Nudibranchia), including additional Moluccan material. *Zool Med Leiden* 75, no. 1: 1–50.

Yonow, N. 2008. *Sea slugs of the Red Sea*. Sofia/Moscow: Pensoft Publications, 304 p.

Yonow, N. 2012. Opisthobranchs from the Western Indian Ocean localities, with descriptions of two new species and eleven new records (Mollusca: Gastropoda). *ZooKeys* no. 197: 1–129. doi:10.3897/zookeys.197.1728.

Yonow, N. 2015. Sea slugs: Unexpected biodiversity and distribution. In: *The Red Sea: The formation, morphology, oceanography and environment of a young ocean basin*. Eds: N.M.A. Rasul, and I.C.F. Stewart, 531–550. Berlin, Heidelberg: Springer Earth System Sciences.

Yonow, N. 2018. Red Sea Opisthobranchia 5: New species and new records of *Chromodoris* from the Red Sea (Heterobranchia, Nudibranchia, Chromodorididae). *ZooKeys* no. 770: 9–42. doi:10.3897/zookeys.770.26378.

Yonow, N. 2020. Red Sea Opisthobranchia 6: Phyllidiidae and their paradorid mimic: New species and new records (Heterobranchia, Nudipleura, Doridina). *ZooKeys* no. 1006: 1–34. doi:10.3897/zookeys.1006.59732.

Sea slugs of the Red Sea
A 'molecular' update

13

Nathalie Yonow

13.1 INTRODUCTION

Many molecular studies on non–Red Sea species have revised a number of genera that affect Red Sea species, which entailed changes of both genus and species names; few studies included sequences of Red Sea specimens, some of which are new endemic species. Another interesting advancement has been made with underwater macro-photography, meaning that citizen scientists (divers and photographers) are now looking for sea slugs measuring only millimetres in length. Vying with the 'molecular' increase of species numbers, these incredibly detailed photographs have also augmented the species list for the Red Sea. Although no specimens could be collected, some photographs can be identified with some confidence to a species or genus level. This chapter will present cases of several of these new records, many with illustrations, and includes a revised and updated table of species (Table 13.1) recorded in the Red Sea to date. This checklist table is based on my first (Yonow 2008a) and second (Yonow 2015) lists, categorising the species (endemic ones in bold typeface) with their new identifications and listing their updated distributions. Many 'old' endemic species have not been recorded since their original descriptions (now indicated with ****), but they are retained in the current list as, once in a while, an old species is reinstated or rediscovered. For example, *Stiliger ornatus*, only known from the Red Sea, has recently been recorded by Blatterer (2019: p. 393, pl. 207, fig. 5) for the first time since its original description in 1828! Authorities are not given for the Red Sea species mentioned in the text to simplify reading, but all names are provided in full in Table 13.1, following species names in WoRMS (2024) with one exception: *Kabeiro* is used instead of *Doto* for *D. orcha*. Additional distribution records for the Gulf of Oman and the Arabian/Persian Gulf are included in this table since many endemic Red Sea species have now been recorded in these seas comprising their native range (e.g., Hoeksema and Yonow 2021), and the Lessepsian migrant species have also been updated. A full account of the new heterobranch species records in the Arabian/Persian Gulf (APG) is given in Nithyanandan et al. (Chapter 12, this volume), which also includes an analysis of similarities for the APG, Red Sea, Gulf of Oman, and the Arabian Sea.

13.2 CEPHALASPIDEA

Some strides have been made in classification in this loosely termed grouping, mostly within the Aglajidae (Zamora-Silva and Malaquias 2018). Morphological differences have long been observed between species groups within a genus; for example, species of *Philinopsis* and *Chelidonura* have been described as one of two morphological types. Fourteen species were listed in 2008, with six of them not identified to species and only two endemic species. The list today encompasses 17 species, with five of them not identified (e.g., Figures 13.1 and 13.2) and four endemic species. *Philinopsis* sp. 1 is recorded for the third time with several individuals photographed (Figure 13.2). The first photograph was published by Yonow (2008b) as *Chelidonura punctata* Eliot, 1903 in error (see also http://www.seaslugforum.net/find/chelsp7): it is not that species but there were no further records until a photograph taken in Eilat, Israel, in 2014; the more recent photographs taken in Eilat and Dahab last year are much clearer, illustrating a distinctive species with short tails, the slightly raised orange spots, and lack of bristles anteriorly, more likely a species of *Philinopsis*.

Molecular work (Austin et al. 2018) showed that *Phanerophthalmus olivaceous* was indeed endemic to the Red Sea and different from several similar counterparts in the Indo-Pacific, while Oskars and Malaquias (2020) proved exactly the opposite, also using molecular techniques, that the endemic *Haminoea cyanomarginata* (Heller & Thompson, 1983) formed part of the widespread and morphologically variable Indo-Pacific species *Lamprohaminoea ovalis*.

The genus *Bullina* has been recorded from the Red Sea with what is probably three different species: the first living animal is known only through photographs (Figure 13.3) and has a pink shell with a clearly raised spire. There is an older record of a shell with a sunken spire that is also pink (www.seaslugforum.net/find/14192), while the third shell in Blatterer (2019: p. 359, pl. 188, fig. 2) is white and has a clearly raised spire.

DOI: 10.1201/9781003321392-16

TABLE 13.1 Checklist of species recorded from the Red Sea with wider distributions.

KEY: **Endemic species.** ???? unknown distribution. ******** original description only.

ACTEONIMORPHA	
Acteon sp.	????
Pupa affinis (A. Adams, 1855)	W Indian Ocean
Pupa tessellata (Reeve, 1842)	Indo-Pacific
***Pupa* cf. *solidula* (Linnaeus, 1758)**	Arabian/Persian Gulf
Pupa sp.	????
Bullina sp. 1	W Indian Ocean
***Bullina* spp. nov. 2, 3**	
Hydatina physis (Linnaeus, 1758)	Circumtropical
Hydatina zonata (Lightfoot, 1786)	Indo-Pacific
RINGICULIMORPHA	
***Ringicula acuta* Philippi, 1849**	
***Ringicula minuta* H. Adams, 1872**	Lessepsian migrant
CEPHALASPIDEA	
***Roxania lithiensis* (Sturany, 1903)**	
Pyrunculus fourierrii (Audouin, 1826)	Indo-West Pacific (Lessepsian migrant)
Acteocina fusiformis (A. Adams, 1854)	W Indian Ocean
***Acteocina inconspicua* (H. Adams, 1852)**	Arabian/Persian Gulf
Acteocina involuta (Nevill & Nevill, 1871)	W Indian Ocean
***Acteocina mucronata* (Philippi, 1849)**	Lessepsian migrant
***Acteocina simplex* (A. Adams, 1850)**	
***Cylichna villersii* Audouin, 1826**	Lessepsian migrant
***Retusa desgenettii* (Audouin, 1826)**	Lessepsian migrant
Retusa tarutana Smythe, 1979	W Indian Ocean
Mnestia girardi (Audouin, 1826)	W Pacific Ocean (Lessepsian migrant)
Colpodaspis thompsoni Brown, 1979	Indo-Pacific
Biuve fulvipunctata (Baba, 1938)	Indo-Pacific (Lessepsian migrant)
***Chelidonura flavolobata* Heller & Thompson, 1983**	
Chelidonura livida Yonow, 1994	W Indian Ocean
Mariaglaja inornata (Baba, 1949)	W Pacific Ocean
Mariaglaja tsurugensis (Baba & Abe, 1959)	W Indian Ocean
Odontoglaja mosaica Gosliner, 2011	W Indian Ocean
Philinopsis ctenophoraphaga Gosliner, 2011	W Pacific Ocean
Philinopsis speciosa (Pease, 1860)	Indo-West Pacific
***Philinopsis* sp. nov. 1**	
Philinopsis sp. 2	????
Spinoaglaja cf. *orientalis* (Baba, 1949)	W Indian Ocean
'Spinophallus' falciphallus Gosliner, 2011	W Pacific Ocean
Tubulophilinopsis pilsbryi (Eliot, 1900)	Indo-Pacific
Tubulophilinopsis reticulata (Eliot, 1904)	W Indian Ocean
Noalda sp.	Indo-West Pacific
Chelidonurid spp. nov. 1, 2	
Chelidonurid sp. 3	????
Sagaminopteron cf. *nigropunctatum* Carlson & Hoff, 1973	W Indian Ocean
Sagaminopteron cf. *ornatum* Tokioka & Baba, 1964	W Pacific Ocean
Sagaminopteron cf. *psychedelicum* Carlson & Hoff, 1974	W Indian Ocean
Siphopteron sp.	Indian Ocean
Aliculastrum cylindricum (Helbling, 1779)	Indo-West Pacific (Lessepsian migrant)
***Atys ehrenbergi* (Issel, 1869)**	
Atys naucum (Linnaeus, 1758)	Indo-West Pacific
Atys semistriatus Pease, 1860	Indo-Pacific
Diniatys dentifer (A. Adams, 1850)	Pacific Ocean + Arabian/Persian Gulf
Diniatys dubius (Schepman, 1913)	Pacific Ocean

TABLE 13.1 (Continued)

Haloa cf. *pemphis* (Philippi, 1847)	Indo-West Pacific
Lamprohaminoea ovalis (Pease, 1868)	Indo-West Pacific (Lessepsian migrant)
Liloa mongii (Audouin, 1826)	Indo-West Pacific
Liloa porcellana (Gould, 1859)	W Pacific Ocean
***Phanerophthalmus albocollaris* Heller & Thompson, 1983**	
***Phanerophthalmus olivaceus* (Ehrenberg, 1828)**	
Smaragdinella calyculata (Broderip & Sowerby I, 1829)	Indo-Pacific
Smaragdinella sieboldi A. Adams, 1864	W Pacific Ocean
Bulla arabica Malaquias & Reid, 2008	NW Indian Ocean (Lessepsian migrant)
Philine sp. 1	NW Indian Ocean
Philine sp. 2	????
APLYSIIDA	
Akera soluta (Gmelin, 1791)	Indo-West Pacific
Aplysia argus Rüppell & Leuckart, 1828	Indo-West Pacific
Aplysia cornigera Sowerby, 1869	Indo-West Pacific
Aplysia ghanimii Golestani et al., 2019	W Indian Ocean (+ Atlantic Ocean)
Aplysia nigrocincta von Martens, 1880	Indo-West Pacific
Aplysia oculifera Adams & Reeve, 1850	Indo-Pacific
Bursatella leachii Blainville, 1817	Circumtropical (? Lessepsian migrant)
Dolabella auricularia (Lightfoot, 1786)	Indo-Pacific
Dolabrifera dolabrifera (Cuvier, 1817)	Circumtropical
Notarchus indicus Schweigger, 1820	Circumtropical
Petalifera petalifera Rang, 1828	Mediterranean
Petalifera ramosa Baba, 1959	Pacific Ocean
***Phyllaplysia* sp.**	
Stylocheilus longicauda (Quoy & Gaimard, 1824)	Circumtropical
Stylocheilus striatus (Quoy & Gaimard, 1832)	Indo-Pacific
Syphonota geographica (Adams & Reeve, 1850)	Circumtropical
SACOGLOSSA	
Ascobulla sp.	????
Costasiella kuroshimae Ichikawa, 1993	Indo-West Pacific
Costasiella cf. *kuroshimae* Ichikawa, 1993	Indo-West Pacific
Costasiella cf. *usagi* Ichikawa, 1993	W Pacific Ocean
Costasiella sp. 1	????
Costasiella sp. 2	Indo-West Pacific
Costasiella sp. 3	W Pacific Ocean
***Cyerce* sp. nov., cf. *elegans* Bergh, 1870**	
Cyerce sp. nov.	W Pacific Ocean
Elysia mercieri Pruvot-Fol, 1930	W Pacific Ocean + Arabian/Persian Gulf + Gulf of Oman
Elysia pusilla (Bergh, 1872)	Indo-West Pacific
Elysia tomentosa Jensen, 1997 complex	Indo-Pacific (Lessepsian migrant)
***Elysia trilobata* Heller & Thompson, 1983**	
Elysia cf. *thompsoni* Jensen, 1993	Indo-West Pacific
***Elysia* sp. nov. 1, *grandifolia-marginata* complex**	Arabian/Persian Gulf + Gulf of Oman (Lessepsian migrant)
***Elysia* sp. nov. 2**	
Elysia sp. 3	????
Elysia sp. 4 (on *Tydemannia*)	Pacific Ocean
***Elysia* sp. 34 of Gosliner et al. 2015**	
Julia exquisita Gould, 1862	Indo-Pacific
Julia zebra Kawaguchi, 1981	Indo-Pacific
Lobiger viridis Pease, 1863	Indo-West Pacific (Lessepsian migrant)
Oxynoe jordani Krug et al., 2018	Indo-West Pacific (Lessepsian migrant)
Panderevela sp.	Indian Ocean
Placida cremoniana (Trinchese, 1892)	Mediterranean (+ circumtropical)

(Continued)

TABLE 13.1 (Continued)

Plakobranchus ocellatus van Hasselt, 1824	Indo-Pacific
Plakobranchus cf. *ocellatus* van Hasselt, 1824	W Pacific Ocean
Plakobranchus sp.	????
Polybranchia burni Medrano et al., 2018	Indo-Pacific
Polybranchia jensenae Medrano et al., 2018	Indo-Pacific
Polybranchia rubicunda (Bergh, 1871)	Indo-West Pacific
Polybranchia samanthae Medrano et al., 2018	Indo-Pacific (Lessepsian migrant)
***Stiliger ornatus* Ehrenberg, 1828**	
***Thuridilla decorata* (Heller & Thompson, 1983)**	Gulf of Oman
Thuridilla cf. *indopacifica* Gosliner, 1995	W Indian Ocean
***Thuridilla* cf. *albopustulosa* Gosliner, 1995**	
PLEUROBRANCHIDA	
Berthella martensi (Pilsbry, 1896)	Indo-Pacific
Berthella stellata (Risso, 1826)	Mediterranean (+ circumtropical)
Berthella sp.	????
***Berthellina citrina* (Rüppell & Leuckart, 1828)**	Lessepsian migrant
Berthellina delicata (Pease, 1861)	Indo-Pacific
Euselenops luniceps (Cuvier, 1817)	Indo-Pacific
Pleurobranchus albiguttatus (Bergh, 1905)	Indo-Pacific
Pleurobranchus forskalii Rüppell & Leuckart, 1828	Indo-Pacific (Lessepsian migrant)
Pleurobranchus grandis Pease, 1868	Pacific Ocean
***Pleurobranchus* cf. *grandis* Pease, 1868**	
Pleurobranchus mamillatus Quoy & Gaimard, 1832	Indo-West Pacific
SYSTELLOMMATOPHORA	
***Peronia anomala* Labbé, 1934**	****
Peronia verruculata (Cuvier, 1830)	Indo-Pacific
***Onchidium durum* Labbé, 1934**	****
***Paraperonia gondwanae* Labbé, 1934**	****
***Paraperonia jousseaumei* Labbé, 1934**	****
NUDIBRANCHIA	
Doridina – Phanerobranchs	
Aegires citrinus Pruvot-Fol, 1930	Indo-Pacific
Aegires villosus Farran, 1905	Indo-West Pacific
***Aegires* sp. nov.**	
Goniodoridella savignyi Pruvot-Fol, 1933	Indo-West Pacific
Gymnodoris alba (Bergh, 1877)	Pacific Ocean
Gymnodoris ceylonica (Kelaart, 1858)	Indo-West Pacific
Gymnodoris citrina (Bergh, 1877)	Indo-Pacific
Gymnodoris impudica (Rüppell & Leuckart, 1828)	Indo-Pacific
Gymnodoris inornata (Bergh, 1880)	Indo-West Pacific
Gymnodoris cf. *nigricolor* Baba, 1960	W Pacific Ocean
Gymnodoris okinawae Baba, 1936	Indo-Pacific
Gymnodoris striata (Eliot, 1908)	Indo-West Pacific
Gymnodoris sp. nov.	Indo-West Pacific
Gymnodoris sp. 1	Indo-West Pacific
Gymnodoris sp. 2	????
***Gymnodoris* spp. nov. 3, 4, 5**	
***Hexabranchus sanguineus* (Rüppell & Leuckart, 1830)**	
Hexabranchus giganteus Tibiriçá et al., 2023	Indo-Pacific
Kaloplocamus ramosus (Cantraine, 1835)	Mediterranean (Indo-Pacific + E Atlantic)
Martadoris limaciformis (Eliot, 1908)	Indo-West Pacific
***Nembrotha megalocera* Yonow, 1990**	
Notodoris citrina Bergh, 1875	Indo-West Pacific
Okenia rhinorma Rudman, 2007	Indo-West Pacific
Pelagella castanea (Alder & Hancock, 1845)	Circumglobal

TABLE 13.1 (Continued)

Pelagella joubini (Risbec, 1928)	Pacific Ocean
Plocamopherus indicus Bergh, 1890	Indo-West Pacific
***Plocamopherus ocellatus* Rüppell & Leuckart, 1828**	Arabian/Persian Gulf + Gulf of Oman (Lessepsian migrant)
Plocamopherus tilesii Bergh, 1877	W Pacific Ocean
Polycera melanosticta Miller, 1996	Indo-West Pacific
Polycera risbeci Odhner, 1941	Indo-West Pacific
Trapania euryeia Gosliner & Fahey, 2008	Indo-Pacific
Trapania palmula Gosliner & Fahey, 2008	W Pacific Ocean
Trapania toddi Rudman, 1987	W Pacific Ocean
Trapania lemnalioides Smirnoff et al., 2022	W Pacific Ocean
Tambja affinis (Eliot, 1904)	W Indian Ocean
Thecacera pacifica (Bergh, 1883)	Indo-Pacific
Doridina – Cryptobranchs	
***Actinocyclus sordida* (Rüppell & Leuckart, 1830)**	****
Actinocyclus verrucosus Ehrenberg, 1831	Indo-Pacific
Aldisa cf. *zavorensis* Tibiriçá et al., 2017	W Indian Ocean
Aldisa sp.	W Indian Ocean
Alloiodoris sp.	????
***Archidoris vayssierea* O'Donoghue, 1929**	****
Archidoris sp.	????
***Artachaea clavata* Eliot, 1907**	****
***Artachaea verrucosa* Eliot, 1908**	****
Asteronotus cespitosus (van Hasselt, 1824)	Indo-Pacific
***Asteronotus denticulata* (Eliot, 1908)**	****
***Asteronotus markaensis* Donohoo & Gosliner, 2020**	
Asteronotus mimeticus Gosliner & Valdés, 2002	W Pacific Ocean
***Asteronotus namuro* Donohoo & Gosliner, 2020**	
Asteronotus spongicolus Gosliner & Valdés, 2002	Indo-West Pacific
Atagema ornata (Ehrenberg, 1831)	Indo-Pacific
Atagema spongiosa (Kelaart, 1858)	Indo-West Pacific
***Atagema* spp. 1, 2**	
Atagema sp. 3	????
Carminodoris estrelyado (Gosliner & Behrens, 1998)	Indo-West Pacific
Carminodoris flammea (Fahey & Gosliner, 2003)	W Pacific Ocean
Carminodoris grandiflora (Pease, 1860)	Indo-Pacific
Carminodoris pustulata (Abraham, 1877)	Indo-West Pacific
Carminodoris cf. *armata* Baba, 1993	W Pacific Ocean
Discodoris cebuensis Bergh, 1877	Indo-Pacific
Discodoris coerulescens (Bergh, 1888)	Indo-West Pacific
Discodoris cf. *coerulescens* (Bergh, 1888)	Indo-West Pacific
Doriopsis granulosa Pease, 1860	Indo-Pacific
Doris immonda (Risbec, 1928)	Indo-Pacific
Doris spp. 1, 2, 3, 4, 5	????
***Halgerda biqiea* Donohoo & Gosliner, 2023**	
***Halgerda paulayi* Donohoo & Gosliner, 2023**	
Hallaxa indecora (Bergh, 1905)	Indo-West Pacific
***Hallaxa* sp.**	
Hallaxa/Geitodoris sp.	????
Jorunna funebris (Kelaart, 1858)	Indo-West Pacific
Jorunna labialis (Eliot, 1908)	W Indian Ocean
Jorunna parva (Baba, 1938)	Indo-West Pacific
Jorunna cf. *parva* (Baba, 1938)	Indo-West Pacific
Jorunna sp. 1	Indo-West Pacific
Jorunna sp. 2	????
Jorunna/Rostanga sp.	????

(Continued)

TABLE 13.1 (Continued)

***Paradoris erythraeensis* (Vayssière, 1912)**	Arabian/Persian Gulf
***Paradoris hypocrita* Yonow, 2020**	
***Paradoris* spp. nov. 1, 2**	
Peltodoris murrea (Abraham, 1877)	Indo-West Pacific
Peltodoris rubra Bergh, 1905	Indian Ocean
Platydoris inframaculata (Abraham, 1877)	Indo-West Pacific
Platydoris pulchra Eliot, 1904	Indian Ocean
Platydoris scabra (Cuvier, 1804)	Indo-West Pacific
Platydoris striata (Kelaart, 1858)	W Indian Ocean
Rostanga lutescens (Bergh, 1905)	Indo-Pacific
Rostanga sp.	????
Sclerodoris apiculata (Alder & Hancock, 1864)	Indo-West Pacific
Sclerodoris tuberculata Eliot, 1904	Indo-West Pacific
***Sclerodoris rugosa* (Vayssière, 1912)**	****
Sclerodoris cf. *rugosa* (Vayssière, 1912)	????
Sclerodoris spp. 1, 2, 3	????
Sebadoris fragilis (Alder & Hancock, 1864)	Indo-Pacific (Lessepsian migrant)
Sebadoris nubilosa (Pease, 1871)	Indo-Pacific
***Spongiodoris* sp.**	
Thordisa tahala Chan & Gosliner, 2007	Indo-West Pacific
Ardeadoris electra (Rudman, 1990)	W Pacific Ocean
Ardeadoris egretta Rudman, 1984	Indo-West Pacific
Ardeadoris symmetrica (Rudman, 1990)	Indo-West Pacific
Cadlinella ornatissima (Risbec, 1928)	Indo-West Pacific
Cadlinella subornatissima Baba, 1996	W Pacific Ocean
***Cadlinella* sp. nov.**	
Ceratosoma tenue Abraham, 1876	Indo-Pacific
Ceratosoma trilobatum (Gray, 1827)	Indo-West Pacific
Chromodoris africana Eliot, 1904	W Indian Ocean
***Chromodoris baqe* Bonomo & Gosliner, 2020**	
Chromodoris quadricolor (Rüppell & Leuckart, 1830)	W Indian Ocean (Lessepsian migrant)
Chromodoris strigata Rudman, 1982	Indo-West Pacific
Diversidoris aurantionodulosa Rudman, 1987	Indo-West Pacific
Diversidoris flava (Eliot, 1904)	Indo-Pacific
Doriprismatica atromarginata (Cuvier, 1804)	Indo-Pacific
***Doriprismatica kyanomarginata* Yonow, 2018**	Arabian/Persian Gulf
Doriprismatica plumbea (Pagenstecher, 1877)	W Indian Ocean
***Doriprismatica rossi* Matsuda & Gosliner, 2018**	
***Glossodoris andersonae* Matsuda & Gosliner, 2018**	
Glossodoris cincta (Bergh, 1888) complex	Indo-West Pacific
Glossodoris hikuerensis (Pruvot-Fol, 1954)	Indo-West Pacific
***Glossodoris kahlbrocki* Yonow, 2018**	
Glossodoris pallida (Rüppell & Leuckart, 1830)	W Indian Ocean
***Glossodoris* cf. rufomarginata (Bergh, 1890)**	(****)
Goniobranchus annulatus (Eliot, 1904)	W Indian Ocean (Lessepsian migrant)
***Goniobranchus charlottae* (Schrödl, 1999)**	
Goniobranchus fidelis (Kelaart, 1858)	Indo-West Pacific
Goniobranchus geminus (Rudman, 1987)	Indian Ocean
Goniodoris hintuanensis (Gosliner & Behrens, 1998)	W Pacific Ocean
***Goniobranchus obsoletus* (Rüppell & Leuckart, 1830)**	Arabian/Persian Gulf + Gulf of Oman (Lessepsian migrant)
***Goniobranchus pseudodecorus* Yonow, 2018**	Arabian/Persian Gulf + Gulf of Oman
Goniobranchus tennentanus (Kelaart, 1859)	W Indian Ocean
Goniobranchus tinctorius (Rüppell & Leuckart, 1830)	W Indian Ocean
Goniobranchus verrieri (Crosse, 1875)	Indo-West Pacific

TABLE 13.1 (Continued)

***Goniobranchus* cf. *verrieri* (Crosse, 1875)**	Gulf of Oman
***Hypselodoris alburtuqali* Gosliner & Johnson, 2018**	
Hypselodoris dollfusi (Pruvot-Fol, 1833)	NW Indian Ocean
***Hypselodoris ghardaqana* (Gohar & Aboul-Ela, 1957)**	Gulf of Oman
Hypselodoris infucata (Rüppell & Leuckart, 1830)	Indo-Pacific (Lessepsian migrant)
Hypselodoris maridadilus Rudman, 1977	Indo-Pacific
Hypselodoris nigrostriata (Eliot, 1904)	W Indian Ocean
Hypselodoris pulchella (Rüppell & Leuckart, 1830)	W Indian Ocean
Hypselodoris yarae Gosliner & Johnson, 2018	W Indian Ocean
***Hypselodoris* sp. nov.**	
Mexichromis katalexis Yonow, 2001	Indo-West Pacific
Mexichromis lemniscata (Quoy & Gaimard, 1832)	Indo-West Pacific
Mexichromis pusilla (Bergh, 1874)	Indo-West Pacific
Miamira magnifica Eliot, 1910	W Indian Ocean
Miamira miamirana (Bergh, 1875)	Indo-West Pacific
Thorunna africana Rudman, 1984	W Indian Ocean
Verconia alboannulata (Rudman, 1986)	W Pacific Ocean
Verconia norba (Er. Marcus & Ev. Marcus, 1970)	Indo-West Pacific
Verconia simplex (Pease, 1871)	Pacific Ocean
***Verconia sudanica* (Rudman, 1985)**	
Verconia sp.	????
Doridina – Phyllidioidea (= Porostomata)	
Dendrodoris albopurpura Burn, 1957	Indo-Pacific
Dendrodoris coronata Kay & Young, 1969	Pacific Ocean
***Dendrodoris cuprea* (Ehrenberg, 1831)**	****
Dendrodoris elongata Baba, 1936	Indo-Pacific
Dendrodoris fumata (Rüppell & Leuckart, 1830)	Indo-Pacific
***Dendrodoris immaculata* (Audouin, 1826)**	****
***Dendrodoris jousseaumei* (Vayssière, 1912)**	****
Dendrodoris krusensternii (Gray, 1850)	Indo-Pacific
***Dendrodoris leptopus* Ehrenberg, 1831**	****
***Dendrodoris lugubris* Ehrenberg, 1831**	****
Dendrodoris nigra (Stimpson, 1955)	Indo-Pacific
***Dendrodoris nigropunctata* (Vayssière, 1912)**	****
***Dendrodoris tigrina* (Audouin, 1826)**	****
Dendrodoris tuberculosa (Quoy & Gaimard, 1832)	Indo-Pacific
Doriopsilla sp.	????
***Phyllidia monacha* Yonow, 1986**	
***Phyllidia multifaria* Yonow, 1986**	
***Phyllidia rueppelii* (Bergh, 1869)**	Arabian/Persian Gulf + Gulf of Oman
***Phyllidia schupporum* Fahrner & Schrödl, 2000**	Arabian/Persian Gulf + Gulf of Oman
'*Phyllidia*' *sudanensis* Heller & Thompson, 1983	****
Phyllidia undula Yonow, 1986	W Indian Ocean
Phyllidia varicosa Lamarck, 1801	Indo-West Pacific
Phyllidia sp.	????
***Phyllidiella amphitrite* Yonow, 2020**	
***Phyllidiella melanocera* Yonow, 1986**	
Phyllidiella pustulosa (Cuvier, 1804) complex	Indo-West Pacific
Phyllidiella zeylanica (Kelaart, 1858)	W Indian Ocean
Phyllidiopsis cardinalis Bergh, 1873	Indo-West Pacific
***Phyllidiopsis dautzenbergi* (Vayssière, 1912)**	Arabian/Persian Gulf
***Phyllidiopsis sinaiensis* (Yonow, 1988)**	Arabian/Persian Gulf
***Phyllidiopsis* spp. nov. 1, 2**	
Dendronotoidea	
***Marionia glama* (Rüppell & Leuckart, 1828)**	
***Marionia rubra* (Rüppell & Leuckart, 1828)**	

(Continued)

TABLE 13.1 (Continued)

Marionia sp.	????
***Marioniopsis cyanobranchiata* (Rüppell & Leuckart, 1828)**	
Marioniopsis levis (Eliot, 1904)	W Indian Ocean
Marioniopsis viridescens Eliot, 1904	W Indian Ocean
Tritonia bollandi Smith & Gosliner, 2003	W Pacific Ocean + Arabian/Persian Gulf
***Tritonia* sp. nov.**	
Tritoniopsis elegans (Audouin, 1826)	Indo-West Pacific
Tritoniopsis cf. *frydis* Er. Marcus & Ev. Marcus, 1970	Indo-West Pacific
Tritoniopsis gravieri Vayssière, 1912	Indo-West Pacific
Tritoniopsis sp.	Indo-West Pacific
Tritoniid gen. et sp. nov.	
Bornella stellifera (Adams & Reeve in Adams, 1848)	Indo-Pacific
Bornella cf. *johnsonorum* Pola et al., 2009	Indo-Pacific
***Bornellopsis kabretiana* O'Donoghue, 1929**	****
***Doto* cf. *kya* Er. Marcus, 1961**	Arabian/Persian Gulf
***Kabeiro orcha* (Yonow, 2000)**	
Kabeiro rubroreticulata Shipman & Gosliner, 2015	W Pacific Ocean + Gulf of Oman
Lomanotus vermiformis Eliot, 1908	Pacific Ocean
***Lomanotus* sp. nov. 1**	
Lomanotus spp. 2, 3	W Pacific Ocean
Scyllaea fulva Quoy & Gaimard, 1824	Indo-West Pacific
Crosslandia viridis Eliot, 1902	Indo-West Pacific
Notobryon wardi Odhner, 1936	Circumtropical
Melibe bucephala Bergh, 1902	Indo-West Pacific
Melibe engeli Risbec, 1937	Indo-Pacific
Melibe viridis (Kelaart, 1858)	Indo-West Pacific (Lessepsian migrant)
Arminoidea	
Armina magna Baba, 1955	W Pacific Ocean
Armina semperi (Bergh, 1866)	W Pacific Ocean
***Armina* sp. nov.**	
Dermatobranchus albus (Eliot, 1904)	W Indian Ocean
Dermatobranchus caeruleomaculatus Gosliner & Fahey, 2011	Indo-West Pacific
Dermatobranchus dendronephthyphagus Gosliner & Fahey, 2011	W Pacific Ocean
Dermatobranchus cf. *fortunatus* (Bergh, 1888)	Indo-West Pacific
***Dermatobranchus glabrus* (Eliot, 1908)**	****
Dermatobranchus gonatophorus van Hasselt, 1824	W Pacific Ocean
Dermatobranchus cf. *leoni* Gosliner & Fahey, 2011	W Pacific Ocean
Dermatobranchus ornatus (Bergh, 1874)	W Pacific Ocean
Dermatobranchus pustulosus van Hasselt, 1824	Indo-West Pacific
Dermatobranchus striatus van Hasselt, 1824	W Pacific Ocean
Dermatobranchus tuberculatus Gosliner & Fahey, 2011	W Pacific Ocean + Gulf of Oman
Dermatobranchus spp. 1, 2, 3, 4, 5, 6	????
Madrella ferruginosa Alder & Hancock, 1864	Indo-West Pacific
Madrella gloriosa Baba, 1949	Indo-West Pacific
Janolus flavoannulatus Pola & Gosliner, 2019	Indo-West Pacific
Janolus mirabilis Baba & Abe, 1970	Indo-West Pacific
Janolus savinkini Martynov & Korshunova, 2012	Indo-West Pacific
Janolus sp. 1	Indo-West Pacific
Janolus sp. 2	????
Aeolidioidea	
***Aeolidia habessinica* Ehrenberg, 1828**	****
Aeolidiopsis cf. *ransoni* Pruvot-Fol, 1956	W Pacific Ocean
Anteaeolidiella indica (Bergh, 1888)	Circumtropical (Lessepsian migrant)
Baeolidia moebii Bergh, 1888	W Pacific Ocean + Gulf of Oman (Lessepsian migrant)

TABLE 13.1 (Continued)

Bulbaeolidia alba (Risbec, 1928)	Indo-Pacific (Lessepsian migrant)
Caloria indica (Bergh, 1896)	Indo-West Pacific (Lessepsian migrant)
***Caloria* sp. nov.**	
***Catriona susa* Ev. Marcus & Er. Marcus, 1960**	****
Cerberilla annulata (Quoy & Gaimard, 1832)	Indo-West Pacific
Cerberilla sp.	W Pacific Ocean
***Coryphellina rubrolineata* O'Donoghue, 1929**	Arabian/Persian Gulf + Gulf of Oman (Lessepsian migrant)
Cratena aff. *affinis* (Baba, 1949)	Indo-West Pacific
Embletonia gracilis Risbec, 1928	Indo-Pacific
Eubranchus ocellatus (Alder & Hancock, 1864)	W Indian Ocean
Eubranchus rubropunctatus Edmunds, 1969	Indo-West Pacific
***Eubranchus* spp. nov. 1, 2**	
Facelina lineata Eliot, 1905	W Indian Ocean
***Facelina* spp. nov. 1, 2, 3**	
'*Facelina*' cf. *rhodopos* Yonow, 2000	
Favorinus japonicus Baba, 1949	Indo-Pacific
Favorinus mirabilis Baba, 1955	Indo-West Pacific
Favorinus tsuruganus Baba & Abe, 1964	W Pacific Ocean
Glaucilla marginata Reinhardt & Bergh, 1868	Circumtropical
Herviella albida Baba, 1966	Indo-West Pacific
Limenandra barnosii Carmona et al., 2013	W Pacific Ocean + Gulf of Oman
Limenandra confusa Carmona et al., 2013	Indo-Pacific
Limenandra fusiformis (Baba, 1949)	W Pacific Ocean + Gulf of Oman
Limenandra nodosa Haefelfinger & Stamm, 1959	Mediterranean (+ circumtropical)
Limenandra rosanae Carmona et al., 2013	Pacific Ocean
Phestilla goniophaga Hu et al., 2020	W Pacific Ocean + Gulf of Oman
Phestilla lugubris (Bergh, 1870)	Indo-Pacific
Phestilla melanobrachia Bergh, 1874	Indo-Pacific
Phidiana anulifera (Baba, 1949)	Indo-West Pacific
Phidiana bouraili (Risbec, 1928)	Indo-West Pacific
Phyllodesmium colemani Rudman, 1991	W Pacific Ocean
Phyllodesmium crypticum Rudman, 1981	Indo-West Pacific
***Phyllodesmium fastuosum* Ehrenberg, 1831**	****
***Phyllodesmium hyalinum* Ehrenberg, 1831**	
Phyllodesmium koehleri Burghardt et al., 2008	W Indian Ocean
Phyllodesmium magnum Rudman, 1991	Indo-West Pacific
Pleurolidia juliae Burn, 1966	Indo-West Pacific
Protaeolidiella atra Baba, 1955	Pacific Ocean
Pruvotfolia rhodopos (Yonow, 2000)	W Indian Ocean
Pteraeolidia semperi (Bergh, 1870) complex	Indo-Pacific
***Pteraeolidia* sp. nov. 1**	
Pteraeolidia sp. nov. 2	????
Samla bicolor (Kelaart, 1858)	Indo-Pacific
Samla bilas (Gosliner & Willan, 1991)	W Pacific Ocean
'*Samla*' sp. nov.	
Spurilla braziliana MacFarland, 1909	Circumtropical
Trinchesia sibogae (Bergh, 1905)	Indo-West Pacific
Trinchesia yamasui (Hamatani, 1993)	Indo-West Pacific
***Trinchesia* spp. 1, 2, 3, 4, 5**	
Trinchesia spp. 6, 7, 8, 9	Indo-West Pacific
Trinchesia sp. 10	W Indian Ocean

13.3 APLYSIIDA

Within the Aplysiida, several genera have received attention in recent years. The species complex of *Aplysia parvula* (e.g., *A. parvula* and *Aplysia* sp. in Yonow 2008b, pp. 98–100) was more complicated and resolved into several species, some with narrow distributions and others with broad distributions by Golestani et al. (2019). Of the two species occurring in the Red Sea, the pale form was described as a new species, *A. ghanimii*, having an Atlantic and Western Indian Ocean distribution, very unusual for heterobranchs. The dark form was found to have an old available name, *A. nigrocincta*, and is found throughout the Indo-West Pacific.

Bursatella leachii in all its forms is now shown to be a single, circumtropical, genetically identical species with morphological variations, and *Stylocheilus striatus* is now restricted to the Indo-Pacific; it is unfortunate that Bazzicalupo et al. (2020) could not include specimens of the pelagic yellow '*Stylocheilus longicauda*' in their study, which has been considered the same species in the past. Finally, the rare *Syphonota geographica* has been photographed several times in the last decade, and an image is included in this work as none was available when Yonow (2008, 2015) were published (Figure 13.4).

13.4 SACOGLOSSA

Amazingly, the list of sacoglossans has expanded from 16 species (three endemic species; five species without names) to 36 species (including eight endemic species and 18 without names). This is mostly due to macro-photography of small animals, such as the tiny species of *Costasiella* (Figures 13.5 to 13.8) and a species of *Panderevela* sp. (K.R. Jensen, pers. comm.; Figure 13.9);

FIGURE 13.1 *Philinopsis* sp. 2 (photograph by Hsini Lin). The pale zone in front of the eyes is reminiscent of *Mariaglaja tsurugensis*, but that species does not have orange and blue marks on the front of the head shield, not the blue edging to the parapodia.

FIGURE 13.2 *Philinopsis* sp. 1 (photograph by R. Amar) showing the short tails and slightly raised yellow/orange spots on the 'face' that distinguish it from *C. punctata*.

FIGURE 13.3 *Bullina* sp. (photograph by A. Ryanskiy) is similar to one recorded in La Réunion (https://www.facebook.com/photo/?fbid=647360719945422&set=pcb.3143988612547622), and the first record of a living animal.

FIGURE 13.4 *Syphonota geographica* (photograph by R. Amar) has been very rarely recorded in the Red Sea and, to my knowledge, never illustrated from the Red Sea with a photograph.

FIGURE 13.5 *Costasiella* cf. *kuroshimae* (photograph by Hsini Lin) differs from the Japanese species that was poorly described. In fact, several other species have been recorded in the Red Sea that also cannot be identified further.

FIGURE 13.7 *Costasiella* sp. 3 (photograph by A. Ryanskiy) appears similar to an unnamed species that is rare in the western Indian Ocean (https://www.facebook.com/photo/?fbid=2577618139011468&set=a.652464771526824) and more common in the western Pacific.

FIGURE 13.6 *Costasiella* sp. 2 (photograph by Hsini Lin).

FIGURE 13.8 *Costasiella* sp. 1 (photograph by Hsini Lin).

this is the first time this latter genus has been recorded in the Red Sea.

Molecular work (Krug et al. 2018) has identified the *Oxynoe viridis* (Pease, 1861) illustrated by Yonow (2008b: p. 118) as a new species, *O. jordani*, that has migrated into the Mediterranean. Of the colourful genus *Thuridilla*, an apparently endemic species has been photographed from Eilat; it resembles *Th. albopustulosa* from the Indo-West Pacific but with significant colour differences regardless of its size (Figure 13.10). Despite the numbers of commonly recorded but unidentified species of *Elysia* in the Red Sea (e.g., Figure 13.11), work has just begun on the Indo-Pacific species, and many endemic species remain unnamed. Of the widely distributed *Elysia*, *E. ornata* (Swainson, 1840) was separated from the *E. grandifolia-marginata* complex and restricted to the Caribbean; the complex currently contains four Indo-Pacific candidate species based on molecular analysis of Pacific specimens only (Krug et al. 2013). For now, the endemic Red Sea species is called *Elysia* sp. *grandifolia-marginata* complex

FIGURE 13.9 *Panderevela* sp. (photograph by Hsini Lin). The genus, only defined in 2015, is related to *Costasiella* and currently contains only four species; this is the third species from the Indo-Pacific, with the first (similar to this one) described from Thailand in 2021.

FIGURE 13.10 *Thuridilla* cf. *albopustulosa* (photograph by R. Amar) is only the third species of *Thuridilla* to be recorded in the Red Sea.

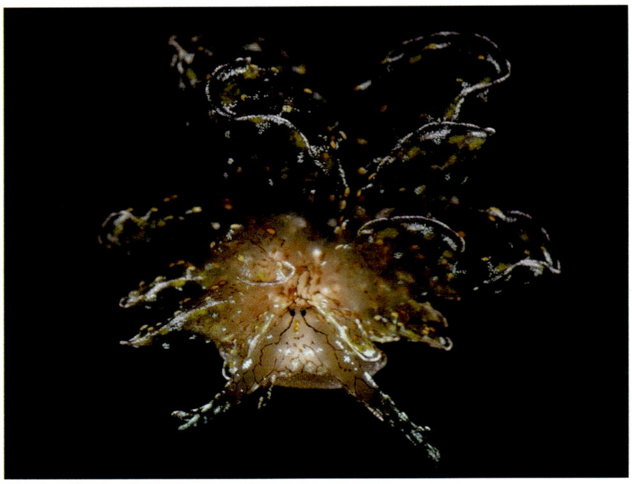

FIGURE 13.12 *Cyerce* sp. nov. (photograph by S. Kahlbrock) is one of two new species occurring in the Red Sea. Although *C.* cf. *elegans* is endemic, this species is recorded from the Philippines.

FIGURE 13.11 *Elysia* cf. *thompsoni* (photograph by A. Ryanskiy) is one of the many species of un-identified *Elysia* in the Red Sea. Whilst it is tempting to match them to known species, some are certainly new and endemic species.

FIGURE 13.13 *Plakobranchus* cf. *ocellatus* (photograph by R. Amar) clearly showing the head, which is lacking the row of ocellated spots typical of *P. ocellatus*.

as the name has bearing for studies on the migrant species recorded in the Mediterranean, and it was illustrated by Yonow (2008b: pp. 120–121) as *Elysia* sp. 1.

The genus *Cyerce* has been subject to molecular work and included Red Sea material: Morena (2020) showed that *Cyerce elegans* is a complex of species and that the Red Sea species is genetically different from the nominal Indo-Pacific type species. At present, this Red Sea species is presumed endemic in the checklist table. Surprisingly, three records of another unnamed species were photographed in the Red Sea in the spring of 2022 (Figure 13.12), but this species is also known from the Philippines (www.medslugs.de/E/Pac-W/Cyerce_sp_03/Selection.htm).

The genus *Plakobranchus* is clearly a complex of species (e.g., Yonow and Jensen 2018), but it was with great surprise that a species differing from *P. ocellatus* was photographed from Eilat in the last few years (Figure 13.13). It does not bear the row of large ocellated spots along the front of the head that are diagnostic of *P. ocellatus*, and closely resembles *P.* cf. *ocellatus* from Ambon, Indonesia, with paler spots, comprising pinkish or yellowish centres and brownish or rust-coloured rings, and violet-tipped rhinophores, described and illustrated in Yonow and Jensen (2018). As a result of this observation, a search through my photographic archives revealed two records of this as well as a third species, photographed by E. Pridgen off Jeddah, Saudi Arabia, in the 1980s, possibly as early as the 1970s (Figure 13.14). This third species was again photographed by A. Ryanskiy in Egypt in March 2023 (https://www.facebook.com/photo/?fbid=10221751648252757&set=gm.995557008492234&idorvanity=982126079835327).

FIGURE 13.14 *Plakobranchus* sp. (photograph by E. Pridgen) is the third *Plakobranchus* species recorded from the Red Sea, with slightly raised white bumps, many with a brown spot, especially along the edges of the parapodia. The rhinophores, edges of oral veil, and tail all have a violet margin.

FIGURE 13.15 *Berthellina delicata* (photograph by E. Pridgen) has subtly different colouration to the other species recorded in the Red Sea.

13.5 PLEUROBRANCHIDA

Few changes have occurred in this order for the Red Sea; taxonomically, the large *Pleurobranchus* (Yonow 2000: pp. 98–99 as *P. forskalii*; Yonow 2008b: p. 115) remains without an identification. However, the problematic situation of *Berthellina* was partially resolved by Moustafa et al. (2018) in the first molecular study to specifically examine Red Sea material. I originally thought *Berthellina saidensis* (O'Donoghue, 1929) was endemic, but the type locality is Port Said, located on the Mediterranean coastline of Egypt and not the Red Sea, and these authors correctly synonymised it with the Mediterranean/Atlantic species *B. edwardsi* Vayssière, 1897; it is hereby removed from the Red Sea species list. They also recognised *Berthellina citrina* in the Red Sea as at least two species, *B. citrina* (the common orange or yellow one) and *B. delicata* (the rarely recorded red species); the latter is illustrated for the first time in Figure 13.15. This image is the only individual of *B. delicata* in my archives, taken near Jeddah during 1970–1980. Blatterer (2019: p. 363, pl. 189, fig. 3) noted the similarity of his ~ 2 mm shells with those of *Berthellina delicata* and not with those of *B. citrina*, but he found no living material of *B. delicata*. He did illustrate the two colour forms of *B. citrina* but had no shells, and it is interesting that only the shells of *B. delicata* persevered in sediments. Moustafa et al. (2018) additionally suggested that *B. citrina* may not be as geographically widespread as previously thought, and that the identity of the specimens outside the Red Sea needs confirmation; hence, the species is now recorded as endemic in the Red Sea checklist table until further data from other locations prove otherwise.

Berthella sp. is recorded for the first time from the Red Sea (Figure 13.16). It has great similarities to both *B. caledonica* (Risbec, 1928) from the western Pacific and *B. africana* (Pruvot-Fol, 1953) from the Atlantic coasts of north Africa. It may be that the two species are in fact the same, but until collections from both regions are made and compared, the debate continues.

FIGURE 13.16 *Berthella* sp. (photograph by S. Kahlbrock) is recorded for the first time in the Red Sea.

13.5.1 Nudibranchia – Doridina

There are no recent publications involving Red Sea species of phanerobranch dorids at this point, somewhat surprisingly. However, there are two photographic records in Eilat of the distinctive *Polycera risbeci* (Figure 13.17), both found on *Padina* (one of two genera of calcified brown macroalgae). This is a new record for the Red Sea, a tiny species with a wide Indo-West Pacific distribution (pers. obs.). Another new record is *Trapania lemnalioides* (Figure 13.18) with a single photograph, and originally described from the Philippines. The most speciose genus in the Red Sea is *Gymnodoris*, with at least 14 species recorded, of which five remain unnamed and three appear endemic. This large Indo-Pacific genus is in dire need of revision, including the establishment of diagnostic morphological characters. Even the commonly recorded species can be problematic to identify, and some are certainly species complexes.

FIGURE 13.17 *Polycera risbeci* (photograph by R. Amar) was originally described from New Caledonia, and subsequent records were published from Australia, Singapore, and Indonesia.

FIGURE 13.18 *Trapania lemnalioides* (photograph by R. Amar) was recently named and described from the Philippines, and this is the first record since the type description.

FIGURE 13.19 *Glossodoris andersonae* (photograph by Hsini Lin), one of the three endemic species belonging to the *G. cincta* complex.

FIGURE 13.20 *Goniobranchus pseudodecorus* (photograph by S. Kahlbrock). This species has long been known and is fairly common in the Red Sea.

On the other hand, the cryptobranch dorids have received much attention in the years since the last update (Yonow 2015), especially the chromodorids, and several publications include new species from the Red Sea and tease out species pairs and complexes. As a result, many of the 'slightly different' Red Sea species have now been distinguished from their Indo-Pacific sibling species and received new names, all designated endemic species. *Glossodoris* has been split into two genera, *Glossodoris* and *Doriprismatica*, both occurring in the Red Sea, with the *cincta* complex of species spanning both genera. Three new and endemic species have been described in this group (Matsuda and Gosliner 2018; Yonow 2018; Figure 13.19), and *G. cincta* is redefined as a species complex in the Red Sea checklist and the wider Indo-Pacific region until the true *G. cincta* is recovered and the other species untangled. The Red Sea variant of *Chromodoris aspersa* (A. Gould, 1852) is now the endemic Red Sea *Chromodoris baqe* based on molecular and morphological differences (Bonomo and Gosliner 2020). Similarly, the colour variations of the Red Sea *Chromodoris* sp. (Yonow 2008b: p. 166) and *H. maculosa* (Pease, 1871) (Yonow 2008b: p. 195 CT photographs) are now two endemic species (Epstein et al. 2018; Yonow 2018) placed in two different genera despite very similar colour patterns, *Goniobranchus pseudodecorus* (Figure 13.20) and *Hypselodoris alburtuqali* (Figure 13.21), and *H. yarae* with a Western Indian Ocean distribution (Yonow 2008b: p. 195, non CT photographs).

A new species, probably a *Hypselodoris*, is recorded only from photographs of two individuals by different photographers in different regions of the Red Sea, and resembles nothing in the published literature or online (Figure 13.22). Three photographs of *Halgerda biqiea* (Figure 13.23) from the northern half of the Red Sea extend the range northwards of this recently described endemic species whose type locality is Djibouti (Donohoo et al. 2023). That paper also established that *H. willeyi* (Eliot, 1904) from the Red Sea is the new endemic species *H. paulayi*. Finally, there is a single photographic record of *Cadlinella subornatissima* (Figure 13.24), a species only recorded from the Western Pacific.

FIGURE 13.21 *Hypselodoris alburtuqali* (photograph by S. Kahlbrock) is rare and was first recorded by the Suakin Expedition in the spring of 1980.

FIGURE 13.22 *Hypselodoris* sp. nov. (photograph by H. Sjoeholm), one of the two records of a new species that is most probably endemic to the Red Sea.

FIGURE 13.23 *Halgerda biqiea* (photograph by S. Kahlbrock) was originally thought to be a form of *H. brunneomaculata* (Carlson & Hoff, 1993) from the western Pacific but is in fact yet another endemic species.

FIGURE 13.24 *Cadlinella subornatissima* (photograph by B. & S. Koretz) was photographed once in the Gulf of Eilat in 2009 and has never been seen again. It was described from Japan, a considerable range extension, and is not established in the Red Sea.

13.5.2 Phyllidioidea (= Porostomata)

The taxonomy of dendrodorids continues to pose problems in the Indo-Pacific realm, with no recent work on these species. One paper was published in Farsi providing a molecular analysis of *Dendrodoris fumata* (Fatemi et al. 2021). Their tree included two specimens from the Persian Gulf and one specimen each from the Gulf of Oman, the Red Sea, and Australia, as well as many sequences of *D. rubra* (Kelaart, 1858) (without stating where these sequences extracted from GenBank originated). What is clear from their trees is that *D. fumata* from the Red Sea, Persian Gulf, and Gulf of Oman are the same species and this species is different from the Australian '*D. fumata*' and all other species recorded as '*D. rubra*'. More work is needed to ascertain that *D. fumata* (type locality Red Sea) is endemic in its native range.

13.5.3 Dendronotoidea

The dendronotids have thrown up two surprises: *Tritonia bollandi* was described from the western Pacific, assuming this record from the Red Sea is the same species (Figure 13.25). There are photographic records also from the APG (see Nithyanandan et al., Chapter 12, this volume), so it appears to be establishing itself in these seas, feeding on the same red soft coral, *Verrucella*. The other surprise is the first record of the genus *Doto* in the Red Sea, with an unidentified species (Figure 13.26) most similar to *D. kya* from the East Pacific; despite some similarities, pigmentation differences in combination with this highly improbable distribution render this identification unlikely, although the same species is also recorded from the APG by Nithyanandan et al. (Chapter 12, this volume). The genus *Doto* is understudied in the tropics, with only

FIGURE 13.25 *Tritonia bollandi* (photograph by Silke Baron) is also recorded from the APG. It is from the Western Pacific, like *Cadlinella subornatissima*, but appears to be recently well established in the Arabian seas.

FIGURE 13.26 *Doto* cf. *kya* (photograph by A. Ryanskiy) is new to both the Red Sea and the APG (see previous chapter) faunas, photographed during the last few years, and the only species of *Doto* occurring in both areas.

a limited number of very distinctive species currently named. For now, this species is considered endemic, simply because the only other Red Sea dotoid, *Kabeiro orcha*, is endemic. Whilst *K. orcha* is listed in WoRMS as *Doto*, Shipman and Gosliner (2015) hinted at the change of genus for this species, and its morphology is certainly more allied to *Kabeiro* than *Doto. Kabeiro rubroreticulata* is newly recorded from the Red Sea and APG (Nithyanandan et al., Chapter 12, this volume).

13.5.4 Arminoidea

A number of species have been photographed in recent years, of which more than half can be tentatively identified, and the species numbers recorded in the Red Sea have increased from 18 to 26, with the most speciose genus being *Dermatobranchus*. This genus has a rich fauna in the Indo-West Pacific tropical oceans, but many species remain unidentified. The first species illustrated here (Figure 13.27) resembles its known counterpart but with significant external differences and is listed as cf. in the species list; the second is probably correctly identified as *D. dendronephthyphagus* (Figure 13.28) and is also recorded from the Gulf of Oman; it is a west Pacific species. Until specimens are collected, however, we may never be sure of their identities. A further five species are numbered simply as spp. 1–5 (e.g., Figure 13.29), as they appear different from both named and unnamed species, so their distributions are listed as unknown. They have only been recorded once (or twice) each from Eilat, which is a large shipping port, so they may simply have arrived via ballast tanks.

13.5.5 Aeolidioidea

The final superfamily discussed in this work with many recent changes in taxonomy are the aeolids. Photographers

FIGURE 13.27 *Dermatobranchus* cf. *fortunatus* (photograph by R. Amar).

FIGURE 13.28 *Dermatobranchus dendronephthyphagus* (photograph by S. Kahlbrock).

FIGURE 13.29 *Dermatobranchus* sp. 3 (photograph by R. Amar). This is the second record of this distinctive species and both are from coarse sandy substrates in the Gulf of Eilat.

FIGURE 13.30 *Eubranchus* sp. nov. 1 (photograph by R. Amar), one of two unidentified endemic species photographed several times in the northern Red Sea.

have exceeded expectations and have added more than a dozen unidentified species to the list (e.g., Figures 13.30 to 13.33). As aeolids are predators that lay their eggs on, in, or near their prey, they may have arrived with these invasive prey items or as larvae in ballast tanks, but time will tell if they manage to establish viable breeding populations. Some of them appear similar to individuals photographed from other regions, so some distributions can be attributed, but the group still needs basic taxonomic work to identify species accurately.

One interesting story involves the aeolid commonly known as the dragon sea slug, *Pteraeolidia*. It was thought to be a single species with several morphs until a molecular study (Wilson and Burghardt 2015) based on 42 Pacific specimens (no Indian Ocean or Red Sea material) named only two species, the endemic Australian *P. ianthina* (Angas, 1864) and the widely distributed Indo-Pacific *P. semperi* complex. They supposed that *P. scolopendrella* Risbec, 1928 (type locality New Caledonia) was a synonym of *P. semperi* but left it as an available name for future studies. However, a fourth named species was described in the last century from the Saya de Malha Bank, NE Madagascar/SE Seychelles in the western Indian Ocean, *Pteraeolidia annulata* Eliot, 1910, which has been completely ignored in the literature except for a mention by Yorifuji et al. (2012): using the nuclear markers 18S and 5.8S rRNA genes, these authors found at least two different species in Japan, the Philippines, and Guam, with both species occurring sympatrically in the Ryukyu Islands and the Philippines. Eliot (1910) separated his western Indian Ocean species on the basis of radular differences (70 teeth in the radula of a 35–40 mm long specimen compared to 15–36 teeth in ≤ 160 mm *semperi* specimens and 12–27 teeth in ≤ 100 mm *ianthina*), but he describes 11 groups of cerata and two purple rings on the oral tentacles: the latter is a characteristic of the Australian *P. ianthina*, as *P. semperi* has more than three purple rings on its oral tentacles. It is worth noting that Marcus and Marcus (1960), in describing a preserved Red Sea specimen 25 mm long, mentioned only two purple bands on the oral tentacles (but 18 pairs of cerata). In this excellent photograph of an individual from near Jeddah, Saudi Arabia

FIGURE 13.31 *Trinchesia* sp. (photograph by Hsini Lin).

FIGURE 13.32 *Trinchesia* sp. (photograph by Hsini Lin).

FIGURE 13.33 *Trinchesia* sp. (photograph by A. Ryanskiy).

FIGURE 13.34 *Pteraeolidia 'semperi'* (photograph by E. Pridgen).

(Figure 13.34), there are clearly only two purple bands on the oral tentacles (diagnostic of *ianthina*), a white mark on the head (supposedly only occurring in *semperi*), and 23 pairs of ceratal clusters. Clearly there is much work to be done: a recent search of public sequences in BOLD and GenBank revealed 61 available sequences [accessed November 2022]. All are from Pacific localities except four, which are from Dahab, Egypt, in the Red Sea; there are no Indian Ocean sequences. A preliminary analysis of these sequences using the BOLD platform shows that there are at least six and maybe as many as ten species, and two different species in the Red Sea, with two specimens of each. It is interesting to note that the Red Sea material clusters separately from all the other branches, and it is probable that these two species are endemic.

13.6 DISCUSSION

It has been a valuable exercise revising and updating the Red Sea species list, from both taxonomic and biogeographic points of view. Many of the endemic species are found only in the Red Sea, while others have 'spread' to the nearby APG and Gulf of Oman, termed the native range of a Red Sea endemic species (e.g., Hoeksema and Yonow 2021). It is not known if this change in status (indicated in Table 13.1) is due to an increased expansion of distribution since the original species list 16 years ago (Yonow 2008a) or, an increased interest in heterobranch photography by divers and photographers living in these areas recording them. The Red Sea has had a very long history of sea slug research, beginning with two expeditions in the second half of the 18th century (see Yonow 2008b), while the earliest work known on the Persian Gulf is very recent (Smythe 1972), and there has been no baseline species list until now (see Nithyanandan et al., Chapter 12, this volume).

Bowen et al. (2016) identified the Red Sea as an 'endemism hotspot', with an interesting discussion on evolving ideas following recent research. Originally, peripheral endemism hotspots, such as the Red Sea and Hawaii, had been regarded as evolutionary dead ends, whereas the current thinking now is that, in fact, they can export novel biodiversity. DiBattista et al. (2015) concluded that the unique conditions in the Red Sea, in addition to characteristics of the Arabian Peninsula region as a whole, drove the divergence of populations through a combination of isolation and selection events. Interestingly, through their work on fishes, they found that "most phylogeographical studies of broadly distributed species indicate some level of genetic distinction of Red Sea populations." It seems that many endemic taxa (or lineages) diverged from their Indian Ocean counterparts long before the most recent glaciations, lending support to my early suggestions of high endemism in the Red Sea of sea slugs, despite their not having a strong fossil record.

Many of the 'old' endemic records in Yonow (2008a) have not been found again nor been validated by type specimens, but remain on the list (24 species, mostly dorids and including the onchidiids, indicated by four asterisks in the distribution column). The 'new' endemic species resulting from recent research comparing species complexes often confirmed the morphological differences observed by workers on Red Sea material in the 1980s through to the 2000s. The percentage of endemism today is little higher than it was 15 years ago, 24% in Yonow (2008b) compared to 27% today; 257 species were recorded by Yonow (2008b), but there are 442 species on the list today (excluding the five species of Systellommatophora that were not included in the first list)!

Wider geographical distributions have also changed, with a reduction from 31% to 23% of the species being found also in the Indo-West Pacific; for the western Indian Ocean, the percentage similarity has increased from 8% to 10%, confirming the idea of sister species. Note that these are simply comparisons of numbers, and they have not been statistically analysed. In Nithyanandan et al. (Chapter 12, this volume), data was collated for the APG, Gulf of Oman, the Arabian Sea, and the Red Sea, and similarity indices were calculated to provide further insights to the native range of Red Sea species, which are discussed therein.

ACKNOWLEDGEMENTS

I would like to sincerely thank the photographers for their generosity, new (Rafi Amar, Silke Baron, Hsini Lin, Andrey Ryanskiy) and 'old' (Sven Kahlbrock, B. & S. Koretz, Elwood Pridgen, Hans Sjoeholm). Hubert Blatterer provided interesting discussions while preparing his book *Mollusca of the Dahab Region*. I cannot thank Woody Pridgen enough for spending hours scanning his slides, providing a wonderful record of species that were present in the Jeddah area more than 50 years ago. In fact, just as proofs were being prepared, Woody scanned a photograph of Phyllidiella zeylanica taken in the 1980s, quashing my previous hypothesis that it was a recently introduced western Indian Ocean species (https://zookeys.pensoft.net/article/59732/). I am grateful to Dr. Argyro Zenetos for discussions on Lessepsian invasive species, which have been updated in this list. I am also very grateful to my dear friend Karin Fletcher for her constructive comments and insights.

I dedicate this chapter to Issam Sharabati, who died 16 April 2023. He and his wife Doreen were instrumental in setting me off on my love affair with Red Sea sea slugs, and our trip circumnavigating the Red Sea in the *Ibn Batuta* is forever etched in my memory. Rest in peace, Issam.

REFERENCES

Austin, J., Gosliner, T., and M.A.E. Malaquias. 2018. Systematic revision, diversity patterns and trophic ecology of the tropical Indo-West Pacific sea slug genus *Phanerophthalmus* A. Adams, 1850 (Cephalaspidea, Haminoeidae). *Invertebr Syst* 32, no. 6: 1336–1387. doi:10.1071/IS17086.

Bazzicalupo, E., Crocetta, F., Gosliner, T.M., et al. 2020. Molecular and morphological systematics of *Bursatella leachii* de Blainville, 1817 and *Stylocheilus striatus* Quoy & Gaimard, 1832 reveal cryptic diversity in pantropically distributed taxa (Mollusca: Gastropoda: Heterobranchia). *Invertebr Syst* 34, no. 5: 535–568. doi:10.1071/IS19056.

Blatterer, H. 2019. Mollusca of the Dahab region. *Denisia (Linz)* 43: 1–480. ISSN 1608–8700.

Bonomo, L.J., and T.M. Gosliner. 2020. Adding stars to the *Chromodoris* (Nudibranchia, Chromodorididae) galaxy with the description of four new species. *Zootaxa* 4819, no. 3: 401–435. doi:10.11646/zootaxa.4819.3.1.

Bowen, B.W., Gaither, M.R., DiBattista, J.D., et al. 2016. Comparative phylogeography of the ocean planet. *Proc Nat Acad Sci* 112, no. 29: 7962–7969. doi:10.1073/pnas.1602404l1.

DiBattista, J.D., Choat, J.H., Gaither, M.R., et al. 2015. On the origin of endemic species in the Red Sea. *J Biogeogr* 43: 13–30. https://onlinelibrary.wiley.com/doi/pdf/10.1111/jbi.12631.

Donohoo, S.A., Villalobos, S.G., Hallas, J.M., and T.M. Gosliner. 2023. Hyperdiversity of the genus *Halgerda* Bergh, 1880 (Nudibranchia: Discodorididae) with descriptions of fourteen new species. *Mar Biodiv* 53: 42. doi:10.1007/s12526-022-01334-9.

Eliot, C.N.E. 1910. Nudibranchs collected by Mr. Stanley Gardiner from the Indian Ocean in H.M.S. Sealark. *Trans Linnean Soc London* 13: 411–438. https://scholar.archive.org/work/bgtyvcefnzhgnkwt5kz54bijqa/access/ia_file/crossref-pre-1923-scholarly-works/10.1111%252Fj.1095-8312.1909.tb01116.x.zip/10.1111%252Fj.1096-3642.1910.tb00082.x.pdf.

Epstein, H.E., Hallas, J.M., Johnson, R.F., et al. 2018. Reading between the lines: Revealing cryptic species diversity and colour patterns in *Hypselodoris* nudibranchs (Mollusca: Heterobranchia: Chromodorididae). *Zool J Linn Soc* 186, no. 1: 1–74. doi:10.1093/zoolinnean/zly048.

Fatemi, Y., Taheri, M.R., Esmaeili, H.R., and A. Shahdadi. 2021. Morphological and molecular analysis of *Dendrodoris fumata* (Rüppell & Leuckart, 1830) (Gastropoda: Nudibranchia) from the Persian Gulf. *J Aquatic Ecol* 10, no. 4: 49–57. http://jae.hormozgan.ac.ir/article-1-965-en.html.

Golestani, H., Crocetta, F., Padula, V., et al. 2019. The little *Aplysia* coming of age: From one species to a complex of species complexes in *Aplysia parvula* (Mollusca: Gastropoda: Heterobranchia). *Zool J Linn Soc* 187, no. 2: 279–330. doi:10.1093/zoolinnean/zlz028.

Hoeksema, B.W., and N. Yonow. 2021. Rarity in the native range of the Lessepsian migrant *Plocamopherus ocellatus* (Nudibranchia): Fact or artifact? *Ecology* 102, no. 11: e03481. doi:10.1002%2Fecy.3481.

Krug, P.J., Vendetti, J.E., Rodriguez, A.K., Retana, J.N., Hirano, Y.M., and C.D. Trowbridge. 2013. Integrative species delimitation in photosynthetic sea slugs reveals twenty candidate species in three nominal taxa studied for drug discovery, plastid symbiosis or biological control. *Mol Phylogenet Evol* 69: 1101-1119. https://doi.org/10.1016/j.ympev.2013.07.009.

Krug, P.J., Berriman, J.S., and A. Valdés. 2018. Phylogenetic systematics of the shelled sea slug genus *Oxynoe* Rafinesque, 1814 (Heterobranchia: Sacoglossa), with integrative descriptions of seven new species. *Invertebr Syst* 32: 950–1003. doi:10.1071/IS17080.

Marcus, E., and E. Marcus. 1960. Opisthobranchia aus dem Roten Meer und von den Malediven. Abhandl. *Akademie der Wissenschaften und der Literatur in Mainz*, 1959, no. 12: 871–934, 86 figs.

Matsuda, S.B., and T.M. Gosliner. 2018. Glossing over cryptic species: Descriptions of four new species of *Glossodoris* and three new species of *Doriprismatica* (Nudibranchia: Chromodorididae). *Zootaxa* 4444, no. 5: 501–529. Doi:10.11646/zootaxa.4444.5.1.

Morena, K. 2020. *Phylogenetic systematics of the sea slug genus Cyerce Bergh, 1871 using molecular and morphological data*. M.Sc. Thesis, California State Polytechnic University, Pomona. https://scholarworks.calstate.edu/downloads/0z708z56p.

Moustafa, A.Y., Golestani, H., and A. Valdés. 2018. A review of the genus *Berthellina* (Mollusca: Heterobranchia: Pleurobranchidae) in the Red Sea based on molecular and anatomical data. *Mar Biodiv* 48: 1121–1134. Doi:10.1007/s12526-016-0558-2.

Oskars, T.R., and M. Malaquias. 2020. Systematic revision of the Indo-West Pacific colourful bubble-snails of the genus *Lamprohaminoea* Habe, 1952 (Cephalaspidea: Haminoeidae). *Invertebr Syst* 34: 727–756. doi:10.1071/IS20026.

Shipman, C., and T. Gosliner. 2015. Molecular and morphological systematics of *Doto* Oken, 1851 (Gastropoda: Heterobranchia), with descriptions of five new species and a new genus. *Zootaxa* 3973, no. 1: 57–101. doi:10.11646/zootaxa.3973.1.2.

Smythe, K.R. 1972. Marine Mollusca from Bahrain Island, Persian Gulf. *J Conchol London* 27: 491–496.

Wilson, N.G., and I. Burghardt. 2015. Here be dragons – phylogeography of *Pteraeolidia ianthina* (Angas, 1864) reveals multiple species of photosynthetic nudibranchs (Aeolidina: Nudibranchia). *Zool J Linn Soc* 175: 119–133. https://academic.oup.com/zoolinnean/article/175/1/119/2449830.

WoRMS Editorial Board. 2024. World Register of Marine Species. VLIZ. Available from https://www.marinespecies.org.

Yonow, N. 2000. Red Sea Opisthobranchia 4: The orders Cephalaspidea, Anaspidea, Notaspidea, and Nudibranchia:

Dendronotacea and Aeolidacea. *Fauna of Arabia* 18: 87–131. www.researchgate.net/profile/Nathalie-Yonow/publication/236259266_Red_Sea_Opisthobranchia_4_The_Orders_Cephalaspidea_Anaspidea_Notaspidea_and_Nudibranchia_Dendronotacea_and_Aeolidacea/data/00b7d517681de78d90000000/RSO-4-Ceph-Anasp-Notasp-Aeol-Dendro-2000.pdf.

Yonow, N. 2008a. Opisthobranchs of the Gulf of Eilat and the Red Sea: An account of similarities and differences. In: *Aqaba – Eilat, the improbable Gulf – Environment, biodiversity and preservation*. Ed: F. Por, 177–196. Jerusalem, Israel: The Hebrew University Magnes Press.

Yonow, N. 2008b. *Sea slugs of the Red Sea*. Sofia/Moscow: Pensoft Publications. ISSN 1312-0174.

Yonow, N. 2015. Sea Slugs: Unexpected biodiversity and distribution. In: *The Red Sea: The formation, morphology, oceanography and environment of a young ocean basin*. Eds: N.M.A. Rasul, and I.C.F. Stewart, 531–550. Berlin/Heidelberg, Germany: Springer Earth System Sciences.

Yonow, N. 2018. Red Sea Opisthobranchia 5: New species and new records of chromodorids from the Red Sea (Heterobranchia; Nudibranchia; Chromodorididae). *ZooKeys* 770: 9–42. https://zookeys.pensoft.net/articles.php?id=26378.

Yonow, N., and K.R. Jensen. 2018. Results of the Rumphius Biohistorical Expedition to Ambon. Part 17: The Cephalaspidea, Anaspidea, Sacoglossa, and Pleurobranchida (Mollusca, Gastropoda, Heterobranchia). *Archiv für Molluskenkunde* 147, no. 1: 1–48. www.researchgate.net/profile/Nathalie-Yonow/publication/326067784_Results_of_the_Rumphius_Biohistorical_Expedition_to_Ambon_1990_Part_17_The_Cephalaspidea_Anaspidea_Pleurobranchida_and_Sacoglossa_Mollusca_Gastropoda_Heterobranchia/links/5b415263aca2728a0d618d79/Results-of-the-Rumphius-Biohistorical-Expedition-to-Ambon-1990-Part-17-The-Cephalaspidea-Anaspidea-Pleurobranchida-and-Sacoglossa-Mollusca-Gastropoda-Heterobranchia.pdf.

Yorifuji, M., Takeshima, H., Mabuchi, K., et al. 2012. Hidden diversity in a reef-dwelling sea slug, *Pteraeolidia ianthina* (Nudibranchia, Aeolidina), in the Northwestern Pacific. *Zool Sci* 29: 359–367. https://bioone.org/journals/zoological-science/volume-29/issue-6/zsj.29.359/Hidden-Diversity-in-a-Reef-Dwelling-Sea-Slug-Pteraeolidia-ianthina/10.2108/zsj.29.359.short.

Zamora-Silva, A., and M.A.E. Malaquias. 2018. Molecular phylogeny of the Aglajidae head-shield sea slugs (Heterobranchia: Cephalaspidea): New evolutionary lineages revealed and proposal of a new classification. *Zool J Linn Soc* 183: 1–51. doi:10.1093/zoolinnean/zlx064.

PART 4

Benthos

Community structure and spatial variations in the benthic crustaceans of the western Arabian Gulf

14

Seerangan Manokaran, Thadickal V. Joydas, Jeff B.R. Abrogueña, Karuppasamy P. Manikandan, Syed A. Khan and Abdulaziz Al-Suwailem

14.1 INTRODUCTION

Benthic fauna generally have a sedentary nature and a long life. As a result, they are exposed to anthropogenic perturbations for a long time. They have diverse taxa, which can respond to varying degrees to multiple types of stress (Leppakoski 1975; Pearson and Rosenberg 1978; Garaffo et al. 2018; Manokaran et al. 2021). These properties qualify them as good indicators for monitoring the health of the marine ecosystem. Crustaceans represent the second most abundant and diversified group among the benthic fauna and are composed of amphipods, isopods, cumaceans, tanaids, and ostracodes, besides juveniles of decapod crustaceans (crabs, shrimps, hermit crabs, stomatopods, etc.) (Geetha and Nandan 2014; Joydas et al. 2018). Crustaceans are an integral part of the marine food web dynamics and have different feeding types, such as carnivory, filter feeding, omnivory, deposit-feeding, and scavenging (Khanam et al. 2014). Furthermore, being a member of the epifaunal groups, crustaceans have more motile capability. Hence, these groups can react by disappearing from the area or adopting variations in species-level recruitment depending on the environmental changes (Chou et al. 1999).

The Arabian Gulf ('the Gulf') represents an extreme environment because of its location and the influence of natural and anthropogenic stresses. Natural stress factors include hypersaline conditions with temperature extremes combined with the shallow nature, restricted water exchange, and 1–2 m equivalent of evaporation per year (Coles 2003; Sheppard et al. 2010). In addition, oil exploration activities (dredging, construction of causeways and oil platforms, as well as the creation of artificial islands) and urbanization (reclaiming shallow productive marine areas into land for homes, recreation, and industrial facilities) have added tremendous stress to the Gulf ecosystem (Khan 2007; Erdelen 2007).

Generally, oil spills lead to toxic conditions and a reduction in dissolved oxygen in the benthic substratum that can cause massive damage and a drop in the sizes and composition of benthic fauna (Elmgren et al. 1983; Chou et al. 1999). In addition, dredging and trenching activities also impact the benthic substratum, which has repercussions at population and community levels. Several studies have been conducted on the infaunal communities of the Gulf, but no focused research on infaunal crustaceans has been carried out in this region. The previous works mainly dealt with the occurrence of crustaceans such as amphipods, cumaceans, and isopods (Momtazi and Sari 2013; Joydas et al. 2017; Layeghi and Momtazi 2018). This study aims to assess the environmental and ecological characteristics of benthic crustaceans in the Gulf.

14.2 STUDY AREA

The study area includes the Saudi Arabian coastal waters from Khafji to Ras Tanura (latitude 26°40′–28°30′ N and longitude 48°10′–51°00′ E) (Figure 14.1) in the western part of the Gulf. The oceanographic features of this area are similar to those of the general characteristics of the Gulf (Manokaran et al. 2021). Due to the semi-enclosed nature of the Gulf, the tidal regime is rather complex (the tidal range varies between 0.8 and 4 m), with a semi-diurnal pattern in most parts. The variation in sea surface temperature is high, as low as 11°C during winter and over 36°C in summer (Joydas et al. 2017; Manokaran et al. 2021). The salinity is relatively high (around 45 psu in the open waters) due to the high evaporation rate besides the low

DOI: 10.1201/9781003321392-18

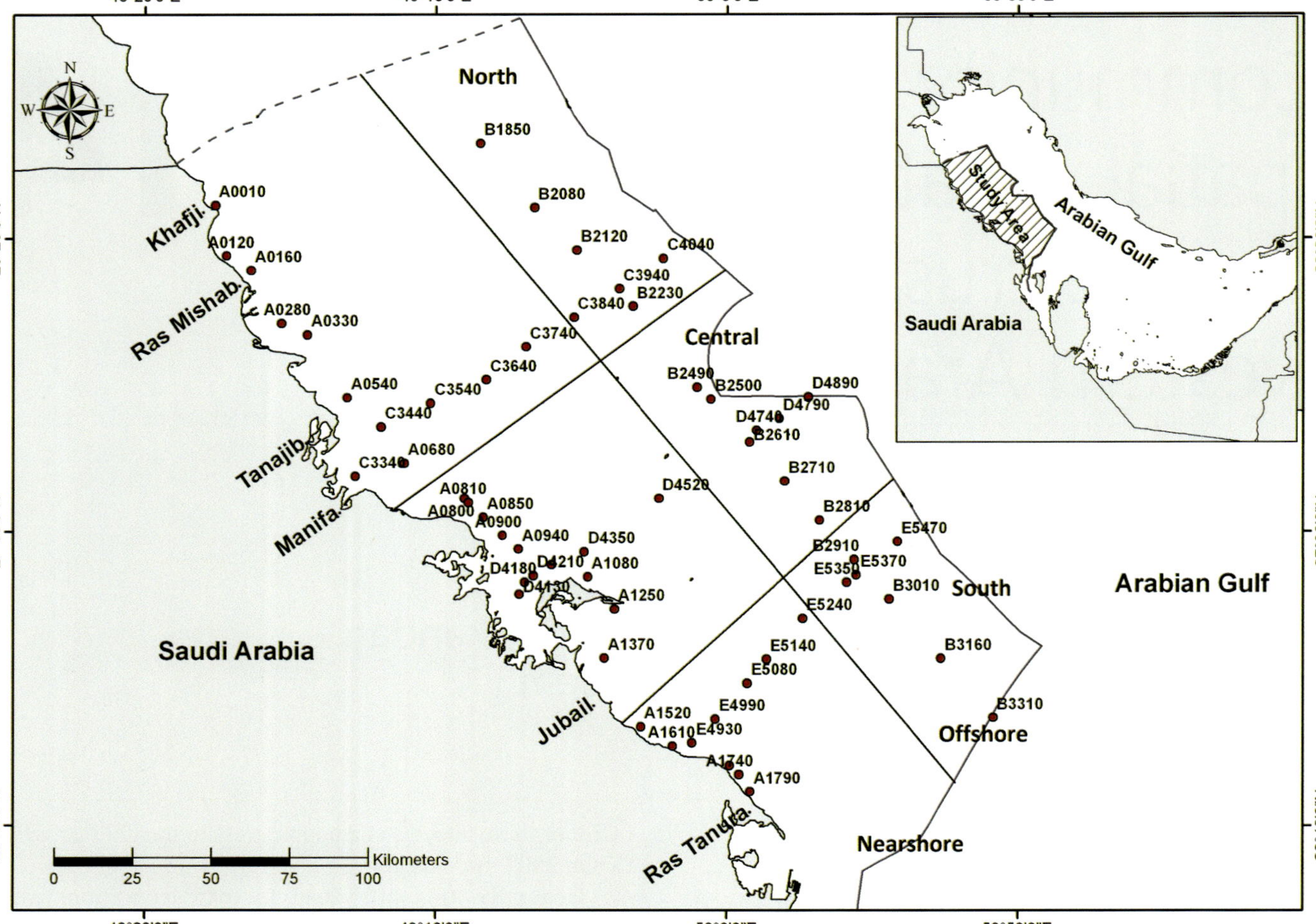

FIGURE 14.1 Map showing the western Arabian Gulf and the sampling locations.

freshwater input. The largest proportion of the seabed in the subtidal region is composed of soft, sandy, or muddy substrate. Sand usually is present from the immediate subtidal levels to a depth of 7–10 m. Below the sandy zone, mud gradually begins to predominate. Seagrass (generally up to 12 m in depth) and coral reefs occur in the subtidal region. The study area was categorized into three regional parts: the north (Khafji to Manifa), central (below Manifa up to Jubail), and south (below Jubail to Ras Tanura); and two depth ranges that include the nearshore (stations at less than 20 m depth) and offshore (stations deeper than 20 m).

14.3 MATERIALS AND METHODS

14.3.1 Sampling

Sediment samples were collected onboard the R/V *Al-Gosaibi*, from November to December 2002. A total of 174 samples from 58 stations (three replicates from each station) (Figure 14.1) were collected using a Van Veen grab (0.1 m^2). The depth of sampling stations ranged from 3 m to 58 m (nearshore – 30 and deeper – 28 stations). Details of the sediment collection and measurement of bottom water quality parameters (bottom water temperature, salinity, pH, and dissolved oxygen) and sediment parameters (sand, silt/clay, mean grain size, Total Organic Carbon (TOC) and Total Petroleum Hydrocarbons (TPH)) are described in Joydas et al. (2017).

14.3.2 Laboratory analysis

The sediment samples were passed through a 0.5 mm sieve with copious filtered seawater. After sieving, the crustaceans were carefully separated, preserved in 5–7% formalin, labeled, and stored for further examination. The samples were rewashed under tap water and 5–7% formalin containing Rose Bengal was added in the laboratory. Identification of crustaceans was made as far as possible up to the species/generic level under a binocular microscope, following Naylor (1972), Jones (1976), Barnard and Drummond (1982), Ruffo (1982, 1989, 1993), and Athersuch et al. (1989).

14.3.3 Data analyses

Univariate biodiversity indices (Shannon Wiener diversity – H' log_2, Pielou's evenness – J', and Simpson dominance index – Lambda') were calculated using PRIMER 6.0 + PERMANOVA (Anderson 2001). PERMANOVA was done to test the latitudinal (northern, central, and southern regions) and bathymetric (nearshore and offshore) differences in density. The distribution pattern in relation to region and depth (fixed ones) was tested using 999 permutations.

Pearson's correlation (corrplot:corrplot; Wei and Simko 2017) analysis was conducted using the R programme to assess pair-wise cross-correlation between environmental variables and the crustacean community structure in the study area. Further, correlations between the distribution and abundance of the most dominant species (19 species) and relevant environmental variables (depth, temperature, salinity, dissolved oxygen (DO), pH, mean grain size, sand, silt/clay, total petroleum hydrocarbons (TPH), and total organic carbon (TOC)) were tested by Canonical Correspondence Analysis (CCA; Ter Braak and Verdonschot 1995) using Canoco Statistical Software (Version 4.5). In CCA, Monte Carlo permutations were used to analyse the statistical significance of the environmental parameters with CCA axes 1 and 2 and to determine the crustacean species assemblage and environment relationships.

14.4 RESULTS

14.4.1 Environmental parameters

A detailed description of the environmental parameters of the study area reported by Joydas et al. (2017) is summarized in Table 14.1. Significant nearshore–offshore variations were found in the bottom temperature, salinity, and DO. These parameters also showed a significant latitudinal gradient with an increase in temperature and a decrease in DO and salinity towards the south. Finer sediment particles increased with depth, and there were no significant latitudinal variations. The TOC also showed a similar trend. The concentration of TPH indicated an unpolluted to moderately polluted nature in the study area. The concentration of TPH also showed significant differences between depths.

14.4.2 Composition of infaunal crustaceans

Overall, 134 species of crustaceans were recorded from the study area (Table 14.2 and Appendix 14.1). Amphipods were found to be a highly dominant group in the samples (44% of the total crustacean species), followed by cumaceans (27%) and ostracods (10%). The remaining taxa ('others') contributed 18% of the total crustacean species. Generally, the same pattern was observed in density; amphipods were most abundant in the samples (41% of the total crustacean density), followed by ostracods (22%) and cumaceans (13%). The 'others' contributed 24% of the total crustaceans.

The number of crustacean species (S) ranged from 10 (station B3310) to 49 m^{-2} (station D4350), with an average ± SE (standard error of the average) of 30 ± 1 m^{-2}. The density of crustaceans ranged from 140 ind. m^{-2} (station B3310) to 15850 ind. m^{-2} (station D4350), with a mean of 1077 ± 270 ind. m^{-2} (Figure 14.2). Results of PERMANOVA revealed significant regional differences in the number of species ($F = 7.61$, $P<0.05$) and depth-driven differences in the density within the regions ($F = 10.1$, $P<0.05$).

Amphipods: 60 amphipod species belonging to 18 families (Table 14.2 and Appendix 14.1) were present in the samples. *Gammaropsis atlantica* (9% of the total macrobenthic crustaceans), *Grandidierella exilis* (4.5%), *Grandidierella bonnieroides* (3%), *Ampelisca scabripes* (2.2%), *Metaprotella sandalensis* (2.2%), and *Ampelisca tulearensis* (2%) were the numerically dominant species in the study area. Gammaridae and Caprellidae (nine species each), Corophidae (eight species), Ampeliscidae, and Amphilochidae (six species each) were the most speciose families. The number of amphipod species in the study area ranged from 4 m^{-2} (stations A0850, A1610, and B3010) to 22 m^{-2} (stations D4350, E4990), and the density from 53 ind. m^{-2} (station C3840) to 11,150 ind. m-2 (station D 4350). The shallower depth stations (655 ± 364 ind. m^{-2}) have more density than deeper stations (216 ± 27 ind. m^{-2}) with a significant difference of $F = 3.41$, $P < 0.05$. Similar variations were observed in the regions; the southern (221 ± 49 ind. m^{-2}) and northern (246 ± 45 ind. m^{-2}) regions have a lower

TABLE 14.1 Summary of the environmental parameters recorded from the western Arabian Gulf.

PARAMETERS	*NORTHERN*	*CENTRAL*	*SOUTHERN*	*NEARSHORE*	*OFFSHORE*
Bot. Temperature (°C)	23.33 ± 0.42	24.30 ± 0.32	24.11 ± 0.37	22.77 ± 0.24	25.17 ± 0.17
Bot. Salinity (psu)	41.32 ± 0.14	41.26 ± 0.23	40.78 ± 0.06	41.64 ± 0.15	40.61 ± 0.04
Bot. DO (mg/l)	5.73 ± 0.11	5.33 ± 0.25	5.04 ± 0.28	5.89 ± 0.07	4.83 ± 0.22
TPH (Sed.) (ppm)	105.95 ± 17.96	70.44 ± 9.99	36.38 ± 11.03	86.66 ± 14.51	56.48 ± 6.96
Mean Grain Size (mm)	0.41 ± 0.06	0.73 ± 0.18	0.73 ± 0.34	0.63 ± 0.10	0.62 ± 0.23
TOC (%)	0.42 ± 0.05	0.39 ± 0.09	0.64 ± 0.16	0.30 ± 0.06	0.66 ± 0.10
Silt/Clay (%)	32.25 ± 6.35	22.60 ± 7	33.42 ± 7.96	11.44 ± 4.16	47.68 ± 5.27
Sand (%)	67.75 ± 6.35	77.40 ± 7	65.58 ± 7.96	88.56 ± 4.16	52.32 ± 5.27
Bot. pH	8.20 ± 0.02	8.14 ± 0.04	8.31 ± 0.06	8.27 ± 0.03	8.14 ± 0.04

Data presented as mean ± SE. Bot. – Bottom, DO – dissolved oxygen; Sed – sediment; TOC – total organic carbon; TPH – total petroleum hydrocarbons.

TABLE 14.2 Number of species and average density (ind. m^{-2}) of crustaceans recorded in the study area.

FAMILY/TAXA	*NO. OF SPECIES*	*DENSITY*	*FAMILY/TAXA*	*NO. OF SPECIES*	*DENSITY*
Amphipods			**Other Crustaceans**		
Ampeliscidae	6	72 ± 10.5	Mysidaceae	3	8 ± 1.2
Amphilochidae	6	75 ± 10.5	Nebaliacea	1	0 ± 0.2
Ampithoidae	4	10 ± 3.4	Stomatopoda	1	1 ± 0.5
Biancolinidae	1	0.1 ± 0.1	Euphausiacea	1	1 ± 0.5
Caprellidae	9	122 ± 83.7	Stenopodidea	1	0 ± 0.1
Corophidae	8	226 ± 133.8	Penaeidea	1	5 ± 0.8
Eusiridae	2	18 ± 8.8	Caridea	1	2 ± 0.5
Gammaridae	9	12 ± 5.0	Scyllaridea	1	0 ± 0.1
Hyalidae	1	4 ± 2.1	Nephropsidea	1	1 ± 0.3
Leucothoidae	4	7 ± 2.3	Thalassinidea	1	2 ± 1.0
Lilgjeborgiidae	1	0 ± 0.2	Paguridea	1	11 ± 3.1
Lysianassidae	2	0 ± 0.2	Hippidea	1	0 ± 0.1
Melphidippidae	1	1 ± 0.6	Brachyura	1	16 ± 2.2
Oedicerotidae	2	2 ± 0.7	Tanaidacea	1	15 ± 3.6
Phoxocephalidae	2	5 ± 2.7	Isopoda	1	49 ± 9.1
Platyischnopidae	1	0 ± 0.1	Cladocera	1	0 ± 0.2
Urothoidae	2	19 ± 3.6	Copepoda	1	69 ± 12.5
Hyperiidea	1	4 ± 2.7	Cirripedia	1	7 ± 7.1
Cumaceans			Mystacocarida	1	0 ± 0.1
Nannastacidae	17	70 ± 18.6			
Bodotriidae	16	52 ± 8.4			
Diastylidae	3	15 ± 4.9			
Pseudocumatidae	1	3 ± 1.2			
Ostracodes					
Polycopidae	1	10 ± 3.4			
Cypridinidae	1	68 ± 9.5			
Cylindroberididae	5	89 ± 26.4			
Philomididae	1	50 ± 8.0			
Cytheridae	1	6 ± 3.1			
Sarsiellidae	1	12 ± 2.7			
Thalassinidea	2	0 ± 0.2			
Healdiidae	1	0 ± 0.1			

density than the central (786 ± 494 ind. m^{-2}) region, and differed significantly ($F = 2.86$, $P < 0.05$). The number of species was similar in both depths (shallower and deeper: 10 ± 1 m^{-2}). However, variations were observed in the regions (northern and southern: 9 ± 1 m^{-2}, central – 12 ± 1 m^{-2}) that were significant ($F = 3.36$, $P < 0.05$).

Cumaceans: 37 cumacean species belonging to four families were observed in the samples (Table 14.2 and Appendix 14.1). The species *Cumella forficula* (3.2% of the total crustaceans) was found to be the dominant one in the samples. Nannastacidae (17 species) and Bodotriidae (16 species) were the most speciose families. The number of species ranged from one (stations A1740, B3310, and C3840) to 19 m^{-2} (station A1250), and the density from 10 ind. m^{-2} (station A1740) to 997 ind. m^{-2} (station C3340). The central region (225 ± 51 ind. m^{-2}) had more density and number of species (9 ± 1m^{-2}) than the northern (density 118 ± 50 ind. m^{-2}, and number of species 7 ± 1 m^{-2}) and southern (density: 56 ± 16 ind. m^{-2}, number of species:4 ± 1 m^{-2}) regions with significant differences ($F = 7.74$, $P<0.05$). However, the density (shallower: 225 ± 47 ind. m^{-2}, and deeper: 49 ± 7 ind. m^{-2}) and the number of species (shallower: 9 ± 1m^{-2}, and deeper: 5 m^{-2}) also significantly differed ($F = 8.6$, $P<0.05$) with respect to depths. There were significant differences in the number of species in relation to the region ($F = 7.72$, $P<0.05$) and depth ($F = 3.35$, $P<0.05$), and also the density with the region ($F = 7.74$, $P<0.05$) and depth ($F = 8.6$, $P<0.05$).

Ostracods: 14 ostracod species belonging to eight families were recorded in the samples (Table 14.2 and Appendix 14.1). The dominant species were *Cypridina* sp. (6.3% of the total crustaceans), *Euphilomedes* sp. (4.6%), *Parasterope* sp. (3.7%), *Cylindroberis* sp. (2.3%), and *Asteropteron* sp. (1.6%). Cylindroberididae (five species) and Cytherellidae (two species) were the most speciose families. Ostracods were present in almost all stations except at station B3310. In the remaining stations, the number of species of ostracods varied from two (stations A1720 and C4040) to 10 m^{-2} (A1250), and the density from 17 ind. m^{-2} (station C4040) to 1773 ind. m^{-2} (station A1610). The southern

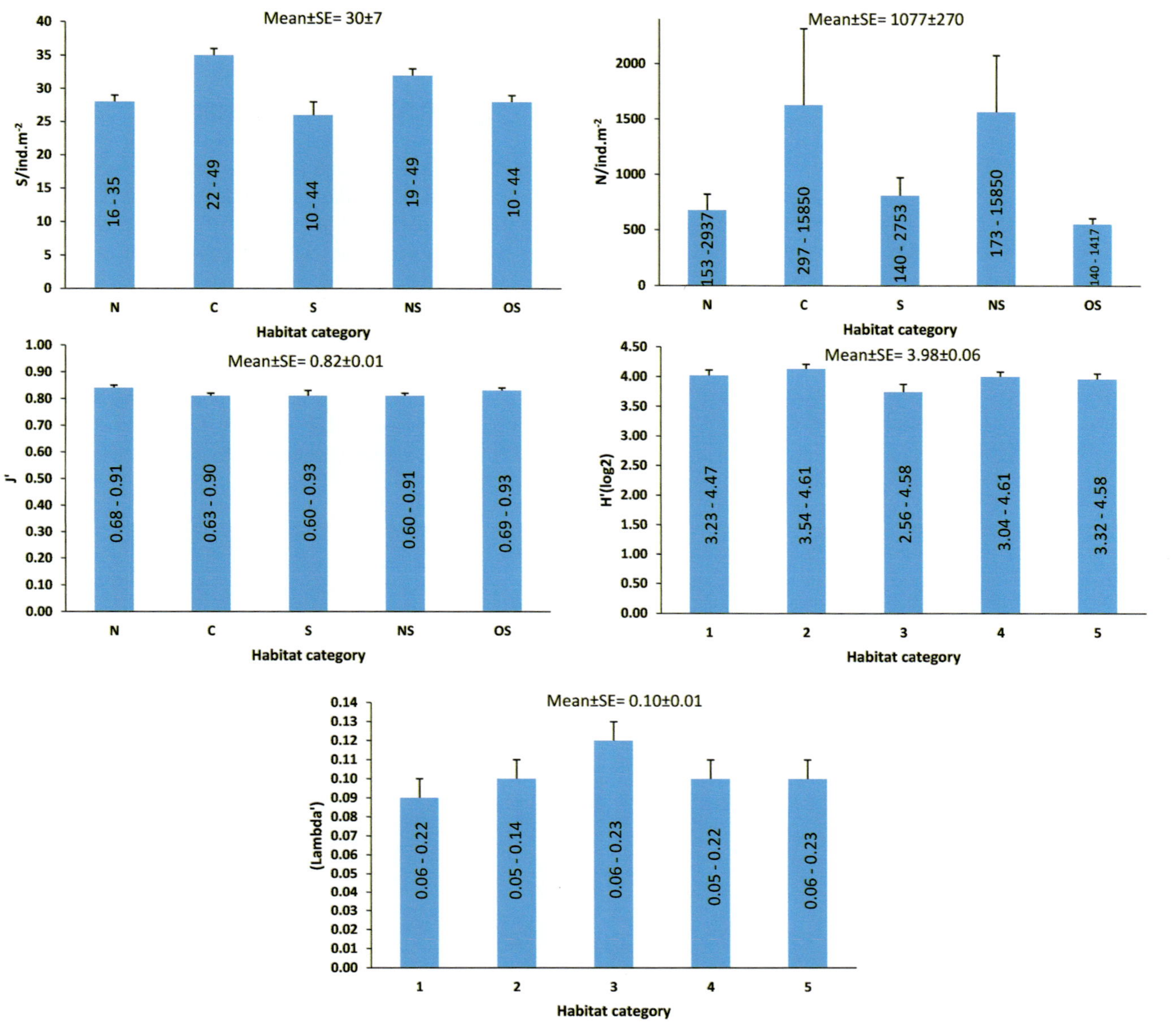

FIGURE 14.2 Number of species and average density (ind. m^{-2}) of crustaceans recorded in the study area.

region (284 ± 104 ind. m^{-2}) had more density than the central (276 ± 52ind. m^{-2}) and northern (142 ± 25ind. m^{-2}) regions with no significant differences ($F = 1.56$, $P>0.05$), whereas, in depth-wise, the density was similar to amphipods and cumaceans, that the shallower depths (338 ± 65 ind. m^{-2}) had more density than deeper depths (123 ± 20 ind. m^{-2}) with significant differences ($F = 4.25$, $P<0.05$). However, the number of species was homogeneous in different regions (northern: 5 m^{-2}, central: 6 m^{-2}, and southern: 5 m^{-2}) and depths (shallower: 6 m^{-2}, and deeper: 5 m^{-2}), and there were no significant differences ($P>0.05$).

Others: 23 species from other taxonomic groups such as mysid (3 species), Nebaliacea, Stomatopoda, Euphausiacea, Stenopodidea, Penaeidea, Caridea, Scyllaridea, Nephropsidea, Thalassinidea, Paguridea, Hippidea, Brachyura, Tanaidacea, Isopoda, Hyperiidea, Caprellidea, Cladocera, Copepoda, Cirripedia, and Mystacocarida (one species each) were present in the samples (Table 14.2 and Appendix 14.1). The dominant taxa were copepods (6.4% of the total crustaceans), caprellidea (6.1%), isopods (4.5%), and brachyura (1.5%). The number of species varied from three (station B3310) to 11 m^{-2} (B2080), and the density from 23 ind. m^{-2} (station A0010) to 3030 ind. m^{-2} (station D4350). The northern region (195 ± 47 ind. m^{-2}) had a lower density than the central (345 ± 131 ind. m^{-2}) and southern (222 ± 33 ind. m^{-2}) regions with no significant differences ($F = 1.47$, $P>0.05$), whereas, in depth-wise, shallower depths (350 ± 98 ind. m^{-2}) had more density than deeper depths (163 ± 23 ind. m^{-2}) with significant differences ($F = 4.28$, $P<0.05$). However, the number of species did not show any variations between regions (northern: 7 m^{-2}, central: 8 m^{-2}, and southern: 7 m^{-2}) and depths (shallower: 7 m^{-2}, and deeper: 8 m^{-2}) significantly ($P>0.05$)).

14.4.3 Diversity

The diversity parameters of microbenthic crustaceans, such as species evenness, diversity, and dominance, are presented in Figure 14.2. The Pielou's evenness ranged from 0.71 (station A1371) to 0.93 (station B3160), with a mean of 0.82 ± 0.01 (regions – northern: 0.84 ± 0.01, central: 0.81 ± 0.01, and southern: 0.81 ± 0.02; depths – shallower: 0.81 ± 0.01, and deeper: 0.83 ± 0.01). The Shannon Wiener diversity ranged from 2.56 (station B3310) to 4.61 (station A1250), with a mean of 3.98 ± 0.06 (regions – northern: 4.02 ± 0.09, central: 4.13 ± 0.08, and southern: 3.74 ± 0.13; depths – shallower: 4 ± 0.08, and deeper: 3.96 ± 0.09). The Simpson dominance index ranged from 0.05 (station A0810) to 0.23 (station B3310) (regions – northern: 0.09 ± 0.01, central: 0.10 ± 0.01, and southern: 0.12 ± 0.01; depths – shallower: 0.10 ± 0.01, and deeper: 0.10 ± 0.01) (Figure 14.2). PERMANOVA results indicated that only the diversity values show significant regional differences (F = 3.84, P<0.05) within a given depth.

14.4.4 Influence of environmental variables on benthic infaunal crustaceans

The Pearson's correlation analysis was done to determine the relationship between the structural parameters of crustaceans and the environmental parameters. The density of cumaceans was positively correlated with salinity ($p<0.01$, $r^2 = 0.38$), TPH ($p<0.05$, $r^2 = 0.27$), and sand ($p<0.01$, $r^2 = 0.34$) and negatively by depth ($p<0.001$, $r^2 = -0.39$) and silt/clay ($p<0.01$, $r^2 = -0.34$). Ostracodes were positively influenced by sand ($p<0.05$, $r^2 = 0.31$) and negatively by depth ($p<0.01$, $r^2 = -0.35$) and silt/clay ($p<0.05$, $r^2 = -0.31$). The number of species ($p<0.01$, $r^2 = -0.36$), density ($p<0.001$, $r^2 = -0.48$), and diversity ($p<0.01$, $r^2 = -0.37$) were negatively influenced by bottom pH (Figure 14.3).

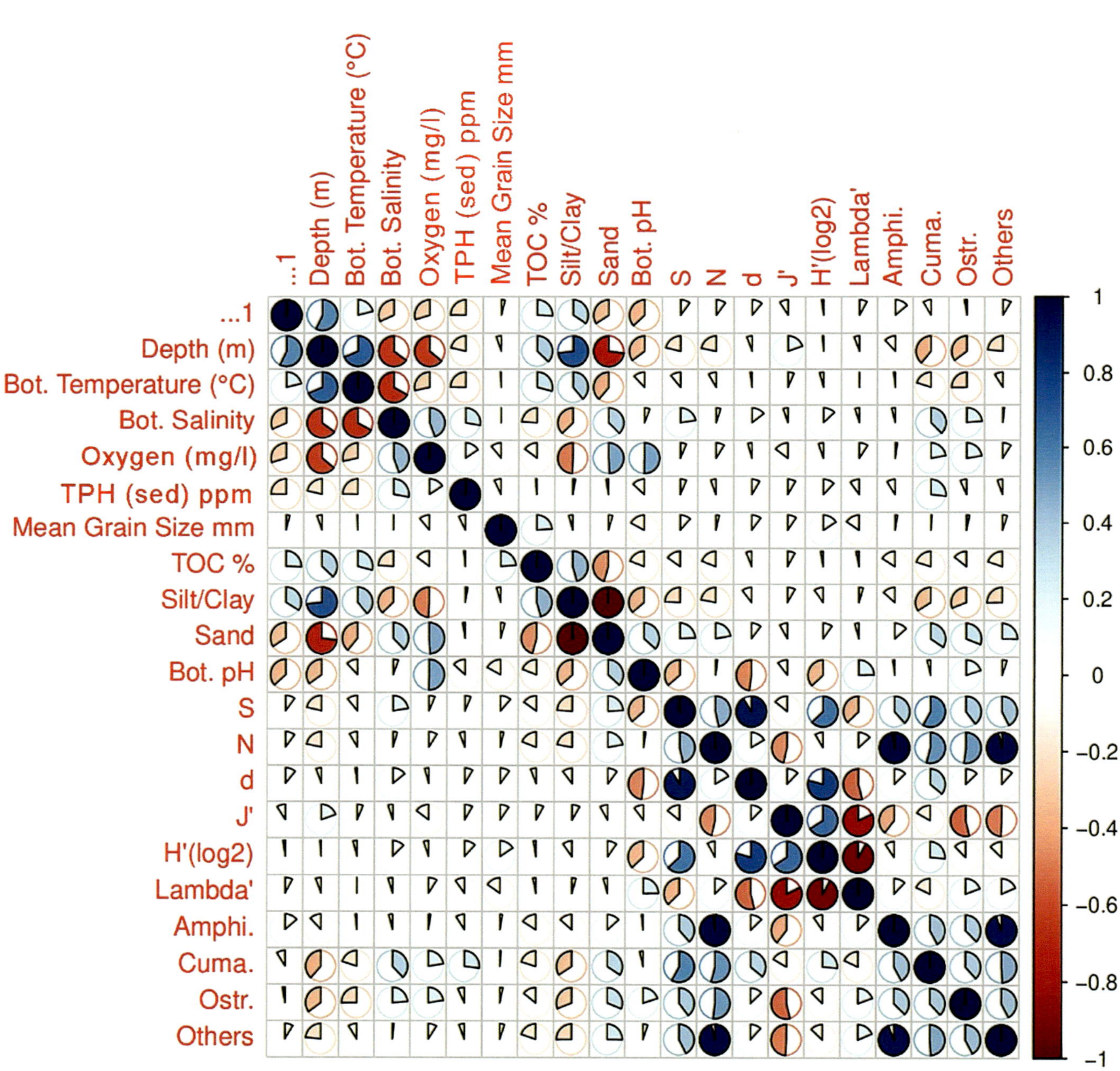

FIGURE 14.3 Pearson's correlation coefficients between environment variables and parameters of benthos.

TABLE 14.3 Results of the CCA: eigenvalues, species-environment correlations, and percentage of variance for infaunal crustaceans.

AXES	*1*	*2*	*3*	*4*	*TOTAL INERTIA*
Eigenvalues	0.112	0.048	0.031	0.012	0.837
Species-environment correlations	0.792	0.599	0.588	0.577	
Cumulative percentage variance					
of species data	13.4	19.1	22.9	24.4	
of species-environment relation	49.9	71.3	85.2	90.7	
Sum of all eigenvalues			0.837		
Sum of all canonical eigenvalues			0.225		

The CCA routine was implemented to link environmental variables with dominant crustaceans density, and this indicated a precise spatial distribution among regions and depths, with axes 1 and 2 explaining 71% of the variation (Table 14.3). The major influencing factors were depth, dissolved oxygen, salinity, sand, sediment TOC, and TPH. The first axis had an eigenvalue of 0.112. A Monte Carlo permutation test (forward selection) was used to identify which environmental variables explained the variance significantly ($p<0.05$). *Ampelisca scabripes* and brachyurans were influenced by depth, *Photis digitata* had an affinity towards sand, *Asteropteron* sp. with bottom water dissolved oxygen, *Grandidierella bonnieroides* and *Cylindroberis* sp. with bottom water salinity, *Cypridina* sp., *Euphilomedes* sp., and isopods with sediment TPH, and *Parasterope* sp. and copepods by TOC (Figure 14.4 a and b).

14.5 DISCUSSION

Soft bottom macrobenthic communities are critical components as indicators of the functioning of coastal and marine ecosystems (Lu 2005). These bring about considerable changes in the physical and chemical composition of sediments, especially in the water-sediment interface (Gaudêncio and Cabral 2007; Shou et al. 2009). The present study showed that the number of species (135) and density (140 – 15,850 ind. m^{-2}), as well as diversity ($H'log_2$–3.98 ±0.06) of crustaceans are comparable to the findings from other world seas. For example, similar densities (9 – 20,686 ind. m^{-2}) of crustaceans were reported from the Southern Ocean and the Weddell Sea (Di Franco et al. 2020), South Sandwich islands (density: 11 – 4123 ind. m^{-2}) (Kaiser et al. 2008), and Southern Ocean (density: 5 – 10,735 ind. m^{-2}) (Brökeland et al. 2007). However, the values recorded from southwestern Taiwan (species richness: 15 – 48; density: 125 – 6235 ind. m^{-2}; diversity: 1.38 – 2.77) (Chou et al. 1999), the northwestern Mediterranean (density: 1 – 123 ind. m^{-2}) (de Juan and Cartes 2011), and the Cochin estuary and Arabian sea (mean density: 125 ind. m^{-2}) (Geetha and Nandan 2014) are lower than those obtained in the present study.

There was a higher species number, density, and diversity (*H' log2*) in the central region than in the other two areas (southern and northern) of the western Arabian Gulf. However, in the structural parameters, a statistically significant regional difference was found only in the case of diversity among the regions. This indicates the spatial variability along the coastal area due to bay structures besides the distribution of several ecosystems, such as mangroves, seagrasses, algal and cyanobacterial mats, and coral reefs. With these biological resources, hydrological and sedimentological characteristics support the essential ecological functions, supplying nutrients and availability of foods, which influence the spatial variations. Moreover, the harsh environmental conditions in the nearshore areas associated with domestic discharges might be the reason for variations in communities in the regions (Joydas et al. 2023). The recorded variations in density and diversity are largely due to the adaptability to the environmental variations in the diverse ecosystems of the Gulf (Joydas et al. 2011; Manokaran et al. 2021).

Amphipods had the highest density compared to other crustaceans, which was found to be similar to the results of previous studies in the Ebro Delta continental shelf in the northwestern Mediterranean (de Juan and Cartes 2011), the Cochin Estuary (Geetha and Nandan 2014), and the Southern Ocean (Brökeland et al. 2007; Kaiser et al. 2008; Di Franco et al. 2020). Amphipods are the leading secondary and tertiary producers among the benthic communities and serve as good indicators of ecosystem health due to their sensitive nature (Manokaran et al. 2021). In addition, the density of the cumaceans was also found to be higher in the study area; their high observed density was due to their preference for fine and silty sediments (Di Franco et al. 2020).

The canonical analysis and the Pearson's correlation indicated a significant relationship between the environmental parameters and the crustacean communities. The first axis of the CCA had a high eigenvalue of 0.112, implying that it has captured more of the variance in the spatial distribution than the other components in the western Arabian Gulf. Monte Carlo permutation tests confirmed the significant association ($p<0.05$) between environmental variables and benthic crustacean species. There was a conspicuous relationship between the dominant species of amphipods and cumaceans with the environmental parameters. The sediment pattern is the main factor that influences the crustaceans, besides other factors such as temperature, organic carbon, salinity, dissolved oxygen, pollution, and food availability (Brandt et al. 2007; Lourido et al. 2008; Di Franco et al. 2020; Stępień et al. 2021). Such an environmental influence on crustacean communities was also observed in the Southern Ocean (Brökeland et al. 2007; Brandt et al. 2007; Kaiser et al. 2008; Di Franco et al. 2020),

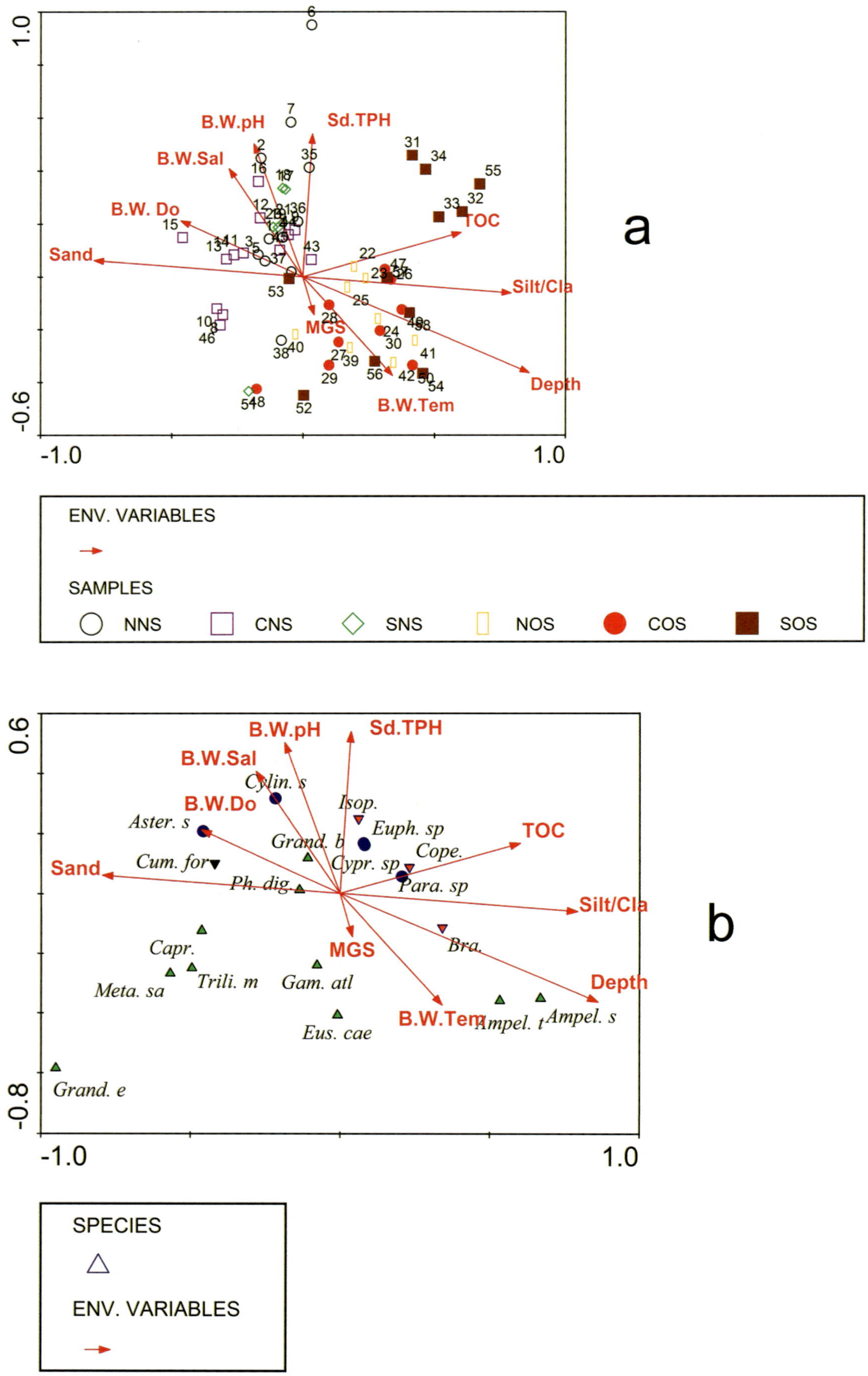

FIGURE 14.4 Correspondence Analysis (CCA) showing the ordination of (a) sampling stations and (b) dominant crustaceans with respect to environmental parameters.

the Turkish coast of the Black and Marmara seas (Mulayim 2021), the South Atlantic Central Water (Cristales and Pires-Vanin 2014), and the Iranian coast of the Arabian/Persian Gulf (Martin et al. 2010).

14.6 CONCLUSION

This study provides detailed data on macrobenthic in faunal crustacean diversity, density, and their relationship with environmental parameters in the Saudi Arabian coastal waters of the Arabian Gulf. The density recorded in the shallower regions (1568 ± 508 ind. m^{-2}) was higher than those recorded from the deeper areas (552 ± 53 ind. m^{-2}). The species composition was found to vary in relation to latitude (northern: 28 ± 1, central: 35 ± 1, and southern: 26 ± 2 m^{-2}), and density with depth (shallower: 32 ± 1, and deeper: 28 ± 1 m^{-2}). The species diversity showed significant differences between the regions within nearshore and offshore areas. The regional variations in the entire coastal area due to several bay structures and depth-related patterns in the communities mirror the patterns of interstitial space, demonstrating the role played by the food supply in determining the distribution of dissolved oxygen in the Gulf. In conclusion, the variations in the benthic crustaceans in terms of region and depth depend on the substratum character and environmental variables, which are influenced by both natural and oil-based industrial activities in the Gulf.

ACKNOWLEDGMENTS

We thank the Center for Environment and Marine Studies (CEMS) at the Research Institute of King Fahd University of Petroleum and Minerals, Dhahran, Saudi Arabia, for providing research facilities and the Marine Studies Section team of ARCEMS for performing the fieldwork and laboratory analyses. We express our sincere thanks to the General Authority of Meteorology and Environmental Protection, Saudi Arabia, and the Saudi Company for Environmental Works Ltd. for their constant support. We also thank M/S *Al-Gosaibi* for providing the services of their Oceanographic Vessel. We are grateful to Drs. Murugesan, Linoy Libini, and Thomas Jacob of Annamalai University, India, and Dr. Sajan Sebastian of the Cochin University of Science and Technology, India, for their valuable support.

REFERENCES

Anderson, M.J. 2001. A new method for non-parametric multivariate analysis of variance. *Austral J Ecol* 26, no. 1: 32–46.

Athersuch, J., Horne, D.J., and J.E. Whittaker. 1989. *Marine and brackish water ostracods*. Synopses British Fauna (New Series) no. 43. Leiden: E. J. Brill, 343 p.

Barnard, J.L., and M.M. Drummond. 1982. Gammaridean amphipoda of Australia, Part V: Superfamily haustorioidea. *Smithson Contrib Zool* 360, nos. i–iv: 1–148.

Brandt, A., Gooday, A.J., Brandao, S.N., Brix, S., Brökeland, W., Cedhagen, T., et al. 2007. First insights into the biodiversity and biogeography of the Southern Ocean deep sea. *Nature* 447, no. 7142: 307–311.

Brökeland, W., Choudhury, M., and A. Brandt. 2007. Composition, abundance and distribution of Peracarida from the Southern Ocean deep sea. *Deep Sea Res Part II: Top Stud Oceanogr* 54, nos. 16–17: 1752–1759.

Chou, W.R., Lai, S.H., and L.S. Fang. 1999. Benthic crustacean communities in waters of southwestern Taiwan and their relationships to environmental characteristics. *Acta Zool Taiwan* 10, no. 1: 25–33.

Coles, S.L. 2003. Coral species diversity and environmental factors in the Arabian Gulf and the Gulf of Oman: A comparison to the Indo-Pacific region. *Atoll Res Bull* 507: 1–19. doi:10.5479/si.00775630.507.1.

Cristales, P.A., and A.M.S. Pires-Vanin. 2014. Structuring factors of the cumacean communities of the continental shelf of southeastern Brazil. *Braz J Oceanogr* 62: 103–116.

de Juan, S., and J.E. Cartes. 2011. Influence of environmental factors on the dynamics of macrobenthic crustaceans on soft-bottoms of the Ebro Delta continental shelf (northwestern Mediterranean). *Sci Mar* 75: 691–700.

Di Franco, D., Linse, K., Griffiths, H.J., Haas, C., Saeedi, H., and A. Brandt. 2020. Abundance and distributional patterns of Benthic Peracarid Crustaceans from the Atlantic sector of the Southern Ocean and Weddell Sea. *Front Mar Sci* 7: 554663.

Elmgren, R., Hansson, S., Larsson, U., Sundelin, B., and P.D. Boehm. 1983. The "Tsesis" oil spill: Acute and long-term impact on the benthos. *Mar Biol* 73, no. 1: 51–65.

Erdelen, W.R. 2007. Foreward. *Aquat Ecosyst Health Manag* 103: 255.

Garaffo, G.V., Jaubet, M.L., Llanos, E.N., Bottero, M.A., and R. Elias. 2018. Assessing functional diversity of macrobenthic assemblages in sewage-affected intertidal shores. *Int Aquat Res* 10, no. 4: 333–347.

Gaudêncio, M.J., and H.N. Cabral. 2007. Trophic structure of macrobenthos in the Tagus estuary and adjacent coastal shelf. *Hydrobiologia* 587, no. 1: 241–251.

Geetha, P.N., and S.B. Nandan. 2014. Ecology, diversity, and abundance of macrobenthic crustaceans in Cochin Estuary, India. *Res J Recent Sci* 3, no. 104: 137–148.

Jones, N.S. 1976. *British cumaceans: Synopses of the British Fauna (New Series) No. 7*. Linnaean Soc London. London and New York: Academic Press.

Joydas, T.V., Krishnakumar, P.K., Qurban, M.A., Ali, S.M., Al-Suwailem, A., and A. Al-Abdulkader. 2011. Status of macrobenthic community of Manifa – Tanajib Bay system of Saudi Arabia based on a once-off sampling event. *Mar Pollut Bull* 62, no. 6: 1249–1260.

Joydas, T.V., Qurban, M.A., Ali, S.M., Albarau, J.F., Rabaoui, L., Manikandan, K.P., and P.K. Krishnakumar. 2018. Macrobenthic community structure in the deep waters of the Red Sea. *Deep Sea Res Part I: Oceanogr Res Pap* 137: 38–56.

Joydas, T.V., Qurban, M.A., Borja, A., Krishnakumar, P.K., and A. Al-Suwailem. 2017. Macrobenthic community structure in the northwestern Arabian Gulf, twelve years after the 1991 oil spill. *Front Mar Sci* 4: 248.

Joydas, T.V., Qurban, M.A., Borja, A., Manokaran, S., Manikandan, K.P., Lotfi, J.R., Garmendia, J.M., Asharaf, T.T.M., Ayranci, K., Shemsi, A.M., Shoeb, M., Basali, A.U., Panickan, P., Nazeer, Z., Lyla, P.S., Khan, S.A., and P.K. Krishnakumar. 2023. Ecological status of macrobenthic communities in the Saudi waters of the western Arabian Gulf. *Reg Stud Mar Sci* 57: 102751.

Kaiser, S., Barnes, D.K., Linse, K., and A. Brandt. 2008. Epibenthic macrofauna associated with the shelf and slope of a young and isolated Southern Ocean island. *Antarct Sci* 20, no. 3: 281–290.

Khan, N.Y. 2007. Multiple stressors and ecosystem-based management in the Gulf. *Aquat Ecosyst Health Manag* 10, no. 3: 259–267.

Khanam, S., Mustaquim, J., and Z. Ayub. 2014. Study of the seasonal abundance and occurrence of macrobenthic crustaceans in mangrove swamps of Sandspit Backwaters, Karachi. *Int J Biol Res* 2, no. 2: 147–152.

Layeghi, Y., and F. Momtazi. 2018. Ampithoe qeshmensis sp. nov. (Amphipoda: Ampithoidae), a new herbivorous amphipod from the Persian Gulf. *J Mar Biol Assoc UK* 98, no. 3: 515–519.

Leppakoski, E. 1975. Assessment of degree of pollution on the basis of macrozoobenthos in marine and brackish-water environments. *Acta Acad Abo Ser B* 30: 1–90.

Lourido, A., Moreira, J., and J.S. Troncoso. 2008. Assemblages of peracarid crustaceans in subtidal sediments from the Ría de Aldán (Galicia, NW Spain). *Helgol Mar Res* 62, no. 4: 289–301.

Lu, L. 2005. The relationship between soft-bottom macrobenthic communities and environmental variables in Singaporean waters. *Mar Pollut Bull* 51, nos. 8–12: 1034–1040.

Manokaran, S., Joydas, T.V., Qurban, M.A., Cheruvathur, L.L., Kariyathil, T.J., Basali, A.U., Khan, S.A., and A. Al-Suwailem. 2021. Baseline patterns of structural and functional diversity of benthic amphipods in the western Arabian Gulf. *Mar Pollut Bull* 164: 112054.

Martin, D., Fernadez-Arcaya, U., Tirado, P., Dutrieux, É., and J. Corbera. 2010. Relationships between shallow-water cumacean assemblages and sediment characteristics facing the Iranian coast of the Persian Gulf. *J Mar Biol Assoc UK* 90, no. 1: 125–134.

Momtazi, F., and A. Sari. 2013. Intertidal caprellids (Crustacea: Amphipoda) of the Persian Gulf and the Gulf of Oman, with description of three new species. *Zootaxa* 3717, no. 2: 195–223.

Mulayim, A. 2021. Soft-bottom crustacean fauna from the Turkish coast of the Black and Marmara seas with new records. *Oceanol Hydrobiol Stud* 50, no. 1: 60–76.

Naylor, I. 1972. *Geology of Olkaria and Eburru: Geothermal prospects*. Report with maps. U.N.D.P./EAPL Geothermal Resources Exploration Project, Report, 52 p.

Pearson, T.H., and R. Rosenberg. 1978. Macrobenthic succession in relation to organic enrichment and pollution of the marine environment. *Oceanogr Mar Biol Ann Rev* 16: 229–311.

Ruffo, S. 1982. *The amphipoda of mediterranean. Part 1: Gammaridea (Acanthonotozomatidae to ammaridae)*. Monaco: Musée Océanographique, Mémoire de l'Institut Océanographique de Monaco 13.

Ruffo, S. 1989. *The amphipoda of the mediterranean. Part 2: Gammaridea (Haustoridae to lysianassidae)*. Monaco: Musée Océanographique, Mémoires de l'Institut Océanographique.

Ruffo, S. 1993. *The amphipoda of the mediterranean. Part 3: Gammaridea (Melphidipiidae to talitridae) – Ingolfiellidea – Caprellidea*. Monaco: Musée Océanographique, Mémoires de l'Institut Océanographique.

Sheppard, C., Al-Husiani, M., Al-Jamali, F., Al-Yamani, F., Baldwin, R., Bishop, J., et al. 2010. The Gulf: A young sea in decline. *Mar Pollut Bull* 60, no. 1: 13–38.

Shou, L., Huang, Y., Zeng, J., Gao, A., Liao, Y., and Q. Chen. 2009. Seasonal changes of macrobenthos distribution and diversity in Zhoushan sea area. *Aquat Ecosyst Health Manag* 12, no. 1: 110–115.

Stępień, A., Pabis, K., Sobczyk, R., and B. Serigstad. 2021. High species richness and extremely low abundance of Cumacean communities along the shelf and slope of the Gulf of Guinea (West Africa). *Front Mar Sci* 8: 703547.

Ter Braak, C.J., and P.F. Verdonschot. 1995. Canonical correspondence analysis and related multivariate methods in aquatic ecology. *Aquatic Sci* 57, no. 3: 255–289.

Wei, T., and V. Simko. 2017. *R package 'corrplot': Visualization of a correlation matrix (Version 0.84)*. https://github.com/taiyun/corrplot.

Appendix 14.1

Benthic infaunal species recorded in the study area

NNS: Northern-Nearshore; CNS: Central-Nearshore; SNS: Southern-Nearshore; NOS: Northern-Offshore; COS: Central-Offshore; SOS: Southern-Offshore

TAXA	*FAMILY*	*SPECIES*	*NNS*	*CNS*	*SNS*	*NOS*	*COS*	*SOS*
Amphipoda	Gammaridae	*Elasmopus pectinicrus*	10	2	3	0	0	2
		Elasmopus rapax	0	0	6	2	0	1
		Maera hemigera	1	1	2	0	1	1
		Maera tenella	0	1	0	0	0	1
		Maera pacifica	0	0	0	0	1	0
		Maera quadrimana	0	0	0	0	0	0
		Maera sp.	0	0	0	0	0	0
		Ceradocus serratous	25	2	2	1	0	0
		Ceradocus rubromaculatus	0	1	0	0	0	0
	Corophidae	*Grandidierella bonnieroides*	18	44	147	10	6	7
		Grandidierella exilis	0	210	6	0	1	1
		Natarajphotis manieni	13	32	16	3	4	3
		Photis longimanus	4	11	0	0	0	0
		Gammaropsis atlantica	13	370	16	12	18	31
		Photis digitata	8	54	14	10	6	8
		Gammaropsis sp.	13	21	21	0	13	10
		Cheirophotis megacheles	0	2	0	0	0	0
	Ampeliscidae	*Ampelisca scabripes*	8	18	11	35	53	20
		Ampelisca insignis	3	1	1	9	33	10
		Ampelisca tulearensis	18	15	24	23	40	15
		Ampelisca brevicornis	2	0	2	8	33	16
		Ampelisca hemigera	0	0	0	0	15	0
		Byblis sp.	0	4	0	1	23	8
	Amphilochidae	*Gitanopsis gouriae*	11	3	2	0	3	0
		Ampilochus sp.	4	5	2	0	3	1
		Ampilochus hemigera	0	0	4	0	0	0
		Ampilochus neopolitanus	1	5	0	0	2	1
	Ampithoidae	*Ampithoe falsa*	2	0	0	3	0	0
		Ampithoe ramondi	8	17	11	3	9	3
		Ampithoe sp.	0	1	0	0	2	0
		Cymadusa filosa	0	0	0	0	0	0

(Continued)

(Continued)

TAXA	*FAMILY*	*SPECIES*	*NNS*	*CNS*	*SNS*	*NOS*	*COS*	*SOS*
	Leucothoidae	*Leucothoella bannwarthi*	0	6	7	0	1	1
		Leucothoella sp.	0	0	0	0	0	2
		Leucothoides sp.	11	0	0	1	4	3
		Leucothoe sp.	2	0	0	0	0	0
	Amphilochidae	*Cyproidea testicauda*	0	6	0	3	4	1
		Gitanopsis sp.	1	0	0	0	0	0
	Urothoidae	*Urothoe* sp.	2	20	21	0	1	6
		Urothoe spinidigitus	8	4	6	31	14	5
	Oedicerotidae	*Perioculoides longimanus*	1	0	2	2	1	3
		Perioculoides pallidus	0	0	0	0	0	1
	Eusiridae	*Eusiroides caesaris*	2	47	26	2	5	6
		Paromoera capensis	1	7	1	1	1	1
	Lysianassidae	*Lysianassa cinghalensis*	0	0	0	0	1	1
		Azotostoma sp.	0	0	0	0	0	1
	Phoxocephalidae	*Eyakia* sp.	14	2	2	0	4	2
		Rhepoxynius sp.	0	0	0	0	2	0
	Platyischnopidae	*Indischnopus herdmani*	0	0	0	0	0	0
	Lilgjeborgiidae	*Idunella* sp.	0	1	0	0	0	0
	Biancolinidae	*Biancolina* sp.	0	0	0	0	0	0
	Melphidippidae	*Hornella incerta*	2	2	0	0	0	0
	Hyalidae	*Parhyale* sp.	16	3	1	0	0	0
	Caprellidae	*Aeginella* sp.	1	1	0	0	0	0
		Aeginellopsis arabica	0	6	0	0	0	0
		Caprella sp.	1	1	0	0	0	0
		Caprella danilevskii	0	8	0	0	0	0
		Metaprotella sandalensis	20	80	6	0	1	3
		Triliropus minutus	9	65	9	2	2	2
		Pseudocaprellina pambanensis	10	33	0	0	0	1
		Caprellidea	50	234	18	6	5	6
		Hyperiidea	10	8	0	0	0	0
Cumacea	Nannastacidae	*Campylaspis* sp.	1	2	1	2	2	1
		Campylaspis glabra	1	4	0	18	9	18
		Cumella forficula	15	127	18	3	1	5
		Cumella hispida	0	34	0	0	0	0
		Cumella sp.	0	0	1	0	0	0
		Nannastacus gurneyi	34	5	0	0	0	0
		Nannastacus lepturus	1	0	3	0	0	0
		Nannastacus inflatus	0	0	0	0	0	0
		Nannastacus reptans	0	5	1	0	0	2
		Nannastacus zimmeri	0	1	0	0	0	0
		Nannastacus gibbosus	28	1	1	0	0	0
		Nannastacus stebbingi	0	0	0	0	0	0
		Nannastacus sp.	0	0	0	1	0	1
		Schizotrema bifrons	0	0	2	0	0	0
		Schizotrema sordidum	1	2	1	0	0	0
		Schizotrema depressum	0	0	0	0	0	0
		Schizotrema sp.	0	0	0	0	0	0
	Bodotriidae	*Eocuma affine*	7	17	7	2	1	0
		Eocuma sarsi	8	16	6	0	0	0
		Eocuma taprobanicum	17	31	7	2	5	2
		Eocuma sp.	0	2	1	0	0	0
		Cyclaspis picta	3	11	4	2	5	7

(Continued)

TAXA	*FAMILY*	*SPECIES*	*NNS*	*CNS*	*SNS*	*NOS*	*COS*	*SOS*
		Cyclaspis hornelli	14	19	4	3	1	4
		Cyclaspis sp.	2	3	1	0	0	0
		Heterocuma andamani	2	2	0	1	0	0
		Heterocuma sp.	0	0	1	0	0	0
		Bodotria similis	1	2	0	0	0	0
		Bodotria siamensis	2	12	3	1	2	0
		Bodotria parva	0	2	0	0	0	0
		Bodotria sublevis	0	0	0	0	0	0
		Iphinoe sp.	0	0	0	17	7	3
		Iphinoe crassipes	0	0	0	4	1	0
		Vaunthompsonia arabica	0	0	0	0	1	0
	Diastylidae	*Paradiastylis* sp.	4	4	1	0	1	0
		Paradiastylis brachyurus	5	41	6	0	0	2
		Gynodiastylis sp.	3	2	0	3	2	1
	Pseudocumatidae	*Pseudosympodomma indicum*	10	3	4	0	1	1
Ostracoda	Polycopidae	*Polycope* sp.	8	28	9	2	1	3
	Cypridinidae	*Cypridina* sp.	54	106	133	40	29	56
	Cylindroberididae	*Parasterope* sp.	18	64	72	23	31	36
		Asteropteron sp.	14	57	5	1	2	1
		Synasterope sp.	2	7	32	0	0	0
		Asteropterygion sp.	1	5	0	1	3	2
		Cylindroberis sp.	3	12	184	2	4	7
	Philomedidae	*Euphilomedes* sp.	52	83	72	27	20	39
	Cyprididae	*Candonna* sp.	8	13	0	2	2	2
		Xestoleberis sp.	0	0	0	0	0	0
	Sarsiellidae	*Eusarsiella* sp.	6	22	13	11	6	9
	Cytherellidae	*Cytherella* sp.	0	1	0	0	0	0
		Cytherelloidea sp.	0	0	0	0	0	0
	Healdiidae	*Hungarella* sp.	0	0	0	0	0	0
Mysid	Mysidaceae	*Siriella brevicaudata*	6	3	6	3	5	2
		Siriella hanseni	0	0	0	1	1	0
		Gastrosaccus kempi	4	5	8	4	4	1
Nebaliacea	Nebaliacea	Nebaliacea	0	1	0	0	1	0
Stomatopoda	Stomatopoda	Stomatopoda	0	3	0	1	1	2
Euphausiacea	Euphausiacea	Euphausiacea	2	1	1	0	0	1
Stenopodidea	Stenopodidea	Stenopodidea	0	0	0	0	0	0
Penaeidea	Penaeidea	Penaeidea	4	6	4	6	9	4
Caridea	Caridea	Caridea	0	1	7	2	0	3
Scyllaridea	Scyllaridea	Scyllaridea	0	0	0	0	0	1
Nephropsidea	Nephropsidea	Nephropsidea	0	1	0	1	1	1
Thalassinidea	Thalassinidea	Thalassinidea	0	0	3	9	3	1
Paguridea	Paguridea	Paguridea	16	5	19	2	16	11
Hippidea	Hippidea	Hippidea	0	0	0	0	0	0
Brachyura	Brachyura	Brachyura	8	16	11	25	20	16
Tanaidacea	Tanaidacea	Tanaidacea	13	27	10	8	7	18
Isopoda	Isopoda	Isopoda	74	53	107	37	19	21
Hyperiidea	Hyperiidea	Hyperiidea	10	8	0	0	0	0
Caprellidea	Caprellidea	Caprellidea	50	229	18	6	5	6
Cladocera	Cladocera	Cladocera	0	1	0	0	0	0
Copepoda	Copepoda	Copepoda	42	105	102	30	27	96
Cirripedia	Cirripedia	Cirripedia	0	0	0	0	47	0
Mystacocarida	Mystacocarida	Mystacocarida	0	0	1	0	0	0

The biology, ecology and life history of Brachyuran crabs of the Arabian Gulf

15

Zainab Al-Wazzan

15.1 INTRODUCTION

Crabs belong to the largest and the most diverse group of decapods called the Brachyura, with more than 9,000 described species according to the World Register of Marine Species (WoRMS 2023). Brachyuran crabs typically have a wide cephalothorax where the abdomen is flapped under the thorax to cover the reproductive appendages in males and to incubate the eggs in females. Their body is segmented, with six head segments that are fused together to form the carapace, eight thoracic segments and six in the abdominal region (Warner 1977; Weis 2012). Crabs have highly specialized appendages. In the head region, they have two pairs of antennae used as sensory organs, a pair of mandibles and two pairs of maxillae that are used for feeding. In the thoracic region, they have eight pairs of appendages, including three pairs of maxillipeds, a pair of chelipeds and four pairs of walking legs. The abdominal appendages have four pairs of pleopods. In females, these appendages are used to hold eggs. In males, two pairs of the pleopods are modified as gonopods used to transfer the sperm during copulation in males (Warner 1977; Weis 2012; Beninger et al. 1988). Brachyuran crabs are dioecious and reproduce sexually by means of internal fertilization (Figure 15.1). Mature males and females reproduce during the breeding season, which varies among species depending on a range of environmental factors (Christy 2003, 2011; Morgan and Christy 1995). Generally, breeding is in spring and summer in temperate regions, in comparison to tropical species, where breeding is often continuous throughout the year (Saigusa et al. 2002; Warner 1977).

The Arabian Gulf is home to more than 200 documented crab species inhabiting a wide range of different very harsh environments (Naderloo 2017). The coastal habitats of this region are subject to a range of ecological and anthropogenic stresses such as habitat loss (Sheppard et al. 2010), pollution (Karam et al. 2019), and intensive gleaning (Al-Wazzan et al. 2020a), in addition to the extreme expected effect of climate change (Ben-Hasan and Christensen 2019). Coastal marine organisms such as crabs have ecological and economical values for coastal communities. They provide biological resources

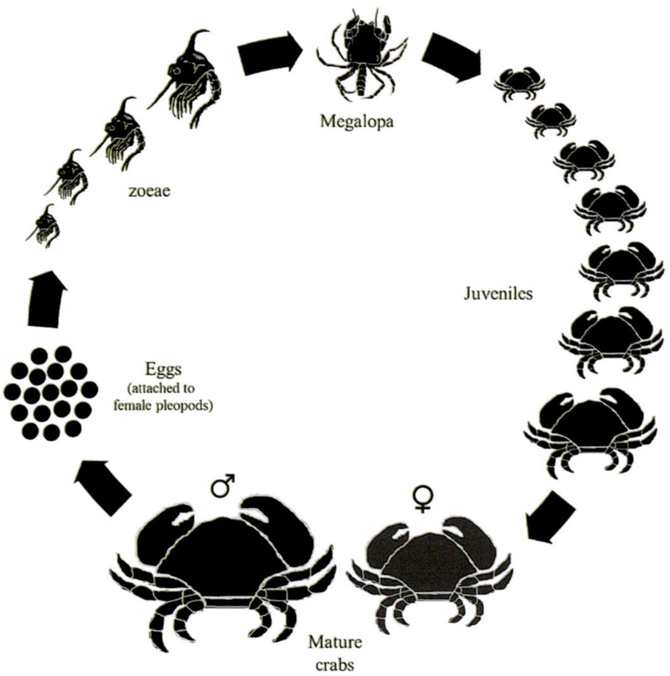

FIGURE 15.1 Schematic diagram of the life cycle of a brachyuran crab (after Al-Wazzan 2017).

such as food sources or bait for fishing (Cunha et al. 2005). In addition, intertidal organisms are essential components in the oceanic and coastal environment (Al-Wazzan et al. 2019) and energy flow between marine to terrestrial habitats (Ellis et al. 2005).

In the literature, there are more than 90 published studies about crabs in the region between 1983–2023; nearly half of them are only related to the taxonomy, whereas the biological and ecological studies represent around 25% (Table 15.1). Moreover, most of these studies were published after 2010. In general, little is known about the biology and ecology of brachyuran crabs in the region, with most of the available literature coming from either commercial species such as the blue crab *Portunus segnis* or a few intertidal crabs such as *Leptochryseus kuwaitense*, *Manningis arabicum* and

DOI: 10.1201/9781003321392-19

TABLE 15.1 Brachyuran crab studies in the literature investigated in the Arabian Gulf.

SUBJECT	*COUNTRY*	*REFERENCES*
Biology and ecology	Bahrain	(Al-Rumaidh 2002; Zainal 2013)
	Iran	(Afkhami et al. 2016; Ehsan et al. 2010)
	Kuwait	(Al-Wazzan 2012, 2017; Al-Wazzan et al. 2018, 2019, 2020a, 2020b, 2021; Clayton 1988; Sattari 2011; Snowden et al. 1991, 1994; Snowden and Clayton 1995; Tahmazian 1998)
	Oman	(Clayton 2001, 2005, 2008)
	Saudi Arabia	(Rabaoui et al. 2021)
	Qatar	(Giraldes et al. 2016; Le Vay and Falamarzi 2009; Al-Khayat et al. 2018; Al-Khayat and Giraldes 2020)
Biodiversity	Iran	(Ebrahimnezhad and Naderloo 2019; Hosseini 2009; Mokhtari et al. 2008)
	Kuwait	(Al Bakri et al. 1997a, 1997b; Al-Zaidan et al. 2003; Clayton 1986; Jones 1986b)
	UAE	(Hornby 1997)
	Qatar	(Al-Khayat and Jones 1999; Galil et al. 2012)
Taxonomy	Iran	(Ghotbeddin and Naderloo 2014; Ebrahimnezhad and Naderloo 2019; Naderloo and Sari 2005, 2007; Naderloo and Türkay 2011, 2012; Naderloo et al. 2010, 2011, 2015, 2016a, 2016b; Shih et al. 2016; Naderloo 2011a, 2011b, 2012, 2013, 2014, 2015a, 2015b; Naderloo and Türkay 2011, 2009; Naderloo and Fatemi 2015; Naderloo and Schubart 2009, 2010; Teng et al. 2016; Parvizi et al. 2019; Shih et al. 2016)
	Kuwait	(Clayton and Jones 1983; Jones and Clayton 1983; Clayton 2001, 2005, 2008; Al-Kandari et al. 2023)
	Oman	(Hogarth 1994; Khvorov et al. 2012)
	Qatar	(Al-Khayat and Jones 1996; Galil et al. 2012; Giraldes et al. 2017)
	UAE	(Hogarth and Beech 2001)
	Arabian Gulf	(Apel and Türkay 1999; Naderloo 2017)
Pollution	Bahrain	(Madany et al. 1996)
	Iran	(Saadati et al. 2020; Monikh et al. 2014; Ghaeni et al. 2015; Nozar et al. 2015)
	Kuwait	(Al-Mohanna and Subrahmanyam 2001; Bu-Olayan and Subrahmanyam 1998; Karam et al. 2023)
	Saudi Arabia	(El Gammal et al. 2016; Barth 2007; Ibrahim et al. 2020; Cormack et al. 2011; Jones et al. 1994)
	Qatar	(Elezz et al. 2022; Al-Qaradawi et al. 2015; Hassan and Al-Maslamani 2015)
Other related studies		(Jones et al. 2002; Malan 1988; Abdul-Salam and Sreelatha 2000)

Leptodius exaratus (Table 15.2). In this chapter, we are highlighting the available knowledge about the biology and ecology of the brachyuran crabs in the Arabian Gulf.

15.2 MATURATION AND GROWTH

Crabs, like other organisms, need to reach sexual maturity to be able to reproduce. There are two main approaches to the determination of maturity: the first is directly related to reproductive activity (functional maturity) and the second is related to indirect assessment of morphometric measurements, as sexual maturity in the brachyurans is often associated with allometric changes in some secondary sexual characteristics relative to their growth – also known as adult ontogeny (Kotb and Hartnoll 2002; Hartnoll 1978). The methods based on functional maturity can determine the size at which an individual is physiologically capable of reproduction, whereas the morphometric methods indicate the changes that can be related to the capability for the individuals to be functionally mature (Muiño et al. 1999; de Lestang et al. 2003).

Maturation in crabs is associated with allometric changes of some body parts relative to their growth (Muiño et al. 1999). The male and female crabs can be differentiated by their abdominal flap, which is V-shaped and slender in males, while in females it is U-shaped and broad (Mariappan et al. 2000; Weis 2012). Generally, it is very common that females mature at a smaller size than males (Afkhami et al. 2016; Lwin et al. 2007; Silva et al. 2014), although a similar size at maturity in both sexes is also present (Flores and Paula 2002). Variation in maturation size is also reported for crabs of the same species inhabiting different habitats (Hines 1989; Henmi 1989). Berrill (1982) has found that geographic variation affects the maturity, growth, and longevity of the crabs. Male crabs in *Leptodius exaratus* from the northeast part of the Arabian Gulf reach their maturity at 14.9 mm in carapace width (CW) (Afkhami et al. 2016), whereas *L. exaratus* from the northwest part of the Gulf reaches maturity at 17.43 ± 0.53 mm CW (Al-Wazzan 2017).

TABLE 15.2 List of brachyuran crabs in the literature investigated in the Arabian Gulf for biological or ecological studies.

SPECIES	*FAMILY*	*HABITAT*	*COUNTRY*	*REFERENCES*
Portunus segnis (Forskål 1775) (*Portunus pelagicus*)	**Portunidae**	Subtidal	Bahrain	(Al-Rumaidh 2002; Zainal 2013)
		Subtidal	Qatar	(Le Vay and Falamarzi 2009; Giraldes et al. 2016)
		Subtidal	Iran	(Monikh et al. 2014; Hosseini et al. 2012; Ehsan et al. 2010)
Scylla serrata (Forskål 1775)		Mangroves	Iran	(Khaksari et al. 2023; Savari et al. 2013)
Leptodius exaratus (H. Milne Edwards 1834)	**Xanthidae**	Rocky shore	Kuwait	(Al-Wazzan et al. 2018, 2020b; Al-Wazzan 2017)
		Rocky shore	Iran	(Afkhami et al. 2016)
Pilumnopeus convexus (Maccagno 1936)	**Pilumnidae**	Rocky shore	Kuwait	(Al-Wazzan et al. 2019)
Opusia indica (Alcock 1900) (*Tylodiplax indica*)	**Camptandriidae**	Mudflats	Kuwait	(Snowden et al. 1994)
Nasima dotilliformis(Alcock1900)		Mudflats	Kuwait	(Snowden and Clayton 1995; Clayton and Jones 1983)
		Salt marshes	KSA	(Barth 2007)
Leptochryseus kuwaitense (D.A. Jones and Clayton 1983)		Mudflats	Kuwait	(Clayton 1988; Jones and Clayton 1983)
Manningis arabicum (DA Jones and Clayton 1983)		Mudflats	Kuwait	(Jones and Clayton 1983)
		Mangroves	Qatar	(Al-Khayat et al. 2018)
Ilyoplax stevensi (Kemp 1919)	**Dotillidae**	Mudflats	Kuwait	(Snowden et al. 1991, 1994)
Macrophthalmus (Mareotis) depressus (Rüppell 1830) *(Macrophthalmus depressus)*	**Macrophthalmidae**	Salt marshes	KSA	(Barth 2007)
		Mudflats	Kuwait	(Al-Wazzan Unpublished data)
Austruca annulipes (H. Milne Edwards 1837) (*Uca lactea annulipes*)	**Ocypodidae**	Mangroves	Iran	(Mokhtari et al. 2008)
Ocypode platytarsis (H. Milne Edwards1852)		Sandy shores	Oman	(Clayton 2001, 2008)
Ocypode jousseaumei (Nobili 1905)		Sandy shores	Oman	(Clayton 2001, 2005)
Ocypode rotundata (Miers 1882)		Sandy shores	Oman	(Clayton 2001)

The growth rate in crabs varies seasonally, between sexes, and between juveniles and adults (Hartnoll 1978; Hartnoll and Bryant 2001; Diele and Koch 2010), and is controlled mainly by temperature and food availability (Allardyce and Linton 2010; Diele and Koch 2010). In Kuwait, a strong relationship between temperature and growth was reported in the rocky shore crab *L. exaratus* (Al-Wazzan 2017), where the growth increased during spring and summer and reduced during winter. Variation in growth rate between males and females is well known, as females invest more in reproduction than in growth (Mantelatto and Martinelli 1999). In contrast, males tend to invest less energy in reproduction and more towards somatic growth (Silva et al. 2014). The growth rate in *L. exaratus* was higher in males than females, resulting from the sexual dimorphism in which males invest low energy in reproduction – unlike females – and direct it toward the growth (Hartnoll et al. 2006). Growth rates have been reported to vary between sexes in several crab species (Diele et al. 2005; Windsland et al. 2013). Although males tend to have higher growth rates than females, some crab species have similar growth rates in both males and females (Diele et al. 2005). Understanding growth rates is important in managing fisheries, being used to determine age from size frequency distributions (and hence estimation of mortality) and to estimate recovery times in highly exploited species.

15.3 SEASONALITY OF REPRODUCTION

Reproduction, like any other physiological process, is affected by interior and exterior factors. Interior or endogenous factors includes hormones that regulate the gonad maturation in crabs (Wu et al. 2010). Exterior or exogenous factors include temperature, photoperiod, and food availability (Cobo and Fransozo 2005; Zapata et al. 2003). Among all environmental factors, temperature is considered to be the main regulating factor that influences crab reproduction (Kumar et al. 2003). Generally, most crab species that inhabit tropical regions reproduce continuously throughout the year (Litulo 2005; Cobo and Fransozo 2005). The pattern of reproduction with a breeding season that starts in spring and extends throughout the summer period has been reported in several subtropical crabs (Pinheiro and Hattori 2003; Carmona-Suárez 2003). However, other patterns of breeding are also reported in these crabs, such as breeding only during summer (Brown and Loveland 1985), winter (Takeda 2005) or throughout the year (Muiño 2002).

The available studies in the Gulf region indicate that the seasonal pattern of reproduction observed in Brachyuran crabs is during the warm season. For example, in Kuwait in the northwest Arabian Gulf, *L. exaratus* are reported to reproduce between March and September (Al-Wazzan 2017). Similarly, ovigerous females of the mud crab *Tylodiplax indica* (Alcock 1900) from Kuwait mudflats were reported during summer between May and September, with a few being found in March (Snowden et al. 1994). An unpublished study by the present author on the reproduction of the mud crab *Macrophthalmus depressus* (Rüppell 1830) in Kuwait using histological examination, as well as the presence of ovigerous females in the population, indicated that the breeding season starts in March and ends in September.

In Bahrain, the breeding season of the blue crab *Portunus segnis* (previously known as *Portunus pelagicus*) is reported to occur between March and November, with ovigerous females being more abundant between June and September (Al-Rumaidh 2002), whereas in Qatar, *P. segnis* is reported to spawn during spring (March-April) with a minor spawning season in the autumn (October-November) (Le Vay and Falamarzi 2009). In contrast, *P. segnis* were found to spawn year-round in the northeast of the Gulf, with the highest gonadosomatic index (GSI) being recorded in December (Ehsan et al. 2010). These variations in the breeding season between the same species across the Gulf could reflect the variation in the annual temperature at different locations. The strongest seasonality in reproduction occurs at the northwest Arabian Gulf, where the annual variation in the temperature between summer and winter is at its maximum.

Other studies used the presence of ovigerous females as an indication of the breeding season. The mud crab *Leptochryseus kuwaitense* was reported to carry eggs in most months, except May and July, with a higher percentage of ovigerous females being present during the winter season from December to March (Sattari 2011). *L. kuwaitense* is an endemic crab species that inhabits the higher intertidal zone of the mudflats in Kuwait (Jones 1986a). Similarly, in Qatar ovigerous females of *Manningis arabicum* were reported between October to May with a peak in December and January (Al-Khayat et al. 2018). Moreover, ovigerous females of another crab from the Kuwait mud flats, *Ilyoplax stevensi*, have been observed throughout the year with a peak between September to March (Snowden et al. 1991). The presence of egg-bearing females during most of the year could indicate a long incubation period over winter rather than a continuous breeding season; however, this cannot be clarified without histological examination. Amongst other crustacea in the Gulf region, the breeding season of the tiger prawn *Penaeus semisulcatus* (De Haan 1844) in Kuwaiti coastal waters is between November and March (Hassan 2002), with spawning reported in March-April in Kuwait waters (Al-Attar and Ikenoue 1974) and December-March with a minor spawning in October on the Iranian coast in the north of the Gulf (Niamaimandi et al. 2008), and November to March-April on the Saudi Arabian coast (Badawi 1975).

15.4 SPAWNING AND INCUBATION PERIOD

Crabs are generally highly productive, with females producing a large number of eggs per brood (Fernández et al. 2000). The number of eggs produced by a female crab in a single batch (fecundity) is variable among species and is basically related to female body size (Sampedro et al. 1999; Kumar et al. 2003). Spawning occurs as the muscular layer of the ovarian wall contracts to extrude mature oocytes, along with the muscles in both cephalothorax and abdomen around the ovary that are involved in forcing the mature oocytes to move toward the oviduct (Adiyodi and Subramoniam 1983; Warner 1977; Saigusa 2000). The eggs have a high accumulation of yolk, which serves as a source of nutrients during embryogenesis. Generally, smaller eggs have a prolonged larval development period with more larval stages. However, larger eggs are associated with a shorter developmental time with fewer stages (Pinheiro and Hattori 2003). The eggs enter an incubation period, which varies depending on species and environmental factors (García-Guerrero and Hendrickx 2004; Green et al. 2014). Female crabs show an active brooding behaviour to provide oxygen to the embryo mass (Ruiz-Tagle et al. 2002; Fernández et al. 2000). The embryonic development takes place during the incubation period after fertilization and the formation of the zygote, and the diploid cell starts cell divisions to form 2, 8, 16, 32 and 64 cells. After that, the embryo continues its development through cleavage, blastula and gastrula stages, and then the embryo organs and other body parts start to develop and grow in size (Bliss 1990) (Figure 15.2). Variations among crabs are observed in the number of embryonic stages. Generally, the embryonic stages are defined based on yolk content (Okazaki and Chang 1991; Bas and Spivak 2000), the embryonic structural development and activities (Simoni et al. 2011; Tuset et al. 2011), embryo size and egg diameter (Nagao et al. 1999).

15.5 LARVAL RELEASE AND HATCHING

Most marine brachyuran crabs release their zoea larvae into the pelagic habitat to complete their development and disperse through the water column. Larval dispersal patterns vary among crab species and depend on their habitat (López-Duarte and Tankersley 2007), larval physiology (Schuh and Diesel 1995; Bas et al. 2009), and larval swimming behaviour (Roegner et al. 2013), in addition to other environmental factors including UV radiation (Hovel and Morgan 1999; Moresino et al. 2011), gravity (Park et al. 2004) and water currents (Tilburg et al. 2007). Queiroga and Blanton (2004)

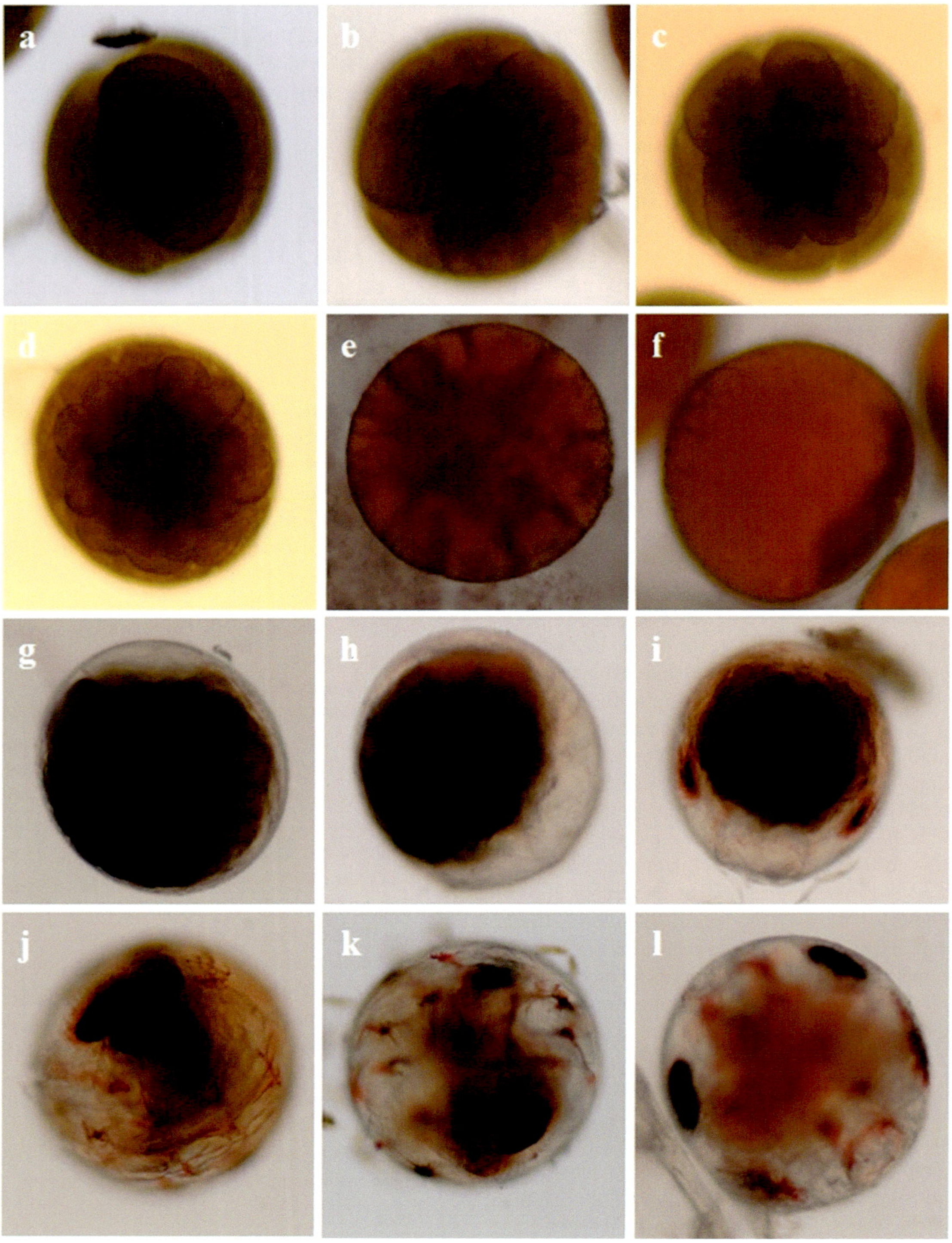

FIGURE 15.2 Embryonic developmental stages (a–l) of *Leptodius exaratus*. Note cell division (a–e), early stage of development with evidence of tissue within the yolk mass (f), formation of the embryonized nauplius "transparent region" (g–h), formation of the zoeal eye (red to black spots), followed by development and growth of the body parts and a reduction in the yolk mass (i–l) (after Al-Wazzan 2017).

described different hatching, dispersal and recruitments patterns in crabs as follows:

1. Hatching in estuaries followed by dispersal in coastal waters, with return back to estuaries for settlement
2. Hatching in coastal waters and then recruitment into estuaries
3. Complete coastal development (hatching, dispersal, and recruitment)
4. Hatching in coastal waters, dispersal in the ocean, and then return to the coastal zone for settlement

In Kuwait, *Leptodius exaratus* release their larvae at night in summer to avoid exposure to the high temperatures during the day (Al-Wazzan et al. 2018). Furthermore, they synchronize their hatching with ebb tide events around neap and spring tides. That could suggest that the larval development occurs in the shallow coastal water where the impact of the warming water is expected to be higher (Bernhardt and Leslie 2013;

Rivera-Ingraham and Lignot 2017). That could mean exposing crab larvae to higher temperatures and higher salinity levels during their development. The increase in water temperature was reported to reduce the duration of the larval cycle (Hamasaki 2003; Nurdiani and Zeng 2007) and the incubation period (Hamasaki 2003). However, if the temperature increases above the natural range, the hatching success decreases (Chou et al. 2019). Additionally, larval survival rate also decreases (Castejón et al. 2018).

15.6 SETTLEMENT AND RECRUITMENT

Crab larvae develop through several zoeal stages, and then they metamorphose into the megalopa stage, which looks much like the adult crab form, except that the abdomen is not flexed under the thorax. The megalopa then undergoes metamorphosis to the first crab instar after settling on a suitable habitat; the most common case is that larvae would settle first and then metamorphose (Diele and Simith 2006; Loher and Armstrong 2000) (Figure 15.1). It has been reported that the megalopa of several crab species actively select their habitat (Pineda et al. 2010; Tapella et al. 2012), as well as choosing the settlement time (Pardo et al. 2007). Settlement and recruitment in crabs are controlled by several factors including the tide (Queiroga et al. 2007), wind (Almeida and Queiroga 2003), tidal stream (Saigusa et al. 2003), food availability (Paula et al. 2003) and habitat (Amaral and Paula 2007; Krimsky and Epifanio 2008). In addition, prenatal or adult cues are also important (O'Connor and Judge 2010), and the effect of the adult cue on metamorphosis is species specific (Steinberg et al. 2007). It has been suggested that megalopae are capable of delaying metamorphosis if the environment is not suitable (Diele and Simith 2006). Furthermore, megalopae are reported to avoid low-oxygen environments (Johnson and Welsh 1985). However, crab juvenile survival rate and body size may be reduced after a delay of metamorphosis of the megalopa (Gebauer et al. 1999). After settlement, crab juveniles undergo a number of moults – varying according to species – to grow in size and reach maturity and join the adult population.

In the Arabian Gulf, the blue crab *P. segnis* has a major recruitment period in the winter season between December to April, with a minor recruitment period during late spring and summer periods in the Bahraini coast (Al-Rumaidh 2002). In Qatar waters, small juveniles of *P. segnis* were present throughout the year, with a major recruitment peak recorded during summer between July and August (Le Vay and Falamarzi 2009). For intertidal mudflats species found in Kuwait, juveniles of *Ilyoplax stevensi* (Snowden et al. 1991), *Opusia indica* (Alcock 1900) –previously known as *Tylodiplax indica* – (Snowden et al. 1994), and *Nasima dotilliformis* (Alcock 1900) (Snowden and Clayton 1995) were recruited during the winter period, with a peak in March. In contrast, *Leptochryseus kuwaitense* recruits throughout most of the year, with peaks in spring/summer between April to July and autumn/winter between September to December, the latter recruitment peak being reported as higher (Sattari 2011).

In the rocky shore crab *Leptodius exaratus*, juvenile crabs recruited during summer between July and September, with some low recruitment levels later in October and November (Al-Wazzan et al. 2020b). These small crabs that are recruited in July are estimated to be two months old, following hatching earlier in the spring and early summer (late April/May) (Al-Wazzan et al. 2018). Furthermore, the low level of juvenile recruitment observed in November is in response to September hatching, as the number of ovigerous females is observed to be less frequent at the end of the breeding season in September in comparison with the summer months (Al-Wazzan et al. 2020b) (Figure 15.3). Furthermore, the latter study indicates that juveniles follow a similar distribution pattern to that of adult crabs, with the abundance of juveniles being directly related to the abundance of the adult crabs and ovigerous females present on each shore (Figure 15.4).

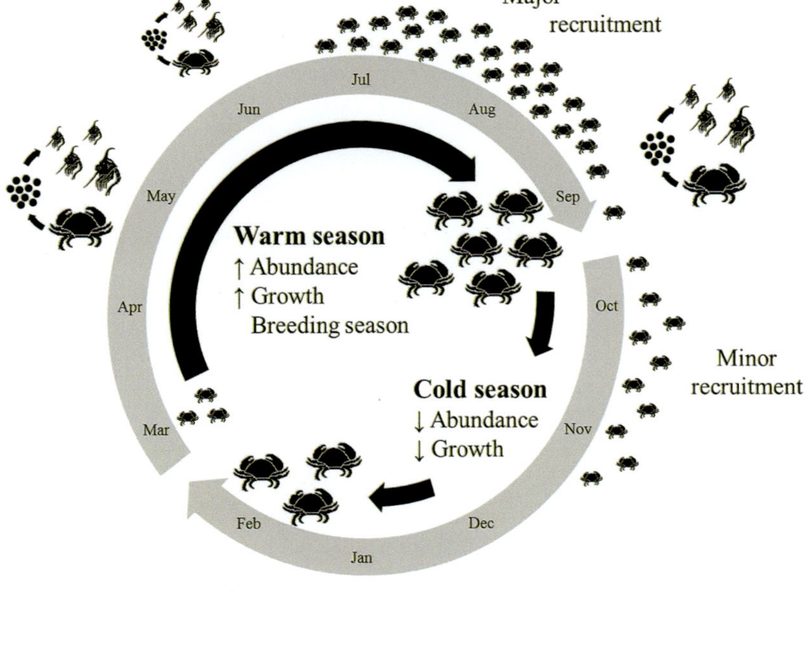

FIGURE 15.3 The main life events of the rocky shore crab *Leptodius exaratus* at different seasons in Kuwait coasts, NW Arabian Gulf (after Al-Wazzan 2017).

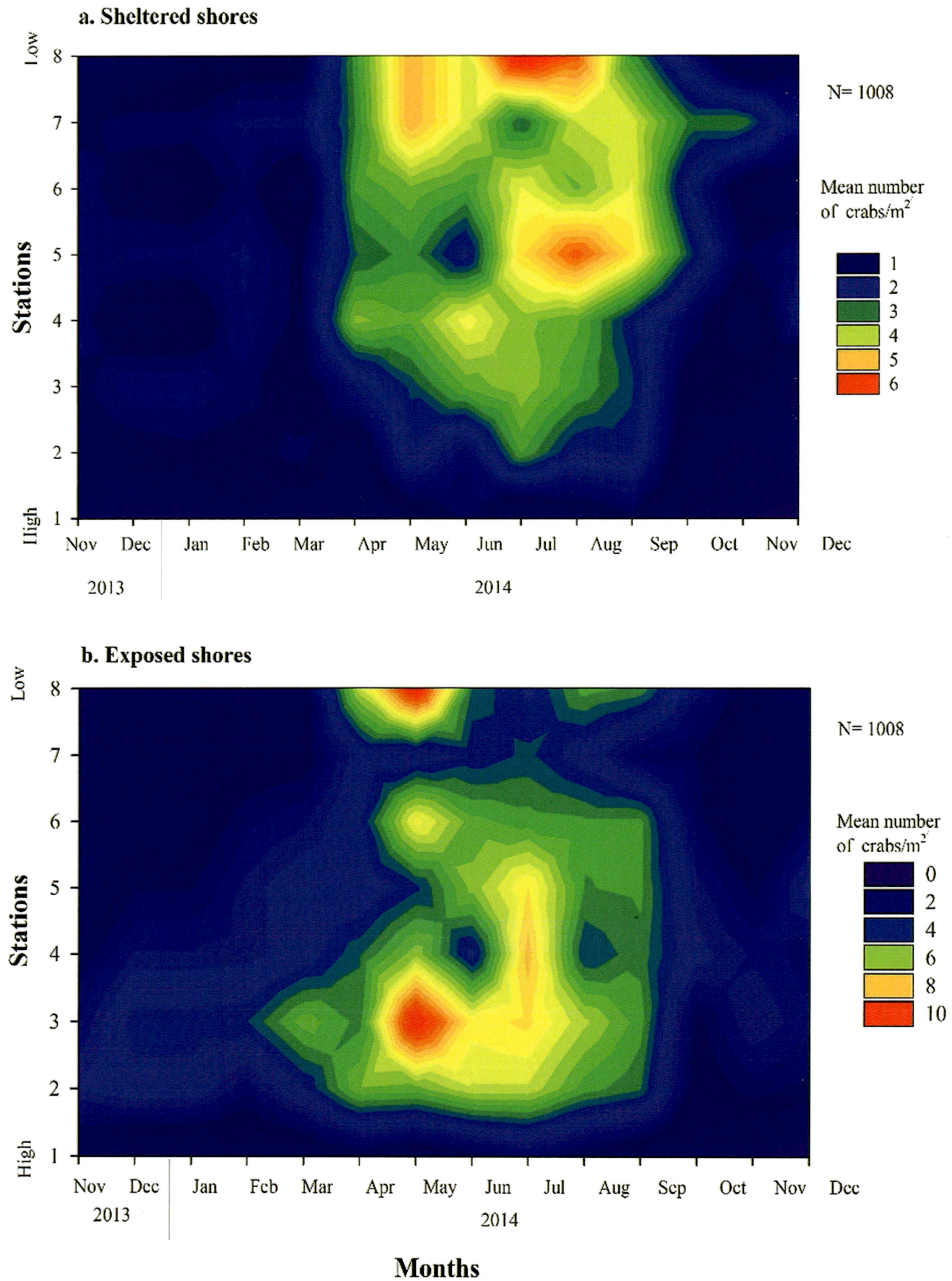

FIGURE 15.4 *Leptodius exaratus* spatial and temporal distribution of abundance along the transect (stations) in sheltered (a) and exposed (b) shores in Kuwait between October 2013 and December 2014. N = total number of quadrats used during sampling period. On the station axis, high/low refer to the location of the station at the intertidal zone; high intertidal, and low intertidal.

15.7 SUMMARY AND CONCLUSION

Under the environmental and anthropogenic influences in the region, coastal marine organisms are facing serious threats that could affect the species population as well as other species in the same habitats (Sheppard et al. 2010). For example, the bait harvesting activity on the crab *L. exaratus* affected not only the crab population but also physical destruction of the rock habitats (Al-Wazzan et al. 2019). These activities, on a larger scale, could potentially result in massive mortality in many sessile faunas, damaging the habitats as a whole. Furthermore, *L. exaratus* plays a very important role in the environmental stability of the rocky shore habitats (Al-Wazzan et al. 2019). Furthermore, crabs are very important macrofauna in the Arabian Gulf due to their significance to the food web of the coastal habitat (Al-Khayat et al. 2018). For that, information such as the size at maturity, fecundity, recruitment, growth, mortality and seasonality of reproduction form the basic essential information needed to develop appropriate conservation and management strategies for economically or ecologically important species (Botter-Carvalho et al. 2007; Rotherham and West 2003). Furthermore, understanding the life history of any species can help underpin population analysis and fisheries stock assessment models (Silva et al. 2014) and also would help in predicting the effects of climate change. For example, environmental factors such as temperature have a major controlling effect on biological activities of marine species (Chou et al. 2019; Yuan et al. 2017). A strong seasonal effect of temperature on abundance, distribution, population structure, growth and reproduction might be expected in this region. Therefore, it is very important to understand the biology and ecology, especially for the key crab species, at different coastal habitats of the Arabian Gulf; we suggest that biological and ecological studies in the area be given more attention.

REFERENCES

Abdul-Salam, J., and B.S. Sreelatha. 2000. Surface morphology of *Probolocoryphe uca* (Sarkisian, 1957) (Digenea: Microphallidae) from Kuwait Bay. *Systematic Parasitol* 46, no. 3: 209–214. doi:10.1023/A:1006394104481/METRICS.

Adiyodi, R.G., and T. Subramoniam. 1983. Arthopoda-crustacea. In: *Reproductive biology of invertebrates: Oogenesis, oviposition and oosorption*. Eds: K.G. Adiyodi, and R.G., vol. 1. Adiyodi. Chichester: John Wiley.

Afkhami, M., Schubart, C.D., and R. Naderloo. 2016. Morphometric differentiation among populations of *Leptodius exaratus* (H. Milne Edwards, 1834) (Brachyura, Xanthidae) from the Persian Gulf and the Gulf of Oman. *Crustaceana* 89, no. 3: 259–271. doi:10.1163/15685403-00003525.

Al-Attar, M.H., and H. Ikenoue. 1974. *Spawning season of shrimp Penaeus semisulcatus in the sea along the coast of Kuwait*. Kuwait: Kuwait Institute for Scientific Research, 16 p.

Al Bakri, D., Behbehani, M., and A. Khuraibet. 1997a. Quantitative assessment of the intertidal environment of Kuwait I: Integrated environmental classification. *J Environ Manage* 51, no. 4: 321–332. doi:10.1006/jema.1997.0152.

Al Bakri, D., Khuraibet, A., and M. Behbehani. 1997b. Quantitative assessment of the intertidal environment of Kuwait II: Controlling factors. *J Environ Manage* 51, no. 4: 333–341. doi:10.1006/jema.1997.0153.

Al-Kandari, M., Anker, A., Khvorov, S., and V. Skryabin. 2023. *Identification guide to the intertidal and shallow-water crustaceans of Kuwait, Cirripeda (Balanomorpha), Decapoda, Stomatopoda*. Kuwait: KISR.

Al-Khayat, J.A., and B.W. Giraldes. 2020. Burrowing crabs in arid mangrove forests on the southwestern Arabian Gulf: Ecological and biogeographical considerations. *Reg Stud Mar Sci* 39: 101416. doi:10.1016/j.rsma.2020.101416.

Al-Khayat, J.A., Jawal, L., and R. Riedel. 2018. Population biology of *Manningis Arabicum* (Jones & Clayton, 1983; Decapoda: Brachyura, Ocypodidae) from Umm SA mangrove swamps, Qatar. *Aquac Fish* 2, no. 2: 1–7. doi:10.24966/AAF-5523/100012.

Al-Khayat, J.A., and D.A. Jones. 1996. Two new genera, *Manningis* and *Leptochryseus* (Decapoda: Camptandriinae), and descriptions of the first Zoea of six Brachyurans from the Arabian Gulf. *J Crustac Biol* 16, no. 4: 797–813. doi:10.2307/1549197.

Al-Khayat, J.A., and D.A. Jones. 1999. A comparison of the macrofauna of natural and replanted mangroves in Qatar. *Estuar Coast Shelf Sci* 49: 55–63. doi:10.1016/S0272-7714(99)80009-2.

Allardyce, B.J., and S.M. Linton. 2010. Functional morphology of the gastric mills of carnivorous, omnivorous, and herbivorous land crabs. *J Morphol* 271, no. 1: 61–72. doi:10.1002/jmor.10781.

Almeida, M.J., and H. Queiroga. 2003. Physical forcing of onshore transport of crab Megalopae in the northern Portuguese upwelling system. *Estuar Coast Shelf Sci* 57, nos. 5–6: 1091–1102. doi:10.1016/S0272-7714(03)00012-X.

Al-Mohanna, S.Y., and M.N.V. Subrahmanyam. 2001. Flux of heavy metal accumulation in various organs of the intertidal marine blue crab, *Portunus pelagicus* (L.) from the Kuwait coast after the Gulf War. *Environ Int* 27, no. 4: 321–326. doi:10.1016/S0160-4120(01)00063-0.

Al-Qaradawi, I., Abdel-Moati, M., Al-Yafei, M.A., Al-Ansari, E., Al-Maslamani, I., Holm, E., Al-Shaikh, I., Mauring, A., Pinto, P.V., Abdulmalik, D., Amir, A., Miller, M., Yigiterhan, O., and B. Persson. 2015. Radioactivity levels in the marine environment along the Exclusive Economic Zone (EEZ) of Qatar. *Mar Pollut Bull* 90, nos. 1–2: 323–329. doi:10.1016/j.marpolbul.2014.10.021.

Al-Rumaidh, M.J. 2002. *The biology, population dynamics and fishery management of the blue swimming crab Portunus pelagicus (Linaeus, 1758) in Bahraini waters*. Ph.D. Thesis, University of Wales, Bangor, 273 p.

Al-Wazzan, Z. 2012. *Distribution, population structure, and intertidal bait fishery of the Xanthid Crab Leptodius exaratus (Brachyura: Xanthidae) in Kuwait's rocky shores*. M.Sc. Thesis, Bangor University, Bangor, UK.

Al-Wazzan, Z. 2017. *The biology and ecology of the rocky shore crab Leptodius exaratus (H. Milne Edwards, 1834) on the coast of Kuwait*. Ph.D. Thesis, Bangor University. http://e.bangor.ac.uk/id/eprint/10249.

Al-Wazzan, Z., Giménez, L., Behbehani, M., and L. Le Vay. 2019. Trophic niche separation in sympatric rocky shore crabs. *J Mar Biol Assoc UK* 99, no. 5: 1171–1180. doi:10.1017/S0025315419000134.

Al-Wazzan, Z., Giménez, L., Behbehani, M., and L. Le Vay. 2020a. Intertidal bait gleaning on rocky shores in Kuwait. *Ocean Coast Manag* 188: 105111. doi:10.1016/j.ocecoaman.2020.105111.

Al-Wazzan, Z., Le Vay, L., Behbehani, M., and L. Giménez. 2018. Hatching rhythm and larval release in *Leptodius exaratus* (H. Milne Edwards, 1834) (Decapoda: Brachyura: Xanthidae) from rocky shores in Kuwait. *J Crustac Biol* 38, no. 6: 804–811. doi:10.1093/jcbiol/ruy080.

Al-Wazzan, Z., Le Vay, L., Behbehani, M., and L. Giménez. 2020b. Scale-dependent spatial and temporal Patterns of abundance and population structure of the xanthid crab *Leptodius exaratus* on rocky shores in Kuwait. *Reg Stud Mar Sci* 37: 101325. doi:10.1016/j.rsma.2020.101325.

Al-Wazzan, Z., Waser, A.M., Glenner, H., Gimenez, L., and D.W. Thieltges. 2021. Temporal and spatial infection patterns of the Rhizocephalan parasite *Parasacculina leptodiae* (Guérin-Ganivet, 1911) in the crab *Leptodius exaratus* along the shores of Kuwait. *Mar Biodiv* 51, no. 6: 90. doi:10.1007/s12526-021-01235-3.

Al-Zaidan, A.S.Y., Jones, D.A., Al-Mohanna, S.Y., and R. Meakins. 2003. Endemic macrofauna of the Sulaibikhat Bay salt marsh and mudflat habitats, Kuwait: Status and need for conservation. *J Arid Environ* 54, no. 1: 115–124. doi:10.1006/jare.2001.0886.

Alcock, A. 1900. Material for a carcinological fauna of India. No. 6: The Brachyura Catometopa, or Grapsoidea. *J Asiat Soc Bengal* 69, no. 2: 279–456. https://biodiversitylibrary.org/page/35612227.

Amaral, V., and J. Paula. 2007. *Carcinus maenas* (Crustacea: Brachyura): Influence of artificial substrate type and patchiness on estimation of megalopae settlement. *J Exp Mar Biol Ecol* 346, nos. 1–2: 21–27. doi:10.1016/j.jembe.2007.01.012.

Apel, M., and M. Türkay. 1999. Taxonomic composition, distribution and zooeoahic relationships of the Grapsid and Ocypodid crab fauna of intertidal soft bottoms in the Arabian Gulf. *Estuar Coast Shelf Sci* 49: 131–142. doi:10.1016/S0272-7714(99)80018-3.

Badawi, H.K. 1975. On maturation and spawning in some penaeid prawns of the Arabian Gulf. *Mar Biol* 32, no. 1: 1–6. doi:10.1007/BF00395155.

Barth, H.-J. 2007. Crab induced salt marsh regeneration after the 1991 Gulf War oil spill. *Aquat Ecosyst Health Manag* 10, no. 3: 327–334. doi:10.1080/14634980701512624.

Bas, C.C., Luppi, T., Spivak, E.D., and L. Schejter. 2009. Larval dispersion of the estuarine crab *Neohelice granulata* in coastal marine waters of the southwest Atlantic. *Estuar Coast Shelf Sci* 83, no. 4: 569–576. doi:10.1016/j.ecss.2009.05.004.

Bas, C.C., and E.D. Spivak. 2000. Effect of salinity on embryos of two southwestern Atlantic estuarine grapsid crab species cultured in vitro. *J Crustac Biol* 20, no. 4: 647–656. doi:10.1163/20021975-99990088.

Ben-Hasan, A., and V. Christensen. 2019. Vulnerability of the marine ecosystem to climate change impacts in the Arabian Gulf – an urgent need for more research. *Glob Ecol Conserv* 17: e00556. doi:10.1016/j.gecco.2019.e00556.

Beninger, P.G., Elner, R.W., Foyle, T.P., and P.H. Odense. 1988. Functional anatomy of the male reproductive system and the female spermatheca in the snow crab *Chionoecetes Opilio* (*O. fabricius*) (Decapoda : Majidae) and a hypothesis for fertilization. *J Crustac Biol* 8, no. 3: 322–332. doi:10.1139/f86-214.

Bernhardt, J.R., and H.M. Leslie. 2013. Resilience to climate change in coastal marine ecosystems. *Ann Rev Mar Sci* 5: 371–392. doi:10.1146/annurev-marine-121211-172411.

Berrill, M. 1982. The life cycle of the green crab *Carcinus maenas* at the northern end of its range. *J Crustac Biol* 2, no. 1: 31–39. doi:10.2307/1548108.

Bliss, D.E. 1990. *Shrimps, lobsters and crabs*. Columbia, USA: Columbia University Press.

Botter-Carvalho, M.L., Santos, P.J.P., and P.V.V.C. Carvalho. 2007. Population dynamics of *Callichirus major* (Say, 1818) (Crustacea, Thalassinidea) on a beach in northeastern Brazil. *Estuar Coast Shelf Sci* 71, nos. 3–4: 508–516. doi:10.1016/j.ecss.2006.09.001.

Brown, S.K., and R.E. Loveland. 1985. A calibration method to generate seasonal hatching profiles for the fiddler crabs *Uca pugnax* (Smith, 1870) and *Uca minax* (LeConte, 1855) (Decapoda : Brachyura). *J Exp Mar Biol Ecol* 90, no. 1: 27–42. doi:10.1016/0022-0981(85)90072-3.

Bu-Olayan, A., and M.N.V. Subrahmanyam. 1998. Trace metal concentrations in the crab *Macrophthalmus depressus* and sediments on the Kuwait coast. *Environ Monit Assess* 53, no. 2: 297–304. doi:10.1023/A:100588242550.

Carmona-Suárez, C.A. 2003. Reproductive biology and relative growth in the spider crab *Maja crispata* (Crustacea: Brachyura: Majidae). *Sci Mar* 67, no. 1: 75–80. doi:10.3989/scimar.2003.67n175.

Castejón, D., Rotllant, G., Giménez, L., Torres, G., and G. Guerao. 2018. Influence of temperature and light regime on the larval development of the common spider crab *Maja brachydactyla* Balss, 1922 (Brachyura: Majidae). *Aquaculture Res* 49, no. 11: 3548–3558. doi:10.1111/are.13820.

Chou, C.C., Head, M.L., and P.R.Y. Backwell. 2019. Effects of temperature on reproductive timing and hatching success in a tropical fiddler crab. *Biol J Linn Soc* 128, no. 4: 817–827. doi:10.1093/biolinnean/blz157.

Christy, J.H. 2003. Reproductive timing and larval dispersal of intertidal crabs: The predator avoidance hypothesis. *Revista Chilena Historia Natural* 76, no. 2: 177–185. doi:10.4067/S0716-078X2003000200005.

Christy, J.H. 2011. Timing of hatching and release of larvae by brachyuran crabs: Patterns, adaptive significance and control. *Integr Comp Biol* 51, no. 1: 62–72. doi:10.1093/icb/icr013.

Clayton, D.A. 1986. Ecology of mudflats with particular reference to those of the north Arabian Gulf. In: *Marine environment and pollution: Proc first Arabian Gulf conference on environment and pollution*. Eds: R. Halwagy, D. Clayton, and M. Behbehani, 83–96. Indiana, USA: Alden Press.

Clayton, D.A. 1988. Hood construction as a spacing mechanism in *Cleistostoma kuwaitense* (Crustacea: Ocypodidae). *Mar Biol* 99, no. 1: 57–61. doi:10.1007/BF00644977.

Clayton, D.A. 2001. Acoustic calling in four species of ghost crabs: *Ocypode jousseaumei, O. platytarsus, O. rotundata* and *O. saratan* (Brachyura: Ocypodidae). *Int J Animal Sound Recording* 12, no. 1: 37–55. doi:10.1080/09524622.2001.9753477.

Clayton, D.A. 2005. Substrate (acoustic/vibrational) communication and ecology of the ghost crab *Ocypode jousseaumei* (Brachyura: Ocypodidae). *Mar Freshw Behav Physiol* 38, no. 1: 53–70. doi:10.1080/10236240500057952.

Clayton, D.A. 2008. Singing and dancing in the ghost crab *Ocypode platytarsus* (Crustacea, Decapoda, Ocypodidae). *J Nat Hist* 42, nos. 3–4: 141–155. doi:10.1080/00222930701840530.

Clayton, D.A., and D.A. Jones. 1983. The systematics and ecology of crabs belonging to the genera *Cleistostoma* De Haan and *Paracleistostoma* De Man on Kuwait mudflats. *Crustaceana* 45, no. 2: 183–199. doi:10.1163/156854083X00613.

Cobo, V.J., and A. Fransozo. 2005. Physiological maturity and relationships of growth and reproduction in the red mangrove crab *Goniopsis cruentata* (Latreille) (Brachyura, Grapsidae) on the coast of São Paulo, Brazil. *Revista Brasileira Zoologia* 22: 219–223. doi:10.1590/S0101-81752005000100027.

Cormack, C.D., Hale, J.A., Gabriel, J.J., and O. Langman. 2011. *Nasima and oil – do they mix? Assessing crab survival in oiled sediments*. Proc 2011 International Oil Spill Conference. doi:10.7901/2169-3358-2011-1-200.

Cunha, T., Hall, A., and H. Queiroga. 2005. Estimation of the *Diopatra neapolitana* annual harvest resulting from digging activity in Canal de Mira, Ria de Aveiro. *Fish Res* 76, no. 1: 56–66. doi:10.1016/j.fishres.2005.05.008.

De Haan, W. 1844. Crustacea. In: *Fauna Japonica sive Descriptio Animalium, quae in Itinere per Japoniam, Jussu et Auspiciis Superiorum, qui Summum in India Batava Imperium Tenent, Suspecto, Annis 1823-1830 Collegit, Notis, Observationibus et Adumbrationibus Illustravit.* von Siebold, P.F. (Ed), i–xxxi, ix–xvi, 1–243, pls. A–J, L–Q, 1–55. Lugduni-Batavorum. (pls. 38, 43–46, 48, 51–55, I–N).

de Lestang, S., Hall, N.G., and I.C. Potter. 2003. Reproductive biology of the blue swimmer crab (*Portunus pelagicus*, Decapoda: Portunidae) in five bodies of water on the west coast of Australia. *Fishery Bull* 101, no. 4: 745–757.

Diele, K., and V. Koch. 2010. Growth and mortality of the exploited mangrove crab *Ucides cordatus* (Ucididae) in N-Brazil. *J*

Exp Mar Biol Ecol 395, nos. 1–2: 171–180. doi:10.1016/j.jembe.2010.08.029.

Diele, K., Koch, V., and U. Saint-Paul. 2005. Population structure, catch composition and CPUE of the artisanally harvested mangrove crab *Ucides cordatus* (Ocypodidae) in the Caeté estuary, North Brazil: Indications for overfishing? *Aquat Living Resour* 18, no. 2: 169–178. doi:10.1051/alr:2005018.

Diele, K., and D.J.B. Simith. 2006. Salinity tolerance of northern Brazilian mangrove crab larvae, *Ucides cordatus* (Ocypodidae): Necessity for larval export? *Estuar Coast Shelf Sci* 68, nos. 3–4: 600–608. doi:10.1016/j.ecss.2006.03.012.

Ebrahimnezhad, S., and R. Naderloo. 2019. Quantitative analyses of decapod crustaceans at a rocky intertidal shore in the Gulf of Oman, Iran: Diversity, zonation and monsoonal fluctuations. *J Mar Biol Assoc UK* 99, no. 2: 445–459. doi:10.1017/S0025315418000310.

Ehsan, K., Nabi, S.A., and M. Yahyavi. 2010. Stock assessment and reproductive biology of the blue swimming crab, *Portunus pelagicus* in Bandar Abbas coastal waters, northern Persian Gulf. *J Persian Gulf* 1, no. 2: 11–22.

Elezz, A.A., Castillo, A., Hassan, H.M., Alsaadi, H.A., and P. Vethamony. 2022. Distribution and environmental geochemical indices of mercury in tar contaminated beaches along the coast of Qatar. *Mar Pollut Bull* 175: 113349. doi:10.1016/j.marpolbul.2022.113349.

El Gammal, M.A., Al-Madan, A., and N. Fita. 2016. Shrimp, crabs and squids as bio-indicators for heavy metals in Arabian Gulf, Saudi Arabia. *Int J Fish Aquatic Studies* 4: 200–207.

Ellis, J.C., Chen, W., O'Keefe, B., Shulman, M.J., and J.D. Witman. 2005. Predation by gulls on crabs in rocky intertidal and shallow subtidal zones of the Gulf of Maine. *J Exp Mar Biol Ecol* 324, no. 1: 31–43. doi:10.1016/j.jembe.2005.04.001.

Fernández, M., Bock, C., and H.-O. Pörtner. 2000. The cost of being a caring mother: The ignored factor in the reproduction of marine invertebrates. *Ecol Lett* 3, no. 6: 487–494. doi:10.1046/j.1461-0248.2000.00172.x.

Flores, A.A.V., and J. Paula. 2002. Sexual maturity, larval release and reproductive output of two brachyuran crabs from a rocky intertidal area in central Portugal. *Invertebr Reprod Dev* 42, no. 1: 21–34. doi:10.1080/07924259.2002.9652506.

Forskål, P. 1775. Descriptiones Animalium, Avium, Amphibiorum, Piscium, Insectorum, Vermium; quae in Itinere Orientali Observavit Petrus Forskål. Post Mortem Auctoris editit Carsten Niebuhr. Adjuncta est materia Medica Kahirina. Mölleri, Hafniae. https://www.biodiversitylibrary.org/page/2088059

Galil, B.S., Ashelby, C.W., and P.F. Clark. 2012. New records of four Leucosiid species (Crustacea: Decapoda: Brachyura: Leucosiidae) from Qatar, Persian Gulf. *Mar Biodiv Rec* 5: e84. doi:10.1017/S1755267212000498.

García-Guerrero, M., and M.E. Hendrickx. 2004. Embryology of decapod crustaceans I: Embryonic development of the mangrove crabs *Goniopsis pulchra* and *Aratus pisonii* (Decapoda: Brachyura). *J Crustac Biol* 24, no. 4: 666–672. doi:10.1651/C-2497.

Gebauer, P., Paschke, K., and K. Anger. 1999. Costs of delayed metamorphosis: Reduced growth and survival in early juveniles of an estuarine grapsid crab, *Chasmagnathus granulata*. *J Exp Mar Biol Ecol* 238, no. 2: 271–281. doi:10.1016/S0022-0981(98)00219-6.

Ghaeni, M., Pour, N.A., and M. Hosseini. 2015. Bioaccumulation of polychlorinated biphenyl (PCB), polycyclic aromatic hydrocarbon (PAH), mercury, methyl mercury, and arsenic in blue crab *Portunus segnis* from Persian Gulf. *Environ Monit Assess* 187, no. 5: 1–9. doi:10.1007/s10661-015-4459-9.

Ghotbeddin, N., and R. Naderloo. 2014. Confirming of the occurrence of three intertidal Xanthid crabs (Crustacea: Decapoda: Brachyura: Xanthidae) in the Gulf of Oman. *Mar Biodiv Rec* 7: e98. doi:10.1017/S1755267214000712.

Giraldes, B.W., Al-Maslamani, I., Al-Ashwel, A., Chatting, M., and D. Smyth. 2016. Basic assessment of Portunus Segnis (Forskål, 1775) – A baseline for stock management in the western Arabian Gulf. *Egypt J Aquat Res* 42, no. 1: 111–119. doi:10.1016/j.ejar.2016.02.001.

Giraldes, B.W., Al-Maslamani, I., and D. Smyth. 2017. A new species of Leucosiid crab (Decapoda: Brachyura: Leucosiidae) from the Arabian Gulf. *Zootaxa* 4250, no. 4: 389. doi:10.11646/zootaxa.4250.4.9.

Green, B.S., Gardner, C., Hochmuth, J.D., and A. Linnane. 2014. Environmental effects on fished lobsters and crabs. *Rev Fish Biol Fish* 24, no. 2: 613–638. doi:10.1007/s11160-014-9350-1.

Hamasaki, K. 2003. Effects of temperature on the egg incubation period, survival and developmental period of larvae of the mud crab *Scylla serrata* (Forskål) (Brachyura: Portunidae) reared in the laboratory. *Aquaculture* 219, nos. 1–4: 561–572. doi:10.1016/S0044-8486(02)00662-2.

Hartnoll, R.G. 1978. The determination of relative growth in crustacea. *Crustaceana* 34, no. 3: 281–293. doi:10.1163/156854078X00844.

Hartnoll, R.G., Baine, M.S.P., Grandas, Y., James, J., and H. Atkin. 2006. Population biology of the black land crab, *Gecarcinus ruricola*, in the San Andres Archipelago, western Caribbean. *J Crustac Biol* 26, no. 3: 316–325. doi:10.1651/C-2640.1.

Hartnoll, R.G., and A.D. Bryant. 2001. Growth to maturity of juveniles of the spider crabs *Hyas coarctatus* Leach and *Inachus dorsettensis* (Pennant) (Brachyura: Majidae). *J Exp Mar Biol Ecol* 263, no. 2: 143–158. doi:10.1016/S0022-0981(01)00303-3.

Hassan, M.A. 2002. *Reproductive biology of the tiger shrimp Penaeus semisulcatus (Decapoda: Penaeidae) in Kuwait, Arabian Gulf.* M.Sc. Thesis, Kuwait University, Kuwait.

Hassan, H.A., and I. Al-Maslamani. 2015. Environmental contamination in mangrove crustaceans inhabiting Qatari waters. In: *Qatar University life science symposium – QULSS 2015 clobal changes: The Arabian Gulf ecosystem*. Hamad bin Khalifa University Press (HBKU Press). doi:10.5339/qproc.2015.qulss2015.26.

Henmi, Y. 1989. Reproductive ecology of three ocypodid crabs II: Incubation sites and egg mortality. *Ecol Res* 4, no. 3: 261–269. doi:10.1007/BF02348447.

Hines, A.H. 1989. Geographic variation in size at maturity in brachyuran crabs. *Bull Mar Sci* 45, no. 2: 356–368.

Hogarth, P.J. 1994. Brachyuran crabs (Xanthoidea: Xanthidae, Pilumnidae, Menippidae and Trapeziidae) of southern Oman. *Trop Zool* 7: 93–108. doi:10.1080/03946975.1994.10539244.

Hogarth, P.J., and M. Beech. 2001. A first modern record of the mangrove crab *Scylla serrata* in the UAE and south-eastern Arabian Gulf. *Tribulus* 11: 30.

Hornby, R.J. 1997. A survey of the habitats, invertebrate fauna and environmental sensitivity of the mainland coast of the UAE, with information on status and distribution of crustaceans. *Tribulus* 7, no. 2: 11–17.

Hosseini, M., Vazirizade, A., Parsa, Y., and A. Mansori. 2012. Sex ratio, size distribution and seasonal abundance of blue swimming crab, *Portunus pelagicus* (Linnaeus, 1758) in Persian Gulf coasts, Iran. *World Applied Sci J* 17, no. 7: 919–925. doi:10.5829/idosi.wasj.2012.19.10.882.

Hosseini, S.H. 2009. The intertidal decapods of Bushehr, northern part of the Persian Gulf. *Iranian J Fish Sci* 8, no. 1: 37–46.

Hovel, K.A., and S.G. Morgan. 1999. Susceptibility of estuarine crab larvae to ultraviolet radiation. *J Exp Mar Biol Ecol* 237, no. 1: 107–125. doi:10.1016/S0022-0981(98)00221-4.

Ibrahim, G.A., Atallah, A.A., Anbtawi, N.M.G., and A. Ben Ahmed. 2020. Interaction of physicochemical parameters and the blue crab *Portunus pelagicus* (Linnaeus, 1758) in the Arabian Gulf. *Int J Oceans Oceanogr* 14: 33–76.

Johnson, D.A., and B.L. Welsh. 1985. Crab larvae. *J Exp Mar Biol Ecol* 86: 73–83. doi:10.1016/0022-0981(85)90043-7.

Jones, D.A. 1986a. *A field guide to the Sea Shores of Kuwait and the Arabian Gulf.* London, UK: Blandford Press and Kuwait: University of Kuwait.

Jones, D.A. 1986b. Ecology of the rocky and sandy shores of Kuwait. In: *Marine environment and pollution: Proc first Arabian Gulf conference on environment and pollution*. Eds: R. Halwagy, D. Clayton, and M. Behbehani, 69–81. Indiana, USA: Alden Press.

Jones, D.A., and D. Clayton. 1983. The systematics and ecology of crabs belonging to the genera Cleistostoma De Haan and *Paracleistostoma* De Man. *Crustaceana* 45, no. 2: 183–199. doi:10.1163/156854083X00613.

Jones, D.A., Price, A.R.G., Al-Yamani, F., and A. Al-Zaidan. 2002. Coastal and marine ecosystem. In: *The Gulf ecosystem; health and sustainability*. Eds: N.Y. Khan, M. Munawar, and A.R.G. Price, 65–103. Leiden, The Netherlands: Ecovision world Monograph Series, Backhuys Publishers.

Jones, D.A., Woodhouse, T.D., and M.D. Richmond. 1994. Intertidal recovery in the Dawhat Ad-Dafi and Dawhat al-Musallamiya region (Saudi Arabia) after the Gulf War oil spill. In: *The status of coastal and marine habitats two years after the Gulf War oil spill*. Eds: A.H. Abuzinada, and F. Krupp, 27–33. Frankfurt: Courier Forsch.-Inst Senckenberg.

Karam, Q., Ali, M., Subrahmanyam, M.N.V., Al-Abdul, K., Bentley, M., and M.U. Beg. 2019. A comparative study on the effect of dispersed and undispersed Kuwait crude oil on egg hatching and larval survival of *Epinephelus coioides*. *J Environ Biol* 40: 192–199. doi:10.22438/jeb/40/2/MRN-836.

Karam, Q., Guermazi, W., Subrahmanyam, M.N.V., Al-Enezi, Y., Ali, M., Leignel, V., and N. Annabi-Trabelsi. 2023. *Portunus pelagicus* (Linnaeus, 1758) as a sentinel species to assess trace metal occurrence: A case study of Kuwait waters (northwestern Arabian Gulf). *Toxics* 11, no. 5: 426. doi:10.3390/toxics11050426.

Kemp, S. 1919. Notes on Crustacea Decapoda in the Indian Museum. XII. Scopimerinae. *Rec Indian Mus* 16, no. 5: 305–348, pls. 12–13.

Khaksari, H., Safaie, M., and A. Salarzadeh. 2023. Population dynamics and reproductive biology of *Scylla serrata* (Forskål, 1775) on the shores overlooking the mangrove forest of the Persian Gulf. *Reg Stud Mar Sci* 57, no. 6: 102758. doi:10.1016/J.RSMA.2022.102758.

Khvorov, S.A., Al-Sinawi, M.N., Al-Kharusi, L.H., and M.N. Al-Mosharfi. 2012. Catalogue of brachyuran crabs (Decapoda) of the Sultanate of Oman: Part I. The Arabian Sea coast (Dhofar, Al-Wusta, southern part of Sharqiya). *Crustaceana* 85: 1735–1744.

Kotb, M.M.A., and R.G. Hartnoll. 2002. Aspects of the growth and reproduction of the coral gall crab *Hapalocarcinus marsupialis*. *J Crustac Biol* 22, no. 3: 558–566. doi:10.1651/0278-0372(2002)022[0558:aotgar]2.0.co;2.

Krimsky, L.S., and C.E. Epifanio. 2008. Multiple cues from multiple habitats: Effect on metamorphosis of the Florida stone crab, *Menippe mercenaria*. *J Exp Mar Biol Ecol* 358, no. 2: 178–184. doi:10.1016/j.jembe.2008.02.010.

Kumar, M.S., Xiao, Y., Venema, S., and G. Hooper. 2003. Reproductive cycle of the blue swimmer crab, *Portunus pelagicus*, off southern Australia. *J Mar Biol Assoc UK* 83, no. 5: 983–994. doi:10.1017/S0025315403008191h.

Le Vay, L., and M. Falamarzi. 2009. *Population biology and inshore fishery assessment for the blue swimming crab*. Doha, Qatar: Portunus Pelagicus.

Litulo, C. 2005. Population structure and reproductive biology of the fiddler crab *Uca inversa* (Hoffman, 1874) (Brachyura: Ocypodidae). *Acta Oecol* 27, no. 3: 135–141. doi:10.1016/j.actao.2004.11.002.

Loher, T., and D.A. Armstrong. 2000. Effects of habitat complexity and relative larval supply on the establishment of early benthic phase red king crab (*Paralithodes camtschaticus* Tilesius, 1815) populations in Auke Bay, Alaska. *J Exp Mar Biol Ecol* 245, no. 1: 83–109. doi:10.1016/S0022-0981(99)00157-4.

López-Duarte, P.C., and R.A. Tankersley. 2007. Circatidal swimming behavior of brachyuran crab zoea larvae: Implications for ebb-tide transport. *Mar Biol* 151, no. 6: 2037–2051. doi:10.1007/s00227-007-0614-3.

Lwin, T.T., Doi, W., Yokota, M., and C.A. Strüssmann. 2007. Juvenile morphology of the xanthid crab *Leptodius exaratus* (H. Milne – Edwards, 1834) (Decapoda: Brachyura), with notes on the appearance of sexual dimorphism. *Invertebr Reprod Dev* 4: 191–201.

Maccagno, T. 1936. Spedizione del Barone Raimondo Franchetti in Doncalia (1928–29). Crostacei di Assab. Decapodi Stomatopodi Anfipodi. *Annali del Museo civico di Storia naturale di Genova* 59: 171–186.

Madany, I.M., Wahab, A.A.A., and Z. Al-Alawi. 1996. Trace metals concentrations in marine organisms from the coastal areas of Bahrain, Arabian Gulf. *Water Air Soil Pollut* 91, nos. 3–4: 233–248. doi:10.1007/BF00666260.

Malan, D.E. 1988. The effects of Qatar light crude oil on the saltmarsh crab *Sesarma caienata* and its implications in the field: Toxicity to adults and larvae. *South African J Marine Sci* 7, no. 1: 37–44. doi:10.2989/025776188784379026.

Mantelatto, F.L.M., and J.M. Martinelli. 1999. Carapace width-weight relationships of *Callinectes ornatus* (Brachyura, Portunidae) from Ubatuba, Bay, Brazil. *Iheringia Serie Zoologia* 87: 111–116.

Mariappan, P., Balasundaram, C., and B. Schmitz. 2000. Decapod crustacean chelipeds: An overview. *J Biosci* 25, no. 3: 301–313. doi:10.1007/BF02703939.

Miers, E.J. 1882. On the species of Ocypoda in the collections of the British museum. *Ann Mag Nat His*, ser. 5. 10(49): 376–388; Pl. XVII

Milne Edwards, H. 1834. *Histoire Naturelle des Crustacés, Comprenant l´Anatomie, la Physiologie et la Classification de ces Animaux*. vol I. i–xxxv, 1–468. Paris: Encyclopédique Roret.

Milne Edwards, H. 1837. *Histoire Naturelle des Crustacés, Comprenant l´Anatomie, la Physiologie et la Classification de ces Animaux*. vol. II. 2: 1–532.Paris: Encyclopédique Roret. http://www.biodiversitylibrary.org/bibliography/16170#/summary

Milne Edwards, H. 1852. Observations sur les affinités zoologiques et la classification naturelle des crustacés, *Ann. sci. nat., Zool.*, ser. 3 18: 109–166, pls. 3–4.

Mokhtari, M., Savari, A., Rezai, H., Kochanian, P., and A. Bitaab. 2008. Population ecology of fiddler crab, *Uca lactea annulipes* (Decapoda: Ocypodidae) in Sirik mangrove estuary, Iran. *Estuar Coast Shelf Sci* 76, no. 2: 273–281. doi:10.1016/J.ECSS.2007.07.010.

Monikh, F.A., Hosseini, M., and S. Rahmanpour. 2014. The effect of size and sex on PCB and PAH concentrations in crab *Portunus pelagicus*. *Environ Monit Assess* 186, no. 3: 1575–1582. doi:10.1007/s10661-013-3475-x.

Moresino, R.D.H., Gonçalves, R.J., and E.W. Helbling. 2011. Sublethal effects of ultraviolet radiation on crab larvae of *Cyrtograpsus altimanus*. *J Exp Mar Biol Ecol* 407, no. 2: 363–369. doi:10.1016/j.jembe.2011.07.019.

Morgan, S.G., and J.H. Christy. 1995. Adaptive significance of the timing of larval release by crabs. *Am Nat* 145, no. 3: 457–479.

Muiño, R. 2002. Fecundity of *Liocarcinus depurator* (Brachyura: Portunidae) in the Ría de Arousa (Galicia, North-West Spain). *J Mar Biol Assoc UK* 82, no. 1: 109–116. doi:10.1017/S0025315402005222.

Muiño, R., Fernández, L., González-Gurriarán, E., Freire, J., and J.A. Vilar. 1999. Size at maturity of *Liocarcinus depurator* (Brachyura: Portunidae): A reproductive and morphometric study. *J Mar Biol Assoc UK* 79, no. 2: 295–303. doi:10.1017/S0025315498000320.

Naderloo, R. 2011a. Grapsoid crabs (Decapoda: Brachyura: Thoracotremata) of the Persian Gulf and the Gulf of Oman. *Zootaxa* 3048: 1–43.

Naderloo, R. 2011b. A new species of the *Macrophthalmus boscii*-group (Decapoda: Brachyura: Macrophthalmidae) from the Persian Gulf with designation of a neotype for M. Boscii Audouin (1826). *Mar Biodiv* 41, no. 4: 503–515. doi:10.1007/s12526-010-0077-5.

Naderloo, R. 2012. Decapod crustaceans of the littoral and shallow sublittoral Iranian coast of the Persian Gulf: Faunistics, biodiversity and zoogeography. *Zootaxa* 3374: 1–67.

Naderloo, R. 2013. The sentinel crabs of the genus *Chaenostoma* (Stimpson, 1858) (Crustacea: Brachyura: Macrophthalmidae), with description of a new species and new records. *J Nat Hist* 47, nos. 45–46: 2835–2848. doi:10.1080/00222933.2013.807948.

Naderloo, R. 2014. Invasive Hepu Mitten crab, *Eriocheir hepuensis* (Crustacea: Decapoda: Brachyura: Varunidae) from the Iranian marshland in the northern Persian Gulf estuarine system. *Mar Biodiv Rec* 7: e23. doi:10.1017/S1755267214000219.

Naderloo, R. 2015a. A new species of decorator crabs, genus *Menaethiops* Alcock, 1895 (Crustacea: Decapoda: Brachyura: Majoidea: Epialthidae), from Abu-Musa Island, Persian Gulf, Iran. *Zootaxa* 3925, no. 3: 431. doi:10.11646/zootaxa.3925.3.7.

Naderloo, R. 2015b. Two new species of *Palapedia* Ng, 1993 (Crustacea, Decapoda, Brachyura, Xanthidae) from the Persian Gulf. *Zootaxa* 3994, no. 2: 265. doi:10.11646/zootaxa.3994.2.6.

Naderloo, R. 2017. *Atlas of crabs of the Persian Gulf*. Cham: Springer. doi:10.1007/978-3-319-49374-9.

Naderloo, R., Ebrahimnejad, S., Dustali, A., and M. Mahdian. 2016a. Five crabs of the families Xanthidae and Pilumnidae (Crustacea: Decapoda: Brachyura) from Abu-Musa Island, Iran; new records for the Persian Gulf. *Mar Biodiv Rec* 9, no. 1: 19. doi:10.1186/s41200-016-0019-4.

Naderloo, R., Ebrahimnezhad, S., and A. Sari. 2015. Annotated checklist of the decapod crustaceans of the Gulf of Oman, northwestern Indian Ocean. *Zootaxa* 4028, no. 3: 397–412. doi:10.11646/zootaxa.4028.3.5.

Naderloo, R., and Y. Fatemi. 2015. New record of the rare alien grapsoid crab *Percnon affine* (H. Milne Edwards, 1853) (Crustacea, Decapoda, Grapsoidea, Plagusidae) from the northern Indian Ocean. *Mar Biodiv Rec* 8: e92. doi:10.1017/S1755267215000603.

Naderloo, R., and A. Sari. 2005. Iranian subtidal Leucosiid crabs (Crustacea: Deacapoda: Brachyura) of the Persian Gulf: Taxonomy and zoogeography. *Iran J Anim Biosyst* 1: 28–43.

Naderloo, R., and A. Sari. 2007. New record of *Hyastenus inermis* (Rathbun, 1911) (Brachyura, Majidae) from the Persian Gulf, Iran. *Crustaceana* 80: 1261–1264.

Naderloo, R., and C.D. Schubart. 2009. Redescription and mitochondrial identification of *Chiromantes boulengeri* (Calman, 1920) (Decapoda: Brachyura: Sesarmidae) based on fresh material from the Persian Gulf, Iran. *Zootaxa* 2128, no. 1: 61–68. doi:10.11646/zootaxa.2128.1.3.

Naderloo, R., and C.D. Schubart. 2010. Description of a new species of *Parasesarma* (Crustacea; Decapoda; Brachyura; Sesarmidae) from the Persian Gulf, based on morphological and genetic characteristics. *Zoologischer Anzeiger – J Comparative Zool* 249, no. 1: 33–43. doi:10.1016/j.jcz.2010.01.003.

Naderloo, R., Schubart, C.D., and H.T. Shih. 2016b. Genetic and morphological separation of Uca Occidentalis, a new East African fiddler crab species, from *Uca annulipes* (H. Milne Edward, 1837) (Crustacea: Decapoda: Brachyura: Ocypodidae). *Zoologischer Anzeiger – J Comparative Zool* 262: 10–19. doi:10.1016/j.jcz.2016.03.010.

Naderloo, R., and M. Türkay. 2009. A new species of the genus *Nanosesarma* (Crustacea: Decapoda: Brachyura: Sesarmidae), and redescription of *Nanosesarma Jousseaumei* (Nobili, 1906), including new records from the Persian Gulf. *J Nat Hist* 43, nos. 47–48: 2911–2923. doi:10.1080/00222930903219996.

Naderloo, R., and M. Türkay. 2011. A new species of the *Macrophthalmus boscii*-group (Decapoda: Brachyura: Macrophthalmidae) from the Persian Gulf with designation of a Neotype for *M. boscii* Audouin (1826). *Mar Biodiv* 41, no. 4: 503–515.

Naderloo, R., and M. Türkay. 2012. Decapod crustaceans of the littoral and shallow sublittoral Iranian coast of the Persian Gulf: Faunistics, biodiversity and zoogeography. *Zootaxa* 3374: 1–67.

Naderloo, R., Türkay, M., and M. Apel. 2011. Brachyuran crabs of the family Macrophthalmidae Dana, 1851 (Decapoda: Brachyura: Macrophthalmidae) of the Persian Gulf. *Zootaxa* 2911: 1–42.

Naderloo, R., Türkay, M., and H.L. Chen. 2010. Taxonomic revision of the wide-front fiddler crabs of the *Uca lactea* group (Crustacea: Decapoda: Brachyura: Ocypodidae) in the Indo-West Pacific. *Zootaxa* 2500, no. 1: 1. doi:10.11646/zootaxa.2500.1.1.

Nagao, J., Munehara, H., and K. Shimazaki. 1999. Embryonic development of the hair crab *Erimacrus isenbeckii*. *J Crustac Biol* 19, no. 1: 77. doi:10.2307/1549549.

Niamaimandi, N., Aziz, A., Khalijah, D.S., Roos, S.C., and B. Kiabi. 2008. Reproductive biology of the green tiger prawn (*Penaeus semisulcatus*) in coastal waters of Bushehr, Persian Gulf. *ICES J Mar Sci* 65, no. 9: 1593–1599. doi:10.1093/icesjms/fsn172.

Nobili, G. 1905. Note sur *Ocypoda fabricii* Milne Edwards. *Bulletin du Muséum d'Histoire naturelle, Paris* 11, no. 4: 229–235.

Nozar, S.L.M., Pauzi, M.Z., Salarpouri, A., Daghooghi, B., and M. Salimizadeh. 2015. Total petroleum hydrocarbons in edible marine biota from northern Persian Gulf. *Environ Monit Assess* 187, no. 4: 1–6. doi:10.1007/s10661-015-4443-4.

Nurdiani, R., and C. Zeng. 2007. Effects of temperature and salinity on the survival and development of mud crab, *Scylla serrata* (Forsskål), larvae. *Aquaculture Res* 38, no. 14: 1529–1538. doi:10.1111/j.1365-2109.2007.01810.x.

O'Connor, N.J., and M.L. Judge. 2010. Molting of megalopae of the Asian shore crab *Hemigrapsus sanguineus* in intertidal habitats of northeastern North America. *J Exp Mar Biol Ecol* 385, nos. 1–2: 92–97. doi:10.1016/j.jembe.2010.01.020.

Okazaki, R.K., and E.S. Chang. 1991. Ecdysteroids in the embryos and sera of the crabs, *Cancer magister* and *C. anthonyi*. *Gen Comp Endocrinol* 81, no. 2: 174–186. doi:10.1016/0016-6480(91)90002-N.

Pardo, L.M., Palma, A.T., Prieto, C., Sepulveda, P., Valdivia, I., and F.P. Ojeda. 2007. Processes regulating early post-settlement habitat use in a subtidal assemblage of brachyuran decapods. *J Exp Mar Biol Ecol* 344, no. 1: 10–22. doi:10.1016/j.jembe.2006.12.024.

Park, S., Epifanio, C.E., and E.K. Grey. 2004. Behavior of larval *Hemigrapsus sanguineus* (de Haan) in response to gravity and pressure. *J Exp Mar Biol Ecol* 307, no. 2: 197–206. doi:10.1016/j.jembe.2004.02.007.

Parvizi, E., Keikhosravi, A., Naderloo, R., Solhjouy-Fard, S., Sheibak, F., and C.D. Schubart. 2019. Phylogeography of *Potamon ibericum* (Brachyura: Potamidae) identifies Quaternary glacial refugia within the Caucasus biodiversity hot spot. *Ecology Evolution* 9, no. 8: 4749–4759. doi:10.1002/ece3.5078.

Paula, J., Dornelas, M., and A.A.V. Flores. 2003. Stratified settlement and moulting competency of brachyuran megalopae in Ponta Rasa mangrove swamp, Inhaca Island (Mozambique). *Estuar Coast Shelf Sci* 56, no. 2: 325–337. doi:10.1016/S0272-7714(02)00165-8.

Pineda, J., Porri, F., Starczak, V., and J. Blythe. 2010. Causes of decoupling between larval supply and settlement and consequences for understanding recruitment and population connectivity. *J Exp Mar Biol Ecol* 392, nos. 1–2: 9–21. doi:10.1016/j.jembe.2010.04.008.

Pinheiro, M.A.A., and G.Y. Hattori. 2003. Embryology of the mangrove crab *Ucides cordatus* (Brachyura: Ocypodidae). *J Crustac Biol* 23, no. 3: 729–737. doi:10.1651/C-2334.

Queiroga, H., and J. Blanton. 2004. Interactions between behaviour and physical forcing in the control of horizontal transport of decapod crustacean larvae. *Adv Mar Biol* 47: 107–214. doi:10.1016/S0065-2881(04)47002-3.

Queiroga, H., Cruz, T., Dos Santos, A., Dubert, J., González-Gordillo, J.I., Paula, J., Peliz, A., Miguel, A., and P. Santos. 2007. Oceanographic and behavioural processes affecting invertebrate larval dispersal and supply in the western Iberia upwelling ecosystem. *Progress Oceanog* 74: 174–191. doi:10.1016/j.pocean.2007.04.007.

Rabaoui, L., Yacoubi, L., Lin, Y.J., Joydas, T.V., Maneja, R.H., Dagoy, J., Qurban, M.A., and R.H. Roa-Ureta. 2021. Distribution, abundance, and life history traits of the blue swimming crab *Portunus segnis* (Forskål, 1775) in the Saudi waters of the Arabian Gulf. *Reg Stud Mar Sci* 46: 101895. doi:10.1016/j.rsma.2021.101895.

Rivera-Ingraham, G.A., and J.H. Lignot. 2017. Osmoregulation, bioenergetics and oxidative stress in coastal marine invertebrates: Raising the questions for future research. *J Exp Biol* 220, no. 10: 1749–1760. doi:10.1242/jeb.135624.

Roegner, G.C., Daly, E.A., and R.D. Brodeur. 2013. Surface distribution of brachyuran megalopae and ichthyoplankton in the Columbia River plume during transition from downwelling to upwelling conditions. *Cont Shelf Res* 60: 70–86. doi:10.1016/j.csr.2013.04.007.

Rotherham, D., and R.J. West. 2003. Comparison of methods for sampling populations of ghost shrimp, *Trypaea australiensis* (Decapoda: Thalassinidea: Callianassidae). *Fish Res* 60, nos. 2–3: 585–591. doi:10.1016/S0165-7836(02)00137-6.

Ruiz-Tagle, N., Fernández, M., and H.O. Pörtner. 2002. Full time mothers: Daily rhythms in brooding and nonbrooding behaviors of brachyuran crabs. *J Exp Mar Biol Ecol* 276, nos. 1–2: 31–47. doi:10.1016/S0022-0981(02)00238-1.

Rüppell, E. (1830). *Beschreibung und Abbildung von 24 Arten Kurzschwänzigen Krabben, als Beitrag zur Naturgeschichte des rothen Meeres*. Frankfurt am Main, Germany: Brönner, 28 pp., 6 pls.

Saadati, M., Soleimani, M., Sadeghsaba, M., and M.R. Hemami. 2020. Bioaccumulation of heavy metals (Hg, Cd and Ni) by sentinel crab (*Macrophthalmus depressus*) from sediments of Mousa Bay, Persian Gulf. *Ecotoxicol Environ Saf* 191: 109986. doi:10.1016/j.ecoenv.2019.109986.

Saigusa, M. 2000. Hatching of an estuarine crab, *Sesarma haematocheir*: Factors affecting the timing of hatching in detached embryos, and enhancement of hatching synchrony by the female. *J Oceanog* 56: 93–102. doi:10.1023/A:1011118726283.

Saigusa, M., Okochi, T., and S. Ikei. 2003. Nocturnal occurrence, and synchrony with tidal and lunar cycles, in the invertebrate assemblage of a subtropical estuary. *Acta Oecol* 24, Suppl. 1: 191–204. doi:10.1016/S1146-609X(03)00034-1.

Saigusa, M., Terajima, M., and M. Yamamoto. 2002. Structure, formation, mechanical properties, and disposal of the embryo attachment system of an estuarine crab, *Sesarma haematocheir*. *Biol Bull* 203, no. 3: 289–306. doi:10.2307/1543572.

Sampedro, M.P., Gonzalez-Gurriarán, E., Freire, J., and R. Muino. 1999. Morphometry and sexual maturity in the spider crab *Maja squinado* (Decapoda: Majidae) in Galicia, Spain. *J Crustac Biol* 9: 578–592. doi:10.2307/1549263.

Sattari, Z.M.H. 2011. *Biological characterises and ecological aspects of the ocypoid crab Leptochryseus kuwaitiansis in intertidal zone of Kuwait*. M.Sc. Thesis, Kuwait University, Kuwait.

Savari, R., Naderloo, R., Kamrani, E., and M. Rezaie-Atagholipour. 2013. Preliminary biological information of *Scylla serrata* (Forskål, 1775) (Brachyura, Portunidae) in the Persian Gulf and Gulf of Oman: A conservation priority. *Crustaceana* 86, no. 3: 322–335. doi:10.1163/15685403-00003174.

Schuh, M., and R. Diesel. 1995. Breeding in a rockpool: Larvae of the semiterrestrial crab *Armases* [=*Sesarma*] *miersii* (Rathbun) (Decapoda: Grapsidae) develop in a highly variable environment. *J Exp Biol Ecol* 185: 109–129. doi:10.1016/0022-0981(94)00139-5.

Sheppard, C., Al-Husiani, M., Al-Jamali, F., Al-Yamani, F., Baldwin, R., Bishop, J., Benzoni, F., et al. 2010. The Gulf: A young sea in decline. *Mar Pollut Bull* 60, no. 1: 13–38. doi:10.1016/j.marpolbul.2009.10.017.

Shih, H.T., Ng, P.K.L., Davie, P.J.F., Schubart, C.D., Türkay, M., Naderloo, R., Jones, D., and M.Y. Liu. 2016. Systematics of the family Ocypodidae rafinesque, 1815 (Crustacea: Brachyura), based on phylogenetic relationships, with a reorganization of subfamily rankings and a review of the taxonomic status of *Uca* Leach, 1814, Sensu lato and its subgenera. *Ruffles Bull Zoology* 64: 139–175.

Silva, T.E., Fumis, P.B., Almeida, A.C., Bertini, G., and V. Fransozo. 2014. Morphometric analysis of the mud crab *Hexapanopeus paulensis* Rathbun, 1930 (Decapoda, Xanthoidea) from the southeastern coast of Brazil. *Lat Am J Aquat Res* 42, no. 3: 588–597.

Simoni, R., Cannicci, S., Anger, K., Pörtner, H.O., and F. Giomi. 2011. Do amphibious crabs have amphibious eggs? A case study of *Armases miersii*. *J Exp Mar Biol Ecol* 409, nos. 1–2: 107–113. doi:10.1016/j.jembe.2011.08.010.

Snowden, D.A., Al-Taher, E.Y., and R.J. Clayton. 1994. Aspects of the ecology of *Tylodiplax indica* Alcock (Brachyura, Ocypodidae) on a Kuwait mudflat. *J Univ Kuwait Sci* 21: 219–228.

Snowden, R.J., and D.A. Clayton. 1995. Aspects of the life history and seasonal ecology of *Nasima dotilliformis* (Alcock) (Brachyura: Ocypodidae) on a Kuwait mudflat. *J Kuwait Univ Sci* 22: 84–94.

Snowden, R.J., Clayton, D.A., and E.Y. Al-Taher. 1991. Population biology of *Ilyoplax-stevensi* (Brachyura, Ocypodidae) on a Kuwait mudflat. *Mar Ecol Prog Ser* 71, no. 3: 219–225.

Steinberg, M.K., Epifanio, C.E., and A. Andon. 2007. A highly specific chemical cue for the metamorphosis of the Asian shore crab, *Hemigrapsus sanguineus*. *J Exp Mar Biol Ecol* 347, nos. 1–2: 1–7. doi:10.1016/j.jembe.2007.02.005.

Tahmazian, A.A. 1998. *The impact of beach modifications on the diversity of the brachyuran crabs of the Kuwait City shoreline*. M.Sc. Thesis, Kuwait University, Kuwait.

Takeda, S. 2005. Sexual differences in behaviour during the breeding season in the soldier crab (*Mictyris brevidactylus*). *J Zool* 266, no. 2: 197–204. doi:10.1017/S0952836905006783.

Tapella, F., Sotelano, M.P., Romero, M.C., and G.A. Lovrich. 2012. Experimental natural substrate preference of southern king crab *Lithodes santolla* larvae. *J Exp Mar Biol Ecol* 411: 70–77. doi:10.1016/j.jembe.2011.11.003.

Teng, S.J., Shih, H.T., Naderloo, R., and L. Corbari. 2016. A review of the *Chaenostoma boscii* species-complex (Decapoda: Brachyura: Macrophthalmidae) from the Indo-West Pacific. *Crustacean Res* 45: 15–27. doi:10.18353/crustacea.45.0_15.

Tilburg, C.E., Dittel, A.I., and C.E. Epifanio. 2007. Retention of crab larvae in a coastal null zone. *Estuar Coast Shelf Sci* 72, no. 4: 570–578. doi:10.1016/j.ecss.2006.11.030.

Tuset, V.M., Espinosa, D.I., García-Mederos, A., Santana, J.I., and J.A. González. 2011. Egg development and fecundity estimation in deep-sea red crab, *Chaceon affinis* (Geryonidae), off the Canary Islands (NE Atlantic). *Fish Res* 109, nos. 2–3: 373–378. doi:10.1016/j.fishres.2011.02.021.

Warner, G.F. 1977. *The biology of crabs*. London, UK: The Gresham Press.

Weis, J.S. 2012. *Walking sideways: The remarkable world of crabs*. New York, USA: Cornell University Press.

Windsland, K., Hvingel, C., Nilssen, E.M., and J.H. Sundet. 2013. Evaluation of von Bertalanffy growth curves for the introduced red king crab (*Paralithodes camtschaticus*) in Norwegian waters. *Fish Res* 145: 15–21. doi:10.1016/j.fishres.2013.03.003.

WoRMS Editorial Board (2023). World Register of Marine Species. VLIZ. https://www.marinespecies.org. doi:10.14284/170

Wu, X., Cheng, Y., Zeng, C., Wang, C., and Z. Cui. 2010. Reproductive performance and offspring quality of the first and the second brood of female swimming crab, *Portunus trituberculatus*. *Aquaculture* 303, nos. 1–4: 94–100. doi:10.1016/j.aquaculture.2010.03.006.

Yuan, Q., Wang, Q., Zhang, T., Li, Z., and J. Liu. 2017. Effects of water temperature on growth, feeding and molting of juvenile Chinese mitten crab *Eriocheir sinensis*. *Aquaculture* 468: 169–174. doi:10.1016/j.aquaculture.2016.10.007.

Zainal, K.A.Y. 2013. Natural food and feeding of the commercial blue swimmer crab, *Portunus pelagicus* (Linnaeus, 1758) along the coastal waters of the Kingdom of Bahrain. *J Assoc Arab Univ Basic Appl Sci* 13, no. 1: 1–7. doi:10.1016/j.jaubas.2012.09.002.

Zapata, V., López Greco, L.S., Medesani, D., and E.M. Rodriguez. 2003. Ovarian growth in the crab *Chasmagnathus granulata* induced by hormones and neuroregulators throughout the year: In vivo and in vitro studies. *Aquaculture* 224, nos. 1–4: 339–352. doi:10.1016/S0044-8486(03)00226-6.

Meiofauna – present status in the Arabian Gulf

16

Gopikrishna Mantha, Manal Al-Kandari, Fathima Thuslim and Faiza Al-Yamani

16.1 INTRODUCTION

Coral reefs are some of the planet's most diverse and productive ecosystems. They provide habitats for countless marine life species, including many essential to human food supplies and economic activities (Sanciangco et al. 2013). Around the Arabian Peninsula, coral reefs are particularly important due to their location in a region of significant geopolitical and economic importance. The Arabian Peninsula is surrounded by the Red Sea to the west and the Arabian Sea to the east. Both bodies of water have extensive coral reef systems, supporting rich and diverse marine communities. In the Red Sea, there are over 600 species of corals and as many as 2000 species of fish and invertebrates in the entire reef ecosystem. The Arabian Sea is also home to many species of corals, including the highly diverse and important fringing reefs off the coast of Oman. Coral reefs serve as essential habitats for a variety of organisms, including those invertebrate animals belonging to the size fraction of meiofauna (Guzmán et al. 1987; Barroso et al. 2018). This diverse group includes small invertebrates among others such as Copepoda, Foraminifera, Gastropoda, Nematoda, Polychaeta, and Ostracoda (Guzmán et al. 1987). These are all important for the functioning of submerged marine ecosystems. These small animals are important food sources for larger marine animals, help to recycle nutrients, and participate in complex microbial interactions that are essential to corals and reef health. Despite their importance, coral reefs and their associated ecosystems around the Arabian Peninsula are under threat from a variety of human activities, including overfishing, pollution, and coral bleaching due to climate change (Coles and McCain 1990; Al-Yamani 2021). These pressures drive declines in reef coral abundance and associated meiofaunal communities. To preserve these vital resources, it is essential to enact strong conservation measures that address the direct and indirect drivers of coral health. By protecting and preserving these ecosystems, we can ensure that the coral reefs and their associated marine fauna including meiofauna around the Arabian Peninsula thrive for generations to come.

Meiobenthic organisms are important metazoan assemblages of the marine realm inhabiting the sub-soil of intertidal and benthic regions. They are defined by their body size (31–1000 μm according to Giere 2009; Baguley et al. 2019; Soto 2020). They are the most diversified groups of marine biota and include around 24 phyla of the 35 animal phyla present today (Balsamo et al. 2010; Baguley et al. 2019; Semprucci and Sandulli 2020), six of which are exclusively meiobenthic, i.e., Gastrotricha, Gnathostomulida, Kinorhyncha, Loricifera, Micrognathozoa, and Tardigrada. These meiobenthic representatives are either temporary or permanent inhabitants of the benthic ecosystem, which fuel other faunal compartments of the ecosystem (Danovaro et al. 2008; Leguerrier et al. 2003; Pascal et al. 2019). In order to avoid predation, they have undergone tremendous structural changes to become meiobenthic and lead a protected life between the sand grains at varying sediment depths and oceanic depths, sometimes even found in the abyssal zones of the deep sea (Giere 2009). Because of the meiobenthic mode of life, these organisms have evolved many life history strategies, suitable for their survival and abundance. As a result of such ventures, several sub-species and species evolved, making the meiobenthic group rich in biodiversity (Giere 2009).

As the evolutionary process is continuous, and meiofauna diversity is less explored, new species continue to be discovered. This warrants continuous monitoring of marine ecosystems and regular studies pertaining to biodiversity, taxonomy, and biogeography of the meiobenthic fauna. Such studies are being carried out throughout the world, which often contribute new information. However, studies on meiobenthic fauna from the Arabian/Persian Gulf (APG) region are scanty, with sporadic accounts here and there. In order to maintain and judiciously utilize our natural resources we should have a comprehensive knowledge of meiobenthic fauna. Our main objective of this chapter is to gather all the available information on meiofauna from the APG and highlight their importance in studying their taxonomy, diversity, biology, and ecology in

DOI: 10.1201/9781003321392-20

order to better understand their role in framing the marine benthic environment of the APG.

16.2 METHODOLOGY

In order to understand the quantity and quality of meiobenthic studies carried out in the APG region, we conducted a systematic literature search using Clarivate Analytics' Web of Science (WoS) and Scopus databases, an international research repository, reported according to PRISMA 2020 guidelines (Page et al. 2021). We searched and used only original peer-reviewed research papers available online; reviews, reports, and commentaries were omitted. The search was not constrained by year, but followed a combination of keywords using Boolean operators "AND" and "OR," and the following search terms used to collect papers that specifically dealt with: "Arabian Gulf" AND "Foraminifera," "Persian Gulf" AND "Foraminifera," and for all the APG countries, namely Bahrain, Iran, Iraq, Kuwait, Saudi Arabia, and the UAE, and then refined this using the keywords "marine" and then "benthic." Although the Arabian Gulf is the same as the Persian Gulf, we retrieved different numbers of records while using the keywords "Arabian Gulf" and "Persian Gulf." Hence, we used the keywords separately and combined the retrieved records. The same pattern was used for all keywords "Ostracod," "meiofauna" OR "meiobenthic," and "nematode." The resulting publications were manually filtered and screened for unrelated literature. All retrieved publications from WoS and Scopus in national or local languages were also included. For nematodes alone, manual screening was done to check whether the publication belonged to benthic or parasitic modes of life. After filtering, we removed certain publications not meeting the specified criteria. The data obtained were used to plot the comparative bar graphs using SigmaPlot version 12.

16.3 RESULTS

The literature review from WoS and Scopus revealed very few articles on the meiobenthic community of the APG region, including its surrounding countries. Except for foraminiferans, studies pertaining to other groups of meiofauna were limited. Scopus retrieved more publications than WoS, where the maximum number of works pertaining to Foraminifera (refined by benthic and marine) were from Iran (158 studies) and the minimum from Qatar (two studies); ostracods were retrieved with a maximum of six studies each from the Arabian/Persian Gulf, whereas Iraq and Saudi Arabia showed no publications; the maximum number of publications retrieved on meiofauna from Iran were seven studies and Qatar showed no publications, while the maximum number of studies retrieved for nematodes were from the Arabian/Persian Gulf (six studies) and Bahrain, Iraq and the UAE (Figure 16.1).

16.3.1 Review of literature from the Arabian/Persian Gulf

16.3.1.1 Arabian/Persian Gulf

As early as 1965, Murray initiated the study of foraminifera from the APG with a series of preliminary but most valuable publications on the diversity and taxonomy in the APG from the Abu-Dhabi region (Murray 1965b, 1970), Halat Al Bahrani region (Murray 1966a), Khor Al Bazam region (Murray 1966b), from the shelf of the Trucial Coast (Murray 1966c), and from parts of the APG (Murray 1965a, 1973). Later, Cherif et al. (1997) identified 98 foraminiferal species from 40 sediment samples collected from different parts of the APG. Different groups of foraminiferans contributed towards the diversity, namely *Textulariina* (15%, agglutinated forms), *Miliolina*

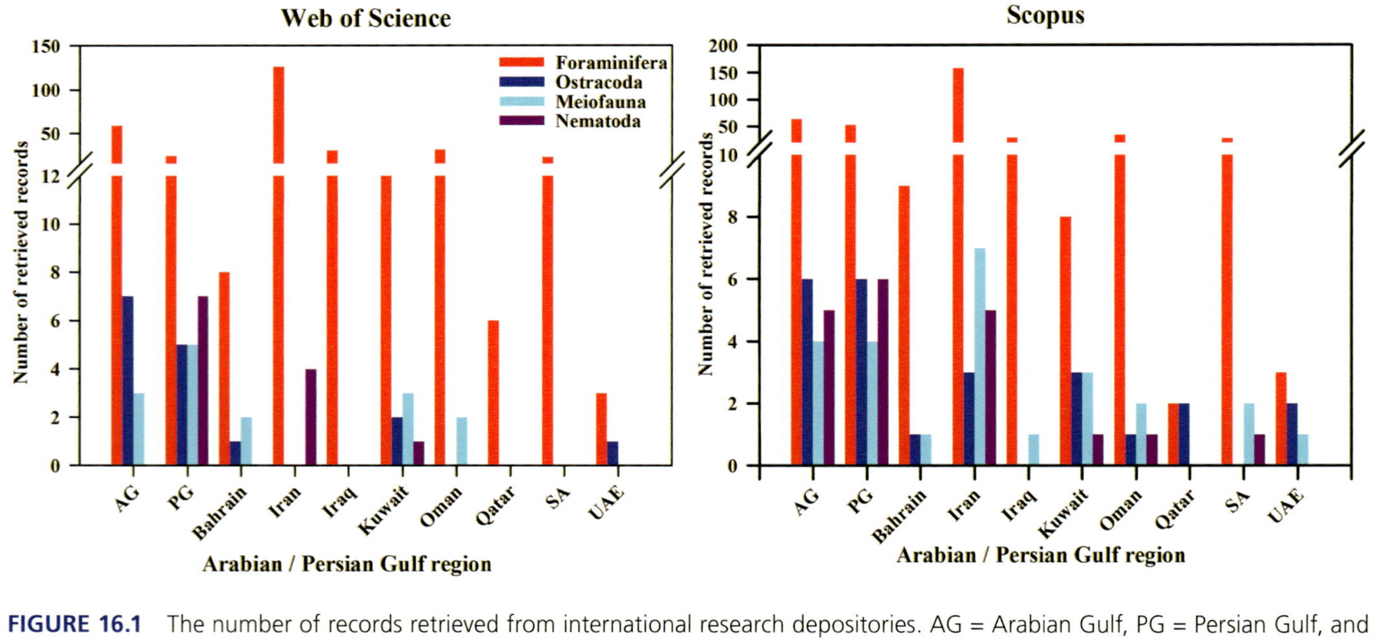

FIGURE 16.1 The number of records retrieved from international research depositories. AG = Arabian Gulf, PG = Persian Gulf, and SA = Saudi Arabia.

(~37%, porcellaneous forms), and *Lagenina* (0.1%, hyaline forms). According to them, the Iranian Shelf, near the mouth of the Shatt Al-Arab, displayed high diversity, as it receives an appreciable amount of nutrients from drainage and river run-offs, whereas the least foraminiferan diversity was shown by the Arabian Shelf, which is arid and practically deprived of active drainage (Al-Yamani 2021). The deepest parts of the APG with more agglutinated foraminiferans were observed near the Strait of Hormuz, with the highest diversities. Cherif et al. (1997) inferred that the amount of nutrients provided in this landlocked basin is an important factor in determining the constitution of the foraminifera of the bottom sediments. Saidova (2010) reported 18 major foraminiferan communities from the APG, where it was noted that less than 80 species are commonly occurring in this region.

Bate (1971) and Bate and Gurney (1981) studied ostracods from the APG region, mainly from the Abu-Dhabi lagoons and the nearby shelf areas and recorded about 47 species, while discovering a new species, *Xestoleberis postangulata* (Bate and Sheppard 1980). Nabavi et al. (2008) inferred that the benthic condition in the APG is appropriate for foraminifera, as they are dominant at all surveyed stations, which could be related to the prevailing favourable conditions, such as sediment grain size, total organic carbon, and water temperature, for their diversity and abundance.

Meiofauna were analysed in three successive cruises in the APG by the Regional Organization for the Protection of the Marine Environment (ROPME) (ROPME 2013), which reported 70 species during Summer 2000, 205 species during Summer 2001, and 281 species during Winter 2006. The highest number of meiofauna recorded during Summer 2000 and Summer 2001 were 39 and 40 species at the middle of the southern part of the ROPME Sea Area (RSA) and for Winter 2006 with 65 species from the Iranian coast of the Sea of Oman. They reported seven meiofauna groups (viz., Foraminifera, Ostracoda, Gastropoda, Bivalvia, Nematoda, Polychaetae larvae, and Copepoda), where Foraminifera, Ostracoda, and small Mollusca were the most dominant taxa. Foraminifera contributed about 90% and 97.05% of the total number, during the three cruises conducted in the RSA, whereby they concluded that meiofauna species diversity and abundance was higher in substrates with a mixture of silt-clay and sandy sediments than in highly silt-clay sediments and were negatively correlated between meiofaunal species distribution and depth, and between meiofauna abundance and salinity (ROPME 2013).

From the available data, Amao et al. (2022) evaluated around 753 benthic foraminiferal species and subspecies belonging to 236 genera, 99 families, and 11 orders, which were considered as high in the extreme conditions of the APG. Although considered high for the land-locked and hypersaline APG, the authors assumed that these species numbers were underestimated, as some areas were poorly surveyed. With their elaborate analysis, Amao et al. (2022) unified and resolved issues related to the use of synonyms and taxonomic-name revisions, which invariably affected species counts and resulted in misaligned datasets. This is a preliminary step in understanding the biodiversity and ecology, and in creating a unified regional database for the benthic foraminiferal assemblage in the APG region.

16.3.1.2 Bahrain

Meiofaunal studies in this region were focussing on foraminiferans rather than on other meiofauna groups. Basson and Murray (1995) carried out a baseline study and recorded the seasonal changes in the benthic foraminiferal diversity, standing crop, and their abundance from the Murray Pool, where they documented seasonal variability among four benthic foraminiferal species, with an increase in diversity and density of *Ammonia*, *Elphidium*, *Brizalina*, and *Nonion*. Amao et al. (2016) conducted an autumn survey of benthic foraminiferal populations from six different sub-environments within the lagoon.

Bhadury et al. (2008) compared the morphological taxonomic identification of the nematode *Terschellingia* sp. with that of molecular taxonomy from the intertidal areas of the sandy beach of Ras al Barr and mangrove vegetation of North Tubli, Bahrain. They found that specimens from North Tubli Bay and Ras al Barr were completely different from the rest of the *Terschellingia longicaudata* sequences, analysed from France, Mexico, and the UK. Ras al Barr sequences clustered with *Sphaerolaimus hirsutus*, whereas North Tubli Bay sequences clustered with *Parodontophora* sp. From this, the authors assumed that the species were misidentified morphologically. Therefore, it is highly recommended that morphological identification of meiofauna should be supported by molecular techniques.

Ali (2014) studied the infauna of northern and southern areas of Bahrain and observed around 186 taxa, wherein the northern area along with the summer season showed higher abundance (105 taxa) than in winter (85 taxa), compared to the southern area, which showed an almost similar abundance during summer (59 taxa) and winter (58 taxa), respectively. Arslan et al. (2016a), by using molecular techniques, confirmed the presence of four groups, i.e., *Ammonia*, *Murrayinella*, *Glabratellina*, and *Elphidium*, that supported the classical taxonomy at the genus level. Arslan et al. (2017) recognized a higher abundance of foraminiferans (viz., *Ammonia*, *Glabratellina*, *Murrayinella*, *Elphidium*, *Brizalina*, miliolids, and peneroplids) in the polluted transects, suggesting the availability of possible nutrients for the benthic Foraminifera to flourish, as they can tolerate eutrophic environments and can withstand fluctuating temperatures, turbidity, and salinities (Melis and Violanti 2006). Moreover, Arslan et al. (2016b) suggested that miliolids can act as pollution proxies for future biomonitoring studies in the region. Garrison (2019) reported an endemic new species of *Agglutinella kaminskii* from the nearshore waters of eastern Bahrain.

Kaminski et al. (2021) implied that due to temperatures reaching as high as 52°C, the upper intertidal zones of "Murray's Pool" are barren of any living meiofauna (e.g., foraminifera, ostracods, gastropods, and worms), and they were observed alive only at low tide in areas that are submerged in water. They also inferred that the meiofaunal organisms in the APG are better acclimatized to high temperatures than in the Mediterranean (Arieli et al. 2011), but no evidential study has yet been carried out to determine their lethal thermal limits in the APG region.

16.3.1.3 *Iran*

Davoodi (1995) surveyed the benthos of Ghazaleh and Ahmadi creeks in the Mahshahr area, Iran, and documented Nematoda, Foraminifera, Copepoda, and Ostracoda as the dominant meiobenthic assemblages and that their respective abundances were higher in winter. While studying the environmental pollution caused by oil spills following the 1991 Gulf War, Mostafawi (2001, 2003) reported around 50 species, of which two were live ostracods showing no response to the contamination. Of these 50 species, seven taxa (*Aglaiocypris pellucida*, *Cytherella bigemina*, *C. retroflexa*, *C. sericea*, *Heinzmalzina ocellata*, *Paijenborchella calcarina*, and *Sagmatocythere nupta*) were observed for the first time when compared with the work carried out by Paik (1977). They suggested that water depth and substrate may be the most important factors controlling the distribution of some ostracod species.

Mirdar et al. (2004) observed four meiofauna taxa, with the observed maximum and minimum abundance from Ostracoda (29%) and Gastropoda (18%), respectively. The number of meiofaunal individuals observed was a maximum during spring (167,603 ind m^{-2}) and a minimum during winter (61,365 ind m^{-2}), respectively. Mollayousefy et al. (2006) identified 54 species from 27 genera of benthic foraminifera from the sediments of Qeshm Island.

Nabavi et al. (2008) reported 34 meiofauna species belonging to nematodes, foraminiferans, copepods, polychaetes, gastropods, oligochaetes, and bivalves from the mangrove vegetation of the Naiband Protected Area, of which Foraminifera were dominant with 51.21%, followed by Crustacea (31.47%), Nematoda, Gastropoda (11.52%), Bivalvia (7.18%); polychaete larvae had the lowest occurrence of 0.62%, suggesting a rich and diverse community that mangroves can inhabit (Nabavi et al. 2005). ROPME (2013) reported *Entzia* sp. for the first time in the region; it is reported from the Gulf of Mexico, the Mediterranean, and the Northeast Atlantic, suggesting that it might have intruded via ballast waters through oil tankers or that it is native to the region but had not been reported.

Dezfouli et al. (2016) studied the meiofauna community structure in the intertidal muddy shores of Sajafi, in the northwest of the APG, and recorded 56 meiofauna species in the warm season and 48 species in the cool season, belonging to 31 genera, where Foraminifera and Nematoda formed the most abundant group quantitatively and qualitatively, along with Ostracoda, Gastropoda, bivalve larvae, larval shellfish, and tube worms. They observed that with increasing depth in the cold season, species richness increased and decreased during the warm season, which can be attributed to a change in water temperature and plankton activity, which provides the main food source for the meiofauna.

Sahraean et al. (2017a) observed 39 genera from 17 families of nematodes and suggested that the entire northern APG area is under substantial stress. About 75% of the nematode abundance was provided by five genera and showed significant differences in species abundance as well as genus diversity between locations, mainly impacted by anthropogenic disturbance. The distance from sources of local pollution significantly affected the nematode diversity. Sahraean et al. (2017b), using molecular tools, investigated the genetic diversity of two species of nematodes, *Ptycholaimellus pandispiculatus* and *Terschellingia longicaudata*.

Mooraki et al. (2018) observed 27 foraminiferan species belonging to 17 genera from 14 families and six orders, where the highest species diversity was shown by the genus *Quinqueloculina*. The genera *Ammonia*, *Elphidium*, *Bolivina*, *Quinqueloculina*, and *Triloculina* were responsible for the similarity of the foraminiferan fauna at other northern and southern coasts of the Arabian/Persian Gulf.

Nazari et al. (2018) studied the meiofauna communities from the Mele Gonze mangrove habitat and observed seven taxa including Foraminifera, Ostracoda, Bivalvia, Gastropoda, Nematodes, Copepoda, and Oligochaeta. Nematoda followed by Foraminifera, Ostracoda, Bivalvia, Copepoda, Gastropoda, and Oligochaeta showed the highest abundance. Fatemeh et al. (2018) did not observe any significant difference between the summer and winter seasons, whereas they observed significant differences between the sampled stations, which was attributed to the silt-clay sediments and total organic carbon of the stations.

Sahraeian et al. (2020) identified 42 genera of free-living nematodes belonging to 17 families from the northern part of the APG. They observed that the genera *Daptonema*, *Ptycholaimellus*, and *Promonhystera* were most abundant, and their assemblage was negatively affected by polychlorinated biphenyls (PCBs) and total organic matter (TOM), with an impact gradient from the polluted Power Plant (PP). Nematode genera *Spirinia*, *Chromadorina*, and *Terschellingia* were resistant to PCBs, while *Oncholaimus* and *Pomponema* were sensitive. Sahraeian et al. (2020) stated that anthropogenic impact was a larger controlling factor in nematode diversity and abundance than seasonal variability at Persian Gulf beaches.

Tigriopus iranicus, a new species from the order Harpacticoida, was reported by Nazari et al. (2021) along with a redescription of *T. raki* (Bradford 1967) from the Iranian coast. Arhami et al. (2021) identified ten species of foraminiferans belonging to eight genera and five families in the Bushehr Province, where *Ammonia beccari* and *Quinqueloculina bicostata* were the dominant species (Mostafawi 2003). They showed that temperature, salinity, and dissolved oxygen influence the density of the foraminiferal communities and that the sediment particles (sand/silt + clay, sand/silt, sand, and calcium carbonate) are important in determining species densities.

According to recent studies, the following number of foraminiferan species were introduced into the waters of the APG: 29 species from Bushehr (Mirdar et al. 2004), nine species from the coasts of Qeshm Island (Mollayousefy et al. 2011), 15 species from the region of Assaluyeh (Mooraki et al. 2013), and 93 species from the northwest APG (Nabavi et al. 2014).

16.3.1.4 *Iraq*

Saad and Arlt (1977) observed around 13 groups, including Coelenterata, Kinorhyncha, Cladocera, and Gastropoda, which were represented by few numbers. Nematoda and Copepoda were found at all stations and generally in much greater numbers than those of the other groups. They concluded that the presence and abundances of meiofauna depend on local environmental conditions of the sediments and the deeper layers of the waters. They observed that Oligochaetae were most abundant at stations where Nematoda decreased in dominance, which was mainly related to the effect of pollution in such localities, which was supported by Laakso (1965), Tulkki

(1968), Reish (1972), and Arlt (1975), as Nematoda are commonly more resistant to pollutants. Elewi and Safawe (1987) studied the foraminifera in southern Iraq and observed that the Miliolidae family is the most dominant group over there.

16.3.1.5 Kuwait

According to the information provided by the Convention on Biological Diversity, Kuwait's Environmental Public Authority is conducting regular monitoring of benthic meiofauna under the Environment Pollution Monitoring protocol (www.cbd.int/countries/profile/?country=kw). They reported around 304 meiofauna species, of which 45 species exist on the shores of Kuwait, although proper authentic publications are scarce. As early as 1973, Clarke and Keij (1973) observed foraminiferans, microgastropods, and ostracods from the Sulaibikhat Bay and Al-Khiran areas. Foraminiferans were abundant and restricted to intertidal areas, which might be due to the dominance of one species, *Ammonia tepida*, which lives on seaweeds and seagrasses that are dominant in these areas, whereas microgastropods, although dominant at Al-Khiran and Sulaibikhat Bay, are least abundant in the main Kuwait Bay and offshore areas, indicating that these faunas are tolerant to high salinities and temperature and prefer to live in shallow, low-energy environments where the sediments are mainly composed of silt. They also observed some microbivalves that were present in very small numbers, indicating their limited tolerance to such areas. By contrast, they show a marked increase in numbers in the deeper waters of the Arabian Gulf, where the environmental factors are much more favourable.

According to Clarke and Keij (1973), ostracods were abundant in the nearshore mud and very poor in the mobile sands of the southern part of Kuwait. Ostracoda species *Loxoconcha (Loxoconcha) indica* and *Xestoleberis rotunda* were abundant in the silty areas of the tidal flats, where there are higher salinities and temperatures, hence indicative of their strong tolerance. Bate (1971) and Clarke and Keij (1973) indicated that *X. rotunda* and *Loxoconcha* were abundant in the shallow coastal lagoons, which are rich in seaweeds. Al Abdul-Razzaq (1991) observed ten ostracod species from the hypersaline waters of Al-Khiran, in southern Kuwait, where they are dependent upon the richness and distribution of algae and the type of sediment. All of these ostracod species were tolerant to high salinity (55–70). According to Al Abdul-Razzaq et al. (1982) and Al Abdul-Razzaq (1991), most ostracod species are found in abundance during summer, when the waters provide a rich diversity of algae.

Al Abdul-Razzaq et al. (1979) studied the meiofauna of tidal flats from Kuwait Bay and Al-Khiran and recorded 42 species of ostracods belonging to 29 genera. Al Abdul-Razzaq et al. (1983b) updated the records with 49 ostracod species belonging to 39 genera in Kuwait Bay. Foraminiferans were the most dominant benthic meiofauna (Al Abdul-Razzaq et al. 1979, 1983a), along with three common species of ostracods, namely *Loxoconcha indica*, *Xestoleberis rotunda*, and *Hemicytheridea paiki*, in the territorial waters of Kuwait. According to Al Abdul-Razzaq et al. (1983b), the density of ostracods increases gradually towards the nearshore areas and then decreases rapidly towards tidal flats, due to turbidity in the northeastern zone and pollution in Sulaibikhat Bay. Moreover, the tidal-flat assemblage differs from that of the nearshore area and from the assemblage living in the restricted area in the western corner of the Bay.

Al Abdul-Razzaq et al. (1983a) observed that microgastropods can tolerate living in these extreme environments, whereas small bivalves are the least abundant. The authors also observed that smaller gastropods and bivalves from Sulaibhikat Bay and Al-Khiran areas are different from Kuwait Bay and the offshore area. The living benthic microfauna are found in low-energy environments of Sulaibhikat Bay and sheltered areas of Al-Khiran, where flats are covered with silt and silty-sand. Benthic microfauna, such as foraminifera and Ostracoda, increase in finer sediments due to the availability of larger amounts of organic material, and these areas are rich in seagrass and seaweeds. No living microfauna is found in high-energy open marine environments, which are characteristics of the southern coast or along the turbid northern coast of Kuwait Bay, where muddy sediments cover the tidal flats (Jacob et al. 1981).

Meiofauna represents the main source of food in the intertidal zone of Kuwait, which provides natural nurseries and rearing grounds for marine biota and a feeding ground for several species of wading birds (Al-Bakri 1984; Al-Bakri et al. 1985; Al-Bakri and Kittaneh 1998). During the early 21st century, the Consortium of International Consultants (CIC) carried out a three-year interim study to investigate the environmental impact of the Gulf War on the marine ecosystem, wherein the Natural History Museum of the United Kingdom (London) studied the intertidal and subtidal meiofauna from Kuwaiti waters (Ferrero et al. 2004). They submitted several reports, but unfortunately, none of them has been published, and most of the information remains confidential. From one of the reports, Ferrero et al. (2004, 2005) reported 12 major meiofaunal taxa, including Archiannelida, Collembola, Copepoda, Foraminifera, Halacarida, Kinorhyncha, Nematoda, Polychaeta, Oligochaeta, Ostracoda, Tardigrada, and Turbellaria.

Ferrero et al. (2004, 2005) observed that the phylum Nematoda (77.56%) was the dominant group, followed by the Copepoda (11.59%). They concluded that there were around 120 copepods and 385 nematode species, of which 101 copepods and 335 nematode species were new to science, including many new genera. They observed that, in general, the abundance was low at the northern muddier stations and higher at the southern coarse sediment habitats. Higher abundance seemed to be associated with a very high variability, indicating spatially heterogeneous sediments. Meiofauna showed strong regional differences in species distribution and significantly higher subtidal diversity (30–60 individuals per sample) compared to intertidal diversity (5–25 individuals per sample). Subtidal inshore stations showed significantly lower diversity compared to offshore stations. For intertidal species, Nematode and copepod abundances were higher in the mid-shore than in the upper shore. Other meiofauna were less abundant and patchy in distribution. Copepoda may be the most sensitive indicator for metal contaminations. The sediment granulometry was the major factor determining the distribution and abundance of meiofauna. Barnes and Ferrero (2009) described two new interstitial nematode species, *Manunema kithara* and *M. kuwaitiensis*, from the intertidal habitats of Kuwait Bay.

Mohamed et al. (2005) observed three different intertidal habitats: Sulaibikhat (muddy), Shuwaikh (muddy, sewage polluted), and Fahaheel (sandy), wherein they observed that meiofauna densities ranged between 878 and 7,271 individuals 100 cm^{-3}. Foraminiferans dominated at Sulaibhikat Bay, foraminiferans and nematodes dominated at Shuwaikh, and nematodes at Fahaheel. They inferred that foraminiferans were dominant in normal or eutrophicated habitats, whereas nematodes were dominant in polluted and oil-contaminated habitats, and suggested that these meiobenthic communities can be assessed and studied in marine pollution assessment and monitoring programs.

El-Sammak et al. (2006) observed that the species diversity of meiofauna ranged from 5 to 24 species m^{-2} in the southern area and 7 to 17 species m^{-2} in the northern regions, suggesting that the northern territorial water of Kuwait's marine environment is different in its nature from the southern region. The diversity and density of meiofauna might be higher in the northern region due to its polluted condition rather than in the southern region, whereas the southern region discriminates by high contents of total petroleum hydrocarbon (TPH) and polycyclic aromatic hydrocarbon (PAH), which are mainly due to the increasing activities related to the oil industry. El-Sammak et al. (2011) reported 25 species of meiofauna from the Az-Zour area, south of Kuwait, and also observed that nematodes, crustaceans, and polychaetes were dominant in the middle zone, which was related to different sediment characteristics such as sediment grain size, total organic carbon, heavy metals, and water temperature. They observed that heavy metal concentrations had little effect on the distribution and abundance of meiofauna.

Al-Marhoun (2013) observed 12 major taxa, dominated by Nematoda, Copepoda, Tardigrada, and Acari from three sandy beaches along the Kuwait coast. Though a clear zonation pattern was visible in meiofaunal distribution and composition, diversity was similar in all the beaches. Meiofaunal densities were within a range of 7.2×10^4 to 3.3×10^5 individuals m^{-2}. Al-Marhoun (2013) showed that the diversity and abundance increased from the high-tidal zone towards the low-tidal zone, where Nematoda and Oligochaeta were dominant in the high-tidal, Nematoda and Tardigrada in the mid-tidal zone, and Copepoda, Nematoda, and Tardigrada in the low-tidal zone, followed by Polychaeta. Al-Marhoun (2013) inferred from her observations that grain size, temperature, and water content are the most influential environmental parameters for the diversity and density of meiofauna at these beaches.

Al-Enezi and Frontalini (2015) evaluated the level of pollution and its effects on the Sulaibhikat Bay by assessing sediment characteristics and benthic foraminifera as bioindicators of environmental quality, while Al-Enezi et al. (2020) documented 451 species belonging to 156 genera, 64 families, and nine orders in Kuwaiti coastal waters. Al-Enezi et al. (2022) attempted to compare the foraminifera from 46 stations using the latest environmental DNA (eDNA) metabarcoding to evaluate the ecological quality status (EcoQS) of Kuwait Bay. Recently, a species of the phylum Gastrotricha was collected from Kuwait and the Adriatic Sea, and their muscular system was comparatively studied with similar populations collected from Italy (Leasi et al. 2006; Leasi and Todaro 2008, 2009), whereas Zotto et al. (2010) used DNA barcoding to obtain taxonomic identities.

16.3.1.6 Oman

Studies have been carried out mainly in the Arabian Sea region of Oman (i.e., Sea of Oman/previously known as Gulf of Oman, Oman Continental shelf, and Indian Ocean margins) (Moghaddasi et al. 2009; Vaziri et al. 2019).

16.3.1.7 Qatar

Dahab (1994) observed that the degradation of petroleum hydrocarbons in marine sediments is caused by the interaction of microfauna, meiofauna, and macrofauna. Al-Hitmi (2000) identified 50 species (viz., seven species of *Textulariina*, one species of *Carterinina*, 32 species of *Miliolina*, and ten species of *Rotaliina*) of Foraminifera from the bottom sediments from the eastern and northern sides of the Qatar Peninsula, where *Miliolina* was frequent and most abundant, contributing about 45.9%, while *Textulariina* and *Rotaliina* represented about 24.3% and 29.8%, respectively, of the total foraminiferans identified.

Estremadura (2021) carried out a preliminary but important study on the meiobenthos of Qatar intertidal areas and reported 74 taxonomic groups belonging to 14 phyla. He observed that Nematoda is the most abundant group with about 51% of mean density, comprising about 43.4% of the entire meiofauna assemblage. Other groups that showed high densities were Tanaididae (9.1%), Harpacticoida (4.3%), *Cypridopus vidua* (3.1%), Hesionidae (3.0%), Bivalves (3.0%), Rotifers (2.1%), Cephalocarida (4.6%), Syncarida (2.7%), *Leptastacus* sp. (2.0%), *Euterpina* sp. (2.0%), and *harpacticoid nauplii* (2.0%). Comparing this with abiotic parameters, Estremadura (2021) found that nematodes and rotifers were positively correlated with silt and clay composition and were negatively correlated with sand composition, interstitial waters, temperature, and salinity, whereas crustaceans were positively correlated with carbonates and total organic matter (TOM). Particularly among crustaceans, copepods were positively correlated with NO_2 and salinity, and negatively correlated with PO_4 and carbonates, and ostracods were negatively correlated with PO_4, CO_3, and TOM, whereas molluscs were negatively correlated with PO_4, CO_3, TOM, and pH. In the same study, flatworms were positively correlated with TOM and negatively correlated with PO_4 and carbonate. Overall, mean densities in the middle intertidal zones were slightly higher when compared with other zones. Finally, Estremadura (2021) concluded that meiofaunal densities and composition were associated with sediment characteristics and total organic matter availability in the area.

16.3.1.8 Saudi Arabia

Along the northern area of the Arabian Gulf off Saudi Arabia, the estimated density of the infauna in the sandy beaches exceeded 400,000 individuals m^{-2} (McCain 1984). Ahmed (1991) identified 43 foraminiferan taxa from Tarut Bay, where *Miliolina* was the most dominant, followed by *Rotaliina*, whereas *Textulariina* were least dominant. *Ammonia beccarii*, *Elphidium advena*, *Quinqueloculina* sp., *Spirolina arietina*, and *Peneroplis planarus* dominated in the intertidal zone, whereas *Spiroloculina* sp. and *Trioculina* sp. become more abundant in the shallow subtidal zone. According to the observations of Ahmed (1991), the *Quinqueloculina* population of Tarut Bay differs completely from the southwest Gulf assemblage, implying that hydrographic conditions and other ecological parameters are responsible for the variation of species.

16.3.1.9 United Arab Emirates (UAE)

El-Deeb (1992) carried out a detailed investigation on the benthic foraminifera of the northern coastline of the United Arab Emirates (UAE) from 150 sediment samples and reported several species, mainly having porcellaneous tests belonging to the genera *Quinqueloculina*, *Triloculina*, *Spiroloculina*, *Sigmoilinita*, *Miliolinella*, *Adelosina*, and *Peneroplis*. *Peneroplis* was abundant in collected seaweed samples; hyaline forms such as *Elphidium*, *Ammonia*, *Spirillina*, *Bolivina*, and *Rosalina* genera were also common in the nearshore shelf areas and beachfront settings, and agglutinated forms such as *Clavulina*, *Textularia*, *Ammobaculites*, and *Reophax* were present with least abundance. Genera *Ammobaculites* and *Reophax* were present only in fine-grained sediments, and were not previously reported in the studied area, whereas coarser sediments did not contain any living Foraminifera, and the dead assemblage was mostly composed of coarse-sized Miliolidae (mainly *Peneroplis* spp.) with fragmented or abraded tests, probably transported from nearby environments.

16.3.2 International literature review emphasizing the importance of meiofauna

Studies on meiobenthic fauna date back to the late 18th century, when Muller presented his preliminary account of the meiobenthic fauna (Duigan 1990). Subsequent methodological improvements in sampling paved the way for a separate field, "meiobenthology"; as its importance was realized, active investigation was carried out on this group since the 1950s (Coull and Giere 1988). The classical studies of McIntyre (1969, 1971), Fenchel (1969), and Fenchel et al. (1967) on meiobenthic fauna indicate that this assemblage comprises nematodes, polychaetes, harpacticoid copepods, amphipods, isopods, and gastropods. These organisms play a key role in the food web and productivity of benthos (Swedmark 1964; Hulings 1971; Hulings and Gray 1971; McIntyre 1971). The ecological role of meiofauna was studied by Gerlach (1971), Coull (1990), Zobrist and Coull (1994), Schratzberger and Ingels (2018), and several others. Zoogeographical distribution studies of these groups indicate a rich biodiversity (Remane 1933, 1940, 1952; Doboutteville 1953; Coull and Herman 1970).

General studies on meiofauna assemblage and interactions with macrofauna were carried out by McIntyre (1969, 1971), Coull (1990), Heip et al. (1988), Warwick and Gee (1984), Zobrist and Coull (1994), and several others. Speciation and the evolutionary trend shown by this group were reported by Coull and Wells (1982), Warwick (2014), Austen and Widdicombe (2006), and Adão (2021), while studies on inter- and intra-specific interactions were reported by Bell (1981), Coull and Herman (1970), and Hicks and Coull (1983).

Meiofauna are known to take part in many ecological functions, e.g., sediment stabilization, biochemical cycling, and food web dynamics (Schratzberger and Ingels 2018). The function of meiofauna in marine benthic systems seems to be much more complex than previously thought and requires detailed investigations to clarify their ecological importance in coastal and benthic ecosystems (Schmid-Araya and Schmid 2000; Moens et al. 2005; Balsamo et al. 2010; Bonaglia et al. 2014). Carbon production of meiofauna is high (~4–29 gC m^{-2} $year^{-1}$) (Chardy and Dauvin 1992; Danovaro et al. 2002), sometimes even surpassing macrofauna (Kuipers et al. 1981), despite their generally low biomass.

Various authors denoted meiobenthic organisms as "pollution markers" or "bioindicators" and studied their role in pollution, toxicology, and environmental impact assessments (Raffaelli and Mason 1981; Amjad and Gray 1983; Shiells and Anderson 1985; Moore and Bett 1989; Garrison et al. 2021; Coull and Chandler 1992; Nabavi 2001; Balsamo et al. 2010; Semprucci and Balsamo 2012; Semprucci et al. 2015, 2018; Zeppilli et al. 2015; Vaziri et al. 2019; di Montanara et al. 2022). Aspects such as bioaccumulation of pollutants (Frontalini and Coccioni 2008), assessment of pollution (Morvan et al. 2004; Moreno et al. 2011), and effects of microplastics (Lagos et al. 2023) were also reported with a view to assessing and managing marine ecosystems. Furthermore, modern studies on meiofauna, such as molecular biomarkers and their effects on organisms (Kovatch et al. 2009), physiological resistance, and genetic markers (Kovatch et al. 2009), were also reported. These studies have thrown new light on our understanding of the role of meiofauna in coastal ecosystems and their importance for the marine ecology of the intertidal zone.

Meiofauna, in general, and harpacticoid copepods, in particular, can serve as excellent live food organisms for commercially important fish and prawn larvae in aqua-hatcheries (Khalaf et al. 1982; Qasim and Jawad 2022). Culture aspects of these groups were studied by Watanabe et al. (1983), Fleeger (2007), Ananthi et al. (2011), Schückel et al. (2013), and Punnarak et al. (2017). The harpacticoid copepods are considered to be an important group of live food organisms, and many have attempted mass cultures of these organisms with varying success (Kraul 1989; Kraul et al. 1992; Stottrup and Norsker 1997).

Ingels et al. (2018) showed that meiofauna are more resilient than microbial communities and macrofauna towards short-term CO_2 exposure, and an elevated temperature experiment observed after 28 days showed that meiofauna, especially nematode communities in muddy sediments, showed a greater response to increased CO_2 concentration and elevated temperature than in sandy sediments. The least studied phyla such as Nemertea and Annelida are good candidate groups for trace markers, which include many burrowing species of very small sized representatives, with body fossils known from the Palaeozoic to the Cenozoic (Valentine 2004).

Recently, van der Heijden et al. (2020) recommended incorporating meiofauna in ecological data monitoring and integrating meiofauna into food web modelling studies; as such they have been poorly considered in the last decades when assessing the functioning of coastal and marine ecosystems (Moens et al. 2011; Schratzberger and Ingels 2018), e.g., roles in trophic processes and energy flow (Leguerrier et al.

2003; Moens et al. 2013) methods for their study through DNA metabarcoding have also been recommended (Gielings et al. 2021)

16.4 CONCLUSION AND RECOMMENDATIONS

Meiofauna are an important component of the marine coastal and benthic environment. However, there is lack of information from the APG, which is evident from the literature records obtained from WoS and Scopus. Although a lot of research has been done on meiofauna from other parts of the world, there is a paucity of information in the APG region, where studies have focused mainly on foraminiferans, while other groups have been poorly studied. Studies such as the one by Clarke and Keij (1973) and most studies on foraminifera have used a lower limit of 75 µm mesh size for filtering the sediments, suggesting that usage of the standard protocol of Somerfield and Warwick (2013) might unearth more groups and species. Even the latest study of Arhami et al. (2021) did not mention the lower mesh size for filtration of the sediment samples for the separation of Foraminifera. Some studies such as that of El-Sammak et al. (2006) have only superficially studied meiofauna, as the main emphasis was to assess the sediment quality. The work of Ferrero et al. (2004) was presented at the International Meiofauna conference held at Ghent in Belgium during 2010, but without giving any reference to the period or location of the samples collected (John et al. 2010).

The majority of works carried out on benthic foraminifera may be inferred because of their role in stratigraphy, paleogeography, the discovery of oil resources, etc. Although several studies have been carried out on benthic foraminifera in the APG, very limited information is available on the impact of natural perturbation, pollutants, chemicals, oil contamination, the impact of dredging, human-related stressors, etc. on this group and is not well documented. If at all, it may be published in the grey literature, in a local journal, or in a regional language.

The Arabian/Pethesian Gulf is a landlocked marginal sea that is under constant stress from anthropogenic and oil industry activities with a heavy inflow of effluents and organic and inorganic pollutants (Sheppard et al. 2010; Al-Yamani 2021), which can be assessed also by studying meiofaunal diversity. The dominant groups such as nematodes, copepods, and foraminiferans and their indices can give a glimpse of the status of pollution prevalent in the APG. The Arabian/Persian Gulf is rich in its endemic diversity of corals and other marine organisms, which account for its endemic meiofaunal diversity.

Overall, studying meiofauna biodiversity is crucial for understanding the ecological, scientific, economic, and conservation value of these small but ecologically significant organisms, especially in the coastal benthic ecosystems of the APG area. Here, we have tried to highlight the work carried out in the APG region and the importance of studying the meiobenthic ecosystem, as it provides background information for future assessment studies in elucidating the role of meiofauna in the APG ecology for sustainable ecosystem management while unearthing species new to science.

REFERENCES

Adão, H. 2021. Metazoan meiofauna: Benthic assemblages for sustainable marine and estuarine ecosystems. In: *Life below water – Encyclopedia of the UN sustainable development goals*. Eds: W.L. Filho, A.M. Azul, L. Brandli, A.L. Salvia, and T. Wall, 1–22. Cham, Switzerland: Springer Nature. doi:10.1007/978-3-319-71064-8_41-1.

Ahmed, M.A.S. 1991. Recent benthic Foraminifers from Tarut Bay, Arabian Gulf coast of Saudi Arabia. *J Micropalenotol* 10, no. 1: 33–38. doi:10.1144/jm.10.1.33.

Al Abdul-Razzaq, S. 1991. The Ostracoda community in hypersaline channels in Al-Khiran, Arabian Gulf. *J Micropalenotol* 10, no. 1: 17–21. doi:10.1144/jm.10.1.17.

Al Abdul-Razzaq, S., Shublaq, W., and Z. Al-Sheikh. 1979. *A preliminary investigation of the marine benthic microfauna of Kuwait*. Kuwait: Kuwait Institute for Scientific Research, 91 p. KISR/PP/1183/EES-RT-R-7915, 111 p.

Al Abdul-Razzaq, S., Shublaq, W., and Z. Al-Sheikh. 1982. Ostracode distribution and ecology of the Sulaibhikat Bay. *Mar Ge¾*46, nos. 3/4: 57–75. doi:10.1016/0025-3227(82)90019-6.

Al Abdul-Razzaq, S., Shublaq, W., and Z. Al-Sheikh. 1983a. The marine benthic microfauna of the tidal flats of Kuwait. *J Univ Kuwait Sci* 10: 101–109.

Al Abdul-Razzaq, S., Shublaq, W., Al-Sheikh, Z., and W. Kittaneh. 1983b. Ecology and distribution of Ostracodes in Kuwait Bay. *J Micropalenotol* 2: 39–45. doi:10.1144/jm.2.1.39.

Al-Bakri, D. 1984. *Environmental assessment of the intertidal zone of Kuwait*. Al-Ahmadi, Kuwait: Kuwait Institute for Scientific Research Report. 429 p.

Al-Bakri, D., Fouda, M., Behbehani, M., et al. 1985. *The environmental assessment of the intertidal zone of Kuwait*. Kuwait: Kuwait Institute for Scientific Research, Report No. KISR 1687, 429 p.

Al-Bakri, D., and W. Kittaneh. 1998. Physicochemical characteristics and pollution indicators in the intertidal zone of Kuwait: Implications for benthic ecology. *Environ Mgmt* 22, no. 3: 415–424. doi:10.1007/s002679900116.

Al-Enezi, E., Francescangeli, F., Balassi, E., et al. 2022. Benthic Foraminifera as proxies for the environmental quality assessment of the Kuwait Bay (Kuwait, Arabian Gulf): Morphological and metabarcoding approaches. *Sci Total Environ* 833: 155093. doi:10.1016/j.scitotenv.2022.155093.

Al-Enezi, E., and F. Frontalini. 2015. Benthic Foraminifera and environmenttheuality: The case study of Sulaibikhat Bay (Kuwait). *Arab J Geosci* 8, no. 10: 8527–8538. doi:10.1007/s12517-015-1812-9.

Al-Enezi, E., Khader, S., Balassi, E., and F. Frontalini. 2020. Modern benthic foraminiferal diversity: An initial insight into the total Foraminiferal diversity along the Kuwait coastal water. *Diversity* 12, no. 4: 142. doi:10.3390/d12040142.

Al-Hitmi, H.H. 2000. Recent benthic Foraminifera from the local water of Qatar, Arabian Gulf. *Qatar Univ Sci* J 20: 167–179. http://hdl.handle.net/10576/9657.

Ali, T.S. 2014. Spatial and temporal variations of marine benthic in-fauna community in northern and southern areas of the Kingdom of Bahrain. *Arab Gulf J Sci Res* 32, no. 1: 80–92. doi:10.51758/AGJSR-01-2014-0010.

Al-Marhoun, A.F. 2013. *Seasonal abundance, distribution and diversity of meiofauna along three coasts of Kuwait*. M.Sc. diss, Kuwait University, Kuwait.

Al-Yamani, F.Y. 2021. *Fathoming the northwestern Arabian Gulf: Oceanography and marine biology*. Kuwait: Kuwait Institute for Scientific Research. ISBN 978-99966-37-40-7.413 p.

Amao, A.O., Kaminski, M.A., Bucci, C., Hallock, P., Al-Enezi, E., Zaky, A.S., and F. Frontalini. 2022. Benthic Foraminifera in the Arabian Gulf: Biodiversity and geographical trends. *Mar Micropaleontol* 176: 102167. doi:10.1016/j.marmicro.2022.102167.

Amao, A.O., Kaminski, M.A., and E. Setoyama. 2016. Diversity of Foraminifera in a shallow restricted lagoon in Bahrain. *J Micropaleontol* 62: 197–211. doi:10.47894/mpal.62.3.01.

Amjad, S., and J.S. Gray. 1983. Use of nematode-copepod ratio as an index of organic pollution. *Mar Pollut Bull* 14: 178–181. doi:10.1016/0025-326X(83)90229-1.

Ananthi, P., Santhanam, P., Nandakumar, R., Ananth, S., Jothiraj, K., Dinesh Kumar, S., Balaji Prasath, B., and T. Jayalakshmi. 2011. Production and utilization of marine copepods as live feed for larval rearing of tiger shrimp *Penaeus monodon* with special emphasis on Astaxanthin enhancement. *Indian J Nat Sci* 11, no. 8: 494–503.

Arhami, M.S., Ardalan, A.A., Manavi, P.N., and B. Moghaddasi. 2021. Identification and distribution of benthic Foraminifera in coastal waters of the Persian Gulf (Bushehr Province). *Iran J Fish Sci* 20, no. 3: 863–878. doi:10.22092/ijfs.2021.124156.

Arieli, R.N., Almogi-Labin, A., Abramovich, S., and B. Herut. 2011. The effect of thermal pollution on benthic Foraminiferal assemblages in the Mediterranean shoreface adjacent to Hadera power plant (Israel). *Mar Pollut Bull* 62: 1002–1012. doi:10.1016/j.marpolbul.2011.02.036.

Arlt, G. 1975. Remarks on indicator organisms (meiofauna) in the coastal waters of the GDR. *Merentutkimusleit Julk/Havsforskninhgsinst. Skr* 239: 272–279.

Arslan, M., Kaminski, M.A., Khalil, A., and M. Holzmann. 2016a. Living behaviors and molecular characterization of benthic Foraminifera in the Arabian Gulf. *Geophys Res Abs* 18: EGU2016–5957. https://ui.adsabs.harvard.edu/abs/2016EGUGA.18.5957A/abstract.

Arslan, M., Kaminski, M.A., Khalil, A., Ilyas, M., and B.S. Tawabini. 2017. Benthic Foraminifera in eastern Bahrain: Relationships with local pollution sources. *Pol J Environ Stud* 26, no. 3: 969–984. doi:10.15244/pjoes/68157.

Arslan, M., Kaminski, M.A., Tawabini, B.S., Ilyas, M., Babalola, L.O., and F. Frontalini. 2016b. Seasonal variations, environmental parameters, and standing crop assessment of benthic Foraminifera in eastern Bahrain, Arabian Gulf. *Geol Quart* 60, no. 1: 24–35. doi:10.7306/gq.1242.

Austen, M.C., and S. Widdicombe. 2006. Comparison of the response of meio- and macrobenthos to disturbance and organic enrichment. *J Exp Mar Biol Ecol* 330, no. 1: 96–104. doi:10.1016/j.jembe.2005.12.019.

Baguley, J.G., Coull, B.C., and G.T. Chandler. 2019. Meiobenthos. In: *Encyclopedia of ocean sciences*. Eds: J.K. Cochran, H.J. Bokuniewicz, and P.L. Yager, 3rd ed., 742–748. Academic Press. doi:10.1016/B978-0-12-409548-9.10800-0.

Balsamo, M., Albertelli, G., Ceccherelli, V.U., et al. 2010. Meiofauna of the Adriatic Sea: Current state of knowledge and future perspective. *Chem Ecol* 26: 45–63. doi:10.1080/02757541003705492.

Barnes, N., and T.J. Ferrero. 2009. Two new species of *Manunema* (Plectida: Peresianidae) from the Arabian Gulf, with notes on the phylogeny of the genus. *Zootaxa* 2053, no. 1: 43–58. doi:10.5281/zenodo.186631.

Barroso, M.S., Da Silva, B.J., Flores Montes, M.J., and P.J.P. Santos. 2018. Anthropogenic impacts on coral reef harpacticoid copepods. *Diversity* 10, no. 2: 32. doi:10.3390/d10020032.

Basson, P.W., and J.W. Murray. 1995. Temporal variations in four species of intertidal Foraminifera, Bahrain, Arabian Gulf. *J Micropaleontol* 41, no. 1: 69–76. doi:10.2307/1485882.

Bate, R.H. 1971. The distribution of recent Ostracoda in the Abu Dhabi lagoon, Persian Gulf. *Bull du Centre de Recher* 5: 239–256.

Bate, R.H., and A. Gurney. 1981. The ostracod genus *Loxoconcha* Sars from Abu Dhabi lagoon and neighbouring near-shore shelf, Persian Gulf. *Bull Br Mus Nat Hist* 41, no. 5: 235–251.

Bate, R.H., and I.M. Sheppard. 1980. On *Xestoleberis postangulata* Bate & Sheppard sp. nov. *Stereo-Atlas Ostracod Shells* 7: 101–106.

Bell, S.S. 1981. Meiofauna-macrofauna interactions in a high salt marsh habitat. *Ecol Monogr* 46: 127–134. doi:10.2307/1942654.

Bhadury, P., Austen, M.C., Bilton, D.T., Lambshead, P.J.D., Rogers, A.D., and G.R. Smerdon. 2008. Evaluation of combined morphological and molecular techniques for marine nematode (*Terschellingia* spp.) identification. *Mar Biol* 154: 509–518. doi:10.1007/s00227-008-0945-8.

Bonaglia, S., Nascimento, F., Bartoli, M., Klawonn, I., and V. Brüchert. 2014. Meiofauna increases bacterial denitrification in marine sediments. *Nat Commun* 5: 5133. doi:10.1038/ncomms6133.

Bradford, J. 1967. The genus *Tigriopus* Norman (Copepoda, Harpacticoida) in New Zealand with a description of a new species. *Trans Royal Soc New Zealand Zoology* 10, no. 6: 51–59.

Chardy, P., and J.C. Dauvin. 1992. Carbon flows in a subtidal fine sand community from the western English Channel: A simulation analysis. *Mar Ecol Prog Ser* 81: 147–161. doi:10.3354/meps081147.

Cherif, O.H., Al-Ghadban, A.N., and I.A. Al-Rifaiy. 1997. Distribution of foraminifera in the Arabian Gulf. *J Micropaleontol* 43, no. 3: 253–280. doi:10.2307/1485827.

Clarke, M.W.H., and A.J. Keij. 1973. Organisms as producers of carbonate sediment and indicators of environment in the southern Persian Gulf. In: *The Persian Gulf*. Ed: B.H. Purser, 33–34. Berlin, Heidelberg: Springer. doi:10.1007/978-3-642-65545-6_3.

Coles, S., and J. McCain. 1990. Environmental factors affecting benthic communities of the western Arabian Gulf. *Mar Environ Res* 29, no. 4: 289–315. doi:10.1016/0141-1136(90)90024-I.

Coull, B.C. 1990. Are members of the meiofauna food for higher trophic levels? *Trans Am Microsc Soc* 109, no. 3: 233–246. doi:10.2307/3226794.

Coull, B.C., and G.T. Chandler. 1992. Pollution and meiofauna: Field, laboratory, and mesocosm studies. *Oceanogr Mar Biol Ann Rev* 30: 191–271. doi:10.1007/bf00815503.

Coull, B.C., and O. Giere. 1988. The history of meiofaunal research. In: *Introduction to the study of meiofauna*. Eds: R.P. Higgins, and H. Thiel, 14–17. Washington, DC: Smithsonian Institution Press. US

Coull, B.C., and S.S. Herman. 1970. Zoogeography and parallel level-bottom communities of the meiobenthic Harpacticoida (Crustacea, Copepoda) of Bermuda. *Oecologia* 5, no. 4: 392–399. doi:10.1007/BF00815503.

Coull, B.C., and J.B.J. Wells. 1982. Refuges from fish predation: Experiments with phytal meiofauna from the New Zealand rocky intertidal. *Ecology* 64: 1599–1609. doi:10.2307/1937513.

Dahab, O.A. 1994. *State of oil pollution along the Qatari coastline*, 70 p. University of Qatar: Scientific and Applied Research Centre.

Danovaro, R., Gambi, C., Dell'Anno, A., Corinaldesi, C., Fraschetti, S., Vanreusel, A., Vincx, M., and A.J. Gooday. 2008. Exponential decline of deep-sea ecosystem functioning linked to benthic biodiversity loss. *Curr Biol* 18, no. 1: 1–8. doi:10.1016/j.cub.2007.11.056.

Danovaro, R., Gambi, C., and S. Mirto. 2002. Meiofaunal production and energy transfer efficiency in seagrass *Posidonia oceanica* bed in the Western Mediterranean. *Mar Ecol Prog Ser* 234: 95–104. doi:10.3354/meps234095.

Davoodi, F. 1995. Survey on benthos of Ghazaleh and Ahmadi creeks in Mahshahr. *Iran Sci Fish J* 3, no. 4: 33–44. http://hdl.handle.net/1834/38312.

Dezfouli, T.T., Nabavi, S.M.B., Ghatromi, E.R., and N. Sajjadi. 2016. Study on the meiofauna community structure in Sajafi shores as the bio-indicator of environmental pollution. *Open J Ecol* 6: 632–644. doi:10.4236/oje.2016.610059.

di Montanara, A.C., Baldrighi, E., Franzo, et al. 2022. Free-living nematodes research: State of the art, prospects, and future directions: A bibliometric analysis approach. *Ecol Inform* 72: 101891. doi:10.1016/j.ecoinf.2022.101891.

Doboutteville, D.C. 1953. Revision des Mystacocarides du genre *Derocheilocaris* Pennak et Zinn. *Vie et Mil*ieu 4: 459–469. https://hal.sorbonne-universite.fr/hal-02559359.

Duigan, C.A. 1990. A historical review of research on Irish Chydoridae (Branchiopoda, Anomopoda) with a checklist of taxa recorded in Ireland. *Irish Nat J* 23, no. 7: 239–246. doi:10.2307/25539494.

El-Deeb, W.Z. 1992. Recent benthonic Foraminifera from the south-eastern coast of the Arabian Gulf, United Arab Emirates. *J Fac Sci UAE. Univ* 4: 181–199.

Elewi, A.H., and N.M. Safawe. 1987. On the distribution of recent Foraminifera from southern lraq. *J Mar Mesopotami—* 2, no. 1: 41–56.

El-Sammak, A., Al-Ghadban, A.N., and M.U. Beg. 2006. Assessment of sediment quality south of Ras Al-Ardh and northern territorial waters, Kuwait, northern Arabian Gulf. *J Coast Res* 39: 1441–1445. www.webofscience.com/wos/WOSCC/full-record/000202961500042.

El-Sammak, A., Mostafa, H.M., and A.N. Al-Ghadban. 2011. Environmental factors controlling the distribution of sediment infauna, Az-Zour area, south of Kuwait. *Arab Gulf J Sci Res* 29, no. 3: 168–182. doi:10.51758/AGJSR-3/4-2011-0016.

Estremadura, D.M.G. 2021. *Meiobenthic assemblages in some intertidal areas around Qatar, Arabian Gulf.* M.Sc. diss, Qatar Univ, Qatar, 65 p.

Fatemeh, N., Amir, V., and F. Ali. 2018. Ecological study on meiofauna communities of Mele Gonze mangrove forest in the north of the Persian Gulf. *J Anim Env* 10, no. 3: 385–396. https://sid.ir/paper/206546/en.

Fenchel, T. 1969. The ecology of marine microbenthos. IV: Structure and function of the benthic ecosystem, its chemical and physical factors and the microfauna communities with special reference to the ciliated Protozoa. *Ophelia* 6: 1–182. doi:10.1080/00785326.1969.10409647.

Fenchel, T., Jansson, B.O., and W. von Thun. 1967. Vertical and horizontal distribution of the metazoan microfauna and of some physical factors in a sandy beach in the northern part of the Oresund. *Ophelia* 4: 227–243. doi:10.1080/00785326.1967.10409622.

Ferrero, T., Barnes, N., Arroyo, N.L., et al. 2005. *A guide to the major meiofaunal taxa of Kuwait's intertidal and subtidal areas.* Consortium Intl Consult, LLC., RSK Gr., and Natl Hist Mus, 92 p.

Ferrero, T., Barnes, N., Huys, R., and R.N. Bamber. 2004. *Environmental monitoring in Kuwait: Marine meiofauna.* Data report on the December 2002/February 2003 intertidal survey. Nat Hist Mus Consul. Report to RSK ENSR.

Fleeger, J.W. 2007. The potential to mass-culture harpacticoid copepods for use as food for larval fish. In: *Copepods in aquaculture.* Eds: C.S. Lee, P.J. O'Bryen, and N.H. Marcus. Amsterdam: Blackwell Publishing. Netherlands. doi:10.1002/9780470277522.ch2.

Frontalini, F., and R. Coccioni. 2008. Benthic foraminifera for heavy metal pollution monitoring: A case study from the central Adriatic Sea coast of Italy. *Estuar Coast Shelf Sci* 76, no. 2: 404–417. doi:10.1016/j.ecss.2007.07.024.

Garrison, T.F. 2019. The microscopic mineral collector of the sea: *Agglutinella kaminskii* n. sp., a new benthic foraminifer from the Arabian Gulf. *Micropaleontology* 65, no. 4: 277–283.

Garrison, T.F., Kaminski, M.A., Tawabini, B., and F. Frontalini. 2021. Sediment oxygen demand and benthic foraminiferal faunas in the Arabian Gulf: A test of the method on a siliciclastic substrate. *Saudi J Biol Sci* 28, no. 5: 2907–2913. doi:10.1016/j.sjbs.2021.02.024.

Gerlach, S.A. 1971. On the importance of marine meiofauna for benthos communities. *Oecologia* 6: 176–190. doi:10.1007/BF00345719.

Gielings, R., Fais, M., Fontaneto, et al. 2021. DNA metabarcoding methods for the study of marine benthic meiofauna: A review. *Front Mar Sci* 8: 730063. doi:10.3389/fmars.2021.730063.

Giere, O. 2009. *Meiobenthology: The microscopic motile fauna of aquatic sediments*, 2nd ed. Berlin-Heidelberg: Springer-Verlag.

Guzmán, H.M., Obando, V.L., and J. Cortés. 1987. Meiofauna associated with a Pacific coral reef in Costa Rica. *Coral Reefs* 6: 107–112. doi:10.1007/BF00301379.

Heip, C., Warwick, R.M., Carr, M.R., et al. 1988. Analysis of community attributes of the benthic meiofauna of Frierfjord, Langesundfjord. *Mar Ecol Prog S*er 46: 171–180. www.int-res.com/articles/meps/46/m046p171.pdf.

Hicks, G.R.F., and B.C. Coull. 1983. The ecology of marine meiobenthic harpacticoid copepods. *Oceanogr Mar Biol Ann Rev* 21: 67–175.

Hulings, N.C. 1971. A quantitative study of the sand beach meiofauna in Tunisia – Preliminary report. *Bull Inst Océanogr Pêche, Salammbô* 2, no. 2: 237–256. http://hdl.handle.net/1834/8533.

Hulings, N.C., and J.S. Gray. 1971. *A manual for the study of meiofauna*, vol. 78, 1–84. Kuwait: Smithsonian Publication.

Ingels, J., dos Santos, G., Hicks, N., et al. 2018. Short-term CO_2 exposure and temperature rise effects on metazoan meiofauna and free-living nematodes in sandy and muddy sediments: Results from a flume experiment. *J Exp Mar Biol Ecol* 502: 211–226. doi:10.1016/j.jembe.2017.07.012.

Jacob, P.G., Anderlini, V., Mohammad, O.S., and M.A. Zarba. 1981. *Physical and chemical oceanography of Kuwait coastal waters.* December 1979-May 1980. KISR 168/EES-11F, 70 p.

John, F., Barnes, N., and R. Huys. 2010. *A regional scale, seasonal study of the intertidal and subtidal meiofaunal major taxa and nematode and copepod biodiversity and species assemblages of Kuwait, Arabian Gulf [abstract].* Fourteenth Intl Meiofauna Conf (FourtIMCo), Ghent, 11–16 July. VLIZ Spl. Publ. 44: 34.

Kaminski, M.A., Amao, A., Babalola, L., et al. 2021. Substrate temperature as a primary control on meiofaunal populations in the intertidal zone: A dead zone attributed to elevated summer temperatures in eastern Bahrain. *Reg Stud Mar Sci* 42: 101611. doi:10.1016/j.rsma.2021.101611.

Khalaf, F.I., Al-Ghadban, A., Al-Saleh, S., and L. Al-Omran. 1982. Sedimentology and mineralogy of Kuwait bay bottom sediments, Kuwait – Arabian Gulf. *Mar Geol* 46, nos. 1–2: 71–99. doi:10.1016/0025-3227(82)90152-9.

Kovatch, C.E., Schizas, N.V., Chandler, G.T., Coull, B.C., and J.M. Quattro. 2009. Tolerance and genetic relatedness of three meiobenthic copepod populations exposed to sediment-associated contaminant mixtures: Role of environmental history. *Env Toxicol Chem* 19, no. 4: 912–919. doi:10.1002/etc.5620190418.

Kraul, S. 1989. Production of live prey for marine fish larvae. *Actes Colloq IFRE*MER 9: 595–607. https://archimer.ifremer.fr/doc/00000/1443/.

Kraul, S., Nelson, A., Brittain, K., Ako, H., and A. Ogasawara. 1992. Evaluation of live feeds for larval and postlarval mahimahi, *Coryphaena hippurus*. *J World Aquac Soc* 23, no. 1: 299–306. doi:10.1111/j.1749-7345.1992.tb00793.x.

Kuipers, B.R., de Wilde, P.A.W.J., and F. Creutzberg. 1981. Energy flow in a tidal flat ecosystem. *Mar Ecol Prog Ser* 5: 215–221. doi:10.3354/meps005215.

Laakso, M. 1965. The bottom fauna in the surroundings of Helsinki. *Ann Zool Fenn* 2: 19–37.

Lagos, A.M., Leon, M.V., Colorado, A., et al. 2023. Effects of microplastics pollution on the abundance and composition of interstitial meiofauna. *Rev Biol Trop* 71: e50031. doi:10.15517/rev.biol.trop.v71i1.50031.

Leasi, F., Rothe, B., Schmidt-Rhaesa, A., and M. Todaro. 2006. The musculature of three species of gastrotrichs surveyed with

Confocal Laser Scanning Microscopy (CLSM). *Acta Zool* 87: 171–180. doi:10.1111/j.1463-6395.2006.00230.x.
Leasi, F., and M. Todaro. 2008. The muscular system of *Musellifer delamarei* (Renaud-Mornant, 1968) and other chaetonotidans with implications for the phylogeny and systematization of the *Paucitubulatina* (Gastrotricha). *Biol J Linn Soc* 94: 379–398. doi:10.1111/j.1095-8312.2008.00974.x.
Leasi, F., and M. Todaro. 2009. Meiofaunal cryptic species revealed by confothemicroscopy: The case of *Xenotrichula intermedia* (Gastrotricha). *Mar Biol* 156: 1335–1346. doi:10.1007/s00227-009-1175-4.
Leguerrier, D., Niquil, N., Boileau, N., et al. 2003. Numerical analysis of the food web of an intertidal mudflat ecosystem on the Atlantic coast of France. *Mar Ecol Prog Ser* 246: 17–37. doi:10.3354/meps246017.
McCain, J.C. 1984. Marine ecology of Saudi Arabia: The intertidal fauna of the sand beaches in the northern area, Arabian Gulf, Saudi Arabia. *Fauna of Saudi Arabia* 6: 53–77.
Mclntyre, A.D. 1969. Ecology of marine meiobenthos. *Biol Rev* 44, no. 2: 245–290. doi:10.1111/j.1469-185X.1969.tb00828.x.
Mclntyre, A.D. 1971. Observations on the status of subtidal meiofauna research. In: *Proc First Int Conf Meiofauna*. Ed: N.C. Hulings, vol. 76, 149–154. Smithsonian Contr Zool. Washington, DC: Smithsonian Institution Press.
Melis, R., and D. Violanti. 2006. Foraminiferal biodiversity and Holocene evolution of the Phetchaburi coastal area (Thailand Gulf). *Mar Micropaleontol* 61, nos. 1–3: 94–115. doi:10.1016/j.marmicro.2006.05.006.
Mirdar, J., Nikoueian, A.R., Owfi, F., and M. Karami. 2004. Study on meiobenthos abundance and their relationship with condition of sediment in northern creek of the Bushehr province. *Iran Sci Fish J* 13, no. 2: 151–162. http://hdl.handle.net/1834/39260.
Moens, T., Braeckman, U., Derycke, S., et al. 2013. Ecology of free-living marine nematodes. In: *Handbook of zoology: Gastrotricha, cycloneuralia and gnathifera 2: Nematoda*. Ed: A. Schmidt-Rhaesa, 109–152. Berlin: De Gruyter. doi:10.1515/9783110274257.109.
Moens, T., Dos Santos, G.A.P., Thompson, F., et al. 2005. Do nematode mucus secretions affect bacterial growth? *Aquat Microb Ecol* 40: 77–83. doi:10.3354/ame040077.
Moens, T., Moodley, L., Steyaert, M., et al. 2011. The structure and functional roles of tidal flat meiobenthos. *Aspects Coast Res Contrib to LOICZ Netherlands Flanders, LOICZ Rep* Stud 38: 171–184. http://hdl.handle.net/1854/LU-3126261.
Moghaddasi, B., Nabavi, S.M.B., Fatemi, S.M.R., and G.H. Vosoughi. 2009. Comparison of species diversity and distribution patterns of benthic Foraminifera in the northern and southern continental shelf of Oman. *J Mar Biol* 6, no. 10: 13–27.
Mohamed, E., Al-Kady, S., Al-Kandari, A., and J. Al-Saffar. 2005. Meiofauna seasonal abundance in three Kuwaiti beaches. *Meiofauna Mar* 14: 115–120.
Mollayousefy, M.S., Tehrani, K.K., and I. Momeni. 2006. Study of benthic foraminifera in mangrove ecosystem of Qeshm Island (Persian Gulf). *Islamic Azad Univ Intl* J Soc Sci 16: 10–19. https://jifro.ir/article-1-4101-en.pdf.
Mollayousefy, S.M., Vaziri, S.H., Sahba, M., Eskandari, S., and R.S. Fard. 2011. Introduction of foraminifera of sedimentary bed of Kish Island coasts to the depth of 30cm. *J Env Geol* 5, no. 14: 76–86.
Mooraki, N., Moghadasi, B., Manoochehri, H., and R. Changizy. 2013. Spatial distribution and assemblage structure of foraminifera in Nayband Bay and Haleh estuary, north-west of the Persian Gulf. *Iran J Fish Sci* 12, no. 3: 654–668.
Mooraki, N., Moghaddasi, B., and R. Nahavandi. 2018. Foraminiferal assemblage structure of Hormoz Island, Persian Gulf. *J Anim Env* 10, no. 1: 369–380. https://dorl.net/dor/20.1001.1.27171388.1397.10.1.49.2.
Moore, C.G., and B.J. Bett. 1989. The use of meiofauna in marine pollution impact assessment. *Zool J Linn Soc* 96, no. 3: 263–280. doi:10.1111/j.1096-3642.1989.tb02260.x.
Moreno, M., Semprucci, F., Vezzulli, L., Balsamo, M., Fabiano, M., and G. Albertelli. 2011. The use of nematodes in assessing ecological quality status in the Mediterranean coastal ecosystems. *Ecol Indicat* 11, no. 2: 328–336. doi:10.1016/j.ecolind.2010.05.011.
Morvan, J., Le Cadre, V., Jorissen, F., and J.P. Debenay. 2004. Foraminifera as potential bioindicators of the 'Erika' oil spill in the Bay of Bourgneuf: Fuel and experiments studies. *Aquat Living Resour* 17, no. 3: 317–322. doi:10.1051/alr:2004034.
Mostafawi, N. 2001. How severely was the Persian Gulf affected by oil spills following the 1991 Gulf War? *Env Geol* 40: 1185–1191. doi:10.1007/s002540100238.
Mostafawi, N. 2003. Recent Ostracods from the Persian Gulf. *Senckenbergiana Maritima* 32, nos. 1–2: 51–75. doi:10.1007/BF03043085.
Murray, J.W. 1965a. The Foraminiferida of the Persian Gulf: Part 1. *Rosalina adhaerens* sp. nov. *Ann Mag Nat Hist Ser* 13, no. 8: 77–79.
Murray, J.W. 1965b. The Foraminiferida of the Persian Gulf. 2: The Abu Dhabi region. *Palaeogeogr Palaeoclimatol Palaeoecol* 1, no. 4: 307–332.
Murray, J.W. 1966a. The Foraminiferida of the Persian Gulf. 3: The Halat Al Bahrani region. *Palaeogeogr Palaeoclimatol Palaeoecol* 2, no. 1: 59–68.
Murray, J.W. 1966b. The Foraminiferida of the Persian Gulf. 4: Khor Al Bazam. *Palaeogeogr Palaeoclimatol Palaeoecol* 2, no. 2: 153–166.
Murray, J.W. 1966c. The Foraminiferida of the Persian Gulf. 5: The shelf of the Trucial coast. *Palaeogeogr Palaeoclimatol Palaeoecol* 2, no. 3: 267–278.
Murray, J.W. 1970. The Foraminifera of the hypersaline Abu Dhabi Lagoon, Persian Gulf. *Lethaia* 3: 51–68.
Murray, J.W. 1973. *Distribution and ecology of living benthonic foraminiferids*. London, UK: Heinemann Educational Books.
Nabavi, S.M.B. 2001. Use of diversity indices as coastal pollution indicators in Khur-e-musa (Mahshahr creeks). *Sci Quart J* 32: 57–61.
Nabavi, S.M.B., Moosapanah, S.G.R., Ghatrami, E.R.Z., Ashrafi, M.G., and S.N. Nabavi. 2014. Distribution, diversity and abundance of benthic Foraminifera of the northwestern Persian Gulf. *J Persian Mar Sci* 5, no. 16: 15–26.
Nabavi, S.M.B., Yavari, V., Kochanian, P., Sabzkabayi, R., Molla, A., and N. Ashrakian. 2005. *Meiofaunal diversity in the naiband marine protected area (Persian Gulf)*. Meknes-Morocco. X Intl Workshop Proc.
Nabavi, S.M.B., Yavari, V., Kochanian, P., Shokat, P., Sabzkabayi, G., Molla, A., and N. Ashraktyan. 2008. Meiofauna of mangrove in the Naiband Protected Area (Persian Gulf) – A preliminary report. *Natl Acad Sci Lett* 31, nos. 7–8: 233–239.
Nazari, F., Amir, V., and F. Ali. 2018. Ecological study on Meiofauna communities of Mele Gonze mangrove forest in the north of the Persian Gulf. *J Anim Env* 10, no. 3: 385–396. https://dorl.net/dor/20.1001.1.27171388.1397.10.3.49.6.
Nazari, F., Mirshamsi, O., and P.M. Arbizu. 2021. *Tigriopus iranicus* sp. nov., a new species of Harpacticidae (Copepoda, Crustacea) from Iran, with a redescription of *T. raki* Bradford, 1967. *ZooKeys* 1035: 115–144. doi:10.3897/zookeys.1035.61584.
Page, M.J., McKenzie, J.E., Bossuyt, P.M., et al. 2021. The PRISMA 2020 statement: An updated guideline for reporting systematic reviews. *BMJ* 372: n71. doi:10.1136/bmj.n71.
Paik, H.K. 1977. Regionale Untersuchungen zur Verteilung der Ostracoden im Persischen Golf und im Golf von Oman: Met forsch deutsche forschungs. *Reihe C Geol Geophys* C28: 37–76. Gebrüder Bornträger, Berlin, Stuttgart. doi:10.1594/PANGAEA.547982.

Pascal, P.Y., Bocher, P., Lefrançois, C., Nguyen, H.T., Chevalier, J., and C. Dupuy. 2019. Meiofauna versus macrofauna as a food resource in a tropical intertidal mudflat. *Mar Biol* 166: 144. doi:10.1007/s00227-019-3588-z.

Punnarak, P., Jarayabhand, P., and A. Piumsomboon. 2017. Cultivation of harpacticoid copepods (families Harpacticidae and Laophontidae) under selected environmental conditions. *Agric Nat Resour* 51, no. 4: 278–285. doi:10.1016/j.anres.2017.05.002.

Qasim, A.M., and L.A. Jawad. 2022. Presence of the Amazon sailfin catfish, *Pterygoplichthys pardalis* (Castelnau, 1855) (Pisces: Loricariidae), in the Shatt al-Arab River, Basrah, Iraq. *Integr Syst* 5, no. 1: 95–103. doi:10.18476/2022.647187.

Raffaelli, D.G., and C.F. Mason. 1981. Pollution monitoring with meiofauna using the ratio of nematodes to copepods. *Mar Pollut Bull* 12, no. 5: 158–163. doi:10.1016/0025-326X(81)90227-7.

Reish, D.J. 1972. *The use of marine invertebrates as indicators of varying degrees of marine pollution: Marine pollution and sea life*. London, UK: FAO Fishing News (Books) Ltd.

Remane, A. 1933. Verteilung und organisation der benthonischen mikrofauna der Kieler Bucht. *Wiss Meeresunter Abt Kiel* 21: 161–221.

Remane, A. 1940. Einfűhrung in die zoologische Ökologie der Nord- und Ostsee. In: *Tierwelt der Nord- und Ostsee*. 238 p. Germany: Becker & Erler.

Remane, A. 1952. Die besiedelung des sandbodens im meere und die bedeutung der lebensformtypen fur die okologie. *Zool Anzeiger Supp* 16: 327–359.

ROPME. 2013. *ROPME oceanographic cruise – winter 2006. Technical report series 10: Meiobenthos in the ROPME Sea area. ROPME/GC-14/7*, 194. Kuwait: Regional Organization for the Protection of the Marine Environment (ROPME).

Saad, M.A.H., and G. Arlt. 1977. Studies on the bottom deposits and the meiofauna of Shatt-Al-Arab and the Arabian Gulf. *Cah Biol Mar* 18: 71–84.

Sahraean, N., Bezerra, T.C., Khanaghah, K.E., Mosallanejad, H., Ranst, E.V., and T. Moens. 2017a. Effects of pollution on nematode assemblage structure and diversity on beaches of the northern Persian Gulf. *Hydrobiologia* 799, no. 1: 349–369. doi:10.1007/s10750-017-3234-z.

Sahraean, N., Campenhout, J.V., Rigaux, A., Mosallanejad, H., Leliaert, F., and T. Moens. 2017b. Lack of population genetic structure in the marine nematodes *Ptycholaimellus pandispiculatus* and *Terschellingia longicaudata* in beaches of the Persian Gulf, Iran. *Mar Ecol* 38, no. 3: e12426. doi:10.1111/maec.12426.

Sahraeian, N., Sahafi, H.H., Mosallanejad, H., Ingels, J., and F. Semprucci. 2020. Temporal and spatial variability of free-living nematodes in a beach system characterized by domestic and industrial impacts (Bandar Abbas, Persian Gulf, Iran). *Ecol Indicat* 118: 106697. doi:10.1016/j.ecolind.2020.106697.

Saidova, K.M. 2010. Benthic foraminiferan communities of the Persian Gulf. *Oceanology* 50, no. 1: 61–66. doi:10.1134/S0001437010010078.

Sanciangco, J.C., Carpenter, K.E., Etnoyer, P.J., and F. Moretzsohn. 2013. Habitat availability and heterogeneity and the Indo-Pacific warm pool as predictors of marine species richness in the tropical Indo-Pacific. *PLoS One* 8, no. 2: e56245. doi:10.1371/journal.pone.0056245.

Schmid-Araya, J.M., and P.E. Schmid. 2000. Trophic relationships: Integrating meiofauna into a realistic benthic food web. *Freshw Biol* 44: 149–163. doi:10.1046/j.1365-2427.2000.00594.x.

Schratzberger, M., and J. Ingels. 2018.theofauna matters: The role of meiofauna in benthic ecosystems. *J Exp Mar Biol Ecol* 502: 12–25. doi:10.1016/j.jembe.2017.01.007.

Schückel, S., Sell, A.F., Kihara, T.C., Koeppen, A., Kröncke, I., and H. Reiss. 2013. Meiofauna as food source for small-sized demersal fish in the southern North Sea. *Helgol Mar Res* 67: 203–218. doi:10.1007/s10152-012-0316-1.

Semprucci, F., and M. Balsamo. 2012. Free-living marine nematodes as bioindicators: Past, present and future perspectives. *Environ Res J* 6: 17–36.

Semprucci, F., Balsamo, M., Appolloni, L., and R. Sandulli. 2018. Assessment of ecological quality status along the Apulian coasts (Eastern Mediterranean Sea) based on meiobenthic and nematode assemblages. *Mar Biodiv* 48: 105–115. doi:10.1007/s12526-017-0745-9.

Semprucci, F., Losi, V., and M. Moreno. 2015. A review of Italian research on freeliving marine nematodes and the future perspectives on their use as Ecol Indicators (EcoInds). *Mediterr Mar Sci* 16, no. 2: 352–365. doi:10.12681/mms.1072.

Semprucci, F., and R. Sandulli. 2020. Meiofauna Biodiversity and Ecology. *Diversity* 12, no. 6: 249. doi:10.3390/d12060249.

Sheppard, C., Al-Husiani, M., Al-Jamali, et al. 2010. The Gulf: A young sea in decline. *Mar Pollut Bull* 60, no. 1: 13–38. doi:10.1016/j.marpolbul.2009.10.017.

Shiells, G.M., and K.J. Anderson. 1985. Pollution monitoring using the nematode/copepod ratio: A practical application. *Mar Pollut Bull* 16: 62–68. doi:10.1016/0025-326X(85)90125-0.

Somerfield, P.J., and R.M. Warwick. 2013. Meiofauna techniques. In: *Methods for the study of marine benthos*· Ed: A. Eleftheriou, 4th ed., 253–284. Chichester, UK: John Wiley & Sons. doi:10.1002/9781118542392.ch6.

Soto, L.A. 2020. The benthic realm. In: *Advances in the studies of the benthic zone*. Ed: L.A. Soto. London, UK: InTechOpen Publ. doi:10.5772/intechopen.81961.

Stottrup, J.G., and N.H. Norsker. 1997. Production and use of copepods in marine fish larviculture. *Aquaculture* 155, nos. 1–4: 231–247. doi:10.1016/S0044-8486(97)00120-8.

Swedmark, B. 1964. The interstitial fauna of marine sand. *Biol Rev* 39, no. 1: 1–42. doi:10.1111/j.1469-185X.1964.tb00948.x.

Tulkki, P. 1968. Effect of pollution on the benthos of Gothenburg. *Helgoländer Wiss Meeresunters* 17: 209–215.

Valentine, J.W. 2004. *On the origin of phyla*. Chicago, IL: University of Chicago Press.

van der Heijden, L.H., Niquil, N., Haraldsson, M., et al. 2020. Quantitative food web modeling unravels the importance of the microphytobenthos-meiofauna pathway for a high trophic transfer by meiofauna in soft-bottom intertidal food webs. *Ecol Modell* 430: 109–129. doi:10.1016/j.ecolmodel.2020.109129.

Vaziri, S.H., Reinhart, E.G., and J.E. Pilarczyk. 2019. Coastal foraminifera from the Iranian coast of Makran, Oman Sea (Chabahar Bay to Gawater Bay) as an indicator of tsunamis. *Geopersia* 9, no. 1: 43–63. doi:10.22059/geope.2018.255122.648378.

Warwick, R.M. 2014. Meiobenthos and macrobenthos are discrete entities and not artefacts of sampling a size continuum: Comment on Bett (2013). *Mar Ecol Prog Ser* 505: 295–298. doi:10.3354/meps10830.

Warwick, R.M., and J.M. Gee. 1984. Community structure of estuarine meiobenthos. *Mar Ecol Prog Ser* 18: 97–111. doi:10.3354/meps018097.

Watanabe, T., Kitajima, C., and S. Fujita. 1983. Nutritional values of live organisms used in Japan for mass propagation of fish: A review. *Aquaculture* 34, nos. 1–2: 115–143. doi:10.1016/0044-8486(83)90296-X.

Zeppilli, D., Sarrazin, J., Leduc, D., et al. 2015. Is the meiofauna a good indicator for climate change and anthropogenic impacts? *Mar Biodiv* 45, no. 3: 505–535. doi:10.1007/s12526-015-0359-z.

Zobrist, E.C., and B.C. Coull. 1994. Meiofaunal effects on growth and survivorship of the polychaete *Streblospio benedicti* Webster and the bivalve *Mercenaria mercenaria* (L.). *J Exp Mar Biol Ecol* 175, no. 2: 167–179. doi:10.1016/0022-0981(94)90024-8.

Zotto, D.M., Ghiviriga, S., Kånneby, T., Jondelius, U., and A. Todaro. 2010. *Probing Gastrotricha taxonomy with DNA barcoding*. Fourteenth Intl Meiofauna Conf (FourtIMCo), Ghent, 11–16 July. Abstracts. VLIZ Special Publ 44: 103.

PART 5

Bacteria, algae and phytoplankton

Bacterial communities and metabolic functions profiled by MG-RAST

An example from Kuwaiti waters

17

Saja A. Fakhraldeen, Nazima Habibi and Sakinah Al-Haddad

17.1 INTRODUCTION

Marine waters are spread over 70% of Earth's surface, provide 97% of the water supply needed by the human population, and account for 90% of habitats (Costello and Chaudhary 2017). Microbes such as bacteria, archaea, fungi, algae, and viruses compete for space and food in coastal environments. These intricate relationships formulate unique niches that play an important role in ocean ecology and planetary health (Tara Ocean Foundation et al. 2022). Microorganisms constitute the majority of the biomass in marine environments (Doytchinov and Dimov 2022). In fact, one millilitre of seawater can contain up to one million microorganisms (Costello and Chaudhary 2017). Microorganisms are vital components of the main biogeochemical cycles, carbon and nitrogen fixation, and marine food webs. These environments are vulnerable to increasing anthropogenic activities, exerting significant pressures on marine biodiversity, which ultimately results in ecological shifts in microbial communities and, by extension, the functions they perform. Such alterations give rise to environmental risk factors that can impart negative consequences on marine productivity, global economies, and public health (Chen et al. 2019; Zhang et al. 2018). Therefore, it is crucial to gather fundamental baseline knowledge about these microorganisms so that any future shifts in their community structures can be adequately documented and evaluated.

Efforts on that front are actively being conducted by many research groups. Studies into marine microbiomes are rapidly evolving, and marine microbiology is one of the hot topics in the field of environmental research (Trevathan-Tackett et al. 2019). 16S ribosomal RNA (rRNA)-based genetic analyses and amplicon sequencing efforts have revealed distinctly recurrent bacterial communities in oceanic environments (Moran 2015). More recently, characterizations of marine archaea, protists, and viruses conducted via a shotgun metagenomic sequencing approach have complemented the taxonomic inventory of the marine microbiome, showing that membership can be assessed across seasons, sea depths, and relative to the characteristics of the surrounding organic matter. Concurrently, the resources available for analysing the vast amounts of data generated from these studies are also continuously being developed and updated (Navgire et al. 2022).

17.2 REGIONAL SETTING

The Arabian Gulf is a shallow sea located between the southwestern part of Iran and the Arabian Peninsula. This marginal water body is an extension of the Indian Ocean and is surrounded by the countries of Oman, United Arab Emirates, Qatar, Bahrain, Saudi Arabia, Kuwait, and Iraq (Al-Zaidan et al. 2006). Extending about 48 km inland from the Arabian Gulf is Kuwait Bay, a semi-estuarine water body located along the northern coast of Kuwait. Kuwait Bay has 130 km of coastline covering an area of 720 km^2 (Pokavanich et al. 2013). The average depth of the bay is 5.2 m below mean sea level. The shallow nature of the bay allows for good mixing of water, and hence no thermocline is observed. The water temperature in the bay reaches up to 35°C during the summer and drops down to 11.5°C during the winter. The salinity of Kuwaiti waters varies between 38 ppt to 42.4 ppt (McEwan

DOI: 10.1201/9781003321392-22

2008). Vast muddy tidal flats within the bay serve as habitats for several species of fishes and shrimps, and they are ideal feeding grounds for many species of shorebirds (Al-Yamani et al. 2004). Importantly, Kuwait Bay houses a vast array of microbial communities.

Over the past several years, the water quality of Kuwait Bay has become heavily compromised. Discharge from the Shatt-Al-Arab to the north brings in a great amount of nutrients, rendering the waters highly eutrophic and resulting in harmful algal blooms, fish kills, and diseases (Al-Yamani et al. 1997, 2004, 2005, 2020). Furthermore, evidence of microplastic contamination in the waters of the Arabian Gulf has also been documented (Uddin et al. 2020, 2021, 2022). The discharge of waste into the bay has introduced pharmaceuticals, drugs, and other pernicious pollutants into the coastal waters and sediments (Gevao et al. 2021, 2022; Habibi et al. 2021b, 2022, 2023). In addition to pollution and acidification, which likely result from anthropogenic activities, Kuwait Bay is also exposed to significantly extreme environmental conditions, including intense fluctuations in temperature, salinity, dissolved oxygen levels, and the global threat of ocean acidification (Al-Yamani et al. 1997, 2004, 2005; Uddin et al. 2011b, 2012).

Furthermore, the shallow nature of the water body (Uddin et al. 2011a) facilitates a reasonably high residence time for contaminants Uddin et al. 2015, 2017a, 2017b). Such phenomena likely place the flora and fauna of Kuwait Bay under a constant state of stress. In view of all these conditions, it is essential to perform a baseline characterization of the microbial communities in terms of structure and function to begin shedding light on how such stressors affect the microbiome within this water body. Advanced molecular methods including shotgun metagenomics play a key role in elucidating these mechanisms.

17.3 SHOTGUN METAGENOMIC SEQUENCING

Advances in high-throughput sequencing have revolutionized the detection of genes and genomes in complex environmental communities, offering a more promising avenue for comprehensive genetic and community profiling. Taxonomic and functional profiles of microbes have been reported from estuaries, rivers, lakes, coral reefs, mangroves, sediments in coastal areas, and deep marine sediments using this approach (Yang et al. 2021, 2022; Zhang et al. 2020; Chen et al. 2013, 2020; Guo et al. 2020; Mootapally et al. 2019; Imchen and Kumavath 2020). Recent research has also included analysis of environmental DNA via whole genome sequencing of sub-surface sediments receiving an influx of pollutants such as metals, oil discharges, and industrial and domestic waste (Catania et al. 2018; Ogaki et al. 2021; Zhang et al. 2018, 2020; Cabello-Yeves et al. 2021; Wang et al. 2020; Yan et al. 2020). Thus, the diverse and impactful utility of this approach has been widely demonstrated by research groups around the world, yet the approach has yet to be applied to the local waters of Kuwait.

17.4 MATERIALS AND METHODS

17.4.1 Sample collection

For the present investigation, 1 L of seawater samples (n = 7) was collected from three sampling stations located in and around Kuwait Bay (Figure 17.1) in pre-sterilized 5 L polypropylene carboys (Azlon® DWK Life Sciences, US). Station K6 is located within Kuwait Bay, Station A is located northeast of Kuwait Bay in a narrow waterway between the Kuwaiti coast and the Kuwaiti island of Bubiyan, and Station 18 is located southeast of Kuwait Bay and is the farthest of the three stations from the Kuwaiti coast. The GPS coordinates of the sampling sites are provided in Table 17.1. Samples were collected between October 2019 (1019) and February 2020 (0220). Four samples were collected from near the surface of the water column (average depth of 1.3 m below the surface), and three samples were collected from the bottom at an average depth of 16.3 m below the surface. Samples were transported to the laboratory on ice and passed through a 0.22 μm polycarbonate membrane filter (Isopore/Millipore) using a vacuum pump (MilliporeSigma). The filter papers were then cut into quarters using a sterile blade and placed in the lysis solution/tube provided in the MO BIO PowerSoil DNA isolation kit (Qiagen) (Habibi et al. 2022; Fakhraldeen et al. 2023). Subsequent DNA purification steps were followed as per the manufacturer's instructions. DNA quality was assessed on a 1.0% agarose gel, and purity and yield were checked on a NanoDrop™ device (ThermoFisher Scientific). High-quality DNA was sent to Admera Health LLC (www.admerahealth.com/) for library preparation and next-generation sequencing. Fluorometric estimation of DNA concentrations and quality were performed by the service provider based on the procedure described by Mardis and McCombie (2017).

17.4.2 Library preparation and sequencing

Isolated whole genomic DNA was used for library preparation. A total of 1 ng of DNA was used for library preparation using the Nextera XT DNA library preparation kit (Ilumina, San Diego) as per the manufacturer's instructions. The final library quantity was assessed using the Qubit 2.0 system (ThermoFisher, MA, USA), and the quality was assessed using the Tape Station 2200 system coupled with high-sensitivity D1000 Screen Tapes (Agilent Technologies, Inc., CA, USA). Equimolar pooling of libraries was performed based on quality control (QC) values, and prepared libraries were subjected to metagenomic sequencing using the HiSeq X platform (Illumina) at a depth of 20 million paired-end reads per sample (10 million for each of the forward and reverse reads) via 2 × 150 bp chemistry. Raw sequences were checked for error rates and low-quality bases via the FastQC tool (Andrews 2010). In-house developed perl scripts were used for merging the paired-end reads.

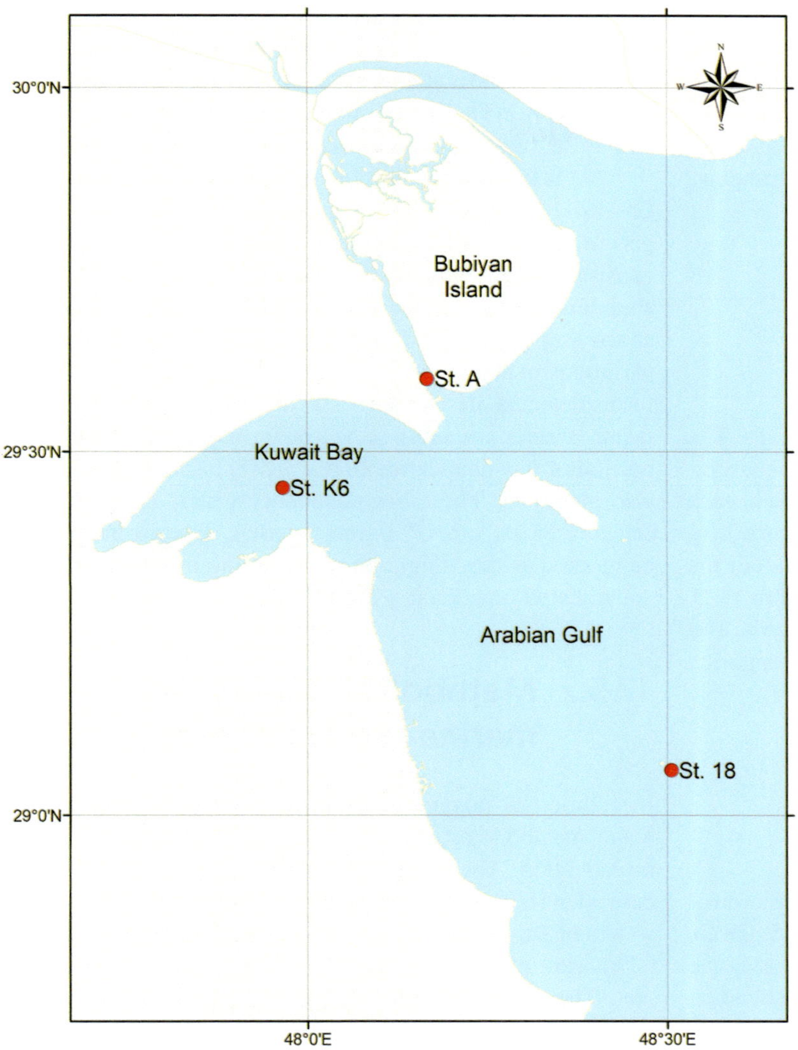

FIGURE 17.1 Map of Kuwait Bay, located within Kuwaiti territorial waters of the northwest Arabian Gulf. Stations used for sample collection in this study (Stations K6, A, and 18) are shown. Map courtesy of T. Yamamoto (Kuwait Institute for Scientific Research).

TABLE 17.1 Sample details and corresponding MG-RAST IDs.

MG-RAST ID	*NAME*	*BASE PAIR COUNT*	*SEQ. COUNT*	*GPS COORDINATES*	*DEPTH (M)*
25db6db8e66d676d343938353236312e33	18_B_0220_R	3,520,970,098	19,738,126	48.3033, 29.0372	20.17
7c152dfdab6d676d343938353333332e33	18_S_0220_R	3,032,566,592	17,081,758	48.3033, 29.0372	1.31
4e796840056d676d343938353238372e33	18_S_1219_R	2,115,891,146	9,871,162	48.3033, 29.0372	1.31
02323280f86d676d343938353236332e33	18_B_1019_R	3,781,834,710	21,011,620	48.3035, 29.0373	18.93
700795af266d676d343938353238382e33	A_S_1019_R	3,954,942,112	21,913,975	48.0999, 29.3599	1.29
f502d66cd86d676d343938353333382e33	K6_B_1219_R	4,016,152,455	22,510,532	47.5799, 29.2698	9.69
378eb29e756d676d343938353332362e33	18_S_0120_R	3,207,931,426	17,987,703	48.3034, 29.0373	1.31

17.4.3 Bioinformatic analysis

The merged (forward and reverse) raw sequences obtained as described earlier were uploaded to the MG-RAST server v 4.0.3 (Keegan et al. 2016). The standard protocol offered by MG-RAST as reported elsewhere (Habibi et al. 2021a) was followed for assessing the quality, removing the adapter sequences and low-quality bases, and thereafter alignment of sequences against the M5RNA database for taxonomic and functional profiling. The initial sequence statistics were computed through DRISEE (Keegan et al. 2012). The adapters were trimmed employing the skewer algorithm (Jiang et al. 2014). The fastq-mcf command line was used for denoising and normalization (Aronesty 2013). The host DNA contamination was removed using Bowtie 2 (Langmead and Salzberg 2012). rRNA gene calling was performed with a cut-off of 70% identity to ribosomal sequences of reduced version of the M5RNA database employing the SortMeRNA algorithm

(Kopylova et al. 2012). Coding regions within the sequences were predicted using FragGeneScan (Rho et al. 2010), and amino acid clustering was performed using CD-HIT (Fu et al. 2012; Li and Godzik 2006). Genes were aligned against the Kyoto encyclopedia of Genes and Genomes (KEGG) database for functional annotation (Kanehisa and Goto 2000; Kanehisa et al. 2021).

17.5 RESULTS

Each sample produced an average of 3.3 Gb of raw reads. The sequence count ranged from 9,871,162 to 22,510,532 sequences. Approximately 20 million reads per sample (10 million in each direction) is considered to have enough depth to perform downstream analyses. Data rarefaction was performed, and all the samples plateaued at a sample size of 100–300 (Figure 17.2). Thus, none of the samples was excluded from the analysis. The dataset can be accessed on the MG-RAST server through the accession IDs provided in Table 17.1.

17.5.1 Taxonomic profiling of bacterial communities

The taxonomic binning of sequences revealed three microbial domains including Bacteria (80–90%), Eukaryotes (0.5–0.8%), and Archaea (0.01–0.1%) (Figure 17.3a), in addition to some unclassified reads. Bacteria were among the most abundant microbial communities. Classification of bacterial genera was performed at the phylum, class, order, family, and genus levels. In total, 400 bacterial genera with a relative abundance (RA) > 0.5% were detected. These were investigated further as they constitute the most abundant microbial biomass in the marine environment. The predominant bacterial genera detected within the study samples are shown in Figure 17.3b.

At the genus level, the predominant forms were *Cytophaga* (21–36%) followed by *Flavobacterium* (8–24%). The top 15 genera are presented in Figure 17.3b. These 15 genera were consistently present in all the tested samples, but with variable abundance. These samples were collected from different sites, at varying depths, and at distinctive time points. Thus, the variations in abundances are not surprising. The predominant forms detected are characteristic of Kuwait's marine environment. Besides the most prevalent forms, a considerable proportion of low-abundance (RA < 0.5%) bacterial genera were also identified. These low-abundance genera are collectively categorized as "others" in most studies. These genera cannot be ignored as they might be a connecting link to many of the documented functional groups.

17.5.2 Metabolic functions in the marine waters of Kuwait

In addition to taxonomic profiling, functional annotation of major metabolic processes was also performed in the present investigation. Alignment of unigenes against the KEGG database identified six prevalent level 1 pathways in the marine waters of Kuwait (Figure 17.4a). The highest abundances (59–62%) were recorded for the metabolism function in all samples. These were followed by genetic information processing (19–21%), environmental information processing (11–12%), and cellular processes (3–5%).

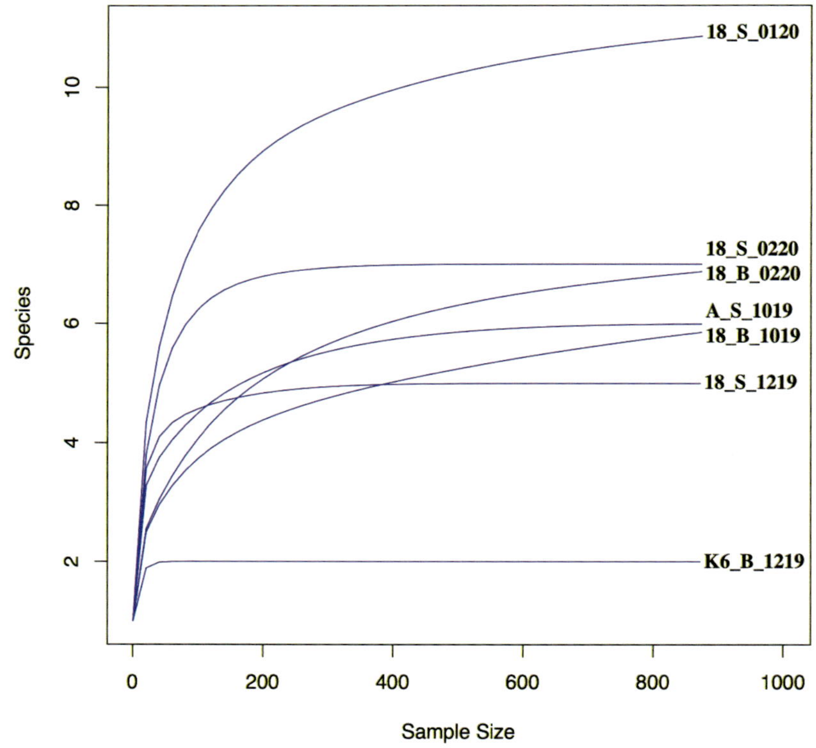

FIGURE 17.2 Rarefaction curves for the seven samples analysed in this study. The y-axis represents species richness, and the x-axis represents sequence sample size.

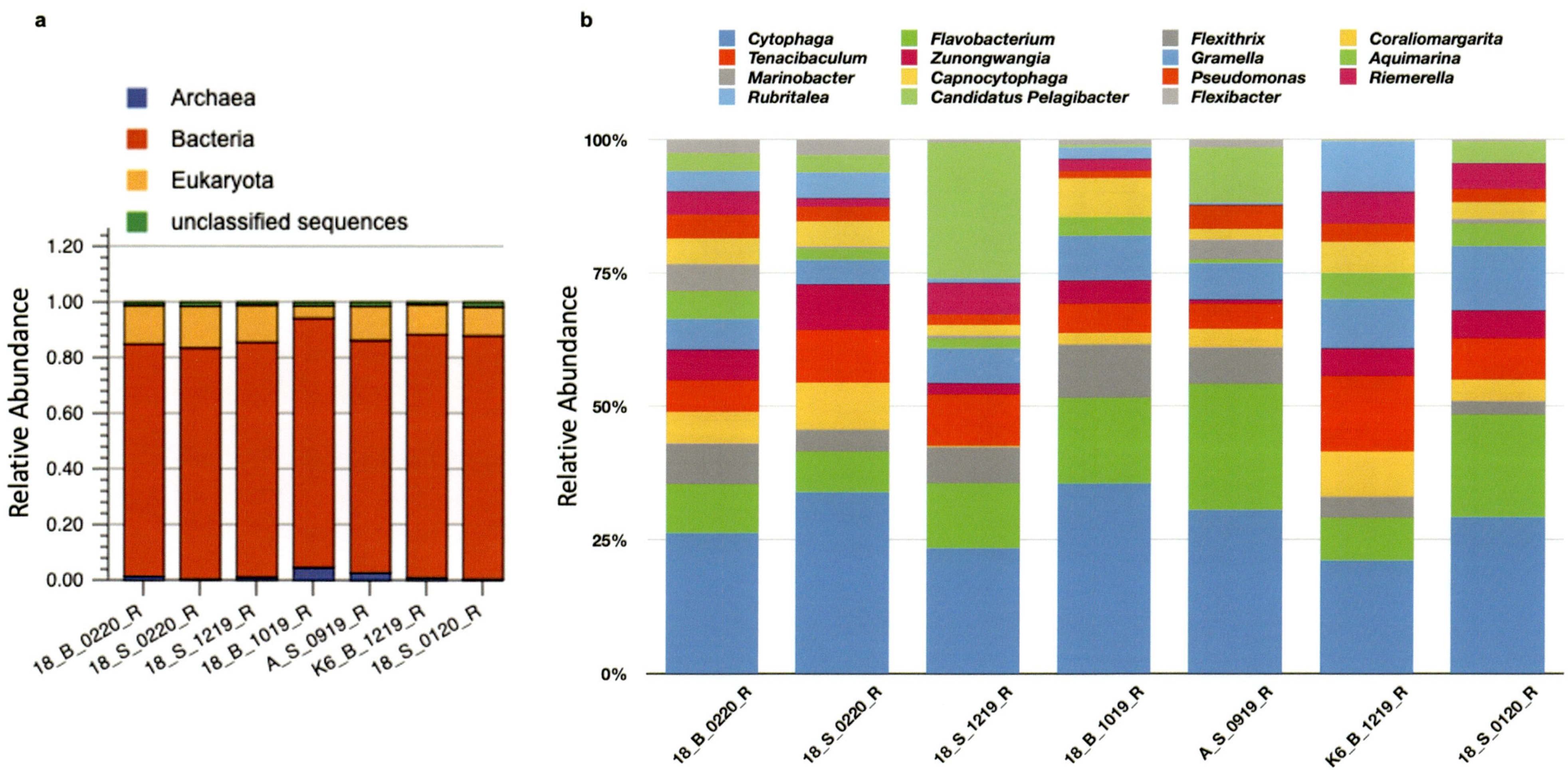

FIGURE 17.3 Taxonomic classification of marine microbial communities: (a) domain-level classification; (b) genus-level classification. The x-axis shows the sample types, and the y-axis shows the relative abundances of the corresponding taxa.

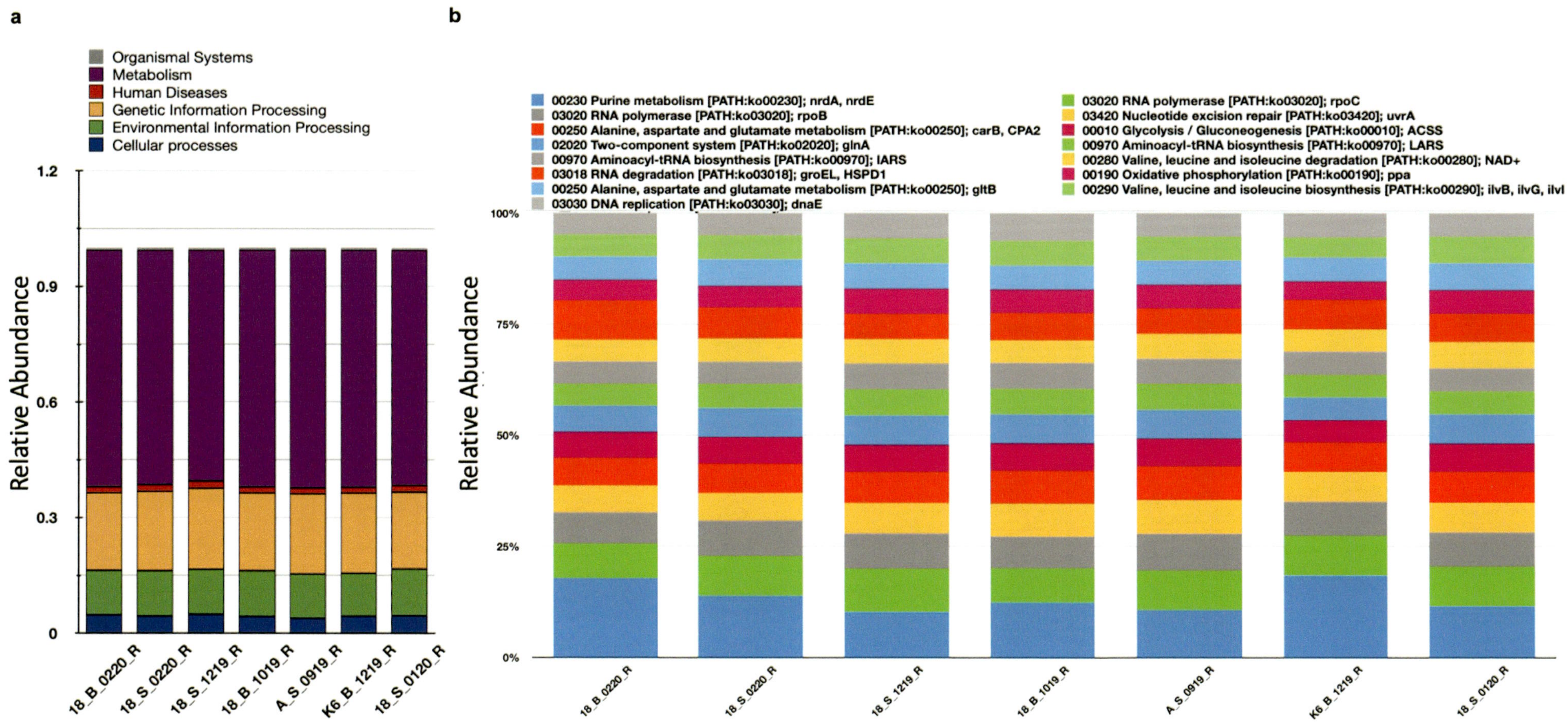

FIGURE 17.4 Functional classification of marine microbial communities: (a) level 1 KEGG pathways; (b) level 3 KEGG pathways and functions. The x-axis shows the sample types, and the y-axis shows the relative abundances of the metabolic pathways.

TABLE 17.2 Alpha diversity indices of marine samples from Kuwait Bay.

SAMPLE	*CHAO1*	*SHANNON*	*SIMPSON*
18_B_0220	14	3.12	3.12
18_S_0220	11	3.25	3.25
18_S_1219	14	2.79	2.79
18_B_1019	9	2.95	2.95
A_S_1019	20	2.90	2.90
K6_B_1219	9	2.52	2.52
18_S_0120	18	1.30	1.30

A closer look at the predominant level 2 KEGG pathways revealed that genes involved in energy production and conversion displayed the highest abundances (23–40%). Other prevalent level 2 pathways detected at a relatively high abundance were lipid transport and metabolism (8–13%), amino acid transport (4–8%), post-translational modification, protein turnover, chaperones (10–13%), replication, recombination, and repair (4–14%), nucleotide transport (3–4%), inorganic ion transport (4–7%), transcription (2–3%), and secondary metabolite biosynthesis (1–2%), among others.

Level 2 KEGG pathways were further classified into level 3 pathways and functions. The dominant functions (n = 15) are shown in Figure 17.4. The most abundant were purine metabolism -nrdA, nrdE (11–18%); RNA polymerase-rpoC (8–10%); RNA polymerase-rpoB (7–8%); nucleotide excision repair-uvrA (6–8%); alanine, aspartate, and glutamate metabolism-carB, CPA2 (6–7%); glycolysis/gluconeogenesis-ACAA (5–6%); two-component system-glnA (6–7%); aminoacyl-tRNA biosynthesis-LARS (5–6%, aminoacyl-tRNA biosynthesis-IARS (5–6%); valine, leucine, and isoleucine degradation-NAD+ (5–6%); RNA degradation-groEL, HSPD1 (6–9%); oxidative phosphorylation-ppa (4–5%); alanine, aspartate, and glutamate metabolism-gltB (5–6%); valine, leucine, and isoleucine biosynthesis-iivB, iivG, iivL (4–6%); and DNA replication-dnaE (5–6%) (Figure 17.4b).

17.5.3 Alpha diversity analysis

Three alpha diversity indices (Chao1, Shannon, and Simpson) were estimated for the study samples. The maximum Chao1 index (20) was noticed in the sample collected from the surface of the seawater at station A in October 2019 (A_S_1019), followed by the sample collected from the surface of the seawater at station 18 in January 2020 (18_S_0120) (18). In the case of Shannon's index, the highest record (3.25) was recorded in the sample collected from the surface of the seawater at station 18 in February 2020 (18_S_0220), and the lowest record (1.30) was recorded in the sample collected from the surface of the seawater at station 18 in January 2020 for (18_S_0120). Similar observations were recorded for the Simpson index (Table 17.2).

17.6 DISCUSSION

This study presents one of the first reports on the taxonomic and metabolic profiles of the microbial communities residing within the marine waters of Kuwait. Some previous studies have attempted similar analyses; however, these were technically limited and only provided broad information pertaining to the microbial and planktonic taxonomies (Almutairi 2015; Chen 2017; Esmaeil et al. 2009; Kumar et al. 2021), with negligible insights into the functions carried out by these communities. Furthermore, the sequencing approach utilized in these studies targeted highly conserved regions within the 16S or 18S rRNA genes, which limits the taxonomic characterization to the genus level. Alternatively, the approach of shotgun metagenomic sequencing utilized in this study and our previous work (REF) can provide a detailed and comprehensive image of the bacterial communities and the dominant functions they carry out, which are of vital importance to the surrounding aquatic environment.

In the present study, the MG-RAST software was utilized to analyse the existing dataset. MG-RAST served the dual purpose of being an analysis tool as well as a public repository in which the data was deposited. The automated workflow allowed for processing of the metagenomic sequencing dataset for quality control, sequence alignment, and functional and phylogenetic assignments through the latest algorithms integrated within its application programming interface (API) (Meyer et al. 2019; Keegan et al. 2012, 2016). Several studies performing similar analyses on datasets from other aquatic bodies utilized the Quantitative Insights Into Microbial Ecology (QIIME) software, which is another popular and freely available online metagenomic analysis tool (Caporaso et al. 2010, 2011, 2012). It is worth noting that variations in taxonomic and functional profiles have been reported when both tools have been used to analyse the same dataset. Such differences in analytical outcomes are likely due to differences in the default databases utilized in the analysis as well as differences in the identity cut-offs utilized in each pipeline. However, in general, both tools have been reported to provide comparable results (D'Argenio et al. 2014; Plummer et al. 2015). Server dependence is a limitation inherent to the MG-RAST software, whereas a high level of bioinformatics skill is required to use QIIME (D'Argenio et al. 2014).

The initial glimpse of Kuwaiti marine waters informs taxa previously reported in aquatic systems (Mootapally et al. 2019; Nathani et al. 2019). Bacteria were the prime component of the marine microbial ecosystem, followed by Eukaryota, Archaea, and other communities, respectively. Similar observations were recorded in marine sediments collected near the storm outlets along the Kuwaiti coastline (Habibi et al. 2023). As in other reports, the phylum Proteobacteria, specifically the classes Alphaproteobacteria and Gammaproteobacteria, were the most abundant (Song et al. 2017; Cabello-Yeves et al. 2021; Nathani et al. 2019). Also, in agreement with results reported in this preliminary study, *Cytophaga* and *Flavobacterium* were recorded in great abundances in several marine and freshwater systems, likely owing to the major roles they play in several biogeochemical cycles (Kirchman 2002). We speculate that these bacteria might have specialized roles in the uptake and degradation of dissolved organic matter. Besides the dominant taxa, a respectable number of low-abundance bacterial genera (RA $< 0.5\%$) were also detected in the present study. Lower-abundance taxa have previously been reported in marine sediments of Kuwait (Habibi et al. 2022, 2023). Sogin et al. (2006) coined the term "rare biosphere" for these low-abundance genera and explained that their presence is of ecological significance. These genera and species are often overlooked but may in fact be the key components of biogeochemical cycling and might represent hidden ecosystem functions (Lynch and Neufeld 2015). They are important contributors to marine microbial diversity. Hence, even modest changes in environmental parameters, which may entirely wipe out their presence, can bring about a shift in the ecological balance (Gobet et al. 2012; Lynch and Neufeld 2015).

We hypothesize that the microbial communities in the present investigation mediated diverse metabolic processes. The active metabolic pathways detected appear to be a pervasive response to the environmental fluctuations during the sampling period. Kültz (2005) demonstrated that the pathways of DNA damage sensing and repair, transcription regulation, lipid metabolism, energy metabolism, and damaged protein degradation are linked to the general stress response. Importantly, the presence of pollutants in the marine environment acts as a stressor for the bacterial population and has been shown to trigger processes involved in energy conversion, carbohydrate metabolism, and transport metabolism (Nowak et al. 1997), all processes that were detected with varying abundance in the samples from the present study.

17.7 CONCLUSIONS

The shotgun metagenomic sequencing approach coupled with bioinformatic analysis through MG-RAST was successfully used to generate a preliminary profile of bacterial community structure and the functional potential of Kuwait's marine waters. A total of seven samples was collected during a span of five months. The dominant bacteria, represented by members of the phylum Proteobacteria, and functional activities, mainly Metabolism, detected are characteristic to the unique waters of the region. The findings described here provide a lead for further investigations at a higher sampling frequency that can provide greater resolution for the dynamic structure and function of microbial communities in the marine environment, particularly in response to fluctuations in various factors, including environmental factors (e.g., temperature, salinity, acidity, etc.), pollutants (e.g., microplastics, various forms of industrial waste, etc.), and others.

REFERENCES

Almutairi, A. 2015. Spatial – Temporal variations and diversity of the bacterioplankton communities in the coastal waters of Kuwait. *Mar Pollut Bull* 100: 699–709.

Al-Yamani, F.A., Al-Samaak, A., Al-Hussaini, M., Al-Ghunaim, A., Al-Rifaie, K., and W. Ismail. 2005. *The current situation of Kuwait's marine environment*. Safat, Kuwait: Kuwait Institute for Scientific Research.

Al-Yamani, F., Bishop, J., Ramadhan, E., Al-Husaini, M., and A.N. Al-Ghadban. 2004. *Oceanographic atlas of Kuwait's waters*. Safat, Kuwait: Kuwait Institute for Scientific Research.

Al-Yamani, F., Durvasula, R., Ismail, W., Al-Rifaie, K., Al-Saeed, T., Al-Yaqout, A., and L. Al-Omran. 1997. *Dynamic oceanography of the northwestern waters of the Arabian Gulf: Ecological significance of the marine food web*. Safat, Kuwait: Kuwait Institute for Scientific Research.

Al-Yamani, F., Polikarpov, I., and M. Saburova. 2020. Marine life mortalities and harmful algal blooms in the northern Arabian Gulf. *Aquat Ecosyst Health Manag* 23: 196–209.

Al-Zaidan, A., Kennedy, H., Jones, D., and S. Al-Mohanna. 2006. Role of microbial mats in Sulaibikhat Bay (Kuwait) mudflat food webs: Evidence from δ13C analysis. *Mar Ecol Prog Ser* 308: 27–36.

Andrews, S. 2010. *FastQC: A quality control tool for high throughput sequence data*. Available online. Retrieved: 17 May 2018.

Aronesty, E. 2013. Comparison of sequencing utility programs. *Open Bioinformatics J* 7: 1–8. doi:10.2174/1875036201307010001.

Cabello-Yeves, P.J., Callieri, C., Picazo, A., Mehrshad, M., Haro-Moreno, J.M., Roda-Garcia, J.J., Dzhembekova, N., Slabakova, V., Slabakova, N., and S. Moncheva. 2021. The microbiome of the Black Sea water column analyzed by shotgun and genome centric metagenomics. *Environ Microbiome* 16: 1–15.

Caporaso, J.G., Kuczynski, J., Stombaugh, J., Bittinger, K., Bushman, F.D., Costello, E.K., Fierer, N., Peña, A.G., Goodrich, J.K., and J.I. Gordon. 2010. QIIME allows analysis of high-throughput community sequencing data. *Nat Methods* 7: 335–336.

Caporaso, J.G., Lauber, C.L., Walters, W.A., Berg-Lyons, D., Huntley, J., Fierer, N., Owens, S.M., Betley, J., Fraser, L., and M. Bauer. 2012. Ultra-high-throughput microbial community analysis on the Illumina HiSeq and MiSeq platforms. *ISME J* 6: 1621–1624.

Caporaso, J.G., Lauber, C.L., Walters, W.A., Berg-Lyons, D., Lozupone, C.A., Turnbaugh, P.J., Fierer, N., and R. Knight. 2011. Global patterns of 16S rRNA diversity at a depth of millions of sequences per sample. *Proc Nat Acad Sci* 108: 4516–4522.

Catania, V., Cappello, S., Di Giorgi, V., Santisi, S., Di Maria, R., Mazzola, A., Vizzini, S., and P. Quatrini. 2018. Microbial communities of polluted sub-surface marine sediments. *Mar Pollut Bull* 131: 396–406.

Chen, B., Yang, Y., Liang, X., Yu, K., Zhang, T., and X. Li. 2013. Metagenomic profiles of Antibiotic Resistance Genes (ARGs) between human impacted estuary and deep ocean sediments. *Environ Sci Technol* 47: 12753–12760.

Chen, K. 2017. Identification of bacterial species in Kuwaiti waters through DNA sequencing. *IOP conference series: Earth and environmental science* 51: 012015. IOP Publishing.

Chen, J., McIlroy, S.E., Archana, A., Baker, D.M., and G. Panagiotou. 2019. A pollution gradient contributes to the taxonomic, functional, and resistome diversity of microbial communities in marine sediments. *Microbiome* 7: 1–12.

Chen, Y.R., Guo, X.P., Niu, Z.S., Lu, D.P., Sun, X.L., Zhao, S., Hou, L.J., Liu, M., and Y. Yang. 2020. Antibiotic Resistance Genes (ARGs) and their associated environmental factors in the Yangtze Estuary, China: From inlet to outlet. *Mar Pollut Bull* 158: 111360.

Costello, M.J., and C. Chaudhary. 2017. Marine biodiversity, biogeography, deep-sea gradients, and conservation. *Curr Biol* 27: R511–R527.

D'Argenio, V., Casaburi, G., Precone, V., and F. Salvatore. 2014. Comparative metagenomic analysis of human gut microbiome composition using two different bioinformatic pipelines. *Bio Med Res Int* 2014: 325340. doi:10.1155/2014/325340.

Doytchinov, V.V., and S.G. Dimov. 2022. Microbial community composition of the Antarctic ecosystems: Review of the bacteria, fungi, and archaea identified through an NGS-based metagenomics approach. *Life (Basel)* 12, no. 6: 916. doi:10.3390/life12060916.

Esmaeil, A.S., Drobiova, H., and C. Obuekwe. 2009. Predominant culturable crude oil-degrading bacteria in the coast of Kuwait. *Int Biodeterior Biodegradation* 63: 400–406.

Fakhraldeen, S.A., Al-Haddad, S., Habibi, N., Alagarsamy, S., Habeebullah, S.F.K., Ali, A.K., and Al-Zakri, W.M. 2023. Diversity and spatiotemporal variations in bacterial and archaeal communities within Kuwaiti territorial waters of the Northwest Arabian Gulf. *Plos One* 18, no. 11: e0291167. doi:https://doi.org/10.1371/journal.pone.0291167.

Fu, L., Niu, B., Zhu, Z., Wu, S., and W. Li. 2012. CD-HIT: Accelerated for clustering the next-generation sequencing data. *Bioinformatics* 28: 3150–3152.

Gevao, B., Uddin, S., and S. Dupont. 2021. Baseline concentrations of pharmaceuticals in Kuwait's coastal marine environment. *Mar Pollut Bull* 173: 113040.

Gevao, B., Uddin, S., Krishnan, D., Rajagopalan, S., and N. Habibi. 2022. Antibiotics in wastewater: Baseline of the influent and effluent streams in Kuwait. *Toxics* 10: 174.

Gobet, A., Böer, S.I., Huse, S.M., Van Beusekom, J.E., Quince, C., Sogin, M.L., Boetius, A., and A. Ramette. 2012. Diversity and dynamics of rare and of resident bacterial populations in coastal sands. *ISME J* 6: 542–553.

Guo, X.P., Zhao, S., Chen, Y.R., Yang, J., Hou, L.J., Liu, M., and Y. Yang. 2020. Antibiotic resistance genes in sediments of the Yangtze Estuary: From 2007 to 2019. *Sci Total Environ* 744: 140713.

Habibi, N., Mustafa, A.S., and M.W. Khan. 2021a. Composition of nasal bacterial community and its seasonal variation in health care workers stationed in a clinical research laboratory. *PLoS One* 16: e0260314.

Habibi, N., Uddin, S., Al-Sarawi, H., Aldhameer, A., Shajan, A., Zakir, F., Abdul Razzack, N., and F. Alam. 2023. Metagenomes from coastal sediments of Kuwait: Insights into the microbiome, metabolic functions and resistome. *Microorganisms* 11: 531.

Habibi, N., Uddin, S., Bottein, M.Y.D., and M. Faizuddin. 2021b. Ciguatera in the Indian Ocean with special insights on the Arabian Sea and adjacent Gulf and seas: A review. *Toxins* 13: 525.

Habibi, N., Uddin, S., Lyons, B., Al-Sarawi, H.A., Behbehani, M., Shajan, A., Razzack, N.A., Zakir, F., and F. Alam. 2022. Antibiotic resistance genes associated with marine surface sediments: A baseline from the shores of Kuwait. *Sustainability* 14: 8029.

Imchen, M., and R. Kumavath. 2020. Shotgun metagenomics reveals a heterogeneous prokaryotic community and a wide array of antibiotic resistance genes in mangrove sediment. *FEMS Microbiol Ecol* 96: fiaa173.

Jiang, H., Lei, R., Ding, S.W., and S. Zhu. 2014. Skewer: A fast and accurate adapter trimmer for next-generation sequencing paired-end reads. *BMC Bioinform* 15: 1–12.

Kanehisa, M., Furumichi, M., Sato, Y., Ishiguro-Watanabe, M., and M. Tanabe. 2021. KEGG: Integrating viruses and cellular organisms. *Nucleic Acids Res* 49: D545–D551.

Kanehisa, M., and S. Goto. 2000. KEGG: Kyoto encyclopedia of genes and genomes. *Nucleic Acids Res* 28: 27–30.

Keegan, K.P., Glass, E.M., and F. Meyer. 2016. MG-RAST, a metagenomics service for analysis of microbial community structure and function. In: *Microbial environmental genomics (MEG)*. Eds: F. Martin, and S. Uroz, 207–233. New York: Springer.

Keegan, K.P., Trimble, W.L., Wilkening, J., Wilke, A., Harrison, T., D'Souza, M., and F. Meyer. 2012. A platform-independent method for detecting errors in metagenomic sequencing data: DRISEE. *PLoS Comput Biol* 8: e1002541.

Kirchman, D.L. 2002. The ecology of Cytophaga – Flavobacteria in aquatic environments. *FEMS Microbiol Ecol* 39: 91–100.

Kopylova, E., Noé, L., and H. Touzet. 2012. SortMeRNA: Fast and accurate filtering of ribosomal RNAs in metatranscriptomic data. *Bioinformatics* 28: 3211–3217.

Kültz, D. 2005. Molecular and evolutionary basis of the cellular stress response. *Ann Rev Physiol* 67: 225–257.

Kumar, V., Al Momin, S., Kumar, V.V., Ahmed, J., Al-Musallam, L., Shajan, A.B., Al-Aqeel, H., Al-Mansour, H., and W.M. Al-Zakri. 2021. Distribution and diversity of eukaryotic microalgae in Kuwait waters assessed using 18S rRNA gene sequencing. *PLoS One* 16: e0250645.

Langmead, B., and S.L. Salzberg. 2012. Fast gapped-read alignment with Bowtie 2. *Nat Methods* 9: 357–359.

Li, W., and A. Godzik. 2006. Cd-hit: A fast program for clustering and comparing large sets of protein or nucleotide sequences. *Bioinformatics* 22: 1658–1659.

Lynch, M.D., and J.D. Neufeld. 2015. Ecology and exploration of the rare biosphere. *Nat Rev Microbiol* 13: 217–229.

Mardis, E., and W.R. McCombie. 2017. Library quantification: Fluorometric quantitation of double-stranded or single-stranded DNA samples using the qubit system. *Cold Spring Harb Protoc* 2017: pdb.prot094730. doi:10.1101/pdb.prot094730.

McEwan, T. 2008. Environmentally tough coral species as candidates for aquarium culture. *Adv Coral Husb Public Aquar* 2: 185–189.

Meyer, F., Bagchi, S., Chaterji, S., Gerlach, W., Grama, A., Harrison, T., Paczian, T., Trimble, W.L., and A. Wilke. 2019. MG-RAST version 4 – Lessons learned from a decade of low-budget ultra-high-throughput metagenome analysis. *Brief Bioinform* 20: 1151–1159.

Mootapally, C., Nathani, N.M., Poriya, P., Beleem, I., Dabhi, J.C., Gadhvi, I.R., and C.G. Joshi. 2019. Antibiotic resistome biomarkers associated to the pelagic sediments of the Gulfs of Kathiawar Peninsula and Arabian Sea. *Sci Rep* 9: 1–12.

Moran, M.A. 2015. The global ocean microbiome. *Science* 350: aac8455.

Nathani, N.M., Mootapally, C., and B.P. Dave. 2019. Antibiotic resistance genes allied to the pelagic sediment microbiome in the Gulf of Khambhat and Arabian Sea. *Sci Total Environ* 653: 446–454.

Navgire, G.S., Goel, N., Sawhney, G., Sharma, M., Kaushik, P., Mohanta, Y.K., Mohanta, T.K., and A. Al-Harrasi. 2022. Analysis and interpretation of metagenomics data: An approach. *Biol Proced Online* 24: 18.

Nowak, M.A., Boerlijst, M.C., Cooke, J., and J.M. Smith. 1997. Evolution of genetic redundancy. *Nature* 388: 167–171.

Ogaki, M.B., Pinto, O.H.B., Vieira, R., Neto, A.A., Convey, P., Carvalho-Silva, M., Rosa, C.A., Câmara, P.E., and L.H. Rosa. 2021. Fungi present in Antarctic deep-sea sediments assessed using DNA metabarcoding. *Microbiol Ecol* 82: 157–164.

Plummer, E., Twin, J., Bulach, D.M., Garland, S.M., and S.N. Tabrizi. 2015. A comparison of three bioinformatics pipelines for the analysis of preterm gut microbiota using 16S rRNA gene sequencing data. *J Proteom Bioinform* 8: 283.

Pokavanich, T., Polikarpov, I., Lennox, A., Al-Hulail, F., Al-Said, T., Al-Enezi, E., Al-Yamani, F., Stokozov, N., and B. Shuhaibar. 2013. Comprehensive investigation of summer hydrodynamic and water quality characteristics of desertic shallow water body: Kuwait Bay. *J Coast Dyn* 12: 1253–1264.

Rho, M., Tang, H., and Y. Ye. 2010. FragGeneScan: Predicting genes in short and error-prone reads. *Nucleic Acids Res* 38: e191.

Sogin, M.L., Morrison, H.G., Huber, J.A., Welch, D.M., Huse, S.M., Neal, P.R., Arrieta, J.M., and G.J. Herndl. 2006. Microbial diversity in the deep sea and the underexplored 'rare biosphere'. *Proc Nat Acad Sci* 103: 12115–12120.

Song, J., Mujahid, A., Lim, P.-T., Samah, A.A., Quack, B., Pfeilsticker, K., Tang, S.-L., Ivanova, E., and M. Müller. 2017. Shotgun metagenomic analysis of microbial communities in the surface waters of the Eastern South China Sea. *Malaysian J Microbiol* 13: 350–362.

Tara Ocean Foundation, Tara Oceans, European Molecular Biology Laboratory (EMBL), European Marine Biological Resource Centre – European Research Infrastructure Consortium (EMBRC-ERIC). 2022. Priorities for ocean microbiome research. *Nat Microbiol* 7: 937–947.

Trevathan-Tackett, S.M., Sherman, C.D., Huggett, M.J., Campbell, A.H., Laverock, B., Hurtado-McCormick, V., Seymour, J.R., Firl, A., Messer, L.F., and T.D. Ainsworth. 2019. A horizon scan of priorities for coastal marine microbiome research. *Nat Ecol Evol* 3: 1509–1520.

Uddin, S., Aba, A., Behbehani, M., Al-Ghadban, A.N., Al-Zekri, W., and H. Al-Shammari. 2017a. Plutonium and cesium baseline concentrations in seawater from northern Arabian Gulf. *Mar Pollut Bull* 120: 396–400.

Uddin, S., Aba, A., Fowler, S.W., Behbehani, M., Ismaeel, A., Al-Shammari, H., Alboloushi, A., Mietelski, J.W., Al-Ghadban, A., Al-Ghunaim, A., Khabbaz, A., and O. Alboloushi. 2015. Radioactivity in the Kuwait marine environment – Baseline measurements and review. *Mar Pollut Bull* 100: 651–661.

Uddin, S., Al Ghadban, A.N., and A. Khabbaz. 2011a. Localized hyper saline waters in Arabian Gulf from desalination activity – An example from South Kuwait. *Environ Monit Assess* 181: 587–594.

Uddin, S., Al Ghadban, A.N., and A. Khabbaz. 2011b. Localized hyper saline waters in Arabian Gulf from desalination activity – An example from South Kuwait. *Environ Monit Assess* 181: 587–594.

Uddin, S., Behbehani, M., Aba, A., and A.N. Al Ghadban. 2017b. Naturally Occurring Radioactive Material (NORM) in seawater of the northern Arabian Gulf – Baseline measurements. *Mar Pollut Bull* 123: 365–372.

Uddin, S., Fowler, S.W., Habibi, N., and M. Behbehani. 2022. Micro-nano plastic in the aquatic environment: Methodological problems and challenges. *Animals* 12: 297.

Uddin, S., Fowler, S.W., and T. Saeed. 2020. Microplastic particles in the Persian/Arabian Gulf – A review on sampling and identification. *Mar Pollut Bull* 154: 111100.

Uddin, S., Fowler, S.W., Uddin, M.F., Behbehani, M., and A. Naji. 2021. A review of microplastic distribution in sediment profiles. *Mar Pollut Bull* 163: 111973.

Uddin, S., Gevao, B., Al-Ghadban, A., Nithyanandan, M., and D. Al-Shamroukh. 2012. Acidification in Arabian Gulf – Insights from pH and temperature measurements. *J Environ Monit* 14: 1479–1482.

Wang, S., Yan, Z., Wang, P., Zheng, X., and J. Fan. 2020. Comparative metagenomics reveals the microbial diversity and metabolic potentials in the sediments and surrounding seawaters of Qinhuangdao mariculture area. *PLoS One* 15: e0234128.

Yan, C., Wang, F., Geng, H., Liu, H., Pu, S., Tian, Z., Chen, H., Zhou, B., Yuan, R., and J. Yao. 2020. Integrating high-throughput sequencing and metagenome analysis to reveal the characteristic and resistance mechanism of microbial community in metal contaminated sediments. *Sci Total Environ* 707: 136116.

Yang, H., Liu, R., Liu, H., Wang, C., Yin, X., Zhang, M., Fang, J., Zhang, T., and L. Ma. 2021. Evidence for long-term anthropogenic pollution: The Hadal Trench as a depository and indicator for dissemination of antibiotic resistance genes. *Environ Sci Technol* 55: 15136–15148.

Yang, X., Yan, L., Yang, Y., Zhou, H., Cao, Y., Wang, S., Xue, B., Li, C., Zhao, C., and X. Zhang. 2022. The occurrence and distribution pattern of antibiotic resistance genes and bacterial community in the Ili River. *Front Environ Sci* 10: 840428. doi:10.3389/fenvs.2022.840428.

Zhang, B., Li, Y., Xiang, S.-Z., Yan, Y., Yang, R., Lin, M.-P., Wang, X.-M., Xue, Y.-L., and X.-Y. Guan. 2020. Sediment microbial communities and their potential role as environmental pollution indicators in Xuande Atoll, South China Sea. *Front Microbiol* 11: 1011.

Zhang, X., Xu, W., Liu, Y., Cai, M., Luo, Z., and M. Li. 2018. Metagenomics reveals microbial diversity and metabolic potentials of seawater and surface sediment from a hadal biosphere at the Yap Trench. *Front Microbiol* 9: 2402.

Harmful algae in the Red Sea 18

Amany Ismael and Abdulsalam Alkawri

18.1 INTRODUCTION

The Red Sea is a narrow oligotrophic tropical sea extending from 12.5°N to 30°N. There are various theories concerning the name of the Red Sea, but the most popular one is that it is due to the seasonal bloom of the harmful cyanobacteria *Trichodesmium erythraeum*, which forms a red tide (Belogorskay 1970). The Red Sea has been characterized as an area sensitive to climate change and classified as a fast-warming large marine ecosystem (Gittings et al. 2018). The temperature increase might stimulate the proliferation of some phytoplankton groups such as cyanobacteria (Chaidez et al. 2017). Although algae are essential for life on Earth and for fisheries, some species may cause harm in various ways. Harmful algal blooms (HABs) include bioaccumulation of toxins in seafood, reaching levels unsafe for human consumption, or causing harm, directly or indirectly to aquatic habitat or their products.

In recent years, HABs have increased worldwide, causing serious problems to marine and public health. Despite its importance, limited studies were carried out along the Red Sea, with a lack of information along its western side except for the Gulf of Suez and Port Sudan. Ninety potentially harmful species (Table 18.1) were recorded from the Red Sea; the causative group mainly included cyanobacteria, diatoms, dinoflagellates and raphidophytes (Mohamed and Mesaad 2007; Alkershi and Nandini Menon 2011; Saburova et al. 2013; Kürten et al. 2015; Alkawri 2016; Alkawri et al. 2016a, 2016b; Banguera-Hinestroza et al. 2016; Devassy et al. 2017; Catania et al. 2017). The Red Sea was classified into three regions: the northern Red Sea (NRS), the central Red Sea (CRS) and the southern Red Sea (SRS). The HAB community structure and distribution from each region was investigated by several workers. Gokul et al. (2020) studied the HAB events along the entire Red Sea using remote sensing. In general, HAB events in the Red Sea appear to decrease northward, with the highest number of HABs reported from the southern region of the Red Sea, followed by the central region, in particular along the Saudi Arabian and Yemeni coasts (Banguera-Hinestroza et al. 2016; Gokul et al. 2020). The increase of HAB species in the Red Sea is associated with increasing nutrients along the coastal areas due to an increase of wastewater discharge. With increasing HAB events, HAB monitoring programs along the Red Sea should be established to avoid health threats and economic loss. This chapter reviews the diversity and distribution of HABs in the Red Sea to improve our knowledge on harmful algae and their impacts on the region and to provide the basis for designing a nationwide monitoring program to protect public health.

18.2 SOUTHERN RED SEA (SRS)

The SRS includes the western coast of Yemen on the eastern side and the coasts of Eritrea and Djibouti on the western side of the Red Sea (Figure 18.1). The eastern side of the SRS exhibited a higher occurrence of HABs than the western side. Although Gokul et al. (2020) confirmed fewer events of high concentrations of chlorophyll on the west coast of the SRS and the open water by using remote sensing algorithms, the species forming these blooms are unknown and could be harmless or harmful. In the past few decades, HABs have been observed and have increased drastically in the SRS, especially in the Yemeni coastal waters (Alkershi and Nandini Menon 2011; Alkawri 2016; Alkawri et al. 2016a, 2016b). This could be due to high nutrient concentrations received from intermediate water inflow through the Gulf of Aden in the summer (Churchill et al. 2014; Dreano et al. 2016), in addition to the discharge of large quantities of nutrients from domestic, industrial and agricultural wastes. Using remote sensing (Gokul et al. 2020), 24 algal bloom events were reported along the southern coasts of the Red Sea, mostly from Yemeni coasts from 2003 to 2017. The first bloom of harmful algae reported in the Yemeni coast of the Red Sea was caused by *Noctiluca scintillans* at Al-Hodeidah coastal water during March 2009 near the fish landing centre, with a maximum abundance of 5.5×10^5 cells l^{-1} (Alkershi and Nandini Menon 2011). A dense bloom of *Blixaea quinquecornis* (*Peridinium/Protoperidinium quinquecorne*) was reported, for the first time, in June 2012 from the coastal water of Al Hodeidah City and Khor Al-Khateeb lagoon (Yemeni coast). The average cell abundance of *Blixaea quinquecornis* during the bloom was 14.3×10^6 cell l^{-1}, contributing more than 99% of the overall phytoplankton population in the area (Alkawri et al. 2016a). Although *Blixaea quinquecornis* is known to be a non-toxic species, the bloom was accompanied by a massive kill of small pelagic fish, mostly *Sardinella* sp., which extended along the surface water of the Al Hodeidah coast due to oxygen depletion (Alkawri et al. 2016a, 2016b). Since then, the bloom has become a recurrent phenomenon in the coastal water of Al Hodeidah (Alkawri 2016; Alkawri et al.

DOI: 10.1201/9781003321392-23

TABLE 18.1 Checklist of the harmful microalgae species reported from the Red Sea.

SPECIES	HABITAT	SOUTHERN RED SEA	CENTRAL RED SEA	NORTHERN RED SEA	REFERENCE
Dinoflagellate					
***Akashiwo sanguinea** (K. Hirasaka) Gert Hansen and Moestrup	P	x	x	x	(Alkawri 2016)
Alexandrium catenella (Whedon & Kofoid) Balech	P		x		(El-Sherbiny et al. 2021)
***Alexandrium minut*um** Halim	P	x	x		(Alkawri 2016; Gomaa et al. 2018)
Alexandrium sp.	P	x	x		(Banguera-Hinestroza et al. 2016)
Alexandrium tamarense (Lebour) Balech	P		x		(Prabowo and Agusti 2019; El-Sherbiny et al. 2021)
Alexandrium tamiyavanichii Balech	P		x		(Al-Amri et al. 2020)
Amphidinium carterae Hulburt	P, B		x		(Prabowo and Agusti 2019; Gomaaet al. 2018)
Amphidinium gibbosum (L. Maranda & Y. Shimizu) Flø Jørgensen and Shauna Murray	P		x		(Prabowo and Agusti 2019)
Amphidinium sp.	P, B		x	x	(Prabowo and Agusti 2019)
*** Blixaea quinquecornis** (F. Stein) Gert Hansen	P	x	x		(Alkawri et al. 2016a, 2016b; Prabowo and Agusti 2019)
***Coolia monotis** Meunier	B		x		(Prabowo and Agusti 2019)
Dinophysis acuminata Claparède and Lachmann	P	x	x		(Alkawri 2016; Alkawri et al. 2016b)
Dinophysis acuta Ehrenberg	P	x	x		(Alkawri 2016; Alkawri et al. 2016b)
Dinophysis caudata Saville-Kent	P	x	x	x	(Alkawri 2016; Alkawri et al. 2016b; Prabowo and Agusti 2019; El-Sherbiny et al. 2021; Gomaa et al. 2018)
Dinophysis fortii Pavillard	P		x		(Prabowo and Agusti 2019)
Dinophysis miles Cleve	P	x	x		(Gomaa et al. 2018)
Dinophysis norvegica Claparède and Lachmann	P		x		(Gomaa et al. 2018)
***Dinophysis schuettii** Murray and Whitting	P		x		(Gomaa et al. 2018)
Dinophysis tripos Gourret	P		x	x	(Ismael 2005, 2015)
Fukuyoa yasumotoi (M.J. Holmes) F. Gómez, D.X. Qiu, R.M. Lopes & Senjie Lin	B			x	(Saburova et al. 2013)
Gambierdiscus belizeanus M. A. Faust	B	x	x	x	(Alkawri et al. 2016b; Prabowo and Agusti 2019; Catania et al. 2017)
Gambierdiscus silvae S.Fraga and F. Rodríguez	B		x		(Prabowo and Agusti 2019)
Gambierdiscus sp.	B			x	(Catania et al. 2017)
Gonyaulax sp.	P		x		(Prabowo and Agusti 2019)
Gonyaulax spinifera (Claparède & Lachmann) Diesing	P				(Ismael 2005, 2015)
Gymnodinium sp.	P, B		x	x	(Gomaa et al. 2018)
***Gyrodinium cf. dominans** Hulbert	B			x	(Prabowo and Agusti 2019)
Gyrodinium sp.	B			x	(Gomaa et al. 2018)
Karenia mikimotoi (Miyake & Kominami ex Oda) Gert Hansen and Moestrup	P		x		(Prabowo and Agusti 2019)

TABLE 18.1 (Continued)

SPECIES	*HABITAT*	*SOUTHERN RED SEA*	*CENTRAL RED SEA*	*NORTHERN RED SEA*	*REFERENCE*
Karenia selliformis A.J. Haywood, K.A. Steidinger & L. MacKenzie	P		x		(Prabowo and Agusti 2019)
****Kryptoperidinium foliaceum*** (F. Stein) Lindemann	P	x			(Alkawri 2016)
Lingulodinium polyedra (F. Stein) J.D. Dodge	P		x		(Prabowo and Agusti 2019)
Margalefidinium polykrikoides ((Margalef) F. Gómez, Richlen & D.M. Anderson	P		x		(Prabowo and Agusti 2019)
****Noctiluca scintillans*** (Macartney) Kofoid and Swezy	P		x		(Alkershi and Nandini Menon 2011; Mohamed and Mesaad 2007)
Ostreopsis lenticularis Fukuyo	B		x		(Prabowo and Agusti 2019)
Ostreopsis ovata Fukuyo	B		x		(Gomaa et al. 2018; Prabowo and Agusti 2019)
Ostreopsis siamensis Johs. Schmidt	B		x		(Prabowo and Agusti 2019)
Ostreopsis sp.	B	x	x	x	(Prabowo and Agusti 2019)
Phalacroma mitra F. Schütt	P		x		(El-Sherbiny et al. 2021)
Phalacroma rotundatum (Claparéde & Lachmann) Kofoid and J.R. Michener	P		x		(Prabowo and Agusti 2019; El-Sherbiny et al. 2021)
Prorocentrum borbonicum L. Ten-Hage et al.	B		x		(Prabowo and Agusti 2019)
Prorocentrum concavum Fukuyo	B	x	x		(Alkawri 2016; Prabowo and Agusti 2019; Gomaa et al. 2018)
Prorocentrum cordatum (Ostenfeld) J.D. Dodge	P		x		(El-Sherbiny et al. 2021)
****Prorocentrum elegans*** M.A. Faust	B			x	(Prabowo and Agusti 2019)
Prorocentrum emarginatum Fukuyo	B		x		(Prabowo and Agusti 2019)
Prorocentrum hoffmannianum M.A.Faust	B		x		(Prabowo and Agusti 2019)
Prorocentrum leve M.A. Faust, M.W. Vandersea, S.R. Kibler, P.A. Tester & R.W. Litaker	B		x		(Prabowo and Agusti 2019)
Prorocentrum lima (Ehrenberg) F. Stein	B	x	x		(Alkawri 2016; Prabowo and Agusti 2019; Gomaa et al. 2018)
Prorocentrum rhathymum Loeblich III, Sherley & Schmidt	B		x		(Gomaa et al. 2018; Prabowo and Agusti 2019)
Prorocentrum sp.	B	x	x	x	(Prabowo and Agusti 2019)
Protoceratium reticulatum (Claparède & Lachmann) Bütschli	P		x		(Prabowo and Agusti 2019)
Pyrodinium bahamense Plate	P	x	x	x	Banguera-Hinestroza et al. 2016; Alkawri et al. 2016b; Prabowo and Agusti 2019)
Pyrodinium bahamense var. compressum (Böhm) Steidinger, Tester & F.J.R. Taylor	P	x		x	(Ismael 2005, 2015)
Diatoms					
Pseudo-nitzschia australis Frenguelli	P		x	x	(Ismael 2015)
Pseudo-nitzschia delicatissima (Cleve) Heiden	P		x	x	(El-Sherbiny et al. 2021)
Pseudo-nitzschia fraudulenta (Cleve) Hasle	P		x	x	(Ismael 2015)

(Continued)

TABLE 18.1 (Continued)

SPECIES	*HABITAT*	*SOUTHERN RED SEA*	*CENTRAL RED SEA*	*NORTHERN RED SEA*	*REFERENCE*
Pseudo-nitzschia granii (G.R. Hasle) G.R. Hasle	P		x	x	(Ismael 2015)
Pseudo-nitzschia pungens (Grunow ex Cleve) G.R. Hasle	P		x	x	(Nassar et al. 2015)
Pseudo-nitzschia seriata (Cleve) H.Peragallo	P		x	x	(El-Sherbin**y** et al. 2021)
Pseudo-nitzschia subcurvata (G.R. Hasle) G.A. Fryxell	P		x	x	(Ismael 2015)
Cyanobacteria					
Anabaena sphaerica Bornet and Flahault	B		x		(Mohamed and Al-Shehri 2015)
Calothrix breviarticulata West and G.S. West	B		x		(Mohamed and Al-Shehri 2015)
Calothrix scopulorum C.Agardh ex Bornet and Flahault	B		x		(Mohamed and Al-Shehri 2015)
Kamptonema chlorinum (Kützing ex Gomont) Strunecký, Komárek & J. Smarda	B		x		(Mohamed and Al-Shehri 2015)
Kamptonema formosum (Bory ex Gomont) Strunecký, Komárek & J. Smarda	B		x	x	(Nassar 2007; Nassar et al. 2015; Mohamed 2018)
Leptolyngbya fragilis (Gomont) Anagnostidis and Komárek	B		x		(Mohamed and Al-Shehri 2015)
Leptolyngbya tenuis (Gomont) Anagnostidis and Komárek	B		x		(Mohamed and Al-Shehri 2015)
Limnoraphis birgei (G.M. Smith) J. Komárek, E. Zapomelová, J. Smarda, J. Kopecký, E. Rejmánková, J. Woodhouse, B.A. Neilan & J. Komárková	B		x		(Mohamed and Al-Shehri 2015)
Lyngbya majuscula Harvey ex Gomont	B		x	x	(Nassar 2007; Nassar et al. 2015; Mohamed 2018)
Lyngbya martensiana Meneghini ex Gomont	B		x		(Mohamed and Al-Shehri 2015)
Oscillatoria curviceps C. Agardh ex Gomont	B		x		(Mohamed and Al-Shehri 2015)
Oscillatoria indica P.C. Silva	B		x		(Mohamed and Al-Shehri 2015)
Oscillatoria sancta Kützing ex Gomont	B		x		(Mohamed and Al-Shehri 2015)
Oscillatoria tenuis C. Agardh ex Gomont	B		x	x	(Mohamed and Al-Shehri 2015)
Oxynema acuminatum (Gomont) Chatchawan, Komárek, Strunecky, Smarda & Peerapornpisal	B		x		(Mohamed and Al-Shehri 2015)
Phormidium ambiguum Gomont	B		x		(Mohamed and Al-Shehri 2015)
Phormidium nigroviride (Thwaites ex Gomont) Anagnostidis and Komárek	B		x		(Mohamed and Al-Shehri 2015)
Phormidium stagninum Anagnostidis	B		x		(Mohamed and Al-Shehri 2015)
Phormidium willei (N.L. Gardner) Anagnostidis and Komárek	B		x		(Mohamed and Al-Shehri 2015)
Planktothrix agardhii (Gomont) Anagnostidis and Komárek				x	(Nassar 2007; Nassar et al. 2015; Mohamed 2018)
Pseudanabaena catenata Lauterborn	B		x		(Mohamed and Al-Shehri 2015)
Pseudanabaena limnetica (Lemmermann) Komárek	B		x	x	(Nassar 2007; Nassar et al. 2015; Mohamed 2018)

TABLE 18.1 (Continued)

SPECIES	HABITAT	SOUTHERN RED SEA	CENTRAL RED SEA	NORTHERN RED SEA	REFERENCE
Scytonematopsis crustacea (Thuret ex Bornet & Flahault) Kovalik and Komárek	B		x		(Mohamed and Al-Shehri 2015)
****Spirulina major*** Kützing ex Gomont	B		x		(Mohamed and Al-Shehri 2015)
****Spirulina subsalsa*** Oersted ex Gomont	B		x		(Mohamed and Al-Shehri 2015)
****Trichodesmium erythraeum*** Ehrenberg ex Gomont	P	x	x	x	(Madkour et al. 2010; Mohamed 2018; Beraki et al. 2022; Madkour et al. 2010)
****Trichodesmium thiebautii*** Gomont ex Gomont	P		x		(Touliabah et al. 2010)
Raphidophytes					
Chattonella marina (Subrahmanyan) Hara and Chihara	P		x		(Gomaa et al. 2018)
Heterosigma akashiwo (Y. Hada) Y. Hada ex Y.Hara and M. Chihara	P	x			(Mohamed and Al-Shehri 2012)

P: plankton, B: benthic.
* Species showed harmful effect in the Red Sea or elsewhere but not recorded in IOC-UNESCO taxonomic list of harmful microalgae.

2016b). An intensive coastal bloom of the toxic dinoflagellate *Pyrodinium bahamense* was observed during summer 2013 at the coastal water of Yemen, with cell abundances that ranged from 1.6×10^4 to 3.3×10^5 cell l^{-1} (Alkawri et al. 2016b). To the best of our knowledge, the species did not cause a harmful effect until now in the Red Sea, but their potential for harm is very high and should be noted as a dangerous species.

Different potentially harmful dinoflagellates were recorded along the SRS during spring and summer. *Akashiwo sanguinea* showed the highest abundance (3.6×10^3 cells l^{-1}), followed by *Alexandrium minutum* (3.2×10^3 cells l^{-1}), *Prorocentrum concavum* (2.9×10^3 cells l^{-1}) and *Dinophysis* spp. (*Dinophysis acuminata, D. acuta*, and *D. caudata*), with a total abundance of 2.9×10^3 cells l^{-1} (Alkawri 2016; Alkawri et al. 2016b). In addition, a bloom of *Kryptoperidinium foliaceum* was reported with a maximum density during spring and late summer 2013 (2.26×10^6 and 1.4×10^3 cells l^{-1}, respectively) that contributed about 99% of the total dinoflagellate community. The red tide–forming cyanobacteria species *Trichodesmium erythraeum*, with a relative contribution to the total phytoplankton abundance of 14.1%, was the most abundant species at the coastal water of Al Hodeidah in the SRS (Alkawri et al. 2016b). Recently, Beraki et al. (2022) reported a massive fish mortality along the Eritrean coast during summer 2021 due to harmful algae but with no information about the causative species

18.3 CENTRAL RED SEA (CRS)

The CRS includes Saudi Arabia in the eastern side and Sudan in the west (Figure 18.1). Although many potentially harmful species were reported from this region (Mohamed and Al-Shehri 2011, 2012; Kürten et al. 2015; Banguera-Hinestroza et al. 2016; Aynousah 2017; Catania et al. 2017; Devassy et al. 2017; Gomaa et al. 2018; Al-Aidaroos et al. 2019; Prabowo and Agusti 2019; Al-Amri et al. 2020; Gokul et al. 2020; El-Sherbiny et al. 2021), only a few data on HAB events were available from Sudanese coasts.

According to Gokul et al. (2020), a total of 27 blooms was detected from this region using remote sensing over 15 years (2003–2017), with maximum records during summer (12 events), mostly from the Saudi Arabian coast. Touliabah et al. (2010) stated that the high nutrient concentrations due to sewage runoff were the main reason for the dominance of dinoflagellate and diatom species in this area. The first HAB event reported from the Saudi Arabian coast (Al Shuqayq and Gazan) in winter 2004 and 2005 was caused by the heterotrophic dinoflagellate *Noctiluca scintillans*, with a maximum abundance of 3×10^6 cells m^{-3} (Mohamed and Mesaad 2007). The bloom of *Pseudo-nitzschia* cf. *delicatissima* was reported from the coastal water of Jeddah, Saudi Arabia, during spring 2017, representing 97.5% of the total phytoplankton population, with a maximum abundance of 2.98×10^6 m^{-3} due to the anthropogenic impact over a continuous period of time (El-Sherbiny et al. 2021).

An intensive bloom of the raphidophyte *Heterosigma akashiwo* was observed for the first time off the Al Shouqyq coast, southern Saudi Arabia, in May 2010 near a shrimp farm (Mohamed and Al-Shehri 2012). The bloom of the species was restricted to this area, with no other observation of the species along the Red Sea coast, mainly due to the limited research in the area. A bloom of *Pyrodinium bahamense* was reported in a semi-closed lagoon in Saudi Arabian coastal waters, with a maximum abundance that ranged from 10^4–10^5 cells l^{-1} during September to the beginning of December. The bloom was associated with the presence of *Alexandrium* spp., but with a low abundance (Banguera-Hinestroza et al. 2016).

FIGURE 18.1 Map of the Red Sea with the three regions: northern Red Sea (NRS), central Red Sea (CRS) and southern Red Sea (SRS) (modified from: www.worldatlas.com/).

Prabowo and Agusti (2019) identified 47 taxa (Table 18.1) that were considered to be potentially harmful species during 2016–2017 from the CRS off the Saudi Arabian coast. The species recorded are known to be toxin producers: *Dinophysis* spp., *Phalacroma* spp. and *Prorocentrum* spp., which are associated with diarrhetic shellfish poisoning (DSP), *Pyrodinium* spp. and *Alexandrium* spp., associated with paralytic shellfish poisoning (PSP), and *Karenia* spp., which was reported to produce neurotoxic shellfish poisoning (NSP). In addition, some species were associated with ichthyotoxin, such as *Amphidinium* spp., or even resulted in massive blooms leading to hypoxia and discolouration of water, such as *Akashiwo sanguinea*, *Levanderina fissa*, *Noctiluca scintillans* and *Blixaea quinquecornis*.

Furthermore, many studies reported the outbreak of *Trichodesmium* spp. along the Red Sea coasts with more occurrences in the CRS. The summer bloom of the species became a recurrent phenomenon in the area. The bloom of *Trichodesmium erythraeum* and *T. thiebautti* was recorded from the southeastern coast of Jeddah during the summer of 2004 due to the nutrient-rich domestic wastewater discharge along the coastal waters of Jeddah (Touliabah et al. 2010; Gokul et al. 2019). In 2017, *T. erythraeum* dominated the phytoplankton population (average: 399×10^3 filament m^{-3}) along the coastal waters of Jeddah, Saudi Arabia (Al-Amri et al. 2020). An outbreak of *Alexandrium tamiyavanichii* was observed along the coastal waters of Jeddah in 2017, with a maximum abundance of 1.2×10^6 cells m^{-3}, accounting for 6.2% of the total phytoplankton population in the area (Al-Amri et al. 2020).

Because desalination plants along the Saudi Arabian Red Sea coast provide drinking water for millions of people, the HAB events from the period 2014 to 2016 along the coast of Jeddah near desalination plants were studied by Gomaa et al. (2018). They found that the ecosystem near a desalination plant triggers high populations of the detected toxic species but with no harmful effect; *Dinophysis caudata* (1.9×10^5 cells l^{-1}), *Dinophysis miles* (1.67×10^5 cells l^{-1}), *Dinophysis caudata* and *Gymnodinium* sp. were identified at all stations near the desalination plants of Jeddah, Al-Shoaiba and Al-Quonfuduah during September 2016. In addition, many harmful species and potential toxin producers were identified from the study area, including *Prorocentrum lima*, *Ostreopsis ovata*, *Chattonella marina*, *Amphidinium* sp., *Alexandrium minutum*, *Gymnodinium* sp., *Peridinium* sp. and *Goniodoma* sp. (Gomaa et al. 2018). These potentially harmful species can affect future development of desalination plants in the area.

A few studies on the occurrence of harmful species have been conducted in Sudanese waters off the Port Sudan coast. El Hag and Nasr (1989) and Ali (2015) reported high abundances of potentially harmful diatoms, dinoflagellates and cyanobacteria species during the fall of 1986: *Dinophysis* spp., *Gonyaulax* spp., *Peridinium* spp. and *Prorocentrum* spp., with standing crops of 10^5 cells l^{-1}, *Chaetoceros* spp. and *Nitzschia* spp. ranged from 0.68–1.1 10^6 cells l^{-1}, and a bloom of *Trichodesmium erythraeum* reached 3.33×10^5 filament l^{-1}. Recently, 16 potentially harmful species, mainly dinoflagellates, were identified from the coastal water of Sudan with the dominance of *Trichodesmium* spp. throughout the summer season (Osama and Amjed 2020).

18.4 NORTHERN RED SEA (NRS)

The NRS includes the Gulf of Suez, the Gulf of Aqaba and the Egyptian coast (Figure 18.1). Since Ostenfeld and Schmidt (1902), the phytoplankton community in the NRS has been investigated by several authors and revealed variable biodiversity and community structure according to different ecological conditions and different spatial and temporal scales (Nassar and Hamed 2003; Ismael 2005, 2015; Madkour et al. 2010; Nassar et al. 2016; Mohamed 2018). Although the results of these studies also revealed the presence of potentially harmful algal species in the NRS, no HAB events or fish killings were detected from these areas. Using remote sensing, Gokul et al. (2020) reported that HAB events in the Red Sea appear to decrease northward, with no harmful events detected from the northern region throughout the 15-year period 2003 to 2017. This region is characterized by a low diversity of the HAB species with the presence of *Dinophysis* spp., *Gonyaulax* spp., *Noctiluca scintillans* and *T. erythraeum*.

In the Gulf of Suez, *Dinophysis caudata*, *D. tripos*, *Gonyaulax spinifera*, *Noctiluca scintillans*, *Pyrodinium bahamense var. compressum* and *Trichodesmium erythraeum* were reported in low densities during 1999–2001(Ismael 2005, 2015). *Trichodesmium erythraeum* showed a peak with a maximum density of 7×10^3 filament l^{-1} in the Gulf during summer

2008 (Madkour et al. 2010; Mohamed 2018). However, in the summer of 2013, the species was detected in the Gulf with an abundance >200 filament l^{-1} (Nassar et al. 2015). A bloom of *Trichodesmium* spp. that reached >10^6 filament l^{-1} was detected during the fall of 1997 from the Gulf of Aqaba (Post et al. 2002). Al-Najjar et al. (2007) also reported an increase in the species abundance during summer and fall of 1999, reaching 4×10^4 filament l^{-1} from the same area.

Other potentially toxic species of cyanobacteria, *Kamptonema formosum*, *Lyngbya majuscula*, *Planktothrix agardhii*, *O. tenuis* and *Pseudanabaena limnetica* were also reported from both the western and eastern coasts of the Gulf of Suez (Nassar 2007; Nassar et al. 2015; Mohamed 2018). The abundance of such harmful species was correlated with high nutrient concentrations as well as low salinity in this region (Nassar et al. 2014).

In addition to these species, *Pseudanabaena limnetica* was reported with low abundance (15–117 filament l^{-1}) during summer 2001 in mangrove ecosystems of the Red Sea at the southeastern Egyptian coast (Abdel Rahman and Nassar 2005). The potentially harmful species *Dinophysis* spp. and *Gonyaulax* spp. were recorded during spring 2008 in the Red Sea proper at the Hurghada and Sharm El-Sheikh coasts, with densities of 1–2 $\times 10^3$ cells l^{-1} (Madkour et al. 2010; Mohamed 2018). The diatoms *Pseudonitzschia* spp. also dominated the phytoplankton population in these sites during winter 2007, while *T. erythraeum* dominated during summer, with a relatively high density of 3×10^3 filament l^{-1} (Madkour et al. 2010; Mohamed 2018).

18.5 DINOFLAGELLATE CYSTS

Dinoflagellate cysts are a useful tool for understanding the upper water conditions. The composition of dinoflagellate cysts is highly dependent on the biological, physical, chemical and oceanographic conditions of the water and also the sedimentary environment. Time-series analysis of the cyst assemblages can provide information about the history of harmful algal blooms and about possible causes and future risk (Zonneveld et al. 2013). In addition, it can provide information about the past variations of the upper water conditions such as temperature, salinity variations, nutrient content and eutrophication.

In the Red Sea, a total of 54 dinoflagellate cysts representing 28 genera were identified, 24 species of them being potentially harmful. The harmful cyst assemblage comprised 12 genera of the orders Gonyaulacales, Prorocentrales, Gymnodinals and Dinophysiales (Table 18.2). Dinoflagellate cysts in the Red Sea consist of phototrophic (autotrophic and mixotrophic) and heterotrophic species, with a greater abundance of phototrophic species than heterotrophic. The harmful species are more dominant than the harmless species in the sediment assemblages. According to Elshanawany and Zonneveld (2016), the Gulf of Aqaba was dominated by cysts of *Lingulodinium machaerophorum*, *Gonyaulax* spp. and *Spiniferites* spp. This indicates that the distribution and abundance of these cysts might be related to the bloom of *Synechococcus* sp., which dominated the phytoplankton community in the Gulf of Aqaba and was considered to be a food source for the motile cells of these cysts. The Gulf of Suez was dominated mainly by Gonyaulacaceae, *Polysphaeridium zoharyi*, *Spiniferites* spp., *Operculodinium* spp., *Batiacasphaera hirsute*, *Lingulodinium machaerophorum*, *Hystrichokolpoma rigaudiae* and *Tuberculodinium vancampoae*, and the Protoperdinoid cyst *Brigantedinium* spp. (El-Soughier and Mahmoud 2019). The surface sediment of the central and southern Red Sea is dominated by five potentially harmful species: *Alexandrium catenella*, *Alexandrium minutum*, *Margalefidinium (Cochlodinium) polykrikoides*, *Prorocentrum cordatum* and *Scrippsiella trochoidea*. The total abundance of these species ranged from 1816 to 8306 cyst g^{-1} dry wt (Mohamed and Al-Shehri 2011). The dominance of HAB species in the sediment represents a risk factor for early warning for future HABs in the Red Sea.

18.6 BENTHIC HARMFUL MICROALGAE

Little is known about the benthic forms of potentially harmful microalgae from the Red Sea. The BHAB of the Red Sea are dominated by dinoflagellates and cyanobacteria, which are found epiphytically on macroalgae and corals in the intertidal and subtidal zone. In addition, benthic dinoflagellates were recorded from sediment as sand-dwelling dinoflagellates. A total of 48 BHAB species were recorded from different parts of the Red Sea; the highest species richness was recorded from the CRS, while no data are available concerning the SRS.

Benthic dinoflagellate species have been described for the first time from the CRS off Saudi Arabia by Catania (2012), and from the NRS, with first report of the genus *Gambierdiscus* in 2013 (Saburova et al. 2013). The benthic dinoflagellates dominated the BHAB during spring and showed the lowest values during winter (Prabowo and Agusti 2019).

Eight genera of potentially harmful benthic dinoflagellates – *Amphidinium*, *Coolia*, *Gambierdiscus*, *Gymnodinium*, *Gyrodinium*, *Katodinium*, *Ostreopsis* and *Prorocentrum* with 26 species (Table 18.1) – were recorded from the NRS and CRS (Catania 2012; Saburova et al. 2013; El-Semary 2016). Catania (2012) showed that benthic epiphytic dinoflagellates were associated with ciguatera fish poisoning in the CRS. The cell abundance of BHAB species in the Red Sea ranged from <40 to <200 cells g^{-1} wet weight of algae (Catania 2012; Catania et al. 2017). Recently, four different species of *Ostreopsis* were reported during 2016–2017 in the coastal areas of the CRS off Saudi Arabia: *Ostreopsis lenticularis*, *Ostreopsis ovata*, *Ostreopsis siamensis* and *Ostreopsis* spp. (Prabowo and Agusti 2019).

The diversity of cyanobacteria on soil surfaces and mangrove swamps in the Red Sea off the coast of Saudi Arabia was studied by Mohamed and Al-Shehri (2015). A total of 24 cyanobacteria species with three families were recorded: Oscillatoriaceae, Nostocaceae and Rivulariaceae (Table 18.1).

TABLE 18.2 Harmful dinoflagellate cysts recorded from the Red Sea, Gulf of Aqaba (AQ), Gulf of Suez (SG) and Red Sea proper (RSP).

CYST	*MOTILE SPECIES*	*AQ*	*SG*	*RSP*	*REFERENCE*
Cyst of ***Alexandrium catenella*** (Whedon & Kofoid) Balech	*Alexandrium catenella*			x	(Mohamed and Al-Shehri 2011)
Cyst of ***Alexandrium minutum*** Halim	*Alexandrium minutum*			x	(Mohamed and Al-Shehri 2011)
Ataxiodinium choane Reid	*Gonyaulax* spp.	x			(Elshanawany and Zonneveld 2016)
Bitectatodinium tepiliense Wilson	*Gonyaulax* spp.	x			(Elshanawany and Zonneveld 2016)
Cyst of ***Dinophysis acuminata*** Claparède and Lachmann	*Dinophysis acuminata*				(Mohamed and Al-Shehri 2011)
Impagidinium paradoxum (Wall) Stover et Evitt	Gonyaulax spp.	x			(Elshanawany and Zonneveld 2016)
Impagidinium patulum (Wall) Stover et Evitt	Gonyaulax spp.	x			(Elshanawany and Zonneveld 2016)
Impagidinium plicatum Versteegh et Zevenboom	Gonyaulax spp.	x			(Elshanawany and Zonneveld 2016)
Impagidinium sphaericum (Wall) Lentin and Williams	Gonyaulax spp.	x			(Elshanawany and Zonneveld 2016)
Impagidinium spp.	Gonyaulax spp.	x			(Elshanawany and Zonneveld 2016)
Lingulodinium machaerophorum (Deflandre and Cookson) Wall	*Lingulodinium polyedra*	x	x		(Elshanawany and Zonneveld 2016; El-Soughier and Mahmoud 2019)
Margalefidinium polykrikoides (Margalef) F. Gómez, Richlen & D.M. Anderson	*Margalefidinium polykrikoides*			x	(Mohamed and Al-Shehri 2011)
Nematosphaeropsis labyrinthus (Ostenfeld) Reid	*Gonyaulax spinifera*	x			(Elshanawany and Zonneveld 2016)
Operculodinium centrocarpum sensu Wall and Dale	*Protoceratium reticulatum*	x	x		(Elshanawany and Zonneveld 2016)
Operculodinium israelianum (Rossignol) Wall	*Protoceratium* spp.	x			(Elshanawany and Zonneveld 2016)
Atlanticodinium janduchenei (Head, G. Norris & P.J. Mudie) Head & Mantilla-Duran	*Protoceratium* spp.	x			(Elshanawany and Zonneveld 2016)
Operculodinium longispinigerum Matsuoka	*Protoceratium* spp.	x			(El Shanawany and Zonneveld 2016)
Operculodinium spp.	*Protoceratium* spp.		x	x	(Elshanawany and Zonneveld 2016)
Polysphaeridium zoharyi (Rossignol) Bujak et al.	*Pyrodinium bahamense*	x	x		(Elshanawany and Zonneveld 2016)
Cyst of ***Prorocentrum cordatum*** (Ostenfeld) J.D. Dodge	*Prorocentrum cordatum*			x	(Mohamed and Al-Shehri 2011)
Spiniferites membranaceus (Rossignol) Sarjeant	Gonyaulax spp.	x			(Elshanawany and Zonneveld 2016)
Spiniferites mirabilis (M.R. Rossignol) Sarjeant	*Gonyaulax spinifera*	x	x		(Elshanawany and Zonneveld 2016)
Spiniferites ramosus (Ehrenberg) Mantell	*Gonyaulax* spp.	x			(Elshanawany and Zonneveld 2016)
Spiniferites sp.	*Gonyaulax* spp.	x	x		(Elshanawany and Zonneveld 2016; El-Soughier and Mahmoud 2019)

18.7 TOXICITY OF HAB IN THE RED SEA

Several studies related to HAB toxins were carried out along the Red Sea. Most of these studies were carried out off Saudi Arabia, while limited information was related to the African coast (Tamele et al. 2019). Marine toxins (MTs) reported in the Red Sea, including ciguatoxins (CTXs), tetrodotoxins (TTXs) and paralytic shellfish toxins (PSTs), were detected from the puffer fish. It has been noticed that from the limited data available for toxins production, the most reported HAB groups are cyanobacteria, dinoflagellates and raphidophytes. The TTXs recorded from Egyptian waters at the northwestern part of the Red Sea during 1990–1991 and the Gulf of Suez during 2003–2004 had concentrations of 752 MU/g and 3950 MU/g, respectively (El-Semary 2016; Tamele et al. 2019). The first report of CTX was in the coastal waters off Saudi Arabia during 2012–2013 by Catania et al. (2017), due to the presence of the benthic dinoflagellates *Gambierdiscus belizeanus* and

Ostreopsis spp., with maximum toxins content of 6.5 ± 1.14 × 10^{-5} pg P-CTX eq. $cell^{-1}$ (Catania et al. 2017). The toxicity of *Pyrodinium bahamense* was detected in 2012 from a shallow Saudi Arabian lagoon in the CRS (Banguera-Hinestroza et al. 2016). The saxitoxins production due to the proliferation of the species reached 68 ng STX ml^{-1}. In 2012–2014, the production of PSTs in the central Red Sea due to the presence of *Alexandrium* spp. reached >0.4 ng ml^{-1}. Nine cyanobacterial species were isolated and tested for toxicity on the *Artemia salina*. Based on ELISA and HPLC analysis, four species – namely, *Aphanothece elabens*, *Leptolyngbya tenuis*, *Calothrix breviarticulata* and *Oscillatoria tenuis* – were found to produce microcystins with toxins concentration that ranged from 68.3 to 974 μg^{-1}, and three species from the southern coast of Saudi Arabia – *Lyngbya majuscula*, *Leptolyngbya tenuis* and *Oxynema acuminatum* – produced saxitoxins (58.9–93.8 μg^{-1}) (Mohamed and Al-Shehri 2015). In addition to dinoflagellates and cyanobacteria, the toxicity of raphidophyte species *Heterosigma akashiwo* was tested on *Artemia salina* in the laboratory after their bloom in 2010 off the coasts of Saudi Arabia (Mohamed and Al-Shehri 2012). The toxicity of *H. akashiwo* showed a powerful haemolytic activity, with a maximum of $Ec_{50} = 3.64 \times 10^4$ cells ml^{-1}.

18.8 CONCLUSION

The literature data from the Red Sea showed high species diversity of potentially harmful microalgae recorded along the Red Sea. Only a few species are associated with harmful effects in the Red Sea. In addition, the presence of *Alexandrium* spp., *Dinophysis caudata*, *D. miles*, *Gymnodinium* sp., *Heterosigma akashiwo* and *Ostreopsis* in the area can be considered as potentially harmful species that could affect future development of aquaculture and desalination plants. The presence of potentially harmful species in the Red Sea must be considered a threat to marine life. Therefore, more investigations, including establishing long-term monitoring programs of harmful algae, are required in the Red Sea, not only to confirm the above-proposed possibilities, but also to improve our knowledge on harmful algae and their impacts to the region.

REFERENCES

Abdel Rahman, N.S., and M.Z. Nassar. 2005. Preliminary studies on plankton communities associated with the mangrove forest habitats at Halayib-Shalatin Sector, Red Sea, Egypt. *Pakistan J Mar Sci* 14: 133–144.

Al-Aidaroos, A.M., Devassy, R.P., and M.M. El-Sherbiny. 2019. Unusual dominance of harmful microalgae *Pseudo-nitzschia delicatissima* cf. (Cleve) Heiden in the coastal waters of Jeddah, central Red Sea. *Pak J Bot* 51, no. 2: 1–6.

Al-Amri, A.A., Qari, H.A., and M.M. El-Sherbiny. 2020. Distribution and community structure of microphytoplankton in relation to increasing anthropogenic impact along coastal waters of Jeddah, the central Red Sea. *Oceanol Hydrobiol Stud* 49: 193–205.

Ali, M.H.M. 2015. *Ecological study on the phytoplankton in the Sudanese Red Sea coast (Port Sudan Area)*. M.Sc. Thesis, University of Khartoum. Unpublished manuscript.

Alkawri, A. 2016. Seasonal variation in composition and abundance of harmful dinoflagellates in Yemeni waters, southern Red Sea. *Mar Pollut Bull* 112: 225–234.

Alkawri, A., Abker, M., Qutaei, E., Alhag, M., Qutaei, N., and S. Mahdy. 2016b. The first recorded bloom of *Pyrodinium bahamense* var *bahamense* Plate in Yemeni coastal waters off Red Sea, near Al Hodeida City. *Turk J Fish Aquat Sci* 16: 275–282.

Alkawri, A., Al Areeki, M., and K. Alsharaby. 2016a. The first recorded bloom of *Protoperidinium quinquecorne* and its link to a massive fish kill in Yemeni coastal waters, southern Red Sea. *Plankton Benthos Res* 11: 75–78.

Alkershi, A., and N. Nandini Menon. 2011. Phytoplankton in polluted waters of the Red Sea coast of Yemen. *J Mar Biol Assoc India* 53: 1–6.

Al-Najjar, T., Badran, M.I., Richter, C., Meyerhoefer, M., and U. Sommer. 2007. Seasonal dynamics of phytoplankton in the Gulf of Aqaba, Red Sea. *Hydrobiologia* 579: 69–83. doi:10.1997/s10750-006-0365.

Aynousah, A. 2017. *Thermal responses of growth and toxin production in four Prorocentrum species from the central Red Sea*. M.Sc. Thesis, King Abdullah University of Science and Technology, Saudi Arabia. 69 p.

Banguera-Hinestroza, E., Eikrem, W., Mansour, H., Solberg, I., Cúrdia, J., Holtermann, K., Edvardsen, B., and S. Kaartvedt. 2016. Seasonality and toxin production of *Pyrodinium bahamense* in a Red Sea lagoon. *Harmful Algae* 55: 163–171.

Belogorskay, E.V. 1970. Qualitative and quantitative distribution of phytoplankton in the Red Sea and Gulf of Aden in October-November 1963. *Biol Morya (Vladivost)/Mar Biol* 21: 133–152.

Beraki, T., Zerom, H., Arefe, T., Goitom, M., and Y. Yemane. 2022. Report of mass fish mortality along the Eritrean coast and islands. *Acta Entomol Zool* 3, no. 2: 36–39.

Catania, D. 2012. *The prevalence of benthic dinoflagellates associated with ciguatera in the central Red Sea*. M.Sc. Thesis, King Abdullah University of Science and Technology, Saudi Arabia. 40 p.

Catania, D., Richlen, M.L., Mak, Y.L., Morton, S.L., Laban, E.H., Xu, Y., Anderson, D.M., Chan, L.L., and M.L. Berumen. 2017. The prevalence of benthic dinoflagellates associated with ciguatera fish poisoning in the central Red Sea. *Harmful Algae* 68: 206–216. doi:10.1016/j.hal.2017.08.005.

Chaidez, V., Dreano, D., Agusti, C.M., and I. Hotei. 2017. Decadal trends in Red Sea maximum surface temperature. *Sci Rep* 7: 8144.

Churchill, J.H., Bower, A.S., McCorkle, D.C., and Y. Abualnaja. 2014. The transport of nutrient-rich Indian Ocean water through the Red Sea and into coastal reef systems. *J Mar Res* 72: 165–181.

Devassy, R.P., El-Sherbiny, M.M., Al-Sofyani, A.M., and A.M. Al-Aidaroos. 2017. Spatial variation in the phytoplankton standing stock and diversity in relation to the prevailing environmental conditions along the Saudi Arabian coast of the northern Red Sea. *Mar Biodiv* 47, no. 4: 995–1008. doi:10.1007/s12526-017-0693-4.

Dreano, D., Raitsos, D., Gittings, J., Krokos, G., and I. Hoteit. 2016. Southern Red Sea summer phytoplankton blooms regulated by the Gulf of Aden intermediate water intrusion. *PLoS One* 11: e0168440.

El Hag, A.G.D., and D.H. Nasr. 1989. Studies in planktonic populations in Port Sudan coastal area. *Sudan J Sci* 4: 12–26.

El-Semary, N. 2016. Benthic dinoflagellates from Red Sea, Egypt: Early records. *Egypt J Aquat Res* 42: 177–184.

Elshanawany, R., and K. Zonneveld. 2016. Dinoflagellate cyst distribution in the oligotrophic environments of the Gulf of Aqaba and northern Red Sea. *Mar Micropaleontol* 124: 29–44.

El-Sherbiny, M.M., Al-Harbi, M.A., and A.J. Kumar. 2021. Spatio-temporal variation of microphytoplankton communities in

Obhur Creek, the central Red Sea. *Oceanol Hydrobiol Stud* 50: 98–114.

El-Soughier, M.I., and M.S. Mahmoud. 2019. Dinoflagellate cysts stratigraphy and paleoecology from some lower Miocene rocks, GS9-IX well, northern Gulf of Suez, Egypt. *J African Earth Sci* 160: 103650.

Gittings, J.A., Raitsos, D.E., Krokis, G., and I. Hoteit. 2018. Impacts of warming on phytoplankton and phenology in a typical tropical marine ecosystem. *Sci Rep* 8, no. 1: 2240.

Gokul, E.A., Raitsos, D.E., Gittings, J.A., Alkawri, A., and I. Hoteit. 2019. Remotely sensing harmful algal blooms in the Red Sea. *PLoS One* 14: e0215463.

Gokul, E.A., Raitsos, D.E., Gittings, J.A., and I. Hoteit. 2020. Developing an atlas of harmful algal blooms in the Red Sea: Linkages to local aquaculture. *Remote Sens* 12: 3695. doi:10.3390/rs12223695.

Gomaa, M.N., Hannachi, I., Carmichael, W.W., Al-Hazmi, M.A., and A.M. Abou. 2018. Low diversity triggers Harmful Algae Bloom (HAB) occurrence adjacent to desalination plants along the Red Sea. *Desalin Water Treat* 114: 1–12.

Ismael, A.A. 2005. Phytoplankton of the Gulf of Suez and the effect of ship traffic. *J Egypt Acad Soc Environ Dev (D – Environ Studies)* 6, no. 1: 75–92.

Ismael, A.A. 2015. Phytoplankton of the Red Sea. In: *The Red Sea: The formation, morphology, oceanography and environment of a young ocean basin*. Eds: N.M.A. Rasul, and I.C.F. Stewart, 567–583. Berlin Heidelberg: Springer Earth System Sciences.

Kürten, B., Khomayis, H.S., Devassy, R., Audritz, S., Sommer, U., et al. 2015. Ecohydrographic constraints on biodiversity and distribution of phytoplankton and zooplankton in coral reefs of the Red Sea, Saudi Arabia. *Mar Ecol* 36, no. 4: 1195–1214.

Madkour, F.F., El-Sherbiny, M.M., and M.A.A. Aamer. 2010. Phytoplankton population along certain Egyptian coastal regions of the Red Sea. *Egypt J Aquat Biol Fish* 14: 95–109.

Mohamed, Z.A. 2018. Potentially harmful microalgae and algal blooms in the Red Sea: Current knowledge and research needs. *Mar Environ Res* 140: 234–242. doi:10.1016/j.marenvres.2018.06.019.

Mohamed, Z.A., and A.M. Al-Shehri. 2011. Occurrence and germination of dinoflagellate cysts in surface sediments from the Red Sea off the coasts of Saudi Arabia. *Oceanologia* 53, no. 1: 121–136.

Mohamed, Z.A., and A.M. Al-Shehri. 2012. The link between shrimp farm runoff and blooms of toxic *Heterosigma akashiwo* in Red Sea coastal waters. *Oceanologia* 54, no. 2: 287–309.

Mohamed, Z.A., and A.M. Al-Shehri. 2015. Biodiversity and toxin production of cyanobateria in mangrove swamps in the Red Sea off the southern coast of Saudi Arabia. *Bot Mar* 58, no. 1: 23–34.

Mohamed, Z.A., and I. Mesaad. 2007. First report on *Noctiluca scintillans* blooms in the Red Sea off the coasts of Saudi Arabia: Consequences of eutrophication. *Oceanologia* 49: 337–351.

Nassar, M.Z. 2007. Species composition and distribution of phytoplankton in the western coast of Suez Gulf, Egypt. *Egypt J Aquat Res* 33: 113–132.

Nassar, M.Z., Hamdy, R.M., Khiray, H.M., and S.H. Rashedy. 2014. Seasonal fluctuations of phytoplankton community and physico-chemical parameters of the north western part of the Red Sea, Egypt. *Egypt J Aquat Res* 40: 395–403.

Nassar, M.Z., and M.A. Hamed. 2003. Phytoplankton standing crop and species diversity in relation to some water characteristics of Suez Bay (Red Sea), Egypt. *J Aquat Biol Fish* 7, no. 3: 25–48.

Nassar, M.Z., Mohamedin, L.I., and M.A. El-Sawy. 2016. Seasonal variations of phytoplankton and nutrients in the Egyptian Harbors of the northern Red Sea. *Int J Marine Sci* 6, no. 8: 1–17.

Nassar, M.Z., Shams El-Din, N.G., and S.M. Gharib. 2015. Phytoplankton variability in relation to some environmental factors in the eastern coast of Suez Gulf, Egypt. *Environ Monit Assess* 187: 1–18.

Osama, S.S., and A.G. Amjed. 2020. A preliminary study on phytoplankton composition at Bashayer II marine oil terminal, Red Sea, Sudan. *J East China Norm Univ (Natur Sci)* 2020(S1): 89–93. doi:10.3969/j.issn.1000-5641.202092209.

Ostenfeld, C.H., and Schmidt, J. 1902. Plankton fra det Røde Hav og Adenbugten (Plankton from the Red Sea and the Gulf of Aden.). *Videnskabelige Meddelelser fra Dansk Naturhistorisk Forening* 1901: 141–182.

Post, A.F., Dedej, Z., Gottlieb, R., Li, H., Thomas, D.N., El-Absawi, M., El-Naggar, A., El-Gharabawi, M., and U. Sommer. 2002. Spatial and temporal distribution of *Trichodesmium* spp. in the stratified Gulf of Aqaba, Red Sea. *Mar Ecol Prog Ser* 239: 241–250.

Prabowo, D.A., and S. Agusti. 2019. Free-living dinoflagellates of the central Red Sea, Saudi Arabia: Variability, new records and potentially harmful species. *Mar Pollut Bull* 141: 629–648.

Saburova, M., Polikarpov, I., and F. Al-Yamani. 2013. New records of the genus *Gambierdiscus* in marginal seas of the Indian Ocean. *Mar Biodiv Rec* 6: e91. doi:10.1017/S1755267213000675.

Tamele, I.J., Silva, M., and V. Vasconcelos. 2019. The incidence of marine toxins and the associated seafood poisoning episodes in African countries of the Indian Ocean and the Red Sea. *Toxins* 11: 58.

Touliabah, H.E., Abu El-Kheir, W.S., Kuchari, M.G., and N.I.H. Abdulwass. 2010. Phytoplankton composition at Jeddah Coast – Red Sea, Saudi Arabia in relation to some ecological factors. *J King Abdulaziz Univ Sci* 22, no. 1: 115–131.

Zonneveld, K., Marret, F., Versteegh, G.J.M., Bogus, K., Bonnet, S., Bouimetarhan, I., Crouch, E., et al. 2013. Atlas of modern dinoflagellate cyst distribution based on 2405 data points. *Rev Palaeobot Palynol* 191: 1–197.

A comparison of potentially harmful microalgae and phytoplankton blooms in the Arabian Gulf and the Red Sea

19

Faiza Al-Yamani, Maria Saburova, Igor Polikarpov

19.1 INTRODUCTION

The functioning of global marine ecosystems depends on phytoplankton abundance, composition, and size structure, given that they constitute the basis of marine food webs. Microalgal blooms are considered a natural component of marine ecosystems, supporting biological productivity and providing human benefits from the aquatic environment worldwide, including the sustainability of commercial fisheries (e.g., Berdalet et al. 2015; Kudela et al. 2015; Kumar 2021).

In two parts of a single marine province in the northwestern Indian Ocean (REDS according to Longhurst 2007), namely the Red Sea and the Arabian Gulf (also known as the Persian Gulf), large variability in productivity occurs due to the prevailing different hydrographical and biogeochemical regimes. High biological productivity of the coastal waters in the northwestern Arabian Gulf (NWG) is supported by the phytoplankton proliferation, largely due to the influence of nutrient-rich discharge from the Shatt Al-Arab River (e.g., Subba Rao and Al-Yamani 1998; Al-Yamani et al. 2004, 2020; Al-Yamani 2021). In contrast, in the northeastern Arabian Gulf (NEG) and the nearby Sea of Oman, the seasonal phytoplankton succession is predominantly governed by monsoon oscillations and related nutrient availability (e.g., Polikarpov et al. 2016; Piontkovski et al. 2019). Areas of elevated phytoplankton productivity in the Arabian Gulf and the Sea of Oman support the sustainability and size of local fisheries resources (Salarpouri et al. 2018; Piontkovski et al. 2019). Phytoplankton phenology and size structure influence the catch of commercially important fisheries in the Red Sea (Gittings et al. 2021).

Although most of the thousands of microalgal species benefit humans and the environment, a small number cause harmful algal blooms (HABs) under certain environmental conditions through the rapid growth and accumulation of algal biomass and/or production of toxins and harmful metabolites. HABs are a growing problem of global concern affecting aquatic ecosystems, public health, and local economies. HAB impacts are variable in their extent and severity and, depending on the causative species, are often associated with a broad range of ecological consequences, including increased mortalities of wild and farmed fish, displacement of indigenous species, and alterations to marine food webs, which ultimately have a detrimental effect on human societies. Toxic algal blooms produce potent toxins, causing illness or mortality in humans and marine mammals, fish, and other marine life through either direct exposure to the toxin or consumption of bioaccumulated toxin transferred through the food webs to higher trophic levels, e.g., shellfish or fish consumption. Other species cause problems due to the rapid reproduction and accumulation in certain areas because the excessive biomass results in water discolouration, oxygen depletion in the water column and the seabed, marine mortalities, shading of submerged aquatic vegetation, damage to coral reefs, or other adverse ecosystem effects (e.g., Anderson et al. 2002, 2012; Kudela et al. 2015; Al-Yamani et al. 2020).

Recent taxonomic efforts have revealed the high microalgal diversity in the Arabian Gulf (Al-Kandari et al. 2009; Polikarpov et al. 2016; Al-Yamani and Saburova 2019a, 2019b; Attaran-Fariman and Ali Asefi 2022) and the Red Sea (El-Sherif and Aboul Ezz 2000; Nassar and Khairy 2014; Ismael 2015; Prabowo and Agusti 2019). Among this diverse set, a large number of potentially harmful species have been identified based on the capability of the selected microalgae to form toxic and harmful blooms in the studied areas or elsewhere. Regionally based lists were established for the Arabian Gulf (Al-Yamani et al. 2012; Polikarpov et al.

DOI: 10.1201/9781003321392-24

2020; Attaran-Fariman and Ali Asefi 2022) and the Red Sea (Alkawri 2016; Gomaa et al. 2018; Mohamed 2018; Prabowo and Agusti 2019; Osama and Amjed 2020; Saburova 2022). The lists covered a wide range of microalgal taxa for which bloom-related harmful effects were reported. Some examples of the algal blooms, which caused water discolouration and marine mortality in the northwestern Arabian Gulf, are shown in Figure 19.1. Results of surveys and research studies have made an important contribution to the ecological and human health risk assessments on regional and basin-wide scales.

FIGURE 19.1 Examples of harmful algal blooms observed in the northwestern Arabian Gulf, Kuwait's waters (causative species and harmful impact are indicated in insets). A) bloom of the filamentous cyanobacterium *Trichodesmium erythraeum* in Kuwait's southern offshore waters in May 2015; B) intense water discolouration during bloom of diatom *Nitzschia pusilla* in Kuwait Bay in May 2019; C) aerial view of inshore waters in Kuwait Bay (courtesy Yousef Alosairi) showing water discolouration and mortality of bivalves (inset) during bloom of dinoflagellate *Heterocapsa circularisquama* in June-July 2020; D) sea-foam, water discolouration, and demersal fish mortality in the coastal waters of Kuwait Bay during multispecies bloom dominated by dinoflagellate *Karenia papilionacea* in April 2017; E) water discolouration and layer of dead fish (*Sardinella fimbriata*) on the surface of Kuwait's inshore waters during multispecies bloom dominated by dinoflagellate *Gonyaulax verior* in July 2011; F) bloom of the prymnesiophyte algae *Phaeocystis globosa* in Kuwait's coastal waters in November 2006; G) intense water discolouration in Kuwait Bay during bloom of the photosynthetic ciliate *Mesodinium rubrum* in November 2014.

Along with long-term taxonomic surveys, the phytoplankton blooms have been widely monitored using satellite-derived chlorophyll-*a* concentrations as a proxy of the phytoplankton biomass over the past decades throughout both the Arabian Gulf (Nezlin et al. 2010; Hamzehei et al. 2013; Olalekan and Malik 2015; Zhao et al. 2015; Ghanea et al. 2016; Polikarpov et al. 2019, 2021) and the Red Sea (Raitsos et al. 2013; Dreano et al. 2016; Gokul et al. 2019; Kheireddine et al. 2021). Although satellite monitoring and assessment of phytoplankton blooms have some advantages, including rapidity, simplicity, efficiency, and wide coverage, they are inferior in accuracy compared to on-site observations and are limited in the identification of bloom-forming species (e.g., Zhao and Ghedira 2014; Gokul et al. 2019; Polikarpov et al. 2019). However, on-site observations and reports on phytoplankton blooms are relatively scarce, often restricted to a single bloom event, scattered geographically and localized, and frequently published in difficult-to-access sources. Hence, establishing a collaborative database on HABs with broad spatial and temporal coverage is essential for assessing regional environmental and social risks and sustainable coastal development.

Recently, the available HAB records have been summarized and analysed for the NWG (Al-Yamani and Saburova 2019a; Polikarpov et al. 2020), the NEG (Esmaeili et al. 2021), and the Red Sea (Mohamed 2018). These historical accounts provide details on the regional HAB status and are highly comparable due to the shared coverage time (roughly over the past three decades) and because they are based on *in situ* observations of described bloom events. Thus, these datasets give an opportunity to compare HABs between different areas of the Arabian Gulf and the Red Sea to provide a present-day pattern of HAB diversity, spatial distribution, seasonality, and long-term trends, and to identify the most ecologically relevant microalgal species, which may impact coastal areas, fishery resources, and public health along geographically distant and environmentally distinct regions.

19.2 DATA COLLECTION AND ANALYSIS

Data on blooms frequency, bloom-forming species identity, location, date, bloom timing, and cell density (when available), as well as bloom-associated harmful effects (water discolouration, hypoxia, marine mortality, strain toxicity, etc.), were gathered from the published reports and overviews. In the NWG (Figure 19.2A, inset 1), the dataset was partly based on available published reports and mainly on data from long-term phytoplankton monitoring of Kuwait's coastal waters by the Kuwait Institute for Scientific Research (KISR), and covered 29 years between 1988 and 2017 (Polikarpov et al. 2020). This dataset was updated with recent observations in 2018–2021 (Polikarpov et al. 2021; Saburova et al. 2022; KISR internal reports). In the NEG (Figure 19.2A, inset 2), a historical account of HAB events along the coast of the Hormozgan Province in Iran was based on long-term phytoplankton monitoring by the Persian Gulf and the Sea of Oman Ecological Research Institute (PGSOERI). It covered 35 years from 1980 to 2015 (Esmaeili et al. 2021). In the Red Sea (Figure 19.2A, inset 3), the available data were assembled from the published reports and reviews (Saburova et al. 2013; Gomaa et al. 2018; Mohamed 2018 and references therein) and covered 29 years from 1989 to 2018. In addition, the observation of Ehrenberg on a *Trichodesmium* bloom in 1823 on the coast of the Sinai Peninsula (Ehrenberg 1830), long before the phytoplankton systematic monitoring began, was accounted as the first historical report of a phytoplankton bloom in the Red Sea and included in the analysis. Thus, all analysed datasets fell into roughly the same three-decadal period from the 1980s to the present.

There are no easy generalizations about the cell concentrations that could lead to blooms, as the abundance of blooming species causing harmful impacts can vary significantly worldwide being regionally and species-specific (Anderson et al. 2017). Some highly toxic dinoflagellates causing paralytic shellfish poisoning (PSP), e.g., *Alexandrium* spp., *Pyrodinium bahamense*, can have a dramatic effect even at low concentrations of 10^2–10^3 cells/l (Hallegraeff 2003; Sombrito et al. 2007). The species of the dinoflagellate genus *Dinophysis*, responsible for diarrhetic shellfish poisoning (DSP), rarely reach concentrations higher than 10^4–10^5 cells/l (Anderson et al. 2017). In contrast, the highly motile dinoflagellate *Margalefidinium* (*Cochlodinium*) *polykrikoides* and the photosynthetic ciliate *Mesodinium rubrum* (*Myrionecta rubra*) can exceed 10^6–10^7 cells/l in dense patches at the water surface (Kudela and Gobler 2012; Polikarpov et al. 2020). The large dinoflagellate *Noctiluca scintillans* forms less dense blooms, causing intense water discolouration even at a concentration of 10^4 cells/l (Baliarsingh et al. 2016).

Phytoplankton abundance varies significantly among the considered sub-basins. Overall, phytoplankton concentration in the NWG is higher than in the NEG and over an order of magnitude above that in the Red Sea (Subba Rao and Al-Yamani 1998; Kürten et al. 2015; Polikarpov et al. 2016; Devassy et al. 2017). Therefore, the same phytoplankton concentration threshold cannot be used to categorize a bloom in different sub-basins of the Arabian Gulf and the Red Sea. Here, we define a bloom as a phytoplankton outbreak when a much higher than sub-basin averaged concentration of cells was observed. Bloom-forming species in highly productive waters of the NWG were defined based on phytoplankton outbreaks exceeding a density of 10^6 cells/l (Polikarpov et al. 2020). For the NEG and the Red Sea, we refer to a bloom as a phytoplankton cell concentration greater than 10^4 cells/l (Mohamed 2018; Esmaeili et al. 2021).

The characterization of algal species as potentially toxic to humans or marine biota was based on published reports of their toxicity wherever they occurred (e.g., Lundholm et al. 2009 onwards; Lassus et al. 2016; Śliwińska-Wilczewska et al. 2018, and references therein). Prior to analysis, the available lists of phytoplankton species from the Arabian Gulf and the Red Sea were verified according to AlgaeBase (Guiry and Guiry 2022) to standardize the high-level classification and currently accepted taxonomic names of microalgal taxa encountered. All the recorded algal bloom events were mapped across the considered sub-basins, NWG, NEG, and the entire Red Sea, to visually represent the distribution pattern. Bloom-affected sites were denoted according to maps provided in the published reports. Phytoplankton outbreaks were then classified

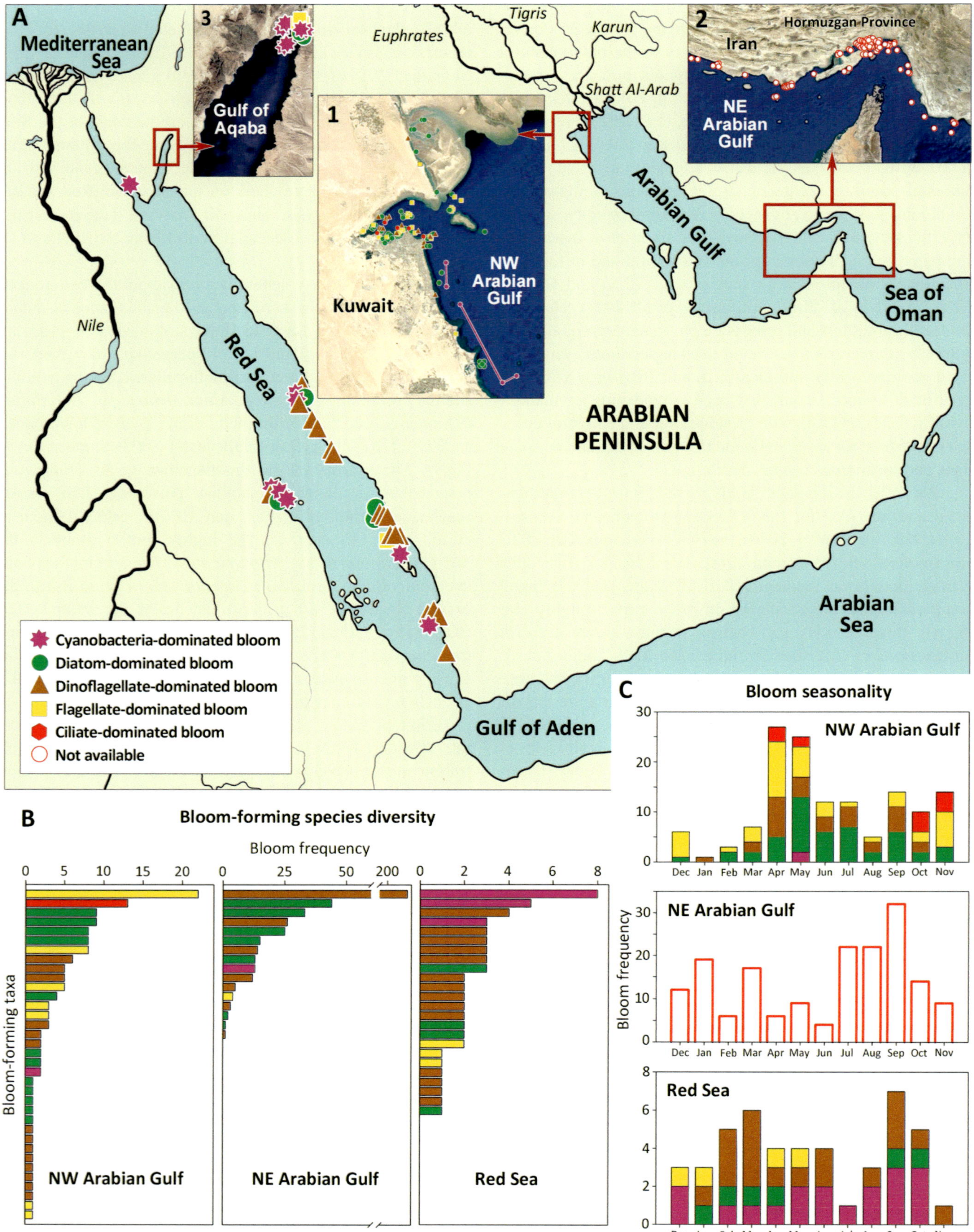

FIGURE 19.2 Distribution, frequency, and seasonality of algal bloom events in the Arabian Gulf and the Red Sea. A) spatial distribution of algal bloom events dominated by different taxonomic groups in the NW Arabian Gulf along Kuwait's shore (inset 1; 1988–2021), the NE Arabian Gulf along the Hormozgan Province of Iran (inset 2; 1980–2015), and throughout the Red Sea (1823*, 1986–2016; inset 3, the northern part of the Gulf of Aqaba); B) frequency of algal blooms dominated by different phytoplankton taxa in the NW and NE Arabian Gulf and in the Red Sea (for species names of bloom-forming taxa see Table 19.1); C) monthly summarized frequency of algal bloom events dominated by different taxonomic groups in the NW Arabian Gulf and in the Red Sea and monthly distribution of total algal blooms frequency in the NE Arabian Gulf (bloom events of *Margalefidinium polykrikoides* not shown due to atypical seasonality).

* Ehrenberg 1830.

based on the dominant phytoplankton group as diatom-, dino-flagellate-, flagellate-, cyanobacteria- and ciliate-dominated blooms. Algal bloom seasonality was calculated by summing all the bloom events that occurred within a given month over the considered period to show a general annual pattern and for each phytoplankton group separately. Microalgal taxa of ecological concern (i.e., potentially dangerous to the marine environment and human activities in the coastal zone) were identified based on their capacity to form recurring blooms, their potential or confirmed toxicity, and bloom-associated harmful effects reported from at least one sub-basin under consideration in the Arabian Gulf and the Red Sea. To assess the risk of HABs in a certain sub-basin, the presence of the species in the phytoplankton community composition, the occurrence, and frequency of its blooms, its known toxicity, as well as the local hydrographic and hydrochemical characteristics were considered.

19.3 BACKGROUND: OCEANOGRAPHIC SETTINGS

The northern Indian Ocean is an exceptional region from the biogeographic perspective (e.g., Longhurst 2007), composed of different basins with distinct climatic and hydrological characteristics. The Arabian Gulf and the Red Sea on either side of the Arabian Peninsula are semi-enclosed marginal seas with a limited oceanic exchange, linked to the northern Indian Ocean through narrow, shallow straits (Figure 19.2A). Both basins are subtropical seas surrounded by a large, arid land mass. Evaporation dramatically exceeds freshwater inputs from precipitation and runoff, resulting in hypersaline conditions and making both basins among the most saline in the world's oceans. To compensate for the negative hydrologic budget, basin-wide water circulation is anti-estuarine, with the low-salinity surface seawater influx from the Indian Ocean through narrow straits and dense, saline water flowing out of the basins into the Indian Ocean in the opposite direction.

In both basins, surface waters are characterized by a broad temperature gradient with low seasonal variability. The lowest temperatures occur in the north, gradually increasing southward. Water temperatures in the Red Sea range from a winter minimum of 22°C in the northern Gulf of Aqaba to a summer maximum of >31°C (locally up to 34°C) in the southern sub-basin. Extreme water temperatures also characterize the Arabian Gulf, ranging from <10°C in winter in the northern waters to >35–36°C in summer in this hottest sea on Earth. Salinity generally follows an opposite pattern in that the lowest values of 36–37 occur near the inlet straits in the waters connected to the Indian Ocean and increase with distance from the inlets toward the interior to 40–41 in the northern Gulf of Suez in the Red Sea, and up to 40–44 in the northwestern Arabian Gulf, with more than 50–70 in some embayments, e.g., in the Gulf of Salwah off Qatar in the western Arabian Gulf (e.g., John et al. 1990; Reynolds 1993; Sofianos and Johns 2002; Sheppard et al. 2010; Rasul et al. 2015; Rasul and Stewart 2019; Neelamani et al. 2021).

Despite the proximity of these basins and some oceanographic similarities, the Arabian Gulf and the Red Sea differ in terms of their geographical settings, hydrography, and biogeochemistry. The seasonally reversing Arabian monsoon winds strongly influence atmospheric conditions over this region. The monsoon forcing is most intense over the coastal Arabian Sea and the southern Red Sea. At the same time, continental patterns become more dominant over the northern Red Sea and the Arabian Gulf, with strong winds influencing water circulation and mixing processes (Johns et al. 1999).

The long, narrow Red Sea, with an area of about 440,300 km^2, extends over 2,000 km between 12°N and 30°N latitude and reaches its maximum width of 350 km at the Eritrean coast. A large part of the Red Sea is shallow, with depths of less than 100 m, whereas in the central deep-water part, bathymetric depressions filled with hot, hypersaline, acidic, and anoxic water occur at depths of 1,500–2,800 m (Rasul et al. 2015; Rasul and Stewart 2019). The Arabian Gulf is much smaller and shallower than the Red Sea. It is located on the other side of the Arabian Peninsula between 24°–30°N and 48°–57°E and covers about 250,000 km^2. Its width varies from 338 km inside the Gulf to as narrow as 56 km at the Strait of Hormuz, with a length of approximately 1,000 km from the Strait to the northern end. The depth of the Gulf is less than 100 m with an average of 36 m, and the northwestern and western parts are shallower than the rest of the area (Reynolds 1993; Sheppard et al. 2010).

The rainfall over the Red Sea ranges from a few mm per year along the northern part, gradually increasing southward to 180 mm/yr in Sudan, averaging 6 mm/yr (Edwards 1987). Along the coast of Saudi Arabia, the northern area receives considerably less precipitation, with an average of 25 mm/yr compared to the southern coast, which has an average of 132 mm/yr (Hasanean and Almazroui 2015). In long-term perspective, along with a sharp warming of the Red Sea waters by 0.7°C since 1994 (Raitsos et al. 2011), an increasing trend in precipitation was revealed along the Saudi Arabian coast during 2000–2009 compared to 1980–1989, whilst rainfall has decreased over most mainland areas (Hasanean and Almazroui 2015). The intense air-sea interactions over the Red Sea, the scarcity of rainfall, and the lack of a major source of freshwater to the sea resulted in one of the highest evaporation rates globally, averaging 2.06 ± 0.22 m per year with peaks exceeding 5 m/yr in the central part (Sofianos et al. 2002). Like the Red Sea, the average annual evaporation of the Arabian Gulf is approximately 2 m. However, in contrast to the Red Sea with negligible runoff input, the freshwater budget of the Arabian Gulf is contributed by river discharge rather than precipitation, mainly in the northern end through the Shatt Al-Arab River, which joins three rivers, namely the Tigris, Euphrates, and Karun (Reynolds 1993; Al-Yamani et al. 2004, 2021; Sheppard et al. 2010).

19.4 HABS OCCURRENCE IN THE ARABIAN GULF AND THE RED SEA

Algal blooms in the Arabian Gulf have been a common phenomenon, particularly during the past few decades (e.g., Subba

Rao and Al-Yamani 1998; Al-Yamani et al. 2004; Sheppard et al. 2010; Al-Yamani and Saburova 2019a; Polikarpov et al. 2020; Esmaeili et al. 2021). In 1999, extensive phytoplankton blooms resulted in massive fish mortality in the coastal waters of Iran and Kuwait in the northern Gulf. Fish kill in Kuwait Bay, affecting about 150 tons of sobaity seabream (*Sparidentex hata*) from the fish farm cages and 60 tons of wild grey mullet (*Liza macrolepis*), was attributed to a massive bloom of ichthyotoxic dinoflagellate *Karenia selliformis* and resulted in the closure of shellfish and finfish markets with an economic loss to the region of US $7 million (Al-Yamani et al. 2000, 2004; Heil et al. 2001). An unprecedentedly severe and widespread bloom of the dinoflagellate *Margalefidinium polykrikoides* in the Arabian Gulf and the Sea of Oman in 2008–2009 lasted more than eight months and killed thousands of tons of fish and limited traditional fishery operations, damaged coral reefs, impacted coastal tourism, and forced the closure of desalination plants in the region (Richlen et al. 2010). In general, the oligotrophic nature of the Red Sea provides a less suitable environment for the development of algal blooms compared to the more productive waters of the Arabian Gulf. The first scientific report on phytoplankton blooms in the Red Sea dates back to 1830, when Christian Gottfried Ehrenberg published his observations on the bloody colouration of the marine coastal waters in December 1823 near Tor (El-Tor) on the Sinai Peninsula in Egypt (Ehrenberg 1830). The blooming cyanobacterium that caused the red colouration of the seawater was so unusual that he raised it to its own genus, which he named *Trichodesmium*.

Moreover, a conventional hypothesis about the name of the Red Sea attributed its origin to the water colour, which is thought to be due to the frequent red tides of *Trichodesmium* in that sea (e.g., Ehrenberg 1830; Hoyt 1912). Table 19.1 provides an overview of phytoplankton bloom events recorded in the Arabian Gulf and the Red Sea over the past three decades and details the main bloom-forming species, their harmful impact, and the frequency of blooms occurring in different sub-basins. These blooms are mapped in Figure 19.2A to visualize their spatial dispersion.

Spatially, both the western and eastern parts of the Arabian Gulf experience HAB events (Figure 19.2A; Table 19.1), whereas a few HAB reports are available for the central coastal waters of the Gulf, mostly off Qatar, with cyanobacterium *Trichodesmium erythraeum* as a main bloom-forming species (Al-Musawi 2012; not mapped in Figure 19.2A). A total of 136 phytoplankton bloom events were recorded in the NW Arabian Gulf along Kuwait's coast between 1988 and 2021. In contrast, a similar monitoring period yielded 436 phytoplankton outbreaks in the NE Gulf in the Hormozgan Province of Iran, over half of them *Margalefidinium polykrikoides* during its persistent bloom in the Arabian Gulf in 2008–2009. Conversely, the diversity of bloom-forming species was much poorer in the NE Gulf (16 taxa) than that of either the NW Gulf (35 taxa) or the Red Sea (24 taxa) (Figure 19.2B). In contrast to the Arabian Gulf, phytoplankton blooms were about an order of magnitude less frequent in the Red Sea, with 58 outbreaks recorded from 1989 to 2018, mostly occurring in the central and southern sub-basins (Figure 19.2A; Table 19.1).

The biological productivity of the northern Arabian Sea region is highly variable and closely limited to the nutrient distribution throughout the area that, in turn, is predominantly governed by seasonally reversing monsoonal winds and associated physical alterations in the water column (e.g., Wiggert et al. 2005; Lotliker et al. 2020; Lakshmi et al. 2021). The high productivity and frequent high biomass phytoplankton blooms in the Sea of Oman are essentially promoted by the monsoon cycle and related upwelling that enriches the surface layer with nutrients (e.g., Jones et al. 2002; Polikarpov et al. 2016). In the adjacent Arabian Gulf, the lack of favourable upwelling conditions results in nutrient limitation in most of the offshore waters (Jones et al. 2002), which, in turn, determines this basin's low to moderate overall productivity. The chlorophyll climatology pattern depicts the inflow of highly productive waters from the Sea of Oman through the Strait of Hormuz into the NE Arabian Gulf along the Iranian coast. Westward from the Strait of Hormuz, the marginal coastal and shelf areas, particularly river mouths with very high chlorophyll concentrations, are clearly distinguished from the central regions with lower chlorophyll levels (e.g., Nezlin et al. 2010; Polikarpov et al. 2019). An upward trend from north to south in the biological productivity of the Red Sea is largely shaped by the basin-wide nutrient balance, where maximum values in the south are due to the influx of nutrient-rich water from the upwelling areas of the Arabian Sea through the Bab al Mandab Strait, particularly during winter monsoons (Raitsos et al. 2015; Qurban et al. 2019).

Regional differences in the occurrence of phytoplankton blooms reflect local meteorological and oceanographic conditions governing the frequency, severity, composition, and seasonality of the phytoplankton outbreaks in the different sub-basins of the Arabian Gulf and the Red Sea. A wide range of potentially harmful species (seed populations) encountered within local phytoplankton assemblages across these marginal seas respond differently to local environmental settings and constitute unique HAB complexes that differ numerically and in composition and shape the characteristic HAB status of each sub-basin considered. In the following sections, we review these local HAB characters from area to area, based on comprehensive datasets covering roughly the past three decades, and on examples of most characteristic blooms, to delineate the HAB status for each sub-basin and to assess other risks of HABs occurrence, their possible composition, and impact on the local marine environment and coastal human activity.

19.4.1 NW Arabian Gulf

The northwestern extreme of the Arabian Gulf is greatly influenced by large runoff from the Shatt Al-Arab River, which is the main source of freshwater and nutrients inputs in this area and sustains a much higher phytoplankton biomass, productivity, and bloom occurrence compared to the Gulf's open waters. Interactions between the river discharge and the Gulf water mass result in distinct environmental gradients across the NW Gulf waters. These provide a wide range of niche opportunities for phytoplankton development, including a diverse complex of bloom-forming species (Subba Rao and Al-Yamani 1998; Jones et al. 2002; Subba Rao et al. 2003; Polikarpov et al. 2009, 2016, 2020; Sheppard et al. 2010; Devlin et al. 2015; Al-Yamani and Saburova 2019a).

TABLE 19.1 Bloom-forming microalgal species, their bloom frequency and harmful impact reported from the Arabian Gulf and the Red Sea.

BLOOM-FORMING TAXON[a]	*KNOWN HARMFUL EFFECT SYNDROME/ TOXIN*	*BLOOM FREQUENCY*			*REFERENCE*
		RED SEA	*ARABIAN GULF*		
			NWG	*NEG*	
Cyanobacteria					
Prochlorococcus sp.	Neurotoxicity/ BMAA	5	–	–	RS [1–4]
Trichodesmium spp.	Cytotoxicity/STXs	8	2*	13	RS [4–11]; NWG [12–14]; NEG [15]
Synechococcus sp.	Hepatotoxicity/ MCs	3	–	–	RS [1,2,8]
Diatoms					
Chaetoceros spp.	HB/Hypoxia	3	8*	25	RS [3,4,6,8]; NWG [13,14,16,17]; NEG [15]
Coscinodiscus sp.	HB	–	–	2	NEG [15]
Cyclotella spp.	HB/Hypoxia	–	1*	–	NWG [13,14,17]
Cylindrotheca closterium	HB	–	1	–	NWG [13,14,17]
Eucampia zodiacus	HB	–	2	–	NWG [13,14,17]
Guinardia flaccida	HB	–	1	–	NWG [13,14,17]
Leptocylindrus sp.	HB	–	1	44	NWG [18]; NEG [15]
Minutocellus polymorphus	HB/Hypoxia	–	2*	–	NWG [19]
Navicula spp.	HB	–	–	33	NEG [15]
Nitzschia spp.	HB/Hypoxia	2	9*	13	RS [4,6,20]; NWG [13,14,17,21]; NEG [15]
Pseudo-nitzschia spp.	ASP/Domoic Acid	–	4	–	NWG [13,14,18]
Rhizosolenia spp.	HB	1	–	1	RS [4,6]; NEG [15]
Skeletonema grevillei	HB/Hypoxia	–	9*	–	NWG [13,14]
Streptotheca sp.	HB	–	–	15	NEG [15]
Thalassionema nitzschioides	HB	–	1	–	NWG [37]
Thalassiosira spp.	HB/Hypoxia	2	8*	–	RS [4,22]; NWG [13,14]
Dinoflagellates					
Alexandrium spp.	PSP/STXs, Ichthyotoxicity/ Goniodomin A	–	1	–	NWG [23]
Blixaea quinquecornis	HB/Hypoxia	1*	2*	–	RS [4,24]; NWG [13,14,17]
Dinophysis acuminata/ovum	DSP/Okadaic acid, DTXs	–	1	–	NWG [13,14]
Dinophysis caudata	DSP/Okadaic acid, DTXs	4	–	–	RS [20]
Dinophysis miles	DSP/Okadaic acid, DTXs	2	–	–	RS [20]
Gonyaulax polygramma	HB	3	–	–	RS [20]
Gonyaulax spinifera	DSP/Yessotoxin	3	–	–	RS [20]
Gonyaulax verior	HB/Hypoxia	–	1*	–	NWG [13,14]
Gonyaulax sp.	HB	2	–	–	RS [4,6,25]
Gymnodinium spp.	HB, Ichthyotoxicity	3	5	14	RS [20]; NWG [13,14,17,18,23]; NEG [15]
Heterocapsa circularisquama	Toxicity to shellfish/ Haemolytic toxins	–	3*	–	NWG [19]
Karenia papilionacea	NSP/Brevetoxin	–	6*;**	–	NWG [13,14,37]
Karenia selliformis	Ichthyotoxicity/ Gymnodimines	–	2*	–	NWG [26]
Karlodinium/Takayama spp.	Ichthyotoxicity/ Karlotoxins	–	1*	–	NWG [13,14]
Kryptoperidinium foliaceum	HB	1	1	–	RS [4,27]; NWG [28]
Margalefidinium polykrikoides	Ichthyotoxicity/ Haemolytic toxins, fatty acids	–	–	225*	NEG [15,29]
Noctiluca scintillans	HB/Hypoxia	3	1	26	RS [4,22,30]; NWG [13,14,31]; NEG [15]

(Continued)

TABLE 19.1 (Continued)

BLOOM-FORMING TAXON[a]	KNOWN HARMFUL EFFECT SYNDROME/ TOXIN	BLOOM FREQUENCY			REFERENCE
		RED SEA	ARABIAN GULF		
			NWG	NEG	
Ostreopsis spp.	CPT, AST/ Palytoxin	1	–	–	RS [4,32]
Peridinium sp.	HB	1	–	5	RS [4,6]; NEG [15]
Prorocentrum lima	DSP/Okadaic acid, DTXs	–	–	3	NEG [15]
Prorocentrum spp.	DSP/Okadaic acid, DTXs, HB/Hypoxia	2	5*	–	RS [4,6,8]; NWG [13,14,17]
Protoperidinium spp.	–	–	–	12	NEG [15]
Pyrodinium bahamense	PSP/STXs	2**	1	–	RS [4,10,33]; NWG [23]
Tripos furca	HB/Hypoxia	–	1*	–	NWG [23]
Tripos spp.	HB/Hypoxia	2	–	1	RS [4,6,8]; NEG [15]
Flagellates					
Dunaliella salina	HB	–	3	–	NWG [34,35]
Chrysochromulina sp.	Ichthyotoxicity/ Haemolytic toxins	–	5*	–	NWG [13,14,37]
Phaeocystis globosa	Ichthyotoxicity/ Haemolytic toxins	–	8	–	NWG [13,14,17,36]
Eutreptiella sp.	HB/Hypoxia	–	1*	–	NWG [13,14]
Chattonella sp.	Ichthyotoxicity/ Haemolytic, neurotoxins, ROS	–	1*;**	–	NWG [37]
Heterosigma akashiwo	Ichthyotoxicity/ Haemolytic toxins, ROS	1**	3	–	RS [4,25]; NWG [13,14,17]
Chlorophyceae	–	2	–	–	RS [3]
Cryptophyceae	–	1	22	–	RS [3]; NWG [13,14]
Prasinophyceae	–	–	–	4	NEG [15]
Photosynthetic ciliate					
Mesodinium rubrum	HB	–	13	–	NWG [12–14,17,26,37]

NWG refers to the northwestern Gulf, NEG refers to the northeastern Gulf, and RS refers to the Red Sea.
[a] The taxon given here may deviate from that given in the publication; for changes in taxonomy, see Guiry and Guiry (2022).
Phycotoxin and syndrome specifications are based on the latest comprehensive reviews (Lassus et al. 2016; Śliwińska-Wilczewska et al. 2018) and constantly updating IOC-UNESCO Taxonomic Reference List of Harmful Micro Algae (Lundholm et al. 2009 onwards) and abbreviated as follows: AST – Aerosol Toxicity; ASP – Amnesic Shellfish Poisoning; CPT – Clupeotoxism; DSP – Diarrhetic Shellfish Poisoning; NSP – Neurotoxic Shellfish Poisoning; PSP – Paralytic Shellfish Poisoning; BMAA – neurotoxic nonprotein amino acid, b-N-methylamino-L-alanine; MCs – microcystin cyanobacterial toxins; DTXs – dinophysistoxins; ROS – reactive oxygen species; STXs – saxitoxins.
Harmful impacts including high biomass bloom (HB), bloom-associated hypoxia and fish kills (*), and toxicity (**) indicate based on published reports as follows: ([1] Lindell and Post 1995; [2] Stambler 2005; [3] Al-Najjar et al. 2007; [4] Mohamed 2018; [5] Ehrenberg 1830; [6] El Hag and Nasr 1989; [7] Post et al. 2002; [8] Touliabah et al. 2010; [9] Kürten et al. 2015; [10] Alkawri et al. 2016a; [11] Osama and Amjed 2020; [12] Subba Rao et al. 2003; [13] Al-Yamani and Saburova 2019a; [14] Polikarpov et al. 2020; [15] Esmaeili et al. 2021; [16] Ismail et al. 2007; [17] Al-Yamani et al. 2012; [18] Subba Rao et al. 2003; [19] Saburova et al. 2022; [20] Gomaa et al. 2018; [21] Subba Rao et al. 1999; [22] Mohamed and Mesaad 2007; [23] Glibert et al. 2002; [24] Alkawri et al. 2016b; [25] Mohamed and Al-Shehri 2012; [26] Heil et al. 2001; [27] Alkawri 2016; [28] Saburova et al. 2012; [29] Richlen et al. 2010; [30] Alkershi and Menon 2011; [31] Al-Yamani et al. 2012; [32] Catania et al. 2017; [33] Banguera-Hinestroza et al. 2016; [34] Al-Kandari et al. 2009; [35] Al-Yamani and Saburova 2010; [36] Al-Hassan et al. 1990; [37] recent unpublished data (internal KISR reports)).

Phytoplankton blooms in the NW Gulf are distinctly dominated by cryptophycean flagellates, photosynthetic ciliates, and diatoms (Figure 19.2B). Small-sized autotrophic cryptophycean flagellates are a group of high abundance in Kuwait's waters, with frequent outbreaks of up to 10^6–10^7 cells/l, particularly in the coastal nutrient-rich waters of Kuwait Bay, with *Teleaulax* spp. as the predominant taxa. Cryptophycean blooms occurred most frequently during spring and fall, and no harmful effect on the marine environment was recorded (Al-Yamani and Saburova 2019a).

The ciliate *Mesodinium rubrum* containing endosymbiotic phycoerythrin-rich plastids derived from its principal food, the photosynthetic cryptophytes (Johnson et al. 2006), can be considered a functional phototroph because of its ability to utilize dissolved inorganic nutrients (Wilkerson and Grunseich 1990) and its high photosynthetic rates associated with bloom events (Stoecker et al. 1991). Blooms of *M. rubrum* are common in Kuwait's waters (Figure 19.2B). Its autumnal blooms in October-November have been recorded since 1995, and likely are a yearly phenomenon. Since 2017, outbreaks of this ciliate have also been observed in spring. Like cryptophyceans, *Mesodinium* blooms seem harmless to Kuwait's marine environment, except for short-term water discolouration. In bloom conditions, this ciliate typically forms reddish-brown patches on the sea surface (Figure 19.1G) in which its abundance reaches 10^6–10^7 cells/l. However, in large proportions, *M. rubrum* represents an important trophic link in the phytoplankton community, feeding on cryptophycean flagellates to acquire plastids and, in turn, being consumed by kleptoplastic dinoflagellates *Dinophysis* spp. (Park et al. 2006). In Kuwait's waters, the blooms of *M. rubrum* closely followed the seasonal peaks of cryptophytes (Figure 19.2C). Its spring bloom in April 2017 in Kuwait Bay preceded the extensive outbreak of DSP-related *Dinophysis ovum* (Polikarpov et al. 2020).

Small-sized diatoms *Nitzschia* spp., *Skeletonema grevillei*, *Thalassiosira* spp., and *Chaetoceros* spp. are among the frequently bloomed taxa in Kuwait's waters (Figure 19.2B; Table 19.1). These diatoms form recurrent high-biomass blooms causing water discolouration (Figure 19.1B) and are sometimes implicated in fish kills in Kuwait Bay. Fish mortality events happened during the hot period from June to September when the water was warm and poorly mixed. High water temperatures coupled with organic carbon supplied in plenty by the decomposition of phytoplankton cells, and water stratification in shallow coastal waters resulted in significant oxygen depletion in the water column due to eventual accelerated decay of organic matter. Extremely high phytoplankton abundance could cause hypoxia in surrounding waters, leading to sublethal to lethal effects on marine biota (Al-Yamani et al. 2020; Polikarpov et al. 2020).

Less frequent blooming species in Kuwait's waters are the dinoflagellates *Karenia papilionacea*, *Gymnodinium* spp., and *Prorocentrum* spp., and two prymnesiophycean flagellates, colony-forming *Phaeocystis globosa* and unicellular *Chrysochromulina* sp. (Figure 19.2B; Table 19.1). Recurrent spring blooms of toxic *Karenia papilionacea* (potentially producing neurotoxins) in Kuwait's coastal waters resulted in intense water discolouration (Figure 19.1D) when its abundance attained >10^6–10^7 cells/l within bloom patches. In bloom conditions, this species often co-occurred with *Prorocentrum* spp. (potentially producing DSP toxins), and the ichthyotoxic *Karenia selliformis* and *Chrysochromulina* sp. During the recent bloom of *K. papilionacea* in 2020, seawater samples and cultural strains isolated from the bloom area were positive for brevetoxin (Maria Saburova and Sabeena F.K. Habeebullah, pers. comm.). Seasonal blooms of *Phaeocystis globosa* occurred in fall, winter, and spring, covering primarily northern Kuwait's waters. At the terminal bloom stage, extraordinarily large colonies of this flagellate (up to 7 cm in diameter) were washed ashore (Figure 19.1F), forming a foam and contaminating the beaches with mucilage (Al-Yamani and Saburova 2019a).

The dinoflagellate *Heterocapsa circularisquama*, although not so frequently blooming, merits special mention here due to its recent invasion in the NW Arabian Gulf from the Asia Pacific region and its specific toxicity to bivalve mollusks. Blooms of this non-indigenous species with exceptionally high concentrations of up to >10^8 cells/l and bloom-associated shellfish mortality (Figure 19.1C) were detected for the first time in Kuwait's waters in summer 2020 (Saburova et al. 2022), and its reappearance in bloom proportion in the summer of 2021 (Saburova et al., in prep.) points to a successful establishment of this invasive toxic species in the phytoplankton assemblage.

Microalgal blooms occur in the NW Arabian Gulf year-round but most commonly during the spring season in April-May (Figure 19.2C), coinciding with the annual flow peak of the Shatt Al-Arab River in spring (Aqrawi 2001; Al-Yamani et al. 2004) and seasonal rise in seawater temperature to >20°C. This seasonal pattern of HAB occurrence emphasizes the nutrient input from the river flows as the prime supply for phytoplankton proliferation in the NW Gulf.

Over the past decade, an increased algal bloom frequency in the NW Arabian Gulf is considered to be a eutrophication-induced phenomenon due to growing anthropogenic nutrient loading in the area (Devlin et al. 2015; Al-Yamani and Naqvi 2019; Polikarpov et al. 2020; Al-Yamani 2021). Recent alterations in nutrient supply driven by the rising coastal urbanization of Kuwait Bay and the decline of Shatt Al-Arab River runoff shifted a nutrient balance, particularly the nitrogen-to-phosphorus stoichiometric ratio, towards favourable conditions for phytoplankton proliferation (Polikarpov et al. 2020). Both overall elevated nutrient availability and modified nutrient ratios largely contribute to the phytoplankton bloom development and to increasing HABs frequency in Kuwait's waters.

19.4.2 NE Arabian Gulf

The oceanographic conditions of the eastern extremum of the Arabian Gulf are greatly influenced by water exchange with the Sea of Oman through the Strait of Hormuz. The Sea of Oman is among the most productive basins of the World Ocean, in which the productivity regime is predominantly governed by seasonally reversing monsoonal winds. Upwelling during the summer southwest monsoon and convective mixing during the northeast winter monsoon elevate nutrient levels in the surface water of the Sea of Oman and consequently increase the phytoplankton biomass and primary production (e.g., Piontkovski et al. 2012, 2019; Al-Hashmi et al. 2015). Phytoplankton blooms from the Sea of Oman often spread into the adjacent Arabian Gulf due to the inflow of the nutrient-enriched waters through the Strait of Hormuz, resulting in high productivity of the northeastern part of the Arabian Gulf along the Iranian coast (e.g., Polikarpov et al. 2016; Piontkovski et al. 2019).

Unlike the NW Gulf, the phytoplankton blooms in the NE part are largely contributed by the ichthyotoxic dinoflagellate *Margalefidinium polykrikoides* (Figure 19.2B). A total of 225 outbreaks of this species were reported in the coastal waters along the Hormozgan Province in Iran in 2008–2009 (Esmaeili et al. 2021). The bloom of *M. polykrikoides* originated in the Sea of Oman in 2008 and expanded into the Arabian Gulf with the Iranian Current over more than 300 km of its coastline (Richlen et al. 2010; Zhao and Ghedira 2014; Polikarpov et al. 2019). This bloom caused catastrophic impacts on the environment and human economic activities in the region, including massive mortalities of wild and farmed fish and shrimp, coral reef damage, and the closure of desalination plants (e.g., Richlen et al. 2010; Esmaeili et al. 2021).

Apart from the *M. polykrikoides*, native diatoms *Leptocylindrus* sp., *Navicula* sp., *Chaetoceros* sp.,

Streptotheca sp., and *Nitzschia* spp., dinoflagellates *Noctiluca scintillans*, *Gymnodinium* sp., and *Protoperidinium* sp., and cyanobacterium *Trichodesmium erythraeum* were among the most frequently blooming phytoplankton taxa in the NE Gulf (Figure 19.2B; Table 19.1). Commonly, short-term sporadic blooms caused by native phytoplankton species with the summer-fall peak in July-September (Figure 19.2C) did not lead to strong adverse impacts on the environment of the NE Gulf (Esmaeili et al. 2021). However, the sudden emergence and extensive long-lasting harmful bloom of *M. polykrikoides* in 2008–2009 has supplanted the commonly occurring algal blooms in the Sea of Oman and the Arabian Gulf (Al-Azri et al. 2014; Esmaeili et al. 2021) due to high competitiveness of this species. Tolerance of this dinoflagellate to a wide range of temperatures and salinities (e.g., Kim et al. 2004) supplemented by mixotrophic capabilities and the existence of a cyst stage in its life cycle likely contributed to the progression of its bloom and rapid expansion in the warm-water and hypersaline environment of the Arabian Gulf. A range of natural meteorological and oceanographic forcing (strong summer upwelling along the Iranian and northern Omani coasts followed by unusually warm coastal plume water along the coast of Oman) along with an elevated nutrient load and warmer than normal temperatures were considered as possible reasons for the occurrence and persistence of high densities of *M. polykrikoides* in the Sea of Oman and the Arabian Gulf in 2008–2009 (Richlen et al. 2010; Al-Azri et al. 2014).

Similarly to the NW part of the Gulf, the occurrence and high frequency of phytoplankton blooms in the NE part have been attributed to eutrophication, mainly of anthropogenic origin. Along with the natural nutrient availability in these waters, untreated industrial and municipal wastewaters are discharged into the sea along the shore of Hormozgan Province, particularly in the coastal waters of the highly urbanized Bandar Abbas area and in shallow waters stretching to Hormuz Island and the northern coasts of Qeshm Island. Over the past decade, nutrient pollution combined with high temperatures and restricted water exchange through the Strait of Hormuz have stimulated the excess growth of algae and caused a sharp increase in phytoplankton bloom occurrence in this area (Esmaeili et al. 2021).

19.4.3 Red Sea

Most of the Red Sea is an oligotrophic and highly stratified system, with the main nutrient influx coming from the upwelling areas of the Arabian Sea and the Gulf of Aden through the Bab al Mandab Strait in the south. Inflowing nutrient-enriched water is subjected to mixing, evaporation, and nutrient depletion along its path from the southern Red Sea to the north (Rasul et al. 2015; Yao and Hoteit 2015; Rasul and Stewart 2019). Fluctuations in phytoplankton biomass and productivity are closely related to surface circulation, which is governed by the wind system, and, in turn, shapes the distribution of nutrients across the Red Sea. Based on the basin-wide chlorophyll distribution pattern, the southern Red Sea is more productive than the central and northern parts (Raitsos et al. 2013; Brewin et al. 2015; Qurban et al. 2019; Kheireddine et al. 2021).

In contrast to eutrophic areas of the Arabian Gulf discussed previously, the Red Sea is less attractive to algal growth due to nutrient limitation, which could be the primary reason for the scarcity of phytoplankton blooms in this basin. The proliferation of algal blooms in the Red Sea follows an overall pattern of its productivity. Blooms occur throughout the Red Sea but increase in frequency southward (Figure 19.2A).

Cyanobacteria lead the list of bloom-forming species, including filamentous *Trichodesmium* and picoplanktonic *Prochlorococcus* and *Synechococcus*. *Trichodesmium erythraeum* is the most frequently blooming phytoplankton species of the Red Sea. Recurrent *Trichodesmium* blooms occur across the Red Sea but more often in its extremely oligotrophic northern part, including the Egyptian coast and the northern tip of the Gulf of Aqaba (Figure 19.2A, inset 3). Blooms of this diazotrophic species often indicate stratification and nitrogen depletion in the sea surface layer. Its known ability to fix atmospheric nitrogen allows this alga to thrive under low-nutrient oceanic conditions in sub-tropical and tropical waters worldwide (Capone et al. 2005; Fernández et al. 2010). Along with another diazotroph, diatom-associated cyanobacterium *Richelia intracellularis*, *Trichodesmium* is commonly recorded in the phytoplankton composition in the Red Sea year-round (Ismael 2015; Kürten et al. 2015; Devassy et al. 2017), pointing to prevailing oligotrophic conditions in this basin.

Picoplankton (mainly cyanobacteria *Prochlorococcus* and *Synechococcus*) is the dominant fraction in the Gulf of Aqaba and the northern Red Sea for much of the year, contributing significantly to the primary productivity of these waters under scarce nutrient supply, typically below the detection limit (Lindell and Post 1995; Al-Najjar et al. 2007; Ismael 2015). The tiny size (<2 μm) of these picocyanobacteria and the resulting high surface-to-volume ratio are adaptive advantages for efficient nutrient uptake (Chisholm 1992), allowing them to potentially outcompete larger species in oligotrophic conditions. The abundant and frequent cyanobacteria blooms can be considered a characteristic feature of the oligotrophic waters of the Red Sea. The dominance of cyanobacteria may be attributable to the unique adaptations of these organisms to the oligotrophic environment, allowing their blooms to occur in nutrient-poor waters. At the same time, other phytoplankton groups replace them under mesotrophic conditions.

Dinoflagellates comprise the second dominant group in terms of bloom frequency in the Red Sea (Figure 19.2B). It includes *Dinophysis* spp., *Noctiluca scintillans*, *Gonyaulax* spp., *Gymnodinium* sp., *Pyrodinium bahamense*, *Tripos* spp., and *Prorocentrum* spp. (Table 19.1). Dinoflagellate-dominated blooms are restricted to mesotrophic conditions in the central and southern Red Sea. In contrast, such blooms have not been reported in the northern Red Sea (Figure 19.2A). Rare diatom-dominated blooms are caused by species belonging to the genera *Chaetoceros*, *Thalassiosira*, *Nitzschia*, and *Rhizosolenia*, and mainly occur in the central Red Sea (Figure 19.2A; Table 19.1).

Although harmless phytoplankton blooms have prevailed in the Red Sea, a few HAB events have recently been observed in the coastal waters of its central and southern sub-basins, which were induced mainly by human-mediated activity. In May 2010, a bloom of the ichthyotoxic raphidophycean flagellate *Heterosigma akashiwo* occurred in the southern Red Sea off Saudi Arabia in the coastal waters influenced by runoff from a nearby shrimp farm (Mohamed and Al-Shehri 2012). During bloom conditions, *H. akashiwo* attained concentrations $>10^7$

cells/l. Although no marine mortality events were observed at bloom sites, bloom samples and cultural strains of bloom-forming species were toxic to brine shrimp (*Artemia salina*) and exhibited haemolytic activity against rabbit erythrocytes. In June 2012, a high-biomass bloom caused by the dinoflagellate *Blixaea quinquecornis* (*Peridinium/Protoperidinium quinquecorne*) occurred in a semi-enclosed lagoon in Yemeni coastal waters in the southern Red Sea (Alkawri et al. 2016b). The bloom stretched for several km along the coast of the highly urbanized Al Hodeidah city, reduced water quality due to oxygen depletion, and caused massive mortality of small pelagic sardines. In 2012–2014, the persistent presence of toxic PSP-producing dinoflagellate *Pyrodinium bahamense* in the phytoplankton composition and its recurrent outbreaks of up to 10^3–10^5 cells/l were observed in the central Red Sea off Saudi Arabia (Banguera-Hinestroza et al. 2016) and in the southern Red Sea in Yemeni coastal waters (Alkawri et al. 2016a). The toxicity of the *P. bahamense* strain isolated from the bloom samples and the net haul sample collected from the bloom-affected area in Saudi Arabia were confirmed by a positive test for saxitoxins (Banguera-Hinestroza et al. 2016).

The annual cycle of phytoplankton blooms in the Red Sea follows a bimodal pattern, with a winter peak in February-March and an autumn peak in September-October when cyanobacterial blooms are more frequent (Figure 19.2C). This suggests a close relationship between phytoplankton productivity and conditions resulting from monsoonal events. In general, blooms are more frequent between December and June.

19.5 RISK ASSESSMENT OF HABS IN THE ARABIAN GULF AND THE RED SEA

Based on the historical analysis and statistics of HAB events over the past three decades, we identified a range of phytoplankton taxa as the most frequent HAB species reported in the Arabian Gulf and the Red Sea. These species are listed in Table 19.2 and illustrated in Figures 19.3–19.6.

19.5.1 HAB-associated risks to public health

A diverse complex of toxigenic algal species inhabits the coastal waters of the Arabian Gulf and the Red Sea (Table 19.2; Figures 19.3 and 19.4). Their presence in the phytoplankton composition with recurrent outbreaks poses a health hazard to the local human populations through various shellfish and fish poisonings. Toxicological surveys of the marine environment on algal toxins' presence throughout the Arabian Gulf and the Red Sea are scarce and localized (Glibert et al. 2002; Al Muftah et al. 2016; Banguera-Hinestroza et al. 2016; Mohebbi-Nozar et al. 2022). However, even these limited outcomes are already enough to conclude that these waters are at high risk of contamination by algal toxins of various origins.

The dinoflagellate *Pyrodinium bahamense* (Figure 19.3 G-I) is the most toxic and frequently detected saxitoxin-producer among the algal species responsible for paralytic shellfish poisoning (PSP). Its recurrent outbreaks in the NW Gulf (Glibert et al. 2002; Polikarpov et al. 2020), central Red Sea (Banguera-Hinestroza et al. 2016), and southern Red Sea (Alkawri et al. 2016a), detection of saxitoxin-positive bivalves (1.26 µg STX eq. 100 g^{-1} tissue) in the NW Gulf (Glibert et al. 2002), and toxicity of local strains documented in the central Red Sea (Banguera-Hinestroza et al. 2016) all contribute to the high risk of PSP and pose a threat to human health in the listed sub-basins. As to neurotoxic shellfish poisoning (NSP), the dinoflagellate *Karenia papilionacea* (Figure 19.3L) is the main risk species in the NW Gulf due to its recurrent highly abundant blooms (Polikarpov et al. 2020; Figure 19.1D) and recently detected brevetoxin in bloom samples and isolated strains of this species.

Sporadic outbreaks of known producers of okadaic acid and dinophysistoxins, *Dinophysis* and *Prorocentrum* species (Figure 19.3 M-U), may facilitate the high risk of diarrhetic shellfish poisoning (DSP) in the NW and NE Gulf, as well as in the central Red Sea. *Dinophysis ovum* was observed in bloom proportion (>10^6 cell/l) in the NW Gulf (Polikarpov et al. 2020), *D. caudata* and *D. miles* attained concentrations >10^5 cells/l in the central Red Sea (Gomaa et al. 2018), and several blooms of *Prorocentrum lima* (>10^6 cells/l) were recorded in the NE Gulf (Esmaeili et al. 2021). Moreover, the seasonal blooms of the photosynthetic ciliate *Mesodinium rubrum* in the NW Gulf (Figure 19.1G), being harmless themselves to the marine environment, need to be mentioned in terms of DSP risks, as the availability of *M. rubrum* is an essential pre-condition for the harmful blooms of *Dinophysis* spp., due to the specific trophic link between plastid-containing *Mesodinium* and *Dinophysis*, feeding on ciliates to obtain plastids (see section 19.4.1). In the NE Gulf, toxic dinoflagellates *D. caudata* and *P. lima* were identified as the main source of detected contamination of edible bivalves by okadaic acid in the Hormozgan Province (Mohebbi-Nozar et al. 2022). Concentrations of okadaic acid in bivalves, measured at densities of causative algal species of 10^2–10^3 cells/l, were lower than the regulatory limit for human consumption. However, higher concentrations of DSP toxins during blooms of these species can cause clinical DSP symptoms in bloom-affected areas.

Ciguatera fish poisoning or Ciguatera poisoning (CFP/CP) is the most frequently reported seafood-toxin illness globally. Ciguatera results from consuming fish or other seafood (sea urchins, giant clam, trochus) contaminated with bioaccumulated potent neurotoxins produced by epibenthic dinoflagellates in the genera *Gambierdiscus* and *Fukuyoa* (e.g., Friedman et al. 2008; Chinain et al. 2020). CFP/CP is considered a major public health problem in the Pacific and Caribbean tropical areas, where its occurrence and effects of ciguatoxins are well described (Chinain et al. 2020). In contrast, this syndrome and causative algae are poorly documented in the northern Indian Ocean. They likely are underestimated in both the Arabian Gulf and the Red Sea. The genus *Fukuyoa* (Figure 19.4 E-H) was recorded in the NW Gulf and the northern Red Sea (Saburova et al. 2013). *Gambierdiscus* species (Figure 19.4 A-D) were documented in the northern and central Red Sea (Saburova et al. 2013; Catania et al. 2017; Prabowo and Agusti 2019). The toxicity of *Gambierdiscus belizeanus* was further confirmed for the strains isolated from the central Red Sea (Catania et al. 2017). A single CFP/CP event was diagnosed on the Egyptian

TABLE 19.2 Regionally assessed hazard of HABs occurrence in the Arabian Gulf and the Red Sea.

VECTOR/HARMFUL EFFECT/TAXON	*ARABIAN GULF*		*RED SEA*	
	NWG	*NEG*	*NRS*	*CRS-SRS*
HUMAN HEALTH:				
Amnesic Shellfish Poisoning				
Pseudo-nitzschia spp.	++	++	+	++
Paralytic Shellfish Poisoning				
Alexandrium spp.	+++	++	+	++
Pyrodinium bahamense	+++	++	+	+++
Trichodesmium erythraeum	++	+++	+++	+++
Neurotoxic Shellfish Poisoning/Aerosol Toxicity				
Karenia papilionacea	+++	++	+	+
Diarrhetic Shellfish Poisoning				
Dinophysis spp.	+++	++	+	+++
Gonyaulax spinifera	++	++	+	+++
Prorocentrum spp.	+++	+++	++	+++
Ciguatera Fish Poisoning/Clupeotoxism/Aerosol Toxicity				
Gambierdiscus spp.	+	+	+++	+++
Fukuyoa spp.	++	+	++	++
Ostreopsis spp.	++	+	++	+++
MARINE MORTALITY:				
Ichthyotoxicity				
Heterocapsa circularisquama	+++	++	+	+
Karenia selliformis	+++	++	+	++
Karlodinium/Takayama spp.	+++	+	+	+
Margalefidinium polykrikoides	++	+++	+	++
Chrysochromulina sp.	+++	++	+	+
Chattonella sp.	+++	+++	+	++
Heterosigma akashiwo	+++	++	+	+++
Phaeocystis globosa	++	+	+	+
High biomass bloom/Hypoxia				
Trichodesmium erythraeum	++	+++	+++	+++
Chaetoceros spp.	+++	+++	+	++
Minutocellus polymorphus	+++	+	+	+
Nitzschia spp.	+++	+++	+	+++
Skeletonema spp.	+++	++	+	+
Thalassiosira spp.	+++	+	+	+++
Gonyaulax verior	+++	+	+	+++
Kryptoperidinium foliaceum	+++	+	+	+++
Noctiluca scintillans	++	+++	+	+++
Blixaea quinquecornis	+++	++	+	+++
Tripos spp.	+++	++	+	+++
Eutreptiella sp.	+++	+	+	+
Phaeocystis globosa	++	+	+	+
Mesodinium rubrum	+++	++	+	+

Risk of HABs occurrence is indicated as follows: +, low; ++, moderate; +++, high; NWG refers to the northwestern Gulf, NEG refers to the northeastern Gulf, NRS refers to the northern Red Sea, CRS refers to the central Red Sea, and SRS refers to the southern Red Sea.

Red Sea coast (de Haro et al. 2003), but the local origin of the fish involved was not confirmed. Scarce taxonomic surveys of epibenthic algae revealed low *Gambierdiscus* and *Fukuyoa* densities along with comparatively low toxicity that likely resulted in underreporting of CFP/CP throughout the area (Catania et al. 2017). However, the presence of toxigenic Ciguatera-implicated algal species and bioaccumulation of their toxins through marine food webs contribute to the occurrence of Ciguatera poisoning, and particularly the coastal Red Sea waters are areas with a probability of high risk. Along with *Gambierdiscus* and *Fukuyoa* species, other benthic dinoflagellates may be involved in human seafood poisonings, including

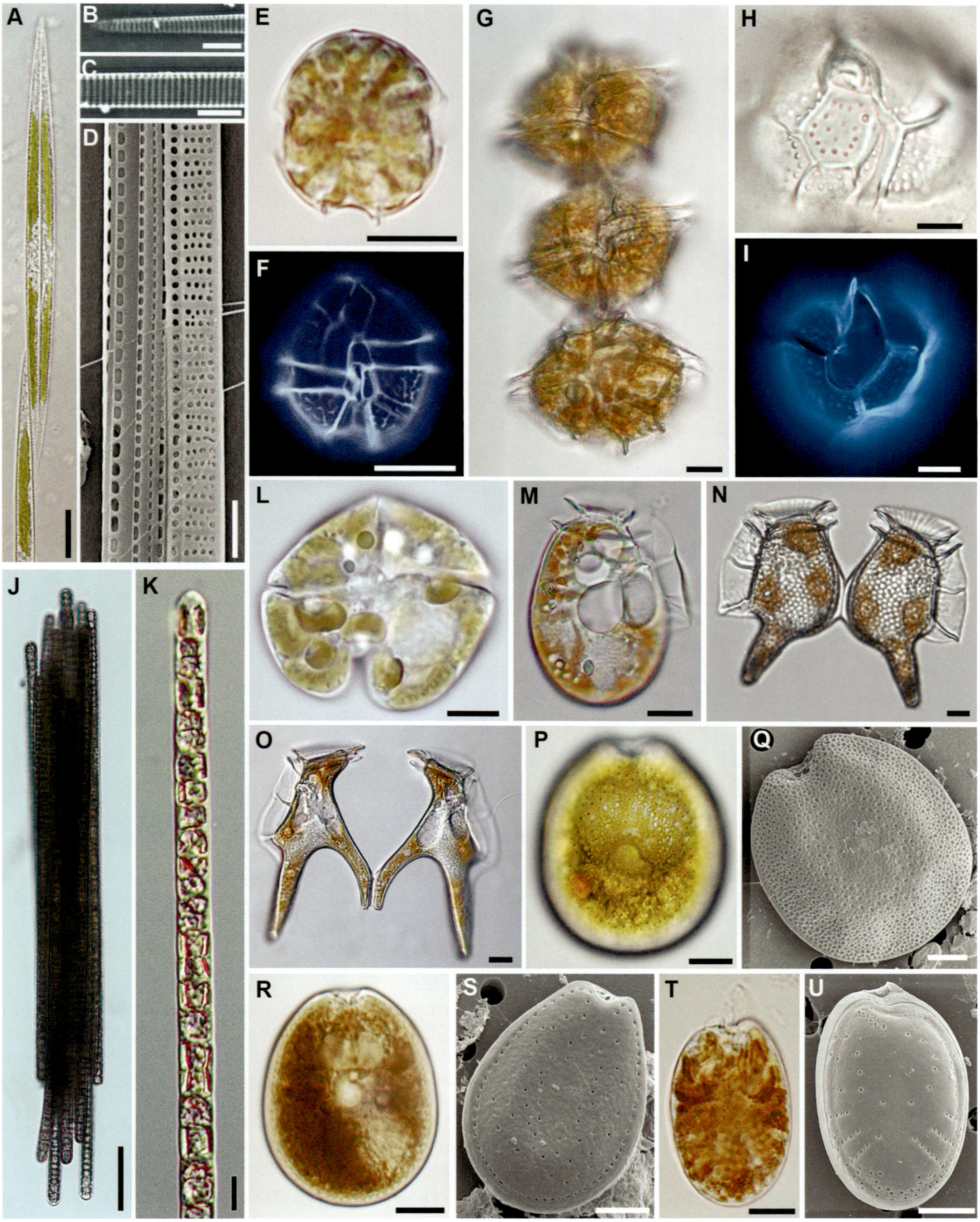

FIGURE 19.3 Toxic microalgae that occur in the Arabian Gulf and the Red Sea. A-D) *Pseudo-nitzschia pungens* (ASP); E, F) *Alexandrium minutum* (PSP); G-I) *Pyrodinium bahamense* var. *compressum* (PSP); J, K) *Trichodesmium erythraeum* (PSP); L) *Karenia papilionacea* (NSP); M) *Dinophysis ovum* (DSP); N) *Dinophysis caudata* (DSP); O) *Dinophysis miles* (DSP); P, Q) *Prorocentrum concavum* (DSP); R, S) *Prorocentrum lima* (DSP); T, U) *Prorocentrum rhathymum* (DSP). A-O – isolated from the NW Arabian Gulf, Kuwait; P-U – isolated from the subtidal sediments in the northern Red Sea; A-C, E-P, R, T – LM; D, Q, S, U – SEM; scale bars: A-C – 5 µm; D – 2 µm; E-I, K-U – 10 µm; J – 50 µm.

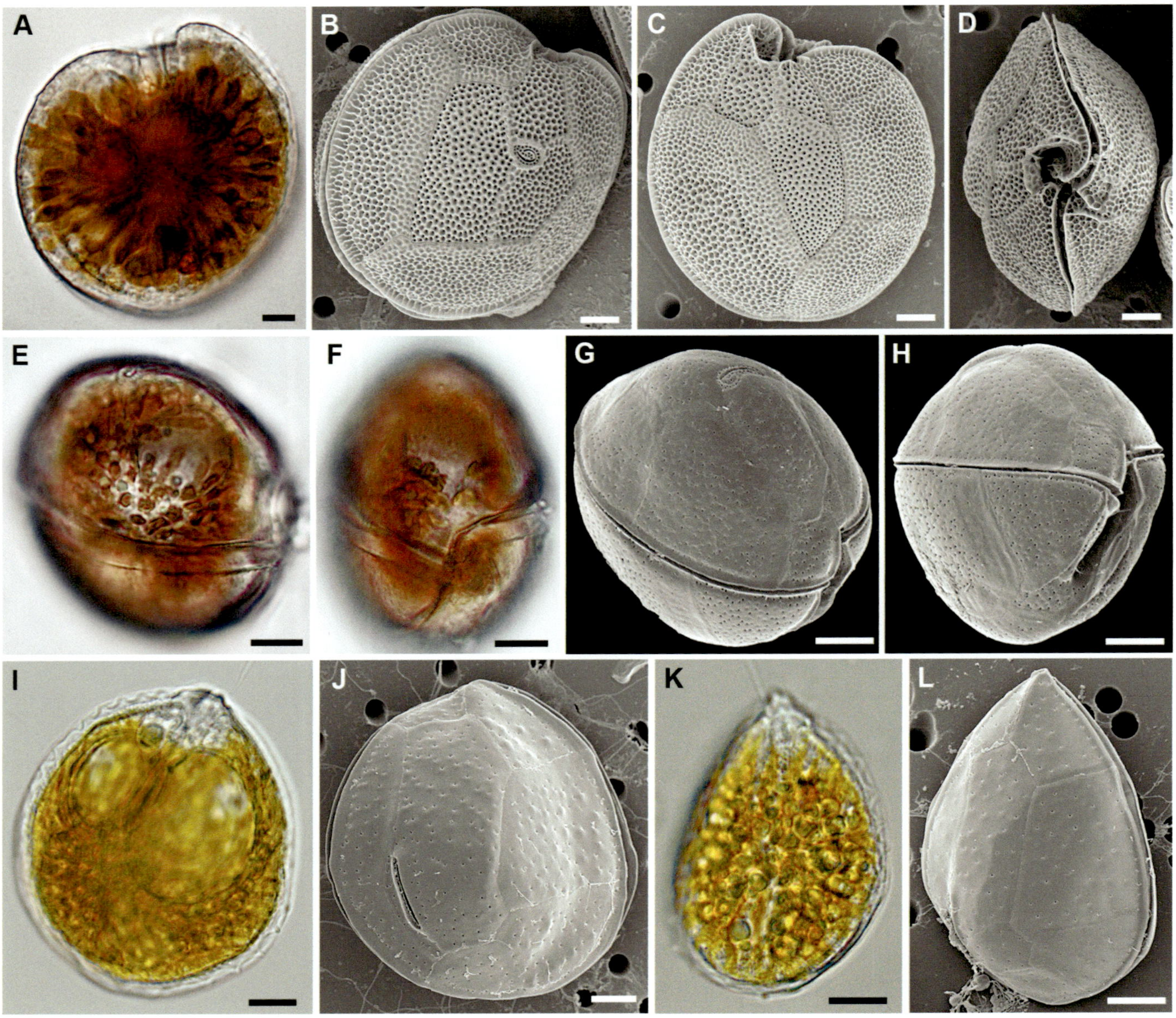

FIGURE 19.4 Toxic microalgae that occur in the Arabian Gulf and the Red Sea. A-D) *Gambierdiscus belizeanus* (CFP/CP); E-H) *Fukuyoa yasumotoi* (CFP/CP); I, J) *Ostreopsis* cf. *siamensis* (CPT); K, L) *Ostreopsis* cf. *ovata* (CPT). A-L – isolated from the subtidal sediments in the northern Red Sea; A, E, F, I, K – LM; B-D, G, H, J, L – SEM; scale bars: 10 μm.

Ostreopsis species (Figure 19.4 I-L) producing palytoxin analogues (e.g., Pavaux et al. 2020) and toxic *Prorocentrum* species (Figure 19.3 P-U).

19.5.2 HAB-associated risks to marine biota

A complex of bloom-forming ichthyotoxic algal species identified from the Arabian Gulf and the Red Sea consists of dinoflagellates, as well as raphidophycean and prymnesiophycean flagellates (Table 19.2; Figure 19.5). Among them, the dinoflagellate *Margalefidinium polykrikoides* (Figure 19.5 H, I) is of particular concern because of its well-documented capacity to form huge and long-lasting blooms, known from the NE Gulf and the Sea of Oman, and its catastrophic impact on the marine environment and industry in bloom-affected areas (Richlen et al. 2010; Esmaeili et al. 2021). The presence of this species in the phytoplankton composition in the NW Gulf (Al-Yamani and Saburova 2019a) and in the Red Sea (Prabowo and Agusti 2019) could potentially contribute to new HAB occurrences in these areas in response to nutrient surplus and an increase of seawater temperatures, particularly in areas with low water exchange rates.

Another dinoflagellate species of particular concern is the ichthyotoxic *Karenia selliformis* (Figure 19.5 C, D). Its intense bloom in Kuwait Bay in 1999 resulted in massive fish kills and considerable economic losses (Heil et al. 2001). Although no blooms of this species have been reported since then, the NW Gulf remains the area most at risk of further *K. selliformis* blooms. In Kuwait's coastal waters, this species co-occurs with *K. papilionacea*, often in large proportions (Polikarpov et al. 2020), and may, under certain environmental conditions, be a hazard to the marine environment at bloom concentrations. Further, the dinoflagellate *Heterocapsa circularisquama* (Figure 19.5 A, B) is a species of concern in terms of HAB risk. The successful establishment of this non-indigenous invasive toxic species into the phytoplankton community composition

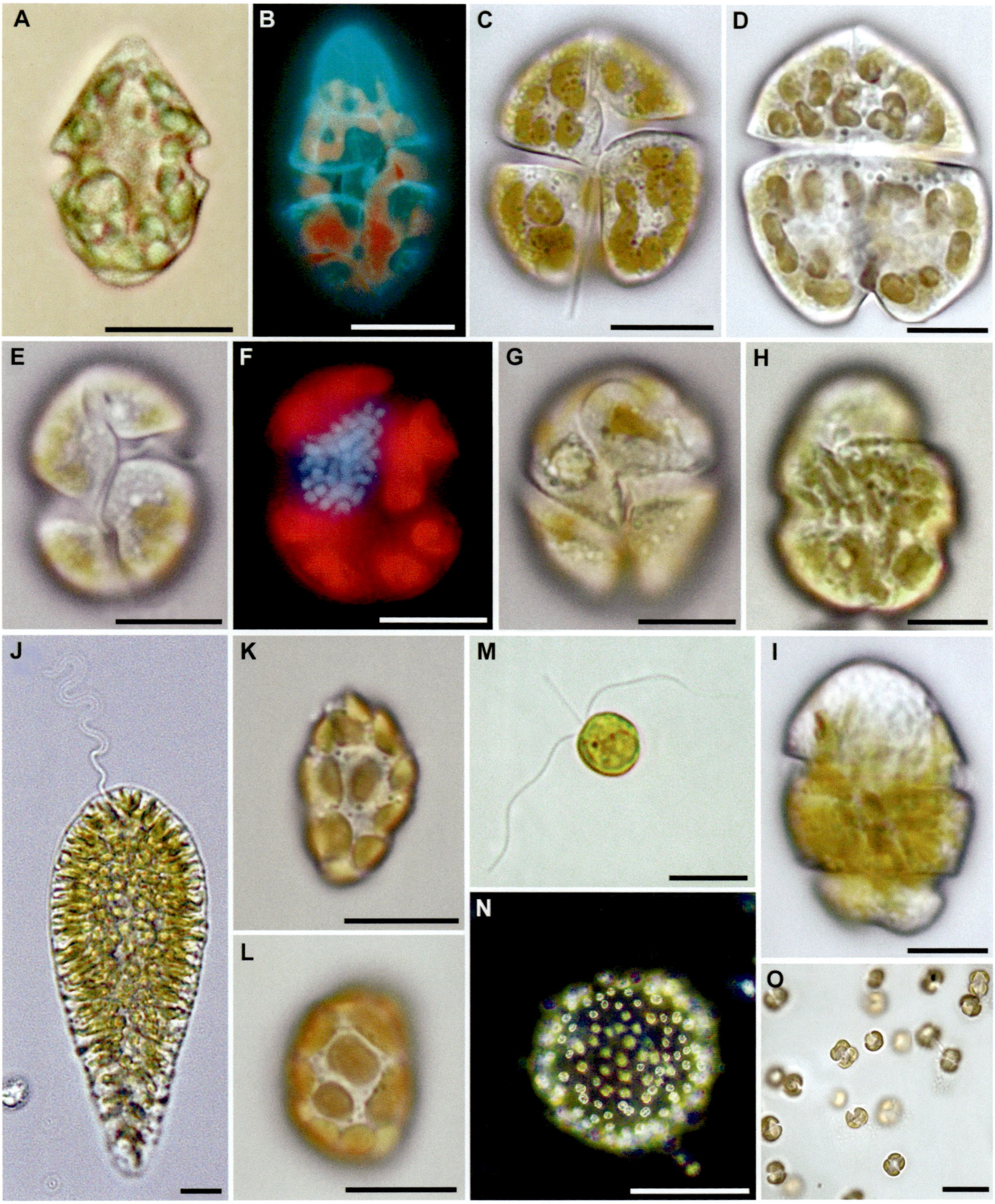

FIGURE 19.5 Ichthyotoxic microalgae that occur in the Arabian Gulf and the Red Sea. A, B) *Heterocapsa circularisquama*; C, D) *Karenia selliformis*, E, F) *Karlodinium* sp., G) *Takayama tasmanica*; H, I) *Margalefidinium polykrikoides*; J) *Chattonella marina var. antiqua*; K, L) *Heterosigma akashiwo*; M) *Chrysochromulina* sp., N, O) *Phaeocystis globosa*. A-O – isolated from the NW Arabian Gulf, Kuwait, LM; scale bars: 10 µm.

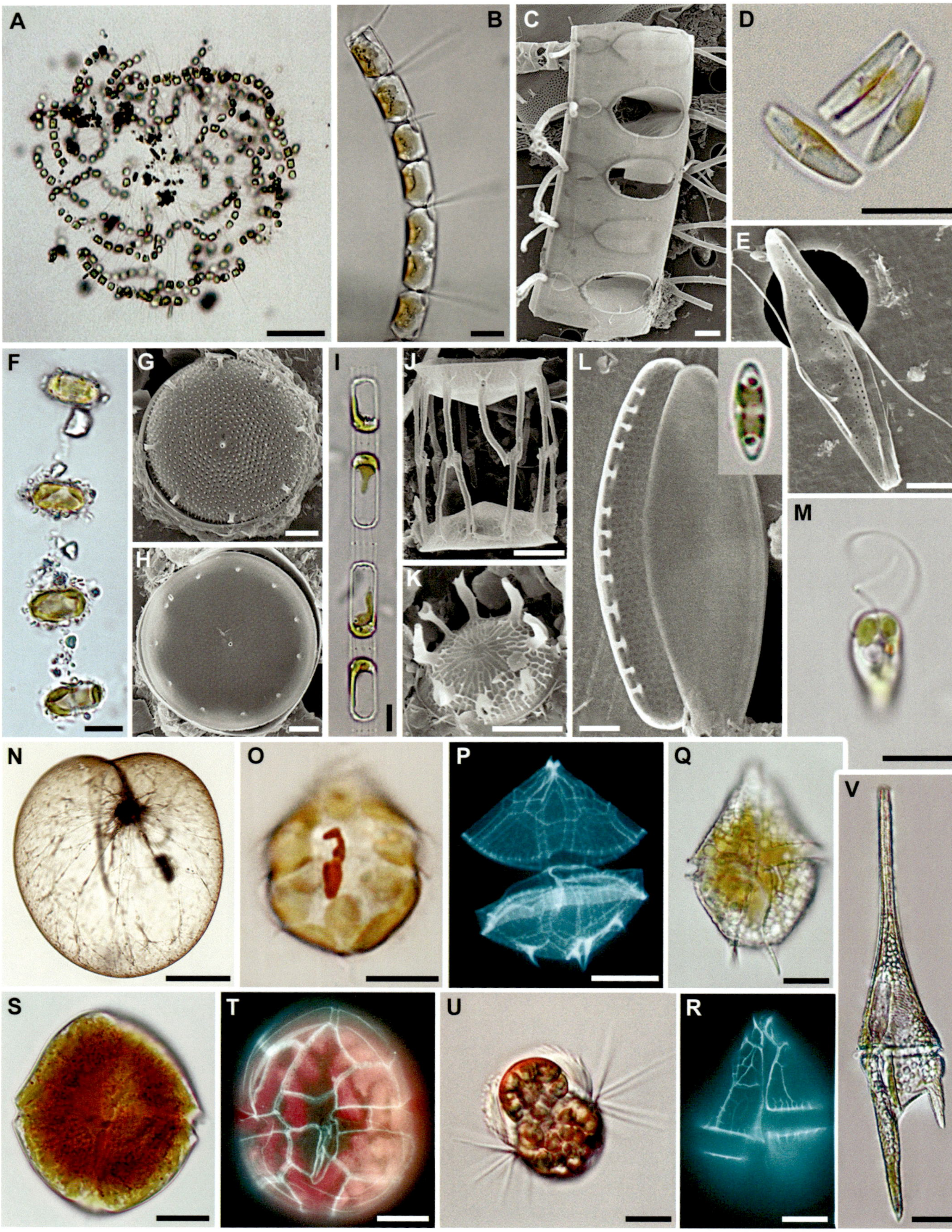

FIGURE 19.6 Bloom-forming microalgal and ciliate species that occur in the Arabian Gulf and the Red Sea. A) *Chaetoceros socialis*; B, C) *Chaetoceros pseudocurvisetus*; D, E) *Minutocellus polymorphus*; F-H) *Thalassiosira concaviuscula*; I-K) *Skeletonema grevillei*; L) *Nitzschia pusilla*; M) *Eutreptiella* sp.; N) *Noctiluca scintillans*; O, P) *Blixaea quinquecornis*; Q, R) *Gonyaulax verior*; S, T) *Kryptoperidinium foliaceum*; U) *Mesodinium rubrum*; V) *Tripos furca*. A-V – isolated from the NW Arabian Gulf, Kuwait; A, B, D, F, I, M-V – LM; C, E, G, H, J-L – SEM; scale bars: A – 50 µm; B – 20 µm; C, F, M-V – 10 µm; D, I – 5 µm; E, G, H, J-L – 2 µm.

in Kuwait (Saburova et al. 2022) suggests a potential threat of recurrent blooms in the NW Gulf and the risk of further spread of this harmful dinoflagellate along the coastal waters of the Arabian Gulf. The mortality of bivalves caused by toxic blooms of *H. circularisquama* (Figure 19.1C) has far-reaching impacts on the shellfish farming industry and natural ecosystems, adversely affecting macrofaunal abundance and food stocks of shellfish-eating fish and birds.

The NW Gulf largely experiences flagellate-dominated blooms, among which the bloom-forming raphidophycean flagellates *Chattonella marina var. antiqua* and *Heterosigma akashiwo* (Figure 19.5 J-L) attract special attention due to their known ichthyotoxicity and common bloom-associated marine mortality (e.g., Lassus et al. 2016). In the Red Sea, the southern part is most at risk from *H. akashiwo* blooms, with higher expected marine mortality due to this flagellate's confirmed toxicity under bloom conditions (Mohamed and Al-Shehri 2012).

19.5.3 HAB-associated risks to the marine environment

The list of species found to produce high biomass blooms in the Arabian Gulf and the Red Sea has a great variety of microalgae and includes cyanobacteria, diatoms, dinoflagellates, flagellates, and photosynthetic ciliate (Table 19.2; Figure 19.6), whose outbreaks can cause damage to ecosystems, fishery resources, recreational areas, and commercial facilities such as desalination power plants due to the biomass of accumulated algae.

The NW Gulf is a high-risk area for diatom-dominated blooms. Small-sized species in the genera *Chaetoceros*, *Minutocellus*, *Nitzschia*, *Skeletonema*, and *Thalassiosira* (Figure 19.6 A-L) are among the most frequently blooming taxa. During hot, calm weather conditions, diatom-dominated blooms often cause intense water discolouration (Figure 19.1B) and fish kills in the shallow coastal waters of Kuwait Bay due to reduced water quality and oxygen depletion through the water column (Al-Yamani et al. 2020; Polikarpov et al. 2020). Frequent and intense diatom blooms in the NW Gulf are largely supported by high silicate concentrations brought through the Shatt Al-Arab River discharge (Al-Yamani et al. 2004; Polikarpov et al. 2009).

High biomass dinoflagellate-dominated blooms are typical in the NW Gulf and the central and southern Red Sea. Dense dinoflagellate blooms are sporadically associated with fish kills, such as during blooms caused by *Gonyaulax verior* (Figures 19.1E, and 19.6 Q, R) in the NW Gulf (Polikarpov et al. 2020), and as in the case of *Blixaea quinquecornis* bloom (Figure 19.6 O, P) in the southern Red Sea (Alkawri et al. 2016b).

The NE Gulf and the Red Sea are the areas of the most frequent cyanobacterial blooms (Table 19.2). Conversely, sporadic blooms of *Trichodesmium erythraeum* (Figures 19.1A, and 19.3 J, K) occurred in the NW Gulf, only in nutrient-depleted offshore waters (Polikarpov et al. 2020), in line with the general trend of this diazotrophic organism to bloom in oligotrophic conditions. Under bloom proportions, its trichomes with positive buoyancy accumulate on the sea surface, forming dense mats arranged in extensive several-kilometre-long patches and colouring the water pinkish red. Decaying *Trichodesmium* blooms can lead to anoxic conditions and reduced water quality. Furthermore, these cyanobacteria need to be mentioned in the context of potentially harmful algae, as these species have been reported to produce neurotoxins (Lassus et al. 2016). However, no harmful effects of *Trichodesmium* bloom on humans or marine life have been reported so far, either in the Arabian Gulf or the Red Sea.

19.6 CONCLUSION

In the two marginal seas around the Arabian Peninsula, namely the Arabian Gulf and the Red Sea, extensive algal blooms are common phenomena. The formation of blooms is a function of the environmental conditions, both natural and anthropogenically driven, and the resource requirements of the different algal species. Alterations of these factors lead to qualitative and quantitative changes in the structure of planktonic communities. Recently, anthropogenic stresses have been cited as one of the biggest challenges facing the Arabian Gulf and the Red Sea marine environment. The increase in bloom frequency and severity strongly confirms the role of anthropogenic activity forcing in HAB development in this particular area.

Even a single HAB occurrence in a particular sea area warns of risk from further proliferation of this bloom-forming species to adjacent waters or area(s) with suitable environmental conditions, and recurrent bloom events contribute to this probability. This trend makes evident the necessity of large-scale phytoplankton monitoring programs in the coastal areas of all countries bordering the Arabian Gulf and the Red Sea. The presence of the toxigenic microalgae in a great variety, even at low concentrations, requires a systematic watch of phycotoxins for comprehensive regional-based assessments of HABs risk and their expected impacts on the marine environment, public health, and human coastal activities.

ACKNOWLEDGEMENTS

The authors thank Dr. Najeeb M.A. Rasul for the invitation to participate in this book project. We also wish to thank our reviewers, Elisa Berdalet, Huda Mahmoud, and Jacob Larsen, for their thorough, constructive, and encouraging comments. This work is partly supported by the Kuwait Institute for Scientific Research (project FM001O).

REFERENCES

Al-Azri, A., Piontkovski, S., Al-Hashmi, K., et al. 2014. Mesoscale and nutrient conditions associated with the massive 2008 *Cochlodinium polykrikoides* bloom in the Sea of Oman/Arabian Gulf. *Estuaries Coast* 37: 325–338. doi:10.1007/s12237-013-9693-1.

Al-Hashmi, K., Smith, S., Claereboudt, M., et al. 2015. Dynamics of potentially harmful phytoplankton in a semi-enclosed bay in the Sea of Oman. *Bull Mar Sci* 91: 141–166. doi:10.5343/bms.2014.1041.

Al-Hassan, R.H., Ali, A.M., and S.S. Radwan. 1990. Lipids and their constituent fatty acids of *Phaeocystis* sp. from the Arabian Gulf. *Mar Biol* 105: 9–14. doi:10.1007/BF01344265.

Al-Kandari, M., Al-Yamani, F., and K. Al-Rifaie. 2009. *Marine phytoplankton atlas of Kuwait's waters*. Kuwait: Kuwait Institute for Scientific Research, ISBN:99906-41-24-2.

Alkawri, A. 2016. Seasonal variation in composition and abundance of harmful dinoflagellates in Yemeni waters, southern Red Sea. *Mar Pollut Bull* 112: 225–234. doi:10.1016/j.marpolbul.2016.08.015.

Alkawri, A., Abker, M., Qutaei, E., et al. 2016a. The first recorded bloom of *Pyrodinium bahamense* var. *bahamense* Plate in Yemeni coastal waters off Red Sea, near Al Hodeida City. *Turk J Fish Aquat Sci* 16: 275–282. doi:10.4194/1303-2712-v16_2_07.

Alkawri, A., Al Areeki, M., and K. Alsharaby. 2016b. The first recorded bloom of *Protoperidinium quinquecorne* and its link to a massive fish kill in Yemeni coastal waters, southern Red Sea. *Plankton Benthos Res* 11: 75–78. doi:10.3800/pbr.11.75.

Alkershi, A., and N. Nandini Menon. 2011. Phytoplankton in polluted waters of the Red Sea coast of Yemen. *J Mar Biol Assoc India* 53: 1–6. doi:10.6024/jmbai.2011.53.2.01630-01.

Al Muftah, A., Selwood, A.I., Foss, A.J., et al. 2016. Algal toxins and producers in the marine waters of Qatar, Arabian Gulf. *Toxicon* 122: 54–66. doi:10.1016/j.toxicon.2016.09.016.

Al-Musawi, L. 2012. *Significant algal blooms observed in the ROPME Sea Area (2005–2012)*. In: ICES. Report of the ICES – IOC Working Group on Harmful Algal Bloom Dynamics (WGHABD), 24–27 April 2012, 15–16. Oban, Scotland, UK. www.ices.dk/sites/pub/Publication%20Reports/Expert%20Group%20Report/SSGHIE/2012/WGHABD12.pdf.

Al-Najjar, D., Badran, M.I., Richter, C., et al. 2007. Seasonal dynamics of phytoplankton in the Gulf of Aqaba, Red Sea. *Hydrobiologia* 579: 69–83. doi:10.1007/s10750-006-0365-z.

Al-Yamani, F.Y. 2021. *Fathoming the Northwestern Arabian Gulf: Oceanography and marine biology*. Kuwait: Kuwait Institute for Scientific Research, ISBN:978-99966-37-40-7.

Al-Yamani, F.Y., Al-Ghunaim, A., Subba Rao, D.V., et al. 2000. *Fish kills, red tides, Kuwait's marine environment*. Internal Publication. Kuwait: Kuwait Institute for Scientific Research.

Al-Yamani, F.Y., Bishop, J., Ramadhan, E., et al. 2004. *Oceanographic atlas of Kuwait waters*. Kuwait: Kuwait Institute for Scientific Research, ISBN:99906-41-19-6.

Al-Yamani, F.Y., and S.W.A. Naqvi. 2019. Chemical oceanography of the Arabian Gulf. *Deep Sea Res Part II: Top Stud Oceanogr* 161: 72–80. doi:10.1016/j.dsr2.2018.10.003.

Al-Yamani, F.Y., Polikarpov, I., and M. Saburova. 2020. Marine life mortalities and harmful algal blooms in the northern Arabian Gulf. *Aquat Ecosyst Health Manag* 23: 196–209. doi:10.1080/14634988.2020.1798157.

Al-Yamani, F.Y., Polikarpov, I., and M. Saburova. 2021. Northern Gulf marine biodiversity in relevance to the river discharge. In: *Southern Iraq's marshes: Their environment and conservation*. Ed: L.A. Jawad, 379–438. Coastal Research Library 37, Springer Nature Switzerland AG. doi:10.1007/978-3-030-66238-7_20.

Al-Yamani, F.Y., and M. Saburova. 2010. *Illustrated guide on the flagellates of Kuwait's intertidal soft sediments*. Kuwait: Kuwait Institute for Scientific Research, ISBN:978-99906-95-04-5.

Al-Yamani, F.Y., and M. Saburova. 2019a. *Marine phytoplankton of Kuwait's waters. Vol. 1: Cyanobacteria, dinoflagellates, flagellates*. Kuwait: Kuwait Institute for Scientific Research, ISBN:978-99966-37-20-9.

Al-Yamani, F.Y., and M. Saburova. 2019b. *Marine phytoplankton of Kuwait's waters. Vol. 2: Diatoms*. Kuwait: Kuwait Institute for Scientific Research, ISBN:978-99966-37-20-9.

Al-Yamani, F.Y., Saburova, M., and I. Polikarpov. 2012. A preliminary assessment of harmful algal blooms in Kuwait's marine environment. *Aquat Ecosyst Health Manag* 15, Suppl. 1: 64–72. doi:10.1080/14634988.2012.679450.

Anderson, D.M., Boerlage, S.F.E., and M.B. Dixon (Eds.). 2017. *Harmful algal blooms (HABs) and desalination: A guide to impacts, monitoring and management*. IOC/2017/MG/78. Paris: Intergovernmental Oceanographic Commission of UNESCO.

Anderson, D.M., Cembella, A.D., and G.M. Hallegraeff. 2012. Progress in understanding harmful algal blooms: Paradigm shifts and new technologies for research, monitoring, and management. *Annu Rev Mar Sci* 4: 143–176. doi:10.1146/annurev-marine-120308-081121.

Anderson, D.M., Glibert, P.M., and J.M. Burkholder. 2002. Harmful algal blooms and eutrophication: Nutrient sources, composition, and consequences. *Estuaries* 25: 704–726. doi:10.1007/BF02804901.

Aqrawi, A.A.M. 2001. Stratigraphic signatures of climatic change during the Holocene evolution of the Tigris-Euphrates delta, Lower Mesopotamia. *Glob Planet Change* 28: 267–283. doi:10.1016/S0921-8181(00)00078-3.

Attaran-Fariman, G., and M. Ali Asefi. 2022. Checklist of phytoplankton of the tropical Persian Gulf and Sea of Oman. *Nova Hedwigia* 114: 251–301. doi:10.1127/nova_hedwigia/2022/0687.

Baliarsingh, S.K., Lotliker, A.A., Trainer, V.L., et al. 2016. Environmental dynamics of red *Noctiluca scintillans* bloom in tropical coastal waters. *Mar Pollut Bull* 111, nos. 1–2: 277–286. doi:10.1016/j.marpolbul.2016.06.103.

Banguera-Hinestroza, E., Eikrem, W., Mansour, H., et al. 2016. Seasonality and toxin production of *Pyrodinium bahamense* in a Red Sea lagoon. *Harmful Algae* 55: 163–171. doi:10.1016/j.hal.2016.03.002.

Berdalet, E., Fleming, L.E., Gowen, R., et al. 2015. Marine harmful algal blooms, human health and wellbeing: Challenges and opportunities in the 21st century. *J Mar Biol Assoc UK* 96: 61–91. doi:10.1017/s0025315415001733.

Brewin, R.J.W., Raitsos, D.E., Dall'Olmo, G., et al. 2015. Regional ocean-colour chlorophyll algorithms for the Red Sea. *Remote Sens Environ* 165: 64. doi:10.1016/j.rse.2015.04.024.

Capone, D.G., Burns, J.A., Montoya, J.P., et al. 2005. Nitrogen fixation by *Trichodesmium* spp.: An important source of new nitrogen to the tropical and subtropical North Atlantic Ocean. *Glob Biogeochem Cy* 19: GB2024. doi:10.1029/2004GB002331.

Catania, D., Richlen, M.L., Mak, Y.L., et al. 2017. The prevalence of benthic dinoflagellates associated with ciguatera fish poisoning in the central Red Sea. *Harmful Algae* 68: 206–216. doi:10.1016/j.hal.2017.08.005.

Chinain, M., Gatti, C., Darius, H., et al. 2020. Ciguatera poisonings: A global review of occurrences and trends. *Harmful Algae* 102: 101873. doi:10.1016/j.hal.2020.101873.

Chisholm, S.W. 1992. Phytoplankton size. In: *Primary productivity and biogeochemical cycles in the sea*. Eds: P.G. Falkowski, and A.D. Woodhead, 213–237. New York: Plenum Press. doi:10.1007/978-1-4899-0762-2_12.

De Haro, L., Pommier, P., and M. Valli. 2003. Emergence of imported ciguatera in Europe: Report of 18 cases at the poison control centre of Marseille. *J Toxicol Clin Toxicol* 41: 927–930. doi:10.1081/CLT-120026512.

Devassy, R.P., El-Sherbiny, M.M., Al-Sofyani, A.M., and A.M. Al-Aidaroos. 2017. Spatial variation in the phytoplankton standing stock and diversity in relation to the prevailing environmental conditions along the Saudi Arabian coast of the northern Red Sea. *Mar Biodiv* 47: 995–1008. doi:10.1007/s12526-017-0693-4.

Devlin, M.J., Massoud, M.S., Hamid, S.A., et al. 2015. Changes in the water quality conditions of Kuwait's marine waters: Long term impacts of nutrient enrichment. *Mar Pollut Bull* 100: 607–620. doi:10.1016/j.marpolbul.2015.10.022.

Dreano, D., Raitsos, D.E., Gittings, J., et al. 2016. The Gulf of Aden intermediate water intrusion regulates the southern Red Sea summer phytoplankton blooms. *PLoS One* 11: e0168440. doi:10.1371/journal.pone.0168440.

Edwards, F.J. 1987. Climate and oceanography. In: *Key environments: Red Sea*. Eds: A.J. Edwards, and S.M. Head, 45–70. Oxford: Pergamon Press.

Ehrenberg, C. 1830. Neue Beobachlungen über blutartige Erscheinungen in Aegypten, Arabien und Sibirien, nebst einer Uebersicht und Kritik der früher bekannnten. *Ann Phys* 94: 477–514.

El Hag, A.G.D., and D.H. Nasr. 1989. Studies in planktonic populations in Port Sudan coastal area. *Sudan J Sci* 4: 12–26.

El-Sherif, Z.M., and S.M. Aboul Ezz. 2000. Check list of plankton of the northern Red Sea. *Pakistan J Mar Sci* 9: 61–78.

Esmaeili, F.M., Mortazavi, M.S., Banadaki, A.D., et al. 2021. Algal blooms historical outbreaks in the northern coastal waters of the Persian Gulf and Oman Sea (1980–2015). *Environ Monit Assess* 193: 648. doi:10.1007/s10661-021-09413-3.

Fernández, A., Mouriño-Carballido, B., Bode, A., et al. 2010. Latitudinal distribution of *Trichodesmium* spp. and N_2 fixation in the Atlantic Ocean. *Biogeosci* 7: 3167–3176. doi:10.5194/bg-7-3167-2010.

Friedman, M., Fleming, L., Fernandez, M., et al. 2008. Ciguatera fish poisoning: Treatment, prevention and management. *Mar Drugs* 6: 456–479. doi:10.3390%2Fmd20080022.

Ghanea, M., Moradi, M., and K. Kabiri. 2016. A novel method for characterizing harmful algal blooms in the Persian Gulf using MODIS measurements. *Adv Space Res* 58: 1348–1361. doi:10.1016/j.asr.2016.06.005.

Gittings, J.A., Raitsos, D.E., Brewin, R.J.W., et al. 2021. Links between phenology of large phytoplankton and fisheries in the northern and central Red Sea. *Remote Sens* 13: 231. doi:10.3390/rs13020231.

Glibert, P.M., Landsberg, J.H., Evans, J.J., et al. 2002. A fish kill of massive proportion in Kuwait Bay, Arabian Gulf, 2001: The roles of bacterial disease, harmful algae and eutrophication. *Harmful Algae* 1: 215–231. doi:10.1016/S1568-9883(02)00013-6.

Gokul, E.A., Raitsos, D.E., Gittings, J.A., et al. 2019. Remotely sensing harmful algal blooms in the Red Sea. *PLoS One* 14: e0215463. doi:10.1371%2Fjournal.pone.0215463.

Gomaa, M.N., Hannachic, I., Carmichaeld, W.W., et al. 2018. Low diversity triggers harmful algae bloom (HAB) occurrence adjacent to desalination plants along the Red Sea. *Desalin Water Treat* 114: 1–12. doi:10.5004/dwt.2018.22323.

Guiry, M.D., and G.M. Guiry. 2022. *AlgaeBase*. World-wide electronic publication. Galway: National University of Ireland. www.algaebase.org; downloaded 24 August 2022.

Hallegraeff, G.M. 2003. Harmful algal blooms: A global overview. In: *Manual on harmful marine microalgae*. Eds: G.M. Hallegraeff, D.M. Anderson, and A.D. Cembella, vol. 11, 2nd ed., 25–49. Paris: IOC-UNESCO.

Hamzehei, S., Bidokhti, A.A., Mortazavi, M.S., et al. 2013. Red tide monitoring in the Persian Gulf and Gulf of Oman using MODIS sensor data. *Tech J Eng Appl Sci* 3: 1100–1107.

Hasanean, H., and M. Almazroui. 2015. Rainfall: Features and variations over Saudi Arabia, a review. *Climate* 3, no. 3: 578–626. doi:10.3390/cli3030578.

Heil, C.A., Glibert, P.M., Al-Sarawi, M.A., et al. 2001. First record of a fish-killing *Gymnodinium* sp. bloom in Kuwait Bay, Arabian Sea: Chronology and potential causes. *Mar Ecol Prog Ser* 214: 15–23. doi:10.3354/meps214015.

Hoyt, S.F. 1912. The name of the Red Sea. *J Am Orient Soc* 32: 115–119.

Ismael, A.A. 2015. Phytoplankton of the Red Sea. In: *The Red Sea: The formation, morphology, oceanography and environment of a young ocean basin*. Eds: N.M.A. Rasul, and I.C.F. Stewart, 567–583. Berlin Heidelberg: Springer Earth System Sciences. doi:10.1007/978-3-662-45201-1_32.

Ismail, W.A., Al-Yamani, F.Y., and K.S. Al-Rifaie. 2007. Field surveys and a perturbation experiment in testing the role of eutrophication in initiating red tides in Kuwait Bay. *Int J Oceans Oceanogr* 2: 187–211.

John, V.C., Coles, S.L., and A.I. Abozed. 1990. Seasonal cycle of temperature, salinity and water masses of the western Arabian Gulf. *Oceanol Acta* 13: 273–281.

Johns, W.E., Jacobs, G.A., Kindle, J.C., et al. 1999. *Arabian marginal seas and gulfs: Report of a workshop held at Stennis space center*. Technical Report 2000–01. 11–13 May. Mississippi: University of Miami RSMAS.

Johnson, M.D., Tengs, T., Oldach, D., et al. 2006. Sequestration, performance, and functional control of cryptophyte plastids in the ciliate *Myrionecta rubra* (Ciliophora). *J Phycol* 42: 1235–1246. doi:10.1111/j.1529-8817.2006.00275.x.

Jones, D.A., Price, A.R.G., Al-Yamani, F., and A. Al-Zaidan. 2002. Coastal and marine ecology. In: *The Gulf ecosystem: Health and sustainability*. Eds: N.Y. Khan, M. Munawar, and A.R.G. Price, 65–104. Leiden, the Netherlands: Backhuys Publishers.

Kheireddine, M., Mayot, N., Ouhssain, M., et al. 2021. Regionalization of the Red Sea based on phytoplankton phenology: A satellite analysis. *J Geophys Res Oceans* 126: e2021JC017486. doi:10.1029/2021JC017486.

Kim, D.-I., Matsuyama, Y., Nagasoe, S., et al. 2004. Effects of temperature, salinity and irradiance on the growth of the harmful red tide dinoflagellate *Cochlodinium polykrikoides* Margalef (Dinophyceae). *J Plankton Res* 26: 61–66. doi:10.1093/plankt/fbh001.

Kudela, R.M., Berdalet, E., Bernard, S., et al. 2015. *Harmful algal blooms. A scientific summary for policy makers*. Paris: IOC/UNESCO, IOC/INF-1320.

Kudela, R.M., and C.J. Gobler. 2012. Harmful dinoflagellate blooms caused by *Cochlodinium* sp.: Global expansion and ecological strategies facilitating bloom formation. *Harmful Algae* 14: 71–86. doi:10.1016/j.hal.2011.10.015.

Kumar, P.S. 2021. *Modern treatment strategies for marine pollution*. Oxford: Elsevier (Academic Press). doi:10.1016/C2019-0-04371-3.

Kürten, B., Khomayis, H.S., Devassy, R., et al. 2015. Ecohydrographic constraints on biodiversity and distribution of phytoplankton and zooplankton in coral reefs of the Red Sea, Saudi Arabia. *Mar Ecol* 36: 1195–1214. doi:10.1111/maec.12224.

Lakshmi, R.S., Prakash, S., Lotliker, A.A., et al. 2021. Physicochemical controls on the initiation of phytoplankton bloom during the winter monsoon in the Arabian Sea. *Sci Rep* 11: 13448. doi:10.1038/s41598-021-92897-3.

Lassus, P., Chomérat, N., Hess, P., and E. Nézan. 2016. *Toxic and harmful microalgae of the World Ocean/Micro-algues toxiques et nuisibles de l'océan mondial*, vol. 68. International Society for the Study of Harmful Algae/Intergovernmental Oceanographic Commission of UNESCO. Denmark: IOC Manuals and Guides.

Lindell, D., and A.F. Post. 1995. Ultraphytoplankton succession is triggered by deep winter mixing in the Gulf of Aqaba (Eilat), Red Sea. *Limnol Oceanogr* 40: 1130–1141. doi:10.4319/lo.1995.40.6.1130.

Longhurst, A. 2007. *Ecological geography of the sea*. London: Academic Press.

Lotliker, A.A., Baliarsingh, S.K., Samanta, A., et al. 2020. Growth and decay of high-biomass algal bloom in the northern Arabian Sea. *J Indian Soc Remote Sens* 48: 465–471. doi:10.4319/lo.1995.40.6.1130.

Lundholm, N., Churro, C., Fraga, S., Hoppenrath, M., Iwataki, M., Larsen, J., Mertens, K., Moestrup, Ø, Murray, S., Tillmann, U., and Zingone, A. (Eds.). 2009 onwards. *IOC-UNESCO taxonomic reference list of harmful micro algae*. www.marinespecies.org/hab on 2024–03–05. doi:10.14284/362.

Mohamed, Z.A. 2018. Potentially harmful microalgae and algal blooms in the Red Sea: Current knowledge and research needs. *Mar Environ Res* 140: 234–242. doi:10.1016/j.marenvres.2018.06.019.

Mohamed, Z.A., and A.M. Al-Shehri. 2012. The link between shrimp farm runoff and blooms of toxic *Heterosigma akashiwo* in Red Sea coastal waters. *Oceanologia* 54: 287–309. doi:10.5697/oc.54-2.287.

Mohamed, Z.A., and I. Mesaad. 2007. First report on *Noctiluca scintillans* blooms in the Red Sea off the coasts of Saudi Arabia: Consequences of eutrophication. *Oceanologia* 49: 337–351.

Mohebbi-Nozar, S.L., Mortazavi, M.S., Seraji, F., and P. Bahreini. 2022. Health risk assessment of okadaic acid and domoic acid in some edible bivalves from Hormozgan Province in the north of Persian Gulf. *Pollution* 8, no. 3: 934–947. doi:10.22059/POLL.2022.339519.1379.

Nassar, M.Z.A., and H.M. Khairy. 2014. Checklist of phytoplankton species in the Egyptian waters of the Red Sea and some surrounding habitats (1990–2010). *Ann Res Rev Biol* 4: 3566–3585. doi:10.9734/ARRB/2014/10866.

Neelamani, S., Al-Osairi, Y., Al-Salem, K., et al. 2021. Some physical oceanographic aspects of Kuwait and Arabian Gulf marine environment. In: *The Arabian seas: Biodiversity, environmental challenges and conservation measures*. Ed: L.A. Jawad, 99–119. Cham, Switzerland: Springer. doi:10.1007/978-3-030-51506-5_5.

Nezlin, N.P., Polikarpov, I.G., Al-Yamani, F.Y., et al. 2010. Satellite monitoring of climatic factors regulating phytoplankton variability in the Arabian (Persian) Gulf. *J Mar Syst* 82: 47–60. doi:10.1016/j.jmarsys.2010.03.003.

Olalekan, A.A., and K. Malik. 2015. Application of Giovanni for rapid assessment of harmful algal blooms in the Arabian Gulf. *Arab J Geosci* 8: 8767–8775. doi:10.1007/s12517-015-1826-3.

Osama, S.S., and A.G. Amjed. 2020. A preliminary study on phytoplankton composition at Bashayer II marine oil terminal, Red Sea, Sudan. *J East China Norm Univ Natur Sci Ed* S1: 89–93. doi:10.3969/j.issn.1000-5641.202092209.

Park, M.G., Kim, S., Kim, H.S., et al. 2006. First successful culture of the marine dinoflagellate *Dinophysis acuminata*. *Aquat Microb Ecol* 45: 101–106. doi:10.3354/ame045101.

Pavaux, A.-S., Berdalet, E., and R. Lemée. 2020. Chemical ecology of the benthic dinoflagellate genus *Ostreopsis*: Review of progress and future directions. *Front Mar Sci* 7: 498. doi:10.3389/fmars.2020.00498.

Piontkovski, S.A., Al-Gheilani, H.M.H., Jupp, B.P., et al. 2012. Interannual changes in the Sea of Oman ecosystem. *Open Mar Biol J* 6: 38–52. doi:10.2174/1874450801206010038.

Piontkovski, S.A., Hamza, W.M., Al-Abri, N.M., et al. 2019. A comparison of seasonal variability of Arabian Gulf and the Sea of Oman pelagic ecosystems. *Aquat Ecosyst Health Manag* 22: 108–130. doi:10.1080/14634988.2019.1621133.

Polikarpov, I., Al-Yamani, F.Y., Petrov, P., et al. 2021. Phytoplankton bloom detection during the COVID-19 lockdown with remote sensing data: Using Copernicus Sentinel-3 for north-western Arabian/Persian Gulf case study. *Mar Pollut Bull* 171: 112734. doi:10.1016/j.marpolbul.2021.112734.

Polikarpov, I., Al-Yamani, F.Y., and M. Saburova. 2009. Space-time variability of phytoplankton structure and diversity in the north-western part of the Arabian Gulf (Kuwait's waters). *BioRisk* 3: 83–96. doi:10.3897/biorisk.3.8.

Polikarpov, I., Al-Yamani, F.Y., and M. Saburova. 2019. Remote sensing of phytoplankton variability in the Arabian/Persian Gulf. In: *Remote sensing of the Asian seas*. Eds: V. Barale, and M. Gade, 485–501. Cham, Switzerland: Springer. doi:10.1007/978-3-319-94067-0_27.

Polikarpov, I., Saburova, M., and F. Al-Yamani. 2016. Diversity and distribution of winter phytoplankton in the Arabian Gulf and the Sea of Oman. *Cont Shelf Res* 119: 85–99. doi:10.1016/j.csr.2016.03.009.

Polikarpov, I., Saburova, M., and F. Al-Yamani. 2020. Decadal changes in diversity and occurrence of microalgal blooms in the NW Arabian/Persian Gulf. *Deep Sea Res Part II: Top Stud Oceanogr* 179: 104810. doi:10.1016/j.dsr2.2020.104810.

Post, A.F., Dedej, Z., Gottlieb, R., et al. 2002. Spatial and temporal distribution of *Trichodesmium* spp. in the stratified Gulf of Aqaba, Red Sea. *Mar Ecol Prog Ser* 239: 241–250. doi:10.3354/meps239241.

Prabowo, D.A., and S. Agusti. 2019. Free-living dinoflagellates of the central Red Sea, Saudi Arabia: Variability, new records and potentially harmful species. *Mar Pollut Bull* 141: 629–648. doi:10.1016/j.marpolbul.2019.03.012.

Qurban, M., Wafar, M., and M. Heinle. 2019. Phytoplankton and primary production in the Red Sea. In: *Oceanographic and biological aspects of the Red Sea*. Eds: N.M.A. Rasul, and I.C.F. Stewart, 491–506. Cham, Switzerland: Springer Nature. doi:10.1007/978-3-319-99417-8_27.

Raitsos, D.E., Hoteit, I., Prihartato, P.K., et al. 2011. Abrupt warming of the Red Sea. *Geophys Res Lett* 38: L14601. doi:10.1029/2011GL047984.

Raitsos, D.E., Pradhan, Y., Brewin, R.J., et al. 2013. Remote sensing the phytoplankton seasonal succession of the Red Sea. *PLoS One* 8: e64909. doi:10.1371/journal.pone.0064909.

Raitsos, D.E., Xing, Y., Platt, T., et al. 2015. Monsoon oscillations regulate fertility of the Red Sea. *Geophys Res Lett* 42: 855–862. doi:10.1002/2014GL062882.

Rasul, N.M.A., and I.C.F. Stewart (Eds.). 2019. *Oceanographic and biological aspects of the Red Sea*. Cham, Switzerland: Springer Nature. doi:10.1007/978-3-319-99417-8.

Rasul, N.M.A., Stewart, I.C.F., and Z.A. Nawab. 2015. Introduction to the Red Sea: Its origin, structure, and environment. In: *The Red Sea: The formation, morphology, oceanography and environment of a young ocean basin*. Eds: N.M.A. Rasul, and I.C.F. Stewart, 1–28. Berlin Heidelberg: Springer Earth System Sciences. doi:10.1007/978-3-662-45201-1_1.

Reynolds, M. 1993. Physical oceanography of the Gulf, Strait of Hormuz, and the Gulf of Oman: Results from the Mt Mitchell expedition. *Mar Pollut Bull* 27: 35–39. doi:10.1016/0025-326X(93)90007-7.

Richlen, M.L., Morton, S.L., Jamali, E.A., et al. 2010. The catastrophic 2008–2009 red tide in the Arabian Gulf region, with observations on the identification and phylogeny of the fish-killing dinoflagellate *Cochlodinium polykrikoides*. *Harmful Algae* 9: 163–172. doi:10.1016/j.hal.2009.08.013.

Saburova, M. 2022. *Biodiversity of benthic dinoflagellates in the northern Red Sea*. 12th International Conference on Modern and Fossil Dinoflagellates, 4–8 July 2022, Las Palmas de Gran Canaria, Spain, Abstracts.

Saburova, M., Al-Kandari, M., Polikarpov, I., et al. 2022. Alien toxic dinoflagellate *Heterocapsa circularisquama* from the western pacific in Kuwait, NW Indian Ocean. *Deep Sea Res Part II: Top Stud Oceanogr* 196: 105027. doi:10.1016/j.dsr2.2022.105027.

Saburova, M., Polikarpov, I., and F. Al-Yamani. 2012. First record of *Kryptoperidinium foliaceum* (Dinophyceae, Peridiniales) from a hypersaline environment in Kuwait, northwestern Arabian Gulf. *Mar Biodiv Rec* 5: e104. doi:10.1017/S1755267212000838.

Saburova, M., Polikarpov, I., and F. Al-Yamani. 2013. New records of the genus *Gambierdiscus* in marginal seas of the Indian Ocean. *Mar Biodiv Rec* 6: e91. doi:10.1017/S1755267213000675.

Salarpouri, A., Kamrani, E., Kaymaram, F., et al. 2018. Essential fish habitats (EFH) of small pelagic fishes in the north of the Persian Gulf and Oman Sea, Iran. *Iran J Fish Sci* 17: 74–94.

Sheppard, C., Al-Husiani, M., Al-Jamali, F., et al. 2010. The Gulf: A young sea in decline. *Mar Pollut Bull* 60: 13–38. doi:10.1016/j.marpolbul.2009.10.017.

Śliwińska-Wilczewska, S., Maculewicz, J., Felpeto, A.B., et al. 2018. Allelopathic and bloom-forming picocyanobacteria in a changing world. *Toxins* 10: 48. doi:10.3390/toxins10010048.

Sofianos, S.S., and W.E. Johns. 2002. An Oceanic General Circulation Model (OGCM) investigation of the Red Sea circulation. 1. Exchange between the Red Sea and the Indian Ocean. *J Geophys Res Oceans* 107, no. C11: 3196. doi:10.1029/2001JC001184.

Sofianos, S.S., Johns, W.E., and S.P. Murray. 2002. Heat and freshwater budgets in the Red Sea from direct observations at Bab el Mandeb. *Deep Sea Res Part II: Top Stud Oceanogr* 49: 1323–1340. doi:10.1016/S0967-0645(01)00164-3.

Sombrito, E.Z., Honrado, M.C.V., de Vera, A., et al. 2007. Use of *Perna viridis* as a bioindicator of paralytic shellfish toxins at low *Pyrodinium bahamense* var. *compressum* density using a radioreceptor assay. *Environ Bioindic* 2, no. 4: 264–272. doi:10.1080/15555270701693612.

Stambler, N. 2005. Bio-optical properties of the northern Red Sea and the Gulf of Eilat (Aqaba) during winter 1999. *J Sea Res* 54: 186–203. doi:10.1016/j.seares.2005.04.006.

Stoecker, D.K., Putt, M., Davis, L.H., et al. 1991. Photosynthesis in *Mesodinium rubrum*: Species-specific measurements and comparison to community rates. *Mar Ecol Prog Ser* 73: 245–252. doi:10.3354/MEPS073245.

Subba Rao, D.V., Al-Hassan, J.M., Al-Yamani, F., et al. 2003. Elusive red tides in Kuwait coastal waters. *Harmful Algae News* 24: 10–13.

Subba Rao, D.V., and F. Al-Yamani. 1998. Phytoplankton ecology in the water between Shatt Al-Arab and the Straits of Hormuz, Arabian Gulf: A review. *Plankton Biol Ecol* 45: 101–116.

Subba Rao, D.V., Al-Yamani, F., Lennox, A., et al. 1999. Biomass and production characteristics of the first red tide noticed in Kuwait Bay, Arabian Gulf. *J Plankton Res* 21, no. 4: 805–810. doi:10.1093/plankt/21.4.805.

Touliabah, H.E., Abu El-Kheir, W.S., Kuchari, M.G., and N.I.H. Abdulwassi. 2010. Phytoplankton composition at Jeddah coast – Red Sea, Saudi Arabia in relation to some ecological factors. *J King Abdulaziz Univ Sci* 22: 115–131. doi:10.4197/SCI.22-1.9.

Wiggert, J., Hood, R., Banse, K., et al. 2005. Monsoon-driven biogeochemical processes in the Arabian Sea. *Prog Oceanogr* 65: 176–213. doi:10.1016/j.pocean.2005.03.008.

Wilkerson, F.P., and G. Grunseich. 1990. Formation of blooms by the symbiotic ciliate *Mesodinium rubrum*: The significance of nitrogen uptake. *J Plankton Res* 12: 973–989. doi:10.1093/plankt/12.5.973.

Yao, F., and I. Hoteit. 2015. Thermocline regulated seasonal evolution of surface chlorophyll in the Gulf of Aden. *PLoS One* 10, no. 3: e0119951. doi:10.1371/journal.pone.0119951.

Zhao, J., and H. Ghedira. 2014. Monitoring red tide with satellite imagery and numerical models: A case study in the Arabian Gulf. *Mar Pollut Bull* 79: 305–313. doi:10.1016/j.marpolbul.2013.10.057.

Zhao, J., Temimi, M., and H. Ghedira. 2015. Characterization of harmful algal blooms (HABs) in the Arabian Gulf and the Sea of Oman using MERIS fluorescence data. *ISPRS J Photogramm Remote Sens* 101: 125–136. doi:10.1016/j.isprsjprs.2014.12.010.

Phytoplankton community dynamics in the western Arabian Gulf

20

Spatial and seasonal trends

Kolchar M. Rajaneesh, Thadickal V. Joydas, Moritz Heinle, Shamseer Pavithratha, Mohamed Asharaf, Karuppasamy P. Manikandan, Mohammed A. Qurban and Abdulaziz Al-Suwailem

20.1 INTRODUCTION

Planktonic components are vital in the transformation of energy in aquatic ecosystems (Bar-On and Milo 2019). Among these planktonic components, phytoplankton are believed to sustain the health of the ecosystem as they are the primary producers and serve as food to higher trophic levels (Azam et al. 1983). Among the phytoplankton, dinoflagellates and diatoms, belonging to Kingdom Protista, are the dominant groups in marine environments. Some dinoflagellates possess an interesting characteristic in their mode of nutrition, as they can switch between autotrophy and heterotrophy, depending on the environmental conditions (Tomas 1997). These mixotrophic dinoflagellates can survive in habitats with low nutrient concentrations by switching to phagotrophy whereas, under conditions of low availability of particulate food, they survive by switching to photosynthesis (Jost et al. 2004). Diatoms are either centric or pennate forms, depending on their symmetry. Centric diatoms have radial symmetry, whereas pennate diatoms have bilateral symmetry (Tomas 1997). The third important group of phytoplankton are cyanobacteria, referred to as the "blue-green algae." They played a pivotal role in the evolution of life on earth by creating an oxygenic atmosphere through photosynthesis (Bendall et al. 2008). All phytoplankton are controlled by the prevailing environmental conditions such as temperature, salinity, nutrient availability, the flow of currents, precipitation, wind patterns, etc. (Cloern 1979; Trombetta et al. 2019). These parameters can alter the phytoplankton community structure in any given ecosystem.

The Arabian Gulf (AG) is a semi-enclosed marginal sea with a unique geographical setting and highly arid climate among all marginal seas. This inland sea is a 251,000 km^2 shallow basin with an average depth of 35–40 m and a maximum depth of 90 m. The AG is connected to the Gulf of Oman and the Arabian Sea through the narrow bottleneck of the Strait of Hormuz (Sheppard 2016). The AG experiences wide ranges in some environmental conditions, such as temperature (<12 °C to 36 °C) and salinity (<30‰ at the Shatt al-Arab River (SAR) mouth to >60‰ in the Gulf of Salwah) (Al-Yamani et al. 2004; Riegl and Purkis 2012; Devlin et al. 2015). It is also under severe anthropogenic pressure. The AG receives a freshwater discharge in the north via the SAR estuary, with a peak during March–July. However, hydrological interventions along the SAR and its precursors, the Euphrates and Tigris, in the recent past have considerably reduced the amount of freshwater that reaches the NW AG (Isave and Mikhailova 2009). The widespread activities associated with the enormous oil and gas exploration and processing industry apply immense pressure on the Gulf's marine ecosystem. The world's largest oil spill (1991 Oil Spill) occurred in the AG. The oil spill affected the coastal areas located between Ras-Al-Khafji in the north and Abu Ali to the south on the Saudi Arabian coast (Figure 20.1; Fowler et al. 1993; Readman et al. 1996; Price 1998). Offshore areas claimed for urbanization are a major threat to the marine environment (Loughland et al. 2012) in the AG.

Studies reported from the western AG (e.g., Al-Kaisi 1977; Al-Kandari et al. 2009; Al-Yamani et al. 2004; Azis et al. 2003; Nour El-Din and Al-Khayat 2005; Quigg et al. 2013) revealed the importance of environmental parameters, such as temperature, salinity, nutrients, dust deposition, wind and water currents, in governing the distribution of phytoplankton communities in the region. The seasonal and inter-annual variabilities in phytoplankton are correlated with periods of low precipitation, associated with elevated levels of dust deposition over the water surface, also causing phytoplankton blooms in the open waters (Singh et al. 2008; Nezlin et al. 2010). Although

DOI: 10.1201/9781003321392-25

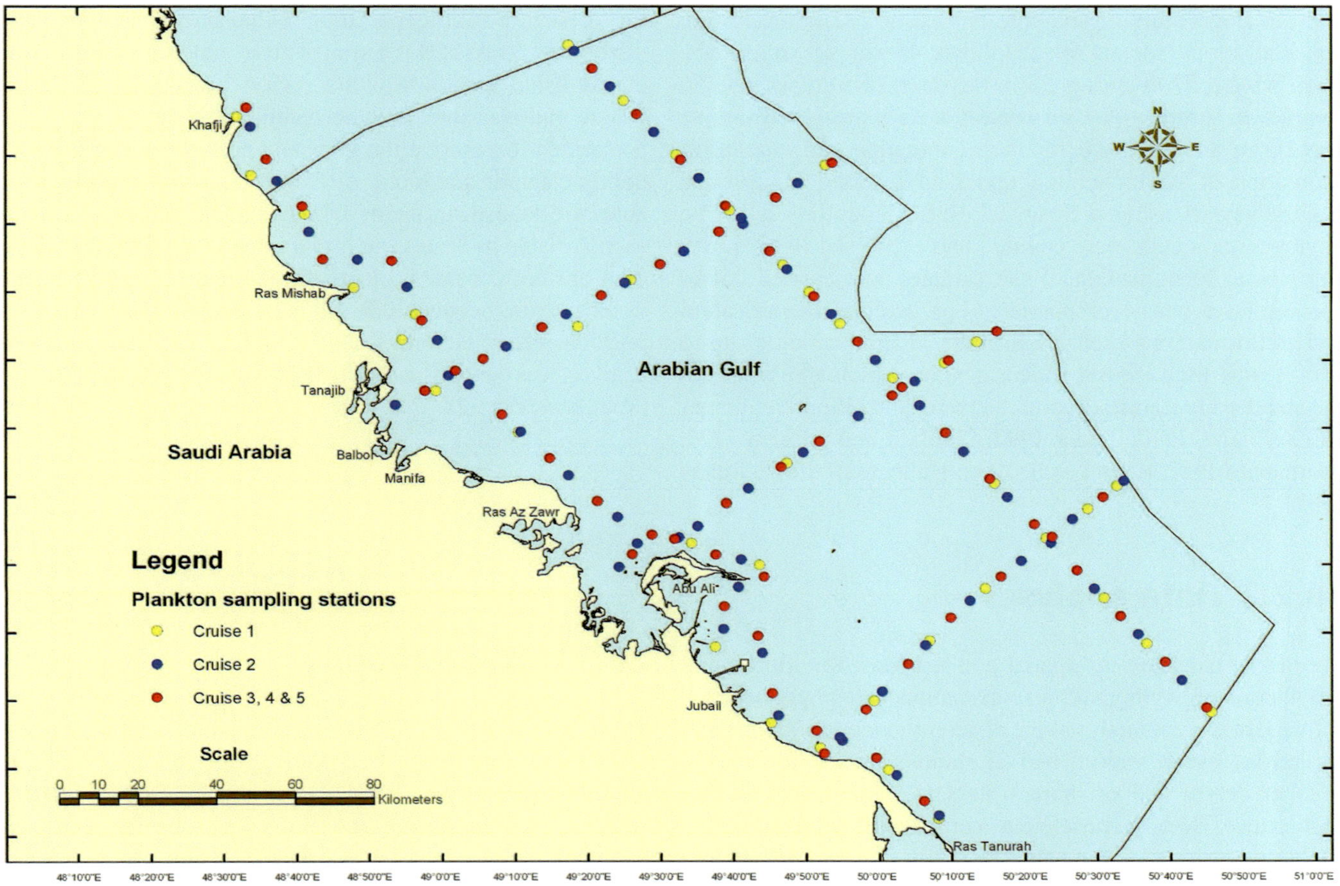

FIGURE 20.1 Location of sampling stations in the western Arabian Gulf.

several phytoplankton studies were conducted in the western AG, these studies are infrequent, geographically restricted, and insufficient to give a broad picture of phytoplankton abundance and composition in the AG, especially in Saudi Arabian waters. The scarce knowledge of the primary producers limits our precise understanding of biological productivity, diversity, and trophic successions both qualitatively and quantitatively for this area.

In the present study, a detailed survey of phytoplankton community structure and composition on a large spatial and temporal scale was carried out for the first time in the Saudi Arabian waters of the AG, and the results were compared with earlier studies to assess the long-term impact of the 1991 oil spill, if any. This study also provides essential new information on the structure and functioning of phytoplankton communities and how they are driven by natural gradients of environmental conditions in this region. Moreover, knowing potential harmful algal species is necessary to correctly plan and implement mitigation and/or response measures.

20.2 MATERIALS AND METHODS

20.2.1 Sampling strategy

Sampling was carried out in the Saudi Arabian waters of the AG between latitudes 26°40′N and 28°40′N and longitudes 48°40′E and 50°45′E (between Al-Khafji and Ras-Tanurah; Figure 20.1). Five oceanographic cruises were conducted using the research vessel *Al Gosaibi-I* during the months of November-December 2002 (late fall), January 2003 (winter), April-May 2003 (spring), June-July 2003 (onset of summer), and August 2003 (summer).

Temperature, salinity, dissolved oxygen (DO), and pH profiles were taken with a pre-calibrated CTD (Conductivity, Temperature, and Depth meter). Water samples for nutrient determination (NO_2, NO_3, NH_3, and PO_4) were analysed using a Skalar San++ SA 1074 Continuous Flow Auto-Analyzer by following the Instrument Manual and Methods of Seawater Analysis (Grasshoff et al. 1999). Analytical Quality Control calibration was carried out every day (or more frequently if needed) to check the performance of the instrument. The phytoplankton biomass (standing crop) was estimated using a CTD-mounted fluorometer (SCUFA-Self-Contained Underwater Fluorescence Apparatus, Turner Design). The minimum detection limit is 0.02 mg m^{-3} for chlorophyll *a* (chl *a*). Total petroleum hydrocarbons (TPH) in seawater samples were extracted twice with 50 mL mixtures of dichloromethane (DCM)/hexane (1:1 v/v) for 5 min, and the extracts were combined and concentrated to 10 mL using a Kuderna Danish concentrator. This was then further concentrated to 1 mL with purging nitrogen and replacing the solvent with hexane. TPH concentrations in this extract were measured in a gas chromatograph (HP Agilent 6890N) equipped with a flame ionization detector (GC/FID) following US EPA Method 8015 (2000).

Phytoplankton samples were collected by vertical hauls using a collection net. The haul duration varied depending upon the depth sampled. The phytoplankton samples were fixed in Lugol's Iodine solution. A total of 272 samples were collected for phytoplankton as follows: 45 samples in late fall, 56 in winter, 57 in spring, 58 in the onset of summer, and 56 samples in summer. A 1 ml aliquot of each preserved sample was taken and analysed for the quantitative and qualitative estimation of phytoplankton under an inverted microscope (Leitz-Diavert) using a Sedgwick-Rafter counting cell. The phytoplankton cells were counted and expressed in cells m^{-3}. Taxonomic identification of the samples was carried out to species level whenever possible. For detailed identification and conformation of species, a higher magnification of up to 1000X was used. Species identification was made based on the standard literature (Subrahmanyan 1946, 1968; Steidinger and Williams 1970; Taylor 1992; Guillard and Kilham 1977; Santhanam 1987; Round et al. 1990; Hallegreaff 2003; Tomas 1997).

20.2.2 Data analyses

In order to elucidate the spatial and temporal variation in the distribution of phytoplankton, correlations with physicochemical variables were analysed for different modes: 1) all stations pooled for every season; and 2) stations pooled for coastal (<25 m depth) and offshore waters (>25 m depth), during each cruise. Both univariate and non-parametric multivariate analyses were carried out using PRIMER software (Plymouth Routines in Multivariate Ecological Research) (version 6.1.5) (Clarke et al. 2001) to study the spatial and seasonal variations in the phytoplankton community. The Q-mode (site similarity) analysis using the Bray-Curtis similarity was applied to double square-root-transformed species-abundance data to group samples with similar community composition or ecological conditions, following the procedure described by Clarke and Warwick (1994). One-way analysis of similarity (ANOSIM-1) (Clarke and Ainsworth 1993) was applied to assess the significance of differences in the composition of biological assemblages between the areas/regions over time. The similarity percentages program (SIMPER) was used to determine species characterizing clusters of stations identified in the cluster analysis and Multidimensional Scaling (MDS) ordinations. Indicator/discriminating species were determined based on consistency in their distribution between the groups.

The species diversity of each cluster was examined using univariate diversity indices such as Shannon-Weiner (H′), Simpson (λ′), Margalef (d), Pielou's evenness (J′), and V statistics (Caswell's neutral model), using the sub-routine DIVERSE of PRIMER. Multivariate diversity indices (based on Linnaean classification), as suggested by Clarke et al. (2001), using taxonomic relatedness of the species (taxonomic diversity (Δ), average taxonomic distinctness (Δ+), and variation in taxonomic distinctness (λ+)), were also used to delineate the status of different biological communities. Average taxonomic distinctness is the mean path length through the taxonomic tree, connecting every pair of species in the list. Variation in taxonomic distinctness is the variance of these pairwise path lengths and reflects the unevenness of the taxonomic tree. These two taxonomic diversity indices have the advantage of a lack of dependence on its mean value on the sample size. This feature is not shared by indices related to species richness. As with AvTD (Δ+), the VarTD (λ+) statistic for any local study can be tested for 'departure from expectation' by constructing a simulation distribution (say 95% CI range) from random subsets of the master list of species from that region (Clarke et al. 2001).

A multivariable analytic technique, factor analysis, was performed to describe the interrelationships among the set of multiple abiotic and biotic variables. The most important variables are loaded on factor 1 (x-axis). The value obtained for each variable indicates the importance of the variable. Values <0.4 are not considered, between 0.4 and 0.5 are considered weak factors, between 0.5 and 0.75 as moderate factors, and >0.75 as strong factors (Liu et al. 2003). The aforementioned analysis was done using SPSS Multi-variate Statistical Package (Windows Ver. 22).

20.3 RESULTS

20.3.1 Hydrography

There was a 14°C difference in temperature between winter (20°C) and summer (~34°C). During late fall and winter, the offshore waters were 2°C warmer than the coastal waters, whereas during spring, the onset of summer, and summer, coastal waters were warmer than the offshore waters (Figure 20.2a). The temperature did not vary much between the north and south during the study period. The salinity did not vary much between the seasons (Figure 20.2b). However, salinity was higher in the coastal waters (~42‰) than in the offshore waters (~40.8‰). The salinity decreased from the northern waters to southern waters. DO concentration was highest during winter (6.20 ± 0.24 mg l^{-1}) and decreased in the following seasons (Figure 20.2c). During summer, the DO concentration was lowest (4.69 ± 0.39 mg l^{-1}). Spatial variation was not observed in DO concentration during the study period. TPH concentration (112 ± 44 ppm) was highest during late fall, especially in the coastal and offshore waters (Figure 20.2d). During the rest of the season, the TPH concentration was 82 ± 21 ppm, with a peak at coastal stations (111 ± 44 ppm) during the onset of summer.

NO_2 concentrations did not vary significantly between the seasons and ranged from 0.28 to 24.37 ppb (Figure 20.2e). The lowest concentration was observed in coastal waters and the highest in offshore waters. NO_3 concentrations were lowest (65 ± 35 ppb) during late fall and winter (Figure 20.2f). It increased (101 ± 25 ppb) during spring and the onset of summer, followed by a decrease in summer (96 ± 41 ppb). There was no significant variation between coastal and offshore waters during the study period except in summer, where offshore water concentration was high compared to the coastal waters. NO_3 concentration was high in the northern waters during the onset of summer and spring. PO_4 concentrations were highest (17 ± 6 ppb) during winter and decreased in the following seasons, except in central waters during summer (Figure 20.2g). In the coastal waters, PO_4 concentration decreased from late fall to summer, whereas in the offshore waters an opposite trend was observed. NH_4 concentrations were low during winter (146 ± 65 ppb) and increased in the following seasons (Figure 20.2h). The highest NH_4 concentration

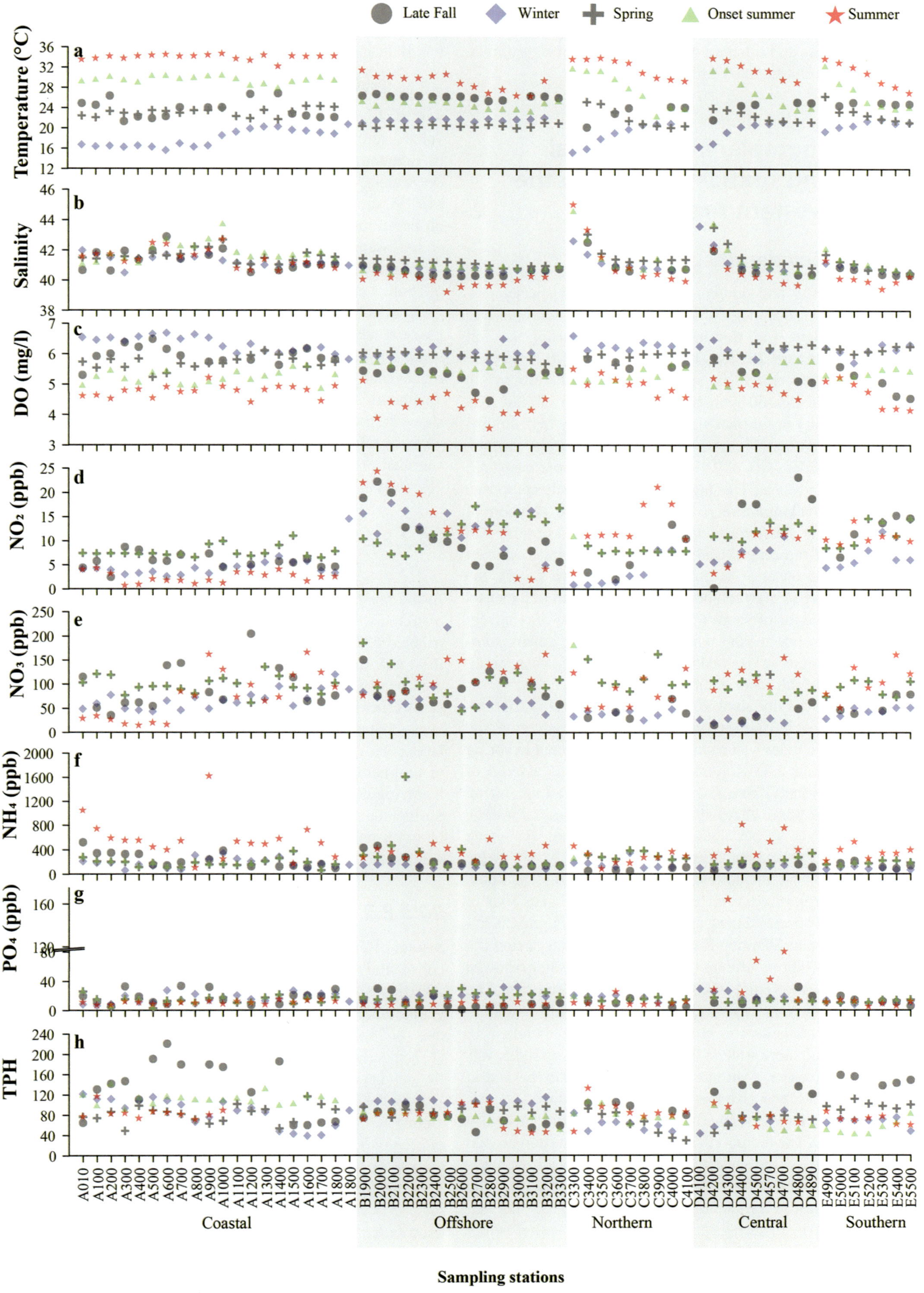

FIGURE 20.2 Variation in (a) temperature (°C), (b) salinity, (c) DO (mg/l), (d) NO_2 (ppb), (e) NO3 (ppb), (f) NH_4 (ppb), (g) PO_4 (ppb), and (h) TPH concentrations during different seasons (late fall, winter, spring, onset summer, and summer) at different sampling locations (coastal, offshore, northern, central southern) in the western Arabian Gulf.

(428 ± 245 ppb) was observed in summer. During spring and the onset of summer, the NH_4 concentration was high in the northern waters and offshore waters, whereas during summer, it was high in coastal and central waters.

20.3.2 Phytoplankton seasonal and spatial variation in the western Arabian Gulf

20.3.2.1 Cell density

The cell density and number of phytoplankton species collected from the oceanographic cruises showed seasonal and spatial variations. Cell density was very low during late fall and spring (Figure 20.3a). The mean cell density increased from 1.70 × 10^6 cells m^{-3} in spring to 8.32 × 10^6 cells m^{-3} in summer, but decreased again to 1.41× 10^6 cells m^{-3} in fall. The high cell density in summer was not supported by high chlorophyll concentrations (Figure 20.3a and 20.3b). During summer, very high cell densities of 7.49 × 10^7 cells m^{-3} were measured in the southern waters. This high density was mostly composed of *Alexandrium monilatum*, a toxic dinoflagellate. A bloom of the cyanophyte *Trichodesmium erythraeum* was observed in the surface of the central transect and offshore waters during spring and the onset of summer. In the southern waters, the coastal waters had high cell densities from late fall to the onset of summer, and the offshore waters had two peaks of relatively high densities, one during winter and the other during summer. The cell density was relatively low in spring, when strong stratification of the water column was apparent.

Species diversity showed seasonal and spatial variations and was highest during the spring. Across the study area, the mean Shannon index value was less than 2.5, except in spring when it reached 2.88 (Figure 20.3d). There was a trend of increasing diversity from coastal to offshore stations during the onset of the summer season. In the northern coastal waters, the species diversity per station was generally highest during spring except in some locations, i.e., Manifa and Jubail. These two locations had the lowest diversity during the summer. Similarly, the northern and central waters were also poor in the number of species during spring to summer (Figure 20.3d). The lowest mean Shannon diversity was recorded during the late fall (1.85). The central and southern coastal waters showed higher indices (>3) from winter to the onset of summer. The Shannon index was low (0.75) in summer south of Jubail, where the dinoflagellate *A. monilatum* was most abundant. In the southern offshore waters, the Shannon diversity index was low during the onset of summer and summer, and in the central waters, the index was also low (<1.0) during late fall.

Pielou's evenness showed spatial and temporal variations (Figure 20.3c). The temporal variations were highest in coastal waters throughout the seasons. A latitudinal species evenness trend was observed in the study area, i.e., high evenness in the northern and low in the southern transects during summer that marginally changed during spring, with relatively lower values in the other regions of the study site. However, Pielou's evenness increased from north to south along the coastal waters. Low values (<0.3) were observed near Abu Ali during late fall and summer. In the southern coastal waters, high evenness was recorded during spring and the onset of summer, whereas in the offshore waters, the southern transect showed low values (<0.3) during the onset of summer and summer.

The offshore waters had higher Average Taxonomic Distinctness (Delta+) than coastal waters in all seasons except winter (Figure 20.3e). The Average Taxonomic Distinctness values were low in winter and high in summer. The Variation in Taxonomic Distinctness (Lambda+) was high in the northern waters, particularly during spring and the onset of summer (Figure 20.3f). A north (low) to south (high) gradient was observed for Average Taxonomic Distinctness (Delta+) in the coastal waters during winter. Low values (<77) were observed off Manifa during winter through the onset of summer. In the offshore waters, a north (low) to south (high) gradient (Delta+ <80) was observed during the late fall, but in winter a reverse gradient with relatively high values in the north and low in the south was noted. In other seasons, most Delta+ values were higher than 80 but showed no spatial trend. The Variation in Taxonomic Distinctness (Lambda+) along the coastal waters showed a positive southward gradient during winter and spring, whereas a northward trend was observed during the onset of summer. Off Manifa and Abu Ali relatively low values were observed in all the seasons.

20.3.2.2 Species composition and community structure

In the present investigation, a total of 490 species of phytoplankton were identified along the AG, of which diatoms contributed 63% (32% of centric and 31% of pennate diatoms) to the total number of species. Dinoflagellate species contributed nearly 33% to the total species, and the rest was made up by other groups comprising silicoflagellates, flagellates, and cyanophytes. Generally, the number of diatom and dinoflagellate species was highest during spring, followed by the onset of summer (June-July 2003), winter (January 2003), and summer (August 2003). The lowest species numbers were observed during the late fall (Figure 20.3g). There were 43 potential bloom-forming phytoplankton species identified during the study period (Figure 20.3g).

20.3.2.3 Late fall 2002

A total of 213 species were recorded during the late fall in the entire study area. Diatoms constituted 62% (40% centric and 22% pennate diatoms) of the total species, followed by dinoflagellates (35%). The coastal waters had a higher number of species (177 species belonging to 67 genera) than offshore waters (137 species belonging to 61 genera). A decreasing trend in the number of species was also observed from north to south. During the study period, 11 toxic algal species were observed among 23 potential bloom-forming species. *Thalassionema nitzschioides*, a potent algal bloom species, was observed among the dominant species during this season.

Stations along the coastal waters showed a 26% similarity in species composition, mainly due to *Proboscia alata*, which contributed more than 70% to the total similarity (Table 20.1). Other species that contributed to the total similarity were *Chaetoceros curvisetus* (5%) and *Chaetoceros diversus* (4%). The samples along the offshore waters showed a similarity of 41%, mainly represented by *Chaetoceros curvisetus* (59%),

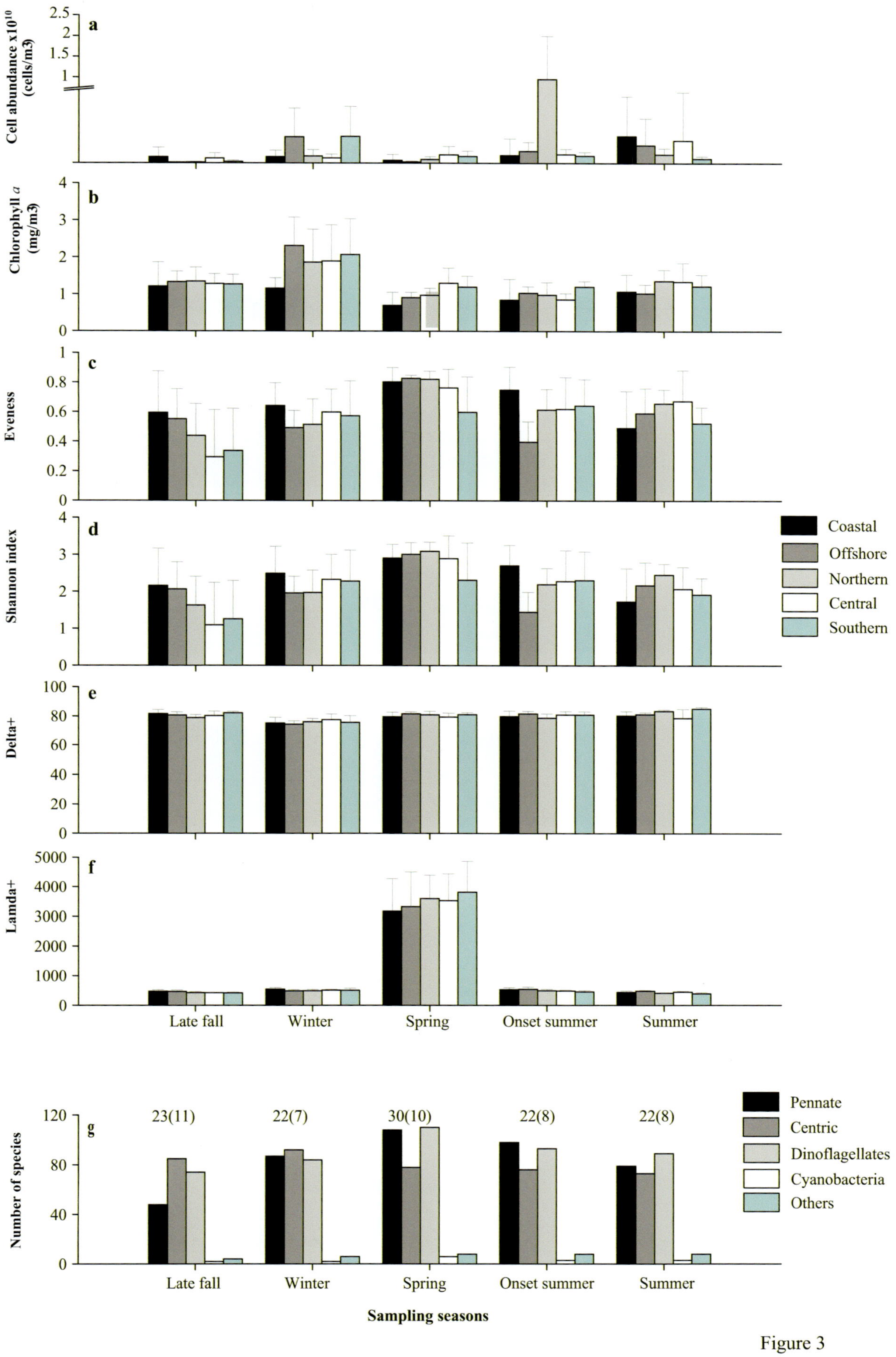

FIGURE 20.3 Variation in phytoplankton (a) cell abundance (x1010 cells/m³), (b) chlorophyll *a* (mg/m³), (c) evenness, (d) Shannon index, (e) Delta+, (f) Lamda+ during different seasons at different sampling locations in the western Arabian Gulf. (g) The number of species (number denotes total bloom-causing species, and in brackets, total toxic species) during different seasons at different sampling locations in the western Arabian Gulf.

TABLE 20.1 ANOSIM-derived spatially dominant five phytoplankton species during five different seasons in the western Arabian Gulf.

AREA	LATE FALL		WINTER		SPRING		ONSET OF SUMMER		SUMMER	
	DOMINANT SPECIES	%	DOMINANT SPECIES	%	DOMINANT SPECIES	%	DOMINANT SPECIES	%	DOMINANT SPECIES	%
Coastal	*Rhizosolenia alata*	70.89	*Rhizosolenia alata*	29.03	*Chaetoceros affinis*	30.32	*Chaetoceros affinis*	45.63	*Chaetoceros curvisetus*	67.66
	Chaetoceros curvisetus	4.59	*Thalassionema nitzschioides*	28.91	*Thalassionema frauenfeldii*	15.57	*Skeletonema costatum*	6.39	*Chaetoceros coarctatus*	5.71
	Chaetoceros diversus	3.71	*Rhizosolenia calcar-avis*	7.68	*Ceratium furca*	5.99	*Bacteriastrum cosmosum*	4.61	*Amphora ovalis*	5.22
	Bacteriastrum cosmosum	3.1	*Rhizosolenia bergoni*	2.51	*Climacodium fraunfeldium*	3.29	*Thalassiothrix frauenfeldii*	4.58	*Synedra sp*	4.92
	Thalassiothrix longissima	2.85	*Thalassionema frauenfeldii*	2.21	*Ceratium fusus*	2.86	*Thalassionema nitzschioides*	3.89	*Bacteriastrum delicatulam*	2.8
Offshore	*Chaetoceros curvisetus*	58.84	*Chaetoceros curvisetus*	21.3	*Climacodium fraunfeldium*	9.38	*Trichodesmium erythraeum*	62.05	*Gonyaulax monilata*	39.5
	Nitzschia lineola	8.11	*Lauderia anulata*	19.16	*Trichodesmium erythraeum*	9.29	*Thalassionema nitzschioides*	19.54	*Triceratium reticulatum*	31.62
	Thalassionema nitzschioides	3.17	*Thalassiosira subtilis*	18.36	*Thalassionema nitzschioides*	8.49	*Pleurosigma angulatum*	2.71	*Synedra sp*	5.83
	Thalassionema frauenfeldii	3.06	*Thalassionema nitzschioides*	16.16	*Ceratium fusus*	7.41	*Climacodium fraunfeldium*	1.54	*Chaetoceros curvisetus*	2.95
	Coscinodiscus marginatus	2.3	*Lauderia borealis*	4.86	*Ceratium furca*	3.67	*Ceratium fusus*	1.26	*Ceratium fusus*	1.59
Northern	*Thalassiosira subtilis*	58.57	*Lauderia anulata*	55.55	*Thalassionema nitzschioides*	16.57	*Trichodesmium erythraeum*	45.73	*Bacteriastrum cosmosum*	14.71
	Thalassionema nitzschioides	9.1	*Thalassionema nitzschioides*	15.07	*Rhizosolenia alata*	8	*Thalassionema nitzschioides*	24.67	*Bacteriastrum delicatulam*	14.26
	Thalassionema frauenfeldii	6.35	*Rhizosolenia alata*	3.52	*Rhizosolenia calcar-avis*	5.26	*Pleurosigma angulatum*	5.9	*Triceratium reticulatum*	13.43
	Nitzschia lineola	5.92	*Nitzschia longissima*	1.56	*Pleurosigma angulatum*	5.18	*Ceratium fusus*	3.28	*Gonyaulax monilata*	12.07
	Guinardia striata	5.68	*Nitzschia sigma*	1.46	*Ceratium furca*	5.03	*Nitzschia closterium*	2.6	*Synedra sp*	7.81
Central	*Thalassiosira subtilis*	90.47	*Thalassiosira subtilis*	28.14	*Thalassiothrix frauenfeldii*	13.67	*Trichodesmium erythraeum*	39.76	*Chaetoceros curvisetus*	40.87
	Chaetoceros curvisetus	3.96	*Thalassionema nitzschioides*	16.15	*Psedonitzschia fraudulenta*	10.2	*Thalassionema nitzschioides*	18.17	*Gonyaulax monilata*	19.38
	Odentella sinensis	1.28	*Lauderia anulata*	12.17	*Coscinodiscus gigas*	5.5	*Skeletonema costatum*	5.34	*Triceratium reticulatum*	15.16
	Odentella mobiliensis	1.19	*Rhizosolenia alata*	10.14	*Peridinium pellucidum*	5.5	*Pleurosigma angulatum*	3.42	*Synedra sp*	2.88
	Coscinodiscus marginatus	0.31	*Nitzschia sigma*	3.55	*Gonyaulax spinifera*	5.35	*Chaetoceros affinis*	2.45	*Chaetoceros affinis*	2.37
Southern	*Thalassiosira subtilis*	83.64	*Lauderia anulata*	34.08	*Thalassiothrix frauenfeldii*	49.55	*Trichodesmium erythraeum*	31.19	*Gonyaulax monilata*	46.3
	Chaetoceros curvisetus	7.54	*Thalassiosira subtilis*	24.02	*Pleurosigma simonsenii*	14.41	*Thalassionema nitzschioides*	23.74	*Triceratium reticulatum*	22.2
	Chaetoceros affinis	1.09	*Thalassionema nitzschioides*	20.73	*Psedonitzschia fraudulenta*	1.96	*Chaetoceros affinis*	5.68	*Chaetoceros dedymus*	4.09
	Thalassionema frauenfeldii	0.89	*Chaetoceros curvisetus*	6.57	*Navicula distans*	1.88	*Rhizosolenia setigera*	2.65	*Chaetoceros curvisetus*	3.7
	Coscinodiscus marginatus	0.65	*Chaetoceros lorenzianus*	1.42	*Peridinium nipponicum*	1.75	*Rhizosolenia styliformis*	2.17	*Ceratium furca*	2.85

Pseudonitzschia lineola (8%), *Thalassionema frauenfeldii* (3%), *Thalassionema nitszchioides* (3%), and *Coscinodiscus marginatus* (3%).

Generally, the species composition and community structure in the study area varied considerably among the regions. This differentiation in the species structure of the coastal (depth <25 m) and offshore (depth >25 m) transects are clearly simulated in the MDS plot (Figure 20.4a). The similarity in species composition among the regions was lowest between coastal and central waters. *Thalassiosira subtilis* was the most widely distributed species, accounting for more than 50% of total numerical abundance, and in central waters they accounted for 90% of total numerical abundance. The phytoplankton community in the northern and offshore waters had low dominance (Figure 20.4a). The funnel plot for Average Taxonomic Distinctness (Delta+) showed that the Delta+ value for stations offshore of Khafji was outside the expected range (probability limit 95%), which probably implies that in this station the biotic community may be under stress (Figure 20.4b). However, in other regions, the values fell well within the expected range. The funnel plot for variation in Taxonomic Distinctness (Lambda+) showed that offshore stations north of Abu Ali had higher variance.

20.3.2.4 Winter 2003

A total of 271 species was recorded during winter, a value higher than late fall. Diatoms were the predominant group (179) and constituted 66% (34% centric and 32% pennate diatoms) of the total species identified, followed by dinoflagellates (84), which constituted about 31%. The dominant species along each transect differed significantly however, in general, *Thalassionema nitzschioides* and *Lauderia annulata* were the most abundant species during winter. The species abundance in coastal and offshore waters was dominated by *Proboscia alata* (29%) and *Chaetoceros curvisetus* (21%), respectively. In the northern transect, the abundance of *Lauderia annulata* was highest (56%), while along the central one, *Thalassiosira subtilis* contributed 28% of the total abundance. In the southern transect, *Lauderia annulata* was found to be most abundant (34%). Coastal waters showed an average similarity of 40% for species composition that was mainly contributed by the diatom *Proboscia alata* (16% to the average similarity). An average similarity of 44% along the offshore waters was estimated, of which *Chaetoceros curvisetus* alone contributed 15%. The similarity among the stations in the northern, central, and southern regions was relatively low. The northern waters showed an average similarity of 35.6%, mostly represented by *Thalassionema nitzschioides* (13.1%). Along central waters, a similarity of 35% was recorded, mainly due to the contribution of *P. alata* (12 %). An average similarity of 34% was estimated for stations along the southern waters, and the species *Thalassionema nitzschioides* contributed 11% to the total similarity. The five most dominant species of each area during winter are presented in Table 20.1. Seven toxic algae species were observed among 22 potential bloom-forming species. *Thalassionema nitzschioides* and *Chaetoceros lorenzianus* are potent algal bloom species that were observed among the dominant species.

The difference in species diversity and abundance between the coastal (depth <25 m) and offshore (depth >25 m) samples were shown through MDS plots and ANOSIM R-values. Two separate domains were visible in the MDS plot for nearshore and offshore, which shows a significant difference in the species composition with respect to depth (Figure 20.4c). A significantly higher ANOSIM R-value (>0.25), observed between coastal and offshore waters, indicated a clear difference in the species composition and abundance among these regions. The cumulative species dominance plot clearly showed that the dominance was highest along the coastal waters (Figure 20.4d). The funnel plot of the offshore stations for Average Taxonomic Distinctness (Delta+) showed stations B2140, B2540, B2640, B2840, B2940, 3040, and B3140 to be outside the 95% confidence limit. The funnel plot for variation in Taxonomic Distinctness (Lambda+) showed the stations lying nearshore of Abu Ali, offshore of Tanajib, Jubail, and Ras Tanura with a confidence limit above 95%, while the nearshore stations off Khafji were found below the 95% confidence limits.

20.3.2.5 Spring 2003

A total of 310 species were identified in spring, which was higher than in late fall and winter. Diatoms were the most diverse group (186 sp.), comprising about 60% (25% centric and 35% pennate diatoms) of the total species identified, followed by dinoflagellates (110 sp., 35%). A noteworthy feature in spring was the increase in dinoflagellates species. A total of 30 potential bloom-forming species with ten toxic species were observed during this period. *Tripos furca*, *Tripos fusus*, *Trichodesmium erythraeum*, and *Thalassionema nitzschioides* were potent algal bloom species, and *Gonyaulax spinifera*, *Pseudonitzschia fraudulenta* were toxic species observed among the dominant species.

The stations in coastal waters showed an average similarity of 28% in species composition, and the dinoflagellates *Tripos furca* and *CT fusus* together contributed 21% to the average similarity. The offshore stations showed an average similarity of 32%; *Climacodium frauenfeldianum* and *Trichodesmium erythraeum* contributed 18% to the total similarity. The similarity among the stations in northern, central, and southern waters were 42%, 32%, and 35%, respectively. The important species contributing to the similarities in northern waters were *Thalassionema nitzschiodes* and *T. furca*, which contributed 19% of the total similarity index. Along the central waters, the species *T. nitzschioides* and *T. erythraeum* contributed 21% to the total similarity. *T. erythraeum* and *C. frauenfeldianum*, which contributed 26% to total similarity, were the major species in the southern waters (Table 20.1).

A *Trichodesmium* bloom was found in the surface waters of offshore stations. *Trichodesmium erythraeum* and *Thalassionema frauenfeldii* were the dominant species along the offshore transect. Along the shallow coastal stations, the diatoms *Chaetoceros affinis* and *Thalassionema frauenfeldii* were dominant, whereas *C. frauenfeldianum* and *T. erythraeum* were the dominant species along the offshore transect. In northern waters, *Thalassionema nitzschiodes* and *Proboscia alata* were found in high numbers, while in the central waters, a predominance of *Thalassionema frauenfeldii* and *Pseudonitzschia fraudulenta* was seen. The southern waters showed a predominance of *Thalassionema frauenfeldii* and *Pleurosigma simonsenii*.

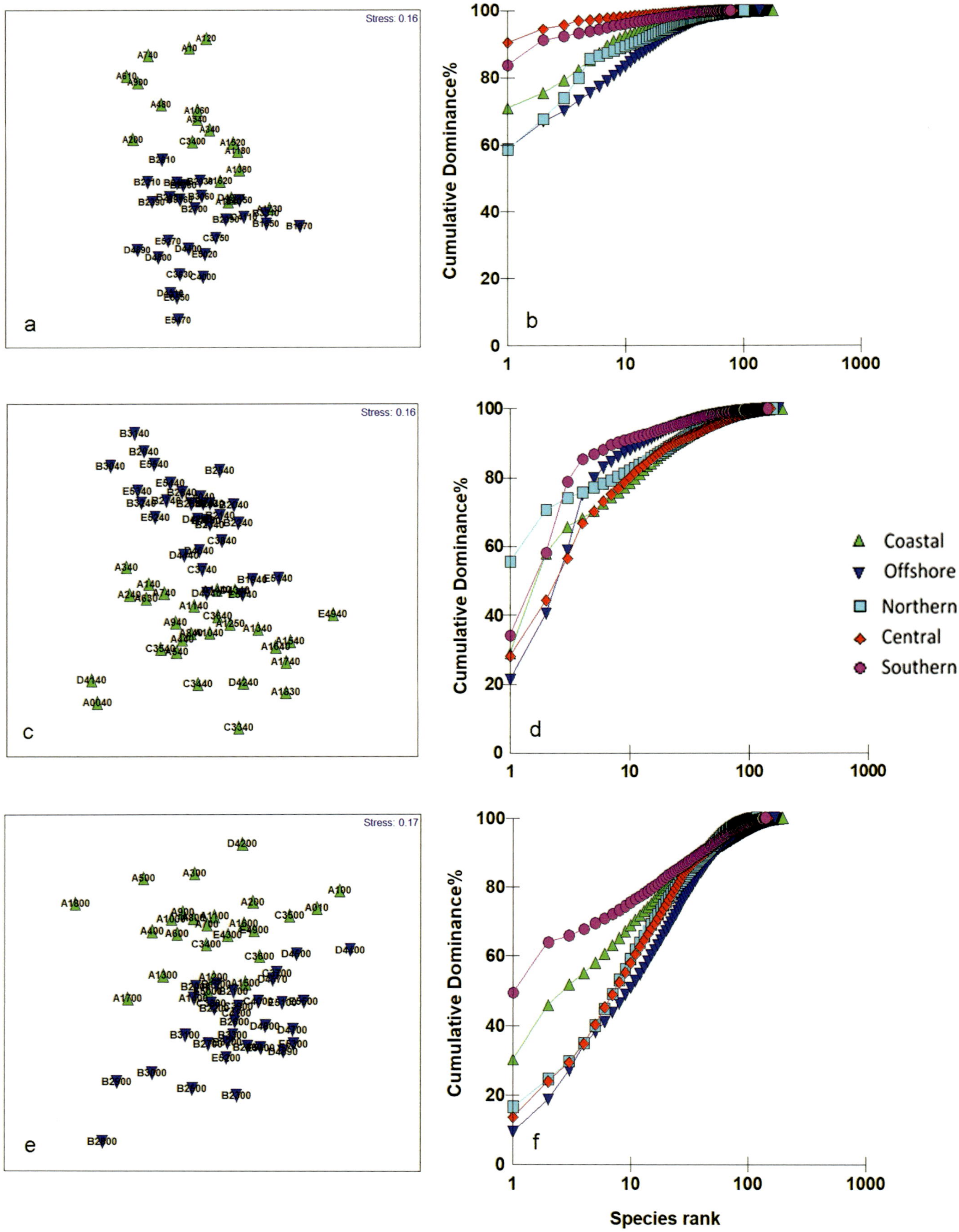

FIGURE 20.4 MDS for nearshore-offshore stations and dominance plot for the study area during (a, b) late fall; (c, d) winter; (e, f) spring; (g, h) onset of summer; and (i, j) summer.

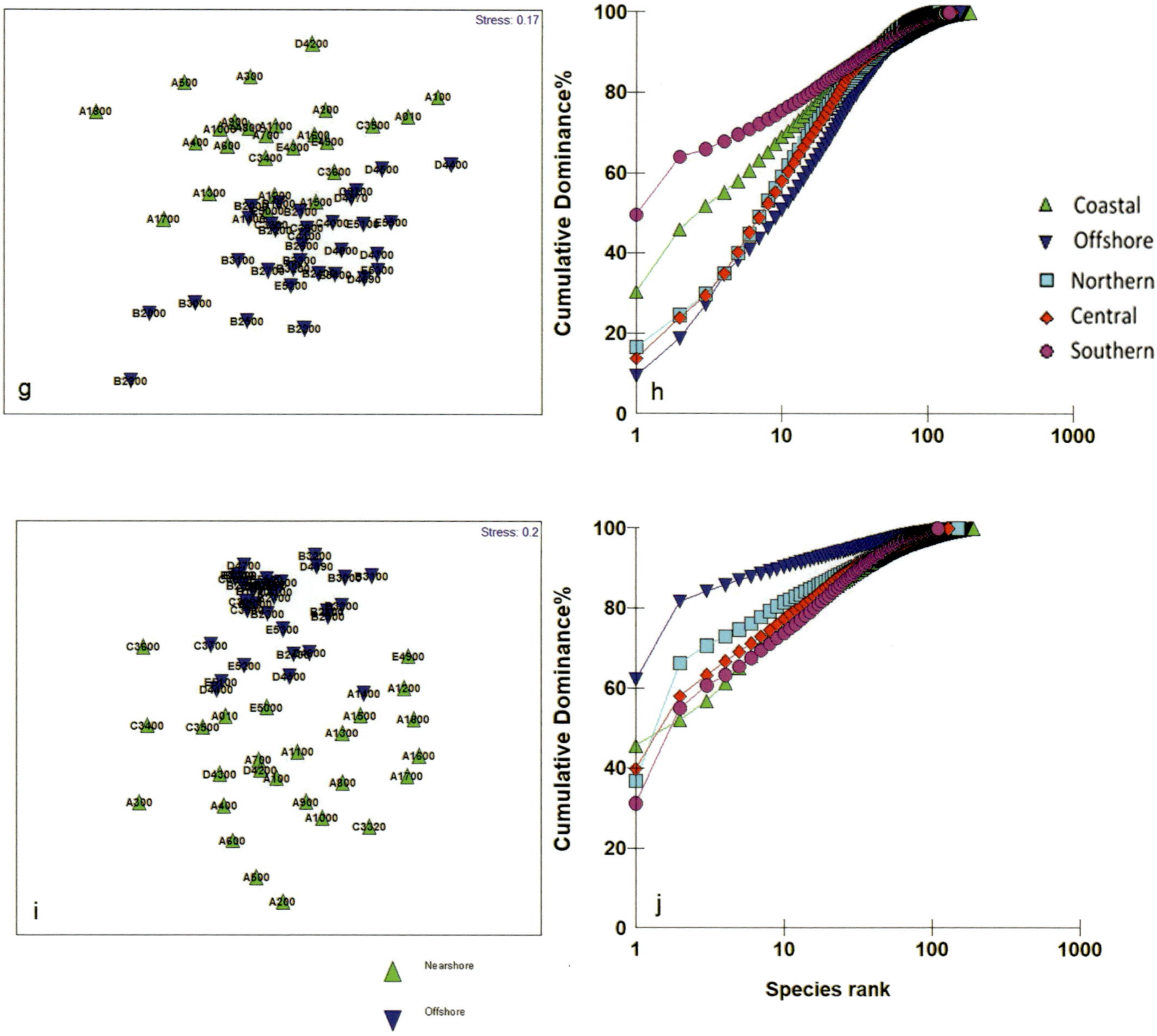

FIGURE 20.4 (Continued)

The MDS plot and ANOSIM R-values showed the difference in the species abundance and diversity between nearshore (<25 m) and offshore (>25 m) stations. Two distinct groups or clusters were found in the MDS, and this indicated the existence of different species with respect to depth (Figure 20.4e). The global R-values were less than 0.5, which indicated that the species composition between each transect was overlapping and barely separable. The cumulative dominance plot described the dominance of *Thalassionema frauenfeldii* along southern waters. The dominance plot showed that southern waters had the highest dominance (Figure 20.4f). The funnel plot for Average Taxonomic Distinctness (Delta+) simulated for the five transects indicated that, except for few stations lying close to the coast off Ras Mishab and Ras Tanurah, all other areas had a value within the 95% confidence limits. The funnel plot for the Lambda+ showed the stations A1600, B2700, and B2900 to be above the 95% confidence level, which could be due to some confined disturbances, while most of the other stations were within the simulated limits.

20.3.2.6 Onset of summer 2003

A total of 278 species was recorded during the onset of summer, which was lower than in spring and winter. Diatoms contributed 63% (27% centric and 35% pennate diatoms) to the total number of species along the offshore waters, followed by dinoflagellates (33%). In most of the offshore stations, the dominant species was the cyanophyte *Trichodesmium erythraeum*. The surface bloom of *Trichodesmium* was also observed in spring. The dominant species along the coastal waters were *Chaetoceros affinis* and *Skeletonema costatum*. *Trichodesmium erythraeum* and *Thalassionema nitzschioides* constituted the dominant group in all the other transects. Eight toxic algae species were observed among 22 potential bloom-forming species. *Skeletonema costatum*, *Sundstroemia setigera*, *Tripos fusus*, *Trichodesmium erythraeum*, and *Thalassionema nitzschioides* were potent algal bloom species observed among the dominant species.

The stations in coastal waters showed an average similarity of 28% in the phytoplankton population, and it was mostly contributed by the diatom *Chaetoceros affinis* and *T. erythraeum*, which together accounted for nearly 20% of the average similarity. The stations along the offshore showed an average similarity of 45%, *T. erythraeum* and *T. nitzschioides* being the major contributors. The similarity among the stations in the northern, central, and southern regions was 32%, 34%, and 40%, respectively. The important species contributing to the similarities in northern stations were *Thalassionema nitzschiodes* and *Pleurosigma angulatum*, which contributed 40% to the total similarity, while along the central and southern regions, the predominant species were *T. nitzschioides* and *T. erythraeum* (35% and 44%, respectively; Table 20.1).

The MDS plot and ANOSIM-R values showed a clear difference between nearshore-offshore transects both for species composition and abundance. The stations along the coastal transect were much more scattered. The offshore stations showed an R-value greater than 0.5 (R = 0.601), which indicates that the population was different between these two transects. In the northern, central, and southern waters, the R-value was less than 0.1 and indicated no significant differences between these regions. The MDS plot for near-offshore stations (25 m cut-off depth) placed the stations from both depths in two different domains, which suggests an influence of depth on phytoplankton population (Figure 20.4g). The cumulative dominance plot showed that offshore waters had the highest dominance, followed by the coastal stations (Figure 20.4h). The dominance along the northern, central, and southern areas was less than 40%, while northern waters showed the least cumulative dominance.

The funnel plot for Delta+ showed stations A200, A300, and A500 to be below the 95% confidence limit, while stations A1300 and B3300 were above the 95% confidence limit. The funnel plot for Lambda+ indicated stations A500, A600, A1400, B2700, and B2900 to be above the 95% confidence limit. The stations deviating from the expected normal, and placed above or below the confidence level, could be in stress, or some environmental disturbances may be influencing community structure. The funnel plot for Delta+ showed the stations along the perpendicular transects (Transect C, D and E) did not deviate from the simulated 95% confidence limit, while the funnel plot for Lambda+ indicated that stations C4000 and E5200 are above the simulated 95% confidence level, which implies a deviation from the neighbouring community structure.

20.3.2.7 Summer 2003

A total of 252 species were identified during the summer cruise, and a notable increase in the total number of dinoflagellate species was observed. Diatoms constituted the major group (152 species), followed by dinoflagellates (89 species). Diatoms contributed 60% (29% centric and 31% pennate diatoms) to the total number of species along the offshore waters, followed by dinoflagellates (33%). Eight toxic algae species were observed among 22 potential bloom-forming species. *Tripos fusus*, *Tripos furca*, and *Alexandrium monilatum* were potent algal bloom species observed among the dominant species.

The stations along the coastal waters showed an average similarity of 37% in species composition, which was contributed mainly by the diatom *Chaetoceros curvisetus* (23%). Along the offshore, northern, and central waters, a similarity of 44%, 39%, and 35%, respectively, was observed among the stations. *Triceratium reticulatum* in these transects contributed 20%, 13%, and 18% to the total similarity, respectively. An average similarity of 49% was estimated for the stations along the southern waters, and the dinoflagellate *Alexandrium monilatum* contributed 21% to this total similarity. The dominant species of each area during summer are listed in Table 20.1.

The abundance of the toxic dinoflagellate *Alexandrium monilatum*, especially in the offshore waters, was noteworthy. In the offshore waters, *Alexandrium monilatum* and *Triceratium reticulatum* were the dominant species, among which the former contributed about 40% to the total cell density. Along the coastal waters, the dominant species were the diatoms *Chaetoceros curvisetus* and *Chaetoceros coarctatus*, both accounting for nearly 68% of the total abundance. In the northern waters, the diatoms *Bacteriastrum comosum* and *Bacteriastrum delicatulum* were dominant, contributing 30% of the total cell density. Along central waters, the species *Chaetoceros curvisetus* and *Alexandrium monilatum* were predominant and, among them, *C. curvisetus* alone contributed 41% to the total cell density. *Alexandrium monilatum* and *Triceratium reticulatum* were common in the southern transect. The toxic dinoflagellate *Alexandrium monilatum* in this transect contributed more than 46% to the total cell density.

There was a clear difference between coastal and offshore waters in species composition, abundance, and diversity, which was evident in the MDS plot and ANOSIM R-values. Two distinct groups of clusters were found in the MDS, which indicated different species composition with respect to depth (Figure 20.4i). The cumulative dominance plot describes the high dominance along coastal waters (Figure 20.4j).

The funnel plot for the distributions of Average Taxonomic Distinctness (Delta+) showed that most of the stations, except B3300, A400, A1100, and A1200, were below the 95% simulated confidence limit, while the funnel plot for the Variation of Taxonomic Distinctness (Lambda+) showed stations A600 and A1800 above the 95% confidence limit. The stations falling out of the 95% confidence contours are considered imbalanced, and probably some environmental disturbances affected the phytoplankton diversity at these stations. The funnel plot for Average Taxonomic Distinctness (Delta+), simulated for the northern, central, and southern waters, showed most of the station within the 95% confidence limits, except for station D4800.

20.3.3 Ecological relationships

During the late fall, none of the variables other than TPH and nitrite influenced the chl *a* concentration significantly. TPH and nitrite showed a significant positive correlation with chl *a* concentration (r = 0.383 and r = 0.629, $P<0.05$) along the coastal waters. Another noteworthy observation was the positive correlation of TPH with cell density (r = 0.555, $P<0.05$), indicating that TPH had a profound influence on phytoplankton abundance during the late fall. During winter, across the study area, chl *a* showed a significant positive correlation with depth, temperature, and pH and a negative correlation with salinity. The cell density along the coastal waters showed a significant positive correlation with TPH ($r = 0.498$, $P<0.05$), while temperature exhibited a significant negative correlation ($r = -0.421$, $P<0.05$). In the offshore waters, cell density showed a significant positive correlation with phosphate ($r = 0.454$, $P<0.05$), temperature, and DO. During spring, none of the hydrographical parameters showed a significant correlation with chl *a* in shallow waters. In deep waters, temperature, DO, and pH exhibited a significant positive correlation with chl *a* ($r = 0.380$, $P<0.05$). During the onset of summer, chl *a* showed a significant positive correlation with depth and a negative correlation with salinity and temperature when stations were pooled together. During summer, for all the stations, chl *a* showed a significant correlation with depth and pH. None of the variables significantly influenced the phytoplankton abundance along the coastal waters and other regions. The phytoplankton abundance during summer seems to have been influenced by multiple parameters rather than by any single parameter.

The correlation for all the seasonal means showed that chl *a* values had a significant negative correlation with nitrite and nitrate ($r>-0.953$, $P<0.05$), but there was no significant correlation with cell density and any of the other variables throughout the study period. On the other hand, the seasonal means of cell density for all the seasons showed a significant positive correlation with phosphate ($r = 0.822$, $P<0.05$). Likewise, diatoms also showed a significant positive correlation with phosphate ($r = 0.844$, $P<0.05$), which indicates that diatoms flourished with phosphate enrichment during winter mixing.

In a factor analysis, two main factors explained 26% of the total variance among the environmental variables and 35 dominant phytoplankton species (Figure 20.5a). *Synedra sp., Alexandrium monitalum, Amphora ovalis, Triceratium reticulatum*, and *Pseudonitzschia fraudulenta* showed a strong positive correlation with temperature and NH_4 and were negatively correlated with DO. *Bacteriastrum comosum, B. delicatulam, Chaetoceros affinis, C. coarctatus, C. curvisetus, C. didymus, C. diversus, Climacodium frauenfeldianum, Thalassiosira eccentrica, C. gigas, Guinardia striata, Lauderia annulata, Proboscia alata, R. calcar-avis, Sundstroemia setigera*, and *R. styliformis* were positively correlated with temperature and NH_4 and negatively correlated with DO. *Thalassionema nitzschiodes* was positively correlated with DO and negatively correlated with temperature and NH_4. *Climacodium frauenfeldianum, Tripos furca*, and *Trichodesmium erythraeum* were positively correlated with NO_3 and negatively correlated with TPH. *Pseudonitzschia lineola, Thalassionema frauenfeldii, Thalassiothrix longissima, Guinardia striata, Lauderia annulata, Thalassiosira subtilis, Chaetoceros diversus, Thalassiosira eccentrica, Proboscia alata*, and chl *a* were positively correlated with TPH and negatively correlated with NO_3.

Factor analysis for the environmental variables and potential bloom-causing phytoplankton explained 15% of total variance between the two main factors (Figure 20.5b). *Alexandrium monitalum, Lingulodinium polyedra, Karenia mikimotoi, Dinophysis miles*, and *Pseudonitzschia fraudulenta* showed a positive correlation with temperature, NH_4, and PO_4, and were negatively correlated with DO. *Trichodesmium erythraeum, Prorocentrum gracile, Prorocentrum mexicanum*, and *Tripos furca* were positively correlated with NO_3. *Trichodesmium erythraeum* and *Prorocentrum gracile* were closely associated with salinity. *Thalassionema nitzschioides* was positively correlated with DO and negatively correlated with temperature, PO_4, and NH_4. *Dinophysis caudata, Dinophysis tripos, Thalassiosira decipiens, Chaetoceros lorenzianus, Tripos muelleri, Pseudonitzschia multiseries, Pseudonitzschia seriata, Sundstroemia setigera*, and chl *a* were negatively correlated with NO_3.

20.4 DISCUSSION

20.4.1 Spatial and temporal variation of the phytoplankton community

A wide variation of environmental parameters showed a seasonal cycle in the study area, especially temperature. This variation was well reflected in the phytoplankton abundance and community dynamics. Diatoms constituted the most significant component, contributing 65% to the total identified species. This is consistent with earlier studies (Al-Kandari et al. 2009; Al-Yamani et al. 2004; Husain and Ibrahim 1998). Among the diatoms, pennate species dominated in abundance over centric, suggesting better tolerance against environmental variables. Similar observations were reported from the Indian sector of the Southern Ocean (Gomi et al. 2007) and Turkish waters (Cetin and Sen 2004). Dinoflagellates followed diatoms, and they constituted nearly 30% of the total species. In general, variation in phytoplankton biomass, abundance, and diversity was statistically significant between transects, and there existed an apparent spatial trend, which changed with seasons. It has been suggested that the ratio between diatoms and dinoflagellates may provide a good indicator for regional environmental changes, as these two groups show a consistent pattern of ecological succession and distinct pollution responses (Wasmund et al. 2017). During this investigation, a seasonal shift in the number of diatom and dinoflagellate species was observed. The maximum number of species was identified during the onset of summer, while it was least during late fall 2002. The diatom to dinoflagellate ratio calculated for the entire study area was nearly 2 for most of the study period (Table 20.2). Comparing this diatom to dinoflagellate ratio with those previously reported from the AG by Al-Kaisi (1976) and Al-Saadi and Hadi (1987), the ratio seems to have reduced by nearly three-fold. There was variation in the abundance of both high surface-to-volume ratios (*Navicula* sp, *Pleurosigma* sp, *Pinnularia* sp, and *Synedra* sp) and low surface-to-volume ratios (*Chaetoceros* spp) on a seasonal basis. This variation could be attributed to the seasonal changes in nutrient concentration that may have affected the competition between diatom species as they differ in their nutrient-absorbing capabilities (Achary et al. 2014).

Total cell density in the study area displayed both spatial and temporal variations. The mean cell density was a maximum during summer and minimum during late fall and spring, and ranged from 1.24×10^4 cells/m^3 to 7.49×10^6 cells/m^3. A post-spill study along the ROPME sea area by Husain and

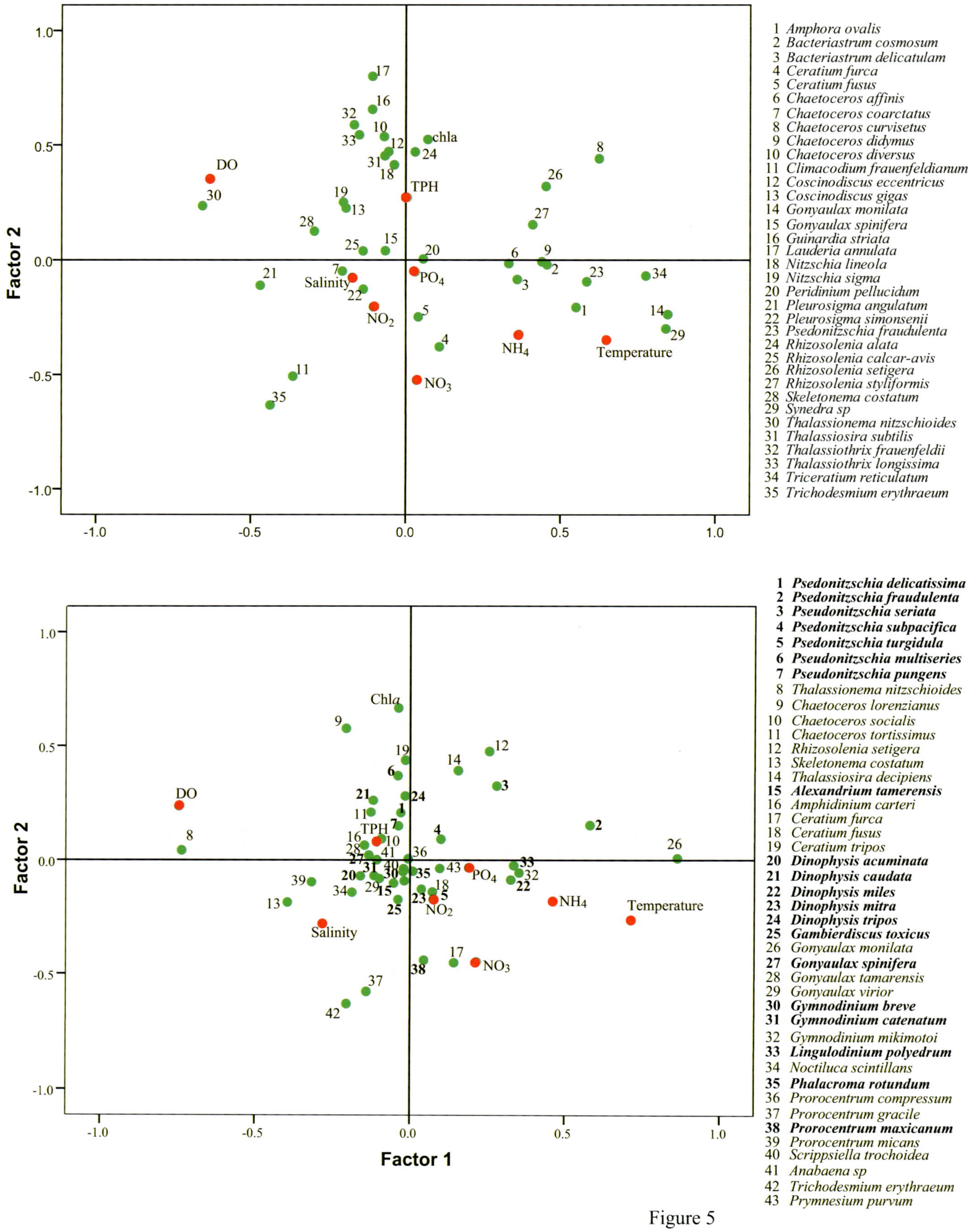

FIGURE 20.5 Factor analysis to find relationship between different abiotic parameters of seawater (temperature, salinity, DO, NO_2 NO_3 NH_4 PO_4 and TPH) and different species of phytoplankton in the western Arabian Gulf. a) dominant species, b) potential bloom-forming phytoplankton species.

TABLE 20.2 Diatom to dinoflagellates ratio during the study period.

SEASON	*DIATOMS/DINOFLAGELLATES*
Late Fall	2.01
Winter	2.22
Spring	1.77
Onset of Summer	1.86
Summer	1.80

Ibrahim (1998) during *Umitaka-Maru* cruises (1993–1994) also showed that the Saudi Arabian waters had relatively low cell abundances. They reported a cell density of 2.26×10^6 to 3.03×10^6 cells/m^3 during December 1993 along the Saudi Arabian territorial waters. Higher cell densities were observed at stations in the south, which were in UAE and Omanian waters, and ranged from 3.21 xx 10^6 to 4.35×10^6 cells/m^3, respectively, during December 1993. Habashi et al. (1992) estimated a cell density of 5.40×10^6 cells/m^3 in the Arabian Gulf. Estimates in the present study on phytoplankton abundance were close to the range of values in the pre-spill period reports of Al-Kaisi (1976). Chl *a* also showed the same trend, but it was highest during winter (1.75 mg m^{-3}) and lowest during spring and the onset of summer (0.93 and 0.95 mg m^{-3}, respectively; Figure 20.3b). In the northern coastal waters from fall through the onset of summer, a high cell density was observed, where the coastal circulation of nutrient-rich water was high. Low cell densities were generally recorded in the central waters and northern offshore waters. The oceanographic process, such as eddies, deep-sea storms, secondary circulation, internal waves and tsunamis, rogue waves and cyclone waves, in this area is weak and would not have perhaps favoured higher primary production.

In a review of phytoplankton ecology in the Arabian Gulf, Subba Rao and Al-Yamani (1998) describe a north-to-south gradient in phytoplankton species diversity, characterizing the northern Gulf as an area of low diversity, dominated by diatoms, and the open Arabian Gulf and the Gulf of Oman as an area of high phytoplankton biodiversity, comprising primarily tropical species. Temperature is a very important factor and was found to be positively and strongly related to most of the dominant species in the study area. Community changes of phytoplankton have been reported in relation to temperature (Goldman and Mann 1980), light intensity (Ryther 1956), and chemical conditions (Sanders and Riedel 1987). Since phytoplankton species have different tolerances to environmental changes, the patterns of seasonal succession may be a consequence of these changes (Guillard and Kilham 1977). Generally, the dominant phytoplankton species and the community structure along the Arabian Gulf were not rigid and changed with space and time along the Gulf. An apparent onshore-offshore distinction was always present. All of the dominant species observed in this study were also reported in earlier studies in the Gulf. However, some exotic species, mostly flagellates (e.g., *Chrysochromulina* sp. and *Dinobryon* sp.), recorded in the present investigation have not been recorded in the reviewed literature.

Phytoplankton production in the AG is supported by the external supply of nutrients from the land, atmosphere, river input, and internal nutrient regeneration by decomposition and demineralization processes. A lack of long-term observational data precludes direct comparison of nutrient and phytoplankton abundance from the present study area. Cell density and chlorophyll were high in winter, when the phosphate levels were also relatively high. A similar trend was visible during summer, but it was not prominent during winter. A relatively high biomass and cell density were visible to the north of the Abu Ali region, where nutrient concentrations, particularly nitrate and phosphate, were high. The low cell density during spring could be due to the depletion of nutrients after the winter bloom and the disturbance in the phytoplankton succession during this transition period. A correlation of cell density with any of the nutrient or environmental variables for the seasons was not visible in the correlation matrix. Still, it showed a significant positive correlation ($P<0.05$) with TPH during late fall and winter along the coastal waters. This could be due to the different hydrographical parameters and species that prevailed in these waters.

The Pielou's evenness values were relatively high and uniform along the study region during spring, while a coastward trend was visible during the onset of summer, which was further substantiated by the dominance of *Trichodesmium* and *A. monilatum* in offshore waters. A north-to-south gradient was visible during summer, and a southern inshore trend was observed during winter. Two separate domains for the north and south with low evenness in the central waters were the peculiarity for the late fall. This suggests the existence of shifting patchiness and dominance with respect to the changing seasons. Shannon Index (H') values less than 2.5 are consistent with polluted or stressed environments, or upwelling areas (Margalef 1967; Moss 1973; Heip et al. 1985; Ismael and Dorgham 2003). Applying this criterion to phytoplankton diversity in the present study, it could be assumed that most of the offshore, central, and some pockets along the coastal waters were under stress throughout the year, except during spring. The changes in the environmental and hydrographical parameters were well reflected in the diversities. The offshore low values during the warmer period could be due to a strong thermocline that arrests the nutrient dynamics by acting as a barrier between nutrient-poor surface water and nutrient-rich bottom waters of the Arabian Gulf. Another possible reason could be the slow response of the offshore waters to any atmospheric processes. Particular correlation with any of the nutrient or environmental variables for all the seasons was not visible in the correlation matrix, but H′ showed a significant ($P<0.05$) negative correlation with TPH during late fall and winter and a positive correlation during the onset of summer ($r = 0.399$, $P<0.05$). The inverse correlation between TPH and H′ could be due to the growth stimulation, low grazing pressure due to avoidance, modulation of light during summer, and patchiness.

Average Taxonomic Distinctness (Delta+) and Variation in Taxonomic Distinctness (Lambda+) showed, in general, overall low indices during late fall and winter. Low values for all of the above indices indicate a loss or change in the biodiversity status (Austen and Warwick 1995). Perhaps the decrease in indices could have been due to the changes in the hydrographical components rather than any effect of the 1991 oil spill. It may be important to remember here that no detailed work with phytoplankton taxonomic diversity and distinctness using the software PRIMER was done previously. Hence, a comparison with the scenario prior to or soon after the 1991 oil spill is difficult. The ANOSIM-R values for pooled data for biodiversity showed that there existed seasonal differences in biodiversity. The R-values ranged between 0.064 (onset of summer and summer) to 0.453 between winter and summer. R-values more than 0.25 indicated a considerable difference in species composition and abundance, which was evident from the succession of diverse dominant species during different seasons. The onshore-offshore difference was minimal during transitional periods (late fall and spring).

20.4.2 Potential bloom-forming species in the Arabian Gulf

The occurrence and frequency of algal blooms in the Arabian Gulf are common, and most of them are caused by potentially toxic phytoplankton (Subba Rao and Al-Yamani 1998; El Gammal et al. 2017; Rajan and Al Abdessalaam 2008). In the present study, 43 potential bloom-forming or harmful algal taxa have been reported, which is higher than the earlier reported numbers, i.e., 38 bloom-forming species. The most common species reported earlier were *Dinophysis caudata*, *D. miles*, *Prorocentrum cordatum*, *Pyxidicula compressa var. compressa*, *Tripos furca*, and *Pyrodinium bahamense* (Subba Rao and Al-Yamani 1998; Rajan and Al Abdessalaam 2008). In the present study, nine species among the dominant species were bloom-forming species, suggesting that the AG ecosystem is at high risk due to the consequences of algal bloom.

A *Trichodesmium* bloom was observed during the warmer period when the sea was calm and had strong stratification; these conditions could have favoured the growth of trichomes. This genus is commonly responsible for blooms in the AG (Thangaraja et al. 2007; Miller and Wheeler 2012). The optimum temperature for *Trichodesmium* growth is reported to be 20°C to 34°C, and the maximum growth rate and nitrogen-fixing rate were found at the temperature range of 24–30°C (Agawin et al. 2007). Therefore, the increasing temperature during the spring season and accumulated nutrients during winter could have triggered the *Trichosdesmium* bloom. It can provide new nutrients for other blooms, once initiated (Lenes et al. 2001; Walsh and Steidinger 2001; Mulholland et al. 2006; Lenes and Heil 2010). The dominance of the toxic dinoflagellate *A. monilatum* along the southern waters during summer was driven by temperature. This was also supported by factor analysis. *Margalefidinium polykrikoides*, a species that caused a major toxic algal bloom incident in Kuwait waters during 2008–2009 and killed a large number of fish, was not found in the present study location. The primary factor for the occurrence of the *M. polykrikoides* bloom was nutrient enrichment in the epizootic in Kuwait Bay (Glibert et al. 2002). The coastal waters of the Arabian Gulf are exposed to different types of environmental stresses, such as pollutants, dust storms, construction, and industrialization, etc., in the coastal belt. Therefore, knowledge of certain fundamental biological aspects of this taxon, such as its life cycle and toxicity mechanisms, is needed to understand the causes, evidence, and operational guidelines for future decision-making mechanisms and emergency response actions.

20.4.3 Impact of 1991 oil spill on phytoplankton

The cell density observed was similar to previous pre-spill estimations (Basson et al. 1977). The diatom to dinoflagellate ratio calculated for the entire study area was nearly 2 for most of the study period (Table 20.2). When comparing this diatom to dinoflagellate ratio with those reported by Al-Kaisi (1976) and Al-Saadi and Hadi (1987), the ratio seems to have reduced by nearly three-fold. The present shift in the ratio cannot be directly related to the 1991 oil spill since many environmental variables are found to be influencing the phytoplankton community. Moreover, different techniques were used between these studies for collection and enumeration. A major difficulty in assessing the impacts of the 1991 oil spill is related to the theoretical and practical difficulties associated with highly variable spatial and temporal scales over which marine processes and species operate. However, after analysing the data obtained through this study, it can be concluded with a certain amount of confidence that, although sporadically distributed, the phytoplankton population in the Saudi Arabian waters was ecologically healthy at the time of sampling.

A lack of data over the 11 years after the spill made it difficult to ascertain whether the perceived changes in diversity, especially the shift in the diatom to dinoflagellate ratio, were due to the aftermath of the 1991 oil spill. However, the data generated, covering five seasons (late fall to summer), conclusively showed that there had been a considerable influence of temporal and seasonal variables on the phytoplankton community. Many factors simultaneously influence the phytoplankton variables in the Arabian Gulf, including TPH, but the present study did not consider the atmospheric and other inputs of nutrients and trace elements into the Arabian Gulf.

20.5 CONCLUSION

During the present investigation, 490 phytoplankton species were recorded, in which diatoms constituted the major component in all seasons and contributed more than 65% of the total identified species. Dinoflagellates constituted nearly 30% of the total species. In general, phytoplankton species diversity of the Arabian Gulf was rich and indicated an ecologically healthy community. Phytoplankton biomass was evenly distributed in the water column during winter. However, the chl *a* maxima depth shifted gradually with the rise in ambient temperature and water stratification. The seasonal changes in the water circulation pattern along the region were reflected on the phytoplankton standing stock, and its signature was more perceptible during spring. Generally, the central waters were less productive because of the weak oceanographic process, such as eddies, deep-sea storms, secondary circulation, internal waves and tsunamis, rogue waves and cyclone waves, while northern coastal waters were more productive. Spatial domination of a particular species of phytoplankton with changing seasons was observed, and their numerical abundance contributed more than 90% to the total density in certain spells. Surface blooms of the cyanophyte *Trichodesmium erythraeum* were observed during spring and the onset of summer along the offshore waters, while a high abundance of the toxic dinoflagellate *Alexandrium monilatum* was noted during summer. Spatial and temporal variations in phytoplankton abundance and biomass were observed during the study, and the above variables deviate marginally from pre- and post-oil spill studies. This could be due to the difference in sampling methods used by the different studies. The present investigation observed changes in the diatom to dinoflagellate ratio and a reduction in diatom contribution compared to the pre-oil spill reports. Changes observed in the diatom to dinoflagellate ratio in the present study could not be directly attributed to the 1991 oil spill alone, as multiple abiotic factors would have influenced the phytoplankton distribution along the Arabian Gulf.

ACKNOWLEDGMENTS

We thank the Center for Environment and Marine Studies (CEMS) at the Research Institute of the King Fahd University of Petroleum and Minerals, Dhahran, Saudi Arabia, for providing research facilities and the Marine Studies Section team of CEMS for performing the fieldwork and laboratory analyses. We express our sincere thanks to the General Authority of Meteorology and Environmental Protection, Saudi Arabia, and the Saudi Company for Environmental Works Ltd. for their constant support. We also thank M/S *Al-Gosaibi* for providing services of their Oceanographic Vessel.

REFERENCES

Achary, M.S., Panigrahi, S., Satpathy, K.K., Sahu, G., Mohanty, A.K., Selvanayagam, M., and R.C. Panigrahy. 2014. Nutrient dynamics and seasonal variation of phytoplankton assemblages in the coastal waters of southwest Bay of Bengal. *Environ Monit Assess* 186, no. 9: 5681–5695.

Agawin, N.S., Rabouille, S., Veldhuis, M.J., Servatius, L., Hol, S., van Overzee, H.M., and J. Huisman. 2007. Competition and facilitation between unicellular nitrogen-fixing cyanobacteria and non-nitrogen-fixing phytoplankton species. *Limnol Oceanogr* 52, no. 5: 2233–2248.

Al-Kaisi, K.A. 1976. *On the phytoplankton of the Arabian Gulf.* 2nd Joint Oceanographic Assembly, Edinburgh, Scotland. Abstracts, p. 125.

Al-Kaisi, K.A. 1977. Qualitative and quantitative assessment of phytoplankton in the sea water of Kuwait. Third Seminar on Environment and Pollution, Kuwait.

Al-Kandari, M., Al-Yamani, F., and K. Al-Rifaie. 2009. *Marine phytoplankton atlas of Kuwait's waters.* Kuwait: Kuwait Institute for Scientific Research, 351 p.

Al-Saadi, H.A., and R.A. Hadi. 1987. Ecological and taxonomical studies on phytoplankton of Arab Gulf. *J Biol Sci Res* 18, no. 3: 7–31.

Al-Yamani, F.Y., Bishop, J.M., Ramadhan, E., Al-Husaini, M., and A. Al-Ghadban. 2004. *Oceanographic atlas of Kuwait's waters.* Kuwait: Kuwait Institute for Scientific Research (Publisher), 202 p.

Austen, M.C., and R.M. Warwick. 1995. Effects of manipulation of food supply on estuarine meiobenthos. *Hydrobiologia* 311, no. 1: 175–184.

Azam, F., Fenchel, T., Field, J.G., Gray, J.S., Meyer-Reil, L.A., and F. Thingstad. 1983. The ecological role of water-column microbes in the sea. *Mar Ecol Prog Ser* 91: 257–263.

Azis, P.A., Al-Tisan, I.A., Daili, M.A., Green, T.N., Dalvi, A.G.I., and M.A. Javeed. 2003. Chlorophyll and plankton of the Gulf coastal waters of Saudi Arabia bordering a desalination plant. *Desalination* 154, no. 3: 291–302.

Bar-On, Y.M., and R. Milo. 2019. The biomass composition of the oceans: A blueprint of our blue planet. *Cell* 179, no. 7: 1451–1454.

Basson, P.W., Burchard, J.E., Hardy, J.T., and A.R. Price. 1977. *Biotopes of the Western Arabian Gulf: Marine life and environments of Saudi Arabia.* Dhahran, Saudi Arabia: Arabian American Oil Co. (Aramco), 284 p.

Bendall, D.S., Howe, C.J., Nisbet, E.G., and R.E. Nisbet. 2008. Photosynthetic and atmospheric evolution. Introduction. *Phil Trans R Soc Lond, B, Biol Sci* 363, no. 1504: 2625–2628.

Cetin, A.K., and B. Sen. 2004. Seasonal distribution of phytoplankton in Orduzu dam lake (Malatya, Turkey). *Turk J Botany* 28, no. 3: 279–285.

Clarke, K.R., and M. Ainsworth. 1993. A method of linking multivariate community structure to environmental variables. *Mar Ecol Prog Ser* 92: 205–205.

Clarke, K.R., Gorleya, R.N., Somerfield, P.J., and R.M. Warwick. 2001. *Change in marine communities. An approach to statistical analysis and interpretation*, 2nd ed. Plymouth, UK: PRIMER-E Ltd., 168 p.

Clarke, K.R., and R.M. Warwick. 1994. Similarity-based testing for community pattern: The two-way layout with no replication. *Mar Biol* 118, no. 1: 167–176.

Cloern, J.E. 1979. *Phytoplankton ecology of the San Francisco Bay system – the status of our current understanding.* San Francisco, California, USA: Am Assoc Advance Sci – Pacific Division.

Devlin, M.J., Le Quesne, W.J., and B.P. Lyons. 2015. The marine environment of Kuwait – emerging issues in a rapidly changing environment. *Mar Pollut Bull* 2: 593–596.

El Gammal, M.A.M., Nageeb, M., and S. Al-Sabeb. 2017. Phytoplankton abundance in relation to the quality of the coastal water – Arabian Gulf, Saudi Arabia. *Egypt J Aquat Res* 43, no. 4: 275–282.

Fowler, S.W., Readman, J.W., Oregioni, B.J.P.V., Villeneuve, J.P., and K. McKay. 1993. Petroleum hydrocarbons and trace metals in nearshore Gulf sediments and biota before and after the 1991 war: An assessment of temporal and spatial trends. *Mar Pollut Bull* 27: 171–182.

Glibert, P.M., Landsberg, J.H., Evans, J.J., Al-Sarawi, M.A., Faraj, M., Al-Jarallah, M.A., Haywood, A., Ibrahem, S., Klesius, P., Powell, C., and C. Shoemaker. 2002. A fish kill of massive proportion in Kuwait Bay, Arabian Gulf, 2001: The roles of bacterial disease, harmful algae, and eutrophication. *Harmful Algae* 1, no. 2: 215–231.

Goldman, J.C., and R. Mann. 1980. Temperature-influenced variations in speciation and chemical composition of marine phytoplankton in outdoor mass cultures. *J Exp Mar Biol Ecol* 46, no. 1: 29–39.

Gomi, Y., Taniguchi, A., and M. Fukuchi. 2007. Temporal and spatial variation of the phytoplankton assemblage in the eastern Indian sector of the Southern Ocean in summer 2001/2002. *Polar Biol* 30, no. 7: 817–827.

Grasshoff, K., Kremling, K., and M. Ehrhardt (Eds.). 1999. *Methods of seawater analysis.* Weinheim, Germany: John Wiley & Sons.

Guillard, R., and P. Kilham. 1977. The ecology of marine planktonic diatoms. *Biology of diatoms* 13: 372–469.

Habashi, B.B., Nageeb, F., and M. Faraj. 1992. *Distribution of phytoplankton cell abundance and chlorophyll with certain environmental factor in the ROPME Sea Area*. ROPME/LOC(UNESCO)/UNEP/NOAA Scientific workshop on results of R/V Mt. Mitchell Cruise in the ROPME Sea Area, Kuwait, pp. 1–23.
Hallegreaff, G.M. 2003. Harmful algal blooms: A global overview. In: *Manual on harmful marine microalgae*. Eds: G.M. Hallegreaff, D.M. Anderson, and A.D. Cembella, 1–22. IOC Manual and Guides No. 33. Paris, France: UNESCO.
Heip, C., Vincx, M., and G. Vranken. 1985. The ecology of marine nematodes. *Oceanogr Mar Biol* 23: 399–489.
Husain, M., and S.H. Ibrahim. 1998. *Study phytoplankton in the ROPME Sea Area. Offshore environment of the ROPME sea area after the war-related oil spill-results of the 1993–94 Umitaka-Maru Cruises*, 281–301. Tokyo: Terra Scientific Publishing Company (TERRAPUB).
Isave, V.A., and M.V. Mikhailova. 2009. The hydrography, evaluation, and hydrological regime of the mouth area of the Shatt al-Arab river. *Water Res* 36: 402–417.
Ismael, A.A., and M.M. Dorgham. 2003. Ecological indices as a tool for assessing pollution in El-Dekhaila Harbour (Alexandria, Egypt). *Oceanologia* 45, no. 1: 121–131.
Jost, C., Lawrence, C.A., Campolongo, F., Van De Bund, W., Hill, S., and D.L. DeAngelis. 2004. The effects of mixotrophy on the stability and dynamics of a simple planktonic food web model. *Theor Popul Biol* 66, no. 1: 37–51.
Lenes, J.M., Darrow, B.P., Cattrall, C., Heil, C.A., Callahan, M., Vargo, G.A., Byrne, R.H., Prospero, J.M., Bates, D.E., Fanning, K.A., and J.J. Walsh. 2001. Iron fertilization and the *Trichodesmium* response on the West Florida shelf. *Limnol Oceanogr* 46, no. 6: 1261–1277.
Lenes, J.M., and C.A. Heil. 2010. A historical analysis of the potential nutrient supply from the N2 fixing marine cyanobacterium *Trichodesmium* spp. to *Karenia brevis* blooms in the eastern Gulf of Mexico. *J Plankton Res* 32, no. 10: 1421–1431.
Liu, C.W., Lin, K.H., and Y.M. Kuo. 2003. Application of factor analysis in the assessment of groundwater quality in a Blackfoot disease area in Taiwan. *Sci Total Environ* 313: 77–89.
Loughland, R.A., Al-Abdulkader, K.A., Wyllie, A., and B.O. Burwell. 2012. Anthropogenic induced geomorphological change along the Western Arabian Gulf coast. In: *Studies on environmental and applied geomorphology*. Eds: T. Piacentini, and E. Miccadei, 191–218. IntechOpen. doi:10.5772/28330.
Margalef, R. 1967. Some concepts relative to the organization of plankton. *Oceanogr Mar Biol Ann Rev* 5: 257–289.
Miller, C.B., and P.A. Wheeler. 2012. *Biological oceanography*. Chichester, UK: John Wiley & Sons.
Moss, B. 1973. Diversity in fresh-water phytoplankton. *Am Midl Nat* 90, no. 2: 341–355.
Mulholland, M.R., Bernhardt, P.W., Heil, C.A., Bronk, D.A., and J.M. O'Neil. 2006. Nitrogen fixation and release of fixed nitrogen by *Trichodesmium* spp. in the Gulf of Mexico. *Limnol Oceanogr* 51, no. 4: 1762–1776.
Nezlin, N.P., Polikarpov, I.G., Al-Yamani, F.Y., Subba Rao, D.V., and A.M. Ignatov. 2010. Satellite monitoring of climatic factors regulating phytoplankton variability in the Arabian (Persian) Gulf. *J Mar Syst* 82, nos. 1–2: 47–60.
Nour El-Din, N.M., and J.A. Al-Khayat. 2005. Phytoplankton – zooplankton relations in three inland seas along the Qatari coast (Arabian Gulf). *Int J Environ Sci* 62, no. 4: 375–390.
Price, A.R.G. 1998. Impact of the 1991 Gulf War on the coastal environment and ecosystems: Current status and future prospects. *Environ Int* 24, nos. 1–2: 91–96.
Quigg, A., Al-Ansi, M., Al Din, N.N., Wei, C.L., Nunnally, C.C., Al-Ansari, I.S., Rowe, G.T., Soliman, Y., Al-Maslamani, I., Mahmoud, I., and N. Youssef. 2013. Phytoplankton along the coastal shelf of an oligotrophic hypersaline environment in a semi-enclosed marginal sea: Qatar (Arabian Gulf). *Cont Shelf Res* 60: 1–16.
Rajan, A., and T.Z. Al-Abdessalaam. 2008. *Harmful algal blooms eutrophication: Nutrient sources, composition and consequences in the Arabian Gulf bordering Abu Dhabi Emirate*. Proc 12th International Conference on Harmful Algae, Abu Dhabi, UAE, pp. 226–229.
Readman, J.W., Bartocci, J., Tolosa, I., Fowler, S.W., Oregioni, B., and M.Y. Abdulraheem. 1996. Recovery of the coastal marine environment in the Gulf following the 1991 war-related oil spills. *Mar Pollut Bull* 32, no. 6: 493–498.
Riegl, B.M., and S.J. Purkis. 2012. Coral reefs of the Gulf: Adaptation to climatic extremes in the world's hottest sea. In: *Coral reefs of the Gulf*, 1–4. Dordrecht: Springer.
Round, F.E., Crawford, R.M., and D.G. Mann. 1990. *Diatoms: Biology and morphology of the genera*. Cambridge: Cambridge University Press.
Ryther, J.H. 1956. Photosynthesis in the ocean as a function of light intensity. *Limnol Oceanogr* 1, no. 1: 61–70.
Sanders, J.G., and G.F. Riedel. 1987. Control of trace element toxicity by phytoplankton. In: *Phytochemical effects of environmental compounds*. Eds: J.A. Saunders, L. Kosak-Channing, and E.E. Conn, 131–149. Boston, MA: Springer.
Santhanam, R. 1987. *Phytoplankton of the Indian seas*. New Delhi, India: Daya Publishing House.
Sheppard, C. 2016. Coral reefs in the Gulf are mostly dead now, but can we do anything about it? *Mar Pollut Bull* 105, no. 2: 593–598.
Singh, R.P., Prasad, A.K., Kayetha, V.K., and M. Kafatos. 2008. Enhancement of oceanic parameters associated with dust storms using satellite data. *J Geophys Res Oceans* 113: C11008. doi:10.1029/2008JC004815.
Steidinger, K.A., and J. Williams. 1970. *Dinoflagellates*. St. Petersburg, FL: Florida Dept Natural Resources, Marine Research Laboratory.
Subba Rao, D.V., and F. Al-Yamani. 1998. Phytoplankton ecology in the waters between Shatt Al-Arab and Straits of Hormuz, Arabian Gulf: A review. *Plankton Biol Ecol* 45, no. 2: 101–116.
Subrahmanyan, R. 1946. A systematic account of the marine plankton diatoms of the Madras coast. *Proc Indian Acad Sc, Section B* 24, no. 4: 85–197.
Subrahmanyan, R. 1968. *MBAI memoir no. 2: The Dinophyceae of the Indian seas part I, genus Ceratium schrank*. Cochin: Marine Biological Association of India.
Taylor, F.J.R. 1992. The taxonomy of harmful marine phytoplankton. *Giornale Botanico Italiano* 126, no. 2: 209–219.
Thangaraja, M., Al-Aisry, A., and L. Al-Kharusi. 2007. Harmful algal blooms and their impacts in the middle and outer ROPME sea area. *Int J Oceans Oceanogr* 2, no. 1: 85–98.
Tomas, C.R. (Ed.). 1997. *Identifying marine phytoplankton*. San Diego, California: Elsevier.
Trombetta, T., Vidussi, F., Mas, S., Parin, D., Simier, M., and B. Mostajir. 2019. Water temperature drives phytoplankton blooms in coastal waters. *PLoS One* 14, no. 4: 0214933.
US EPA. 2000. Method 8015C (SW-846): Nonhalogenated Organics Using GC/FID, Rev. 3. Washington, DC.
Walsh, J.J., and K.A. Steidinger. 2001. Saharan dust and Florida red tides: The cyanophyte connection. *J Geophys Res Oceans* 106, no. C6: 11597–11612.
Wasmund, N., Kownacka, J., Göbel, J., Jaanus, A., Johansen, M., Jurgensone, I., Lehtinen, S., and M. Powilleit. 2017. The diatom/dinoflagellate index as an indicator of ecosystem changes in the Baltic Sea 1. Principle and handling instruction. *Front Mar Sci* 4, no. 22: 1–13. doi:10.3389/fmars.2017.00022.

PART 6

Marine reptiles

Natural history and biometrics of sea turtle populations nesting in the Saudi Arabian waters of the Arabian Gulf

21

Rommel H. Maneja, Jeffrey D. Miller,
Ace V.B. Flandez, Jinoy V. Gopalan, Yasir Y. Asiri,
Muaadh A. Alnuwairah, and Diego F. Lozano-Cortés

21.1 INTRODUCTION

21.1.1 Marine turtle biology and ecology

Marine turtles are part of a dynamic, integrated marine ecosystem that supports not only their needs but also those of other species (Bjorndal and Jackson 2003). In the marine ecosystem, the various species of marine turtles have different ecological roles and use different habitats during their lifetime (Musick and Limpus 1997; Miller 1997; Bolten 2003; Heppell et al. 2003), each of which is vital to their survival. In addition, marine turtle species differ in regard to their biological characteristics (van Buskirk and Crowder 1994). For example, they differ in size, reproductive aspects (i.e., size and number of eggs), as well as diet (van Buskirk and Crowder 1994; Bjorndal 1997; Miller 1997; Jones and Seminoff 2013; Heithaus 2013). However, all species of marine turtles share a general life history pattern (NRC 1990; Limpus and Miller 1993; Miller 1997; Musick and Limpus 1997). Turtles lay their eggs in beaches that support embryonic development and that are adjacent to marine habitats that facilitate dispersal of the hatchlings (Georges et al. 1993). Very young turtles in their first years of life typically forage in pelagic gyres near the surface of the ocean (Musick and Limpus 1997; Bolten 2003). As older immature turtles and as adults, they utilize a variety of foraging habitats associated with their food preferences including benthic, epipelagic, and pelagic zones (Bjorndal 1997; Bjorndal and Jackson 2003; Bolten 2003; Plotkin 2003; Eckert et al. 2012; Saba 2013; Mansfield and Putman 2013; Jones and Seminoff 2013), where they forage and develop towards adulthood and reproduction (Miller 1997; Miller et al. 2003). During their reproductive years, they migrate in varying distances between their foraging area and their nesting site, often passing through multiple marine areas (Luschi et al. 2003; Plotkin 2003).

21.1.2 Threats to marine turtles

Around the world, marine turtles complete their life cycles in marine ecosystems that are currently impacted by numerous threats, including over-fishing by commercial and artisanal fishers, pollution from many sources, loss of foraging areas and nesting beach habitats from human activities (Lutcavage et al. 1997), and the impacts of climate change (Hawkes et al. 2009; Hamann et al. 2003; Naser 2014; WMO 2017; IPCC 2018; Lincoln et al. 2021; Patrício et al. 2021). The threatening processes and changes in the marine environment impact the functioning of the ecosystems by disrupting the interconnectedness of their parts (Wabnitz et al. 2018). Understanding the complex interactions among marine turtles, as well as other marine species, and their environment is essential to good conservation management (Frazier 2005; Miller et al. 2019; Wallace et al. 2010).

Studies conducted elsewhere (Lewison and Crowder 2007; Lewison et al. 2008; Finkbeiner et al. 2011; Jennings and Kaiser 1998; Donlan et al. 2010; Williams et al. 2019; NRC 1990; Lutcavage et al. 1997; Fuentes et al. 2011; Ditmer and Stapleton 2012; Hamann et al. 2013) help to identify which threatening processes may be acting in the Arabian (Persian)

DOI: 10.1201/9781003321392-27

Gulf (the Gulf) but do not identify the level of impact they may have in the Gulf, with a few exceptions. For example, artisanal shrimp trawl-fishing operations have been identified as having important impacts (86.3% of 4726 captures) on marine turtle populations in Saudi Arabian territorial waters of the northwestern Arabian Gulf (Abdulqader 2000; Abdulqader et al. 2017). As a result, conservation management in the Gulf requires more regionally relevant information (Naser 2014). In addition to a lack of information on threatening processes, detailed information concerning the biology of the species and their habitats is missing or has only been partially defined (Phillott and Rees 2021; Al Ameri et al. 2022; Mobaraki et al. 2022; Maneja et al. 2023). Although connectedness between important foraging areas and nesting areas has been defined inside and outside the Gulf using satellite tracking (Pilcher et al. 2014, 2020; Rees et al. 2012a, 2012b, 2013, 2019), the details of interconnectedness of foraging and nesting areas within the Gulf are only partially known (Rees et al. 2013, 2019) Also, the genetic relatedness among nesting populations is only partially defined (D'aloia and Al-Ghais 2000; Jensen et al. 2019; Fitzsimmons and Limpus 2014; Tabib et al. 2011, 2014; Zolgharnein et al. 2011; Al-Mohanna et al. 2014). A better understanding of such information is important for effective conservation management in the Gulf region.

21.1.3 Studies on marine turtles in the Arabian Gulf

In recent years, research and/or monitoring on marine turtles in the Gulf has expanded and now includes information from every country where nesting occurs (Al Ameri et al. 2022; Phillott and Rees 2021; Mobaraki et al. 2022; IOSEA 2019a, 2019b, 2019c, 2019d, 2019e).

Of the five species of marine turtles known to inhabit the Gulf (Gasperetti et al. 1993), three species have been recorded nesting on the beaches within the Gulf (Phillott and Rees 2021; Al Ameri et al. 2022): the olive ridley (*Lepidochelys olivacea*), the hawksbill turtle (*Eretmochelys imbricata*), and the green turtle (*Chelonia mydas*). The olive ridley is listed in the IUCN Red List Category and Criteria as Vulnerable A2bd ver 3.1 (Abreu-Grobois and Plotkin 2008). Although hawksbill and green turtle nesting has been known for decades (Gasperetti et al. 1993), until recently nesting by olive ridley turtles was unknown within the Gulf. Olive ridley turtles have been tracked using satellite transmitters from nesting sites in Oman into the southern basin of the Gulf (Rees et al. 2012b), and a few stranded specimens have been reported in the northern basin (Abdulqader and Miller 2012). Recently, olive ridley turtles have been identified nesting at a few places along the Iranian coast and nearshore islands (Tollab et al. 2015). Knowledge of the locations of nesting sites used by hawksbill turtles and green turtles has improved along with data on the biological characteristics of the nesting species (Al Ameri et al. 2022; Phillott and Rees 2021; Mobaraki et al. 2022). The green and hawksbill turtles have widespread nesting in the Gulf (Gasperetti et al. 1993; Phillott and Rees 2021; Al Ameri et al. 2022). The hawksbill is listed by the IUCN as being Critically Endangered globally (CR A2bd, ver. 3.1) (Mortimer and Donnelly 2008), and the north Indian Ocean subpopulation of the green turtle is listed as being Vulnerable (A2b, ver. 3.1) (Mancini et al. 2019). Multiple management units have been defined for marine turtles around the world using multiple research techniques and analytical tools (Wallace et al. 2010). The Gulf is located in the northwest Indian Ocean regional management unit (RMU) for hawksbill turtles and green turtles (Wallace et al. 2010; see also Phillott and Rees 2021, Vargas et al. 2015).

The purpose of this chapter is to present a synopsis of the results of the studies concerning the nesting turtles on Jana and Karan Islands located in Saudi Arabian territory of the northern Gulf basin.

21.2 THE ARABIAN GULF AND OFFSHORE NESTING ISLANDS

21.2.1 Brief description of the Arabian Gulf

The Gulf is a semi-enclosed, shallow, subtropical sea surrounded by hyper-arid terrestrial environs fluctuating between hot summers (June to September) and cool winters (December to March) (Vaughan et al. 2019; Krishnakumar et al. 2019; Beni et al. 2021). The Gulf extends approximately 1,000 km along a NNW to SSE axis, and the width varies from about 56 km to 340 km (Sheppard 1993) (Figure 21.1). The mean depth of the Gulf is approximately 35 m; it has a maximum depth of approximately 100 m in the Strait of Hormuz. It covers an area of roughly 240,000 km^2 (Price et al. 1993; Krishnakumar et al. 2019; Vaughan et al. 2019; Sheppard et al. 2010; Barth and Khan 2008). The Qatar Peninsula divides the Gulf into northern and southern basins that slope from the shallow western and southwestern sides toward the deeper trench that extends along the eastern side (Johns et al. 1999; Reynolds 1993; Krishnakumar et al. 2019; Vaughan et al. 2019; Sheppard et al. 2010). The Saudi Arabian part of the northwestern Gulf includes an area of approximately 27,050 km^2 (Krishnakumar et al. 2019).

21.2.2 Main offshore nesting islands of Saudi Arabia

Jana Island (27°22′6.85″N, 49°53′50.85″E; dimensions: 1,105 m × 300 m; circumference: 2.6 km) and Karan Island (27°43′5.05″N, 49°49′28.91″E; dimensions: 2,024 m × 632 m; circumference: 5.3 km) are the two largest islands in the northwestern portion of the Gulf within the territorial waters of Saudi Arabia (Basyoni 1999; Miller 2011a, 2011b; Miller et al. 2019). Although nesting does occur on the other islands, these two islands comprise the major breeding sites for turtles and seabirds in the area (Miller et al. 2019; Figure 21.1).

The islands are characterized by coral sand augmented by vegetation, debris, and bird guano. The islands are low and flat and have no fresh water. The central areas of the islands are covered by low shrubs (*Suaeda* sp., *Salsola* sp.) (Maneja et al. 2020). Turtle nesting activity helps to keep the outer margin of the island free of vegetation. Each island is situated on the

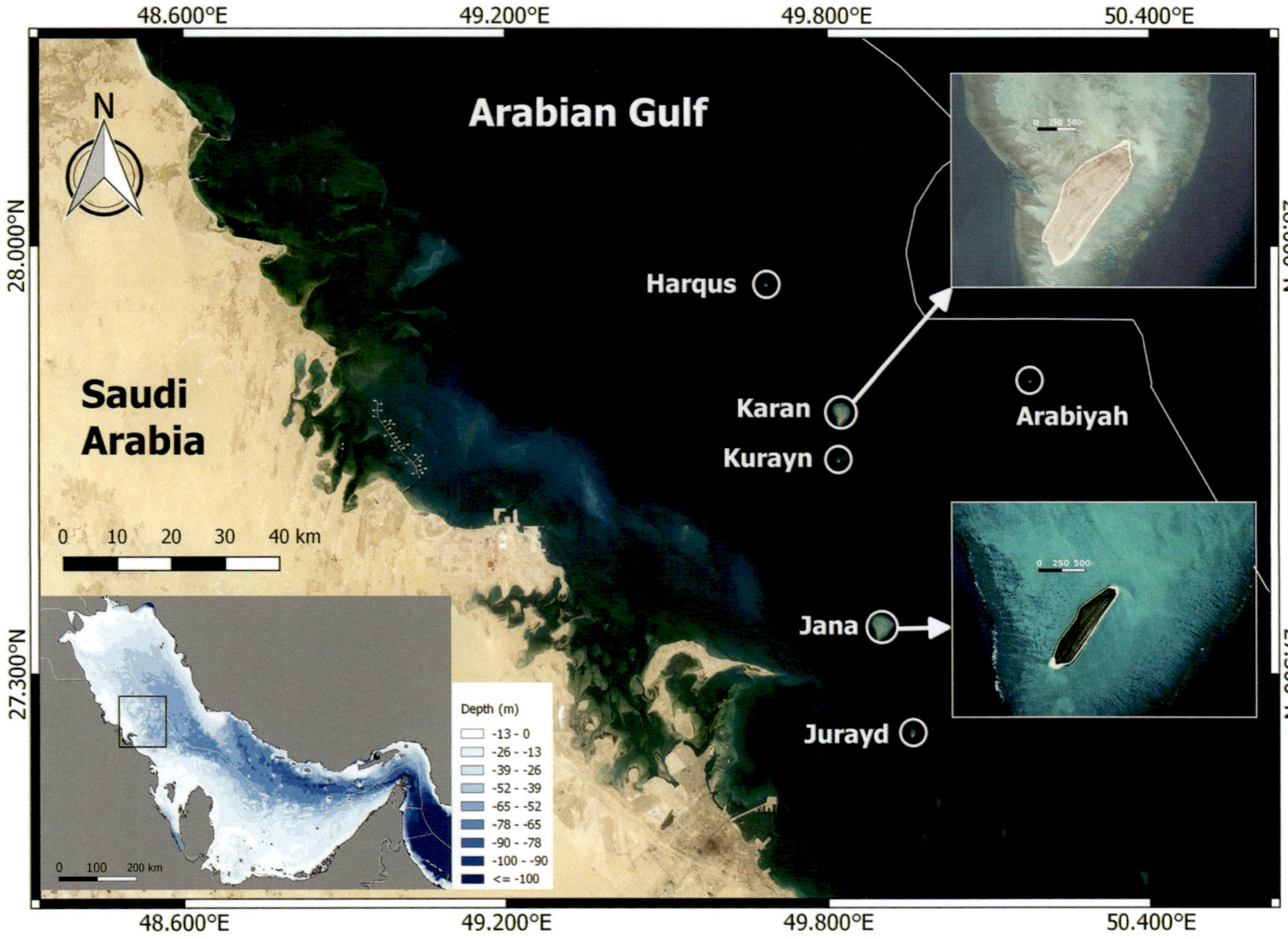

FIGURE 21.1 Satellite image base map showing the location of Karan and Jana Islands (shown by white arrows) in the Saudi Arabian waters of the Arabian Gulf. White circles represent the six offshore Saudi Arabian islands. Area is shown in lower left inset map of the whole Arabian Gulf with 30-sec GEBCO bathymetry (source: The GEBCO_2014 Grid, version 20150318, www.gebco.net).

southern portion of a shallow reef flat (Basson et al. 1977). Both islands have exposed beach rock and a storm berm along the northwestern side. As a result of the prevailing winds, the beach rock along the northwestern side is coated with tar and is heavily littered with debris from ships (e.g., wood, Styrofoam, plastic, light bulbs, etc.) (Maneja et al. 2021). In contrast, the sand of the southeastern side has noticeably less tar content and is comparatively clear of debris.

21.3 MARINE TURTLES IN THE SAUDI ARABIAN PORTION OF THE WESTERN GULF

21.3.1 Nesting marine turtles in the Saudi Arabian waters of the Gulf

21.3.1.1 Nesting locations

Both the green turtle and the hawksbill turtle nest on these Saudi Arabian offshore islands (Miller 1989, 2021; Pilcher 1999, 2000; Al-Merghani et al. 2000; Al-Mansi 2016) (Table 21.1). Hawksbill turtles nest on both Jana and Karan Islands, but the majority nest on Jana Island, with a small percentage of individuals moving between the islands (Miller 1989). In contrast, green turtles nest on both Karan and Jana Islands, but the majority utilize Karan. Some nesting by green turtles and hawksbill turtles occurs on Kurayn Island and Jurayd Island in low (albeit unquantified) numbers (Miller 1989). No major nesting areas occur on the mainland, although minor nesting occurs at Ras Tannurah (Miller 1989).

21.3.1.2 Nesting seasonality

Throughout the Gulf, hawksbill and green turtles nest seasonally and produce multiple clutches within a nesting season (Witzell 1983; Miller 1989; Hirth 1997; Al-Merghani et al. 1996, 2000; Al-Ghais 2009; Pilcher 1999, 2000; Al Ameri et al. 2022; Mobaraki et al. 2022) (Table 21.1). In the western portion of the Gulf, hawksbill turtles nest earlier in the spring (end of April, May, mid-June) than green turtles (mid-June, July, late August), with some overlap during the middle of June (Miller 1989; Al-Merghani et al. 1996, 2000; Pilcher 1999, 2000). In other areas of the Gulf, hawksbill turtles nest during April, May, and June as well (Qatar, Chatting et al. 2018; Iran, Mobaraki et al. 2004b, 2004c; Al Ameri et al.

TABLE 21.1 General biological characteristics of hawksbill and green turtles that nest on Jana and Karan Islands in the western Arabian Gulf (Miller 1989, 2011b; Miller et al. 2019; Pilcher 1999, 2000).

	HAWKSBILL TURTLES	*GREEN TURTLES*
BREEDING SEASON in the northern Gulf		
Mating	Commences in the spring and reaches a peak in late April.	Commences in late May and reaches a peak in early June.
Nesting	Commences in May, reaches a peak in late May, and ends by late June.	Commences in early June, reaches a peak in mid to late July, and ends in mid-September.
Hatchlings	Emerge from early July through mid-August, with a peak of hatching in late June.	Emerge from late July through mid-October, with a peak of hatching in late August to mid-September.
NESTING CYCLE: Mean, range [sample size]		
Incubation Period	55–60 days	59–65 days
Renesting Interval (days)	≅14	≅13
Remigration Interval (years)	3.9, 2–8 [14]	3.8, 1–7 [90]
Number of Clutches	1.7, 1–4	1.9, 1–9
KNOWN NESTING AREAS		
No important nesting areas occur on the mainland, although occasional nesting may occur during some years.	Jana Island, Karan Island, Kurayn Island, and Jurayd Island	Karan Island, Kurayn Island, Jana Island, and Jurayd Island
KNOWN FORAGING AREAS		
Both species may be found in low numbers in suitable habitats.	Around coral reefs and rocky reef out crops North of Abu 'Ali Island	Dawhat Abu 'Ali, around reefs, and shallow seagrass areas along the coast

2022). However, in the southern basin, hawksbill turtles begin nesting in March and end during June (UAE, Miller and Loughland 2006; UAE, Al-Ghais 2009; Iran, Mobaraki 2004b, 2021, Al Ameri et al. 2022). In the northern basin, green turtles begin nesting during the middle period of June and continue to nest into early September (Kuwait, Al-Mohanna et al. 2014; Iran, Mobaraki 2004a, 2004c; Sinaei et al. 2018; Al Ameri et al. 2022). In the southern basin, green turtles begin nesting in July and may continue through November (Iran: Mobaraki 2004c, 2021; Mohammadizadeh and Soltanpour 2014; Sinaei et al. 2018; UAE: Miller and Loughland 2006; Al Ameri et al. 2022).

21.3.1.3 Nesting population investigations

Determining the numbers of nesting turtles and whether the population is decreasing or increasing can be difficult because this requires long-term datasets that have been collected using consistent methods (i.e., consistent tagging interval) (Chaloupka and Musick 1997). Between 1987 and 1997, a total of 448 hawksbill turtles were tagged on Jana Island; an additional 16 hawksbill turtles were tagged in 1986 on Karan Island (Al Merghani et al. 2000; Al-Mansi 2016). Between 1986 and 2014, a total of 714 hawksbill turtles were tagged (Al-Mansi 2016). Over the years the tagging interval has varied from 7 days in 1987 to 70 days in 1991, and in recent years the interval has been about 15 to 21 days. Between 1987 and 1997, a total of 4784 green turtles were tagged while nesting at Karan Island (Al-Merghani et al. 2000; Al-Mansi 2016). The tagging effort varied from 15 days in 1997 to 93 days in 1986.

21.3.1.4 Adult turtle foraging sites

The foraging areas used by the nesting turtles during their non-reproductive periods are poorly defined in Saudi Arabia. However, by satellite tracking post-nesting turtles, Pilcher et al. (2014) identified the nearshore area located to the north of Abu Ali Island as an important hawksbill turtle foraging area. Rees et al. (2019) reported two post-nesting hawksbill turtles from Kuwait inhabiting foraging areas in the offshore islands and coastal areas in the central Saudi Arabian waters of the Gulf. The majority of the green turtles tracked by satellite from nesting in Kuwait stayed in Failaka Island in Kuwait, but a few individuals went to foraging areas located in the shallow water along the northern and central Saudi Arabian coast as well as one reaching the foraging sites in the UAE (Rees et al. 2013, 2018). In addition, aerial and watercraft-based surveys identified Dawhat Abu Ali as a foraging area for green turtles (Miller 1989). Incidental sightings of turtles in the shallow coastal waters indicate that both species forage in areas around reefs and seagrass meadows along the coast and around the offshore islands, at least in low numbers (Miller 1989). This is similar to the situation in the shallow offshore area of the UAE, where foraging by both species is widespread (Miller et al. 2004; Pilcher et al. 2021).

21.3.2 Nesting turtles and their eggs

The current analysis of turtles nesting on Jana Island and Karan Island is based on four summarized datasets for each species collected between 1986–2022. These studies were focused on defining the morphological and reproductive characteristics of the nesting turtles. Miller (1989) conducted field work on the islands in 1986 and 1987. Pilcher reported summarized data on hawksbill turtles (1999) and green turtles (2000) collected during the 1991 and 1992 nesting seasons. Al-Mansi (2016) presented data collected between 2001 and 2014. KFUPM/RI (2020) reported data collected on both species in 2017 through 2022.

These four summarized studies included determining curved carapace lengths and weights of nesting turtles, along with clutch size and egg size measurements following the methods used previously (Miller 1989; Pilcher 1999, 2000; Al-Mansi 2016). Turtles were measured and weighed after oviposition was completed. Eggs were measured, weighed, and returned to their nest chamber within 1 hour (Limpus et al. 1979). The location of each nest was marked by triangulation from marker posts in the dunes and by GPS coordinates. Assessment of hatching/emergence success followed standard methods (Miller 1999). Hatching success refers to the number of hatchlings that successfully leave the eggshell within the nest chamber, and emergence success refers to the number of hatchlings leaving the nest and emerging onto the beach surface (Miller 1999). Hatching success was assessed for egg clutches laid in the western and southern parts of Karan Island (Maneja et al. 2023). Temperatures in the middle of the clutches were recorded at 1-hour intervals using HOBO 8k data loggers (Onsetcomp.com, MA). Briefly, monitored clutches were allowed to incubate for a minimum of 60 days and excavated afterwards. The contents of nests were separated into categories. Each empty eggshell that was greater than 50% intact was considered to represent a hatched egg (Miller 1999). Unhatched eggs were opened and their contents categorized as 'No sign of embryonic development or embryo unpigmented' or 'late embryonic death' (i.e., carapace pigmented embryo) (Miller et al. 2017). Because the contents of the nest were examined after the hatchlings had emerged and because no live or dead hatchlings were recovered from the nests, hatching success and emergence success were equal (Miller 1999).

For hawksbill turtles nesting on Jana Island between 2017 and 2022, the mean curved carapace length ($CCL_{standard}$; Bolten 1999) was 71.2 cm (SD = 1.76 cm, Min = 67.0 cm, Max = 75.5 cm, n = 33) with a mean weight of 40.2 kg (SD = 2.13 kg, Min = 35.0 kg, Max = 47.6 kg, n = 20). The clutches oviposited by these turtles contained a mean of 68.7 eggs (SD = 11.21 eggs, Min = 29 eggs, Max = 84 eggs, n = 20 clutches) and a mean of 19.6 yolkless eggs (SD = 7.88 yolkless eggs, Min = 4 yolkless eggs, Max = 37 yolkless eggs, n = 20 clutches) (Table 21.2).

Comparison among the four datasets for hawksbill turtles showed there was no significant difference among the summarized data tested for curved carapace length (ANOVA, $F_{3,692} = 0.0103$, $p = 0.999$) (Table 21.3). However, there was a significant difference among the weights of the nesting turtles (ANOVA, $F_{3,180} = 2.67$, $p = 0.049$). There was no significant difference among the numbers of eggs in the clutches oviposited by these turtles (ANOVA, $F_{3,104} = 0.1629$, p = 0.9211) or among the numbers of yolkless eggs contained in these clutches (ANOVA, $F_{3,104} = 0.7289$, p = 0.5369).

Based on the pooled data, the hawksbill turtles nesting on Jana and Karan islands had a mean CCL of 71.2 cm (SD = 3.49 cm, Min = 59 cm, Max = 93 cm, n = 696 turtles) and mean weight of 38.7 kg (SD = 6.12 kg, Min = 20 kg, Max = 64 kg, n = 184 turtles (Table 21.3). The pooled clutches contained a mean of 70.0 eggs (SD = 17.9 eggs, Min = 29 eggs, Max = 127 eggs, n = 115 clutches). These clutches contained a mean of 17.2 yolkless eggs (i.e., without any propensity for development) (Min = 2 yolkless eggs, Max = 42 yolkless eggs, n = 115 clutches). The eggs had a mean diameter of 37.5 mm (SD = 1.96 mm, Min = 22 mm, Max = 49.8 mm, n = 1077 eggs from 102 clutches) and a mean weight of 29.3 g (SD = 3.05 g, Min = 16 g, Max = 62 g, n = 1077 eggs from 102 clutches). There was no significant difference among the groups concerning the number of eggs in the clutches sampled (ANOVA, $F_{3,104} = 0.1629$, $p = 0.9211$). However, both the diameters and the weights of these eggs varied significantly (diameter: ANOVA, $F_{3,1074} = 10.34$, $p = 0.0000$; weight: ANOVA, $F_{3,1074} = 72.19$, $p = 0.0000$).

Comparison among the four summarized datasets from hawksbill turtles nesting at Jana and Karan Islands to the summarized data collected from hawksbill turtles nesting at four different sites in Iran (Sheedvar Is., Hendourabi Is., Nakhiloo Is., Ommolkaram Is.; Mobaraki et al. 2022) showed no significant difference among the CCL measurements and among the numbers of eggs in the clutches (Table 21.4). However, significant differences did occur among the diameters and weights of the eggs. Because the biological meaning of these results remains unknown, the summary data are accepted as describing the general reproductive characteristics of the nesting hawksbill turtles.

For green turtles nesting on Karan Island between 2017 and 2022, the mean curved carapace length ($CCL_{standard}$; Bolten 1999) of green turtles was 99.4 cm (SD = 4.60 cm, Min = 90.0 cm, Max = 109 cm, n = 69) with a mean weight of 108.0 kg (SD = 14.64 kg, Min = 74.8 kg, Max = 136.3 kg, n = 55) (Table 21.5). The clutches oviposited by these turtles contained a mean of 90.6 eggs (SD = 22.65 eggs, Min = 59 eggs, Max = 134 eggs, n = 29 clutches) with additional yolkless eggs (mean = 5.5 yolkless eggs, Min = 1 yolkless egg, Max = 15 yolkless eggs, n = 29 clutches).

When data from the previous and current studies were compared, the pooled mean CCL measurements of green turtles nesting on Karan Island was 97.9 cm (SD = 4.702 kg, Min = 72.3 kg, Max = 114 kg, n = 2098 turtles), and they had a pooled mean weight of 107.9 kg (SD = 14.7 kg, Min = 54 kg, Max = 168 kg, n = 758). There was a significant difference among CCL measurements of turtles from the data summaries tested (ANOVA, $F_{3,2094} = 5.016$, $p = 0.0018$) but there was no significant difference among the weights of these turtles (ANOVA, $F_{3,754} = 2.589$, $p = 0.0519$) (Table 21.6).

The pooled data indicated that green turtles nesting on Karan Island produced clutches that contained a mean of 88.4 eggs (SD = 19.83, Min =33 eggs, Max = 146 eggs, n = 142 clutches). The eggs had a mean diameter of 42.9 mm (SD = 1.75 mm, Min = 23.6 mm, Max = 50.4 mm, n = 1186 eggs) and a mean weight of 44.8 g (SD = 5.7 g, Min = 30 g, Max = 60 g, n = 1027 eggs) (Table 21.6). In addition, the pooled mean number of yolkless eggs was 10.1 (SD = 8.30, Min = 0, Max = 36, n = 189 clutches). There was no significant difference among the groups concerning the number of eggs laid in the clutches sampled (ANOVA, $F_{3,138} = 0.2043$, p = 0.8932). However, there was a significant difference among the diameters of eggs (ANOVA, $F_{3,1182} = 21.813$, p = 0.0000) and

TABLE 21.2 Summarized reproductive characteristics of hawksbill turtles (***Eretmochelys imbricata***) nesting at Karan and Jana Islands in the northwestern Gulf.[1] Based on summary data from Miller (1989), Pilcher (1999), Al-Mansi (2016), and KFUPM between 2017 and 2022 (KFUPM/RI 2020).

CHARACTERISTIC	*MEAN*	*SD*	*MIN*	*MAX*	*N*	*PERIOD OF DATA*	*SOURCE*
Adult CCL (cm)	71.3	3.60	60.5	78.0	35	1986–1987	(Miller 1989)
	71.2	4.10	59.0	93.0	240	1991–1992	(Pilcher 1999)
	71.2	3.16	63.0	78.5	388	2010–2014	(Al-Mansi 2016)
	71.2	1.76	67.0	75.5	33	2017–2022	(KFUPM/RI 2020)
Adult Weight (kg)	38.4	4.29	30.5	45.0	28	1986–1987	(Miller 1989)
	39.1	6.70	30.0	64.0	123	1991–1992	(Pilcher 1999)
	34.5	7.50	20.0	46.0	13	2010–2014	(Al-Mansi 2016)
	40.2	2.13	35.0	47.6	20	2017–2022	(KFUPM/RI 2020)
# Eggs in Clutch	68.6	15.72	30	95	20	1986–1987	(Miller 1989)
	70.2	19.28	36	127	63	1991–1992	(Pilcher 1999)
	72.8	19.80	44	121	12	2010–2014	(Al-Mansi 2016)
	68.7	11.21	29	84	20	2017–2022	(KFUPM/RI 2020)
# Yolkless Eggs[2]	16.9	–	2	36	20	1986–1987	(Miller 1989)
	17.1	–	4	42	63	1991–1992	(Pilcher 1999)
	14.9	–	4	37	12	2010–2014	(Al-Mansi 2016)
	19.6	–	4	37	20	2017–2022	(KFUPM/RI 2020)
Diameter (mm)	37.6	0.2	35.0	39.8	140	1986–1987	(Miller 1989)
	37	2.3	22.0	49.8	724	1991–1992	(Pilcher 1999)
	38.1	0.1	34.8	40.6	84	2010–2014	(Al-Mansi 2016)
	37.2	1.3	34.8	42.2	130	2017–2022	(KFUPM/RI 2020)
Egg Weight (g)	28.6	3.63	19.4	36.9	130	1986–1987	(Miller 1989)
	31.9	4.36	16.0	62.0	724	1991–1992	(Pilcher 1999)
	29.5	1.56	27.0	35.0	84	2010–2014	(Al-Mansi 2016)
	27.2	2.67	22.1	31.7	130	2017–2022	(KFUPM/RI 2020)
Hatchling SCL (mm)	39.2	0.78	36.6	41.4	19	1986–1987	(Miller 1989)
	38.0	1.89	28.8	41.9	261	1991–1992	(Pilcher 1999)
	37.3	0.19	32.0	41.6	25	2010–2014	(Al-Mansi 2016)
Hatchling Weight (g)	14.1	1.34	12.1	16.1	19	1986–1987	(Miller 1989)
	14.3	1.96	10.0	20.1	261	1991–1992	(Pilcher 1999)
	12.6	0.24	12.0	13.3	25	2010–2014	(Al-Mansi 2016)

[1]A few tagged turtles were recorded within and between study periods.
[2]Yolkless eggs were not normally distributed, but there is no non-parametric test for summarized data. Therefore, the mean, minimum, and maximum are presented.

among the groups for egg weights (ANOVA, $F_{3,1023} = 24.629$, p = 0.0000).

Comparison among the datasets for reproductive characteristics of green turtles nesting at Karan Island and elsewhere in the Gulf and Sea of Oman indicated significant differences exist among the nesting groups (Table 21.7). For example, the CCL measurements of green turtles nesting at Karan Island (Table 21.6) were similar to those of green turtles nesting in Kuwait (mean = 98.2 cm, SD = 4 cm, Min-Max = 93–105 cm, n = 10; Rees et al. 2013) and at Ras Al Hadd in Oman (mean = 99.7 cm, SD = 6.53 cm, Min-Max = 91.1–109.1 cm, n = 6; Pilcher et al. 2021). However, those CCL measurements were smaller than the CCL measurements from green turtles nesting at Chabahar beach (Mean = 106.3 cm, SD = 6 cm, Min-Max = 100–120 cm, n = 6; Sinaei et al. 2018), Ras Al Jinz near Ras Al Hadd, Oman (Mean = 102.7 cm, SD = 1.87 cm, Min-Max = 89–116 cm, n = 36; Mendonca et al. 2010), and along the northern coast of the Sea of Oman (Mean = 111 cm, SD = 4 cm, Min-Max = 103–122 cm, n = 18; Sinaei et al. 2019). Because the genetic relationships among these nesting groups are unclear, the biological meaning of these results remains unknown and the summary data are accepted as describing the general reproductive characteristics of the nesting green turtles in the region.

Looking at the biological characteristics of sea turtle species, the combination of carapace length and total body mass (weight) are commonly used to define the size of nesting sea turtles, along with the numbers of eggs produced in a clutch (Bolten 1999; Miller 1999). Although significant differences were found for some of the variables among the datasets and not others, the similarities of the means, variance, and minimum-maximum within the groups suggest that the differences are not biologically definitive. Another source of potential difference occurs among the summarized datasets resulting from the style

TABLE 21.3 Pooled data on the reproductive characteristics of hawksbill turtles (***Eretmochelys imbricata***) nesting at Karan and Jana Islands in the northwestern Gulf. Based on summary data from Miller (1989), Pilcher (1999, 2000), Al-Mansi (2016), and KFUPM/RI (2020).

	MEAN	*SD*	*MIN*	*MAX*	*N*	*F =*	*DF*	*P =*	*
Nesting Females									
CCL (cm)	71.2	3.488	59.0	93.0	696	0.0103	3, 692	0.9986	
Weight (kg)	38.7	6.116	20.0	64.0	184	2.6690	3, 180	0.0491	*
Eggs									
#Eggs in Clutch	70.0	17.960	29.0	127.0	115	0.1629	3, 104	0.9211	
Diameter (mm)	37.5	1.960	22.0	49.8	1077	10.336	3, 1074	0.0000	*
Weight (g)	29.3	3.054	16.0	62.0	1077	72.199	3, 1074	0.0000	*
Yolkless Eggs1	17.2	–	2.0	42.0	115	–	–	–	
Hatchlings									
SCL (mm)	38.2	1.770	28.8	41.9	305	6.345	2, 302	0.0019	*
Weight (g)	14.15	1.852	10.0	20.1	305	9.3139	2, 302	0.0001	*

* = ANOVA indicated significant differences among data sets at alpha 0.05.
[1]Yolkless eggs were not normally distributed, but there is no non-parametric test for summarized data. Therefore, the mean, minimum and maximum are presented.

TABLE 21.4 Pooled data on the reproductive characteristics of hawksbill turtles (***Eretmochelys imbricata***) nesting at Karan and Jana Islands in the northwestern Gulf and in Iran. Based on summary data from Miller (1989), Pilcher (1999, 2000), Al-Mansi (2016), KFUPM (between 2017 and 2022) (KFUPM/RI 2020), and Mobaraki et al. (2022).

	MEAN	*SD*	*MIN*	*MAX*	*N*	*F =*	*DF*	*P =*	*
Nesting Females									
CCL (cm)	71.4	3.489	58	93	1124	0.8563	7, 1116	0.5408	
Weight (kg)	41.5	5.643	20	64	498	11.509	7, 490	0.0000	*
Eggs									
#Eggs in Clutch	73.5	17.710	29	131	281	1.396	7, 273	0.2067	
Diameter (mm)	37.6	1.70	22	49.8	273	2.399	7, 265	0.0214	*
Egg Weight (g)	30.2	3.74	16	62	273	4.244	7, 265	0.0002	*

* = ANOVA indicated significant differences among data sets at alpha 0.05.

of carapace measurement (Bolten 1999), which was not defined in all of the studies. Although producing similar values in adult green turtles, measuring the curved carapace length from the boundary between the nuchal scale and the skin of the neck to the anterior of the notch between the supracaudal scales does not yield the same measurement as measuring to the posterior of the supracaudal scales (see Bolten 1999). The difference in the two measurements tends to be greater in adult hawksbill turtles. In addition, because hawksbill and green turtles do not nest each year, some of the variation in the sizes of turtles may derive from different cohorts of turtles nesting each year in different ratios from various foraging areas. In this context, we consider the current data as characterizing the morphometrics of hawksbill and green turtles nesting at Jana and Karan Islands.

21.3.3 Brief comparison with other nesting populations in the Gulf

The hawksbill turtles nesting in the western Gulf are similar in size to those nesting along the eastern coast of the Gulf and in the southern basin (Al Ameri et al. 2022). For example, the range in mean CCL measurements of hawksbill turtles nesting in Kuwait was 69–78.5 cm (Mean = 73.3 cm, SD = 4.2 cm, n = 5) (Rees et al. 2019). The range in the CCL measurements for hawksbill turtles nesting along the coast of Qatar was 60.5–80.0 cm (mean = 70.8 cm, SD = 2.9 cm, n = 428) (Chatting et al. 2018). Mobaraki et al. (2022) summarized morphometric data of hawksbill turtles nesting at four sites along the Iranian coast and found no significant difference among the data sets for CCL, weight, and number of eggs per clutch, as well as the diameter and weight of the eggs. They reported a combined mean CCL of 71.6 cm (SD = 3.5, range = 58–82, n = 428). These nesting hawksbill turtles had a mean weight of 43.3 kg (SD = 5.5, range = 28–60, n = 314).

When the summary data from the four Iranian sites were compared to the four datasets from Jana and Karan islands in the western Gulf, there was no significant difference among the data (ANOVA, $F_{(7,1116)} = 0.856$, $p = 0.541$). The mean CCL of the combined data was 71.4 cm (SD = 3.49, Min = 58, Max = 93, n = 1124). However, the mass of hawksbill turtles nesting on the islands in the western Gulf was significantly less than those turtles nesting in Iran (ANOVA $F_{(7,490)} = 11.5$, $p = 0.000$).

TABLE 21.5 Summarized data on reproductive characteristics of green turtles (***Chelonia mydas***) nesting at Karan and Jana Islands in the northwestern Gulf.[1] Based on summary data from Miller (1989), Pilcher (2000), Al-Mansi (2016), and KFUPM between 2017 and 2022 (KFUPM/RI 2020).

CHARACTERISTIC	*MEAN*	*SD*	*MIN*	*MAX*	*N*	*PERIOD OF DATA*	*SOURCE*
Adult CCL (cm)	97.8	4.21	89.5	108.5	102	1986–1987	(Miller 1989)
	97.7	4.73	72.3	114.0	1391	1991–1992	(Pilcher 2000)
	98.4	4.73	83.5	113.5	536	2010–2014	(Al-Mansi 2016)
	99.4	4.604	90.0	109.0	69	2017–2022	(KFUPM/RI 2020)
Adult Weight (kg)	103.3	13.25	74.0	144.0	69	1986–1987	(Miller 1989)
	108.5	14.63	54.0	168.0	619	1991–1992	(Pilcher 1999)
	108.1	22.64	71.0	163.0	15	2010–2014	(Al-Mansi 2016)
	108.0	14.64	74.8	136.3	55	2017–2022	(KFUPM/RI 2020)
# Eggs in Clutch	88.5	15.81	68	114	58	1986–1987	(Miller 1989)
	86.9	22.63	33	146	42	1991–1992	(Pilcher 2000)
	87.8	19.40	45	125	12	2010–2014	(Al-Mansi 2016)
	90.6	22.65	59	134	29	2017–2022	(KFUPM/RI 2020)
# Yolkless Eggs	9.9	6.84	0	28	580	1986–1987	(Miller 1989)
	5	5.24	0	22	422	1991–1992	(Pilcher 1999)
	16	12.18	0	36	60	2010–2014	(Al-Mansi 2016)
	5.5	2.34	1	15	29	2017–2022	(KFUPM/RI 2020)
Egg Dia (mm)	43.2	0.18	33.8	49.0	420	1986–1987	(Miller 1989)
	42.7	2.61	35.0	50.4	422	1991–1992	(Pilcher 2000)
	43.6	0.23	40.0	49.0	60	2010–2014	(Al-Mansi 2016)
	42.0	2.41	23.6	44.7	124	2017–2022	(KFUPM/RI 2020)
Egg Weight (g)	44.13	4.48	33.2	56.3	580	1986–1987	(Miller 1989)
	46.3	7.41	30.0	60.0	423	1991–1992	(Pilcher 1999)
	46.2	3.91	40.0	53.0	60	2010–2014	(Al-Mansi 2016)
	41.7	2.57	36.46	46.84	124	2017–2022	(KFUPM/RI 2020)
Hatchling SCL (mm)	47.9	0.12	45.0	52.1	120	1986–197	(Miller 1989)
	47.7	1.78	41.4	51.4	162	1991–1992	(Pilcher 2000)
	48.7	0.20	44.0	52.	90	2010–2014	(Al-Mansi 2016)
Hatchling Weight (g)	21.9	1.33	18.2	25.0	120	1986–1987	(Miller 1989)
	22.3	2.15	15.0	29.0	162	1991–1992	(Pilcher 1999)
	22.6	3.54	19.0	27.5	90.	2010–2014	(Al-Mansi 2016)

[1]A few tagged turtles were recorded within and between study periods.

TABLE 21.6 Pooled data on the reproductive characteristics of green turtles (***Chelonia mydas***) nesting at Karan and Jana Islands in the northwestern Gulf. Based on summary data from Miller (1989), Pilcher (2000), Al-Mansi (2016), and KFUPM/RI (2020).

NESTING FEMALES	*MEAN*	*SD*	*MIN*	*MAX*	*N*	*F =*	*DF*	*P =*	
CCL (cm)	97.9	4.702	72.3	114.0	2098	5.0163	3, 2094	0.0018	*
Weight (kg)	107.9	14.702	54.0	168.0	758	2.5890	3, 754	0.0519	
# Eggs in Clutch	88.4	19.830	33	146.0	142	0.204	3, 138	0.8933	
Diameter (mm)	42.9	1.75	23.6	50.4	1186	21.813	3, 1182	0.0000	*
Yolkless Eggs[1]	10.1	–	0.0	36.0	189	–	–	–	
Eggs:									
Weight (g)	44.8	5.707	30.0	60.0	1027	24.6291	3, 1023	0.0000	*
Hatchlings									
SCL (mm)	48.0	1.183	41.4	52.1	372	21.433	2, 369	0.0000	*
Weight (g)	22.2	2.372	15.0	29.0	372	2.3265	2, 369	0.0991	

* = ANOVA showed significant differences among data sets at alpha 0.05.

[1]Yolkless eggs were not normally distributed, but there is no non-parametric test for summarized data. Therefore, the mean, minimum and maximum are presented.

TABLE 21.7 Pooled data on the reproductive characteristics of green turtles (***Chelonia mydas***) nesting at Karan and Jana Islands in the northwestern Gulf and elsewhere in the Gulf and Sea of Oman. Based on summary data from Miller (1989), Pilcher (2000), Al-Mansi (2016), KFUPM/RI (2020), Sinaei et al. (2018, 2019), Al-Mohanna et al. (2014), Rees et al. (2013), Mendonca et al. (2010), and Pilcher et al. (2021).

NESTING FEMALES	*MEAN*	*SD*	*MIN*	*MAX*	*N*	*F =*	*DF*	*P =*	
CCL (cm)	97.2	4.59	72.3	122	2047	24.861	9, 2237	0.0000	*
Weight (kg)	107.8	14.0	54.0	168.0	838	2.860	5, 832	0.0144	*
# Eggs in Clutch	91.8	25.1	33	157	183	2.753	5, 177	0.0202	*
Diameter (mm)	42.9	1.74	23.6	54.4	1201	34.122	4,1196	0.0000	*

* = ANOVA showed significant differences among data sets at alpha 0.05.

Comparison of data from nesting populations inside the Gulf region to worldwide data shows that hawksbill turtles nesting in the Gulf are among the smallest nesting hawksbill turtles in the world (Witzell 1983; van Buskirk and Crowder 1994; Pilcher 1999; Chatting et al. 2018; Al Ameri et al. 2022; Phillott and Rees 2021) (Figure 21.2). In addition, the turtles nesting in the Gulf region lay fewer eggs than those produced by the species nesting elsewhere (Witzell 1983; Van Buskirk and Crowder 1994; Pilcher 1999; Al Ameri et al. 2022). van Buskirk and Crowder (1994) reviewed data from 17 hawksbill turtle populations and reported a mean CCL of 78.6 cm (SD = 1.6, Min = 66.0 cm, Max = 96.0 cm, n = 17) and a mean number of eggs oviposited of 130 (SD = 6.8, Min = 73.2, Max = 150 cm, n = 17). The biological causes and life history impacts, if any, of these differences are unknown. It is possible that the extreme temperature fluctuation in the Gulf may impact food production or nutritional uptake and hence growth (Pilcher 1999; Chatting et al. 2018).

Green turtles nest in very low density at a few locations in the Gulf, including the UAE (Al Suweidi et al. 2012; Hebbelmann et al. 2016; Al Ameri et al. 2020), Iran (Sheedvar, Hendourabi and Khargoo Islands; Mobaraki 2004d; Mostafavi et al. 2010), and Kuwait (Rees et al. 2013, 2019), as well as on the Oman Sea coast (Mobaraki 2004c, 2021).

Green turtles nesting in the western Gulf were similar in size to those nesting elsewhere in the Gulf and in the Sea of Oman (Sinaei et al. 2018, 2019; Al-Mohanna et al. 2014; Mendonca et al. 2010; Pilcher et al. 2021), although there was a significant difference among the nesting groups ($ANOVA_{9,2237}$ = 24.860, p = 00000). The pooled mean CCL was 98.2 cm (SD = 4.59, Min = 72.3 cm, Max = 122 cm, n = 2247). Until genetic research clarifies the interrelationships among the foraging and nesting populations in the Gulf, this value is accepted as an approximation of the size of nesting green turtles in the region.

The typical CCL measurements of green turtles nesting in the Arabian Gulf are in the lower range (but not the lowest) for the species (Hirth 1997; Pilcher 2000; Al Ameri et al. 2022) and about the same size as those found nesting in Oman (Ross and Barwani 1982) (Figure 21.2). van Buskirk and Crowder

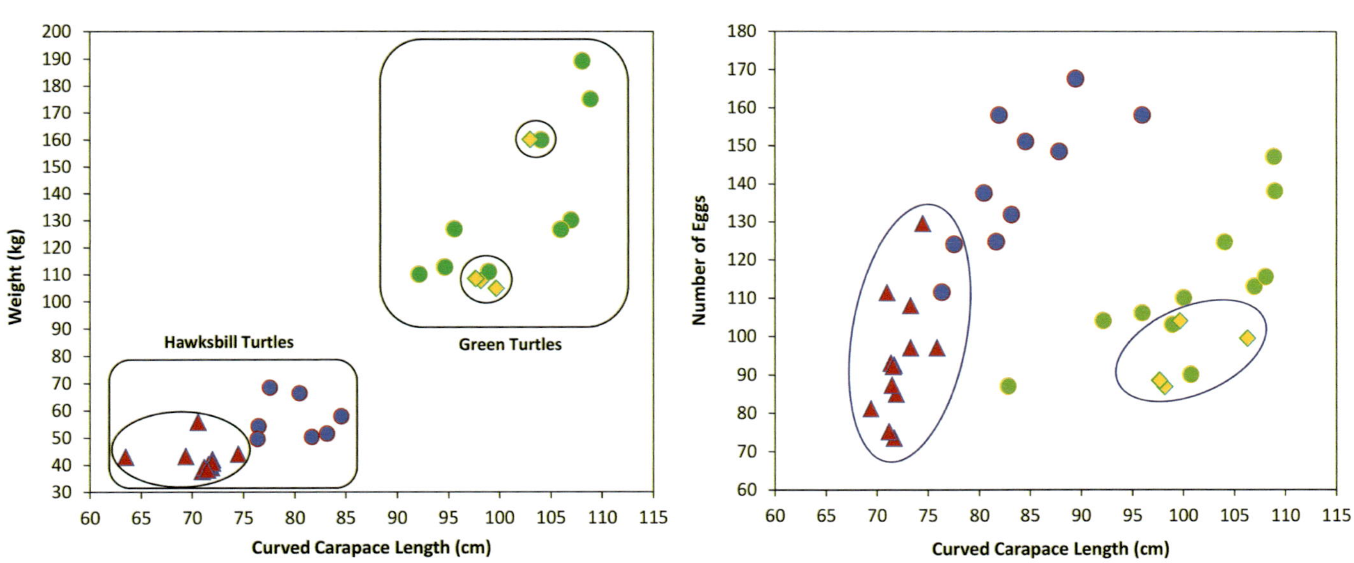

FIGURE 21.2 Comparison of the curved carapace lengths of hawksbill turtles and green turtles relative to their weights (left) and numbers of eggs oviposited (right) from nesting populations inside the Gulf and in the Indian Ocean, Pacific Ocean, and Atlantic Ocean regions. Red triangles indicate hawksbill turtle populations from within the Gulf; blue circles indicate hawksbill turtle populations from outside the Gulf region; yellow diamonds indicate green turtle populations from within the Gulf; green circles indicate green turtle populations from outside the Gulf region. Data from Witzell (1983), Hirth (1997), van Buskirk and Crowder (1994), Pilcher (1999, 2000), and Al Ameri et al. (2022).

(1994) reviewed data from 22 green turtle populations and reported a mean CCL of 99.1 cm (SD = 1.9, Min = 77.3 cm, Max = 112.9 cm, n = 22), which is only slightly larger than the pooled mean CCL in the Gulf region.

21.3.4 Hatching success

In 2017 on Karan Island, the mean hatching success of green turtle eggs was 26.6% (SD = ±38.8%, Min = 0%, Max = 83.3%, n = 4). In 2019, the mean hatching success was 38.8% (SD = 25.2%, Min = 2.5%, Max = 75%, n = 14). The mean hatching success for the combined nests was 36.1% (SD = 27.9%, Min = 0%, Max = 83.3%, n = 18) (Maneja et al. 2023). There was no significant difference in the hatching success of eggs that incubated in the two years ($F_{1,19}$ = 0.3247, p = 0.5754). Eggs were oviposited in two areas on the southeastern (n = 7) and western (n = 6) parts of Karan Island. There was no significant difference in the hatching success of eggs that incubated in nests located in two of the clusters (t_{11} = 1.175, p = 0.265), although the southeastern area had a higher mean hatching success (mean = 48.5%, SD = 22.5%, n = 7) compared to the western area (mean = 32.9%, SD = 25.5%, n = 6) (Maneja et al. 2023).

The unhatched eggs comprised a mean of 61.2% of the total number of eggs in the clutches. Of the total unhatched eggs, 32.5% were classified as "no signs of development" and the remainder were classified as late embryonic death. The percentage of eggs that showed "no signs of development" had a mean of 19.8% (Min = 3.3%, Max = 64.2%), and the mean percentage of eggs that contained pigmented embryos was 41.4% (Min = 4.8, Max = 92.6%).

Higher temperatures in nests during the incubation period resulted in lower hatching success (Maneja et al. 2023). At 50 days in the monitored nests, noon temperatures increased by 5.4°C from a mean of 31.2°C (SD = 1.0°C, Min = 29°C, Max = 33°C, n = 13) to a mean of 36.6°C (SD = 1.9°C, Min = 33.7 °C, Max = 38.8°C, n = 13). Hatching success showed an inverse correlation with the duration of exposure above temperatures of 34°C, 35°C, 36°C, and 37°C (Maneja et al. 2023). The highest inverse correlation was for exposure above 36°C (R^2 = 0.633, p = 0.0104).

Hatching and emergence success vary among species and nesting locations used by sea turtles (Hirth 1997; Witzell 1983; van Buskirk and Crowder 1994). They vary as a result of nest site selection by the gravid turtle, which determines the conditions experienced by the developing embryos in the context of the stochasticity of the environment. The primary abiotic factors that determine the conditions experienced by the embryos include: temperature (Matsuzawa et al. 2002; Segura and Cajade 2010), gas availability (Ackerman 1980), moisture (Ackerman 1997; Mortimer 1982), and salinity (Foley et al. 2006; Wood et al. 2000) as filtered through the sand structure and composition (Mortimer 1990). These variables synergistically impact development. In addition, biotic factors such as depredation of eggs and hatchlings by a variety of predators, including humans, can impact hatching and emergence success (Stancyk 1982; Witzell 1983; Hirth 1997). As a consequence of differences in the microclimate experienced by the developing embryos in different nesting beaches and in different places in a beach, hatching and/or emergence success varies within and among the nesting sites (Witzell 1983; Hirth 1997; Bentley et al. 2020). For example, emergence success of hawksbill turtles varied from 46.7% to 96.9% (mean = 72.3%, SD = 18.2%, n = 13 reports) (Witzell 1983). Emergence success of green sea turtle clutches varied from 1.9% to 88% with a mean of 67.6% (SD = 16.7%, n = 47 reports) (Hirth 1997). Turkozan et al. (2011) reported hatching success in vegetated areas between 39.8–47.9% and between 62.4–69.5% in non-vegetated areas. The mean hatching success was 75% in Northern Cyprus (Özdemir and Türkozan 2006) and 46% in the Galapagos Islands (Zarate et al. 2013). The temperature experienced by the developing embryos is the primary factor, but its impact is in the context of the other environmental conditions. The result is that some embryos in the clutch may survive even when the recorded incubation temperatures approach 34°C (hatching success: 25–57% at 33.4°C, Weber et al. 2012; 71% mean temperature of 33°C, Tilley et al. 2019). The mean temperature in clutches at Karan Island during the last third of the incubation period was 36.0°C and the mean hatching success was 38.8%. These conditions may be a selective factor in determining the thermal tolerance in the surviving hatchlings (Gatto et al. 2021).

In the Gulf region, hatching and emergence success of hawksbill and green turtle eggs and hatchlings are not well documented, although some data do exist (Al-Ghais 2009; Phillott and Rees 2021; Al Ameri et al. 2022; Hamann et al. 2022). For example, the range in hatching success of hawksbill turtle eggs in the UAE was reported to be between about 80% and 15%, decreasing through the season (Al-Ghais 2009). In Kuwait, Rees et al. (2020) reported a mean hatching success for 16 hawksbill turtle nests of 57.7% (SD = 26.3%, min-max = 5.5–95.5%). The mean hatching success from seven locations in Iran was 68.3%, with the lowest value being 22.7% (Al Ameri et al. 2022), and the hatching success ranged from 86% to 73.3% from Qatar (Al Ameri et al. 2022).

The range in hatching success for green turtles at Karan Island in 1986 and 1987 was 60.9–95% (mean = 81.7%, 4 clutches, Miller 1989) and in the early 1990s was reported to be in the range of 53.7–96.8% (mean of 84.7%, 21 clutches, Pilcher 2000). The mean hatching success for 72 green turtle eggs from three clutches (15, 28, 29 eggs, respectively) incubated in an artificial nest in Karan Island sand was 93.1% (Al-Mansi et al. 1999). The combined data from Kuwait, Iran, and Saudi Arabia between 1985 and 2005 show a range in hatching success of 27.6–96.8% with a mean of ~69% (n = 211) (Al Ameri et al. 2022). The mean hatching success (36.1%) of 18 green turtle nests recorded at Karan Island (2017: mean = 26.6%, n = 4; 2019: mean = 38.8%, n = 14) was lower relative to the previously reported hatching success for Karan Island and for green turtles nesting at various sites in the world (Hirth 1997; Zarate et al. 2013). The reason(s) for the low hatching success are likely related to the temperatures experienced by the embryos during incubation (Maneja et al. 2023), but more assessment of temperatures within nests needs to be done.

Finding emerged nest locations can be difficult when the surface sand is dry. Usually nest locations are found when hatchling tracks are followed from the beach to the nest location. Consequently, hatching and emergence success are recorded for nests that produce hatchlings whose tracks are found but not for the others. This overestimates hatchling production because the assumption is that the emerged nests that are found are representative of the entire beach. Unfortunately, this assumption precludes calculating emergence success of nests from which hatchlings failed to emerge or their tracks are not found.

As a result of differences in the microenvironments in the sand surrounding a clutch of eggs, the incubation success will vary among different locations, as well as within and among nesting seasons (Ackerman 1997). Emergence success will be lower than hatching success if hatchlings die in the nest or while digging to the surface (Miller 1999). Given the lower hatching success found for green turtle eggs incubating at Karan Island, the emergence success is likely to be considerably lower than previously suggested.

21.4 SUMMARY AND RECOMMENDATIONS

Although statistical significance was demonstrated among some variables and not others (Tables 21.3, 21.4, 21.6, 21.7), the biological importance of these determinations is unknown (Johnson 1999; EFSA Scientific Committee 2011). The data collected at Karan and Jana islands for the curved carapace length of nesting hawksbill turtles was not significantly different among the cohorts of the total population during the study years (Table 21.3). This suggests no biological reason to consider the annual cohorts as being different, even though the annual nesting cohort is likely comprised of turtles in different ratios from different foraging areas in the Gulf. In contrast, the weight of these turtles was significantly different (Table 21.3). The reverse situation occurred with green turtles nesting at Karan Island. The CCL measurements from the four study periods were significantly different, and the weights were not (Table 21.6). Although the species forage on different food items (Witzell 1983; Hirth 1997; Jones and Seminoff 2013), the environmental conditions in the Gulf are likely to affect the quantity and quality of food available in the foraging areas, which may contribute to the differences in growth as expressed by CCL and weight differences found among the cohorts. How quantity and quality of food available in the foraging areas impact the functioning of the total population in the Gulf requires longer-term studies of nesting turtles and an expanded understanding of the genetic similarities and differences among the groups of turtles nesting among years and at different locations (Jensen et al. 2019; Karaman et al. 2022; Komoroske et al. 2017). In addition, mixed-stock analysis of turtle populations in the foraging areas that support breeding in the Gulf should be conducted and include males to enhance the understanding of the total population (Mobaraki et al. 2020; Read et al. 2015; Vargas et al. 2016).

The average hawksbill turtle nesting in the northern part of the Gulf is smaller than nesting populations elsewhere in the world. They produce clutches that contain fewer eggs as well. The same general statements are true for green turtles. In addition, the hatching success for green turtles at Karan Island, Saudi Arabia, is lower than at other green turtle nesting sites around the world. Less is known about the hatchling production at Jana Island. The cause(s) of this situation remain unknown. However, if this level of low hatching success represents the long-term situation, recruitment to the nesting population is likely not sufficient for the survival of the population. Because Jana and Karan islands host the largest aggregations of nesting hawksbill and green turtles, respectively, in the northern Gulf, the monitoring program should be expanded to include assessment of the hatching and emergence success, as well as to determine the microenvironment of the beach and reproductive characteristics of nesting turtles.

Although longer-term datasets would be useful, most of the data required for the development of regional conservation management have been compiled (Phillott and Rees 2021; IOSEA 2019a, 2019b, 2019c, 2019d, 2019e). These reports identify the current status of marine turtles, as well as current conservation and management measures. In addition, Al Ameri et al. (2022) summarized the existing information from the published literature for the turtles in the Gulf, and Mobaraki et al. (2022) reviewed hawksbill nesting in Iran in detail. Based on these resources, the Regional Organization for Protection of the Marine Environment (ROPME), which comprises the eight countries surrounding the Arabian Gulf and is tasked to coordinate efforts towards protection of the marine and coastal environment in the region, could develop a cooperative regional research and monitoring program for marine resources, including the management of marine turtles in the Gulf region (van Lavieren and Klaus 2013), similar to the one developed by the Regional Organization for the Conservation of the Environment of the Red Sea and Gulf of Aden (PERSGA 2007). Also, ROPME could recommend adoption of the PERSGA (2004, 2019) research methods and/or use of techniques from the Marine Turtle Specialist Group methods manual (Eckert et al. 1999) to standardize the methods used to collect data and facilitate information exchange among research groups for different countries around the Gulf.

In addition to monitoring the reproductive characteristics of hawksbill and green turtles nesting in the Gulf region, assessment of the hatching and emergence success is required to understand the impact of the reproductive effort of the turtles to future generations. This research needs to be conducted consistently by sampling at multiple places on the nesting beaches over several years to gather the data necessary to define the current status of the production of hatchlings and to guide conservation decision-making because of within-site and between-site variations. For example, at Jana and Karan Islands, the research program would include marking a large number of nests in the early, middle, and late portions of the nesting season and at multiple places around the islands, followed by an assessment after a period of 7 to 9 weeks. Temperature data loggers would have to be inserted among the eggs in multiple nests during oviposition to record the temperatures experienced by the embryos. Collecting these data would require that the research team return multiple times to the islands to examine the material remaining in the marked nests, to retrieve the temperature data loggers, and to determine how many hatchlings emerged from the nest (*sensu* Miller 1999).

ETHICS STATEMENT

The KFUPM (unpublished) study was reviewed and approved by the National Center for Wildlife (NCW) (formally Saudi Wildlife Authority) Projects Committee and complies with all applicable rules and regulations of the NCW for research on animals in the Kingdom of Saudi Arabia.

FUNDING

Support and funding for this study was provided by the King Fahd University of Petroleum and Minerals (KFUPM) – Saudi Aramco Environmental Sustaining Research – Phase VI Project (CEW02428) and Environmental Sustaining Research – Phase VII Project (CEM02700).

ACKNOWLEDGEMENTS

We would like to acknowledge the Center for Environment and Marine Studies of KFUPM for providing the logistics and administrative support to the research team, including the additional field, logistics, and administrative support team comprising J.F.A. Alcaria, A.U.B. Basali, T. Vasanthi Duraisamy, Jeff B.R. Abrogueña, A.A. Bawazier, P.B. Das, A. Jukhdar, M. Al-Jedani, I. Alzhogiby, Abdaljabae Al-Saddah, E. Santiago, N.M. Mudasir, Mohamed Asharaf, Karuppasamy P. Manikandan, and Mohammed Ali B. Qurban. We also acknowledge the crew of *Selbar-1325* led by Mr. Bader Abuainain for supporting the fieldwork logistics at the offshore islands. We would also like to acknowledge the support of the National Center for Wildlife and the Saudi Border Patrol for providing the permits to conduct research in the Saudi Arabian offshore islands in the Arabian Gulf.

REFERENCES

Abdulqader, E.A.A. 2000. Turtle incidences in shrimp trawl nets during the 1997–1998 shrimp season in Bahrain waters. *Asian Fish Sci* 13: 287–296.

Abdulqader, E.A.A., and J. Miller. 2012. Marine turtle mortalities in Bahrain territorial waters. *Chelonian Conserv Biol* 11: 133–138.

Abdulqader, E.A.A., Miller, J., Al-Mansi, A., et al. 2017. Turtles and other marine megafauna bycatch in artisanal fisheries in the Saudi waters of the Arabian Gulf. *Fish Res* 196: 75–84.

Abreu-Grobois, A., and P. Plotkin (IUCN SSC Marine Turtle Specialist Group). 2008. *Lepidochelys olivacea*. The IUCN Red List of Threatened Species 2008: e.T11534A3292503. doi:10.2305/IUCN.UK.2008.RLTS.T11534A3292503.en.

Ackerman, R. 1997. The nest environment and the embryonic development of sea turtles. In: *The biology of sea turtles*. Eds: P.L. Lutz, and J.A. Musick, vol. 1, 83–106. Boca Raton: CRC Press.

Ackerman, R.A. 1980. Physiological and ecological aspects of gas exchange by sea turtle eggs. *Am Zool* 20: 575–584.

Al Ameri, H.M., Al Harthi, S., Al Kiyumi, A., et al. 2022. Biology and conservation of marine turtles in the North-Western Indian Ocean: A review. *Endang Species Res* 48: 67–86. doi:10.3354/esr01183.

Al Ameri, H.M., Das, H.S., Rodriguez-Zarate, C.J., and M. Antonopoulou. 2020. *United Arab Emirates*. MTSG Regional Report Middle East South Asia 2020–2.

Al-Ghais, S.M. 2009. Nesting of hawksbill turtles, *E Imbricata*, on the islands of the Arabian Gulf (Reptilia: Cheloniidae). *Zool Middle East* 48: 43–48.

Al-Mansi, A.M. 2016. *Sea turtles in the Red Sea and the Arabian Gulf, Saudi Arabia*. Report. Riyadh: Saudi Wildlife Authority, 43 p.

Al-Mansi, A.M.A., Khushaim, O.A., and M.M.H. Al-Marghani. 1999. On the effect of substrate on nesting success of the green turtle, *Chelonia mydas*, in the Arabian Gulf. *Zool Middle East* 19: 5–11.

Al-Merghani, M., Miller, J.D., Al-Mansi, A., Khushaim, O., and N.J. Pilcher. 1996. The marine turtles of the Arabian Gulf: NCWCD studies 1991–1994. In: *A marine wildlife sanctuary for the Arabian Gulf: Environmental research and conservation following the 1991 Gulf War oil spill*. Eds: F. Krupp, A.H. Abuzinada, and L.A. Nader, 351–359. Frankfurt: Senckenberg Research Institute.

Al-Merghani, M., Miller, J.D., Pilcher, N.J., and A. Al-Mansi. 2000. The green and hawksbill turtles in the Kingdom of Saudi Arabia: Synopsis of nesting studies 1986–1997. *Fauna of Arabia* 18: 369–384.

Al-Mohanna, S.Y., Al-Zaidan, A.S.Y., and P. George. 2014. Green turtles (*Chelonia mydas*) of the north-western Arabian Gulf, Kuwait: The need for conservation. *Aquat Conserv Mar Freshw Ecosyst* 24, no. 2: 166–178. doi:10.1002/aqc.2371.

Al Suweidi, A.S., Wilson, K.D.P., Healy, T., and L. Vanneyre. 2012. First contemporary record of green turtle (*Chelonia mydas*) nesting in the United Arab Emirates. *Marine Turtle Newsletter* 133: 16.

Barth, H.-J., and N.Y. Khan. 2008. Biogeophysical setting of the Gulf. In: *Protecting the Gulf's marine ecosystems from pollution*. Eds: A.H. Abuzinada, H.-J. Barth, F. Krupp, B. Böer, and T.Z. Al Abdessalaam, 1–21. Basel: Birkhäuser.

Basson, P.W., Burchard, J., Hardy, J.T., and A.R. Price. 1977. *Biotopes of the Western Arabian Gulf: Marine life and environments of Saudi Arabia*. Dhahran: ARAMCO Department of Loss Prevention and Environmental Affairs.

Basyoni, M.H. 1999. Sedimentology and geochemistry of the Karan Island sediments, Arabian Gulf, Saudi Arabia. *J KAU Mar Sci* 10: 89–106.

Beni, A.N., Marriner, N., Sharifi, A., et al. 2021. Climate change: A driver of future conflicts in the Persian Gulf Region? *Heliyon* 7, no. 2: e06288. doi:10.1016/j.heliyon.2021.e06288.

Bentley, B.P., Kearney, M.R., Whiting, S.D., and N.J. Mitchell. 2020. Microclimate modelling of beach sand temperatures reveals high spatial and temporal variation at sea turtle rookeries. *J Therm Biol* 88: 102522. doi:10.1016/j.jtherbio.2020.102522.

Bjorndal, K.A. 1997. Foraging ecology and nutrition of sea turtles. In: *The biology of sea turtles*. Eds: P.L. Lutz, and J.A. Musick, 199–231. New York: CRC Press.

Bjorndal, K.A., and J.B.C. Jackson. 2003. Roles of sea turtles in marine ecosystems: Reconstructing the past. In: *The biology of sea turtles*. Eds: P.L. Lutz, J.A. Musick, and J. Wyneken, vol. 2, 259–274. Boca Raton: CRC Press.

Bolten, A.B. 1999. Techniques for measuring sea turtles. In: *Research and management techniques for the conservation of sea turtles*. Eds: K.L. Eckert, K.A. Bjorndal, F.A. Abreu-Grobois, and M. Donnelly, 110–114. IUCN/SSCMarine Turtle Specialist Group Publication No. 4. Washington, DC: IUCN/SSC Marine Turtle Specialist Group.

Bolten, A.B. 2003. Variation in sea turtle life history patterns: Neritic vs. oceanic developmental stages. In: *The biology of sea turtles*.

Eds: P.L. Lutz, J. Musick, and J. Wyneken, vol. 2, 243–257. Boca Raton, FL: CRC Press.

Chaloupka, M.Y., and J.A. Musick. 1997. Age, growth, and population dynamics. In: *The biology of sea turtles*. Eds: P.L. Lutz, and J.A. Musick, 233–276. Boca Raton, FL: CRC Press.

Chatting, M., Smyth, D., Al-Maslamani, I., et al. 2018. Nesting ecology of hawksbill turtles, *E Imbricata*, in an extreme environmental setting. *PLoS One* 13, no. 9: e0203257. doi:10.1371/journal.pone.0203257.

D'aloia, M.A., and S.M. Al-Ghais. 2000. Preliminary genetic analysis of the green turtle, *Chelonia mydas*, in the Arabian Gulf using mitochondrial DNA. *Zool Middle East* 21: 47–54.

Ditmer, M.A., and S.P. Stapleton. 2012. Factors affecting hatch success of hawksbill sea turtles on Long Island, Antigua, West Indies. *PLoS One* 7: e38472.

Donlan, C.J., Wingfield, D.K., Crowder, L.B., and C. Wilcox. 2010. Using expert opinion surveys to rank threats to endangered species: A case study with sea turtles. *Conserv Biol* 24, no. 6: 1586–1595.

Eckert, K.L., Bjorndal, K.A., Abreu-Grobois, F.A., and M. Donnelly. 1999. *Research and management techniques for the conservation of sea turtles*, vol. 4. Washington, DC: IUCN/SSC Marine Turtle Specialist Group, 245 p.

Eckert, K.L., Wallace, B.P., Frazier, J.G., Eckert, S.A., and P.C.H. Pritchard. 2012. *Synopsis of the biological data on the leatherback sea turtle (Dermochelys coriacea)*. Washington, DC: U.S. Department of Interior, Fish and Wildlife Service. 158 p.

EFSA Scientific Committee. 2011. Statistical significance and biological relevance. European Food Safety Authority (EFSA), Parma, Italy. *EFSA J* 9, no. 9: 2372.

Finkbeiner, E.M., Wallace, B.P., Moore, J.E., Lewison, R.L., Crowder, L.B., and A.J. Read. 2011. Cumulative estimates of sea turtle bycatch and mortality in USA fisheries between 1990 and 2007. *Biol Conserv* 144: 2719–2727.

Fitzsimmons, N.N., and C.J. Limpus. 2014. Marine turtle genetic stocks of the Indo-Pacific: Identifying boundaries and knowledge gaps. *Indian Ocean Marine Turtle Newslett* 20: 2–12.

Foley, A.M., Peck, S.A., and G.R. Harman. 2006. Effects of sand characteristics and inundation on the hatching success of loggerhead sea turtle (*Caretta caretta*) clutches on low-relief mangrove islands in southwest Florida. *Chelonian Conserv Biol* 5: 32–41.

Frazier, J. 2005. Marine turtles: The role of flagship species in interactions between people and the sea conservation. *Mast* 3, no. 2, and 4, no. 1: 5–38.

Fuentes, M.M.P.B., Limpus, C.J., and M. Hamann. 2011. Vulnerability of sea turtle nesting grounds to climate change. *Glob Change Biol* 17: 140–153.

Gasperetti, J., Stimson, A., Miller, J.D., Ross, J.P., and P. Gasperetti. 1993. Turtles of Arabia. *Fauna Saudi Arabia* 13: 170–367.

Gatto, C.R., Matthews, B., and R.D. Reina. 2021. Role of incubation environment in determining thermal tolerance of sea turtle hatchlings. *Endang Species Res* 44: 397–408. doi:10.3354/esr01111.

Georges, A., Limpus, C.J., and C.J. Parmenter. 1993. Natural history of the Chelonia. In: *Fauna of Australia, Volume 2: Amphibia, reptilia, aves*. Ed: J. Godsell, 120–128. Canberra: Australian Biological Resources Study, DASETT.

Hamann, M., Flavell, F., Frazier, J., Limpus, C.J., Miller, J.D., and J.A. Mortimer. 2022. *Assessment of the conservation status of the hawksbill turtle in the Indian Ocean and South-East Asia*. IOSEA Species Assess. Bonn, Germany: CMS, 113 p.

Hamann, M., Fuentes, M.M.P.B., Ban, N., and V. Mocellin. 2013. Climate change and marine turtles. In: *The biology of sea turtles*. Eds: J. Wyneken, K.J. Lohmann, and J.A. Musick, 353–378. Boca Raton, FL: CRC Press.

Hamann, M., Limpus, C.J., and D.W. Owens. 2003. Reproductive cycles of males and females. In: *Biology of sea turtles*. Eds: P. Lutz, J. Musick, and J. Wyneken, vol. 2, 135–161. Boca Raton, FL: CRC Press.

Hawkes, L.A., Broderick, A.C., Godfrey, M.H., and B.J. Godley. 2009. Climate change and marine turtles. *Endang Species Res* 7: 137–154.

Hebbelmann, L., Pereira, J., Yaghmour, F., and A. Al-Ali. 2016. New records of sea turtle nesting at Al Qurm Wa Lehhfaiiah Protected Area beach after a 30-year absence. *Marine Turtle Newsletter* 150: 7–9.

Heithaus, M.R. 2013. Predators, prey, and ecological roles of sea turtles. In: *Biology of sea turtles*. Eds: J. Wyneken, J.K. Lohman, and J.A. Musick, vol. 3, 249–285. Boca Raton: CRC Press.

Heppell, S.S., Snover, M.L., and L.B. Crowder. 2003. Sea turtle population ecology. In: *The biology of sea turtles*. Eds: P.L. Lutz, J.A. Musick, and J. Wyneken, vol. 2, 275–306. Boca Raton: CRC.

Hirth, H.F. 1997. Synopsis of the biological data on the green turtle *Chelonia mydas* (LINNAEUS 1758). *Biol Rep* 97, no. 1: 97.

IOSEA. 2019a. *Iran national report*. IOSEA. www.cms.int/iosea-turtles/en/documents/national-reports.

IOSEA. 2019b. *United Arab emirates national report*. IOSEA. www.cms.int/iosea-turtles/en/documents/national-reports.

IOSEA. 2019c. *Oman national report*. IOSEA. www.cms.int/iosea-turtles/en/documents/national-reports.

IOSEA. 2019d. *Saudi Arabia national report*. IOSEA. www.cms.int/iosea-turtles/en/documents/national-reports.

IOSEA. 2019e. *Bahrain national report*. IOSEA. www.cms.int/iosea-turtles/en/documents/national-reports.

IPCC. 2018. Summary for policymakers. In: *Global warming of 1.5°C. an IPCC special report on the impacts of global warming of 1.5°C above pre-industrial levels and related global greenhouse gas emission pathways, in the context of strengthening the global response to the threat of climate change, sustainable development, and efforts to eradicate poverty*. Eds: V. Masson-Delmotte, P. Zhai, H.-O. Pörtner, D. Roberts, J. Skea, P.R. Shukla, A. Pirani, W. Moufouma-Okia, C. Péan, R. Pidcock, S. Connors, J.B.R. Matthews, Y. Chen, X. Zhou, M.I. Gomis, E. Lonnoy, T. Maycock, M. Tignor, and T. Waterfield, 3–24. Cambridge, UK and New York, NY: Cambridge University Press.

Jennings, S., and M. Kaiser. 1998. The effects of fishing on marine ecosystems. *Adv Mar Biol* 34: 201–352.

Jensen, M.P., Miller, J., FitzSimmons, N.N., and M. Al-Merghani. 2019. Identification of *Chelonia mydas* populations in the Kingdom of Saudi Arabia through regional genetic analyses. *Marine Turtle Newsletter* 156: 16–20.

Johns, W.E., Jacobs, G.A., Kindle, J.C., Murray, S.P., and M. Carron. 1999. *Arabian marginal seas and Gulfs: Report of a workshop*. Report prepared by the Workshop Organizing Committee: RSMAS Technical Report 2000–01. Naval Research Lab Stennis Space Center MS Oceanography Div.

Johnson, D.H. 1999. The insignificance of statistical significance testing. *J Wildl Manag* 63, no. 4: 763–772.

Jones, T.T., and J.A. Seminoff. 2013. Feeding biology: Advances from field-based observations, physiological studies, and molecular techniques. In: *Biology of sea turtles*. Eds: J. Wyneken, K. Lohmann, and J. Musick, vol. 3, 211–247. Boca Raton: CRC Press.

Karaman, S., Turkozan, O., Carreras, C., et al. 2022. Population genetic diversity of green turtles, *Chelonia mydas*, in the Mediterranean revisited. *Mar Biol* 169: 77. doi:10.1007/s00227-022-04068-1.

KFUPM/RI. 2020. *Saudi Aramco/KFUPM-RI Sustaining Research Project Marine Environmental Studies – Phase VI*. Final report. Prepared for Saudi Aramco by the Center for Environment and Water, Research Institute, King Fahd University of Petroleum and Minerals, Dhahran, Saudi Arabia, CEW02428.

Komoroske, L.M., Jensen, M.P., Stewart, K.R., Shamblin, B.M., and P.H. Dutton. 2017. Advances in the application of genetics in marine turtle biology and conservation. *Front Mar Sci* 4: 156. Doi:10.3389/fmars.2017.00156.

Krishnakumar, P.K., Ashraf, M., Sheppard, C., and J.D. Miller. 2019. The physical environment of the western Arabian Gulf. In: *Ecosystems and biodiversity of the Arabian Gulf: Saudi Arabian waters, fifty years of scientific research*. Eds: K.A. Al-Abdulkader, R.A. Loughland, and M.A.B. Qurban, 19–59. Dhahran, KSA: ARAMCO & KFUPM.

Lewison, R.L., and L.B. Crowder. 2007. Putting longline bycatch of sea turtles into perspective. *Conserv Biol* 21: 79–86.

Lewison, R.L., Hooker, S., Hodgson, D., et al. 2008. Fisheries bycatch: Problems and solutions. *Endang Species Res* 5, nos. 2–3.

Limpus, C.J., Baker, V., and J.D. Miller. 1979. Movement induced mortality of loggerhead eggs. *Herpetologica* 35, no. 4: 335–338. www.jstor.org/stable/3891966.

Limpus, C.J., and J.D. Miller. 1993. Family Cheloniidae. In: *Fauna of Australia, Volume 2A: Amphibia and reptilia*. Eds: C.J. Glasby, G.J.B. Ross, and P.L. Beesley, 113–138. Canberra: Australian Government Publishing Service.

Lincoln, S., Buckley, P., Howes, E.L., et al. 2021. A regional review of marine and coastal impacts of climate change on the ROPME Sea Area. *Sustainability* 13: 13810.

Luschi, P., Hays, G.C., and F. Papi. 2003. A review of long-distance movements by marine turtles, and the possible role of ocean currents. *Oikos* 103: 293–302.

Lutcavage, M.E., Plotkin, P., Witherington, B.E., and P.L. Lutz. 1997. Human impacts on sea turtle survival. In: *Biology of sea turtles*. Eds: P.L. Lutz, and J.A. Musick, 387–409. Boca Raton, FL: CRC Press.

Mancini, A., Phillott, A.D., and A.F. Rees. 2019. *Chelonia mydas* (North Indian Ocean subpopulation) (errata version published in 2019). The IUCN Red List of Threatened Species 2019: e.T142121108A154845002.

Maneja, R.H., Miller, J.D., Flint, J.B., et al. 2023. Extreme conditions reduce hatching success of green turtles (*Chelonia mydas* L.) at Karan Island, the major nesting site in the Arabian Gulf. *Mar Pollut Bull* 190: 114801.

Maneja, R.H., Miller, J.D., Li, W., et al. 2020. Long-term NDVI and recent vegetation cover profiles of major offshore island nesting sites of sea turtles in Saudi waters of the northern Arabian Gulf. *Ecol Indic* 117: 106612.

Maneja, R.H., Thomas, R., Miller, J.D., et al. 2021. *Marine litter survey at the major sea turtle nesting islands in the Arabian Gulf using in-situ and remote sensing methods*. 2021 IEEE International Geoscience and Remote Sensing Symposium, IGARSS 2021, Brussels, Belgium, July 11–16, pp. 8582–8585. IEEE. https://IGARSS2021.com. doi:10.1109/IGRSS47720.2021.

Mansfield, K., and N. Putman. 2013. Oceanic habits and habitats – *Caretta caretta*. In: *The biology of sea turtles*. Eds: P.L. Lutz, J. Musick, and J. Wyneken, vol. 3, 189–210. Boca Raton, FL: CRC Press.

Matsuzawa, Y., Sato, K., Sakamoto, W., and K.A. Bjorndal. 2002. Seasonal fluctuations in sand temperature: Effects on the incubation period and mortality of loggerhead sea turtle (*Caretta caretta*) pre-emergent hatchlings in Minabe, Japan. *Mar Biol* 140: 639–646.

Mendonca, V.M., Saady, S.A., Kiyumi, A.A., and K. Erzini. 2010. Interactions between green turtles (*Chelonia mydas*) and foxes (*Vulpes vulpes arabica*, *V. rueppellii sabaea*, and *V. cana*) on turtle nesting grounds in the Northwestern Indian Ocean: Impacts of the fox community on the behavior of nesting sea turtles at the Ras Al Hadd Turtle Reserve, Oman. *Zool Stud* 49, no. 4: 437–452.

Miller, J.D. 1989. *Marine turtles Volume 1: An assessment of the conservation status of marine turtles in the Kingdom of Saudi Arabia*. Technical Report, MEPA Coastal and Marine Managemental Series, Report No. 9, 209 p. MEPA, Jeddah.

Miller, J.D. 1997. Reproduction in sea turtles. In: *The biology of sea turtles*. Eds: P. Lutz, and J.A. Musick, 51–81. Boca Raton, FL: CRC Press.

Miller, J.D. 1999. Determining clutch size and hatching success. In: *Research and management techniques for the conservation of sea turtles*. Eds: K.L. Eckert, K.A. Bjorndal, F.A. Abreu-Grobois, and M. Donnelly. 124–129. IUCN/SSC Marine Turtle Specialist Group Publication No. 4. Washington, DC: IUCN/SSC Marine Turtle Specialist Group.

Miller, J.D. 2011a. The coastal and offshore island ecosystems. In: *Marine atlas of the Western Arabian Gulf*. Eds: R.A. Loughland, and K. Al-Abdulkader, 158–185. Dhahran: Saudi ARAMCO Environmental Protection Pubs.

Miller, J.D. 2011b. Marine and coastal reptiles. In: *Marine atlas of the Western Arabian Gulf*. Eds: R.A. Loughland, and K. Al-Abdulkader, 264–295. Dhahran: Saudi ARAMCO Environmental Protection Pubs.

Miller, J.D. 2021. Saudi Arabia. In: *Sea turtles in the Middle East and South Asia Region MTSG Annual Regional Report 2021*. Eds: A.D. Phillott, and A.F. Rees, 265–310. Washington, DC: IUCN-SSC Marine Turtle Specialist Group.

Miller, J.D., Abdulqader, E.A.A., Heatwole, H., Al Mansi, A., and M. Pope. 2019. Marine reptiles, marine mammals and marine birds. In: *Ecosystems and biodiversity of the Arabian Gulf: Saudi Arabian waters, fifty years of scientific research*. Eds: K.A. Al-Abdulkader, R.A. Loughland, and M.A.B. Qurban, 409–449. Dhahran, KSA: ARAMCO & KFUPM.

Miller, J.D., Limpus, C.J., and M. Godfrey. 2003. Nest site selection, oviposition, eggs, development, hatching and emergence of loggerhead turtles. In: *The biology of the loggerhead turtle*. Eds: A. Bolten, and B. Witherington, 125–143. Washington, DC: Smithsonian Institution Press.

Miller, J.D., and R. Loughland. 2006. *Distribution and relative abundance of sea turtle nesting in the United Arab Emirates*. Compilers: M. Frick, A Panagopoulou, A. Rees, K. Williams, 307. 26th Annual Conference on Sea Turtle Conservation and Biology, April, Crete, Greece. ISTS, Athens, Greece.

Miller, J.D., Mortimer, J.A., and C.J. Limpus. 2017. A field key to the developmental stages of marine turtles (Cheloniidae) with notes on the development of *Dermochelys*. *Chelonian Conserv Biol* 16, no. 2: 111–122. doi:10.2744/CCB-1261.1.

Miller, J.D., Preen, A., Loughland, R., Youssif, A., and A. Darwish. 2004. Marine turtles and sea snakes in Abu Dhabi Emirate. In: *Marine atlas of Abu Dhabi*. Eds: R. Loughland, F. Al Muhairi, S. Fadel, A. Almehdi, and P. Hellyer, 184–201. Abu Dhabi, UAE: Emirates Heritage Club.

Mobaraki, A. 2004a. Marine turtles in Iran: Results from 2002. *Mar Turtle Newsl* 104: 13.

Mobaraki, A. 2004b. Nesting of the hawksbill turtle at Shidvar Island, Hormozgan Province, Iran. *Mar Turtle Newsl* 103: 13.

Mobaraki, A. 2004c. Green turtle nesting on the Gulf of Oman coastline of the Islamic Republic of Iran. *Mar Turtle Newsl* 104: 11.

Mobaraki, A. 2004d. Green sea turtle nesting on the Gulf of Oman coastline of the I.R. Iran. *Marine Turtle Newsletter* 104: 11.

Mobaraki, A. 2021. Iran. In: *Sea turtles in the Middle East and South Asia Region MTSG annual regional report 2021*. Eds: A.D. Phillott, and A.F. Rees, 133–145. Washington, DC: IUCN-SSC Marine Turtle Specialist Group.

Mobaraki, A., Phillott, A.D., Erfant, M., Ghasem, M., and H. Jafar. 2022. Inferred impacts of extreme environments on hawksbill turtle (*Eretmochelys imbricata*) body size and reproductive output. *Chelonian Conserv Biol* 21: 187–198.

Mobaraki, A., Rastegar-Pouyani, E., Kami, H.G., and N. Khorasani. 2020. Population study of foraging green sea turtles (*Chelonia mydas*) in the northern Persian Gulf and Oman Sea, Iran. *Reg Stud Mar Sci* 39: 101433.

Mohammadizadeh, M., and N. Soltanpour. 2014. Identification and prioritizing important nesting sites of green turtle in Iranian beaches of Oman Sea during 2008–2010. *Bull Env Pharmacol Life Sci* 3: 81–86.

Mortimer, J.A. 1982. Factors influencing beach selection by nesting sea turtles. In: *Biology and conservation of sea turtles*. Ed: K.A. Bjorndal, 45–51. Washington: Smithsonian Institution Press.

Mortimer, J.A. 1990. The influence of beach sand characteristics on the nesting behavior and clutch survival of green turtles (*Chelonia mydas*). *Copeia* 3: 802–817. doi:10.2307/1446446.

Mortimer, J.A., and M. Donnelly (IUCN SSC Marine Turtle Specialist Group). 2008. *Eretmochelys imbricata*. The IUCN Red List of Threatened Species 2008: e.T8005A12881238. http://dx.doi.org/10.2305/IUCN.UK.2008.RLTS.T8005A12881238.en.

Mostafavi, H., Kiabi, B., Liagati, H., Abdoli, A., et al. 2010. *Management plan of Kharkoo Wildlife refuge*. Iran: Shahid Beheshti University and Department of Environment. Unpublished.

Musick, J.A., and C.J. Limpus. 1997. Habitat utilization and migration in juvenile sea turtles. In: *The biology of sea turtles*. Eds: P.L. Lutz, and J.A. Musick, vol. 1, 137–163. Boca Raton, FL: CRC Press.

Naser, H.A. 2014. Marine ecosystem diversity in the Arabian Gulf: Threats and conservation. In: *Biodiversity – The dynamic balance of the planet*. Ed: O. Grillo, 297–327. Rijeka: InTech Open Access Publisher.

NRC. 1990. *Decline of the sea turtles: Causes and prevention*. Washington, DC: National Research Council.

Özdemir, B., and O. Türkozan. 2006. Hatching success of original and hatchery nests of the green turtle, *Chelonia mydas*, in Northern Cyprus. *Turk J Zool* 30, no. 4: 377–381. https://journals.tubitak.gov.tr/zoology/vol30/iss4/6.

Patrício, A.R., Hawkes, L.A., Monsinjon, J.R., Godley, B.J., and M.M.P.B. Fuentes. 2021. Climate change and marine turtles: Recent advances and future directions. *Endang Species Res* 44: 363–395.

PERSGA. 2004. *Standard survey methods for key habitats and key species in the Red Sea and Gulf of Aden*, 1st ed. PERSGA Technical Series no. 10. PERSGA, Jeddah, 320 p.

PERSGA. 2007. Regional action plans for the conservation of marine turtles, seabirds and mangroves in the Red Sea and Gulf of Aden. PERSGA Technical Series no. 12. PERSGA, Jeddah, 149 p.

PERSGA. 2019. Standard survey methods for key habitats and key species in the Red Sea and Gulf of Aden, 2nd ed. PERSGA Technical Series no. 19, PERSGA, Jeddah, 344 p.

Phillott, A.D., and A.F. Rees (Eds.). 2021. *Sea turtles in the Middle East and South Asia Region MTSG annual regional report 2021*. Washington, DC: IUCN-SSC Marine Turtle Specialist Group. 412 p.

Pilcher, N.J. 1999. The hawksbill turtle, *Eretmochelys imbricata*, in the Arabian Gulf. *Chelonian Conserv Biol* 3, no. 2: 312–317.

Pilcher, N.J. 2000. The green turtle, *Chelonia mydas*, in the Saudi Arabian Gulf. *Chelonian Conserv Biol* 3, no. 4: 730–734.

Pilcher, N.J., Antonopoulou, M.A., Perry, L., et al. 2014. Identification of important sea turtle areas (ITAs) for hawksbill turtles in the Arabian Region. *J Exp Mar Biol Ecol* 460: 89–99.

Pilcher, N.J., Antonopoulou, M.A., Rodriguez-Zarate, C.J., et al. 2021. Movements of green turtles from foraging areas of the United Arab Emirates: Regional habitat connectivity and use of marine protected areas. *Mar Biol* 168: 1–15.

Pilcher, N.J., Rodriguez-Zarate, C.J., Antonopoulou, M.A., Mateos-Molina, D., Das, H.S., and I.A. Bugla. 2020. Combining laparoscopy and satellite tracking: Successful round-trip tracking of female green turtles from feeding areas to nesting grounds and back. *Glob Ecol Conserv* 23: e01169.

Plotkin, P.T. 2003. Adult migrations and habitat use. In: *The biology of sea turtles*. Eds: P.L. Lutz, J.A. Musick, and J. Wyneken, vol. 2, 225–241. Boca Raton, FL: CRC Press.

Price, A.R.G., Sheppard, C.R.C., and C.M. Roberts. 1993. The Gulf: Its biological setting. *Mar Pollut Bull* 27: 9–15.

Read, T.C., Fitzsimmons, N.N., Wantiez, L., et al. 2015. Mixed stock analysis of a resident green turtle, *Chelonia mydas*, population in New Caledonia links rookeries in the South Pacific. *Wildlife Res* 42: 488–499.

Rees, A.F., Al Hafez, A., Lloyd, J.R., Papathansopoulou, N., and B.J. Godley. 2013. Green turtles, *Chelonia mydas*, in Kuwait: Nesting and movements. *Chelonian Conserv Biol* 12: 157–163.

Rees, A., Al-Kiyumi, A., Broderick, A., Papathanasopoulou, N., and B.J. Godley. 2012a. Each to their own: Inter-specific differences in migrations of Masirah Island turtles. *Chelonian Conserv Biol* 11: 243–248.

Rees, A.F., Al-Kiyumi, A., Broderick, A.C., Papathanasopoulou, N., and B.J. Godley. 2012b. Conservation related insights into the behaviour of the olive ridley sea turtle, *Lepidochelys olivacea*, nesting in Oman. *Mar Ecol Prog Ser* 450: 195–205.

Rees, A.F., Llyod, J.R., and N. Papathanasopoulou. 2020. Hawksbill nest and hatchling data from Kuwait. *Indian Ocean Turtle Newsletter* 31: 13–16.

Rees, A.F., Papathanasopoulou, N., and B.J. Godley. 2018. Tracking hawksbill and green sea turtles in Kuwait reveals variability in migratory and residency strategies. *Indian Ocean Turtle Newsletter* 28: 23–26.

Rees, A.F., Papathanasopoulou, N., and B.J. Godley. 2019. Tracking hawksbills in Kuwait: Contributions to regional and behavioral insights. *Chelonian Conserv Biol* 18: 86–90.

Reynolds, M. 1993. Physical oceanography of the Gulf, Strait of Hormuz, and the Gulf of Oman: Results from the Mt. Mitchell Expedition. *Mar Pollut Bull* 27: 35–59.

Ross, J.P., and M.A. Barwani. 1982. Review of sea turtles in the Arabian area. In: *Biology and conservation of sea turtles*. Ed: K.A. Bjorndal, 373–383. Washington, DC: Smithsonian Institution Press.

Saba, V.S. 2013. Oceanic habits and habitats: *Dermochelys coriacea*. In: *The biology of sea turtles*. Eds: P.L. Lutz, J. Musick, and J. Wyneken, vol. 3, 163–188. Boca Raton, FL: CRC Press.

Segura, L.N., and R. Cajade. 2010. The effects of sand temperature on the pre-emergent green sea turtle hatchlings. *Herpetol Conserv Biol* 5, no. 2: 196–206.

Sheppard, C.R.C. 1993. Physical environment of the Gulf relevant to marine pollution: An overview. *Mar Pollut Bull* 27: 3–8.

Sheppard, C.R.C., Al-Husiani, M., Al-Jamali, F., et al. 2010. The Gulf: A young sea in decline. *Mar Pollut Bull* 60, no. 1: 13–38. doi:10.1016/j.marpolbul.2009.10.017.

Sinaei, M., Bolouki, M., Ghorbanzadeh, S., and M. Talebi Matin. 2019. Evaluation of hematological and plasma biochemical parameters in green sea turtle (*Chelonia mydas* Linnaeus, 1758) from nesting colonies of the northern coast the Sea of Oman. *Iranian J Fish Sci* 18, no. 4: 891–902.

Sinaei, M., Bolouki, M., Ghorbqanzadeh-Zaferani, G., Matin, M.T., Alimoradi, M., and S. Dalir. 2018. On a poorly known rookery of green turtles (*Chelonia mydas*) nesting at the Chabahar Beach, Northeastern Gulf of Oman. *Russian J Mar Biol* 44, no. 3: 254–261.

Stancyk, S.E. 1982. Non-human predators of sea turtles and their control. In: *Biology and conservation of sea turtles*. Ed: K.A. Bjorndal, 139–152. Washington: Smithsonian Institution Press.

Tabib, M., Frootan, F., and M.A. Hesni. 2014. Genetic diversity and phylogeography of hawksbill turtle in the Persian Gulf. *J Biodivers Environ Sci* 4: 51–57.

Tabib, M., Zolgharnein, M., Mohammadi, M.A., et al. 2011. mtDNA variation of the critically endangered hawksbill turtle (*Eretmochelys imbricata*) nesting on Iranian islands of the Persian Gulf. *Genetics Molecular Res* 10, no. 3: 1499–1503.

Tilley, D., Ball, S., Ellick, J., et al. 2019. No evidence of fine scale thermal adaptation in green turtles. *J Exp Mar Biol Ecol* 514–515: 110–117. doi:10.1016/j.jembe.2019.04.001.

Tollab, M.A., Dakhteh, M.H., Zaferani, G.G., et al. 2015. The olive ridley turtle, *Lepidochelys olivacea*, in the Persian Gulf: A review of the observations, including the first nesting of the species in the area. *Chelonian Conserv Biol* 14: 192–196.

Turkozan, O., Yamamoto, K., and C. Yılmaz. 2011. Nest site preference and hatching success of green (*Chelonia mydas*) and loggerhead (*Caretta caretta*) sea turtles at Akyatan Beach, Turkey. *Chelonian Conserv Biol* 10, no. 2: 270–275. doi:10.2744/ccb-0861.1.

van Buskirk, J., and L.B. Crowder. 1994. Life-history variation in marine turtles. *Copeia* 1: 66–81.

Van Lavieren, H., and R. Klaus. 2013. An effective regional Marine Protected Area network for the ROPME Sea Area: Unrealistic vision or realistic possibility? *Mar Pollut Bull* 72, no. 2: 389–405.

Vargas, S.M., Jensen, M.P., Ho, S.Y., et al. 2015. Phylogeography, genetic diversity, and management units of hawksbill turtles in the Indo-Pacific. *J Hered* 107, no. 3: 199–213.

Vargas, S.M., Jensen, M.P., Ho, S.Y., et al. 2016. Phylogeography, genetic diversity, and management units of hawksbill turtles in the Indo-Pacific. *J Hered* 107: 199–213.

Vaughan, G.O., Al-Mansoori, N., and J.A. Burt. 2019. The Arabian Gulf. In: *World Seas: An environmental evaluation. The Indian Ocean to the Pacific*. Ed: C. Sheppard, vol. 2, 1–23. New York: Academic Press.

Wabnitz, C.C.C., Lam, V.W.Y., Reygondeau, G., et al. 2018. Climate change impacts on marine biodiversity, fisheries and society in the Arabian Gulf. *PLoS One* 13, no. 5: e0194537.

Wallace, B.P., Dimatteo, A.D., Hurley, B.J., et al. 2010. Regional management units for marine turtles: A novel framework for prioritizing conservation and research across multiple scales. *PLoS One* 5, no. 12: e15465. doi:10.1371/journal.pone.0015465.

Weber, S.B., Broderick, A.C., Groothuis, T.G., Ellick, J., Godley, B.J., and J.D. Blount. 2012. Fine-scale thermal adaptation in a green turtle nesting population. *Proc Biol Sci Roy Soc* 279, no. 1731: 1077–1084. doi:10.1098/rspb.2011.1238.

Williams, J.L., Pierce, S.J., Hamann, M., and M.M.P.B. Fuentes. 2019. Using expert opinion to identify and determine the relative impact of threats to sea turtles in Mozambique. *Aquat Conserv Mar Freshw Ecosyst* 29: 1936–1948.

Witzell, W.N. 1983. Synopsis of biological data on the hawksbill turtle, *E Imbricata* Linnaeus, 1766. *FOA Fisheries Synopsis* 137: 78.

WMO (World Meteorological Organization). 2017. *Statement on the state of the global climate in 2017*. https://public.wmo.int/en/wmo-statement-state-of-global-climate-2017; downloaded 18 January 2019.

Wood, D.W., Bjorndal, K.A., and S.T. Ross. 2000. Relation of temperature, moisture, salinity, and slope to nest site selection in loggerhead sea turtles. *Copeia*, 2000: 119–128.

Zarate, P., Fernie, A., and P. Dutton. 2013. First results of the East Pacific green turtle, *Chelonia mydas*, nesting population assessment in the Galapagos Islands. *NOAA Technical Memorandum NMFS-SEFSC* 503: 70–73.

Zolgharnein, H., Salari-Aliabadi, M.A., Forougmand, A.M., and S. Roshani. 2011. Genetic population structure of hawksbill turtle (*Eretmochelys imbricata*) using microsatellite analysis. *Iranian J Biotechnol* 9: 56–62.

Status of green sea turtles (*Chelonia mydas*) at Zabargad Island, southern Egyptian Red Sea

22

Ashraf S. Mohammad, Hussein N.M. Hussein, Ahmed Ghallab, Aldoushy Mahdy and Hashem Madkour

22.1 INTRODUCTION

Zabargad Island in the southern Egyptian Red Sea (Figures 22.1 and 22.2) is 4.5 km^2 in area, mostly made of rocks, and is distinguished by distinctive geological features. It is thought to have been a piece of the mainland that split apart with the formation of the Red Sea. There are several hills on the sides that were formed by fossilized coral reefs, and it is surrounded by a fringing reef that is used for diving activities. Very fine sand along the shoreline makes it ideal for sea turtle nesting (El-Sadek et al. 2016). The precious stone peridot that is found on the island gave it the name Zabargad. Magnesium carbonate, magnetite, talc, serpentine, quartz, gold, calcite, and other raw materials are abundant in the hills of Zabargad Island, as well as valuable gems like peridot (Styles and Gerdes 1983; Brueckner et al. 1996; Thoresen and Harrell 2014).

The island is distinguished by the presence of a sandy tongue that extends from the southern side and measures three kilometres in length and 50 metres on average in width. According to Hanafy (2012), Attum et al. (2014), and El-Sadek et al. (2016), this region is one of the most significant nesting locations for green sea turtles (*Chelonia mydas*) in the Red Sea. Since the sooty falcon (*Falco concolor*) inhabits the highest peaks on the island's southern and northern sides, where it is challenging to reach its nesting locations, the height of the island's hills also creates a suitable habitat for nesting seabirds, particularly the sooty falcon, to nest and reproduce. The inability to constantly monitor Zabargad Island to stop fishermen and tourist boats from landing and taking its natural resources, such as turtles during the breeding season and migratory birds, poses one of the island's greatest risks due to its distance from the coast.

Seven species of sea turtles are well-known worldwide and belong to two taxonomic families. The Cheloniidae has six species: the green (*Chelonia mydas*), the loggerhead (*Caretta caretta*), the Kemp's ridley (*Lepidochelys kempii*), the hawksbill (*Eretmochelys imbricate*), the flatback sea turtle (*Natator depressus*), and the leatherback turtle (*Dermochelys coriacea*). The number and types of scutes on the carapace, the pattern of scalation on the plastron, and other morphological characteristics can be used to distinguish among species (Meylan et al. 2011; Sönmez 2019).

According to the IUCN Red List (Godley et al. 2008; IUCN 2010), green sea turtles (*Chelonia mydas*) are an endangered species due to factors such as delayed maturation age, ontogenetic habitat use, and the long distance for migration that makes them vulnerable to predators and loss in the sea (Seminoff and Shanker 2008). Due to numerous natural factors, such as birds that prey on young turtles and crustaceans that eat turtle eggs, green turtle populations are constrained.

This is in addition to the detrimental effects of human intervention, such as the general lack of environmental awareness and the overfishing of turtles and birds of all ages, collecting eggs, or eradicating their feeding sources and nesting locations as a result of the recent and unexpected boom in tourism development (Short et al. 2011).

Sea turtles lay their eggs and hatch them on sandy beaches. They then spend their juvenile, sub-adult, and adult years swimming in the ocean. The baby turtles emerge from the nest and scurry towards the water. Before becoming an adult, the animal's biology is unknown (Bowen and Karl 2007). Marine turtles migrate over long distances, sometimes hundreds of kilometres, between their feeding grounds and nesting grounds (Casale et al. 2012). Marine turtles are creatures that spend almost their entire lives in the ocean; it takes them decades to reach sexual maturity. These species come ashore during the nesting season and lay numerous clutches of eggs on sandy beaches (El-Sadek et al. 2016). Sea turtles have a high level of site loyalty to their birth beaches (Carr and Hirth 1962).

DOI: 10.1201/9781003321392-28

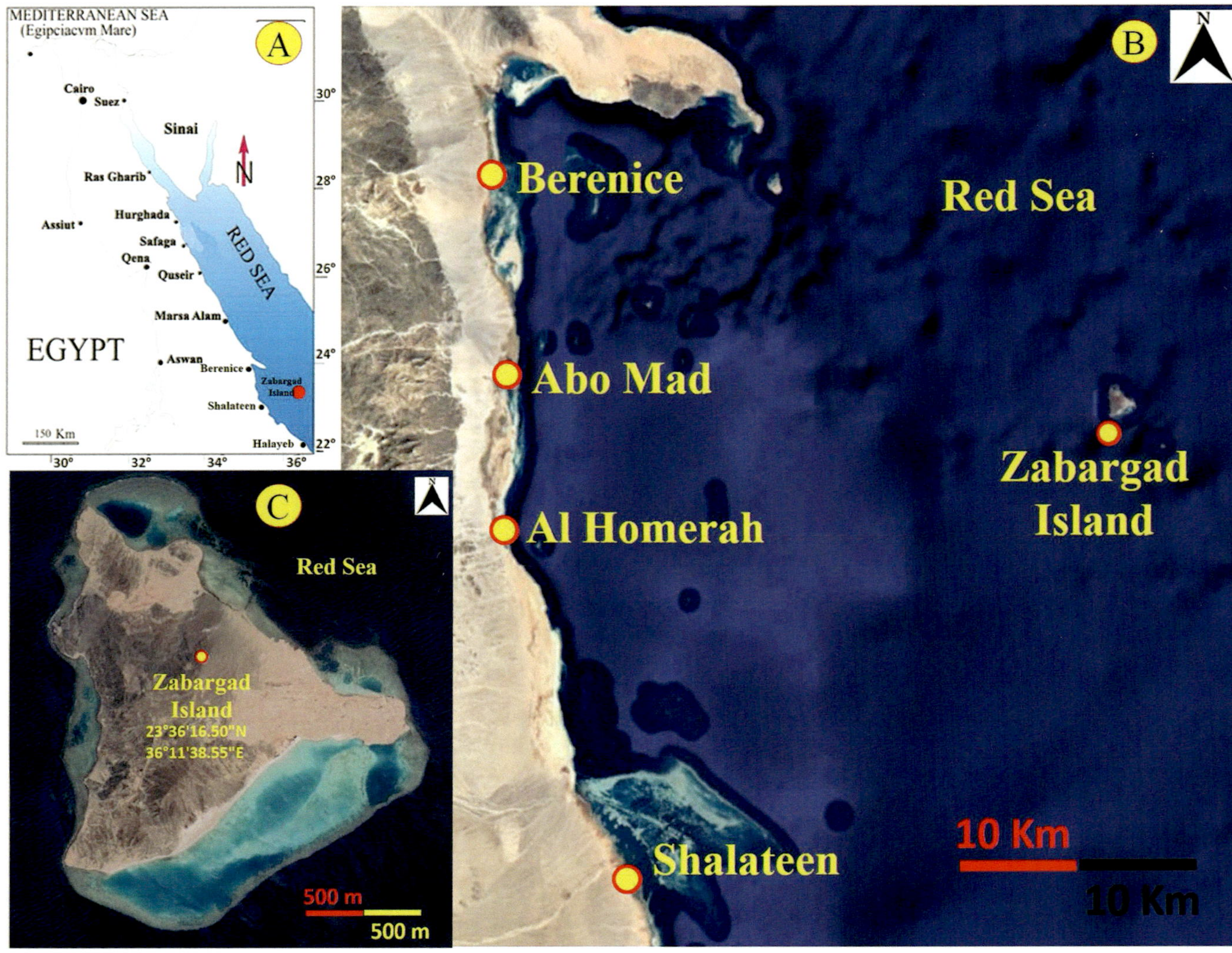

FIGURE 22.1 (A) Location map, (B, C) Google Earth images of Zabargad Island and mainland.

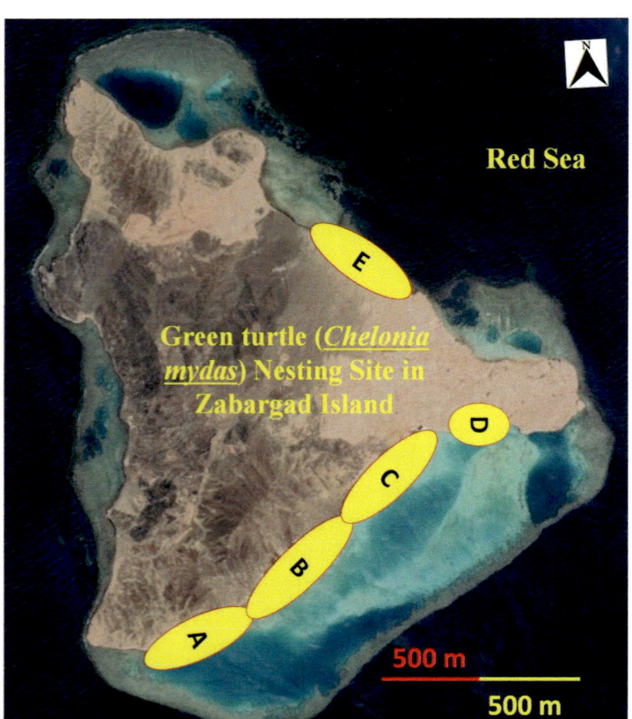

FIGURE 22.2 Google Earth image showing the nesting sites for the green turtle (*Chelonia mydas*) on Zabargad Island.

Every two to four years, adult female sea turtles go back to the beach to nest, where they lay their eggs. Mature females scrounge on the sandy beach at night to locate a suitable location for the nest, which they then construct by removing sand with their flippers. Depending on the species and size, female turtles lay between 50 and 350 soft-shelled eggs from the cloaca, one at a time or twice. Once the eggs are laid, the female turtle begins to carefully cover the nest with sand before diving back into the water (Carr 1987; Pilcher and Ali 1999; El-Sadek et al. 2016).

Before the chicks hatch, the eggs must be incubated for 50–60 days in the nest. Once they emerge from the egg, they instinctively move towards the sea. Once there, they go right into the water, where they must contend with a variety of predators. Until now, studies on marine turtles in the Red Sea are scarce (Hanafy and Sallam 2003; PERSGA 2004; Hanafy 2012; Rouphael et al. 2013; Attum et al. 2014; Mancini et al. 2015; El-Sadek et al. 2016).

Due to the Egyptian Environmental Affairs Agency's (EEAA) and environmental researchers' limited capacity to cover the remote Red Sea islands, many of the islands are exposed to environmental risks. On Zabargad Island, green turtle nesting activities were estimated from 2001 to 2012 (Hanafy 2012) and from 2009 to 2014 (El-Sadek et al. 2016). Prior research focused on the midseason nesting period, which lasted only three nights during the nesting season. Therefore, it was essential that the relevant authorities take actionable measures to lessen the adverse effects. In this situation, the EEAA, represented by Elba National Park, is in the process of preparing environmental studies as well as monitoring, protecting, and preserving all of the offshore islands that correspond to the reserve and the creatures on them from the risk of extinction.

In order to prepare for recurring follow-up field trips by concerned and research authorities to prevent environmental violations, the current study's objectives included monitoring local green sea turtles (*Chelonia mydas*) and estimating the island's environmental value. The current study concentrated on giving a general description of these areas in order to make further research and the identification of appropriate locations for nesting and egg-laying on the island easier. This included determining the factors that are harmful to the spread of green turtles and choosing the best ways to lessen these influences. In collaboration with the Tourism Development Authority and the Egyptian Border Guard Intelligence, recommendations should be made to safeguard nesting areas on distant islands.

22.2 MATERIALS AND METHODS

22.2.1 Study area

Zabargad Island is located 71 km southeast of Ras Banas at latitude 23°36′4.43″N, longitude 36°11′53.89″E, and is a volcanic cone-shaped island. It is about 70 km from Al Homerah village, which is in Elba National Park (Figures 22.1 and 22.2). Zabargad Island is the only island from which both the Egyptian and Saudi Arabian coasts can be seen. It has the form of an equilateral pentagon, the length of each side being 2.5 km, and its total area is 4.5 km^2, with a circumference of about 10 km.

Zabargad Island is made up of basic and metamorphic rocks with dams of basalt and dolerite interspersed throughout. Old coral barriers (skins) that rise about 60 metres above sea level are atop layers of sedimentary rocks (limestone and black clay) (Styles and Gerdes 1983). Since ancient times, mines in the northern peaks have been used to extract valuable ores, particularly Zabargad mineral (a precious gemstone), and ores such as magnetite, quartz, gold, talc, serpentine, and calcite are dispersed nearby (Styles and Gerdes 1983). The island's coast is surrounded by coral reefs and rises from water 500 to 1000 m deep.

Zabargad Island has a rocky, outdated pier. The island, which is situated in the middle of the sea close to the navigational route, has been known to sailors since ancient times. Green sea turtles (*Chelonia mydas*), which fishermen and miners visit for the purpose of hunting, have made the island famous since antiquity. Ancient nesting activity can still be seen in the northeastern region of the island, where saline-adapted plants such as *Zygophyllum album*, Z. simplex, *Cyperus* spp., *Halocnemum strobilaceum*, and *Arthrocnemum glaucum* are common. There is a large sandy tongue at the base of the hills on the south side and a small, narrow tongue on the north. These areas are suitable for sea turtles to come ashore. Additionally, they are suitable for nesting by a variety of seagulls, martens, falcons, and eagles. Rocky Island, a small island just above sea level to the south, is encircled by coral reefs on all sides. Its lack of a sandy beach renders it unsuitable for green turtle nesting.

Access to the low-lying southern beaches is a little challenging due to the underwater reefs south of the island. It is not unusual for the Red Sea to be difficult to navigate due to the abundance of hazardous reefs and strong northerly winds. According to the shore's habitat and how far it is from the high tide, Zabargad Island's shore was divided into five habitats: A, B, C, D, and E (Tables 22.1 and 22.2 and Figures 22.1 and 22.2). As can be seen on the map, Site C is the widest area along the shore.

TABLE 22.1 Locations of turtle nesting sites on Zabargad Island (A, B, C, D, E in Figure 22.2).

NESTING SITE	*LATITUDE*	*LONGITUDE*
A	23°35′48.57″N	36°11′22.97″E
B	23°35′58.84″N	36°11′49.73″E
C	23°36′14.54″N	36°12′4.18″E
D	23°36′23.38″N	36°12′30.90″E
E	23°36′57.00″N	36°12′4.20″E

TABLE 22.2 Measurements of turtle nest widths and lengths (cm) and distance to beach (m) in sites A, B, C, D, and E on Zabargad Island.

		NEST W. CM	*NEST L. CM*	*DIST. (NEST TO BEACH) M*
NESTING SITE	*NO. NEST*	*AVE (MIN-MAX) ± S.D.*	*AVE (MIN-MAX) ± S.D.*	*AVE (MIN-MAX) ± S.D.*
A	191	111.5 (80–160) ± 25.7	132.3 (90–200) ± 35.3	14.7 (5–70) ± 14.7
B	69	104.2 (70–150) ± 25.8	134.3 (90–160) ± 20.9	28.5 (6–80) ± 29.3
C	453	104.4 (70–160) ± 23.6	128.2 (75–200) ± 29.6	11.9 (2.5–80) ± 10.2
D	53	109.3 (70–150) ± 23.3	124.3 (85–200) ± 28	4.8 (3–7.5) ± 1.2
E	9	90 (70–110) ± 28.3	96.5 (75–118) ± 30.4	6.5 (5–8) ± 2.1
Total	775	105.9 (70–160) ± 23.9	128.2 (75–200) ± 29.6	12.5 (2.5–80) ± 13.7

22.2.2 Green sea turtles survey methods

For marine turtles, nesting surveys are used to assess the status of populations (Schroeder and Murphy 1999). The surveys of nesting beaches for green sea turtles in Zabargad Island refers to each time a female turtle emerges from the water to nest, creating a distinctive set of tracks on the sand. To describe the locations surveyed and identify the qualities of the most ideal areas for green turtle nesting, Zabargad Island was divided into various sites, utilizing GPS to give the locations of those places. The most valuable nesting shores on Zabargad Island (A, B, C, D, and E) were surveyed over a period of ten days in the nesting season, from July 1–10, 2006. Based on the presence of the nests and their crawls or tracks on the surveyed shores, marine turtle nesting activities were identified. The measurements of each nest's width, length, and the distance from the shore in metres were taken. Turtle tracks were measured for track length and track width (internal and external) (Figure 22.3). Using the shape of the track (symmetric for green turtles; Pritchard and Mortimer 1999), nesting turtle species were identified.

22.3 RESULTS

In the current study's 10-night survey, the turtle nest at each Zabargad Island location was unique. Table 22.2 details the measurements of the turtle nests and the number of tracks that have been counted on Zabargad Island's various shores (A, B, C, D, and E) (average, minimum, maximum, and standard deviation).

There were a total of 775 turtle nests, with site C having the highest number (453 nests) and site E having the lowest number (9 nests) (Table 22.2). With a minimum of 70 cm and a maximum of 160 cm, the average turtle nest width was 105.9 ± 23.9 cm. Site A had the widest average nest, measuring 111.5 ± 25.7 cm, while site E had the narrowest, measuring 90 ± 23.9 cm (Table 22.2). The nests ranged in length from 75 to 200 cm, with an average length of 128.2 ± 29.6 cm. Site B had the highest average across all sites at 134.3 ± 20.9 cm, while site E had the lowest average at 96.5 ± 30.4 cm (Table 22.2).

With a minimum of 2.5 m and a maximum of 80 m, the average distance of a turtle nest from the shore was 12.5 ± 13.6 m. The average for site B was the highest of all the sites (28.5 ± 29.3 m), and the average for site E was the lowest (6.5 ± 2.1 m). Table 22.3 shows the number of turtle tracks at different sites on Zabargad Island. The total number of turtle tracks was 57, with the highest number being 27 at site C and the lowest being 2 at site E (Table 22.3). The averages of internal track widths were much closer together, with a total average of 48.4 ± 7.5 cm and a range of 40–64 cm, with the highest average of 49.9 cm in site A and the lowest of 47.5 cm in site E (Table 22.3).

The total average of external turtle track widths was 94.4 ± 8.3 cm, with the highest average being 95.2 ± 8.2 cm in site C and the lowest 90 ± 28.3 cm in site E (Table 22.3). Finally, the total average length of turtle tracks was 79.4 ± 52.8 m with a range of 70–110 m, with the highest average being 117.8 ± 40.3 m in site C and the lowest 13.5 ± 3.1 m in site D (Table 22.3).

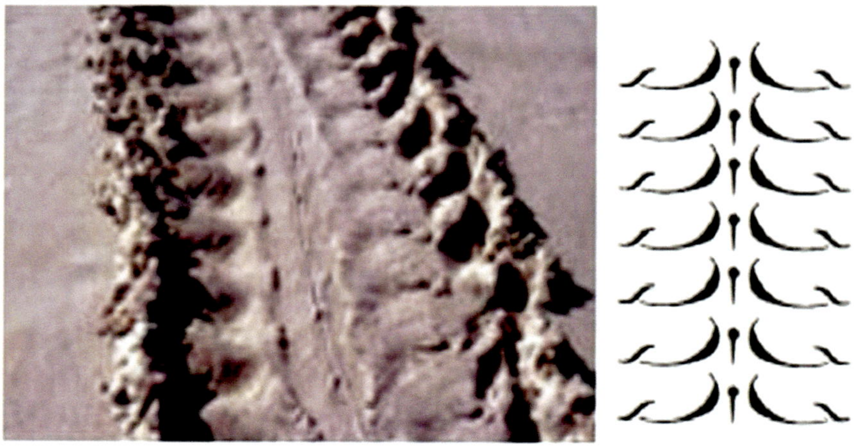

FIGURE 22.3 Shapes of green turtle track at nesting shore.

TABLE 22.3 Measurements of the turtle track widths (cm), internal (int.) and external (ext.), and turtle track lengths (m) in sites A, B, C, D, and E on Zabargad Island.

		TRACK W. INT.	*TRACK W. EXT.*	*TRACK L. M*
TRACK SITE	*NO. TRACKS*	*AVE (MIN-MAX) ± S.D.*	*AVE (MIN-MAX) ± S.D.*	*AVE (MIN-MAX) ± S.D.*
A	9	49.9 (40–64) ± 8.1	94.6 (90–110) ± 7.2	57.9 (30–120) ± 26.3
B	7	48.4 (40–64) ± 8.7	94.7 (90–110) ± 8.3	81.9 (38–135) ± 39.8
C	27	48.2 (40–64) ± 7.4	95.2 (90–110) ± 8.2	117.8 (20–200) ± 40.3
D	14	48 (40–64) ± 7.6	93.2 (90–110) ± 6.3	13.5 (10–20) ± 3.1
E	2	47.5 (40–55) ± 10.6	90 (70–110) ± 28.3	96.5 (75–118) ± 30.4
Total	57	48.4 (40–64) ± 7.5	94.4 (70–110) ± 8.3	79.4 (10–200) ± 52.8

FIGURE 22.4 Photographs of activities while observing green turtles on Zabargad Island. A) Turtle searching for food; B) turtles mating; C) measuring the length of the dorsal shell; D) measuring the width of the abdominal shell; E) turtle digging a place to lay eggs; F) tag on one of a turtle's limbs.

Figure 22.4 shows photographs of different activities while observing sea turtles on Zabargad Island. The island has been one of the most successful locations for saving female marine green turtles that are nesting because the island is remote from many anthropogenic activities and development. However, there are some recorded threats to individual females, which may have been eaten by local fishermen or others on the island.

22.4 DISCUSSION

One species of sea turtle, *Chelonia mydas*, is seen nesting on Zabargad Island, as also observed by Hanafy (2012) and El-Sadek et al. (2016); however, other studies on different Red Sea islands have found nesting activities for green turtles and hawksbill sea turtles on Shedwan Island and the nearby Giftun Islands (Hanafy and Sallam 2003).

Frazier and Salas (1984) note that *Chelonia mydas* are rare in the Red Sea, whereas they are most common in the western Indian Ocean (Frazier 1979) and further into the Mediterranean (Geldiay et al. 1982). They also estimate that there are only a few hundred nesting females in the entire Red Sea territory (Frazier and Salas 1984). Green turtles are the next most frequent species of marine turtles to feed and lay their eggs in this region, along with hawksbill sea turtles, and Farasan Bank is regarded as a good feeding ground. Hawksbill sea turtles are the most frequently sighted species on the eastern side of the Red Sea (Al-Mansi 2016). Assuming that the population of the nesting green turtles at Zabargad Island has recovered after a decline for the years prior to 2014, the nesting success shows no trend during the previous years at 44% in 2014, 42% in 2015, 41% in 2016, and 44% in 2017 (unpublished data).

Globally, variations in the number of nesting females between years were noted in the nesting population of green turtles. Old studies indicate that there were only 200 nesting females in the Red Sea (Frazier and Salas 1984; PERSGA 2004), but Hanafy (2012) demonstrates that over an eight-year period, the number of green turtle nesting females in Egypt fluctuated between 175 and 610. Follow-up research by El-Sadek et al. (2016) reveals that over the course of a five-year study from 2009 to 2014, the number of nesting females on Zabargad Island fluctuated between 37 and 362. The highest estimated number of green sea turtles has been estimated to be over 150 green turtles in one site according to the availability of food and the size of the site. It was noted that in the national park, where there is no human stress or limited stress, the estimated abundance of green turtles is better than in sites outside of the national park (Mancini et al. 2015; El-Sadek et al. 2016).

According to the IUCN Red List assessment and its inclusion in Appendix 1 of the CITES agreement, which shows the distribution of marine turtles around the world, *Chelonia mydas* is recognised as being endangered. They migrate widely and intricately through habitats that are sparsely populated geographically. Nesting takes place in more than 80 nations around the world (Hirth 1997), and the coastal waters of about 140 countries are home to them (Groombridge and Luxmoore 1989).

Hatchlings are thought to enter the oceanic phase after leaving the nesting beach, floating passively in the main current systems that serve as their development grounds (Carr and Meylan 1980). The turtles return to the neritic area, which is rich in seagrass and/or marine algae, after spending a few years in the ocean, where they forage and develop until maturity (Musick and Limpus 1997). Green turtles begins their annual breeding migration between their foraging grounds and nesting grounds when they reach adulthood (Hirth 1997). Both sexes migrate, and they frequently cover thousands of kilometres (Carr 1987). Adults live in the foraging ground, which serves as the juveniles' habitat for development at some point during the non-breeding season (Seminoff et al. 2003). From the egg to the adult stage of development, green turtles are more susceptible to anthropogenic impacts. The main human threats are egg and adult harvesting from nesting beaches (Fretey 2001).

The main factors contributing to the decline of the green turtle population globally are bycatch in marine fisheries, habitat degradation for the nesting beaches and feeding grounds, disease, entanglement in fishing nets, and fishing hooks (Lutcavage and Lutz 1997). These factors may alter the quantity and quality of nesting areas available to females directly through the loss of beach habitat or indirectly through changing thermal profiles and increasing erosion. They may also cause changes in the natural behaviour of adults and hatchlings (Ackerman 1997).

The adult females' behaviour is changed by lights at the beach or near nesting shores, which draws the hatcheries to the light source and drowns them out at sea (Ackerman 1997). The health of the coastal marine ecosystem was impacted by an increase in contamination, coastal development, the building of marinas, increased boat traffic, and the harvesting of marine algae resources, which in turn had an effect on green turtles (George 1997). Green turtles after hatching and arriving at the sea begin to recruit to neritic development foraging habitats, such as mangrove bays and creeks, hard bottom habitats, seagrass habitats, and coral reef habitats (Musick and Limpus 1997). As they mature, they engage in a number of habitat development activities, growing larger in these habitats (van Dam and Diez 1997; Bowen et al. 2007). When they reach sexual maturity, they travel during periods of several years between the habitat of the feeding grounds and breeding areas (Mortimer and Bresson 1999).

Genetic research has shown that female sea turtles return to their natal rookery to breed (Bowen and Karl 1997). Others may settle into foraging habitat close to their shores or places of origin, while some may find foraging habitats or development habitats hundreds or thousands of kilometres from their natal beach (Bowen et al. 2007).

22.5 RECOMMENDATIONS

In order to gain more experience through outside organizations, researchers must attend training sessions on green turtle monitoring. A programme should be set up to inform residents and fishermen about the importance of turtles, how to protect them, and what to do if one is entangled in a net. Nest destruction and other activities that prevent turtles from laying eggs there should be discouraged, including creating notice boards for turtle nesting areas. The use of traditional fishing methods and those used by local fishermen, which stand out for their lack of negative effects on marine life, including green turtles, should be made mandatory, and patrols to observe fishing operations in these areas should also be conducted.

REFERENCES

Ackerman, R. 1997. The nest environment and the embryonic development of sea turtles. In: *The biology of sea turtles*. Eds: P.L. Lutz, and J.A. Musick, 83–106. Boca Raton, FL: CRC Press.

Al-Mansi, A.M. 2016. *Sea turtles in the Red Sea and the Arabian Gulf, Saudi Arabia*. Saudi Wildlife Authority, Riyadh, Report, p. 1–43.

Attum, O., Kramer, A., Mahmoud, T., and M. Fouda. 2014. Post-nesting migrations patterns of green sea turtles (*Chelonia mydas*) from the Egyptian Red Sea. *Zool Middle East* 60, no. 4: 299–305.

Bowen, B.W., Grant, W.S., Hillis-Starr, Z., Shaver, D.J., Bjorndal, K.A., Bolten, A.B., and A.L. Bass. 2007. Mixed stock analysis reveals the migrations of juvenile hawksbill turtles (*Eretmochelys imbricata*) in the Caribbean Sea. *Mol Ecol* 16: 49–60.

Bowen, B.W., and S.A. Karl. 1997. Population genetics, phylogeography, and molecular evolution. In: *The biology of sea turtles*. Eds: P.L. Lutz, and J.A. Musick, 29–50. Boca Raton, FL: CRC Press.

Bowen, B.W., and S.A. Karl. 2007. Population genetics and phylogeography of sea turtles. *Mol Ecol* 16: 4886–4907. doi:10.1111/j.1365-294X.2007.03542.x.

Brueckner, H.K., Bonatti, E., Elhaddad, M.A., Hamelin, B., Kröner, A., Reisberg, L., and M. Seyler. 1996. A Nd, Sr, Pb, and Os study of the gneisses and ultramafic rocks of Zabargad Island, Red Sea: Miocene Moho or Pan African peridotites. *J Geophys Res* 100, no. B11: 22283–22297.

Carr, A. 1987. New perspectives on the pelagic stage of sea turtle development. *Conserv Biol* 1: 103–121. doi:10.1111/j.1523-1739.1987.tb00020.x.

Carr, A.F., and H.F. Hirth. 1962. The ecology and migrations of sea turtles. 5, Comparative features of isolated green turtle colonies. *Am Mus Novit* no. 2091.

Carr, A.F., and A.B. Meylan. 1980. Extinction or rescue for the hawksbill? *Oryx* 15, no. 5: 449–450.

Casale, P., Affronte, M., Scaravelli, D., Lazar, B., Vallini, C., and P. Luschi. 2012. Foraging grounds, movement patterns and habitat connectivity of juvenile loggerhead turtles (*Caretta caretta*) tracked from the Adriatic Sea. *Mar Biol* 159: 1527–1535. doi:10.1007/s00227-012-1937-2.

El-Sadek, I., Ahmed, M.I., Aamer, M.A., Mancini, A., and M.H. Hanafy. 2016. Nesting activities of Green sea turtles (*Chelonia mydas*) on the beaches of Zabargad Island, Southern Egyptian Red Sea, Egypt. *J Aquat Biol Fish* 20, no. 4: 29–37.

Frazier, J. 1979. Marine turtle management in Seychelles: A case-study. *Environ Conserv* 6, no. 3: 225–230.

Frazier, J., and S. Salas. 1984. The status of marine turtles in the Egyptian Red Sea. *Biol Conserv* 30, no. 1: 41–67.

Fretey, J. 2001. *Biogeography and conservation of marine turtles of the Atlantic coast of Africa/Biogographie et conservation des tortues marines de la cote atlantique de lAfrique*. CMS Technical Series Publication, no. 6, Bonn, Germany: UNEP/CMS Secretariat, 429 p.

Geldiay, R., Koraly, T., and S. Balik. 1982. Status of sea turtle populations (*Caretta caretta* and *Chelonia mydas*) in the Northern Mediterranean Sea, Turkey. In: *Biology and conservation of sea turtles*. Ed: K.A. Bjorndal, 425–434. Washington, DC: Smithsonian Institution Press.

George, R.H. 1997. Health problems and diseases of sea turtles. In: *The biology of sea turtles*. Eds: L. Lutz, and J.A. Musick, 363–386. Boca Raton, FL: CRC Press.

Godley, B.J., Blumenthal, J.M., Broderick, A.C., Coyne, M.S., Godfrey, M.H., Hawkes, L.A., and M.J. Witt. 2008. Satellite tracking of sea turtles: Where have we been and where do we go next? *Endang Species Res* 4: 3–22.

Groombridge, B., and R. Luxmoore. 1989. *The green turtle and hawksbill (Reptilia: Cheloniidae): World status, exploitation, and trade*. Lausanne, Switzerland: CITES Secretariat, 601 p.

Hanafy, M.H. 2012. Nesting of marine turtles on the Egyptian beaches of the Red Sea. *Egypt J Aquat Biol Fish* 16: 59–71.

Hanafy, M.H., and A. Sallam. 2003. *Status of marine turtles nesting on the Egyptian beaches of the Red Sea*. National Report to PERSGA, 45 p.

Hirth, H.F. 1997. Synopsis of the biological data on the green turtle *Chelonia mydas* (Linnaeus, 1758). *Biol Rep* 97, no. 1, US Fish and Wildlife Service, US Department of the Interior, Washington, DC.

IUCN (International Union for the Conservation of Nature). 2010. *IUCN Red list of threatened species*. Version 2010.4.8, December. www.iucnredlist.org.

Lutcavage, M., and P.L. Lutz. 1997. Diving physiology. In: *The biology of sea turtles*. Eds: P.L. Lutz, and J.A. Musick, vol. 1, 277–296. Boca Raton, FL: CRC Press.

Mancini, A., Elsadek, I., and M. Benedicte. 2015. When simple is better: Comparing two sampling methods to estimate green turtles' abundance at coastal feeding grounds. *J Exp Mar Biol Ecol* 465: 113–120.

Meylan, P.A., Meylan, A.B., and J.A. Grey. 2011. The ecology and migrations of sea turtles: Tests of the developmental habitat hypothesis. *Bull Am Mus Nat Hist* 357: 1–70.

Mortimer, J., and R. Bresson. 1999. Temporal distribution and periodicity in hawksbill turtles (*Eretmochelys imbricata*) nesting at Cousin Island, Republic of Seychelles, 1971–1997. *Chelonian Conserv Biol* 3, no. 2: 318–325.

Musick, J.A., and C.J. Limpus. 1997. Habitat utilization and migration in juvenile sea turtles. In: *The biology of sea turtles*. Eds: P.L. Lutz, and J.A. Musick, 137–163. Boca Raton, FL: CRC Press.

PERSGA (Regional Organization for the Conservation of the Environment of the Red Sea Gulf of Aden). 2004. *Regional action plan for the conservation of marine turtles and their habitats in the Red Sea and Gulf of Aden*. PERSGA Technical Report, Jeddah, Saudi Arabia.

Pilcher, N.J., and L. Ali. 1999. Reproductive biology of the hawksbill turtle, *Eretmochelys imbricata*, in Sabah, Malaysia. *Chelonian Conserv Biol* 3, no. 2: 330–336.

Pritchard, P.C., and J.A. Mortimer. 1999. Taxonomy, external morphology and species identification. In: *Research and management techniques for the conservation of sea turtles*. Eds: K.L. Eckert, K.A. Bjorndal, F.A. Abreu-Grobois, and M. Donnelly. IUCN/SSC Marine Turtle Specialist Group Publication no. 4. Washington, DC.

Rouphael, T., Abdulla, A., Attum, O., Marshall, N., and U. Ghazali. 2013. Do marine protected areas in the Red Sea afford protection to dugongs and sea turtles? *J Biodivers Endanger Species* 1: 1–6. doi:10.4172/jbes.1000102.

Schroeder, B., and S. Murphy. 1999. Population surveys (ground and aerial) on nesting beaches. In: *Research and management techniques for the conservation of sea turtles*. Eds: K.L. Eckert, K.A. Bjorndal, F.A. Abreu-Grobois, and M. Donnelly, 45–55. Washington, DC: IUCN/SSC Marine Turtle Specialist Group Publication no. 4.

Seminoff, J.A., Jones, T.T., Resendiz, A., Nichols, W.J., and M.Y. Chaloupka. 2003. Monitoring green sea turtles (*Chelonia mydas*) at a coastal foraging area in Baja California, Mexico: Using multiple indices to describe population status. *J Mar Biol Assoc UK* 83: 1355–1362.

Seminoff, J.A., and K. Shanker. 2008. Marine turtles and IUCN Red Listing: A review of the process, the pitfalls, and novel assessment approaches. *J Exp Mar Biol Ecol* 356, nos. 1–2: 52–68.

Short, F.T., Polidoro, B., Livingstone, S.R., Carpenter, K.E., Bandeira, S., Bujang, J.S., and J.C. Zieman. 2011. Extinction risk assessment of the World's seagrass species. *Biol Conserv* 144: 1961–1971.

Sönmez, B. 2019. Head and plastron scalation patterns of the green turtle, *Chelonia mydas*, hatchlings in natural and relocated nests on Samandağ Beach. *J Black Sea/Mediterr Environ* 25, no. 3: 280–293.

Styles, P., and K.D. Gerdes. 1983. St. John's Island (Red Sea): A new geophysical model and its implications for the emplacement of ultramafic rocks in fracture zones and at the continental margin. *Earth Planet Sci Lett* 66: 353–368.

Thoresen, L., and J.A. Harrell. 2014. *Discovery of the Red Sea source of Topazos (ancient gem peridot) on Zabargad Island, Egypt (by James A. Harrell); Archaeogemology of Peridot*. Twelfth Annual Sinkankas Symposium Peridot and Uncommon Green Gem Minerals, The Robert Mouawad Campus, Carlsbad, CA.

van Dam, R.P., and C.E. Diez. 1997. Diving behavior of immature hawksbills (*Eretmochelys imbricata*) in a Caribbean reef habitat. *Coral Reefs* 16: 133–138.

Sea turtle nesting beach conditions in the western Arabian Gulf

23

Rommel H. Maneja, Jeffrey D. Miller, Ace V.B. Flandez, Yasir Y. Asiri, Jinoy V. Gopalan, Ibrahim A. Alzoghiby and Ali Qasem

23.1 INTRODUCTION

Beaches used for nesting by marine turtles share a general set of characteristics, including being accessible from the ocean, having stability through time, and permitting nest construction as well as providing the conditions necessary for successful incubation of the eggs and dispersal of the hatchlings (Mortimer 1990; Georges et al. 1993; Santos et al. 2016; Patrício et al. 2018; Wood and Bjorndal 2000; Türkozan et al. 2011). The selection of a site for oviposition on a beach may be based on specific criteria, including beach elevation, temperature, and presence of vegetation (i.e., discriminating or non-random nesters) (Mellanby et al. 1998; Hays and Speakman 1993; Hays et al. 1995; Pfaller et al. 2009), or it may be random (Mrosovsky 1983; Eckert 1987; Kelly et al. 2017). Additionally, the nesting process may be interrupted by the presence of obstacles (e.g., debris) on or buried in the beach that may cause the turtle to move to a different place, even after starting to dig the nest chamber, indicating that nesting turtles respond to their surroundings while attempting to nest (Fujisaki and Lamont 2016). The location of a nest site must provide the appropriate thermal, hydric, and gaseous environment to support successful embryonic development (Georges et al. 1993; Miller 1997; Ackerman 1997; Wood and Bjorndal 2000; Miller et al. 2003; Suss et al. 2012; Santos et al. 2016; Gatto and Reina 2022). This requirement is usually met by a combination of physical characteristics, including slope, aspect, sand grain size, and environmental conditions, including moisture content and temperature within the beach (Mortimer 1990; Foley et al. 2006; Fuentes et al. 2010; Zare et al. 2012; Booth et al. 2020). Successful development is affected by stochastic events that impact the beach, such as the inundation of nests by storms (Caut et al. 2010; Booth et al. 2022). In addition, after incubating successfully for 50–60 days, emerged hatchlings may be subjected to depredation as they cross the beach and the reef flat (National Research Council 1990; Georges et al. 1993; Heithaus 2013; Santos et al. 2016). As a consequence, beach and nest site selection have a major influence on the survival of embryos, the production of hatchlings to the beach surface and, thus, the long-term functioning of the population (Maurer et al. 2021).

The size distribution of the sand grains provides the matrix through which the embryos experience the environmental conditions within the beach. The physical characteristics of the sand impact the interactions among the gaseous, hydric, and thermal environment surrounding the developing embryos (Ackerman 1997; Yalçın-Özdilek et al. 2007; Fuentes et al. 2010; Salleh et al. 2018; Stewart et al. 2019; Tezak et al. 2020). For example, the size, shape, and roughness of the sand create spaces among the particles, and their surface roughness holds water via surface tension. The thermal properties of water buffer the rate of temperature change in the surrounding substrate. Also, the water contains dissolved salts and organic material that contribute, albeit slightly, to the osmotic gradient between the embryo and the surrounding substrate (Ackerman 1991, 1997; Smith et al. 2021; Tezak et al. 2020). In addition, if the amount of fluid water held among the particles is too high, the exchange of gas between the environment and the embryo may be impeded.

Concentrations of oxygen and carbon dioxide in marine turtle nests must be at appropriate levels for the gas gradient pressure to support embryonic development (Ackerman 1997; Chen et al. 2010; Stubbs and Mitchell 2018; Booth et al. 2020). The energy assimilation and growth by the embryos are regulated by the hydric conditions and, as a result, impact hatchling fitness (Fleming et al. 2020; Cavallo et al. 2015). Within the thermal tolerance range of about 24–34°C, sand temperatures facilitate normal metabolic activity and embryonic development (Miller 1997; Bladow and Milton 2019; Calderón Peña and Azanza Ricardo 2021; Türkozan et al. 2021). However, because warmer temperatures increase the developmental rate and cooler temperatures slow it (Bentley et al. 2020a), excursions outside this temperature range may cause metabolic stress and impact fitness (Booth 2017; Booth and Astill 2001a), depending on the actual temperature experienced and the duration of exposure. Plus, the thermal environment determines the sex of the hatchlings (Wibbels 2003; Sim et al. 2015; Schwanz and Georges 2021).

DOI: 10.1201/9781003321392-29

Emergence success of hatchlings, defined as the proportion of eggs laid that result in hatchlings that emerge from the sand, is an important measure of the outcome of incubation. However, because conservation management needs a broad understanding of the beach with hatching success, it is not sufficient to examine just the emergence success of the hatchlings. It is necessary to define the environmental conditions in the area that surrounds the incubating eggs.

Three of the five species that occur in the Arabian (Persian) Gulf (the Gulf) are known to nest on the beaches within the Gulf (Tollab et al. 2015; Phillott and Rees 2021; Al Ameri et al. 2022). The two most common nesting species are hawksbill (*Eretmochelys imbricata*) and green (*Chelonia mydas*) turtles. Both species are distributed circumglobally in tropical and subtropical marine environments (Witzell 1983; Hirth 1997; Seminoff 2004), including in the Arabian Gulf (Phillott and Rees 2021; Al Ameri et al. 2022). Globally, the hawksbill turtle is classified as *Critically Endangered* (IUCN: CR A2bd, ver. 3.1) (Mortimer and Donnelly 2008). A hawksbill turtle sub-population has not been defined in the Gulf region, and at present it is considered as part of the northwest Indian Ocean regional management unit (Wallace et al. 2010). The green turtle is classified globally as *Endangered* (IUCN: A2bd; Seminoff 2004) and as being *Vulnerable* (IUCN: A2b, ver. 3.1; Mancini et al. 2019) in the Gulf region. The green turtle population in the Gulf is part of the northwest Indian Ocean for the green turtle management unit (Wallace et al. 2010).

Regionally, large nesting aggregations of hawksbill turtles occur in Oman (Masirah Island, Daymaniat Islands), Iran (multiple coastal islands), and Saudi Arabia (offshore islands) (Willson et al. 2021; Mobaraki 2021; Miller 2021), and the largest nesting aggregation of green turtles occurs at the mainland beach at Ras al Hadd (Willson et al. 2021). Within the northern basin of the Gulf, the largest aggregation of nesting by hawksbill turtles occurs at Jana Island, Saudi Arabia (Miller 1989, 2021; Pilcher 1999; Al-Mansi 2016; Phillott and Rees 2021; Al Ameri et al. 2021, 2022). Multiple small (<100 turtles per year) nesting populations are distributed throughout the Gulf region (Al Ameri et al. 2022; Phillott and Rees 2021; Mobaraki et al. 2022), occurring in every country in the Gulf except Iraq and Bahrain (Phillott and Rees 2021; Al Ameri et al. 2021, 2022; Miller 2021; Mobaraki 2021; Papathanasopoulou and Rees 2021; Willson et al. 2021; Rees 2021).

The largest aggregation of nesting green turtles in the Gulf occurs at Karan Island, Saudi Arabia (Miller 1989, 2021; Pilcher 2000; Al-Mansi 2016; Phillott and Rees 2021; Al Ameri et al. 2022). Multiple small (<100 turtles per year) nesting populations are distributed throughout the Gulf region (Al-Mohanna et al. 2014; Al Ameri et al. 2022; Phillott and Rees 2021; Mobaraki 2021; Papathanasopoulou and Rees 2021; Willson et al. 2021; Rees 2021).

The purpose of this chapter is to provide a baseline definition of the conditions experienced by the developing embryos of hawksbill and green turtles in the Saudi Arabian offshore nesting islands using information from past studies conducted in the area. To achieve this, we examined the shape of Jana and Karan Islands, as well as the size composition of the sand, moisture and salinity content, and temperatures at four reference sites around Jana and Karan Islands, and in the area of Karan Island that supports the highest density of green turtle nesting.

23.2 BRIEF DESCRIPTION OF THE ARABIAN GULF

The Gulf is a semi-enclosed, shallow, subtropical sea situated in an extremely arid terrestrial environment (Krishnakumar et al. 2019; Vaughan et al. 2019; Sheppard et al. 2010). It covers approximately 240,000 km^2 and has a mean depth of about 35 m (Krishnakumar et al. 2019; Vaughan et al. 2019). The length of the Gulf is about 1,000 km along a NNW to SSE axis, and the width varies from about 56 km to 340 km (Sheppard 1993; Krishnakumar et al. 2019; Vaughan et al. 2019; Sheppard et al. 2010; Barth and Khan 2008). The Gulf is divided into northern and southern basins by the Qatar Peninsula. The Saudi Arabian part of the western Gulf includes an area of approximately 27,050 km^2 (Krishnakumar et al. 2019) and slopes toward the deeper trench that extends along the eastern side (Johns et al. 1999; Reynolds 1993; Krishnakumar et al. 2019; Vaughan et al. 2019; Sheppard et al. 2010).

23.2.1 Climate and weather

The general climate of the Gulf region is subtropical. The surrounding area is arid with temperatures that fluctuate seasonally between hot summers (June to September) and cool winters (December to March) (Vaughan et al. 2019; Krishnakumar et al. 2019; Beni et al. 2021). The Gulf loses about 15.8 km^3 of water through evaporation per year (Privett 1959; Hastenrath and Lamb 1979), which increases the salinity to above 55 practical salinity units (psu) in coastal waters during summer (Coles and McCain 1990; Krishnakumar et al. 2019; Beni et al. 2021).

Sea surface temperatures (SST) vary seasonally in the nearshore areas and semi-enclosed bays along the shallow western side (Brewer et al. 1978; John et al. 1990; Vaughan et al. 2019; Krishnakumar et al. 2019). In general, low water temperatures in January (range of 15–18°C) increase to a range of 19–22°C by March, then increase to 31–33°C in July. Temperatures are elevated through September (32–33°C) before they cool (Krishnakumar et al. 2019).

23.2.2 Description of main offshore islands

The offshore area of the northwestern Gulf supports areas of seagrass interspaced with numerous small rocky areas and coral reefs (Sheppard and Borowitzka 2011). Of the six offshore islands, Jana and Karan Islands are the two largest islands in the northwestern territorial waters of Saudi Arabia (Basyoni 1999; Miller 2011a; Miller et al. 2019a) (Figure 23.1). Jana Island (27°22′6.85″N, 49°53′50.85″E) has a circumference of 2.6 km (dimensions: 1105 × 300 m), and Karan Island (27°43′5.05″N, 49°49′28.91″E) has a circumference of 5.3 km (dimensions: 2024 × 632 m) (Basyoni 1999; Miller 2011a; Miller et al. 2019a). Each island is situated on the southern margin of a large shallow reef flat (Basson et al. 1977; Miller 2011a; Miller et al. 2019a). The northwestern sides of both islands have exposed beach rock that is coated with tar

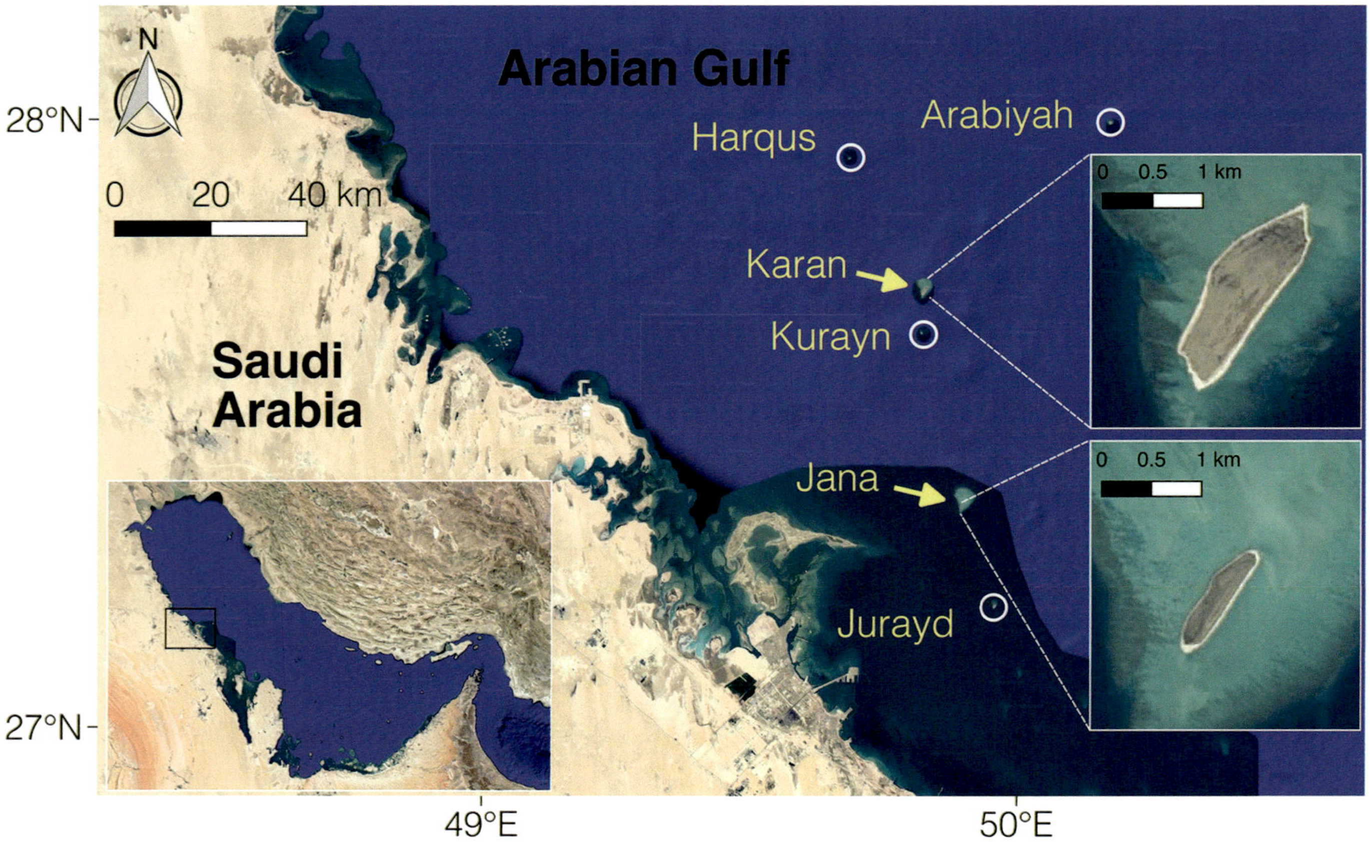

FIGURE 23.1 Satellite image base map showing the location of Karan and Jana Islands (indicated by yellow arrows) in the Saudi Arabian waters of the Arabian Gulf. White circles represent the other four offshore islands in Saudi Arabian waters. Left bottom inset is a map of the whole Arabian Gulf showing image area.

and a storm berm that is heavily littered with debris (wood, Styrofoam, plastic, light bulbs, etc.). In contrast, the southeastern sides of both islands have more exposed sand with less tar content and are comparatively clear of debris. Coral sand soils mixed with vegetation debris and bird guano make up the core areas of the low and flat islands (Miller 2011a, 2011b; Miller et al. 2019a). Although the islands have no freshwater, low shrubs (*Suaeda* sp., *Salsola* sp.) grow in the middle parts of the islands (Maneja et al. 2020a). Turtle nesting activity helps to keep the outer margin of the island free of vegetation. They are the major breeding sites for marine turtles and sea birds in the area (Basson et al. 1977; Miller et al. 2019a), although some nesting does occur on Kurayn and Jurayd Islands.

23.3 METHODS

23.3.1 Island change and reference sites

The potential change in the outline of Karan and Jana Islands was assessed by comparing digitized and georectified aerial photographs and ESRI satellite images taken between 1965 and 2017 using ImageJ software (Schneider et al. 2012) and QGIS (2016) (Maneja et al. 2020b).

Reference sites were established at the northern and southern extremes, as well as approximately halfway between the ends of Jana and Karan Islands (Figure 23.2). These sites are identified using the island name and the direction it faces to the Gulf (e.g., Jana-East, Karan-West). At each site, beach conditions (sand grain size, hydric and salinity conditions, temperature) were assessed at the surface as well as at depths of 30 cm and 50 cm to provide baseline data on the conditions experienced by embryos during incubation. These two depths were selected because they approximate the middle depth of the egg mass of clutches laid by hawksbill (30 cm) and green turtles (50 cm). In addition, the hatching success of a series of green turtle clutches was monitored on Karan Island (Maneja et al. 2023).

23.3.2 Sand grain size

At each reference site, one artificial nest hole was dug landward of the vegetation-beach boundary and one hole was dug seaward of the boundary; the holes were approximately three meters apart. A single sample of surface sand was obtained from the landward site. Triplicate samples of approximately 500g of sand were collected from depths of 30 cm and 50 cm in the artificial nest holes and from the walls of monitored nests. These were analysed separately.

Sub-samples for determining the wet weight were removed from the samples before further processing. The remaining material was oven dried to constant mass (±0.01g) before being sieved through a set of nine graduated sieves (U.S.A. Standard Testing Sieve A.S.T.M. E-11 Specification, VWR Scientific) and a sieve shaker (Ro-Tap Testing Sieve shaker, Model B, CE TYLER, Trademark of Combustion Engineering, Inc.). Sand retained by each sieve and the silt and clay from the pipette analysis were weighed to determine their proportion of the

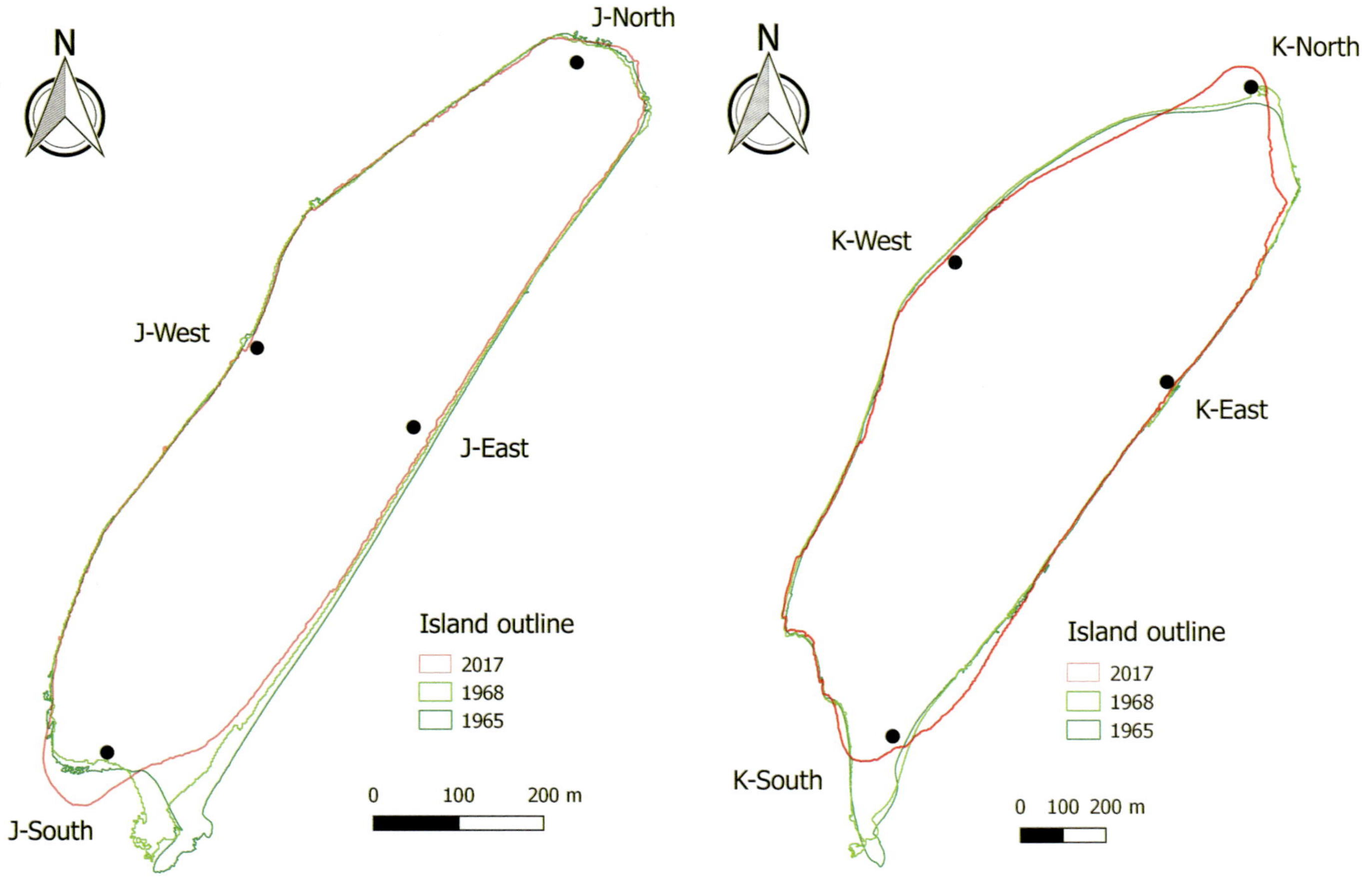

FIGURE 23.2 Outlines of Jana Island (left panel) and Karan Island (right panel) in 1965, 1968, and 2017. Locations of Reference Sites are indicated by black circle symbols. Map modified from Maneja et al. (2020b). (Sources: US Geological Survey (USGS) Earth Resources Observation and Science (EROS) Center; ESRI Satellite ArcGIS/World_Imagery – qms.nextgis.com/geoservices/1300/)

total mass (Carvar 1971). The Wentworth Scale for grain size and nomenclature was used to describe sand grain composition (clay, <0.004 mm; silt, 004–0.063 mm; very fine sand, 0.063–0.125 mm; fine sand, 0.125–0.25 mm; medium sand, 0.25-.5 mm; coarse sand, 0.5–1 mm; very coarse sand, 1–2 mm; granule, 2–4 mm; pebble, >4 mm; Wentworth 1922).

23.3.3 Moisture content

The percentage moisture content of the sand collected at the reference sites and from the nest chambers was determined using an analytical balance (Precisa XT 120A, readability 0.0001g) by calculating the simple difference between the wet mass of the sub samples (approximately 60–75g) and their dry mass after being dried at 70°C to constant weight. The percentage moisture in each sample was calculated on the dry weight of the sand sample.

23.3.4 Salinity

A sub-sample of 50g from the oven-dried samples was mixed with 50g of milliQ water and shaken for one hour, then allowed to stand (Patriquin et al. 1993; Corwin and Lesch 2005; Smith and Doran 1997; Soil Survey Staff 2014). Electrical conductivity (EC) of the fluid extract was measured using a WTW Multi3630 IDS with digital conductivity cell TetraCon®925 (Xylem Analytics, Weilheim, Germany). Salinity was calculated as EC = 1 microSiemens/centimetre [µS/cm], which varies according to the resistance of the concentration of salts dissolved in the water to carry electrical current. Raw values were adjusted for temperature at 25°C using the formula: EC µS/cm @25°C = EC µS/cm/(1 + (0.02*(Temperature of extract-25))). Results were converted to Practical Salinity Units (psu).

23.3.5 Temperature

The temperature (°C) in the sand and nests was monitored using HOBO 8K Pendant loggers (Onsetcomp.com, MA) placed at 30 cm and 50 cm depths at each reference site and in the middle of the eggs of 14 green turtle nests (circa 50 cm), as the eggs were being returned to the nest chamber after being counted. Data loggers were set to record the temperature every hour. Temperatures were summarized at intervals of 10 days to describe the pattern of the thermal environment. Daily noon temperatures were extracted to calculate the duration above threshold temperatures of 34, 35, 36, and 37°C.

23.3.6 Distribution of nesting

The distributions of nesting attempts of green sea turtles at Karan Island were estimated for the 2016, 2017, 2018, and 2019 nesting seasons by counting the accumulated number of nesting tracks on the beach from photographs of the beach while circumnavigating the island in a boat at an average distance of 50 m from the shoreline using a GPS-equipped Canon

PowerShot D30 camera. The geo-tagged images were loaded into the QGIS software (QGIS 2016) to establish their locations around the islands. The images were stacked to create the complete profile of the beach. Only tracks that were in the tide-washed (lower) section of the beach were counted. The total number of tracks was divided by two to estimate the number of nesting attempts to account for the upward and downward tracks of one sea turtle. The locations of the nesting attempts were plotted in QGIS based on the GPS coordinates of the geo-tagged images. To compare the intensities of nesting attempts around the island, the beach was divided into eight sections at equal intervals. A Pearson's chi-squared test was performed to test whether nesting attempts were uniformly distributed among the eight beach sections (Greenwood and Nikulin 1996).

23.3.7 Hatching success

A total of 14 green turtle nests were monitored for hatching success on the southern part of Karan Island (Maneja et al. 2023). After the eggs had been deposited, all handling was completed in one hour (Limpus et al. 1979), and the location of the nest was marked with PVC pipes and the GPS coordinates were obtained. Nests were excavated after allowing a minimum of 60 days of incubation (estimated incubation period: ~62.5 days, Miller 1989; Pilcher 2000), and the contents of each were separated into hatched (empty shells greater than 50% intact), unhatched—early embryonic death (i.e., no visible embryonic material), and unhatched—late embryonic death (i.e., pigmented embryo present) (Miller 1985; Miller et al. 2017). Hatching success refers to the number of hatchlings that exited their eggshells and emergence success refers to the number of hatchlings that emerged onto the beach (Miller 1999). Because no dead-in-nest hatchlings were observed in this study, hatching success and emergence success are the same (Miller 1999).

23.3.8 Statistical analysis

The normality of the data was assessed using the Shapiro-Wilks test. When the null hypothesis (H_o = normal distribution) was rejected, a non-parametric test (e.g., Kruskal-Wallis test (K-W)) was used for analysing data. Otherwise, the corresponding parametric tests (e.g., ANOVA) were used for analysis.

23.4 RESULTS

23.4.1 Beach loss at Jana and Karan Islands

A comparison of images of Jana and Karan Islands taken between 1965 and 2017 revealed changes in the shapes of the islands with significant beach loss (Maneja et al. 2020b). Between 1965 and 1968, Jana Island lost 6940 m^2 of area (2.5%), while Karan Island gained 9098 m^2 of area (0.9%), which could be attributed to the natural dynamics of island erosion and accretion (Maneja et al. 2020b). However, between 1965 and 2017, Jana lost about 14150 m^2 (5.1%) and Karan lost about 16376 m^2 (1.7%) (Maneja et al. 2020b). Most of the changes involved the shape of the sand spits at the southern end of the islands (Figure 23.2), which included part of the area of dense turtle nesting noted in 1986 and 1987 (Miller 1989). In addition, the narrowing of beach width along the middle and northern sections exposed rock outcrops from within the beach, which made accessing the beach more difficult and reduced the area for successful nesting.

23.4.2 Grain size

The contributions of the grain size classes in the samples from the four reference sites on both islands showed considerable variation. There were differences among the reference sites located around each island, and differences between the corresponding sides of the two islands (Figure 23.3). The beaches of both islands contained a majority of coarse and medium-grained sand, with varying contributions from very coarse sand and fine sand.

At the Jana-West reference site at 30 cm, medium sand contributed 43.2%, coarse sand added 27.6%, fine sand made up 26.4%, and very coarse sand provided < 3% to the substrate particle size mixture. In contrast, at 30 cm at the Karan-West reference site, medium sand contributed 52%, coarse sand made up 30.2%, fine sand added 8.5%, and very coarse sand provided < 9.3% to the substrate mixture (Figure 23.3).

Considering the samples from the three depths (surface, 30 cm, 50 cm) from the four reference sites on Jana Island, there was no significant difference among the sand from the sites (K-W, $H_{(1,3)} = 0.2543$, $P = 0.9684$). When the surface sand and the smallest and largest grain size categories were removed from the calculation, there was no significant difference among the very coarse, coarse, and medium size classes (K-W, $H_{(1,3)} = 0.6844$, $P = 0.8769$). The Karan Island samples from the three depths (surface, 30 cm, 50 cm) at four reference sites showed no significant difference in sand from among the sites (K-W, $H_{(1,3)} = 0.0847$, $P = 0.9936$). When the surface sand and the smallest and largest size categories were removed from the calculation, there was no significant difference among the very coarse, coarse, and medium size classes (K-W, $H_{(1,3)} = 0.2300$, $P = 0.9989$).

For the combined reference sites on Jana Island at 30 cm, coarse sand contributed a mean of 47.8% and medium sand contributed a mean of 32.6%. At 50 cm, coarse sand had a mean of 51.7% and medium sand contributed a mean of 29.5% to the total. The total contributions of these two size classes from the combined reference sites on Jana Island was 80.4% at 30 cm and 81.2% at 50 cm. On Karan Island, the pattern of the distribution was reversed. At 30 cm, coarse sand contributed a mean of 28.6% and medium sand contributed 45.6%. At 50 cm, coarse sand comprised 51.7% and medium sand contributed 29.5% to the composition of the sand. At 50 cm, coarse sand made up 27.9% and medium sand contributed 46.4% of the structure of the sand. The total contributions of these two grain size classes to the total for Karan Island was 74.2% at 30 cm and 81% at 50 cm.

The sand surrounding the 14 green turtle egg chambers comprised mostly medium grain material (mean = 33.6%, Min-Max = 12–45%, n = 14) and coarse-grained sand (mean = 44.8%, Min-Max = 30.4–56.9, n = 14) (Maneja et al. 2023).

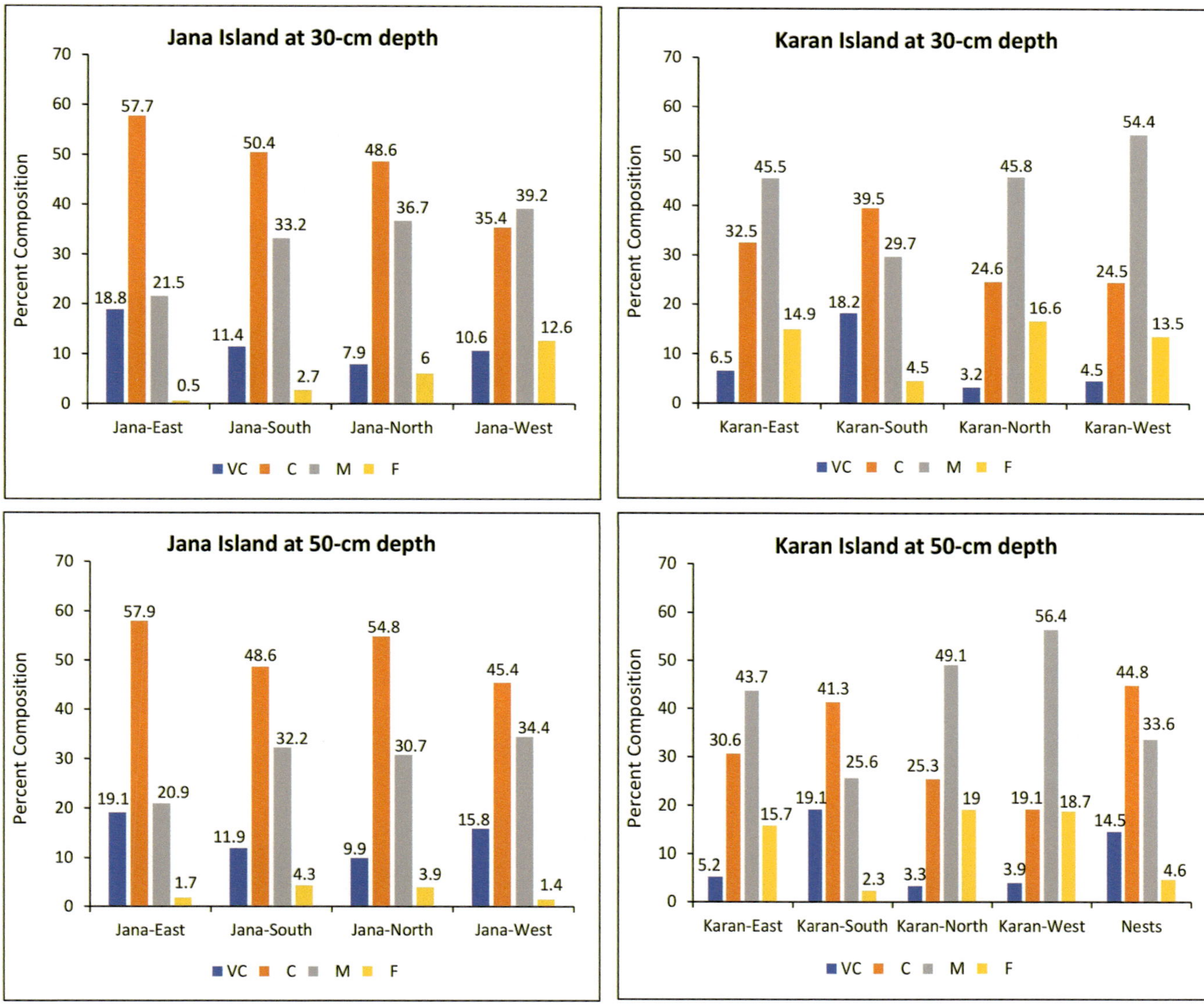

FIGURE 23.3 Mean percent sand grain size distribution at depths of 30 cm and 50 cm at the reference sites on Jana (left) and Karan (right) Islands. VC = Very Coarse (1–2mm), C = Coarse (0.5–1mm), M = Medium (0.25–0.5mm), F = Fine 0.125–0.25mm (Wentworth 1922).

Medium and coarse sand combined contributed 78.4% to the mixture. Very coarse sand added an additional 14.5%, bringing the total contribution by the three size classes to 92.9%.

23.4.3 Sand moisture

The percentage moisture of the sand from the walls of the holes dug at the four reference sites varied among the sites on both islands (Figure 23.4). There was a significant difference in the moisture of the sand among the four reference sites on Jana Island (K-W, $H_{(1,3)}$ = 9.725, P = 0.0211), with the difference occurring between Jana-West and Jana-South (K-W, $H_{(1)}$ = 0.0083, P = 0.0023). Jana-West was the driest of the four reference sites (Jana-West, $\bar{x}$ = 1.44%, Jana-South, $\bar{x}$ = 43.9%, Jana-North, $\bar{x}$ = 3.20%, Jana-East, $\bar{x}$ = 3.11%). In both of these cases, Jana-West was significantly different from Jana-South (K-W, $H_{(1)}$ = 0.0083, P = 0.0022 at 30 cm, and 0.0022 at 50 cm). There was a significant difference in the moisture content of the sand on Karan Island (K-W, $H_{(1,3)}$ = 8.467, P = 0.037), with the difference occurring between Karan-South and Karan-East (K-W, $H_{(1)}$ = 0.0083, P = 0.0048). The southern reference site on Karan was the wettest of the reference sites, with the mean being $\bar{x}$ = 6.61%, and the mean of Karan-South-50 being $\bar{x}$ = 4.89. In both of these cases, Karan-South was significantly different from Karan-East (K-W, $H_{(1)}$ = 0.0083, P = 0.0023 at 30 cm, and 0.0004 at 50 cm).

In comparisons between the islands, there was a significant difference in the moisture content of the sand at 30 cm, but only for Jana-West and Karan-South (K-W, $H_{(1,3)}$ = 0.0018, P = 0.0003 at 30 cm and P = 0.0003 at 50 cm). Jana-West was the driest at both depths (mean = 1.44%) and Karan-South was the wettest at both depths (mean = 5.75%) among the eight reference sites.

The mean moisture content of the sand surrounding the egg chambers dug by green turtles on Karan Island was 4.02% (Min-Max = 3.25–4.89%). If the range values at 50 cm for the reference sites represent the stable moisture environment in the sand during incubation, then two sites offer the best hydric conditions for incubation: Jana-South (mean = 4.39, Min-Max = 4.11–4.81%) and Karan-South (mean = 5.75%, Min-Max = 4.88–7.21%). Among the other sites, Jana-West and Jana-East appear to have insufficient moisture to support development without some water loss from the eggs. The other reference sites have moisture levels near the minimum of that found in the nests.

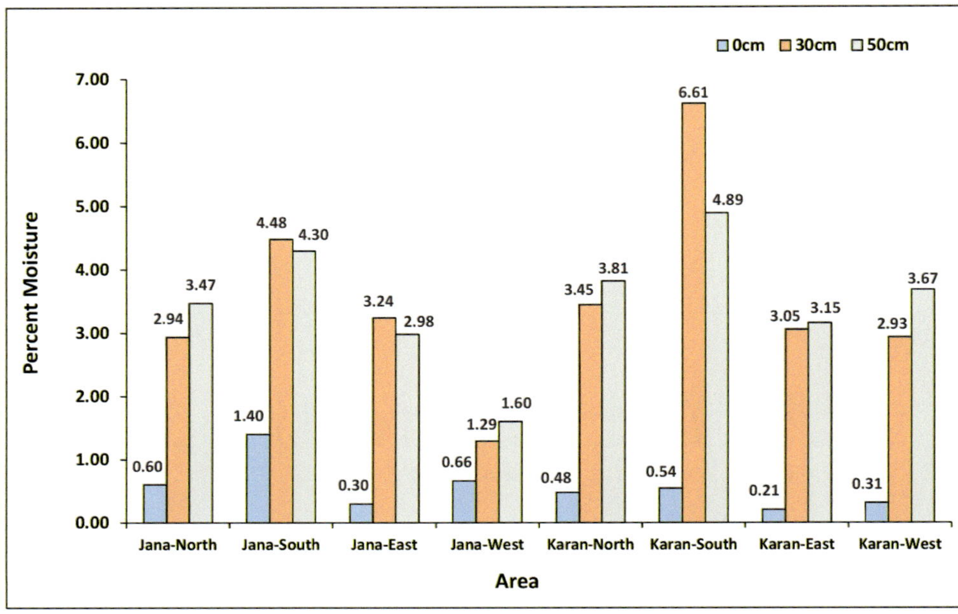

FIGURE 23.4 Mean percentage moisture in sand samples from the surface, 30 cm, and 50 cm collected at the reference sites on Jana and Karan Islands.

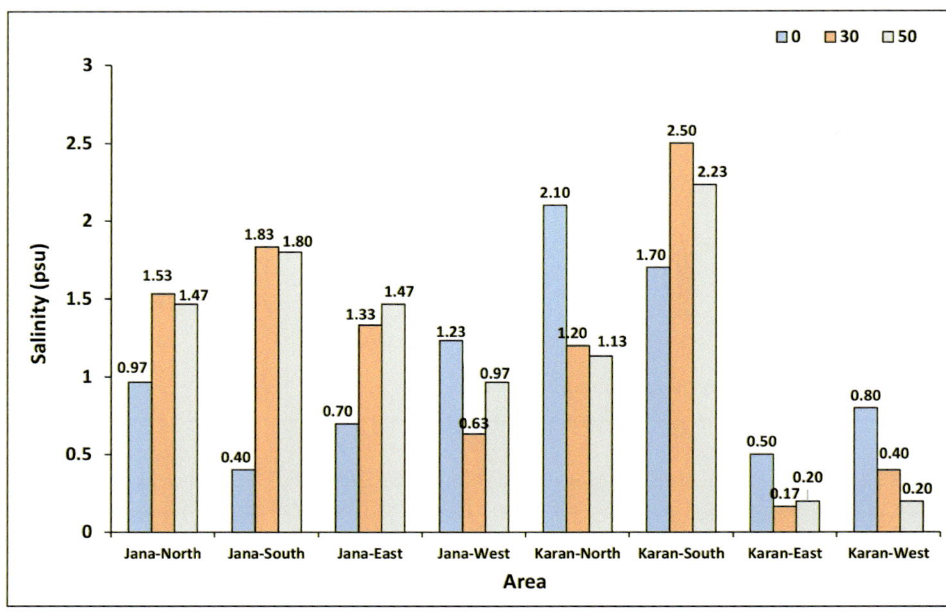

FIGURE 23.5 Mean Practical Salinity Units (psu) in sand samples from the surface, 30 cm, and 50 cm collected at the reference sites on Jana and Karan Islands.

23.4.4 Salinity

The salinity of the sand was not uniform around the islands or by depth (Figure 23.5). The mean salinity of the surface sand varied from 0.4 psu to 1.23 psu (Jana: Mean = 0.825 psu, Min-Max = 0.4–1.23 psu; Karan: Mean = 1.27 psu, Min-Max = 0.50–2.10 psu). The values from Jana Island were slightly less than those from Karan Island but the ranges overlapped. The mean salinity for the sand from 30 cm varied from 0.17 to 2.5 psu (Jana: Mean = 1.33 psu, Min-Max = 0.63–1.83 psu; Karan: Mean = 1.07 psu, Min-Max = 0.17–2.50 psu), and the mean salinity for the sand from 50 cm varied from 0.2 to 2.23 psu (Jana: Mean = 1.43 psu, Min-Max = 0.97–1.80 psu; Karan: Mean = 0.94 psu, Min-Max = 0.20–2.23 psu). There was a significant difference among the reference sites on the islands (K-W, $H_{(1,7)}$ = 15.39, P = 0.0309). However, only the Karan-South and Karan-East were significantly different (K-W, $H_{(1,1)}$ = 0.0018, P = 0.0011).

The mean salinity of the sand surrounding the nests was 0.986 psu (StDev = 0.561 psu, Min-Max = 0.187–1.92 psu, n = 14). Among the reference sites, the range in psu values indicates that salinity is not a major factor impacting hatching success unless it is coupled with low moisture. Jana-West and Karan-West have the lowest minimum psu values.

23.4.5 Temperature

The temperature of the sand at 30 cm and 50 cm increased during the nesting season (Figure 23.6). Hawksbill turtles nest from early May through the middle of June, with a peak period in late May to early June, with a few individuals nesting somewhat before and after this period (Al-Merghani et al. 1996, 2000). Green turtle nesting occurs during the middle of July, with some individuals starting to nest in the middle of June and others nesting into late August (Al-Merghani

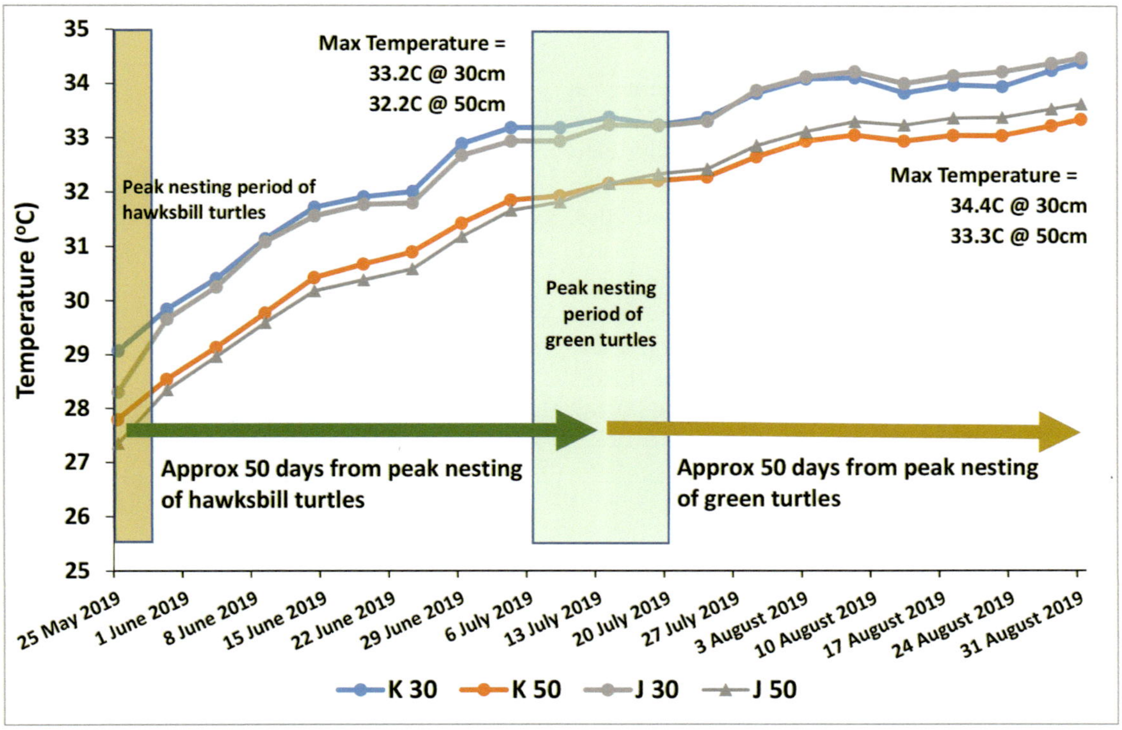

FIGURE 23.6 Mean temperatures recorded 30 cm and 50 cm depths at five-day intervals from the four combined reference sites on Jana and Karan Islands in 2019. Peak nesting for hawksbill turtles occurs during the last half of May; peak nesting of green turtles occurs during the middle of July. Actual dates shift in coordination with the high tides. Arrows indicate approximately 50 days of incubation for each species beginning from the middle of the estimated peak of their nesting period.

et al. 1996, 2000). Both species produce multiple clutches at approximately 14-day intervals (Al-Merghani et al. 1996, 2000). The sand temperatures recorded at the reference sites on Jana and Karan Islands beginning 25 May 2019 were warmer at 30 cm than temperatures at 50 cm by approximately 1°C (30cm: mean = 1.1°C, Min-Max = 0.8–1.5°C; 50 cm: mean = 1.2°C, Min-Max = 09.0–1.5°C). This general pattern continued throughout the incubation period (Figure 23.6).

The temperature data indicated that if a clutch of hawksbill eggs was laid into a nest such that the middle of the eggs was ≈30 cm deep in the peak nesting period (late May), the eggs would start incubation at approximately 28°C (Figure 23.6). Presuming a 50-day incubation period, the temperature would increase ≈5°C to about 33.2°C by the middle of July, which is within the thermal tolerance limits of the embryos. Similarly, if a clutch of green turtle eggs was laid into a nest ≈50 cm to the middle of the egg mass in the middle of July during the peak nesting period, the eggs would start incubation at approximately 32.2°C (Figure 23.6). Assuming a 50-day incubation period, the temperature would increase ≈1°C, to about 33.2°C by the middle of August, which appears to be within the thermal tolerance limits of the embryos. The temperatures in shallower nests dug by hawksbill turtles increased more during incubation than did temperatures in the deeper green turtle nests.

The mean temperature recorded at noon from within the monitored green turtle clutches increased by 2.1°C in 20 days from a mean of 31.2°C (SD = ±1.011°C, range = 29.9–33.0°C, n = 13) starting at 5 days after oviposition to a mean of 33.3°C (SD = ±1.036°C, range = 31.6–35.0°C, n = 13) at 25 days post oviposition (Maneja et al. 2023). By day 40, the mean temperature in the clutches had increased 2.3°C to 35.6°C (SD = ±1.033°C, range = 33.6–36.8°C, n = 13) in 15 days, representing an increase of 4.4°C in 35 days since oviposition. By 50 days of incubation, the mean temperature in the clutches had increased a further 1.0°C in the 10-day interval to 36.6°C (SD = ±1.914 C, range = 33.7–38.8°C, n = 13). The mean temperatures in the clutches increased 5.4°C between 5 and 50 days post-oviposition (Maneja et al. 2023).

The difference between the pooled mean temperatures for reference sites on Jana and Karan Islands ranged from 0.3–0.8°C at 30 cm and from 0.3–0.4°C at 50 cm, indicating that there is little difference in the general thermal conditions experienced by the embryos. In contrast, the eggs in the 14 clutches monitored while incubating on the southern end of Karan Island experienced higher temperatures that ranged from a mean of 31.2°C (Min-Max = 29.8–33.0°C) at 5 days post-oviposition to a mean of 36.4°C (Min-Max = 32.9–39.5°C) after 50 days of incubation, a mean difference of 5.1°C (Min-Max = 2.5–8.6°C). Eight of the 14 clutches of eggs experienced temperatures > 35°C. The temperatures recorded among the eggs of these clutches exceeded 34°C for 44.4% (SD = 25.21%, Min-Max = 9.69–65.3%) of the 50-day incubation period. Eggs in some of these clutches experienced temperatures > 35°C for 36.3% (SD = 15.6%, Min-Max = 7.35–55.5%) of the incubation period. Others were > 36°C for 27.2% (SD = 16.7%, Min-Max = 2.12–47.1%) of the time, and a few experienced temperatures > 37°C for 13.9% (Stdev = 16.7%, Min-Max = 2.12–47.1%) of the time (Maneja et al. 2023). The highest temperatures occurred during the latter part of the incubation period.

23.4.6 Distribution of nesting

The estimated number of green turtles attempting to nest at Karan Island varied from year to year (Figure 23.7). Low nesting events were recorded in 2016 and 2018, with estimated totals of 424 and 507 attempts, respectively. A high number of nesting events were recorded in 2017 and 2019, with totals of 762 and 733 events, respectively. The distribution of nesting attempts by green turtles at Karan Island showed variations among the years and around the island (Figure 23.6). Over the four years, the northern beach sections (A, B, H) had fewer attempts than the middle sectors (C, G) and southern sections (D, E, F) ($X^2_{135.51}$, 0.001>p) (Figure 23.7 and Table 23.1).

23.4.7 Hatching success

No hatching success data were obtained for Jana Island during the study period. The mean hatching success on Karan Island was not different between the two years 2017 and 2019 (t-test: t_{18} = 0.7551, p = 0.4611). In 2017, the hatching success was 26.6% (SD = 38.8%, Min-Max = 0–83.3, n = 4 clutches) and in 2019, the hatching success was 38.8% (SD = ±25.2%, range = 2.5–75%, n = 14 clutches). The combined data yield a mean hatching success of 36.1% (SD = 27.9%, Min-Max = 0–83.3, n = 18) (Maneja et al. 2023).

The nests monitored in 2019 were loosely clustered in two areas, one along the western (F+G, Figure 23.7) and one along the southern-southeastern (D+E, Figure 23.7) portions of Karan Island. Although the mean hatching success in sectors D+E had a higher value (mean = 48.5%, SD = 22.5%, n = 7) than in sectors F+G (mean = 32.9%, SD = 25.5%, n = 6) (Maneja et al. 2023), the difference was not significant (t_{11} = 1.175, p = 0.265). Examination of the contents of the unhatched eggs of clutches oviposited in 2019 (61.2% of eggs) revealed 39% (range = 5.1–95.1%) of embryos had died during early development and 61% (range = 4.9–94.9%) died during late embryonic development (Maneja et al. 2023). There was high variation in hatching success among nests between and within areas of the beach. In addition, the southern sectors (D, E, F) had relatively more early embryonic death while the mid-beach of sectors C and G had relatively more late embryonic deaths (Figure 23.7). No live or dead hatchlings were recovered from within the nests, which means that in this case hatching success equaled emergence success.

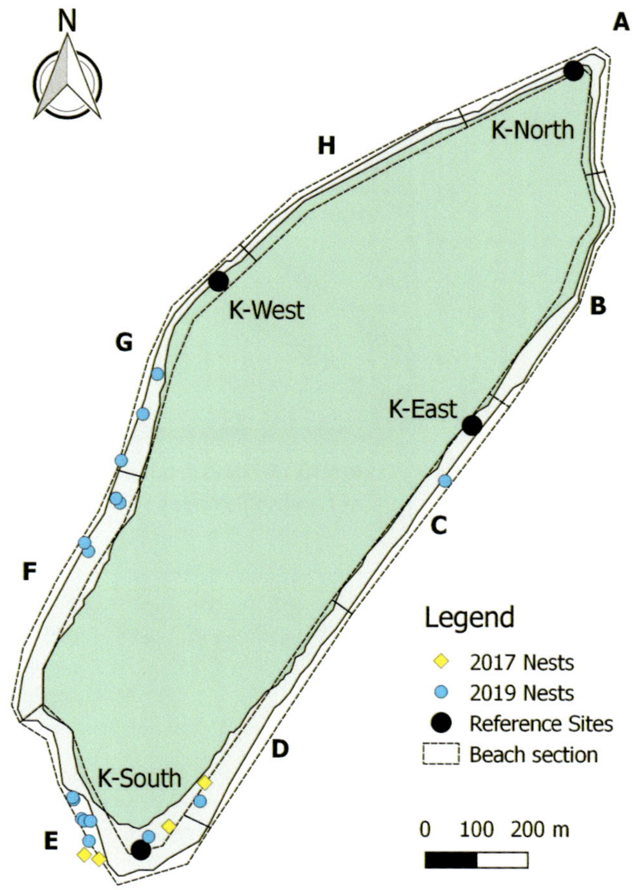

FIGURE 23.7 Beach sections (A-H) with distribution of green turtle (*Chelonia mydas*) nests in 2017 and 2019 at Karan Island.

23.5 DISCUSSION

Marine turtles use historic nesting sites that have been stable through time to support the iteroparous pattern of reproduction (i.e., multiple nestings by individual females) often at irregular intervals (Limpus 1987; Georges et al. 1993; McClenachan et al. 2006). During the nesting season, beaches used for nesting and incubation must be stable enough to support the digging of nest chambers. In addition, beaches must have an appropriate range of temperatures and moisture, and allow adequate gas exchange to support development of the embryos

TABLE 23.1 Number of nesting attempts per beach section and the hatching success by sector on Karan Island.

	NESTING ATTEMPTS				HATCHING SUCCESS OF MONITORED NESTS: % (N, RANGE)	
SECTOR	2016	2017	2018	2019	2017	2019
A	19	38	2	36		
B	25	45	58	108		
C	98	132	109	161		**5.9 (1)**
D	97	139	109	132	**20 (1)**	**47.9 (1)**
E	58	118	97	64	**28.7 (3, 0–83.3)**	**41.7 (7, 0–70.3)**
F	62	125	56	84		**41.7 (5, 1.6–69.9)**
G	35	115	44	81		**18.2 (3, 2.5–43.3)**
H	30	50	32	67		

(Georges et al. 1993; Miller 1997; Ackerman 1997; Suss et al. 2012). If changes to a nesting site are substantial, the beaches may become unavailable or unsuitable for nesting. The abiotic conditions in which marine turtle eggs incubate can impact not only their developmental success but also the fitness of the hatchlings produced (Sim et al. 2015; Cavallo et al. 2015; Stubbs and Mitchell 2018; Bladow and Milton 2019; Wood and Bjorndal 2000; Carpio Camargo et al. 2020; Booth et al. 2020; Fleming et al. 2020; Calderón Peña and Azanza Ricardo 2021; Türkozan et al. 2021; Schwanz and Georges 2021; Gravelle and Wyneken 2022; Gatto and Reina 2020).

Because the Gulf region experiences extreme environmental conditions as a consequence of its physical geography (Sheppard et al. 2010; Krishnakumar et al. 2019; Vaughan et al. 2019), the nesting environment experienced by marine turtle embryos in the Gulf region has the potential to negatively impact their developmental success (Chatting et al. 2018, 2021; Mobaraki et al. 2022). Defining and monitoring the environmental conditions at marine turtle nesting sites in the Gulf contributes to regional and local conservation decision-making because the abiotic conditions at the nest site impact the survival of the embryos, as well as the sex and fitness of the hatchlings (Wibbels 2003; Fuentes and Porter 2013; Fisher et al. 2014; Bentley et al. 2020a). Data on hatchling production are essential to guide conservation efforts in support of healthy marine turtle populations through generations (Blechschmidt et al. 2020). In addition, understanding the impact of abiotic conditions in which marine turtle eggs incubate in the Gulf region contributes to predicting how embryos at locations in less extreme conditions may respond to the projected climate-driven changes in beach conditions at their nesting locations (Maurer et al. 2021; Montero et al. 2018; Wood and Bjorndal 2000; Glen et al. 2003; Marn et al. 2017; Carpio Camargo et al. 2020; Matthews et al. 2021; Gravelle and Wyneken 2022).

23.5.1 Island shape

The images of the Islands revealed significant changes in their outline shape in the 53 years between 1964 and 2017 (Maneja et al. 2020b). The sand movement reduced the area available for nesting at the southern ends and narrowed the width of the beaches along the western and northern sides of the islands. At present, the recorded changes may not be a serious threat to the area available for nesting. However, if the predicted increases in the frequency and severity of storms in the Gulf happen, the islands are likely to be altered more by erosional sand movement (IPCC 2022; Evan and Camargo 2011; Al Senafi 2022; Krishnakumar et al. 2019; Vaughan et al. 2019; Miller et al. 2019a). As a consequence, the availability and suitability of nesting areas to support the incubation of sea turtle embryos may change and reduce hatchling production (Reece et al. 2013; Fujisaki et al. 2018).

23.5.2 Sand grain size

The composition of the beach is critical to successful nesting by adult females and successful incubation of their eggs (Yalçın-Özdilek et al. 2007; Stewart et al. 2019). The grain size, in the context of available moisture, may influence not only the digging of the nest chamber (Karavas et al. 2005; El Daba et al. 2019) but also the success of incubation (Fadini et al. 2011) and emergence success (Foley et al. 2006). The range in size of the sand grains in the beach affects the volume of the spaces between them. If the sand grains are able to fit closely together so that only small spaces are formed between them, then the interstitial volume is restricted. This is important to the developing embryos because the spaces contain air and water vapour. The surface structure (rugosity) of the particles is important because the minute roughness of the surface regulates the amount of water that can be held by surface tension.

Previous analysis of sand from Karan Island demonstrated that medium grain to coarse grain sand were the dominant size classes (Basyoni 1999). Al-Mansi et al. (1999) reported mean sand grain sizes of 0.88 mm (Stdev = ±0.76 mm) from Jana and 0.51 mm (Stdev = ±0.76 mm) from Karan, which are consistent with our results and those for other beaches used by green turtles (Mortimer 1990; El Daba et al. 2019).

Our results show that the distribution of sand grain size varies from place to place on the beach, and the composition of a sample depends on the site at which it was taken (Figure 23.4). It follows that to characterize a beach requires multiple samples and that single samples (or a limited number of samples) will not characterize the range of grain sizes to be found around an island beach.

23.5.3 Hydric environment

The percentage moisture of sand at the level of the eggs in a nesting beach that is required to support normal development and hatching is in the range of 3% to 12% moisture by dry weight (McGehee 1990; Mortimer 1990; Ackerman 1997; Wood and Bjorndal 2000; Sönmez et al. 2013; Erb et al. 2018). If the percentage moisture is lower or higher, then the survival of embryos can be affected (Ragotzkie 1959; Kraemer and Bell 1980; McGehee 1990; Mortimer 1990; Rivas et al. 2018; Kobayashi et al. 2020; Staines et al. 2020). The lower the percentage moisture (e.g., <1%), the more likely the embryos are to experience hydric stress and, possibly, death (Mortimer 1990; Ackerman 1997).

At Jana and Karan Islands, Al-Mansi et al. (1999) reported 3.25% moisture from Karan and 3.6% moisture from Jana from single samples collected at 50 cm depth. These values are similar to values we recorded in nests on Karan (range: 3.2–4.9, at ~50 cm, 14 clutches) and at reference sites on Jana (range: 1.5–4.12%, 50 cm, 4 sites). These values are on the lower end of the range required for successful incubation and suggest the embryos may experience some hydric stress during incubation (McGehee 1990; Mortimer 1990; Ackerman 1997; Wood and Bjorndal 2000; Sönmez et al. 2013; Erb et al. 2018). The percentage moisture content in nests may vary from the beginning of the nesting season (4.1–11.5%) to the end (4.1–6.6%) (Sönmez et al. 2013) depending on external environmental influences and may still support normal development. Also, moisture buffers the rate of temperature change within a sea turtle nest because of its high heat capacity (Mortimer 1990; Santidrian Tomillo et al. 2017; Laloë et al. 2016). However, higher moisture content increases incubation duration and lowers thermal tolerance in green turtle hatchlings, which can potentially lower emergence success (Gatto et al. 2021). Thus, any conservation intervention that will involve watering and shading must first undergo field testing, particularly in the arid environment of the Arabian Gulf.

The mean percentage moisture from sand collected around the islands varied within and among reference sites at each island (Figure 23.4). In addition, the mean percentage moisture varied between the islands. The moisture of sand surrounding the eggs depends on the location on the beach and the depth from which samples were collected. This suggests that using a single or only a few sampling points does not represent the moisture available in the sand surrounding the clutches of eggs at a nesting beach. Given the importance of moisture to the developing embryos, multi-season monitoring and experimentation are necessary to better define the situation and determine what, if any, remediation would be most effective.

23.5.4 Salinity

Beaches receive aerosols of sea water and dissolved salts that are carried ashore by storm winds and tidal surges from the surrounding ocean (Knox 2001). Although sea water in the Gulf may exceed 50 practical salinity units (psu), at least along the shallow western side (Coles and McCain 1990; Jones et al. 1998; Beni et al. 2021), the accumulated salinity in sand is typically less. The salt content in sand from Jana and Karan Islands ranged between 0.1 and 2.7 psu. Although beach sand contains other dissolved ions and organic material that can add to the osmotic gradient to draw water out of the egg, salt content is a more important component of the nest environment (Ackerman 1997; Smith et al. 2021; Booth et al. 2022; Mortimer 1990).

Increased salinity in the sand may add to hydric stress in the developing embryos, particularly when the percentage moisture is low. Using the simple difference in mass of wet and dry sand samples to assess the moisture content of beach sand is not sufficient to understand the dynamic context of the incubation environment (Wood and Bjorndal 2000). At least intermittent monitoring of the total hydric environment, including salinity, should be incorporated into research programs.

23.5.5 Temperature

Embryos of marine turtles require temperatures in the range of 25–27°C to 33–35°C for successful development (Ackerman 1997). Prolonged exposure to temperatures above or below this general range may exceed the embryonic tolerance limits and cause embryonic malformation or death, depending on the amount and duration of deviations (Bladow and Milton 2019). Within the general range of temperatures required for successful development, warmer temperatures result in female hatchlings and cooler temperatures result in male hatchlings (Wibbels 2003).

At Jana and Karan Islands, hawksbill turtles nest about four weeks earlier in the year than green turtles do. The difference between the nesting periods for hawksbill and green turtles is reflected in the temperatures experienced by their eggs at 30 cm and 50 cm (Figure 23.5). For both species, the temperatures during much of the incubation period were similar because the nest depths were dissimilar. As the beach warms in spring, the temperatures at the depth of hawksbill turtle nests (centre of egg mass ≈30 cm) become warmer sooner than the sand deeper in the beach at the level where green turtles deposit their eggs (centre of egg mass ≈50 cm). Consequently, hawksbill turtles begin to nest about a month to five weeks earlier (early May) than green turtles do (late June). The seasonal range of temperatures on the islands at least partially explains the seasonality of nesting by the two turtle species.

Although temperatures and the duration of exposure above 34–35°C increase embryo mortality (Bladow and Milton 2019), some eggs in most of these clutches survived. Not all eggs in a clutch experience the same temperatures (Booth and Astill 2001b), and the susceptibility of the embryos to temperature-induced mortality is likely not the same throughout development (Booth 2017).

There were differences in the temperatures recorded within and among reference sites on each island and between the islands. This is consistent with microclimate modelling that has shown high spatial and temporal variation in beach sand temperatures within and among sea turtle nesting areas (Bentley et al. 2020b).

The nesting beaches at Jana and Karan islands experienced elevated temperatures. In particular, the temperature during July and August 2019 exposed the eggs at 50 cm to temperatures above 33–34°C for different periods of time and at different stages of development. In addition, beach sand temperatures appear to be exposing eggs to elevated temperatures for extended periods of time as the summer progresses, which may exceed the limits of embryonic survival. Because the nest temperatures experienced in the northern Gulf are already higher than most temperatures experienced during incubation, temperatures recorded during incubation in this study are likely contributing to the mortality of the developing embryos and lowering the hatching success.

Equally as important is that the elevated temperatures are likely to be skewing the sex ratio toward the majority of hatchlings being female (Wibbels 2003; Chatting et al. 2021). If the temperatures recorded in nests during 2019 are representative of temperatures in previous nesting seasons, it is likely that Jana and Karan are producing mostly female hatchlings for both species. However, this requires confirmation.

Not only did the reference sites under-predict the temperatures experienced in clutches on the southern beach of Karan, but the eggs in those nests experienced temperatures that likely caused thermal stress and mortality for many of the eggs. The temperatures recorded at the reference sites provide a general picture of the thermal environment and should not be used to estimate the hatchling sex ratio, which must include both the distribution and number of nests around the islands.

23.5.6 Distribution of nesting

The distribution of nesting attempts, such as on Jana and Karan islands, reflects the response of the turtles to the morphology of the nearshore reef flat, the distribution of exposed shore line beach rock, whether they encounter debris on or in the beach while attempting to nest, and other cues for nest site selection (Wood and Bjorndal 2000; Türkozan et al. 2011; Foley et al. 2006; Zare et al. 2012; Kelly et al. 2017; Aguilera et al. 2018;

Booth et al. 2020; Maneja et al. 2021, 2023). The northwestern sides of both Jana and Karan Islands face the extensive reef flats that have exposed beach rock and more accumulated debris compared to the southeastern sides (Maneja et al. 2021). The deepest approach throughout the tidal cycle is along the southeastern to southern portions of the islands, where the sandy beach is widest and most of the nesting occurs.

23.5.7 Hatching success

Understanding hatching and emergence success is a crucial part of conservation management because it represents the success of the reproductive phase of the turtle's life cycle (Ditmer and Stapleton 2012). For green turtles nesting at Karan Island, the current estimate of hatching success (38.8%, Maneja et al. 2023) was a large decrease from the reported hatching success for the species at Karan Island in 1986 and 1987 (81.7%; Miller 1989) and 1993, about 80% (Pilcher and Al-Merghani 1994; Al-Merghani et al. 1996). Both of these reports were based on only a few nests that had been found opportunistically. The mean hatching success of the 14 green turtle nests monitored at Karan Island in 2019 was in the lower part of the range of hatching success at various nesting sites in the world. The mean hatching success for green turtles reported from other nesting populations may range widely (Maneja et al. 2023). The hatching success of green turtle clutches varied from 1.9% to 88% (mean = 67.6%, SD = 16.711%, n = 47 reports; Hirth 1997) and from 47% to 87% (mean = 72.6%, SD = 0.127%, n = 14 reports) in studies from around the world (van Buskirk and Crowder 1994). A similar range in hatching success occurs for hawksbill turtles at nesting sites around the world (mean = 72.3%, SD = 18.191%, min-max = 46.7% to 96.9%, n = 13 studies; Witzell 1983) and from 58% to 91% (mean = 76.5%, SD = 0.111, n = 9 studies; van Buskirk and Crowder 1994). For both species, if the emergence success remains below about 70% for an extended period of years, the incubation systems of the nesting beaches may fail to provide enough hatchlings to support the recruitment into the adult population, which is required to prevent the decline of the populations. Given the similarity of conditions found on the two islands, the hatching success of hawksbill turtle clutches on Jana may be low and needs to be determined.

23.6 CONCLUSIONS

In the short-term on Jana and Karan Islands, the proportions of sand grain sizes at the reference sites and the nesting beach at the southern end of Karan Island were similar and will facilitate development. The moisture contained in the sand at nest depth appears to be in the lower half of the range required for the development of embryos. The amount of sea salt dissolved in the water of the sand is low and is unlikely to increase the osmotic gradient between the surrounding sand and the contents of the eggs. The majority of nesting appears to be on the southern end of Karan Island. The temperatures in the beaches of Jana and Karan Islands during the middle of the summer are high enough to cause stress and potential mortality among the embryos. Temperatures above 36°C were recorded more than 16% of the time. In summary, the temperatures of the nesting beach as experienced by the embryos are likely to be the cause of the low hatching and emergence success.

Although the environmental conditions of the beach in which marine turtle eggs incubate operate synergistically, temperature was identified as the primary variable impacting hatching success. Moisture is the second most important variable to measure, because as weather conditions change into the future and sand temperatures rise, the moisture in the sand is likely to change. Although sand texture impacts both temperature and moisture surrounding the incubating eggs, we suggest analysis of sand grain size might be done occasionally. Determining the salinity of the sand collected from around nests might also be done occasionally. Baseline and long-term monitoring are necessary to identify changes in the beach environment and emergence success.

Marine turtle nesting areas are vulnerable to climatic changes (Hawkes et al. 2009; Fuentes et al. 2011; Hamann et al. 2013; Santos et al. 2017). Conservation management may be able to slow the impacts of temperature change on the nesting beaches by employing actions that will reduce temperatures in the sand surrounding the incubating eggs and, perhaps, slightly increase the moisture while reducing the salt content in the sand. Examples of potential management efforts include the use of seawater or freshwater to cool nests (Jourdan and Fuentes 2015; Smith et al. 2021), translocating eggs to other sites (Ware and Fuentes 2019), and/or shading of nests (Hill et al. 2015; Staines et al. 2020). However, before such action is initiated, a more representative and longer duration study is needed to better define the beach conditions. In addition, a review of the methods currently being trialed at beaches around the world should be made to assess their potential application on Jana and Karan Islands.

Reference sites are useful in allowing environmental variables to be repeatedly measured at specific places on the islands. The reference sites used to monitor the status of environmental variables on Jana and Karan Islands showed that the conditions on the two islands were very similar, yet there were differences. The continuous monitoring of temperatures at the reference sites will provide a baseline data set against which changes in the temperatures around the islands can be evaluated. Although useful in defining the general conditions on the islands, the reference sites located approximately at north, east, south, and west positions on Karan Island did not replicate the specific conditions of the nesting area located on the southern half of the islands (Maneja et al. 2023). A better approach would be to maintain environmental monitoring at the reference sites and to increase the monitoring of conditions at many more nest locations, wherever they occur on the beach.

ETHICS STATEMENT

The animal study was reviewed and approved by the National Center for Wildlife (NCW) (formally Saudi Wildlife Authority) Projects Committee and complies with all applicable rules and regulations of the NCW for research on animals in the Kingdom of Saudi Arabia.

ACKNOWLEDGMENTS

The authors would like to acknowledge the funding of the Sea Turtle Research under the King Fahd University of Petroleum and Minerals (KFUPM) – Saudi Aramco Environmental Sustaining Research – Phase VI Project (CEW02428). We would like to acknowledge the Center for Environment and Marine Studies of KFUPM for providing the logistics and administrative support to the research team comprising J.F.A. Alcaria, A.U.B. Basali, T. Vasanthi Duraisamy, P.B. Das, Muaadh A. Alnuwairah, M. Al-Jedani, E. Santiago, N.M. Mudasir, Mohamed Asharaf, Karuppasamy P. Manikandan, and Mohammed Ali B. Qurban. We also acknowledge the crew (Nasser) of *Selbar-1325* led by Mr. Bader Abuainain for supporting the fieldwork logistics at the offshore islands.

REFERENCES

Ackerman, R.A. 1991. Physical factors affecting the water exchange of buried reptile eggs. In: *Egg incubation: Its effects on embryonic development in birds and reptiles*. Eds: D.C. Deeming, and M.W. Ferguson, 193–212. Cambridge: Cambridge University Press.

Ackerman, R.A. 1997. The nest environment and the embryonic development of sea turtles. In: *The biology of sea turtles*. Eds: P.L. Lutz, and J.A. Musick, 83–106. Boca Raton, FL: CRC Press.

Aguilera, M., Medina-Suárez, M., Pinós, J., LiriaLoza, A., and L. Benejam. 2018. Marine debris as a barrier: Assessing the impacts on sea turtle hatchlings on their way to the ocean. *Mar Pollut Bull* 137: 481–487. doi:10.1016/j.marpolbul.2018.10.054.

Al Ameri, H.M., Al Harthi, S., Al Kiyumi, A., et al. 2022. Biology and conservation of marine turtles in the North-Western Indian Ocean: A review. *Endang Species Res* 48: 67–86. doi:10.3354/esr01183.

Al Ameri, H.M., Das, H.S., Rodriguez-Zarate, C.J., and M. Antonopoulou. 2021. United Arab Emirates. In: *Sea turtles in the Middle East and South Asia Region*. Eds: A.D. Phillott, and A.F. Rees, 277–295. MTSG Annual Regional Report 2021. Report of the IUCN-SSC Marine Turtle Specialist. Washington, DC: IUCN/SSC Marine Turtle Specialist Group.

Al Senafi, F. 2022. Atmosphere-ocean coupled variability in the Arabian/Persian Gulf. *Front Mar Sci* 9: 809355.

Al-Mansi, A.M. 2016. *Sea turtles in the Red Sea and the Arabian Gulf, Saudi Arabia*, 1–43. Report. Riyadh: Saudi Wildlife Authority.

Al-Mansi, A.M.A., Khushaim, O.A., and M.M.H. Al-Marghani. 1999. On the effect of substrate on nesting success of the Green Turtle, *Chelonia mydas*, in the Arabian Gulf. *Zool Middle East* 19: 5–11.

Al-Merghani, M., Miller, J.D., Al-Mansi, A., Khushaim, O., and N.J. Pilcher. 1996. The marine turtles of the Arabian Gulf: NCWCD studies 1991–1994. In: *A marine wildlife sanctuary for the Arabian Gulf: Environmental research and conservation following the 1991 Gulf War oil spill*. Eds: F. Krupp, A.H. Abuzinada, and L.A. Nader, 351–359. Frankfurt: Senckenberg Research Institute.

Al-Merghani, M., Miller, J.D., Pilcher, N.J., and A. Al-Mansi. 2000. The green and hawksbill turtles in the Kingdom of Saudi Arabia: Synopsis of nesting studies 1986–1997. *Fauna Arabia* 18: 369–384.

Al-Mohanna, S.Y., Al-Zaidan, A.S.Y., and P. George. 2014. Green turtles (*Chelonia mydas*) of the north-western Arabian Gulf, Kuwait: The need for conservation. *Aquat Conserv: Mar Freshw Ecosyst* 24, no. 2: 166–178. doi:10.1002/aqc.2371.

Barth, H.J., and N.Y. Khan. 2008. Biogeophysical setting of the Gulf. In: *Protecting the Gulf's marine ecosystems from pollution*. Eds: A.H. Abuzinada, H.J. Barth, F. Krupp, B. Böer, and T.Z. Al Abdessalaam, 1–21. Basel: Birkhäuser. doi:10.1007/978-3-7643-7947-6_1.

Basson, P.W., Burchard, J., Hardy, J.T., and A.R. Price. 1977. *Biotopes of the Western Arabian Gulf: Marine life and environments of Saudi Arabia*, 284 p. Dhahran: ARAMCO Department of Loss Prevention and Environmental Affairs.

Basyoni, M. 1999. Sedimentology and geochemistry of the Karan Island sediments, Arabian Gulf, Saudi Arabia. *J King Abdulaziz Univ Marine Sci* 10: 89–106. doi:10.4197/mar.10-1.6.

Beni, A.N., Marriner, N., Sharifi, A., et al. 2021. Climate change: A driver of future conflicts in the Persian Gulf Region? *Heliyon* 7: e06288.

Bentley, B.P., Kearney, M.R., Whiting, S.D., and N.J. Mitchell. 2020b. Microclimate modelling of beach sand temperatures reveals high spatial and temporal variation at sea turtle rookeries. *J Therm Biol* 88: 102522. doi:10.1016/j.jtherbio.2020.102522.

Bentley, B.P., Stubbs, J.L., Whiting, S.D., and N.J. Mitchell. 2020a. Variation in thermal traits describing sex determination and development in Western Australian sea turtle populations. *Funct Ecol* 34: 2302–2314.

Bladow, R.A., and S.L. Milton. 2019. Embryonic mortality in green (*Chelonia mydas*) and loggerhead (*Caretta caretta*) sea turtle nests increases with cumulative exposure to elevated temperatures. *J Exp Mar Biol Ecol* 518: 151180. doi:10.1016/j.jembe.2019.151180.

Blechschmidt, J., Wittmann, M.J., and C. Blüml. 2020. Climate change and green sea turtle sex ratio – preventing possible extinction. *Genes* 11, no. 5: 588. doi:10.3390/genes11050588.

Booth, D.T. 2017. Influence of incubation temperature on sea turtle hatchling quality. *Integr Zool* 12, no. 5: 352–360. doi:10.1111/1749-4877.12255.

Booth, D.T., Archibald-Binge, A., and C.J. Limpus. 2020. The effect of respiratory gases and incubation temperature on early-stage embryonic development in sea turtles. *PLoS One* 15: e0233580.

Booth, D.T., and K. Astill. 2001a. Incubation temperature, energy expenditure and hatchling size in the green turtle (*Chelonia mydas*), a species with temperature-sensitive sex determination. *Austral J Zool* 49: 389–396. doi:10.1071/ZO01006.

Booth, D.T., and K. Astill. 2001b. Temperature variation within and between nests of the green sea turtle, *Chelonia mydas* (Chelonia: Cheloniidae) on Heron Island, Great Barrier Reef. *Austral J Zool* 49: 71–84.

Booth, D.T., Staines, M.N., and R.D. Reina. 2022. Sand characteristics do not influence hatching success of nests at the world's largest green turtle rookery. *Austral J Zool* 69, no. 4: 113–124. doi:10.1071/ZO21050.

Brewer, P.G., Fleer, A.P., Kadar, S., Shafer, D.K., and C.L. Smith. 1978. *Chemical oceanographic data from the Persian Gulf and the Gulf of Oman*. Cruise Report WHOI-78–37, p 1–14. Woods Hole, MA: Woods Hole Oceanographic Institution.

Calderón Peña, R., and J. Azanza Ricardo. 2021. Incubation temperatures, hatching success and congenital anomalies in green turtle nests from Guanahacabibes Peninsula, Cuba. *Aquat Res* 4, no. 4: 321–330. doi:10.3153/AR21027.

Carpio Camargo, A.J., Álvarez Gutiérrez, Y., Jaramillo Véliz, J., and F. Sánchez Tortosa. 2020. Nesting failure of sea turtles in Ecuador–causes of the loss of sea turtle nests: The role of the tide. *J Coast Conserv* 24, no. 5: 1–10. doi:10.1007/s11852-020-00775-3.

Carvar, R.E. 1971. *Procedures in sedimentary petrology*. New York: Wiley-Interscience, 653 p.

Caut, S., Guirlet, E., and M. Girondot. 2010. Effect of tidal overwash on the embryonic development of leatherback turtles in French Guiana. *Mar Environ Res* 69: 254–261.

Cavallo, C., Dempster, T., Kearney, M.R., et al. 2015. Predicting climate warming effects on green turtle hatchling viability and dispersal performance. *Funct Ecol* 29, no. 6: 768–778. doi:10.1111/1365-2435.12389.

Chatting, M., Hamza, S., Al-Khayat, J., Smyth, D., Husrevoglu, S., and C.D. Marshall. 2021. Feminization of hawksbill turtle hatchlings in the twenty-first century at an important regional nesting aggregation. *Endang Species Res* 44: 149–158.

Chatting, M., Smyth, D., Al-Maslamani, I., et al. 2018. Nesting ecology of hawksbill turtles, *Eretmochelys imbricata*, in an extreme environmental setting. *PLoS One* 13, no. 9: e0203257. doi:10.1371/journal.pone.0203257.

Chen, C.-L., Wang, C.-C., and I.-J. Cheng. 2010. Effects of biotic and abiotic factors on the oxygen content of green sea turtle nests during embryogenesis. *J Comp Physiol B* 180, no. 7: 1045–1055. doi:10.1007/s00360-010-0479-5.

Coles, S.L., and J.C. McCain. 1990. Environmental factors affecting benthic infaunal communities of the western Arabian Gulf. *Mar Environ Res* 29: 289–315.

Corwin, D.L., and S.M. Lesch. 2005. Apparent soil electrical conductivity measurements in agriculture. *Comput Electron Agric* 46, nos. 1–3: 11–43. doi:10.1016/j.compag.2004.10.005.

Ditmer, M.A., and S.P. Stapleton. 2012. Factors affecting hatch success of hawksbill sea turtles on Long Island, Antigua, West Indies. *PloS One* 7: e38472.

Eckert, K.L. 1987. Environmental unpredictability and leatherback sea turtle (*Dermochelys coriacea*) nest loss. *Herpetologica* 43: 315–323.

El Daba, A.E.M., Abd El Wahab, M., and M. Abd El Ghany. 2019. Sedimentary characteristics of some island beaches and their effects on nesting numbers of marine turtles, Red Sea, Egypt. *Egypt J Aquat Biol Fish* 23, no. 1: 185–195. doi:10.21608/ejabf.2019.26931.

Erb, V., Lolavar, A., and J. Wyneken. 2018. The role of sand moisture in shaping loggerhead sea turtle (*Caretta caretta*) neonate growth in southeast Florida. *Chelonian Conserv Biol* 17, no. 2: 245–251. doi:10.2744/CCB-1301.1.

Evan, A.T., and S.J. Camargo. 2011. A climatology of Arabian Sea cyclonic storms. *J Clim* 24: 140–158.

Fadini, L.S., Silva, A.G., and P.D. Ferreira. 2011. Sedimentary characteristics and their effects on hatching success and incubation duration of *Caretta caretta* (Testudines: Cheloniidae) in Espirito Santo, Brazil. *Zoologia* 28, no. 3: 312–320.

Fisher, L.R., Godfrey, M.H., and D.W. Owens. 2014. Incubation temperature effects on hatchling performance in the loggerhead sea turtle (*Caretta caretta*). *PLoS One* 9, no. 12: e114880. doi:10.1371/journal.pone.0114880.

Fleming, K.A., Perrault, J.R., Stacy, N.I., Coppenrath, C.M., and A.M. Gainsbury. 2020. Heat, health and hatchlings: Associations of in situ nest temperatures with morphological and physiological characteristics of loggerhead sea turtle hatchlings from Florida. *Conserv Physiol* 8, no. 1: coaa046. doi:10.1093/conphys/coaa046.

Foley, A.M., Peck, S.A., and G.R. Harman. 2006. Effects of sand characteristics and inundation on the hatching success of loggerhead sea turtle (*Caretta caretta*) clutches on low-relief mangrove islands in Southwest Florida. *Chelonian Conserv Biol* 5: 32–41.

Fuentes, M., Limpus, C., and M. Hamann. 2011. Vulnerability of sea turtle nesting grounds to climate change. *Glob Change Biol* 17: 140–153. doi:10.1111/j.1365-2486.2010.02192.x.

Fuentes, M.M.B., Dawson, J.L., Smithers, S.G., Hamann, M., and C.J. Limpus. 2010. Sedimentological characteristics of key sea turtle rookeries: Potential implications under projected climate change. *Mar Freshw Res* 61, no. 4: 464–473. doi:10.1071/MF09142.

Fuentes, M.M.P.B., and W.P. Porter. 2013. Using a microclimate model to evaluate impacts of climate change on sea turtles. *Ecol Modell* 251: 150–157.

Fujisaki, I., and M.M. Lamont. 2016. The effects of large beach debris on nesting sea turtles. *J Exp Mar Biol Ecol* 482: 33–37.

Fujisaki, I., Lamont, M., and R. Carthy. 2018. Temporal shift of sea turtle nest sites in an eroding barrier island beach. *Ocean Coast Manag* 155: 24–29. doi:10.1016/j.ocecoaman.2017.12.032.

Gatto, C.R., Matthews, B., and R.D. Reina. 2021. Role of incubation environment in determining thermal tolerance of sea turtle hatchlings. *Endang Species Res* 44: 397–408. doi:10.3354/esr01111.

Gatto, C.R., and R.D. Reina. 2020. Sea turtle hatchling locomotor performance: Incubation moisture effects, ontogeny and species-specific patterns. *J Comp Physiol B-Biochem Systems Environ Physiol* 190, no. 6: 779–793. doi:10.1007/s00360-020-01307-z.

Gatto, C.R., and R.D. Reina. 2022. A review of the effects of incubation conditions on hatchling phenotypes in non-squamate reptiles. *J Comp Physiol B* 192: 207–233. doi:10.1007/s00360-021-01415-4.

Georges, A., Limpus, C.J., and C.J. Parmenter. 1993. Natural history of the Chelonia. In: *Fauna of Australia, volume 2: Amphibia, Reptilia, Aves*. Ed: J. Godsell, 120–128. Canberra: Australian Biological Resources Study, DASETT.

Glen, F., Broderick, A.C., Godley, B.J., and G.C. Hays. 2003. Incubation environment affects phenotype of naturally incubated green turtle hatchlings. *J Mar Biol Assoc UK* 83, no. 5: 1183–1186. doi:10.1017/S0025315403008464h.

Gravelle, J., and J. Wyneken. 2022. Resilient eggs: Highly successful loggerhead sea turtle nesting sites vary in their characteristics. *Front Ecol Evol* 10: 853835. doi:10.3389/fevo.2022.853835.

Greenwood, P.E., and M.S. Nikulin. 1996. *A guide to Chi-Squared testing*. New York: John Wiley and Sons, 304 p.

Hamann, M., Fuentes, M.M.P.B., Ban, N., and V. Mocellin. 2013. Climate change and marine turtles. In: *The biology of sea turtles*. Eds: J. Wyneken, K.J. Lohmann, and J.A. Musick, 353–378. Boca Raton, FL: CRC Press.

Hastenrath, S., and P.J. Lamb. 1979. *Climatic atlas of the Indian Ocean: Part I, Surface climate and atmospheric circulation. Part II. The ocean heat budget*. Madison: Univ. of Wisconsin Press, p. 116 and 110.

Hawkes, L.A., Broderick, A.C., Godfrey, M.H., and B.J. Godley. 2009. Climate change and marine turtles. *Endang Species Res* 7: 137–154.

Hays, G.C., Mackay, A., Adams, C.R., Mortimer, J.A., Speakman, J.R., and M. Boerema. 1995. Nest site selection by sea turtles. *J Mar Biol Assoc UK* 75: 667–674.

Hays, G.C., and J.R. Speakman. 1993. Nest placement by loggerhead turtles, *Caretta caretta*. *Anim Behav* 45: 47–53.

Heithaus, M.R. 2013. Predators, prey, and ecological roles of sea turtles. In: *Biology of sea turtles*. Eds: J. Wyneken, J.K. Lohman, and J.A. Musick, vol. 3, 249–285. Boca Raton, FL: CRC Press.

Hill, J.E., Paladino, F.V., Spotila, J.R., and P.S. Tomillo. 2015. Shading and watering as a tool to mitigate the impacts of climate change in sea turtle nests. *PLoS One* 10, no. 6: e0129528. doi:10.1371/journal.pone.0129528.

Hirth, H.F. 1997. *Synopsis of the biological data on the Green Turtle Chelonia mydas (Linnaeus 1758)*. Biological Report 97(1). Fish and Wildlife Service, U.S. Department of the Interior, Washington, DC. 128p. doi:10.1046/j.1365-3008.2001.00168.x.

IPCC. 2022. Summary for policymakers. In: *Climate change 2022: Impacts, adaptation and vulnerability. Contribution of Working Group II to the Sixth Assessment Report of the Intergovernmental Panel on Climate Change*. Eds: H.-O. Pörtner, D.C. Roberts, M. Tignor, E.S. Poloczanska, K. Mintenbeck, A. Alegría, M. Craig, S. Langsdorf, S. Löschke, V. Möller, A. Okem, and B. Rama, 3–33. Cambridge, UK and New York, NY: Cambridge University Press. doi:10.1017/9781009325844.001.

John, V., Coles, S., and A. Abozed. 1990. Seasonal cycles of temperature, salinity and water masses of the western Arabian Gulf. *Oceanologica Acta* 13: 273–281.

Johns, W.E., Jacobs, G.A., Kindle, J.C., Murray, S.P., and M. Carron. 1999. *Arabian marginal seas and Gulfs*. Report of a workshop held at Stennis Space Center, Miss., 11–13 May 1999, prepared by the Workshop Organizing Committee. RSMAS Technical Report 2000–01. Miami, FL: University of Miami RSMAS.

Jones, D.A., Plaza, J., Watt, I., and M. Al Sanei. 1998. Long-term (1991–1995) monitoring of the intertidal biota of Saudi Arabia after the 1991 Gulf War oil spill. *Mar Pollut Bull* 36: 472–489.

Jourdan, J., and M.M.P.B. Fuentes. 2015. Effectiveness of strategies at reducing sand temperature to mitigate potential impacts from changes in environmental temperature on sea turtle reproductive output. *Mitigation Adaptation Strategies Glob Change* 20, no. 1: 121–133. doi:10.1007/s11027-013-9482-y.

Karavas, N., Georghiou, K.A., Arianoutsou, M., and D. Dimopoulos. 2005. Vegetation and sand characteristics influencing nesting activity of *Caretta caretta* on Sekania beach. *Biol Conserv* 121, no. 2: 177–188.

Kelly, I., Leon, J.X., Gilby, B.L., Olds, A.D., and T.A. Schlacher. 2017. Marine turtles are not fussy nesters: A novel test of small-scale nest site selection using structure from motion beach terrain information. *Peer J* 5: e2770. doi:10.7717/peerj.2770.

Knox, G.A. 2001. *The ecology of seashores*. Boca Raton, FL: CRC Press, 557 p.

Kobayashi, S., Endo, D., Kondo, S., et al. 2020. Investigating the effects of nest shading on the green turtle (*Chelonia mydas*) hatchling phenotype in the Ogasawara Islands using a field-based split clutch experiment. *J Experiment Zool A: Ecol Integrative Physiol* 333, no. 9: 629–636. doi:10.1002/jez.2411.

Kraemer, J.E., and R. Bell. 1980. Rain-induced mortality of eggs and hatchlings of loggerhead sea turtles (*Caretta caretta*) on the Georgia coast. *Herpetologica* 36: 72–77. www.jstor.org/stable/3891858.

Krishnakumar, P.K., Ashraf, M., Sheppard, C., and J.D. Miller. 2019. The physical environment of the western Arabian Gulf. In: *Ecosystems and biodiversity of the Arabian Gulf: Saudi Arabian waters. Fifty years of scientific research*. Eds: K.A. Al-Abdulkader, R.A. Loughland, and M.A.B. Qurban, 19–59. Dhahran, KSA: ARAMCO and KFUPM.

Laloë, J.-O., Esteban, N., Berkel, J., and G.C. Hays. 2016. Sand temperatures for nesting sea turtles in the Caribbean: Implications for hatchling sex ratios in the face of climate change. *J Exp Mar Biol Ecol* 474: 92–99. doi:10.1016/j.jembe.2015.09.015.

Limpus, C.J. 1987. A turtle fossil on Raine Island, Great Barrier Reef. *Search* 18: 254–256.

Limpus, C.J., Baker, V., and J.D. Miller. 1979. Movement induced mortality of loggerhead eggs. *Herpetologica* 35, no. 4: 335–338. www.jstor.org/stable/3891966.

Mancini, A., Phillott, A.D., and A.F. Rees. 2019. *Chelonia mydas* (North Indian Ocean subpopulation) (errata version published in 2019). The IUCN Red List of Threatened Species 2019: e.T142121108A154845002. doi:10.2305/IUCN.UK.2019-2.RLTST142121108A154845002.en.

Maneja, R.H., Miller, J.D., Flint, J.B., et al. 2023. Extreme conditions reduce hatching success of green turtles (*Chelonia mydas* L.) at Karan Island, the major nesting site in the Arabian Gulf. *Mar Pollut Bull.* 190: 114801. doi.org/10.1016/j.marpolbul.2023.114801

Maneja, R.H., Miller, J.D., Li, W., et al. 2020a. Long-term NDVI and recent vegetation cover profiles of major offshore island nesting sites of sea turtles in Saudi waters of the northern Arabian Gulf. *Ecol Indic* 117: 106612. doi:10.1016/j.ecolind.2020.106612.

Maneja, R.H., Miller, J.D., Li, W., et al. 2020b. Multidecadal analysis of beach loss at the major offshore sea turtle nesting islands in the northern Arabian Gulf. *Ecol Indic* 121: 107146. doi:10.1016/j.ecolind.2020.107146.

Maneja, R.H., Thomas, R., Miller, J.D., et al. 2021. *Marine litter survey at the major sea turtle nesting islands in the Arabian Gulf using in-situ and remote sensing methods*, 8582–8585. Proc IGARSS 2021.

Marn, N., Jusup, M., Legović, T., Kooijman, S.A.L.M., and T. Klanjšček. 2017. Environmental effects on growth, reproduction, and life-history traits of loggerhead turtles. *Ecol Modell* 360: 163–178. doi:10.1016/j.ecolmodel.2017.07.001.

Matthews, B.L., Gatto, C.R., and R.D. Reina. 2021. Effects of moisture during incubation on green sea turtle (*Chelonia mydas*) development, morphology and performance. *Endang Species Res* 46: 253–268. doi:10.3354/esr01159.

Maurer, A.S., Seminoff, J.A., Layman, C.A., Stapleton, S.P., Godfrey, M.H., and M.O.B. Reiskind. 2021. Population viability of sea turtles in the context of global warming. *BioSci* 71, no. 8: 790–804. doi:10.1093/biosci/biab028.

McClenachan, L., Jackson, J.B.C., and M.J.H. Newman. 2006. Conservation implications of historic sea turtle nesting beach loss. *Front Ecol Environ* 4, no. 6: 290–296.

McGehee, M.A. 1990. Effects of moisture on eggs and hatchlings of loggerhead sea turtles (*Caretta caretta*). *Herpetologica* 46: 251–258.

Mellanby, R.J., Broderick, A.C., and B.J. Godley. 1998. *Nest site selection in Mediterranean marine turtles at Chelones Bay, Northern Cyprus*. Proc 16th Annual Symp on Sea Turtle Biology and Conservation; compilers R. Byles, and Y. Fernandez, 103–104. NOAA Technical Memorandum NMFS-SEFSC-412.

Miller, J.D. 1985. Embryology of marine turtles. In: *Biology of the reptilia*. Eds: C. Gans, F. Billett, and P.F.A. Maderson, 269–328. New York: John Wiley and Sons.

Miller, J.D. 1989. *Marine Turtles Volume 1: An assessment of the conservation status of marine turtles in the Kingdom of Saudi Arabia*. MEPA Coastal and Marine Management Series, Technical Report No. 9, 209 p. MEPA, Jeddah.

Miller, J.D. 1997. Reproduction in sea turtles. In: *The biology of sea turtles*. Eds: P. Lutz, and J.A. Musick, 51–81. Boca Raton, FL: CRC Press.

Miller, J.D. 1999. Determining clutch size and hatching success. In: *Research and management techniques for the conservation of sea turtles*. Eds: K.L. Eckert, K.A. Bjorndal, F.A. Abreu-Grobois, and M. Donnelly, 124–129. IUCN/SSC Marine Turtle Specialist Group Publication No. 4. Washington, DC: IUCN/SSC Marine Turtle Specialist Group.

Miller, J.D. 2011a. The coastal and offshore island ecosystems. In: *Marine atlas of the Western Arabian Gulf*. Eds: R.A. Loughland, and K. Al-Abdulkader, 158–185. Dhahran: Saudi ARAMCO Environmental Protection Pubs.

Miller, J.D. 2011b. Marine and coastal reptiles. In: *Marine atlas of the Western Arabian Gulf*. Eds: R.A. Loughland, and K. Al-Abdulkader, 264–295. Dhahran: Saudi ARAMCO Environmental Protection Pubs.

Miller, J.D. 2021. Saudi Arabia (Kingdom of). In: *Sea turtles in the Middle East and South Asia Region*. Eds: A.D. Phillott, and A.F. Rees, 208–244. MTSG Annual Regional Report 2021. Report of the IUCN-SSC Marine Turtle Specialist Group. Washington, DC: IUCN/SSC Marine Turtle Specialist Group.

Miller, J.D., Hayes, M.O., Michel, J., Krishnakumar, P.K., and R.A. Loughland. 2019a. Coastal beach and island ecosystems. In: *Ecosystems and biodiversity of the Arabian Gulf: Saudi Arabian waters. Fifty years of scientific research*. Eds: K.A. Al-Abdulkader, R.A. Loughland, and M.A.B. Qurban, 68–99. Dhahran, KSA: ARAMCO and KFUPM.

Miller, J.D., Limpus, C.J., and M. Godfrey. 2003. Nest site selection, oviposition, eggs, development, hatching and emergence of loggerhead turtles. In: *The biology of the Loggerhead Turtle*. Eds: A. Bolten and B.W. Witherington, 125–143. Washington, DC: Smithsonian Institution Press.

Miller, J.D., Mortimer, J.A., and C.J. Limpus. 2017. A field key to the developmental stages of marine turtles (Cheloniidae) with notes on the development of *Dermochelys*. *Chelonian Conserv Biol* 16, no. 2: 111–122. doi:10.2744/CCB-1261.1.

Mobaraki, A. 2021. Iran. In: *Sea turtles in the Middle East and South Asia Region*. Eds: A.D. Phillott, and A.F. Rees, 109–118. MTSG Annual Regional Report 2021. Report of the IUCN-SSC Marine Turtle Specialist Group. Washington, DC: IUCN/SSC Marine Turtle Specialist Group.

Mobaraki, A., Phillott, A.D., Erfant, M., Ghasem, M., and H. Jafar. 2022. Inferred impacts of extreme environments on hawksbill turtle (*Eretmochelys imbricata*) body size and reproductive output. *Chelonian Conserv Biol* 21: 187–198.

Montero, N., Dei Marcovaldi, M.A.G., Lopez-Mendilaharsu, M., Santos, A.S., Santos, A.J.B., and M. Fuentes. 2018. Warmer and wetter conditions will reduce offspring production of hawksbill turtles in Brazil under climate change. *PLoS One* 13, no. 11: e0204188. doi:10.1371/journal.pone.0204188.

Mortimer, J.A. 1990. The influence of beach sand characteristics on the nesting behavior and clutch survival of green turtles (*Chelonia mydas*). *Copeia* 1990, no. 3: 802–817. doi:10.2307/1446446.

Mortimer, J.A., and M. Donnelly (IUCN SSC Marine Turtle Specialist Group). 2008. *Eretmochelys imbricata*. The IUCN Red List of Threatened Species 2008: e.T8005A12881238. http://dx.doi.org/10.2305/IUCN.UK.2008.RLTS.T8005A12881238.en

Mrosovsky, N. 1983. Ecology and nest-site selection of leatherback turtles. *Biol Conserv* 26: 47–56.

National Research Council. 1990. *Decline of the sea turtles: Causes and prevention*. Washington, DC: The National Academies Press. doi:10.17226/1536.

Papathanasopoulou, N., and A.F. Rees. 2021. Kuwait. In: *Sea turtles in the Middle East and South Asia region*. Eds: A.D. Phillott, and A.F. Rees, 121–131. MTSG Annual Regional Report 2021. Report of the IUCN-SSC Marine Turtle Specialist Group. Washington, DC: IUCN/SSC Marine Turtle Specialist Group.

Patrício, A.R., Varela, M.R., Barbosa, C., et al. 2018. Nest site selection repeatability of green turtles, *Chelonia mydas*, and consequences for offspring. *Anim Behav* 139: 91–102. doi:10.1016/j.anbehav.2018.03.006.

Patriquin, D.G., Blaikie, H., Patriquin, M.J., and C. Yang. 1993. On-farm measurements of pH, electrical conductivity and nitrate in soil extracts for monitoring coupling and decoupling of nutrient cycles. *Biol Agric Hortic* 9, no. 3: 231–272. doi:10.1080/01448765.1993.9754638.

Pfaller, J.B., Limpus, C.J., and K.A. Bjorndal. 2009. Nest-site selection in individual loggerhead turtles and consequences for doomed-egg relocation. *Conserv Biol* 23, no. 1: 72–80.

Phillott, A.D., and A.F. Rees. (Eds.). 2021. *Sea turtles in the Middle East and South Asia region*. MTSG Annual Regional Report 2021: IUCN-SSC Marine Turtle Specialist Group. 412 p. Washington, DC: IUCN/SSC Marine Turtle Specialist Group.

Pilcher, N.J. 1999. The hawksbill turtle, *Eretmochelys imbricata*, in the Arabian Gulf. *Chelonian Conserv Biol* 3, no. 2: 312–317.

Pilcher, N.J. 2000. The green turtle, *Chelonia mydas*, in the Saudi Arabian Gulf. *Chelonian Conserv Biol* 3, no. 4: 730–734.

Pilcher, N. J. & Al-Merghani, M. 1994. The Marine Turtle Nesting Season, Arabian Gulf, 1992. *In:* Feltkamp, E. & Krupp, F. (eds.) *Establishment of a Marine Habitat and Wildlife Sanctuary for the Gulf Region – Final Report for Phase 2*. pp. 514–533. Jubail and Frankfurt Commission of the European Communities: Brussels, and the National Commission for Wildlife Conservation and Development: Riyadh.

Privett, D.W. 1959. Monthly charts of evaporation from North Indian Ocean, including the Red Sea and the Persian Gulf. *Quart J Royal Meteorol Soc London* 85: 424–428.

QGIS. 2016. QGIS Geographic Information System. Open Source Geospatial Foundation Project. http://qgis.osgeo.org.

Ragotzkie, R.A. 1959. Mortality of loggerhead turtle eggs from excessive rainfall. *Ecology* 40, no. 2: 303–305. doi:10.2307/1930045.

Reece, J.S., Passeri, D., Ehrhart, L., et al. 2013. Sea level rise, land use, and climate change influence the distribution of loggerhead turtle nests at the largest USA rookery (Melbourne Beach, Florida). *Mar Ecol Prog Ser* 493: 259–274. doi:10.3354/meps10531.

Rees, A.F. 2021. Qatar. In: *Sea turtles in the Middle East and South Asia region. MTSG Annual Regional Report 2021*. Eds: A.D. Phillott, and A.F. Rees, 199–207. Report of the IUCN-SSC Marine Turtle Specialist Group. Washington, DC: IUCN/SSC Marine Turtle Specialist Group.

Reynolds, R.M. 1993. Physical oceanography of the Persian Gulf, Strait of Hormuz, and the Gulf of Oman–Results from the Mt. Mitchell Expedition. *Mar Pollut Bull* 27: 35–59.

Rivas, M.L., Spínola, M., Arrieta, H., and M. Faife-Cabrera. 2018. Effect of extreme climatic events resulting in prolonged precipitation on the reproductive output of sea turtles. *Anim Conserv* 21, no. 5: 387–395. doi:10.1111/acv.12404.

Salleh, S.M., Nishizawa, H., Ishihara, T., Sah, S.A.M., and A.J.K. Chowdhury. 2018. Importance of sand particle size and temperature for nesting success of green turtles in Penang Island, Malaysia. *Chelonian Conserv Biol* 17, no. 1: 116–122. doi:10.2744/CCB-1266.1.

Santidrian Tomillo, P., Fonseca, L., Paladino, F.V., Spotila, J.R., and D. Oro. 2017. Are thermal barriers higher in deep sea turtle nests? *PLoS One* 12, no. 5: e0177256. doi:10.1371/journal.pone.0177256.

Santos, A.J.B., Neto, J.X.L., Vieira, D.H.G., et al. 2016. Individual nest site selection in hawksbill turtles within and between nesting seasons. *Chelonian Conserv Biol* 15: 109–114.

Santos, K.C., Livesey, M., Fish, M., and A.C. Lorences. 2017. Climate change implications for the nest site selection process and subsequent hatching success of a green turtle population. *Mitigation Adaptation Strategies Glob Change* 22, no. 1: 121–135. doi:10.1007/s11027-015-9668-6.

Schneider, C.A., Rasband, W.S., and K.W. Eliceiri. 2012. NIH Image to ImageJ: 25 years of image analysis. *Nat Methods* 9, no. 7: 671–675. doi:10.1038/nmeth.2089.

Schwanz, L.E., and A. Georges. 2021. Sexual development and the environment: Conclusions from 40 years of theory. *Sexual Dev* 15, nos. 1–3: 7–22. doi:10.1159/000515221.

Seminoff, J.A. 2004. *Chelonia mydas*. The IUCN Red List of Threatened Species 2004: e.T4615A11037468. doi:10.2305/IUCN.UK.2004.RLTS.T4615A11037468.en; downloaded 2 April 2022.

Sheppard, C.R. 1993. Physical environment of the Gulf relevant to marine pollution: An overview. *Mar Pollut Bull* 27: 3–8.

Sheppard, C., Al-Husiani, M., Al-Jamali, F., et al. 2010. The Gulf: A young sea in decline. *Mar Pollut Bull* 60: 13–39. doi:10.1016/j.marpolbul.2009.10.017.

Sheppard, C., and M. Borowitzka. 2011. Subtidal habitats. In: *Marine atlas of the Western Arabian Gulf*. Eds: R.A. Loughland, and K.A. Al-Abdulkader, 116–157. Dhahran, Saudi Arabia: Saudi Aramco Environmental Protection Pubs.

Sim, E.L., Booth, D.T., and C.J. Limpus. 2015. Incubation temperature, morphology and performance in loggerhead (*Caretta caretta*) turtle hatchlings from Mon Repos, Queensland, Australia. *Biol Open* 4, no. 6: 685–692. doi:10.1242/bio.20148995.

Smith, C.E., Booth, D.T., Crosby, A., et al. 2021. Trialling seawater irrigation to combat the high nest temperature feminisation of green turtle *Chelonia mydas* hatchlings. *Mar Ecol Prog Ser* 667: 177–190. doi:10.3354/meps13721.

Smith, J.L., and J.W. Doran. 1997. Measurement and use of pH and electrical conductivity for soil quality analysis. In: *Methods for assessing soil quality*. Eds: J.W. Doran, and A.J. Jones, vol. 49, 169–185. Madison, WI: Soil Science Society of America Inc.

Soil Survey Staff. 2014. *Soil survey field and laboratory methods manual*. Soil Survey Investigations Report No. 51, Version 2.0. Eds: R. Burt, and Soil. Survey Staff. 1–487. Lincoln, NE: U.S. Department of Agriculture, Natural Resources Conservation Service.

Sönmez, B., Turan, C., and Ş. Yalçın Özdilek. 2013. Comparison of the physical properties of two green turtle (*Chelonia mydas*) nesting beaches (Akyatan and Samandağ) in the eastern Mediterranean (Reptilia: Cheloniidae). *Zool Middle East* 59, no. 1: 30–38. doi:10.1080/09397140.2013.795061.

Staines, M., Booth, D.T., Madden Hof, C.A., and G.C. Hays. 2020. Impact of heavy rainfall events and shading on the temperature of sea turtle nests. *Mar Biol* 167: 190. doi:10.1007/s00227-020-03800-z.

Stewart, T.A., Booth, D.T., and M.U. Rusli. 2019. Influence of sand grain size and nest microenvironment on incubation success, hatchling morphology and locomotion performance of green turtles (*Chelonia mydas*) at the Chagar Hutang Turtle Sanctuary, Redang Island, Malaysia. *Austral J Zool* 66, no. 6: 356–368. doi:10.1071/ZO19025.

Stubbs, J.L., and N.J. Mitchell. 2018. The influence of temperature on embryonic respiration, growth, and sex determination in a Western Australian population of green turtles (*Chelonia mydas*). *Physiol Biochem Zool* 91, no. 6: 1102–1114. doi:10.1086/700433.

Suss, J.S., Patel, S., Neeman, N., et al. 2012. Beach characteristics affect the gas exchange environment for sea turtle nests. *Integr Comp Biol* 52: E335.

Tezak, B., Bentley, B., Arena, M., Mueller, S., Snyder, T., and I. Sifuentes-Romero. 2020. Incubation environment and parental identity affect sea turtle development and hatchling phenotype. *Oecologia* 192, no. 4: 939–951. doi:10.1007/s00442-020-04643-7.

Tollab, M.A., Dakhteh, M.H., Zaferani, G.G., et al. 2015. The olive ridley turtle, *Lepidochelys olivacea*, in the Persian Gulf: A Review of the observations, including the first nesting of the species in the area. *Chelonian Conserv Biol* 14: 192–196.

Türkozan, O., Almpanidou, V., Yilmaz, C., and A.D. Mazaris. 2021. Extreme thermal conditions in sea turtle nests jeopardize reproductive output. *Clim Change* 167, nos. 3–4: 30. doi:10.1007/s10584-021-.03153-6.

Türkozan, O., Yamamoto, K., and C. Yılmaz. 2011. Nest site preference and hatching success of green (*Chelonia mydas*) and loggerhead (*Caretta caretta*) sea turtles at Akyatan Beach, Turkey. *Chelonian Conserv Biol* 10, no. 2: 270–275. doi:10.2744/ccb-0861.1.

van Buskirk, J., and L.B. Crowder. 1994. Life-history variation in marine turtles. *Copeia* 1994, no. 1: 66–81.

Vaughan, G.O., Al-Mansoori, N., and J.A. Burt. 2019. The Arabian Gulf. In: *World seas: An environmental evaluation: The Indian Ocean to the Pacific*. Ed: C. Sheppard, vol. 2, 1–23. New York: Academic Press.

Wallace, B.P., Dimatteo, A.D., Hurley, B.J., et al. 2010. Regional management units for marine turtles: A novel framework for prioritizing conservation and research across multiple scales. *PLoS One* 5, no. 12: e15465. doi:10.1371/journal.pone.0015465.

Ware, M., and M.M.P.B. Fuentes. 2019. Potential for relocation to alter the incubation environment and productivity of sea turtle nests in the Northern Gulf of Mexico. *Chelonian Conserv Biol* 17, no. 2: 252–262. doi:10.2744/CCB-1306.1.

Wentworth, C.K. 1922. A scale of grade and class terms for clastic sediments. *J Geol* 30: 377–392. doi:10.1086/622910.

Wibbels, T. 2003. Critical Approaches to sex determination in sea turtles. In: *The biology of sea turtles*. Eds: P.L. Lutz, J.A. Musick, and J. Wyneken, vol. 2, 103–134. Boca Raton, FL: CRC Press.

Willson, M.S., Phillott, A.D., Al Azri, K., et al. 2021. Oman (Sultanate of). In: *Sea turtles in the Middle East and South Asia region*. Eds: A.D. Phillott, and A.F. Rees, 149–176. MTSG Annual Regional Report 2021. Report of the IUCN-SSC Marine Turtle Specialist Group. Washington, DC: IUCN/SSC Marine Turtle Specialist Group.

Witzell, W.N. 1983. Synopsis of biological data on the hawksbill turtle, *Eretmochelys imbricata* Linnaeus, 1766. *FOA Fisheries Synopsis* 137: 78.

Wood, D.W., and K.A. Bjorndal. 2000. Relation of temperature, moisture, salinity, and slope to nest site selection in Loggerhead sea turtles. *Copeia* 2000, no. 1: 119–128. doi:10.1643/0045-8511(2000)2000[0119:ROTMSA]2.0.CO;2.

Yalçın-Özdilek, Ş., Özdilek, H.G., and F.S. Ozaner. 2007. Possible influence of beach sand characteristics on green turtle nesting activity on Samandağ Beach, Turkey. *J Coast Res* 23, no. 6: 1379–1390. doi:10.2112/06-0630.1.

Zare, R., Vaghefi, M.E., and S.J. Kamel. 2012. Nest location and clutch success of the hawksbill sea turtle (*Eretmochelys imbricata*) at Shidvar Island, Iran. *Chelonian Conserv Biol* 11: 229–234.

PART 7

Marine mammals

Dugong feeding ecology along the Egyptian Red Sea

24

Ahmed M. Shawky

24.1 INTRODUCTION

Large herbivores or mega-grazers such as dugongs play a significant role in the structure of seagrass communities; however, understanding their relative importance has been hindered by the lack of studies (Ebrahim et al. 2014; Bessey et al. 2016). Dugongs have been recorded as changing the distribution, species composition, structure, productivity and nutrient status of plant communities (Beck and Clark 2012; Mizuno et al. 2017). Moore et al. (2017) reported that dugongs can possibly be considered as a keystone species in tropical seagrass ecosystems within the Indo-Pacific region. Several studies documented their ecological role such as bioturbation during feeding on seagrasses (Heinsohn et al. 1977), increased microbial nutrient re-cycling after excavation (Perry and Dennison 1999), influence on seagrass species composition and nutrient content (Preen 1995; Aragones and Marsh 2000; Aragones et al. 2006), promotion of seagrass growth through fertilization by dugong feces and increased seagrass productivity post-grazing (McMahon 2005), and the potential for seed and propagule dispersion (Kendrick et al. 2012).

When dugongs excavate the seagrass, rather than cropping, they create winding feeding pathways along seagrass beds during feeding that are known as feeding trails (Preen 1995; Anand 2012). Those trails are considered to be excellent evidence of dugong feeding and are common in many tropical intertidal regions, including Indonesia (De Iongh et al. 2007), the Andaman Islands and the Nicobar archipelago in India (D'Souza et al. 2015), and the western coast of the Red Sea (Shawky 2019a, 2019b; Nasr et al. 2019). Recently, 13 dugong feeding sites across 27 seagrass meadows were recorded in the northeastern Red Sea in the waters of NEOM, Kingdom of Saudi Arabia, within a range of 4,061 km^2 of offshore and nearshore waters (Khamis et al. 2022). The impact of continued grazing can reduce shoot densities (Preen 1995); however, early successional species might be able to recover rapidly from these reductions.

The aim of this study is, therefore, to obtain a greater understanding of the feeding ecology of dugongs inhabiting the Egyptian Red Sea coast, particularly in the unstudied area of dugong–seagrass interactions (Figure 24.1). The following aspects of dugong feeding were studied:

FIGURE 24.1 Dugong feeding on a seagrass, *Halophila stipulacea*, in Wadi El Gemal National Park, Red Sea, Egypt. Photo credit: Ahmed M. Shawky

1. Estimating the characteristics of feeding sites (seagrass coverage and abundance).
2. Measuring feeding trail dimensions and frequency.
3. Evaluating the grazing rate on seagrass communities.

24.2 MATERIAL AND METHODS

24.2.1 Study sites

The data was collected from eight different sites located in two regions on the western coast of the Egyptian Red Sea at different times at Marsa Alam and Wadi El Gemal National Park (WGNP) (Figure 24.2). Three sites were selected in Marsa Alam: site 1, Marsa Mobarak (late autumn 2016); site 2, Marsa Abou Dabbab (late summer 2017); and site 3, Marsa Assalaya

DOI: 10.1201/9781003321392-31

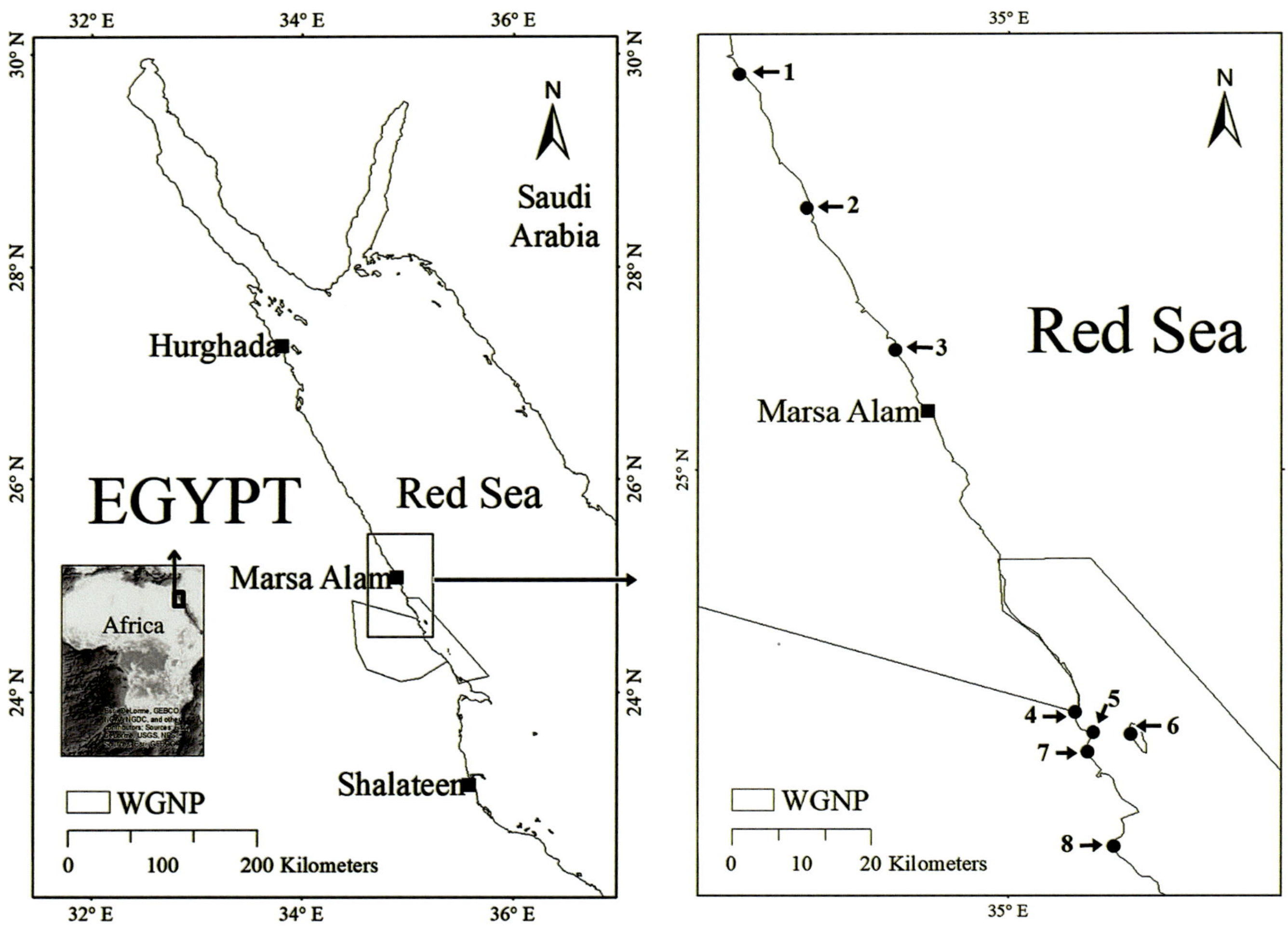

FIGURE 24.2 Map of the Egyptian Red Sea (left) and the details of the study sites (right) in Marsa Alam and Wadi El-Gemal National Park (WGNP): 1, Marsa Mobarak; 2, Marsa Abou Dabbab; 3, Marsa Assalaya; 4, Shams Alam; 5, Ras Baghdady; 6, Wadi El-Gemal Island; 7, Torpha South; 8, Marsa Omm El-Abbas.

(early spring 2016). The other five sites in WGNP were site 4, Shams Alam (late summer 2016); site 5, Ras Baghdady (late autumn 2016); site 6, Wadi El Gemal Island (early spring 2016); site 7, Torpha south (late autumn 2016); and site 8, Marsa Omm El-Abbas (early winter 2016). These sites were chosen due to the information obtained from a questionnaire survey conducted by Shawky et al. (2024) that confirmed the presence of feeding trails in some seagrass sites, and also depended on our availability to visit.

24.2.2 Methodology

24.2.2.1 Seagrass abundance

Following Preen (1992), seagrass abundance was measured and estimated as relative abundance and percentage cover. Two 20 m line transects were constructed in the seagrass area perpendicular to the shoreline. Seagrass abundance was assessed in 10 randomly located quadrats (0.25 m^2) at 2 m intervals along each transect (20 quadrats per site). They were displaced if they fell over an area where a feeding trail occurred. The relative abundance of seagrass in each quadrat was estimated by counting the shoots of each species, and we did not study below ground. The data was collected underwater by means of SCUBA diving. Each quadrat was photographed using an HD digital camera (GoPro Hero 4 Silver Edition set in 4K mode) (Figure 24.3). Photos were later displayed on a laptop screen, zoomed and seagrass shoots were counted. Identification of seagrasses was carried out after El Shaffai (2016).

24.2.2.2 Determination of sediment compaction

Sediment compaction may explain how easily the dugongs can collect the seagrass rhizomes (Preen 1992). It was determined using a steel rod of a length 120 cm and a diameter of 1 cm. In the trail areas, and beside the quadrate location, the rod was randomly pushed vertically into the sediment until it no longer penetrated. Upon withdrawal, the penetrated part was measured in metres using a fibreglass measuring tape. A total of 20 replicate readings were randomly taken at each of the study sites.

24.2.2.3 Feeding trail dimensions

Following Preen (1992) and Amamoto et al. (2009), the width of the feeding trails was measured with a ruler to within a centimetre, and the trail length was measured in metres with a fibreglass tape. A total of 20 feeding trails was measured in each site. The width of the trails sometimes varied along the trail due to the feeding behaviour of the dugong. In our study,

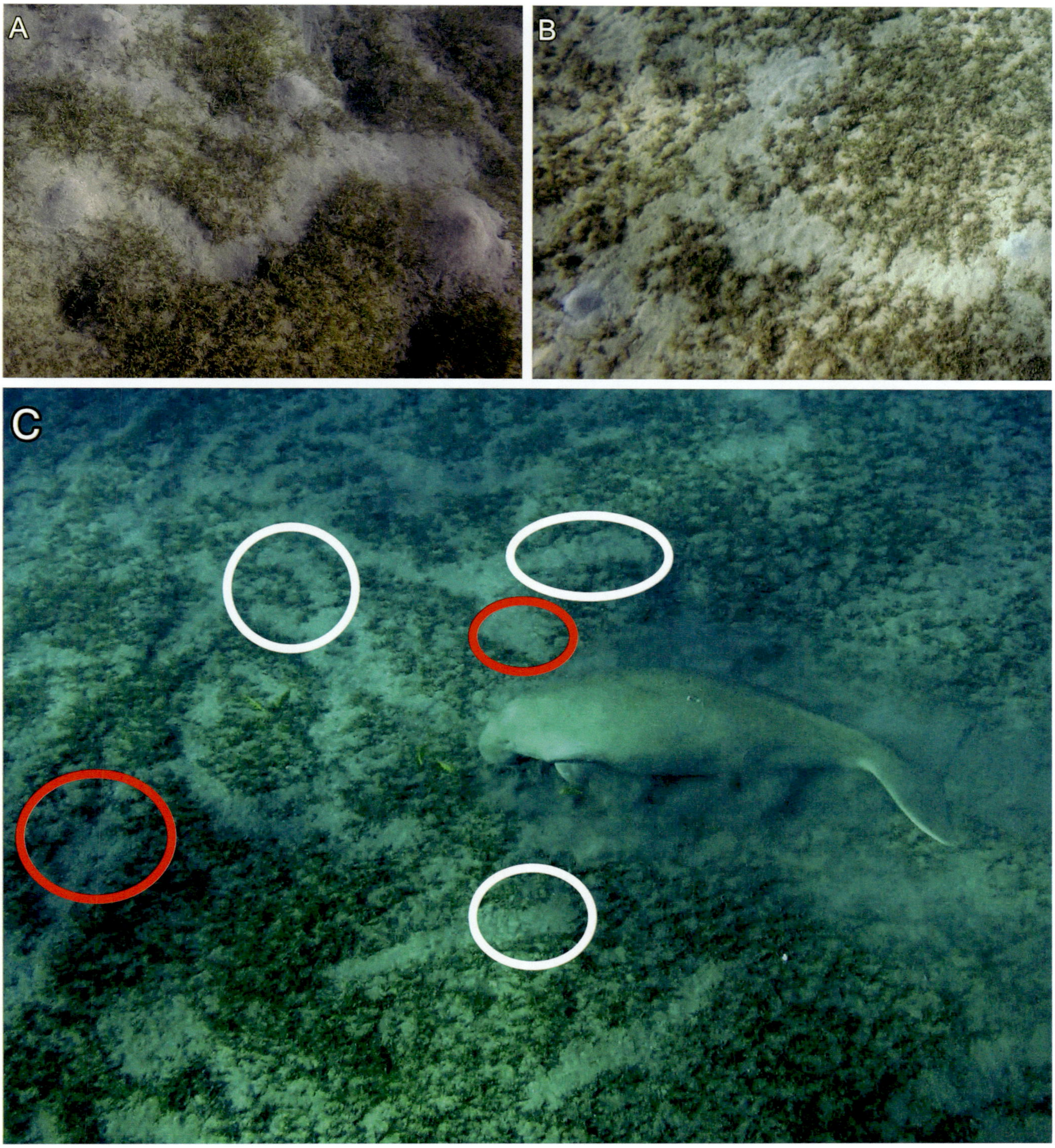

FIGURE 24.3 Examples of feeding trails with regular width (A), irregular width (B), and (C) dugong grazing on the seagrass bed with regular width (white circles) and irregular width (red circles).

we concluded that the width is related to muzzle width and is fixed when the dugong moves forward continuously during feeding, leaving the trail edge as two parallel lines with fixed distance between (Shawky 2019a). In this case only, we measured the regular width between the two parallel edges of the seagrass three times (i.e., the start, middle and end of the trail), then obtained the average, ignoring irregular widths and unclear trails (Figure 24.3). The trail depth was measured in centimetres with a ruler. The grazed area for each trail was calculated by multiplying the length and the width for each trail.

An estimation of a dugong's size is concluded from feeding trail width. Different size categories indicate the number of different sizes of the dugongs. Shawky et al. (2017, 2019) identified the dugongs that visited the study sites previously and are compared to the results of the current study. With more dugong numbers visiting a site, large and small sizes are expected. Calves were expected due to the presence of large trail widths beside small ones, which is evidence for the presence of mothers with their calves.

24.2.2.4 Feeding trail spatial frequency

Feeding trails differed in density among sites. Therefore, 1 m^2 quadrats were placed randomly 20 times over the feeding area, and the number of trails inside was counted. If the quadrat fell in an area of seagrass free of trails, it was relocated again over a trail area.

24.2.2.5 Proportion of seagrass removed from feeding trails

According to Preen (1992), the percentage of seagrass eliminated was measured using 20 quadrats of 0.1 m^2 (10 cm × 10 cm) by comparing the abundance of seagrass inside and outside the feeding trails. This was carried out by randomly locating the quadrats along several feeding trails and another equivalent number of matching quadrats were positioned immediately adjacent to them (2–5 cm away). Each quadrat was photographed using an HD digital camera. Photos were later displayed on a laptop screen, zoomed and seagrass shoots were counted. Seagrasses that were 2–5 cm away from the trail may not always be the same as those inside the feeding trail. Hence, the amount of seagrass removed from the feeding trails was calculated by subtracting the mean abundance of seagrass inside the feeding trails from the mean abundance of those immediately outside the feeding trails.

24.3 RESULTS

24.3.1 Seagrass abundance and species composition

Marsa Alam had a higher density of seagrass shoots per 1 m^2 than did WGNP (2585 ± 1925, SE = 249; 1095 ± 689, SE = 89, N = 20, respectively). In Marsa Alam, the highest density was recorded in Marsa Assalaya (4588 ± 421, SE) while the lowest was in Marsa Mobarak (1464 ± 151) (Table 24.1). The density of seagrass showed highly significant differences among sites ($P = <0.05$). In WGNP, the highest density was recorded in Wadi El Gemal Island (1605 ± 195, SE), while the lowest was in Ras Baghdady (736 ± 100). The density of seagrass showed highly significant differences among sites ($P = 0.00004$).

Four different seagrass species were recorded in the Marsa Alam region, namely *Halophila stipulacea*, *Halophila ovalis*, *Halodule uninervis* and *Cymodocea rotundata*. *H. stipulacea* was most dominant in Marsa Assalaya with a percentage of 52%, but the least abundant was in Marsa Abou Dabbab (2%). *H. ovalis* was the most dominant in Marsa Abou Dabbab (83%) and the least in Marsa Mobarak (29%). *H. uninervis* was the most dominant in Marsa Mobarak (26%) and Marsa Abou Dabbab (15%) but disappeared completely from Marsa Assalaya. *C. rotundata* was only recorded in Marsa Mobarak (5%).

In addition to the previous species, *Syringodium isoetifolium and Thalassia hemprichii* were recorded in WGNP region. *H. stipulacea* was the most dominant in Marsa Omm El-Abbas with a percentage of 97%, but was the least abundant in Wadi El Gemal Island (42%); however, it was absent from Ras Baghdady during the study period. *H. ovalis* was most dominant in Wadi El Gemal Island (47%) and the least in Marsa Omm E-Abbas (3%). *H. uninervis* was the most dominant in Ras Baghdady (75%) and the least in Torpha south (5%), but was not recorded in Omm El-Abbas. *C. rotundata* was only recorded in Shams Alam (7%) and was totally absent in the other sites. *S. isoetifolium* was only recorded in Ras Baghady (3%).

The density of seagrass shoots per 1 m^2 (mean shoot count ±SE) was recorded in the different sites. In Marsa Alam, *H. stipulacea* was the most dominant in Marsa Assalaya (2232 ± 454) and the least in Marsa Mobarak (853 ± 12). *H. ovalis* was the most dominant in Marsa Assalaya (2024 ± 572) and the least in Marsa Abou Dabbab (122 ± 51). *H. uninervis* was most dominant in Marsa Mobarak (288 ± 119) and least in Marsa Assalaya (4 ± 4). *C. rotundata* was only recorded in Marsa Mobarak (61 ± 39) and was completely absent in the other sites. The density of seagrass species among sites differed significantly ($P = 0.0003$).

In WGNP, *H. stipulacea* was the most dominant in Wadi El Gemal Island (682 ± 125), the least in Torpha south (322 ± 60) and absent in Ras Baghdady (Table 24.1). *H. ovalis* was the most dominant species in Wadi El Gemal Island (762 ± 116) and the least in Marsa Omm El-Abbas (11 ± 6). *H. uninervis* was the most dominant in Ras Baghdady (558 ± 58) and least in Torpha south (17 ± 9). *C. rotundata* was the most dominant in Shams Alam (66 ± 20) and the least in Ras Baghdady (3 ± 2.5), and disappeared completely from other sites. The density of seagrass species in WGNP showed highly significant difference among sites ($P = <0.05$).

TABLE 24.1 Seagrass abundance (shoots/m^2) in the study sites.

STUDY SITES	*RANGE*	*MEAN±SE*	*MEDIAN*	*SEAGRASS SPECIES*
Marsa Mobarak	544–2896	1521 ± 134	1504	H.s, H.o, H.u, C.r
Marsa Abou Dabbab	208–3760	1702 ± 255	1520	H.s, H.o, H.u
Marsa Assalaya	2160–8640	4580 ± 411	4320	H.s, H.o, H.u
Shams Alam	240–1888	929 ± 103	912	H.s, H.o, H.u, C.r
Ras Baghdady	272–1200	734 ± 53	760	H.o, H.u, S.i, T.h
Wadi El-Gemal Island	160–4080	1529 ± 216	1464	H.s, H.o, H.u
Torpha South	160–3088	786 ± 151	504	H.s, H.o, H.u
Marsa Omm El-Abbas	320–2000	1143 ± 87	1144	H.s, H.o, H.u

H.s, *Halophila stipulacea*; H.o, *Halophila ovalis*; H.u, *Halodule uninervis*; C.r, *Cymodocea rotundata*; S.i, *Syringodium isoetifolium*; T.h, *Thalassia hemprichii*.

24.3.2 Sediment compaction

The rod penetration (N = 20) was higher in Marsa Alam (0.83±0.33 m) than in WGNP (0.52±34 m). In Marsa Alam, the highest penetration was in Marsa Abou Dabbab (1±0.29 m) and the lowest value was in Marsa Assalaya (0.7±0.1 m) (Table 24.2). In WGNP, the highest penetration was in Torpha south (0.99±0.05 m) and the lowest in Wadi El Gemal Island (0.19±0.008 m).

24.3.3 Feeding trail dimensions

The trail lengths in the Marsa Alam region differed significantly among the three study sites (one-way ANOVA: *df* = 2, *F* = 5.3, *P* = 0.004). Results of the *t*-test showed a significant difference between Marsa Mobarak and Marsa Assalaya (*P* = 0.001) and between Marsa Mobarak and Marsa Abou Dabbab (*P* = 0.001). The longest trail length was in Marsa Abou Dabbab (3.7±0.4 m) and the shortest in Marsa Mobarak (2.4±0.2 m) (Table 24.2). There was no correlation between the mean length of feeding trails and the density of seagrass shoots at the three sites of Marsa Alam (r = −0.16, r = −0.06 and r = −0.02) for Marsa Mobarak, Marsa Abou Dabbab and Marsa Assalaya, respectively. In WGNP, the trail lengths differed significantly among the study sites (one-way ANOVA: *df* = 2, *F* = 20, *P* = <0.05). Results of the *t*-test showed a significant difference between Shams Alam and Wadi El Gemal Island, Shams Alam and Ras Baghdady, and between Wadi El Gemal Island and Ras Baghdady (*P* = <0.05, *P* = <0.05 and *P* = 0.001), respectively. The longest trail length was in Ras Baghdady (4.3±0.34 m) and the shortest in Torpha south (1±0.05 m) (Table 24.2). There was no correlation between the mean length of feeding trails and the density of seagrass shoots at the study sites (r = 0.13, r = −0.34 and r = 0.062) for Shams Alam, Wadi El Gemal Island and Ras Baghdady.

A total of 20 different variations in width of the feeding trails were recorded in all study sites for the feeding trails, ranging from 7 cm to 26 cm. A total of 14 different widths in Marsa Alam and 15 in WGNP were reported (Table 24.2 and Figure 24.4). Trail widths differed significantly among the Marsa Alam sites (one-way ANOVA: *df* = 2, F = 3.5, *P* = 0.033). The biggest width was in Marsa Abou Dabbab (18.6±0.7 cm) and the smallest in Marsa Assalaya (15.4±1 cm) (Table 24.2). Results of the *t*-test showed a significant difference between Marsa Mobarak and Marsa Assalaya (*P* = 0.0128). There was no correlation of the mean width of feeding trails between each study site. Feeding trail widths did not differ significantly among the study sites of WGNP (one-way ANOVA: *df* = 2, *F* = 2.9, *P* = 0.06). The biggest width was in Torpha south (18.9±0.4 cm) and the smallest in Marsa Omm El-Abbas (15.4±0.7 cm). Results of the *t*-test showed a significant difference between Shams Alam and Wadi El Gemal Island (*P* = 0.016) and between Shams Alam and Ras Baghdady (*P* = 0.005). There was low correlation (r = 0.55) between Shams Alam and Wadi El Gemal Island.

24.3.4 Frequency of feeding trails

The number of feeding trails assessed was restricted to an area of seagrass excavated by dugongs. The feeding trail density differed significantly among the Marsa Alam sites (one-way ANOVA: *df* = 2, *F* = 13.4, *P* = <0.05). The density of trails was higher in Marsa Abou Dabbab (2±0.2 trail/m²) and lower in Marsa Mobarak (1.2±0.1) (Table 24.2). Results of a *t*-test showed that Marsa Abou Dabbab was significantly different from Marsa Assalaya (*P* = 0.0001). In WGNP, the feeding trail densities differed significantly among sites (one-way ANOVA: *df* = 2, *F* = 28.2, *P* = <0.05). The density of trails was highest in Shams Alam (3.3±0.3 trail/m²) and lowest in Ras Baghdady (1.2±0.1). From the results of a *t*-test, Shams Alam was significantly different from Wadi El Gemal Island (*P* = 0.007), Shams Alam was significantly different from Ras Baghdady (*P* = <0.05), and Wadi El Gemal Island was significantly different from Ras Baghdady (*P* = <0.05). Three different trails width were recorded at Wadi El Gemal Island at depths of <1.5m.

Feeding trail widths varied among the sites: Wadi El-Gemal Island recorded 10 different sizes of trail width, followed by Marsa Omm El-Abbas, while Marsa Mobarak had only three different widths (Figure 24.4). A few sites recorded small feeding trail widths, where these were related to calving areas. On the other hand, the trails of 18 cm and 20 cm width were dominant in most of the study sites in the different regions.

24.3.5 Expected dugong numbers from feeding trails

The different sizes of trail widths in each site were compared with the number of identified dugongs using the photo identification technique of Shawky et al. (2017, 2019), who identified 30 dugongs in 22 sites. In Marsa Abou Dabbab, seven different sizes of trail width were recorded in this study, and at the same time seven different dugongs were identified in the catalogue using a photo-ID (Table 24.3). In Marsa Mobarak, three different sizes of trail width were recorded (i.e., two large and one small), but only two adults were identified in the catalogue, therefore, one more calf is expected to be visiting the site. In Shams Alam, seven different sizes of trail width were recorded including one small size, but only three large dugongs were photographed, therefore, four more individuals are expected (i.e., three large and one small). A total of 18 different sizes of trail width were recorded, and 34 dugongs were expected at the eight study sites, with an increase of 88%. When applying this rate to all the identified 30 dugongs in 22 sites, 56 dugongs would be expected. Hence an estimate of the population visiting these sites ranges from 30 to 56 dugongs.

24.3.6 Seagrass removed by dugong grazing

In this study, our estimate of the amount of seagrass consumed by dugongs is restricted to the above-ground biomass only. The amount of seagrass removed along the feeding trails varied among sites in both regions. In Marsa Alam, the maximum was recorded in Marsa Assalaya (4110 shoots/m², 92%) and the minimum in Marsa Abou Dabbab (1775 shoots/m², 89%) (Table 24.4). There was a correlation between the amount of seagrass removed from feeding trails and the mean abundance

TABLE 24.2 Feeding trails data in the study sites.

	DIMENSIONS OF FEEDING TRAILS													
	TRAIL DENSITY/M²		*TRAIL LENGTH (M)*		*TRAIL WIDTH (CM)*		*TRAIL DEPTH (CM)*		*GRAZED*	*BOTTOM DEPTH (M)*		*BOTTOM COMPACTION (ROD PENETRATION; M)*	
STUDY SITES	*RANGE*	*MEAN±SE*	*RANGE*	*MEAN±SE*	*RANGE*	*MEAN±SE*	*RANGE*	*MEAN±SE*	*AREA/M₂*	*RANGE*	*MEAN±SE*	*RANGE*	*MEAN±SE*
Marsa Mobarak	1–2	1.1 ± 0.1	1.5–4.1	2.41 ± 0.2	10–20	18.2 ± 0.7	5–8	6.6 ± 0.2	8.98	1–9	4.7 ± 0.6	0.2–1.2	0.7 ± 0.4
Marsa Abou Dabbab	1–3	2 ± 0.2	0.6–10.5	3.97 ± 0.5	9–20	17.05 ± 0.8	2.5–4	3.4 ± 0.1	14.43	1.5–7	4.8 ± 0.8	0.2–1.2	1 ± 0.07
Marsa Assalaya	1–2	1.3 ± 0.1	1.6–5.1	3.4 ± 0.2	11–26	15.4 ± 0.04	1.5–5	2.9 ± 0.2	10.47	1.2–2.8	3 ± 0.1	0.35–1.18	0.7 ± 0.1
Shams Alam	1–6	3.4 ± 0.3	1.2–4.5	2.26 ± 0.2	13–19	16.5 ± 0.4	4–10	6.7 ± 0.3	7.24	2–6	4 ± 0.3	0.8–1.16	0.9 ± 0.2
Ras Baghdady	1–2	1.2 ± 0.08	2.35–8.6	4.2 ± 0.4	7–20	18 ± 0.7	4–8	7.2 ± 0.2	20.3	2.8–5	3.8 ± 0.2	0.37–0.43	0.4 ± 0.004
Wadi El-Gemal Island	1–3	2 ± 0.17	2.5–4.4	3.3 ± 0.1	11–26	18.3 ± 1.1	2–3	2.5 ± 0.1	12	1.1–1.2	1.2 ± 0.007	0.14–0.25	0.19 ± 0.008
Torpha South	1–2	1 ± 0.05	2.8–9	6 ± 0.5	11–23	15.4 ± 0.7	3–6	4.2 ± 0.2	3.08	2.5–8	5.6 ± 0.5	0.57–1.2	0.99 ± 0.05
Marsa Omm El-Abbas	1–2	1.2 ± 0.1	4.5–9	6.9 ± 0.4	17–22	18.9 ± 0.4	1–7	3 ± 0.6	3.77	6–7	6.9 ± 0.06	0.06–0.35	0.2 ± 0.02

CM = centimetre, and M = metre.

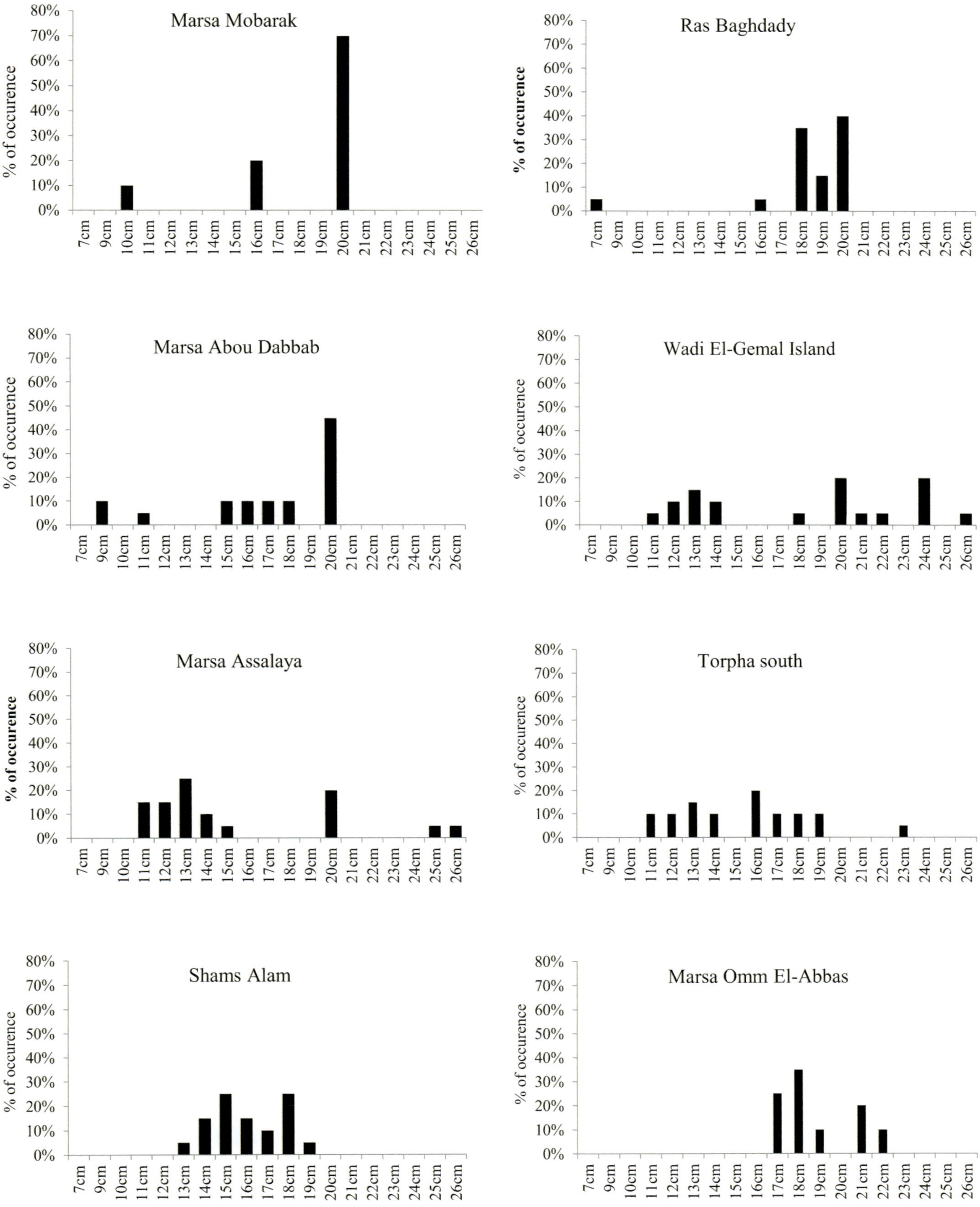

FIGURE 24.4 Percentage of occurrence for different size categories of the feeding trail width (cm) in the study sites (n = 20).

TABLE 24.3 Relationship between the different size categories of feeding trail widths and identified dugongs by photo-ID with the expected ones.

STUDY SITES	TRAIL WIDTH CATEGORIES	NO. OF IDENTIFIED		NO. OF EXPECTED	
		LARGE	SMALL	LARGE	SMALL
Marsa Mobarak	3	2 Male	–	–	1
Marsa Abou Dabbab	7	7 Male	–	–	–
Marsa Assalaya	8	1 Male, 1 Female	–	3	2
Shams Alam	7	3 Male	–	3	1
Ras Baghdady	5	2 Male	–	2	1
Wadi El-Gemal Island	10	–	–	7	3
Torpha South	9	2 Male	–	4	3
Omm El-Abbas	5	1 Male	–	4	–
Total number of dugongs		**18**	**0**	**23**	**11**

Note: Calf is concluded for a trail width less than 14 cm.

TABLE 24.4 Percentage and abundance of grazing rate of dugong on the seagrasses in the study sites.

STUDY SITES	% OF REMOVED SEAGRASS SPECIES					TOTAL SHOOTS REMOVED (MEAN± SE)	TOTAL GRAZING (%±SE)
	H.S	H.O	H.U	C.R	S.I		
Marsa Mobarak	72.4	10.8	16.7	–	–	4035±330	89±1.7
Marsa Abou Dabbab	82	18	–	–	–	1775±104	89.6±2.8
Marsa Assalaya	39.1	60.9	–	–	–	3770±252	90.8±1.4
Shams Alam	92.8	6.2	0.6	0.4	–	4250±297	88.7±2.2
Ras Baghdady	0.2	16.7	80.1	–	3	2820±207	86.6±1.4
Wadi El-Gemal Island	79.5	10.4	10.1	–	–	2925±309	84.5±3.2
Torpha South	99.8	–	0.2	–	–	2095±146	83.8±2
Marsa Omm El-Abbas	45.2	54.8	–	–	–	1560±194	90.5±1.9

H.s, *Halophila stipulacea*; H.o, *Halophila ovalis*; H.u, *Halodule uninervis*; C.r, *Cymodocea rotundata*; S.i, *Syringodium isoetifolium*.

of seagrass adjacent to the feeding trails at the sites ($P = 0.98$, $P = 0.87$ and $P = 0.99$) for Marsa Mobarak, Marsa Abou Dabbab and Marsa Assalaya, respectively. Dugongs removed 94% of shoots of *H. ovalis* at Marsa Mobarak and only 40% from the same species at Marsa Assalaya. In WGNP, the maximum was recorded in Shams Alam (3984 shoots/m^2, 90%) and the minimum in Torpha south (2080 shoots/m^2, 83%). There was a relationship between the amount of seagrass removed from feeding trails and the mean abundance of seagrass adjacent to the feeding trails at the sites ($P = 0.96$, $P = 0.98$ and $P = 0.98$) for Shams Alam, Ras Baghdady and Wadi El Gemal Island, respectively. In Ras Baghdady, dugongs removed 92% and 90% of shoots of *H. stipulacea* and *H. uninervis*, respectively.

24.4 DISCUSSION

Seagrass meadows provide grazing habitats for dugongs throughout their life-cycle (Day 2011; Marsh et al. 2012; Collier et al. 2012; Short et al. 2014; Hossain et al. 2016). In the study sites, dugongs fed mostly on low and moderate seagrass biomass on a sandy bottom in depths that ranged from 1 to 9 m (mean 3±2 m). The eight feeding sites at which seagrass abundance were measured had a median shoot density of 1200 shoots/m^2. However, the estimates of the seagrass varied among regions; the median shoot density was more than double in Marsa Alam (2585 shoots/m^2) than in WGNP. Values in Marsa Alam were similar to those of Preen (1992) in Moreton Bay, Australia (2569 shoots/m^2).

Within the range of the sediment availability in the study sites, substrate compaction appeared to influence the dugongs' choice of feeding sites; the dugongs fed more often in soft sediment sites. Sediment compaction was higher in WGNP than in Marsa Alam. The study sites of WGNP are subdivided into "wadis" (considered to be drainage areas) that mostly drain into the sea (El Shaffai 2016). The deposition of finer sediments is an important factor limiting seagrass distribution in this region. Preen (1992) reported that the amount of shell in the sediment might influence dugong choice for feeding sites. Shells are not easily crushed and may have a sharp edge, and this could impede efficient chewing. The mouthparts of the dugong are highly modified and transport the seagrass backward towards the teeth and throat (Lanyon 1991). The frequent occurrence of hard and sharp shells tangled amongst the seagrass rhizomes and roots may reduce the efficiency of the maceration and transportation processes. In this case, dugongs may choose to avoid areas with a high content of hard shells or modify their grazing mode and process less sediment.

The density of different seagrass species varied among regions. *H. stipulacea* was the most dominant in Marsa Alam with median shoot density of 1064 shoots/m^2. However, the other two species, *H. ovalis* and *H. uninervis*, were most dominant in WGNP, each with a median of 192 shoots m^2 each. Variation in species composition of seagrass may be driven by differences in light availability, hydrodynamics and sediment nutrient characteristics and not spatially explicit information (Manuel et al. 2013; Roelfsema et al. 2014). *H. ovalis* was the dominant species in the high intertidal zone and can be a significant grazing ground for dugong with roots and rhizomes that grow at a depth less than 10 cm (Kaewsrikhaw and Prathep 2014). Rajamani (2009) reported that *H. ovalis* and *H. uninervis* in Banggi Island and Mantanani Island in Malaysia appeared to be mainly at depths between 0.5 to 2 metres. Dugongs are drawn to seagrass meadows dominated by pioneering species, *H. ovalis* and *H. uninervis* (Nakaoka and Aioi 1999). They are often selected by dugongs for their low-fibre and high-nitrogen content (Preen 1995, in Australia; Nakaoka and Aioi 1999, in Thailand; De Iongh et al. 2007, in Indonesia).

During grazing on seagrasses on the sandy bottom in Marsa Alam and WGNP, dugongs left feeding trails in all of the study sites with an average length of 3.3 m. Preen (1992) recorded an average feeding trail length of 2.27 m in Moreton Bay, Australia. Several authors have recorded different feeding trail lengths (2.94 m and 8.05 m, Anderson and Birtles 1978; 1–5 m, Wake 1975; Heinsohn et al. 1977; Thayer et al. 1984: 4.47 m; Wongsuryrat et al. 2011). The dugong possesses a pair of tusks, formed by the first upper incisor teeth in both males and females that are used as cutting instruments for foraging (Nganvongpanit et al. 2017). The older and larger dugongs consumed a larger amount of seagrass than the younger ones and tended to leave more traces of feeding, which are also generally longer. Preen (1992) found no correlation between seagrass density and feeding trail lengths at four sites in Moreton Bay.

The widths of feeding trails in Marsa Alam and WGNP regions were similar (17.1 ± 3.6 cm and 17.3 ± 3.4 cm, respectively). These values appear to be smaller than those reported by other authors (19–25 cm, Wake 1975; Thayer et al. 1984: 23.3 cm; Anderson and Birtles 1978: 22.7 cm; Preen 1992: 19 cm; Wongsuryrat et al. 2011). This change is possibly due to the various differences in bottom type; at some sites, sediments were primarily mud or sandy mud, while others were composed of well-sorted siliceous sand (Preen 1992). When feeding in sandy substrates, dugongs may have to compress their rostral disk to make it easier to push over the sediment. In soft mud, such a response may not be necessary, and the dugongs could flare their upper lip, allowing them to maximize their cropping width. The width of feeding trails recorded by Anderson and Birtles (1978: 23.3 cm) corresponded with the width of the rostral disc of dead dugongs at Townsville (22.3 cm, SE = 1.05 for males; 22.2 cm, SE = 1.24 for females; Spain and Heinsohn 1975).

In the study sites, 88% of the number of shoots was grazed along the feeding trails. Preen (1992) estimated that dugongs removed an average of 86% of seagrass shoots in Moreton Bay, Australia. In Marsa Assalaya, dugongs grazed about 92% of seagrass shoots (mostly dominated by *H. ovalis*), indicating their feeding selectivity and preference. Sheppard (2008) reported that seagrass species possessed the highest digestibility and highest starch concentrations in addition to the lowest fibre content. In other Sirenians such as Florida manatees, the seagrasses were as efficiently removed as by dugongs (83–94% of shoots; Lefebvre and Powell 1990).

Feeding trails give more information about the various sizes of dugongs that visited the area. A complementary study was conducted in Marsa Alam using underwater laser photogrammetry to estimate the relationship between the dugong size and its muzzle width (Shawky 2019b). The author recorded that three individual dugongs with total lengths of 265.5 cm, 277.7 cm and 280 cm created the same trail width of 20 cm while excavating the seagrass. For calf estimation size, a muzzle width of 12.2 cm was measured for a neonatal calf with a size of 120 cm, which stranded in WGNP on 29 September 2015 (Shawky et al. 2016). In October 2015, the author reported small trails with widths of 7 cm and 11 cm in the area. Further study is highly recommended in order to estimate the size of the dugong population via feeding trail widths even if they are not encountered underwater.

The impact of grazing should be examined to determine the preference of individuals for different parts of the seagrass, and management of the preferred species is critical to ensuring the sustainability of dugongs' favourite food. Anderson (1998) and Masini et al. (2001) concluded that dugongs favourably feed on *H. uninervis* to maximize energy intake as its rhizomes are rich in starch. Further study is also required to determine the effect of seasonal variations on the presence of dugongs in their key habitats.

REFERENCES

Amamoto, N., Ichikawa, K., Arai, N., Akamatsu, T., Shinke, T., and K. Adulyanukosol. 2009. The depth of water affects the feeding ground selection by dugongs in dry season. *J Adv Mar Sci Technol Soc* 15: 149–157.

Anand, Y. 2012. First record of feeding trails of dugongs in Gulf of Kachchh (GoK), Gujarat. *Indian For* 138, no. 10: 968–969.

Anderson, P.K. 1998. Shark Bay dugongs (*Dugong dugon*) in summer: II. Foragers in a Halodule-dominated community. *Mammalia* 62: 409–425.

Anderson, P.K., and A. Birtles. 1978. Behaviour and ecology of the dugong, *Dugong dugon* (Sirenia): Observations in Shoalwater and Cleveland Bays, Queensland. *Aust Wildl Res* 5: 1–23.

Aragones, L.V., Lawler, I.R., Foley, W.J., and H. Marsh. 2006. Dugong grazing and turtle cropping: Grazing optimization in tropical seagrass systems? *Oecologia* 149: 635–647.

Aragones, L.V., and H. Marsh. 2000. Impact of dugong grazing and turtle cropping on tropical seagrass communities. *Pacific Conserv Biol* 5: 278–288.

Beck, C.A., and A.M. Clark. 2012. Individual identification of sirenians. In: *Sirenian conservation: Issues and strategies in developing countries*. Eds: E.M. Hines, J.E. Reynolds, L.V. Aragones, A.A. Mignucci-Giannoni, and M. Marmontel, 133–138. Gainesville, FL: University Press of Florida.

Bessey, C., Heithaus, M.R., Fourqurean, J.W., Gastrich, K.R., and D.A. Burkholder. 2016. Importance of teleost macrograzers to seagrass composition in a subtropical ecosystem with abundant populations of megagrazers and predators. *Mar Ecol Prog Ser* 553: 81–92.

Collier, C.J., Waycott, M., and A.G. Ospina. 2012. Responses of four Indo-West Pacific seagrass species to shading. *Mar Pollut Bull* 65: 342–354.

Day, J.C. 2011. Protecting Australia's Great Barrier Reef. *Solutions* 2, no. 1: 56–66.

De Iongh, H.H., Kiswara, W., Kustiawan, W., and P.E. Loth. 2007. A review of research on the interactions between dugongs (*Dugong dugon* Müller 1776) and intertidal seagrass beds in Indonesia. *Hydrobiologia* 591: 73–83.

D'Souza, E., Patankar, V., Arthur, R., Marbà, N., and T. Alcoverro. 2015. Seagrass herbivory levels sustain site-fidelity in a remnant Dugong population. *PLoS One* 10, no. 10: e0141224.

Ebrahim, A., Olds, A.D., Maxwell, P.S., Pitt, K.A., Burfeind, D.D., and R.M. Connolly. 2014. Herbivory in a subtropical seagrass ecosystem: Separating the functional role of different grazers. *Mar Ecol Prog Ser* 511: 83–91.

El Shaffai, A. 2016. *Field guide to seagrasses of the Red Sea*, 2nd ed. Eds: A. Rouphael, and A. Abdulla, viii+56 p. Gland, Switzerland: IUCN and Courbevoie, France: Total Foundation.

Heinsohn, G.E., Wake, J., Marsh, H., and A.V. Spain. 1977. The dugong (*Dugong dugon* [Müller]) in the seagrass system. *Aquaculture* 12: 235–248.

Hossain, M.S., Bujang, S.J., Zakaria, M.H., and M. Hashim. 2016. Marine and human habitat mapping for the Coral Triangle Initiative region of Sabah using Landsat and Google Earth imagery. *Mar Policy* 72: 176–191.

Kaewsrikhaw, R., and A. Prathep. 2014. The effect of habitats, densities and seasons on morphology, anatomy and pigment content of the seagrass *Halophila ovalis* (R. Br.) Hook. f. at Haad Chao Mai National Park, Southern Thailand. *Aquat Bot* 116: 69–75.

Kendrick, G.A., Waycott, M., Carruthers, T.J.B., Cambridge, M.I., Hovey, R., Krauss, S.I., Lavery, P.S., Les, D.H., Lowe, J.R., Vidal, O.M.I., Ooi, J.I.S., Orth, R.J., Rivers, O.D., Montoya, L.R., Sinclair, E.A., Statton, J., Dijk, J.K.V., and J.J. Verduin. 2012. The central role of dispersal in the maintenance and persistence of seagrass populations. *BioSci* 62: 56–65.

Khamis, A., Alcoverro, T., D'Souza, E., Arthur, R., Pagès, J.F., Shah, J., Al-Qahtani, T., and E.A. Abdulla. 2022. Identifying conservation priorities for a widespread dugong population in the Red Sea: Megaherbivore grazing patterns inform management planning. *Mar Environ Res* 181: 105762. doi:10.1016/j.marenvres.2022.105762.

Lanyon, J. 1991. *The nutritional ecology of the dugong (Dugong dugon) in tropical North Queensland*. Ph.D. Thesis, Monash University, Melbourne.

Lefebvre, L.W., and J.A. Powell. 1990. *Manatee grazing impacts on seagrasses in Hobe Sound and Jupiter Sound in southeast Florida during the winter of 1988–89*. Contract Report T62239152 to the Marine Mammal Commission, U.S. Fish Wildlife Service, National Ecology Research Center; Gainesville, Florida.

Manuel, S.A., Coates, K.A., Kenworthy, W.J., and J.W. Fourqurean. 2013. Tropical species at the northern limit of their range: Composition and distribution in Bermuda's benthic habitats in relation to depth and light availability. *Mar Environ Res* 89: 63–75.

Marsh, H., O'Shea, T.J., and J.E. Reynolds III. 2012. *Ecology and conservation of the Sirenia: Dugongs and manatees*. New York, NY: Cambridge University Press.

Masini, R.J., Anderson, P.K., and A.J. McComb. 2001. A Halodule–dominated community in a subtropical embayment: Physical environment productivity, biomass, and impact of dugong grazing. *Aquat Bot* 71: 179–197.

McMahon, K. 2005. *Recovery of subtropical seagrasses from natural disturbance*. Ph.D. Thesis, University of Queensland, Brisbane, QLD.

Mizuno, K., Asada, A., Matsumoto, M., Sugimoto, S., Yamamuro, M., Fortes, M.D., Sarceda, M., and L.A. Jimenez. 2017. A simple and efficient method for making a high-resolution seagrass map and quantification of dugong feeding trail distribution: A field test at Mayo Bay, Philippines. *Ecol Inform* 38: 89–94.

Moore, A.M., Ambo-Rappe, R., and Y. Ali. 2017. "The lost princess (putri duyung)" of the small islands: Dugongs around sulawesi in the anthropocene. *Front Mar Sci* 4: 284.

Nakaoka, M., and K. Aioi. 1999. Growth of seagrass *Halophila ovalis* at dugong trails compared to existing within-patch variation in a Thailand intertidal flat. *Mar Ecol Prog Ser* 184: 97–103.

Nasr, D., Shawky, A.M., and P. Vine. 2019. Status of Red Sea Dugongs. In: *Oceanographic and biological aspects of the Red Sea*. Eds: N.M.A. Rasul, and I.C.F. Stewart, 327–354. Cham, Switzerland: Springer Nature.

Nganvongpanit, K., Buddhachat, K., Piboonl, P., Euppayo, T., Kaewmong, P., Cherdsukjai, P., Kittiwatanawong, K., and C. Thitaram. 2017. Elemental classification of the tusks of dugong (Dugong dugong) by HH-XRF analysis and comparison with other species. *Sci Rep* 7: 46167.

Perry, C.J., and W.C. Dennison. 1999. Microbial nutrient cycling in seagrass sediments. *AGSO J Austr Geol Geophys* 17: 227–231.

Preen, A.R. 1992. *Interactions between dugongs and seagrasses in a subtropical environment*. Ph.D. Thesis. James Cook University of North Queensland, Townsville, Australia.

Preen, A.R. 1995. Impacts of dugong foraging on seagrass habitats: Observational and experimental evidence for cultivation grazing. *Mar Ecol Prog Ser* 124: 201–213.

Rajamani, L. 2009. *The conservation biology of the Dugong (Dugong dugon) and its seagrass habitat in Sabah, Malaysia: A basis for conservation planning*. Ph.D. Thesis, Malaysia Sabah University.

Roelfsema, C.M., Lyons, M., Kovacs, E.M., Maxwell, P., Saunders, M.I., Samper-Villarreal, J., and S.R. Phinn. 2014. Multi-temporal mapping of seagrass cover, species and biomass: A semi-automated object based image analysis approach. *Remote Sens Environ* 150: 172–187.

Shawky, A.M. 2019a. Evidence of the occurrence of a large dugong in the Red Sea, Egypt. *Egypt J Aquat Res* 45: 247–250. doi:10.1016/j.ejar.2019.08.001.

Shawky, A.M. 2019b. *Analysis of feeding trails provides evidence of the number of Dugongs excavating seagrass at key habitats seasonally in the Egyptian Red Sea*. Final Report to Rufford Foundation. doi:10.13140/RG.2.2.15151.97449.

Shawky, A.M., Sallam, W.S., Alwany, M.A., Mohammad, D.A., and S.Z. Mohamed. 2016. Stranding of a neonatal dugong calf in Wadi El Gemal National Park: Implication for dugong conservation in Egypt. *Al Azhar Bull Sci* 27: 1–11.

Shawky, A.M., Sallam, W.S., Alwany, M.A., Mohammad, D.A., and S.Z. Mohamed. 2017. Photo identification of Dugongs in Marsa Alam and Wadi El Gemal National Park, Egyptian coast of the Red Sea. *Al Azhar Bull Sci* 28: 1–10.

Shawky, A.M., Sallam, W.S., Alwany, M.A., Mohammad, D.A., and S.Z. Mohamed. 2019. Photo identification of Dugongs in Marsa Alam and Wadi El Gemal National Park. *Ind J Geo Mar Sci* 48: 1351–1358.

Shawky A.M., Sallam W.S., and S.Z. Mohamed. 2024. Assessing dugong distribution and overlapping threats along the northwestern Red Sea, *Egypt. Egypt J Aquat Res* 28(1): 1455–1488.

Sheppard, J.K. 2008. *The spatial ecology of Dugongs: Applications to conservation management*. Ph.D. Thesis, James Cook University, Townsville, Australia.

Short, F.T., Coles, P., Fortes, M.D., Victor, S., Salik, M., Andrew, J., and A. Seno. 2014. Monitoring in the Western Pacific region shows evidence of seagrass decline in line with global trends. *Mar Pollut Bull* 83: 408–416.

Spain, A.V., and G.E. Heinsohn. 1975. Size and weight allometry in a north Queensland population of Dugong dugon (Muller) (Mammalia: Sirenia). *Austral J Zool* 23: 159–168.

Thayer, G.W., Bjorndal, K.A., Ogden, J.C., Williams, S.L., and J.C. Zieman. 1984. Role of large herbivores in seagrass communities. *Estuaries* 7: 351–376.

Wake, J.A. 1975. *A study of the habitat requirements and feeding biology of the dugong Dugong dugon (Müller)*. Honours Thesis, James Cook University, Townsville.

Wongsuryrat, M., Chunkao, K., Prabuddham, P., and M. Daungsavat. 2011. Distribution, abundance and conservation status of Dugong around Koh Talibong, Trang Province, Thailand. *J Sustain Dev* 3, no. 4: 118–124.

Index

For Product Safety Concerns and Information please contact our EU representative GPSR@taylorandfrancis.com Taylor & Francis Verlag GmbH, Kaufingerstraße 24, 80331 München, Germany

Batch number: 10733477

Printed by Printforce, the Netherlands